3rd edition
CLINICAL ANATOMY

National Medical Series

In the basic sciences

clinical anatomy, 3rd edition
the behavioral sciences in psychiatry, 3rd edition
biochemistry, 3rd edition
clinical epidemiology and biostatistics
genetics
hematology
histology and cell biology, 2nd edition
human developmental anatomy

immunology, 3rd edition
introduction to clinical medicine
microbiology, 2nd edition
neuroanatomy
pathology, 3rd edition
pharmacology, 3rd edition
physiology, 3rd edition
radiographic anatomy

In the clinical sciences

medicine, 3rd edition
obstetrics and gynecology, 3rd edition
pediatrics, 3rd edition
preventive medicine and public health, 2nd edition
psychiatry, 3rd edition
surgery, 3rd edition

In the exam series

review for USMLE Step 1, 3rd edition
review for USMLE Step 2
geriatrics

The National Medical Series for Independent Study

clinical anatomy
3rd edition

Ernest W. April, Ph.D.
Associate Professor of Anatomy and Cell Biology
College of Physicians and Surgeons
Columbia University
New York, New York

Illustrated by
Anne Erickson
Salvatore Montano

Williams & Wilkins
A WAVERLY COMPANY

BALTIMORE • PHILADELPHIA • LONDON • PARIS • BANGKOK
BUENOS AIRES • HONG KONG • MUNICH • SYDNEY • TOKYO • WROCLAW

Editor: Elizabeth A. Nieginski
Managing Editor: Amy G. Dinkel
Development Editors: Catherine Nancarrow, Becky Krumm
Production Coordinator: Cindy Park
Cover Designer: Cathy Cotter
Typesetter: Maryland Composition
Printer: Port City Press
Binder: Port City Press

Copyright © 1997 Williams & Wilkins

351 West Camden Street
Baltimore, Maryland 21201-2436 USA

Rose Tree Corporate Center
1400 North Providence Road
Building II, Suite 5025
Media, Pennsylvania 19063-2043 USA

All rights reserved. This book is protected by copyright. No part of this book may be reproduced in any form or by any means, including photocopying, or utilized by any information storage and retrieval system without written permission from the copyright owner.

Printed in the United States of America

First Edition,

Library of Congress Cataloging-in-Publication Data

April, Ernest W.
   Clinical Anatomy / Ernest W. April : illustrated by Anne Erickson.
Salvatore Montano. —3rd ed.
     p.   cm. —(The National medical series for independent study)
   ISBN 0-683-06199-2.—ISBN 0-683-06199-2
   1. Human anatomy—Outlines, syllabi, etc. 2. Human anatomy—
Examinations, questions, etc. I. Title. II. Series.
   [DNLM: 1. Anatomy—outlines. 2. Anatomy—examination questions.
QS 18.2 A654a   1997]
QM31.A67 1997
611'.002'02—dc20
DNLM/DLC
for Library of Congress   96-1552
                                                                                               CIP

The publishers have made every effort to trace the copyright holders for borrowed material. If they have inadvertently overlooked any, they will be pleased to make the necessary arrangements at the first opportunity.

Call our customer service department at **(800) 638-0672** for catalog information, or fax orders to **(800) 447-8438**. For other book services, including chapter reprints and large-quantity sales, ask for the Special Sales department.

To purchase additional copies of this book or for information concerning American College of Sports Medicine certification and suggested preparatory materials, call **(800) 486-5643**.

Canadian customers should call **(800) 268-4178**, or fax **(905) 470-6780**. For all other calls originating outside of the United States, please call **(410) 528-4223** or fax us at **(410) 528-8550**.

***Visit Williams & Wilkins on the Internet:* http://www.wwilkins.com,** or contact our customer service department at **custserv@wwilkins.com**. Williams & Wilkins customer service representatives are available from 8:30 am to 6:00 pm, EST, Monday through Friday, for telephone access.

                                                                                            97 98 99
                                                                1 2 3 4 5 6 7 8 9 10

Contents

Preface xiii
Acknowledgments xv

SECTION I INTRODUCTORY OVERVIEWS

1 Introduction — 3
    I. Human anatomy 3
    II. Anatomic terminology 3

2 Integument and Mammary Glands — 9
    I. Integument (skin) 9
    II. Fascial layers 11
    III. Mammary gland or breast 12

3 Musculoskeletal System — 15
    I. Skeleton 15
    II. Muscle 20

4 Nervous System — 23
    I. Introduction 23
    II. Central nervous system 24
    III. Peripheral nervous system 26
    IV. Somatic nervous system 29
    V. Visceral nervous system 31

5 Circulatory System — 35
    I. Introduction 35
    II. Heart 35
    III. Arterial system 37
    IV. Capillary beds 38
    V. Venous system 40
    VI. Lymphatic system 41

    Study Questions to Part I 45

SECTION II EXTREMITIES AND BACK
PART IIA UPPER EXTREMITY

6 Shoulder Region — 55
   I. Introduction to the upper extremity 55
   II. Bones of the pectoral girdle 56
   III. Articulations of the shoulder girdle 58
   IV. Muscle function at the shoulder joint 60
   V. Axillary vasculature 65
   VI. Brachial plexus 68

7 Elbow Region — 73
   I. Bones 73
   II. Articulations 74
   III. Muscle function at the elbow joint 76
   IV. Brachial vasculature 78
   V. Brachial innervation 81

8 Extensor Forearm and Posterior Wrist — 85
   I. Bones of the forearm and proximal wrist 85
   II. The forearm extensor compartment 88
   III. Innervation of the extensor compartment 90

9 Flexor Forearm and Anterior Wrist — 93
   I. Bones of the distal wrist 93
   II. The forearm flexor compartment 93
   III. Antebrachial vasculature 97
   IV. Innervation of the flexor compartment 100

10 Hand — 103
   I. Bones and joints 103
   II. Dorsal musculature 105
   III. Palmar musculature 107
   IV. Vasculature 116
   V. Innervation 118
   VI. Function 121
       Study Questions to Part IIA 122

PART IIB BACK

11 Vertebral Column — 131
   I. General structure 131
   II. Structure of typical vertebra 132
   III. Regional modifications of vertebral characteristics 133
   IV. Intervertebral disks 138
   V. Ligaments 140

12 Soft Tissues of the Back — 143
  I. Posterior vertebral muscles   143
  II. Anterior vertebral muscles   144
  III. Functional considerations   148

13 Spinal Cord — 149
  I. General organization   149
  II. Nerve roots   153
  III. Vasculature   154
  IV. Clinical considerations   155
  Study Questions to Part IIB   156

PART IIC   LOWER EXTREMITY

14 Gluteal Region — 161
  I. Introduction   161
  II. Bones of the pelvis   162
  III. Pelvic articulations   164
  IV. Muscle function at the hip joint   167
  V. Vasculature   173
  VI. Innervation   175

15 Thigh and Knee Joint — 179
  I. Bones at the knee   179
  II. Knee joint   182
  III. Fascia and muscles of the thigh   188
  IV. Vasculature of the thigh   193
  V. Innervation of the thigh   196

16 Anterior Leg and Dorsal Foot — 201
  I. Bones of the distal leg and tarsus   201
  II. Ankle and tarsal joints   203
  III. Anterior and lateral crural compartments   206
  IV. Dorsum of the foot   209
  V. Vasculature   211
  VI. Innervation   212

17 Posterior Leg — 215
  I. Musculature   215
  II. Vasculature   218
  III. Innervation   220

18 Plantar Foot — 223
  I. Bones   223
  II. Articulations   223
  III. Musculature   225
  IV. Vasculature   229
  V. Innervation   229
  VI. Ambulation   231
  Study Questions to Part IIC   233

SECTION III TRUNK
PART IIIA THORAX

19 Thoracic Cage — 243
  I. Introduction 243
  II. Rib cage 243
  III. Musculature 247
  IV. Vasculature 252
  V. Innervation 253
  VI. Respiratory mechanics 254

20 Pleural Cavities and Lungs — 259
  I. Introduction 259
  II. Pleural cavities 259
  III. Lungs 263
  IV. Bronchial tree 265
  V. Vasculature 266
  VI. Innervation 268
  VII. Functional anatomy of the respiratory system 269

21 Pericardial Cavity and Heart — 271
  I. Middle mediastinum 271
  II. Pericardial cavity 271
  III. External cardiac structure 273
  IV. Coronary circulation 275
  V. Cardiac wall 278
  VI. Cardiac chambers 279
  VII. Conducting system 283
  VIII. Cardiac innervation 285
  IX. Cardiac dynamics 286
  X. Fetal and early postnatal circulation 290

22 Mediastinum — 293
  I. Anterior mediastinum 293
  II. Superior mediastinum 293
  III. Posterior mediastinum 295
  IV. Vasculature 298
  V. Innervation 300
  Study Questions to Part IIIA 304

PART IIIB ABDOMEN

23 Anterior Abdominal Wall — 313
  I. Introduction 313
  II. Skin and fasciae 314
  III. Musculature 315
  IV. Vasculature 319
  V. Innervation 320
  VI. Clinical considerations 321
  VII. Inguinal region 322

24 Peritoneal Cavity   329
    I. Introduction 329
    II. Development of the peritoneal cavity and viscera 330
    III. Peritoneal relationships 337

25 Gastrointestinal Tract   343
    I. Introduction 343
    II. Foregut 344
    III. Midgut 364
    IV. Hindgut 372
    V. Vasculature 375
    VI. Innervation 380

26 Kidneys and Posterior Abdominal Wall   385
    I. Kidneys 385
    II. Ureters 389
    III. Adrenal glands 391
    IV. Posterior abdominal wall 393
    Study Questions to Part IIIB 399

PART IIIC   PELVIS

27 Pelvis   407
    I. Bones and joints 407
    II. Musculature 411
    III. Fascia and peritoneum 412
    IV. Vasculature 413
    V. Lumbosacral plexus 416

28 Perineum   419
    I. Introduction 419
    II. Development of external genitalia 419
    III. Anal triangle 421
    IV. Male urogenital triangle 425
    V. Female urogenital triangle 435

29 Pelvic Viscera   445
    I. Rectum 445
    II. Pelvic portion of the urinary system 447
    III. Development of urogenital ducts 453
    IV. Male reproductive tract 455
    V. Female reproductive tract 461
    VI. Innervation of pelvic viscera 473
    VII. Functional considerations 475
    Study Questions to Part IIIC 479

SECTION IV   HEAD and NECK
PART IVA   SOMATIC NECK and NEUROCRANIUM

30 Posterior Cervical Triangle — 489
   I. Cervical fascia   489
   II. Somatic neck   491

31 Neurocranium — 501
   I. Introduction   501
   II. Scalp   501
   III. Bones and articulations   503
   IV. Cranial meninges and venous sinuses   513

32 Brain — 519
   I. Introduction   519
   II. Cerebrum   520
   III. Brain stem   525
   IV. Cranial nuclei and associated nerves   527
   V. Ventricular system   537
   VI. Vasculature   538

33 Orbit, Eye, and Ear — 543
   I. Orbit   543
   II. Eyeball (bulbus oculi)   552
   III. Ear   557
   Study Questions to Part IVA   564

PART IVB   FACIAL CRANIUM and VISCERAL NECK

34 Anterior Cervical Triangle — 573
   I. Cervical fascia   573
   II. Hyoid bone   573
   III. Musculature   574
   IV. Thyroid gland and parathyroid glands   577
   V. Vasculature   579
   VI. Innervation   582

35 Facial Skeleton — 585
   I. Introduction   585
   II. Facial cranium   586
   III. Teeth   593

36 Superficial and Deep Face — 597
   I. Introduction   597
   II. Parotid gland   598
   III. Muscles of facial expression   599
   IV. Infratemporal and pterygopalatine fossae   602
   V. Temporomandibular joint and muscles of mastication   604

    VI. Vasculature 607
   VII. Innervation 610

37 Cranial and Cervical Viscera   619
     I. Introduction 619
    II. Nasal cavity 619
   III. Oral cavity 623
   IV. Pharynx 630
    V. Larynx 637
       Study Questions to Part IVB 645

Index   653

Preface

I. PURPOSE. Anatomy constitutes the basis for all clinical practice in the health science professions. It is not a conceptually difficult subject, but the amount of material can be overwhelming. This book places the material in perspective and focuses on important (i.e., clinically relevant) facts and concepts.
  A. The first objective of this book is to **facilitate** organization, study, and review of human anatomy.
    1. It is a **companion and compendium** for human anatomy courses that are part of health sciences curricula. It can be used very successfully as a **course syllabus**.
    2. In **outline** format, it presents an organized, concise, and comprehensive **review** of the essential facts and details of human anatomy for:
      a. In-course examinations in the professional health sciences curricula.
      b. Subsequent courses and clinical clerkships in the health sciences curricula.
      c. Licensure examinations
        (1) USMLE (Part I)
        (2) Specialty board examinations
  B. The second objective is to **maximize use of limited time** available to the student health professional.
    1. The **outline organization** lists topics and items in a logical and easy-to-follow sequence that provides a framework that the student may use to construct a working knowledge of human anatomy. This is achieved by presenting basic anatomy with appropriate illustrations, diagrams, and tables; functional anatomic concepts; and relevant clinical notes.
    2. The emphasis on **structural and functional detail** pertinent to the skillful practice of up-to-date medicine directs the student toward the most fruitful and most easily remembered aspects of anatomic study.
      a. The subject material does not include detailed discussions of every topic in anatomy, nor does it belabor major points of anatomy that are straightforward and easily assimilated.
      b. For the clinician in training, some anatomic structures and concepts are more important than others. This information frequently is emphasized with a clinical note. Effort is made to distinguish between the included minutiae (i.e., clinically important detail) and the excluded trivia (i.e., inconsequential detail of interest only to anatomists and subspecialists).

II. SUGGESTED METHOD OF USE
  A. The subject presentation is regional, which is the method used in most medical and dental programs.
    1. The **section order** is only one of a number of sequences that have proved workable, but one for which I have preference.
    2. Extensive **topic cross-references** enable the user of this outline and study guide to proceed in any order.

B. **To benefit maximally** from this outline and study guide:
   1. **Use it in conjunction with a textbook and an atlas** of human anatomy.
      a. By scanning through the outline points, it can be used to prepare for lectures.
      b. After lectures, it can be used for consolidation and study—taking into consideration those points emphasized by the instructor. When the outline makes sense, it is adequate; when confusion or question persists, the more extensive discussion of a supplementary textbook is necessary.
      c. Because this outline is meant to be used in conjunction with an atlas, the line drawings are primarily for purposes of orientation and review.
   2. Use the **study questions** at the end of each section to determine whether a basic knowledge of anatomy and a working understanding of human anatomy have been attained.
   3. The benefits of working with a **small study group** cannot be overemphasized. The study group format is especially effective with problem-based learning focused on the study questions.

III. **SPECIFIC TO THE THIRD EDITION**
   A. **Reorganization.** In response to student and faculty feedback, numerous changes are incorporated into this edition.
      1. The sections are arranged in a more logical and flexible sequence: Introductory Overviews; Extremities and Back; Trunk; Head and Neck.
      2. Only the most significant outline points have been retained, resulting in a modest reduction in the size of the book.
      3. All major outline points are presented with headings, which are parallel both between and within chapters.
      4. Much of the **illustrative material** has been resized.
      5. The details of individual muscles and nerves are listed in easy-to-use **tables,** whereas group functions and innervations are retained in the text.
      6. **Nomenclature** has been updated.
   B. **Problem-based questions with explanations** are essentially in keeping with the present USMLE format. The number of study questions has been reduced, but more questions involving clinical anatomy and more problem sets centered around case studies have been included.

IV. *ANATOMY INTERACTIVE:* **A Multimedia Tutorial** is available in CD-ROM format to school libraries and computer learning centers. The interactive software program runs under Microsoft Windows (v. 3.1) on stand-alone IBM-compatible personal computers and on local area networks. *NMS Clinical Anatomy* provides convenient hard copy for the electronic tutorial, although a greater depth of information is available, if desired, in the electronic version. All 1600 full-color illustrations (about a quarter are pixel active) are appropriately juxtaposed to the text in split-screen format. Approximately 2500 questions parallel the material, providing the option for an effective anatomic tutorial with a question for each item of anatomic importance. All responses are interactive, with explanations for correct as well as incorrect responses when appropriate. The tutorial method of questioning elicits a clear understanding of human anatomy by focusing attention to appropriate detail. For information regarding this product, contact the author at Columbia University, New York, NY, 10032.

Ernest W. April
New York, New York

Acknowledgments

Because the function of a text or review book is to present concisely the basic information that forms an accepted body of knowledge, only the organization and presentation of those facts and concepts may be original. Therefore, I humbly acknowledge the numerous anatomic reference texts against which the material presented herein has been checked for accuracy. Even more humbly, I acknowledge those uncounted and anonymous deceased individuals, as well as my numerous academic and clinical colleagues, who have contributed to my fund of knowledge over the years. Finally, I am grateful to the student clinicians against whom this anatomic knowledge has been honed. I am very much indebted to Dr. Timothy Chuter, a surgical colleague, for his thorough reading of the first edition manuscript, his lively discussions, his contributions to the section on the extremities, and his assistance in writing much of the section on the head and neck in the first edition.

The anatomic illustrations of Anne Erickson and Salvatore Montano confirm the importance of the visual aspects of anatomy. Much of their work is based upon their own dissections and observations.

For their forbearance, inspiration, and love I dedicate this modest work to Nancy, Jeremy, and Geoffrey.

SECTION 1
INTRODUCTORY OVERVIEWS

Chapter 1

Introduction

I. HUMAN ANATOMY

A. **Definition.** Human anatomy is the study of the macroscopic structure of the human organism, correlated with development, function, and clinical significance.

B. **Origins.** The oldest medical science, human anatomy traces its origins to early Greek civilizations.

1. **Derivation.** The Greek term for anatomy (*anatome*) means cutting apart, as does the Latin term for dissection (*dissecare*).
2. **Offshoots.** Anatomy is the mother science of clinical medicine.
   a. Anatomy gave rise to the daughter medical sciences of pathology (morbid anatomy), physiology, neuroanatomy, histology and cytology (microscopic anatomy), embryology (developmental anatomy), and physical anthropology.
   b. Several of these sciences have produced subdisciplines, such as clinical pathology, hematology, microbiology, immunology, cell biology, molecular biology, cellular biophysics.

C. **Organization**

1. **Systemic (systematic) anatomy** is organized according to functional systems: **integument** (see Chapter 2), **musculoskeletal system** (see Chapter 3), **nervous system** (see Chapter 4), **circulatory system** (see Chapter 5), **respiratory system, digestive system, urinary system, reproductive system,** and **endocrine glands.**
2. **Regional anatomy** is concerned with all systems found in a discrete part of the body, for example, **back and extremities** (see Chapters 6–18), **thorax** (see Chapters 19–22), **abdomen** (see Chapters 23–26), **pelvis** (see Chapters 27–29), and **head and neck** (see Chapters 30–38).
3. **Functional anatomy** concerns the correlation between structure and function. For example, parasympathetic innervation to the male and female external genitalia mediates erection, and the sympathetic innervation mediates ejaculation or uterine contraction.
4. **Clinical anatomy** emphasizes structure and function as it relates to the practice of medicine and other health professions. For example, a displaced fracture of the humerus may injure the adjacent radial nerve, resulting in paralysis of the extensor muscles of the forearm and wrist with loss of sensation over most of the dorsum of the hand.

II. ANATOMIC TERMINOLOGY

A. **Anatomic position.** All structures are described and frequently named with reference to the anatomic position, in which the individual is erect (or lying supine as if erect) with the arms by the sides, palms facing forward, the legs together with the feet directed forward, and the head directed forward (Figure 1-1).

B. **Anatomic adjectives** are arranged as pairs of opposites (see Figure 1-1).

1. **Anterior/posterior**
   a. **Anterior (ventral)** is toward the front aspect of the body.
   b. **Posterior (dorsal)** is toward the back aspect of the body.
   c. **Palmar** is the ventral side of the hand.
   d. **Plantar** is the sole of the foot.

FIGURE 1-1. The body in the anatomic position. Principal anatomic planes and some anatomic adjectives are indicated.

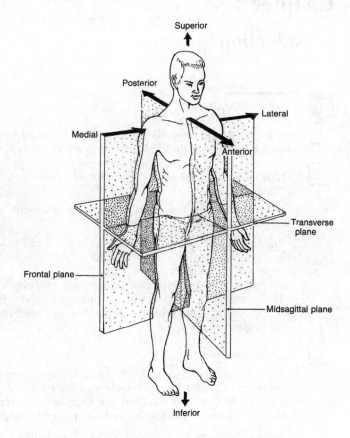

2. **Proximal/distal**
   a. **Proximal** is close to the median or near the origin of a structure.
   b. **Distal** is away from the origin of a structure.
3. **External/internal**
   a. **External (superficial)** is close to the surface of the body.
   b. **Internal (deep)** is closer to the center of the body.
4. **Superior/inferior**
   a. **Superior (cephalad, craniad, cephalic, rostral)** is toward the head.
   b. **Inferior (caudad, caudal)** is toward the tail or base of the spinal column.
5. **Medial/lateral**
   a. **Median** is in the midsagittal plane.
   b. **Medial** is toward the midsagittal plane.
   c. **Lateral** is away from the median.
6. **Central/peripheral**
   a. **Central** is toward the center of the mass of the body.
   b. **Peripheral** is away from the center of body mass.
7. **Prone/supine**
   a. **Prone** is ventral surface down.
   b. **Supine** is ventral surface up.

C. **Anatomic planes** accurately define the location of structures (see Figure 1-1).

1. **The midsagittal plane** is vertical between the anterior midline and the posterior midline, dividing the body into left and right halves.
2. **Parasagittal (paramedian) planes** are parallel to the midsagittal plane.

3. **Coronal planes** are vertical and perpendicular to the midsagittal plane. The **midcoronal (frontal) plane** divides the body into anterior and posterior halves.
4. **Transverse (horizontal) planes** are mutually perpendicular to the midsagittal and coronal planes, dividing the body by cross sections.
5. **An axis** is defined by the intersection of any two planes.
   a. **The vertical axis** is defined by the intersection of the midsagittal and midcoronal planes.
   b. **Anteroposterior axes** are defined by the intersection of the transverse and sagittal planes.
   c. **Bilateral axes** are defined by the intersection of coronal and transverse planes.

D. **Anatomic movements** are usually described as pairs of opposites (Figure 1-2).
   1. **Flexion/extension** usually occurs in the midsagittal or parasagittal planes.
      a. **Flexion** brings primitively ventral surfaces together (e.g., bending the arm at the elbow).
         (1) **Plantar flexion** is downward flexion (true flexion) of the foot at the ankle joint.
         (2) **Dorsiflexion** is upward flexion (extension) of the foot at the ankle joint.
      b. **Extension** is movement away from the ventral surface (e.g., straightening the leg at the knee joint).
   2. **Abduction/adduction** usually occurs in the midcoronal plane.
      a. **Abduction** (lateral flexion) is movement away from the median, away from the middle finger, or away from the second toe (e.g., directing an eye laterally). Radial deviation is abduction of the hand at the wrist joint.
      b. **Adduction** is movement toward the median, toward the middle finger or toward the second toe (e.g., bring one leg adjacent to the other). Ulnar deviation is adduction of the hand at the wrist joint.
   3. **Medial rotation/lateral rotation** usually occurs about a vertical axis.
      a. **Medial rotation** is movement of a ventral surface toward the median (e.g., bringing a flexed arm across the chest).

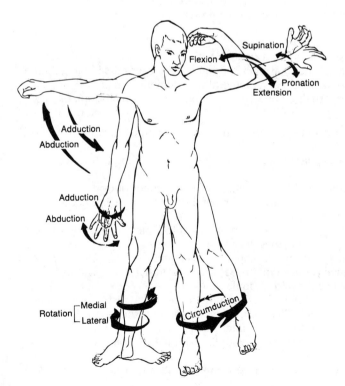

FIGURE 1-2. Principal anatomic movements.

b. Lateral rotation is movement of a ventral surface away from the median (e.g., directing the head toward one side).

4. **Elevation/depression** are generally up and down movements about horizontal axes.
   a. **Elevation** raises or moves a structure cephalad (e.g., shoulder shrug).
   b. **Depression** lowers or moves a structure caudally (e.g., directing the eyes downward).
5. **Protraction/retraction** are generally outward and inward movements.
   a. **Protraction** moves a structure anteriorly (e.g., jutting out the jaw or rolling the shoulder forward).
   b. **Retraction** moves a structure toward the median (e.g., withdrawing a protracted tongue into the oral cavity).
6. **Pronation/supination** generally refers to the arm.
   a. **Pronation,** for example, is a medial rotation so that the palm faces posteriorly.
   b. **Supination** is lateral rotation so that the palm faces anteriorly.
7. **Inversion/eversion** generally refers to the foot.
   a. **Inversion** rotates the plantar surface inward.
   b. **Eversion** rotates the plantar surface laterally.
8. **Specific movements** of particular structures.
   a. **Intorsion/extorsion** of the eye refers to rotation about an axis through the pupil with the top of the eye as the reference.
   b. **Opposition/reposition** of the thumb is a uniquely human characteristic, which refers to a rotation about a resultant axis.
   c. **Circumduction** is a combined movement involving two pairs of movements, such as flexion/extension with abduction/adduction.

E. **Anatomic vocabulary**—the words and language of medicine. Although the vocabulary of anatomy exceeds 5000 terms, usually there is a rationale to each term that makes it easy to remember.

1. **Nomina Anatomica.** Guidelines for the naming of anatomic structures have been established by the International Anatomical Nomenclature Committee.
   a. There shall be only one name for each structure with no alternatives; eponyms shall be discarded.
   b. Names shall be in Latin wherever practical.
   c. Terms shall be short, simple, and informative.
   d. Spatially related structures shall have similar names wherever possible (e.g., femoral artery, femoral vein, femoral nerve, femoral ring, femoral canal).
   e. Differentiating adjectives shall be arranged as opposites (e.g., major–minor; medial–lateral).
2. **Practical nomenclature.** Structures have been named in a number of ways over the past two and one-half millennia.
   a. **In ancient or obsolete languages** (e.g., *esophagus,* Greek; *ileum,* Latin; *liver,* Anglo-Saxon)
      (1) The oldest terms are found in the writings of Hippocrates of Cos.
      (2) Usually the name translates into a meaningful description [e.g., *duodenum,* Latin, 12 (finger breadths) long].
      (3) Many of the original names have been transliterated into modern English (e.g., arteria profunda brachii becomes deep brachial artery).
   b. **By descriptive terms** (e.g., vermiform appendix, meaning wormlike)
   c. **According to the relative position** in the body (e.g., external intercostal muscle)
   d. **According to function** (e.g., levator scapulae muscle)
   e. **By eponyms** associated with mythology (Achilles tendon), the first person to describe the structure (circle of Willis), or the first person to associate the structure with a malformation or disease state (Hunter's canal)
      (1) Most anatomy texts, in agreement with the Nomina Anatomica, avoid eponyms.
      (2) Because the clinical use of many eponyms persists and because eponyms re-

flect the rich history of anatomy and medicine, the more common eponyms are acknowledged herein.
3. Competence with anatomic vocabulary is only the first step in the mastery of anatomy. Nomenclature aside, the study of anatomy is best approached with logic rather than rote memory.
   a. **Comprehension of anatomy** to the point of being able to use it as a medical tool requires integration of structure as it relates to function.
      (1) An understanding of structure often leads to an understanding of function.
      (2) An understanding of function often logically justifies the structure.
   b. **Understanding the development** of a structure often clarifies complex relations (e.g., the innervation of the diaphragm or the basis for referred pain).
   c. **An anatomic principal** is frequently the basis for the diagnosis or choice of treatment for a clinical problem. Anatomy establishes the foundation for clinical observation, physical examination, the interpretation of clinical signs and findings as well as the basis for treatment.

Chapter 2

Integument and Mammary Glands

I. INTEGUMENT (SKIN)

A. **Surface area.** The integument may be considered the largest organ of the body, with a surface area somewhat less than 2 m².

B. **The functions** of the skin are multiple.
1. **Protection**
   a. Preventing fluid loss (the greatest problem in burn patients)
   b. Reducing abrasive trauma
2. **Sensation** mediated by general sensory afferent nerve endings (pain, touch, and temperature)
3. **Secretion**
   a. Sweat glands for temperature regulation mediated by general visceral efferent (sympathetic) nerves
   b. Mammary glands, which are modified sweat glands, under the primary control of endocrine hormones
4. **Thermoregulation** by sweating and control of blood flow

C. **Divisions.** The integument consists of two principal layers.
1. **Epidermis**
   a. The epidermis is a **superficial cellular layer** of stratified epithelium.
   b. It is between 20 and 1400 μm thick, depending on location.
2. **Dermis**
   a. This layer of loose, irregularly arranged connective tissue underlies and supports the epidermis.
   b. It is between 400 and 2500 μm thick, depending on location.
   c. It contains accessory structures, such as hair follicles, sweat glands, mammary glands, blood vessels, lymph vessels, nerves, and special nerve endings.

D. **Cleavage lines** (of Langer) are important surgical considerations (Figure 2-1).
1. **The basis for Langer's lines** is the prevailing directionality of the connective tissue bundles.
   a. The meshwork of collagen fibers (elastic fibers to some extent) provides overall mobility of the skin. Although the connective tissue fibers in the dermis appear randomly oriented, there is a prevailing direction in each area of the body, denoted by the crease, or Langer's, lines.
   b. The elasticity of the collagen fibers creates a slight tension. When connective tissue fibers are severed, they retract so that wounds gape.
      (1) **Incisions made across Langer's lines** retract considerably, resulting in gaping wounds and prominent scar formation.
      (2) **Incisions made parallel to Langer's lines** sever fewer connective tissue fibers, thereby decreasing the tendency to retract and resulting in less unsightly scars.
2. **The direction of Langer's lines on the thorax,** the base of the neck, and the abdominal wall are characteristic.
   a. They tend to be circumferential on the chest and abdomen and longitudinal along the extremities, but are along crease lines at joints. On the female breast, Langer's lines tend to be circumferential to the nipple, an important consideration given the frequency of tumor biopsy and cosmetic mammaplasty.

FIGURE 2-1. Cleavage lines of the skin. *(A)* Anterior surface; *(B)* posterior surface.

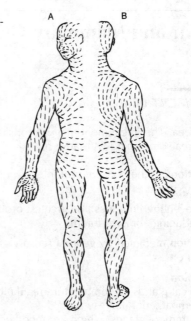

   b. The primary surgical consideration is always adequate exposure; cosmetic considerations are secondary.

E. Innervation

1. **Segmentation.** The skin is innervated in a segmental pattern. All chordates exhibit somatic segmentation (*soma*, G. body) or metamerism. The adult human body retains a segmented vertebral column.
   a. As a first approximation, the body may be thought of as developing from a series of 42–46 identical somites (segments).
   b. With subsequent differential growth and development, segments may lose individual identity, such as in parts of the head, or develop extensively, such as the cervical and lumbosacral contributions to the extremities.
2. **Segmental innervation.** Between each two vertebrae, a left and right mixed (afferent plus efferent) spinal nerve arises to supply a **dermatome** and a **myotome.**
   a. A **dermatome** is a discrete region of skin of the body wall or limb that originates from a segment (see Figure 4-3).
   (1) The superficial branches of each spinal nerve terminate in nerve endings of corresponding dermatomes. Adjacent dermatomes usually overlap somewhat. Understanding the dermatomal arrangement is essential to the interpretation of the findings of a physical examination.
   (2) **Reference landmarks** include the **nipple** and **umbilicus,** which are in the fourth thoracic and tenth thoracic dermatomes, respectively.
   b. A **myotome** is a discrete group of muscles of the body wall or limb that originates from a segment (see Figure 4-3).

F. Clinical significance

1. **General medicine:** manifestations of systemic disease, such as vasoconstriction (cold and clammy skin), vasodilation (flushed skin), eruptions or rash, petechiae, ecchymoses, and edema
2. **General and plastic surgery:** cosmetic incisions, skin grafts, and loss of body fluids in severe burns
3. **Dermatology:** skin disease
4. **Neurology:** manifestations of neurologic dysfunction and disease

II. FASCIAL LAYERS

A. **Fascia** is a plane of loose, irregularly arranged connective tissue composed of fibroblasts, collagen bundles, and some elastic fibers.

1. Fibroblasts secrete tropocollagen, which aggregates into liquid-crystalline collagen bundles.
2. The elasticity of the collagen in fascial planes permits adjacent muscles to shorten and lengthen independently while sliding past each other.

B. **Fascial divisions.** Fascia is divided into superficial and deep layers (Figure 2-2).

1. **Superficial fascia (subcutaneous tissue)** underlies the integument. It is relatively mobile in most regions of the body. Notable exceptions include the palms and soles. Two distinct layers of superficial fascia lie between the dermis and the deep (investing) fascia (see Figure 2-2).
   a. **Superficial layer** (Camper's fascia in the thorax and abdomen, Cruveilhier's fascia in the perineum)
      (1) This predominantly fatty layer does not hold sutures.
      (2) It contains the superficial arteries, veins, lymph vessels, and nerves.
      (3) Its fat content varies depending on genotype, phenotype, and location. It is particularly sensitive to estrogenic hormones. In the abdomen of overweight persons, it forms the **panniculus adiposus.**
   b. **Deep layer** (Scarpa's fascia in the thorax and abdomen, Colles' fascia in the perineum)
      (1) This layer is membranous and relatively thin, but it holds sutures well because of its high collagen content.
      (2) It fuses with the deep fascia.
2. **Deep (investing) fascia** defines fascial planes between muscles. This membranous layer cannot be stripped completely from the structures that it invests (i.e., it becomes continuous with the periosteum, perichondrium, perineurium, perimysium, and other adventitial layers) (see Figure 2-2). With the aid of extracellular fluid, this tissue provides nearly frictionless surfaces for the motion of one muscle over another.
   a. **Organization**
      (1) **Outer investing fascia** overlies the musculature beneath the deep layer of superficial fascia.

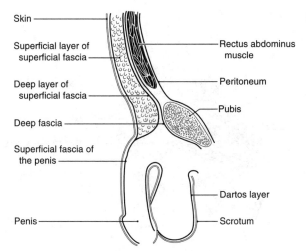

FIGURE 2-2. Fascial layers.

(2) **Inner investing fascia** underlies the musculature of the body and supports the transversalis and endopelvic fasciae.
(3) **Intermediate investing fascia** consists of septa that arise from the outer and inner investing fasciae that course between and around individual muscles as well as neurovascular structures.
   b. **Specialization**
      (1) **Retinacula** are strong fascial bands in the regions of joints that prevent tendons from "bow-stringing" away from the joint.
      (2) **Bursae** are fluid-filled cavities between or within fascial planes that reduce friction between tendons, muscles, ligaments, and bones.
      (3) **Synovial tendon sheaths** are fluid-filled tunnels about the muscle tendons that permit considerable movement and reduce friction.

C. **Clinical considerations**
   1. **Dissection and extravasation.** Fascial planes are opened easily by blunt dissection and by extravasation of fluid, such as blood, edema, pus, or urine.
   2. **Spread of infection** across fascial planes is limited, although these layers are potential pathways for longitudinal spread of infection or extravasation of fluids.
      a. A classic, albeit rare, example is the spread of tuberculosis of the lumbar vertebrae beneath the psoas fascia, resulting in an infection in the femoral triangle.
      b. A more common example is the spread of infection, between fascial layers, from the oropharynx deep into the mediastinum.
      c. Urine from a torn urethra in a male patient can extravasate between fascial planes as far superiorly as the abdominal wall.
      d. Spread of perineal abscesses is limited by planes of perineal fascia.

III. MAMMARY GLAND OR BREAST

A. **Structural considerations.** Because of the contour of the thoracic wall and the teardrop shape of the breast with the axillary tail, about 75% of mammary tissue is lateral to the nipple.
   1. **The breast** consists of 15–20 pyramidal lobes of glandular tissue contained entirely within and considered part of the superficial fascia (Figure 2-3).

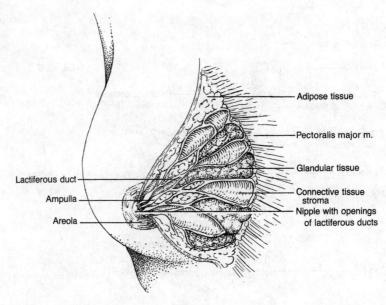

FIGURE 2-3. Partial dissection of the right breast reveals the secretory portion.

a. **Lactiferous ducts** drain the lobes.
b. **Suspensory ligaments** (of Cooper), which are connective tissue septa, define and interconnect the lobes.
   (1) These ligaments attach the breast to the skin and deep layer of the superficial fascia and determine the posture of each breast.
   (2) Tumors (benign or malignant) often displace a suspensory ligament, causing retraction of the breast surface, which is usually a diagnostic sign.
c. **Glandular tissue** of each lobe is surrounded by varying amounts of fat that determine the shape of each breast. Both are under hormonal control.
   (1) The interlobular fat is particularly sensitive to estrogen titers. Thus, the breast undergoes subtle changes during the menstrual cycle.
   (2) Estrogen and progesterone play a role in the proliferation of glandular tissue during pregnancy. Prolactin is involved in secretory activation of glandular tissue.
d. **The lobes** are only slightly developed in the male.

2. **The nipple, or mammary papilla,** receives 15–20 lactiferous ducts and is variable in size.
   a. **Each lactiferous duct** opens separately onto the mammary papilla (nipple).
   b. A large or erected nipple may be mistaken radiographically for a small pulmonary mass.

3. **The areola,** a continuation of pigmented skin beyond the nipple, contains smooth muscle fibers (see Figure 2-3).
   a. Numerous sebaceous glands lubricate the areola. These glands protect it during nursing.
   b. The areola typically darkens irreversibly a few months after the first conception.

4. **The axillary tail** is a normal extension of breast tissue toward or into the axilla.

B. **Blood supply and lymphatic drainage** of the mammary gland are especially important because of the high incidence of malignant tumors of this organ.

1. **Blood supply.** Lateral parts of the breast are supplied by the mammary branch of the **lateral thoracic artery.**
   a. The smaller medial portion of the breast receives blood from anterior intercostal branches of the **internal intercostal artery,** which are branches of the **internal thoracic artery.**
   b. Additional blood is supplied through **posterior intercostal arteries** and the **thoracoacromial trunk.**

2. **Lymphatic drainage.** Lymphatic pathways parallel the vasculature.
   a. Lateral and inferior portions, approximately 75% of the breast, drain along the thoracoacromial and lateral thoracic vessels toward the **axillary nodes.**
   b. Medial portions tend to drain along anterior intercostal vessels toward internal **thoracic (parasternal) nodes.** These nodes lie in a chain along the internal thoracic artery deep to the costal cartilages and parallel to the sternum.
   c. The small superior portion tends to drain toward the **supraclavicular nodes.**
   d. Superficial lymphatics may drain across the midline to the contralateral breast or along the anterior abdominal wall.

C. **Anomalies**

1. **Variation.** Beyond an extensive normal range, the breasts may be over- or underdeveloped, either unilaterally or bilaterally. Such conditions are amenable to mammaplasty.

2. **Accessory breasts (polymastia)** and **supernumerary nipples (polythelia)** may develop along the "milk line" from the axilla to the groin.

3. **Gynecomastia** (abnormal bilateral hypertrophy or hyperplasia of the male breasts) is usually associated with endocrine disorders, impaired liver function, or use of certain drugs.

Chapter 3

Musculoskeletal System

I. SKELETON

A. **Bone,** a calcified connective tissue, forms most of the adult skeleton, which consists of approximately 206 bones. The **axial skeleton** comprises the bones of the head, the bones of the vertebral column, the ribs, and the sternum. The **appendicular skeleton** comprises the bones of the extremities.

1. **Basic functions** include support, mechanical advantage, and a reservoir for ions.
   a. Bone **supports and protects** certain internal organs.
   b. Bones act as **biomechanical levers** on which muscles act to produce motion.
   c. As a tissue, bone is in dynamic equilibrium with its bathing medium and serves as **a reservoir of ions** (Ca^{2+}, PO_4^-, and CO_3^-) in mineral homeostasis.
   d. The bone marrow in the adult is **hemopoietic** (i.e., the source of red blood cells, granular white blood cells, and platelets).

2. **Composition.** Bone is composed of living cells and an organic intercellular matrix with an inorganic component.
   a. Connective tissue cells become **osteocytes** as the collagen meshwork that they secrete undergoes calcification and ossification.
   b. The **collagenous matrix** provides tensile strength. If the mineral content of bone is removed by acid, the remaining collagenous meshwork is flexible but not particularly extensile.
   c. The **mineral content,** a crystalline hydroxyapatite complex of calcium, provides shear strength and compressive strength. If the collagenous matrix is removed by incineration, the remaining inorganic matrix is brittle.

3. **Bone types**
   a. **Long bones,** such as those of the arms and legs (Figure 3-1):
      (1) Develop by replacement of hyaline cartilage
      (2) Usually provide the levers for movement
      (3) Receive blood supply through one or more nutrient foramina
      (4) Have structurally distinct regions (see Figure 3-1)
         (a) **The diaphysis** (shaft) is composed of a thick collar of dense compact bone (cortical bone) beneath which is a thin layer of spongy (trabecular) bone adjacent to the marrow cavity.
            (i) The microstructure of cortical bone is arranged for maximal strength.
            (ii) Growth in thickness occurs by circumferential apposition of bone.
         (b) **The metaphyses** (ends) are composed of a trabecular bony meshwork surrounded by a thinner collar of compact bone. Spicules are arranged along the lines of stress.
         (c) **The epiphyses,** toward the ends of long bones, are separated from the metaphyses in a young person by cartilaginous growth plates, the **epiphyseal disks.**
            (i) Longitudinal growth occurs proximally and distally from the epiphyseal plate by cartilage proliferation, calcification, and remodeling.
            (ii) About the time of puberty, the epiphyseal disks become ossified and longitudinal growth ceases.
   b. **Flat (squamous) bones,** such as ribs and the bones of the cranium:
      (1) Develop by replacement of connective tissue
      (2) Generally serve protective or reinforcement functions
      (3) Consist of two plates of compact bone separated by spongy bone (**diploe**) that bridges the marrow cavity
   c. **Sesamoid bones,** such as the patella and the pisiform bone:
      (1) Develop within tendons

FIGURE 3-1. A tibia, partially cut open, reveals the structure and internal organization of a long bone. The epiphyseal lines represent fusion of the growth plates. The trabeculae of the spongy bone of the metaphyses align along the lines of force and transmit these forces to the compact bone of the diaphyseal shaft. The principal blood supply to the marrow cavity enters via a nutrient foramen.

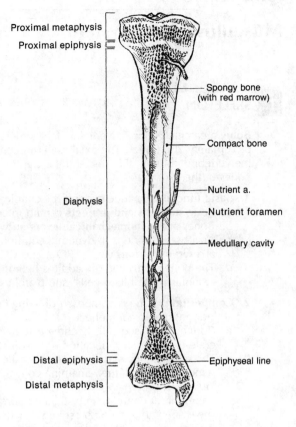

  (2) Reduce attrition on the tendon
  (3) Increase the lever arm of the muscle by moving the tendon away from the fulcrum
4. **Functional and clinical considerations.** As a living tissue, bone is in dynamic equilibrium with the body fluids and reacts to external forces. These two factors interrelate in the process of bone remodeling and healing.
  a. **Remodeling** occurs constantly along the entire length of the bone under the effects of gravity and exertional stress.
    (1) **Mechanism**
      (a) The trabecular meshwork comprising the internal structure of the metaphyses as well as the collagenous substructure of compact bone of the diaphysis develop along the lines of force (compression and stress).
      (b) Lines of force are established by the mass of the body in response to gravitational force and voluntary exertion.
    (2) **Dynamics**
      (a) As the forces acting on bone change over time (e.g., the result of a change in body weight, pregnancy, or exertion), bone undergoes subtle remodeling to maximize intrinsic strength.
      (b) The constant pull of a muscle at its sites of attachment produces ridges, crests, tubercles, and trochanters by the remodeling reaction.
      (c) If bone is not stressed, because of illness, injury, or the weightlessness of outer space, calcium is rapidly resorbed.
  b. **Fractures** are classified according to displacement (nondisplaced or displaced), whether there is compression of the bone fragments (comminution), and whether the skin is torn by displacement (compound fracture).
    (1) Fracture of long bones may result in loss of integrity of the lever arm. Fracture of flat bones results in loss of protective function.

(2) Fractures are visible radiographically. Small fractures with no compression or displacement, however, may not be apparent radiographically until some bone resorption occurs, usually in about 1 week.

(3) Fracture of the cartilaginous epiphyseal plate in a young person is difficult to detect unless there is compression or displacement. Healing of epiphyseal fractures in growing young persons can interfere with subsequent growth.

(4) Fragments of fractured long bones may jeopardize adjacent soft tissues, especially neurovascular bundles.

(5) Cranial fractures may compress or lacerate the nerve or vessels that pass through bony canals and foramina in the flat bones. Such fractures also may result in compression or laceration of the brain.

(6) Aseptic necrosis is a possible complication. If a single major nutrient foramen is involved, a fracture may isolate a significant part of a long bone from its blood supply, with resultant avascular necrosis.

c. **Bone healing** involves three distinct stages.
   (1) **Stage I: Callus formation** occurs within 2 weeks.
      (a) Fibroblasts peripheral to bone in the region of the fracture proliferate and secrete a surrounding collar of collagen, the callus.
      (b) The callus calcifies, providing an internal splint for the fracture.
   (2) **Stage II: Connective tissue calcification** occurs by the sixth week.
      (a) Some bone resorption on either side of the fracture is followed by unorganized proliferation of connective tissue across the fracture.
      (b) The unorganized connective tissue in the fracture calcifies within 6 weeks, thereby healing the fracture.
   (3) **Stage III: Remodeling** requires several months.
      (a) Remodeling occurs in the region of the fracture as well as in adjacent regions.
      (b) In the young person, little evidence remains of the fracture.

B. **Cartilage,** a dense irregular connective tissue, accounts for a small part of the skeleton. Cartilage is formed by living cells and the intercellular matrix that they secrete. It is essentially avascular. The composition of the intercellular matrix determines the type of cartilage.

1. **Hyaline cartilage** has an intercellular matrix especially rich in hyaluronic acid and mucopolysaccharides, which are natural lubricants.
   a. It forms the **chondral (anterior) portion** of most ribs to complete the rib cage and to provide the resiliency necessary for ventilation.
   b. The **articular cartilage** in most joints is of this type.
   c. It provides the anlage for long bone development.

2. **Fibrocartilage** has an intercellular matrix rich in mucopolysaccharides and bundles of **collagenous fibers.**
   a. The osmotic pressure and water content are high as a result of a high concentration of mucopolysaccharides.
   b. This type of cartilage is especially resilient and durable.
   c. Fibrocartilage forms most **symphyses,** such as the pubic symphysis and intervertebral joints, as well as certain joint disks, such as that of the temporomandibular joint.

3. **Elastic cartilage** has an intercellular matrix rich in mucopolysaccharides and bundles of **elastic fibers,** which provide strong, yet flexible, support. It forms the skeletal structure of the external ear and the tip of the nose.

C. **Articulations.** Most bones articulate with each other at joints (Figure 3-2).

1. **Types** of articulations or joints
   a. **Synarthroses,** or fibrous joints, are barely movable or immovable. Examples include:
      (1) **Suture** (e.g., sagittal suture; see Figure 3-2A)
      (2) **Syndesmosis** (e.g., tibiofibular joint, see Figure 3-2B)

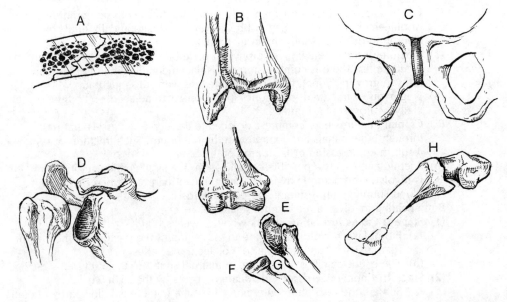

FIGURE 3-2. Articulations. *(A)* Cranial suture; *(B)* tibiofibular syndesmosis; *(C)* pubic symphysis; *(D)* glenohumeral joint (ball-and-socket with three degrees of freedom); *(E)* humeroulnar joint (hinge with one degree of freedom); *(F)* humeroradial joint (ball- and-pivot with two degrees of freedom); *(G)* radioulnar (gliding pivot with one degree of freedom); *(H)* first carpometacarpal joint (saddle with two degrees of freedom).

        (3) **Gomphosis** (tooth joint)
    b. **Amphiarthroses,** or cartilaginous joints, allow limited motion. Examples include the pubic symphysis and intervertebral disks (see Figure 3-2C).
    c. **Diarthroses,** or synovial joints, permit relatively free motion about at least one axis of rotation, as in the shoulder, elbow, and hand (see Figure 3-2D–H).
        (1) **Characteristics** of diarthrodial joints
            (a) A **joint or synovial capsule** reinforces and supports the joint.
            (b) **Articular cartilage** covers the articular surfaces of the bones. Articular cartilage usually is hyaline cartilage but may be fibrous cartilage.
                (i) Because articular cartilage is somewhat fluid, it can change shape so that mechanical forces are distributed over the largest possible surface area within the joint.
                (ii) The degeneration of articular cartilage, **degenerative arthritis,** results in loss of the smooth gliding surface and pain accompanying joint motion.
            (c) The two articular surfaces have reciprocally concave and convex contours.
            (d) **Synovial membrane** encloses the joint and secretes synovial fluid.
                (i) The **synovial cavity** between the synovial membrane and the bone or cartilage is filled with synovial fluid.
                (ii) **Synovial fluid,** rich in hyaluronic acid, lubricates the articular surfaces.
        (2) Diarthrodial joints act as **fulcrums** for bony levers. Muscles act on the lever to produce motion about the fulcrum (Figure 3-3).
            (a) Motions at all diarthroses can be reduced to rotations in one or more mutually perpendicular planes.
            (b) **Paired motions.** Relative movements of bones at common articulations are given pairs of descriptive names to describe motions that occur about a single axis of rotation.
                (i) **Flexion/extension**

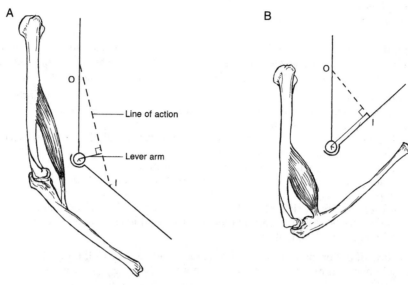

FIGURE 3-3. Musculoskeletal action involving the brachialis muscle. *(A)* Fully stretched. Although intrinsically most powerful in this position, the brachialis has a short lever arm, which provides little mechanical advantage. *(B)* On isotonic contraction. Although intrinsically less powerful when shortened, the muscle has a longer lever arm, which provides greater mechanical advantage.

      (ii) Abduction/adduction, sometimes, elevation/depression
      (iii) Internal rotation/external rotation
      (iv) Circumduction, which is complex movement produced by a combination of flexion/extension and abduction/adduction
  (3) All movement in this type of joint takes place about an **axis of rotation.**
    (a) Joints can have one, two, or three axes of rotation with one degree of freedom about each axis of rotation.
    (b) The pairs of motions that diarthroses permit vary in number. For example, the humeroulnar joint (see Figure 3-2E) and radioulnar joints (see Figure 3-2G) permit only one pair of motions (one degree of freedom), the humeroradial joint (see Figure 3-2F) and the first carpometacarpal joint (see Figure 3-2H) permit two degrees of freedom, and the shoulder joint (see Figure 3-2D) permits three degrees of freedom.
2. **Ligaments** are bands of dense, regularly arranged connective tissue that cross joints and frequently form an articular capsule about the joint. They are embedded in bone on either side of a joint by Sharpey's fibers.
  a. Function
    (1) Connect the bones
    (2) Reinforce the articulations
    (3) Contribute to joint stability
  b. Torn ligaments, sometimes referred to as **sprains,** present difficult clinical problems.
    (1) Because ligaments, unlike bone, are relatively avascular, tears heal slowly.
    (2) When a ligament is detached from bone, its fibers do not grow back into the bone cortex as extensively as before the injury; thus, a healed ligament is usually weaker, predisposing to subsequent re-injury.
    (3) Torn ligaments destabilize the joint and predispose it to dislocation.
      (a) Loss of joint stability is a sign of a torn ligament.
      (b) Although ligaments usually cannot be visualized radiographically, displacement of bones at an articulation is a sign of tearing.

II. MUSCLE

A. Composition.
Muscles are composed of bundles of muscle cells (muscle fibers), which are contractile. **Skeletal muscle** runs between two points of attachment on related bones separated by a considerable distance. About 48% of the body mass is muscle.

1. **Tendons** (bundles) and **aponeuroses** (sheets) are formed by dense, regularly arranged connective tissue into which each end of a muscle inserts, and which, in turn, attach to the outer layer of periosteum.
   a. The more peripheral attachment site of a muscle usually is termed the **origin;** the more distal attachment is the **insertion.**
   b. Tendons may continue into the muscle as **septa.**
   c. A **sesamoid bone** may form in the area of a tendon that is subjected to intense friction.
   d. Muscle pulling at sites of tendinous attachment produce remodeling reaction within the bones. Thus, **attachments sites** are indicated by osseous ridges, crests, tubercles, and trochanters.
2. **The line of action** of a muscle is the line that best describes the mean direction of the muscle between two attachments (see Figure 3-3A).
3. **The lever arm** of a muscle is a line drawn perpendicular to the line of action through the axis of rotation of the joint (see Figure 3-3A).
4. **Arrangement of fascicles** (groups of muscle fibers) within the muscle
   a. **Fusiform.** Fascicles of muscle fibers lie parallel to the line of action along the long axis of the muscle (e.g., the sartorius muscle).
   b. **Pennate.** Fascicles lie at an angle to the long axis of the muscle. Vector analysis demonstrates that the same amount of contraction produces slower, more forceful movement in a pennate muscle.
      (1) **Unipennate.** Fascicles lie at the same angle on one side of the tendon of insertion (e.g., the flexor pollicis longus muscle).
      (2) **Bipennate.** Fascicles lie at an angle on both sides of a single tendinous septum (e.g., the soleus muscle).
      (3) **Multipennate.** Fascicles reach more than one tendinous septa from several directions (e.g., the deltoid muscle).

B. Action

1. **Contraction.** The type of muscle contraction depends on whether or not the joint moves.
   a. **Isometric contraction.** Muscles exert force without producing rotation at the joint (e.g., the elbow flexors trying to lift a weight that is too heavy to move).
   b. **Isotonic contraction.** Muscles shorten to produce motion (e.g., the elbow flexors lifting a manageable weight).
2. **Movement.** By shortening, muscles act on bony levers to produce motion in one or both bones to which they attach (see Figure 3-3A, B).
   a. Principles determining movement
      (1) All muscles exert equal and opposite tension at both attachments.
      (2) Any muscle with a line of action that crosses an unconstrained axis of rotation at a joint must produce movement at that joint.
      (3) Movement at a joint is determined by the sum of the activity of all the muscles the line of action of which cross the axis of rotation.
      (4) The bone that is least stabilized must move.
   b. **Strength** of a muscle is the product of its cross-sectional area (4 kg/cm^2) and mechanical advantage (the length of the lever arm).
3. **Group actions.** Regardless of the specific innervation, some muscles act together as synergists to produce a specific motion; other muscles act together as antagonists to oppose that motion. Synergists and antagonists may act in concert to produce **dynamic stability** at a joint.

a. **Synergistic muscles** cross the *same side* of the axis of rotation; **antagonistic muscles** pass over the *opposite side* of the axis of rotation.
   (1) In most instances, motion at a joint is initiated by one set of synergistic muscles and brought to a close by the antagonists. For example, controlled flexion of the forearm at the elbow joint is initiated by flexor muscles and slowed or stopped at any desired position by extensor muscles.
   (2) Simultaneous contraction of both synergists and antagonists produces maximal joint stability (**dynamic stability**) with little or no movement.
b. Because muscle can only shorten actively, lengthening of a muscle requires contraction of the antagonists on the opposite side of the axis of rotation.

C. Function

1. **Force-generating capacity of muscle** is a function of muscle stretch rather than overall muscle length.
   a. A **force–length relation** has its morphologic basis in the ultrastructure of the muscle fibers.
      (1) **Maximum isometric force** is produced when muscle is stretched to its rest length (i.e., its maximum working length). For a flexor, this point is at full extension, for an extensor, at full flexion.
      (2) **Minimum isometric force** occurs in maximally shortened muscle in situ, which is approximately one-half of the stretched length in fusiform muscles.
   b. A **force–area relation** limits maximal force to 4 kg/cm^2. The greater the cross-sectional area of a muscle, the greater the number of contractile filaments acting in parallel, and the greater the force-generating capacity.
      (1) Muscle use stimulates synthesis of contractile filaments, resulting in muscular **hypertrophy** (enlargement).
      (2) Disuse or paralysis has the opposite effect, resulting in muscular **atrophy.**

2. **The mechanical advantage** of a muscle is a function of the length of the lever arm. As the joint is flexed, the length of the lever arm of a muscle changes (see Figure 3-3A).
   a. On full extension of the elbow, the lever arm of the flexor muscle is relatively short because the joint angle is nearly 180° and the flexor line of action lies close to the axis of rotation. The mechanical advantage of a short lever arm is minimal.
   b. As the elbow is flexed to 90°, the lever arm of the flexor is longer because the line of action of the flexor has moved away from the axis of rotation. The mechanical advantage is maximal.
   c. With additional flexion of the elbow, the lever arm again becomes shorter and the mechanical advantage decreases (see Figure 3-3B).

3. **Effective strength** is a function of pennation, the force–length relation, and the lever arm.
   a. Although the intrinsic force-generating capacity of the flexors is maximal in the fully extended arm, the lever arm of the flexor is minimal. Hence, this muscle is strongest at the point of least mechanical advantage.
   b. As the arm flexes, the force–generating capacity diminishes because of the force–length relationship, but the mechanical advantage increases as the length of the lever arm increases. Hence, as this muscle weakens, its mechanical advantage increases. The net effect is the maintenance of fairly constant strength.
   c. Beyond 90° of flexion, both the force-generating capacity of the flexor muscle and the length of the lever arm diminish. Hence, the muscle becomes weaker and loses its mechanical advantage, with a rapid loss of strength.

D. Nervous control of musculoskeletal movement (see Figure 4-1)

1. **Initiation of movement.** Physiologic recording demonstrates that electrical excitation passes along nerves to the muscles. This event is the basis for **diagnostic nerve conduction studies.**

2. **Excitation–contraction coupling.** Nerve electrical activity causes release of a neurotransmitter at the neuromuscular junctions, initiating electrical excitation along the

muscle fibers and inducing contraction. This sequence of events is the basis for diagnostic **electromyography.**

3. **Quiet phase.** When gravity, friction, and inertia are overcome, the nerve becomes relatively silent, contraction ceases, and motion continues because of inertia.
4. **To halt motion,** the nerves to the antagonistic muscles become active, and the antagonists contract sufficiently to control and cease movement.
5. **Voluntary control** of muscle is from the higher centers of the brain. Reflex control and muscle tone are accomplished by neurons within the spinal cord.

Chapter 4

Nervous System

I. INTRODUCTION

A. **The nervous system** is a complex organ system. It provides a mechanism for monitoring ever-changing external and internal environments.

1. **Sensory.** Signals originating in the sensory receptors are monitored by, processed in, and transmitted through the nervous system.
   a. Sensation is mediated through the **somatic sensory system** and the **visceral sensory system.**
   b. These inputs may reach the conscious sphere or may be used at subconscious and reflex levels.
2. **Somatic motor.** The nervous system controls and integrates various parts of the body (Figure 4-1).
   a. Neural messages modulating and regulating motor activity are processed in and conveyed through the nervous system to **voluntary muscles**.
   b. Output may be voluntary or the result of reflex mechanisms.
3. **Visceral motor.** The nervous system maintains the internal environment within narrow limits mediated through the **autonomic motor system.**
   a. Neural messages modulating and regulating visceral activity are processed in and conveyed through the nervous system to **involuntary muscles** and **glands.**
   b. These outputs may be involuntary or the result of reflex mechanisms.

B. **Neurons (nerve cells),** the principal units of the nervous system, typically have two types of processes.

1. **Dendrites** are afferent processes that typically receive synaptic contact from receptor cells or the axons of other neurons.
2. **The axon** is the single efferent process through which each neuron communicates with other neurons and effectors (i.e., muscles and glands).
   a. **Axons** may be wrapped in **myelin** (multiple layers of membranes), which insulate the axon, thereby promoting conduction.
   b. **Nerves** are formed by groups of axons.

C. **Subdivisions.** The bilaterally symmetric nervous system is subdivided **anatomically** into the central nervous system and the peripheral nervous system as well as **functionally** into the somatic nervous system and the visceral nervous system. Each of these subdivisions has sensory and motor components (see Figure 4-1).

1. **Central nervous system (CNS)**
   a. The CNS is composed of the **brain** and **spinal cord,** which are encapsulated within the skull and vertebral column, respectively.
   b. It is the center of perception and the site of integration of sensory information as well as initiation and coordination of motor activity.
2. **Peripheral nervous system (PNS)**
   a. The PNS includes the **cranial nerves** and **spinal nerves** that arise from the brain and spinal cord, respectively.
   b. It conveys neural impulses:
      (1) To the CNS, as input from the sense organs and sensory receptors of the body
      (2) From the CNS, as output to the muscles and glands of the body
3. **Somatic nervous system**
   a. The **afferent portion** includes the neural structures of the CNS and PNS that are

FIGURE 4-1. Basic organization of the nervous system: somatic sensory input *(above left)*, visceral sensory input *(below left)*, somatic motor output *(above right)*, and visceral motor output *(below right)*.

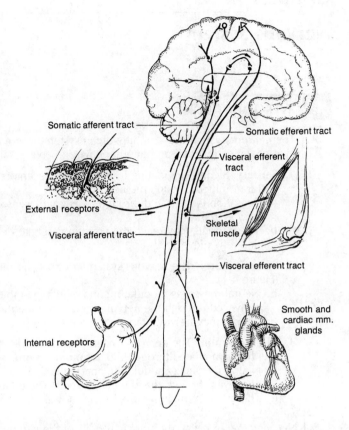

involved in conveying and processing conscious and unconscious sensory information.
   b. The **efferent portion** includes the neural structures of the CNS and PNS that are involved in motor control of voluntary muscle.

4. **Visceral nervous system**
   a. The **afferent portion** includes central and peripheral neural structures that convey sensory information from the splanchnic (visceral) part of the body.
   b. The **efferent portion** constitutes the **autonomic nervous system (ANS).** Although the visceral nervous system is both sensory and motor, the ANS is motor only—a subtle, but important, distinction.
      (1) It is composed of the neural structures of the CNS and PNS that are involved in motor activities influencing the involuntary (smooth) and cardiac musculature and glands of the viscera and skin.
      (2) The ANS is divided into the **sympathetic system** and the **parasympathetic system.**

II. CENTRAL NERVOUS SYSTEM

A. **The brain** is the enlarged, convoluted, and highly developed rostral portion of the central nervous system.

1. **Morphology.** The brain is divided into the **cerebrum, cerebellum,** and **brain stem.**
   a. The average adult human brain weighs about 1400 g, approximately 2% of the total body weight.
   b. The **gray matter** of the brain contains neuron cell bodies that lie principally in

the cortical areas. The **white matter** consists of axons that group to form pathways (tracts) in the central areas.
   c. The surface of the brain is convoluted, which greatly increases its area.
2. **Cranial meninges.** The gelatinous brain is protected by an outer rigid capsule, the bony skull, and is invested by a succession of three connective tissue membranes, the meninges. The **leptomeninx** comprises the softer pia and arachnoid layers; **the pachymenix** is the tough dura.
   a. **Pia mater** is the innermost protective layer.
      (1) This membrane is intimately attached to the brain.
      (2) It contains the small blood vessels that supply the brain and brain stem.
   b. **Arachnoid** is an intermediate layer.
      (1) This thin membranous layer is external to the pia mater and is connected to it by web-like trabeculations, hence its name.
      (2) It supports the large distributing cortical arteries.
      (3) It delimits the **subarachnoid space.**
         (a) The subarachnoid space lies between the arachnoid layer and pia mater.
         (b) It surrounds the brain but does not dip into the sulci.
         (c) This space is filled with **cerebrospinal fluid (CSF)** and may contain blood on hemorrhage of a cerebral artery. The brain floats in CSF, which supports it and acts as a shock absorber when the head moves or is jarred.
   c. **Dura mater** is the external protective layer.
      (1) The dura mater consists of two adherent fibrous membranes (hence its name) external to the arachnoid.
         (a) The **outer dural layer (periosteal dura)** is the periosteum of the cranial vault.
         (b) The **inner dural layer** is the true dura.
         (c) In certain locations, **venous sinuses** (draining blood from the brain) course between the two layers.
      (2) The dura mater defines and delimits the **subdural space.**
         (a) The subdural space is a potential space between the arachnoid and dura.
         (b) It does not contain CSF.
         (c) This space is the location of a **subdural hematoma,** which is usually a low-pressure venous hemorrhage.
      (3) The dura mater also defines the **epidural space.**
         (a) The epidural space is a potential space that lies between the inner and outer layers of the dura mater.
         (b) It contains the meningeal arteries and dural venous sinuses.
         (c) This space is the location of an **epidural hematoma,** which is usually a life-threatening, high-pressure arterial hemorrhage.
3. **Functions.** The brain functions in perception of sensory stimuli, integration and association of stimuli with memory, and neural activity that results in coordinated motor responses to stimuli.

B. **Spinal cord.** This cylindric structure is located in the upper two-thirds of the vertebral canal of the bony vertebral column.

1. **Morphology.** A continuation of the brain stem, the spinal cord extends from the foramen magnum at the base of the skull to its termination as the **conus medullaris,** usually located at the caudal level of the first lumbar vertebra in the adult.
   a. The **gray matter** contains neuron cell bodies and forms the central portion. The **white matter** consists of pathways (tracts) containing the axons and lies peripherally. This relationship is opposite to that found in the brain.
   b. The spinal cord is enlarged in those segments that innervate the extremities.
      (1) **Cervical (brachial) enlargement** extends from spinal levels C5 to T1, the segments that innervate the upper extremities.
      (2) **Lumbosacral enlargement** extends from L3 to S2, the segments that innervate the lower extremities.

c. Law of descent. Because the vertebral column continues to grow after birth and the spinal cord grows little in length, the adult spinal cord (ending at the level of the L1 vertebra) is shorter than the bony vertebral column.
  (1) Spinal nerves emerge from the vertebral column lower than the spinal cord segments from which the corresponding rootlets originate.
  (2) Lumbar and sacral nerves develop long roots that extend from the spinal cord as the **cauda equina** (horse's tail) within the **lumbar cistern** (the dilated caudal termination of the subarachnoid space).

2. **Spinal meninges** are continuous with those of the brain.
   a. **Spinal pia mater** is the most interior protective layer.
      (1) This tissue membrane is intimately attached to the spinal cord and its roots.
      (2) It continues beyond the termination of the spinal cord as the **filum terminale,** which attaches to the sacrum or coccyx.
      (3) It contains the blood supply to the spinal cord.
   b. **Spinal arachnoid** is the intermediate layer.
      (1) This thin membrane is external to the pia mater and is connected to it by web-like trabeculations, hence its name.
      (2) It supports the large distributing vessels.
      (3) It delimits the **subarachnoid space.**
         (a) The subarachnoid space lies between the arachnoid and pia mater. It extends caudally to the level of the second sacral vertebra and is wider between vertebral levels L1 and S1—the **lumbar cistern.**
         (b) It surrounds the spinal cord and its roots.
         (c) It is filled with **CSF.**
            (i) A **lumbar tap** to sample CSF is usually accomplished by inserting a needle in the midline between vertebrae L3 and L4 or L4 and L5. CSF contains blood cerebral artery hemorrhage and many polymorphonuclear leukocytes if bacterial meningitis is present.
            (ii) **Spinal anesthesia** is accomplished by infusing anesthetic about the nerve roots in the lumbar cistern through a needle inserted along the midline between vertebrae L3 and L4 or between L4 and L5.
            (iii) Introduction of a needle above vertebra L3 carries a risk of injury to the spinal cord.
   c. **Spinal dura mater** is the most external protective layer.
      (1) This tough fibrous membrane (hence its name) is external to the arachnoid. Unlike cranial dura, it does not fuse to the vertebral periosteum.
      (2) It delimits the **subdural space.**
         (a) The subdural space is a potential space between the arachnoid and dura.
         (b) It does not contain CSF.
      (3) It defines the **epidural space.**
         (a) The epidural space lies between the dura mater and the periosteum of the vertebral column.
         (b) It contains profuse venous plexuses **(Batson's plexus)** and fat.
         (c) **Epidural (caudal) anesthesia** is accomplished by perfusing the anesthetic agent about the spinal nerves in the epidural space. The approach to the vertebral canal may be between the lower lumbar vertebrae or through the sacral hiatus.

III. PERIPHERAL NERVOUS SYSTEM

A. Definitions

1. **A nerve bundle** or nerve consists of axons. Depending on its location, such a bundle may be called a rootlet, root, trunk, division, cord, ramus, nerve, or branch.
2. **A ganglion** is an aggregation of nerve cells (cell bodies and processes).
3. **A plexus** is a network or interjoining of nerves that may contain one or more ganglia.

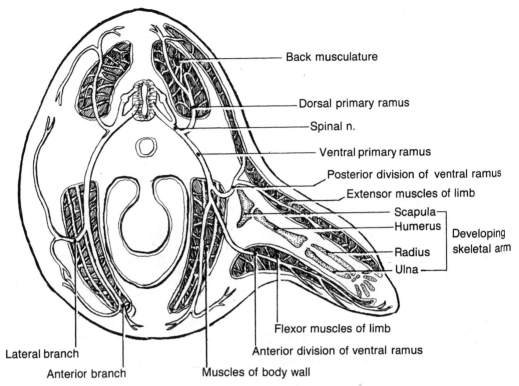

FIGURE 4-2. Organization of the spinal nerve based on development. One side depicts the basic pattern of the spinal nerve with division into dorsal and ventral primary rami and the lateral branch of the ventral ramus. The other side depicts the formation of the brachial or lumbosacral plexus with the lateral branch of the ventral ramus forming the posterior division of the plexus while the continuation of the ventral ramus forms the anterior division of the plexus.

B. Spinal roots. The spinal nerves are formed by **rootlets**.

1. **Rootlets.** Nerve fibers emerge from the spinal cord in a paired, uninterrupted series of dorsal (sensory, afferent) and ventral (motor, efferent) rootlets, which join to form 31 pairs of dorsal and ventral roots (Figure 4-2).

2. **Dorsal (sensory) roots.** Afferent fibers that constitute the dorsal root convey input from the sensory receptors in the body via the spinal nerves to the spinal cord.
   a. The cell bodies of the neurons lie in the **dorsal root ganglion,** which is located within the intervertebral foramen.
   b. Centrally, roots are formed by a series of rootlets that enter the dorsolateral aspect of the spinal cord. A **spinal segment** is that part of the spinal cord that contributes to a spinal nerve.
   c. Peripherally, some fibers of the dorsal root of each spinal nerve supply the sensory innervation to a skin segment known as a **dermatome,** whereas other fibers provide nerve endings to deep structures (Figure 4-3).
      (1) Adjacent dermatomes overlap, and the loss of one dorsal root results in diminished sensation (not a complete loss) in that dermatome.
      (2) There are usually no first cervical (C1) or coccygeal (Cx) dermatomes.
      (3) Anesthesia or paresthesia involving an entire dermatome is indicative of spinal cord injury or root damage. Partial dermatome involvement is indicative of peripheral nerve damage.

3. **Ventral (motor) roots.** The motor nerves, which constitute the ventral root, convey output from the spinal cord to effectors.
   a. **Somatic cell bodies** lie in the **ventral horn** of the spinal cord gray matter.

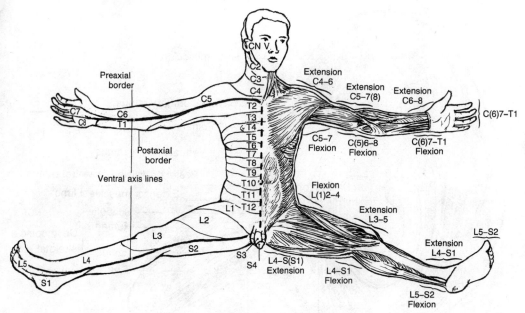

FIGURE 4-3. Cutaneous and muscular nerve distributions in the primitive position; the primitively ventral surfaces are anterior. The basic segmental dermatomal arrangement is depicted with evidence of dermatomal migration into the right extremities during development. The innervating root numbers to the extensor and flexor muscles that act across major joints of the left upper and lower extremities show evidence of myotomal migration.

   (1) Axons innervate voluntary striated muscles (general somatic efferent, GSE).
   (2) The fibers of the ventral root of each spinal nerve supply motor innervation to specific groups of voluntary muscles known as **myotomes** (see Figure 4-3).
      (a) Usually, different myotomes act across opposite sides of a joint. More than one myotome may act across the same side of a joint.
      (b) A central lesion involves all of the muscles of a myotome, whereas a peripheral nerve injury involves only some of these muscles.
   b. **Visceral cell bodies** lie in the **lateral horn** of the spinal cord gray matter.
      (1) These axons synapse with neurons in peripheral ganglia, which, in turn, innervate involuntary smooth muscles and glands (general visceral efferent, GVE).
      (2) In the cranial region, these axons also innervate the striated muscle derived from the branchiomeric (gill) structures (special visceral efferent, SVE).
   c. Some evidence, albeit controversial, suggests that some visceral afferent fibers enter the spinal cord through the ventral root.

C. **Spinal nerves.** In the vicinity of an intervertebral foramen, a dorsal root and a ventral root meet to form a spinal nerve, which supplies the innervation of a segment of the body.

   1. **Organization.** Spinal nerves are numbered in association with vertebral levels.
      a. **Thoracic, lumbar, and sacral nerves** are numbered by the vertebra just rostral to the intervertebral foramen through which they pass (e.g., nerve T4 emerges below vertebra T4).
      b. **Cervical nerves** are numbered by the vertebrae just caudal to them (e.g., nerve C7 is rostral to vertebra C7), except that nerve C8 exits caudal to vertebra C7 and rostral to vertebra T1.
      c. In all, there are 31 pairs of spinal nerves:
         (1) **Eight cervical:** C1–C8. The adult has seven cervical vertebrae, but eight cervical nerves. The first cervical nerve exits above vertebra C1, whereas the eighth cervical nerve exits below vertebra C7.

(2) **Twelve thoracic:** T1–T12
(3) **Five lumbar:** L1–L5
(4) **Five sacral:** S1–S5
(5) **One coccygeal** (occasionally two): Cx1

2. **Subdivisions.** The short mixed spinal nerve divides almost immediately into two primary rami and two secondary rami (see Figure 4-2).
   a. The **dorsal primary ramus** of each spinal nerve arches dorsally to innervate the skin (the posterior portion of the dermatome) and muscles of the back.
   b. The **ventral primary ramus** continues anteriorly to innervate the muscle and skin of the lateral and ventral aspects of the body.
      (1) In the thoracic region, the ventral primary rami are termed **intercostal nerves.**
         (a) The twelfth intercostal nerve is the **subcostal nerve.**
         (b) In the lumbar region, the anterior primary rami form other named nerves (e.g., **iliohypogastric nerve** and **ilioinguinal nerve**).
      (2) When the ventral primary ramus (e.g., intercostal nerve) reaches a point in line with the axilla, it gives off a **lateral cutaneous branch.**
         (a) This branch then penetrates the muscle and divides into **anterior and posterior perforating branches,** which innervate muscle and the lateral portion of the dermatome.
         (b) Homologues of these branches form **posterior divisions** of the brachial plexus and lumbosacral plexus.
      (3) The ventral primary ramus continues anteriorly toward the midline and terminates as the **anterior cutaneous branch.**
         (a) This branch then divides **into lateral and medial perforating branches,** which innervate muscle and the anterior portion of the dermatome.
         (b) Homologues of these branches form the **anterior divisions** of the brachial plexus and lumbosacral plexus.
      (4) The distribution of cutaneous branches of the spinal nerves overlaps so that every region of the body is ensured innervation. The dermatomes also overlap slightly in the anterior and posterior midline.
   c. The **meningeal ramus** is actually the first branch of the spinal nerve.
   d. **Rami communicantes** are connections between the sympathetic division of the autonomic and peripheral nervous systems adjacent to the vertebral column.
      (1) **White rami communicantes** connect the spinal cord and the sympathetic chain between T1 and L2, inclusive (see Figure 4-5).
         (a) They convey sympathetic preganglionic myelinated fibers to the paravertebral and prevertebral ganglia.
         (b) They also convey visceral afferent fibers to the spinal nerve.
      (2) **Gray rami communicantes** connect the sympathetic chain and the spinal nerves at every spinal level (see Figure 4-5).
         (a) They convey sympathetic postsynaptic unmyelinated fibers from the paravertebral ganglia to the spinal nerve to reach the skin.
         (b) They function in vasoconstriction, sweating, and piloerection.

IV. SOMATIC NERVOUS SYSTEM

A. **Somatic afferent nerves.** These nerves convey sensory input from free nerve endings and special receptors to the CNS (Figure 4-4).

1. **Classification**
   a. **General somatic afferent (GSA) nerves** convey pain, temperature, touch, and proprioception from the head, body wall, and extremities to the CNS.
   b. **Special somatic afferent (SSA) nerves** convey vision, hearing, and balance to the CNS.
   c. **General visceral afferent (GVA) nerves** (frequently included in this classification

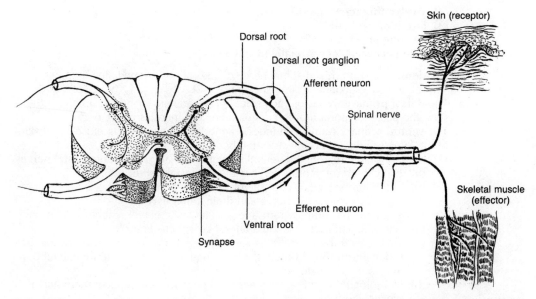

FIGURE 4-4. Somatic afferent and efferent neurons in a reflex arc. Somatic afferent neurons with cell bodies in the dorsal root ganglion convey sensory information from the periphery to the spinal cord. The information is passed through small internuncial neurons in the central gray area to the motor neuron. The efferent neurons with cell bodies in the ventral horn of the spinal cord pass along the spinal nerve to contact effector organs.

for functional reasons) convey sensory information from the visceral organs to the CNS.
   d. **Special visceral afferent (SVA) nerves** (frequently included in this classification for functional reasons) convey smell and taste sensations to the CNS.
2. **Afferent pathway.** One peripheral neuron and at least one central neuron are involved.
   a. **Peripheral neuron**
      (1) The cell bodies of these pseudounipolar sensory neurons lie in the dorsal root ganglia of the spinal nerves or the cranial nerve equivalents.
      (2) The distal end of the axon either is a free nerve ending or makes contact with a specialized receptor; the proximal end of the axon enters the spinal cord through dorsal rootlets.
      (3) The axons may synapse with a central neuron in the gray matter of the spinal cord or run cranially in the white matter to a region of gray matter in the brain stem where synaptic contact is made.
   b. **Central neurons** lie in the gray matter of the spinal cord or brain stem and send axons to higher centers of the brain at which synaptic contact is made.

B. Somatic efferent nerves. These nerves regulate the coordinated muscular activities associated with voluntary motion and the maintenance of posture (see Figure 4-4).
   1. **Classification**
      a. **General somatic efferent (GVE) nerves** supply the muscles of the head, body wall, and extremities, which arise from myotomes of the embryonic somites.
      b. **Special visceral efferent (SVE) nerves** supply the muscles of the head and neck, which arise from branchiomeric (gill) structures.
   2. **Efferent pathway.** Two sets of neurons are involved in producing voluntary somatic motor activity.
      a. **Upper motor neurons** lie in the brain.
         (1) The cell bodies of these neurons lie in the gray matter of the motor areas of the cortex and various nuclei of the brain stem.

(2) The upper motor neuron passes through the cerebrum, brain stem, and white matter of the spinal cord to contact a lower motor neuron.
  b. **Lower motor neurons** lie in the brain stem or spinal cord.
      (1) The cell bodies of these neurons lie in the gray matter of the brain stem for cranial nerves or in the ventral horns of the spinal cord gray matter for the spinal nerves.
      (2) Each somatic lower motor neuron has an axon that courses through a cranial or spinal nerve to make synaptic connections with voluntary muscle fibers at myoneural junctions.
3. **Reflex arcs**
   a. **Simple reflex arcs** are composed of one sensory neuron and one motor neuron.
   b. **Complex reflex arcs** may have one or more interneurons (in the gray matter) intercalated between the sensory and motor neurons.

C. **Visceral afferent (GVA) nerves** are frequently included with somatic afferents, but they are more conveniently discussed in association with the visceral efferent innervation because the pathways are similar.

V. VISCERAL NERVOUS SYSTEM

A. **The autonomic nervous system (ANS)** is composed of the neural structures of the CNS and PNS that are involved with the **motor activities** that influence the involuntary (smooth) and cardiac musculature as well as the glands of the viscera and skin.

1. **Composition.** It is a motor-only system consisting of one upper motor neuron and two lower motor neurons (Figure 4-5).

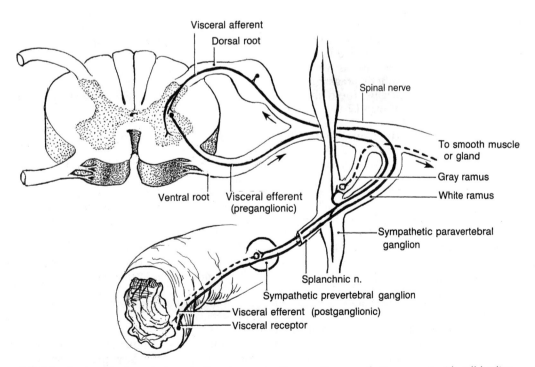

FIGURE 4-5. Autonomic and visceral afferent neurons. Presynaptic sympathetic neurons with cell bodies in the lateral horns of the spinal cord pass along the spinal cord and along a white ramus communicans to reach the sympathetic chain. They then either synapse or pass along a splanchnic nerve to synapse in a pervertebral ganglion. The postganglionic neurons contact the effector organs.

a. The ANS is classified as **general visceral efferent** (GVE) and supplies smooth muscles, cardiac muscle, and glands.
b. It is often called the **visceral (vegetative) motor system** because the effectors are associated with involuntary systems (e.g., cardiovascular, digestive, respiratory, perspiratory, and vasodilatory) over which only minimal direct conscious control can be exerted.

2. Visceral efferent pathway. One central and two peripheral neurons are involved in producing visceral motor activity (see Figure 4-5).
   a. The **upper motor neuron** lies in the brain.
      (1) The cell bodies of these neurons are in the gray matter of the autonomic areas of the brain.
      (2) The axons pass in the descending tracts at the brain stem and spinal cord.
   b. The **preganglionic neuron** lies in the brain stem or in the lateral horn of the spinal cord.
      (1) This neuron, which may be seen as the equivalent of a lower motor neuron, originates in the gray matter of the brain stem or the lateral horn of the spinal cord gray matter.
      (2) Its **myelinated axon** courses through a cranial nerve or spinal nerve.
      (3) It terminates by synapsing with a postganglionic neuron in a ganglion outside the central nervous system.
   c. The **postganglionic neuron** is peripheral and is unique to the ANS.
      (1) The cell body of this neuron is located in an autonomic ganglion.
      (2) The postganglionic neuron has an **unmyelinated axon** that extends peripherally.
      (3) It terminates in endings associated with smooth muscle, cardiac muscle, or glands.

3. Divisions
   a. Sympathetic division. As a rough rule of thumb, the sympathetic system stimulates activities that are mobilized by the organism during emergency and stress situations the so-called **fight, fright, and flight responses.** Other names for this system are the **thoracolumbar system** and the **adrenergic system.**
      (1) Thoracolumbar outflow supplies sympathetic innervation to the entire body. Its functions tend to be excitatory, for example, increasing heart rate and stroke volume as well as sweating.
      (2) Sympathetic preganglionic fibers
         (a) Origins. These fibers originate from cell bodies located in spinal levels **T1 through L2.**
         (b) Course
            (i) Ventral roots, referred to as the thoracolumbar outflow
            (ii) Anterior primary rami of the spinal nerves
            (iii) Myelinated **white rami communicantes** (only spinal levels T1 through L2 have white rami communicantes)
            (iv) Sympathetic (paravertebral) chain
               Preganglionic fibers may synapse with a postganglionic neuron in that paravertebral ganglion associated with the spinal nerve level.
               They may course up or down the chain to synapse with a postganglionic neuron located in a ganglion several levels away.
               They may pass through the chain without synapse to form a splanchnic nerve that leads to a **prevertebral ganglion** from the cervical to coccygeal regions.
         (c) Preganglionic sympathetic neurons are **cholinergic,** because the neurosecretory transmitter **acetylcholine** is released.
      (3) Sympathetic postganglionic fibers
         (a) Fibers from cells in the **paravertebral ganglia** take one of two pathways:
            (i) Back to a spinal nerve along an unmyelinated **gray ramus communicans** from a paravertebral ganglion at every spinal level
               Sympathetic fibers terminate in the dermal sweat glands, the smooth muscles of blood vessels, and hair (arrector pili) muscles of the body wall and extremities.

Every spinal nerve receives a gray ramus from the sympathetic chain.
- (ii) To **splanchnic nerves** and perivascular plexuses, leading to the visceral structures of the head, neck, thorax, abdomen, and pelvis
- (b) Fibers from cells in the **prevertebral ganglia** course in the perivascular plexuses to innervate abdominal and pelvic viscera.
- (c) Sympathetic postganglionic neuron is **adrenergic** because the neurotransmitter usually is **norepinephrine**.
- (d) The cells of the **adrenal medulla** are specialized postganglionic sympathetic neurons.
  - (i) Preganglionic cholinergic fibers stimulate the adrenal chromaffin cells to release both **norepinephrine** and **epinephrine** into the circulatory system, which distributes these neurosecretions throughout the body with a slight time delay.
  - (ii) In conjunction with the immediate norepinephrine release mediated by the sympathetic postganglionic fibers, the adrenal neurosecretions prolong the actions on the receptive tissues.

(4) **Sympathetic ganglia**
- (a) **Paravertebral ganglia** with superior and inferior connectors form the **sympathetic chain.**
- (b) **Prevertebral ganglia** are usually associated with the abdominal viscera.
  - (i) They include the celiac, superior mesenteric, and inferior mesenteric ganglia.
  - (ii) These ganglia receive presynaptic neurons from the splanchnic nerves.
  - (iii) The prevertebral ganglia contain cell bodies of postsynaptic sympathetic neurons.
- (c) The paravertebral and prevertebral ganglia are located some distance from the organ innervated. Therefore, sympathetic preganglionic fibers tend to be short, whereas postganglionic fibers tend to be long.

b. Parasympathetic division. Following the same rough rule of thumb as described for the sympathetic system, the parasympathetic system stimulates those activities associated with conservation and restoration of body resources. The parasympathetic division is also called the **craniosacral system** or the **cholinergic system.**

(1) The **cranial parasympathetic outflow** emerges with cranial nerves III, VII, IX, and X.
- (a) These nerves supply parasympathetic innervation to the head, thorax, and most of the abdominal viscera.
- (b) Parasympathetic function decreases heart rate, increases gastrointestinal activity, and increases secretory activity.

(2) The **sacral parasympathetic outflow** is through sacral spinal levels S2 through S4 as the nervi erigentes, or pelvic splanchnic nerves.
- (a) The sacral spinal cord supplies innervation to lower abdominal and pelvic viscera.
- (b) It is involved with urination, defecation, and sexual function.

(3) **Characteristics**
- (a) Both the preganglionic and postganglionic neurons are **cholinergic** because the neurotransmitter **acetylcholine** is secreted at the terminals.
- (b) **Parasympathetic ganglia.** Postganglionic cell bodies are located close to the organ innervated. Therefore, preganglionic fibers have long axons, whereas postganglionic fibers have short axons.

c. The enteric nervous system comprises the neural networks and plexuses of the gastrointestinal canal, which are considered a distinct division of the ANS.

B. Visceral afferents (GVA). Although part of the visceral nervous system, afferent innervation from the viscera usually is not considered part of the ANS, because, the ANS, by definition, is a visceral motor system.

1. **Afferent pathways.** Afferent fibers from the viscera pass retrograde along autonomic pathways to reach **white rami communicantes** (see Figure 4-5).

a. **In the cervical region,** visceral afferents travel along cervical splanchnic nerves, such as the cardiac accelerator nerves, to reach the sympathetic chain. The neurons then travel down the chain to the white rami communicantes of the upper thoracic (T1–T2) levels to reach the spinal nerves and the upper thoracic levels of the spinal cord.
  b. **In the thorax and abdomen,** they pass along splanchnic nerves to reach the sympathetic chain.
      (1) On reaching the sympathetic chain, the afferents pass through the white rami communicantes to gain access to a spinal nerve.
      (2) If there is no white ramus communicans (above T1, below L2), the afferents course down or up the sympathetic chain until they reach a white ramus that allows access to the spinal cord.
  c. **In the pelvic region,** there are two distinct afferent pathways.
      (1) From upper pelvic viscera, afferent neurons travel along sympathetic pathways to the lumbar splanchnic nerves, then along white rami communicantes to the lumbar spinal nerves that bring the sensory information to the upper lumbar levels of the spinal cord.
      (2) From the lower pelvic viscera, afferent neurons travel along the parasympathetic nervi erigentes (pelvic splanchnic nerves) to reach midsacral (S2–S4) levels of the spinal cord.
2. **Referred pain.** Visceral afferent pathways provide the anatomic basis for referred pain.
    a. Nociceptive input from the viscera at any thoracic, lumbar, or pelvic level of the spinal cord cannot be distinguished from that of somatic origin.
        (1) Pain and pressure afferents travel along thoracic splanchnic (sympathetic) nerves, lumbar splanchnic (sympathetic) nerves, and pelvic splanchnic (parasympathetic) nerves to reach the spinal cord.
        (2) Pain is referred to the **somatic dermatome** associated with the particular spinal level that received the visceral afferent nerve.
    b. The reason that visceral pain is perceived as originating from dermatomes is not understood. Theories range from structural (the final common pathway) to mechanistic (recognition as a function of learning from experience).
    c. Regardless, the patterns of referred pain are well documented and, for the most part, logical. They play a major role in clinical diagnosis.

Chapter 5
Circulatory System

I. INTRODUCTION

A. **The circulatory system** includes the heart and vessels (arteries, capillaries, and veins) that conduct blood through the body (Figure 5-1).

1. **The heart** is a muscular pump that propels blood through the circulatory system. The right side of the heart pumps blood through the **pulmonary circulation;** the left side pumps blood through the **systemic circulation** (see Figure 5-1).
2. **Arteries** conduct blood from the heart to the capillary beds.
3. **Capillaries** in the tissue spaces of the body provide for gaseous diffusion and exchange of nutrients as well as waste products (see Figure 5-2).
4. **Veins** collect blood after its passage through capillary beds and return it to the heart.

B. **The lymphatic system,** a part of the circulatory system, includes the **lymphatic vessels,** a set of channels that begin in the tissue spaces and return excess tissue fluid to the bloodstream (see Figure 5-2).

C. **Vascular patterns**

1. **Development.** Embryonic vessels develop before the initiation of circulation. Subsequent development and lifelong remodeling are functionally regulated.
   a. Remodeling of the embryonic vascular pattern (e.g., aortic arches and cardinal veins) produces the adult patterns with considerable variation.
      (1) An increase in vascular blood pressure leads to an increase in endothelial budding that forms new vessels to accomodate the flow.
      (2) An increase in vascular pressure, which exerts circumferential stress on the walls, results in an increase in the diameter of the vessel and, because the amount of muscular and elastic tissue is a response to stress, an increase in wall thickness.
   b. Vascular malformations result from incomplete remodeling.
2. **Characteristic patterns.** The vascular supply to some organs, such as the kidney, is altered as the organ develops or migrates within the body. Other organs, such as the gonads and the gastrointestinal tract, carry their original vascular supply with them as they migrate during development to their adult position.

II. HEART

A. **Cardiac pump.** The heart is a set of two adjacent folded tubes of cardiac muscle (see Figure 5-1).

1. **Circulation.** Blood flows through two circulatory loops.
   a. **Pulmonary circulation.** The right side of the heart pumps unoxygenated blood through the pulmonary arteries to the capillary beds of the lungs. Oxygenated blood returns to the left side of the heart in the pulmonary veins.
   b. **Systemic circulation.** The left side of the heart pumps oxygenated blood through the aorta to the capillary beds of the tissues of the body. Unoxygenated blood is collected and returned to the right side of the heart by the superior and inferior vena cavae.
2. **Cardiac dynamics.** The work done by the heart muscle is the product of the pressure and the volume of blood pumped.

FIGURE 5-1. The circulatory system. The pulmonary circulation is associated with the right side of the heart, the systemic circulation with the left side of the heart.

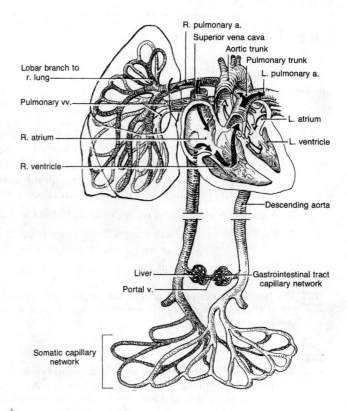

- **a. Relaxation** (diastole) of the myocardium (muscular walls) allows the chambers to fill with blood.
- **b. Contraction** (systole) of the myocardium propels blood out of the chambers.
- **c. Leaflets** guarding the valve lumina close passively (according to pressure differentials) to maintain the blood flow in a fixed direction.

3. **Cardiac control.** The pumping rate is regulated by the autonomic nervous system, which controls an internal pacemaker (the sinoatrial node).
   - a. An internal impulse-conducting system of modified muscle cells (atrioventricular node and bundle) coordinates the contraction in the ventricular chambers of the heart.
   - b. Pathologic change in this impulse initiation and conduction system can lead to heart block or fibrillation (uncoordinated contractions).

B. **Coronary circulation** supplies the myocardium with blood.

1. **Right and left coronary arteries** begin from the sinuses behind the right and left semilunar cusps of the aortic valve. They distribute blood in large part (but not exclusively) to their own half of the heart.
2. **Blood flow in the coronary arteries** is maximal during diastole and minimal in systole.
   - **a. During systole,** no pressure differential exists between the myocardium and left ventricle, so flow is not possible.
   - **b. During diastole,** a pressure differential does exist and the elasticity of the aorta propels blood through the coronary circulation.

C. **Fetal circulation** is different from circulation in the adult. In the fetus, oxygenation of the blood occurs in the placenta.

1. **The right side of the heart** pumps a small amount of blood to the collapsed fetal lungs.

2. **The left side of the heart** pumps blood through the systemic circulation and the placenta.
3. **A system of shunts** operates before birth to bypass partially the lungs and liver.
   a. **The foramen ovale** shunts blood from the right atrium to the left atrium, bypassing the pulmonary circulation.
   b. **The ductus arteriosus** shunts blood from the left pulmonary artery to the aorta, bypassing the pulmonary circulation.
   c. **The ductus venosus** shunts blood from the umbilical vein to the inferior vena cava, bypassing the liver.
4. **Changes at birth** are related to the cessation of placental blood flow and the commencement of pulmonary respiration. Failure of these shunts to close at birth results in defects that may be surgically correctable.
   a. **The foramen ovale** closes to form the **fossa ovalis** in the interatrial septum.
   b. **The ductus arteriosus** constricts and occludes to form the **ligamentum arteriosum,** a fibrous band between the aorta and pulmonary trunk.
   c. **The ductus venosus** also becomes a fibrous band (**ligamentum venosum**), extending the path of the ligamentum teres (former umbilical vein) to the hepatic vein or inferior vena cava.

III. ARTERIAL SYSTEM

A. **Conducting arteries** are the largest vessels of the body (see Figure 5-1). The arterial system begins with the aorta, a single artery with a large diameter (3 cm).
   1. **Structure and major branches.** The conducting and large distributing arteries have generous amounts of intramural elastic tissue. The aorta gives off subclavian arteries to the upper extremities, carotid arteries to the head and neck, as well as numerous arteries to the viscera, kidneys, and trunk wall before bifurcating into iliac arteries to the pelvis and lower extremities.
   2. **Hemodynamics.** Wall structure is related to pressure. The higher the pressure, the greater the amount of elastic tissue relative to muscular tissue.
      a. Elastic tissue in the conducting arteries permits both stretch of the wall during the ejection phase of cardiac systole and a propulsive elastic recoil during ventricular diastole.
         (1) This stretching accommodates the systolic cardiac ejection volume, thereby reducing the potential peak arterial systolic pressure.
         (2) The stored energy of elastic stretch maintains a prolonged diastolic arterial pressure, which drops slowly rather than precipitously.
      b. The aorta gives off distributing arteries with small diameters.

B. **Distributing arteries** account for the remaining named arteries of the body (see Figure 5-1). The distributing arteries progressively divide into arteries with smaller and smaller diameters (although the total cross-sectional area increases).
   1. **Main arterial pathways**
      a. They take the shortest possible course. For example, in the limbs, they:
         (1) Run on flexor surfaces
         (2) Usually do not pass directly through muscles, avoiding compression
         (3) Are somewhat extensible
      b. They branch in particular patterns.
         (1) The angle of branching is related to hemodynamic factors and minimizes pull between parent stem branches.
         (2) Branches of equal size (e.g., the common iliac arteries) make equal angles with the parent stem; the ideal angle is approximately 75°.
         (3) With smaller branches, the diameter of which barely affects the size of the parent stem (e.g., the intercostals), the angle can range from 70° to 90°.

(4) As the arteries decrease in size, intermittent pulsations (from cardiac systole) become steady, continuous flow.

2. **Arterial anastomoses** permit equalization of pressures and alternate channels of supply.
   a. Abundant anastomoses occur in the regions of joints in which movement might temporarily occlude the main channel.
   b. The circle of Willis equalizes the blood supply to brain.
   c. In the gastrointestinal tract, anastomoses are abundant between some regions and sparse between others.
   d. The surgeon must differentiate between potential and functional anastomoses and be aware of variability, which might result in sparse or nonfunctional anastomoses where profuse functional anastomoses are expected.

3. **End-arteries** supply discrete regions of tissue that have no collateral supply. They are found in the heart, kidneys, liver, brain, and organs of the gastrointestinal tract.
   a. Normally, no direct anastomoses occur between end-arteries.
   b. A thrombosis or embolus lodged in an end-artery produces ischemia and necrosis (**infarct**) of the tissue supplied exclusively by that vessel.
   c. Anastomoses may develop between end-arteries in response to a slow-onset pathologic process. Potential anastomoses may not be functional or adequate at a sudden-onset crisis.

C. **Arterioles,** the terminations of the smallest arteries, are nearly as small as capillaries and are not merely conducting channels (see Figure 5-2). These vessels regulate the distribution of blood.

1. **Flow regulation.** Smooth muscle in the walls of arterioles controls the size of the vessel lumen.
2. **Sympathetic control.** Sympathetic innervation of vascular musculature regulates blood flow to the tissues.

D. **Clinical considerations**

1. **An arterial pulse** is palpable in specific places because the elastic aorta and distributing arteries stretch with each cardiac ejection. The **pulse rate** is a measure of heartbeats per minute.
2. **Blood pressure** is measured by auscultation of the **bruit** produced by the intermittent flow of blood through an artery.
   a. When sufficient pressure is applied by a pneumatic tourniquet to compress the artery completely, no bruit is heard.
   b. As pressure is released, an intermittent bruit is heard when **systolic pressure** just exceeds that applied by the tourniquet.
   c. As pressure is further released, **the diastolic pressure** equals that applied by the tourniquet, flow becomes continuous, and the bruit disappears.
3. **Arterial blood samples** are obtained by inserting a needle at a pulse point.
4. **Reflex sympathetic dystrophy** (RSD) occasionally follows afferent nerve injury. Unremitting pain can initiate a reflex in adjacent spinal segments whereby abnormal sympathetic regulation limits blood flow. An entire region, usually an arm or leg, can become ischemic and dystrophic.

IV. CAPILLARY BEDS

A. **Capillaries** are the exchange sites of the circulatory system (Figure 5-2).

1. **Total cross-sectional area** of the capillary bed is approximately 800 times that of the aorta.

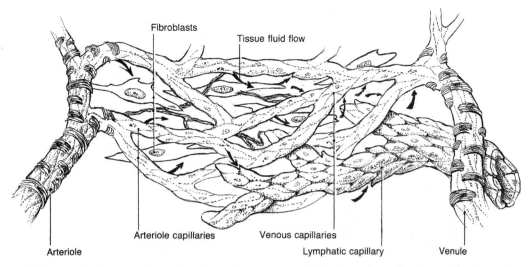

FIGURE 5-2. Capillary bed displaying the continuum from arteriole through capillary to venule. *Arrows* indicate the flow of fluid. Blind lymphatic capillaries drain excess fluid from the tissue spaces.

      a. The diameter of a capillary is about 5 μm, just large enough to enable a red blood cell to squeeze through.

      b. The velocity of circulation changes from 0.5 m/sec in the aorta to 0.5 mm/sec in the capillaries.

    2. Diffusion barrier. Capillary walls are formed by a single layer of endothelial cells. This endothelium and a thin layer of connective tissue fibers (the basal lamina) constitute the diffusion barrier.

      a. Exchange of gas. Gaseous diffusion occurs according to partial pressure gradients.

          (1) **In the peripheral tissues,** oxygen released from arterial (oxygenated) blood diffuses across the endothelium into the tissue spaces. Carbon dioxide diffuses from the tissue spaces into the blood.

          (2) **In the lungs,** carbon dioxide diffuses from the blood into the alveolar air, and oxygen diffuses from the alveolar air to oxygenate the blood.

      b. Exchange of fluid. Fluid movement occurs according to differentials between blood pressure and osmotic pressure.

          (1) **At the arterial end** of the capillary bed, blood pressure exceeds the tissue osmotic pressure.

              (a) An ultrafiltrate of nutrient-rich plasma exudes from the arterial end of the capillaries into the tissue spaces.

              (b) The osmotic pressure of the blood plasma remaining in the capillary increases and the tissue osmotic pressure decreases.

          (2) **At the venous end** of the capillary bed, the blood pressure is less and the plasma osmotic pressure is greater than the tissue osmotic pressure.

              (a) There is osmotic uptake of tissue fluid that is high in metabolic waste products into the venous capillaries.

              (b) The osmotic pressure of the tissue spaces and blood plasma are restored in a constant, finely balanced process.

    3. Density of a capillary bed in a tissue correlates with the metabolic need (e.g., endocrine glands are highly vascular; cartilage is almost avascular).

B. **Sinusoids** substitute for capillaries in some organs, such as the liver, spleen, and red bone marrow, where circulation is slow. Sinusoids often contain phagocytic cells.

    1. Sinusoids are usually wider than capillaries and have gaps between the endothelial cells, which allows underlying parenchyma cells direct access to plasma with its large protein and carbohydrate molecules.

2. The endothelial cells of sinusoids are often phagocytic, especially in the liver and spleen.

C. **Clinical considerations**
1. **Edema** is the collection of excess fluid in the tissue spaces and serous cavities.
   a. **Exudative edema.** When associated with trauma or inflammation, edematous exudate is the result of increased permeability of the walls of the capillary bed. Red blood cells are contained but plasma proteins leak out of the capillaries and raise the tissue osmotic pressure, thus interfering with the hydrostatic-osmotic pressure balance that normally results in fluid uptake at the venous end of the capillary bed.
   b. **Transudative edema.** When associated with elevated venous pressure, edematous transudate is associated with imbalance of the hydrostatic-osmotic pressures that normally result in fluid uptake at the venous end of the capillary bed.
2. **Hematoma** (black-and-blue mark associated with edema) results from a major loss of integrity of capillary walls or the walls of other blood vessels; red blood cells, in addition to plasma, leak into the tissue and cause discoloration.

V. VENOUS SYSTEM

A. **Capillary beds** drain into venules, which come together to form veins.

B. **Veins** return blood to the heart (see Figure 5-1).
1. **Venous blood flow.** The veins must carry the same volume of blood as the arteries but at a lower pressure.
   a. Characteristically, veins have large, somewhat irregular lumina and thin, relatively nonmuscular walls. Thus, these vessels are relatively compressible by external forces, which aids in blood flow.
   b. **Smooth muscle** in venous walls tends to be arranged in a loop near the point of drainage of the tributary to act as a regulator of return flow (e.g., the penile veins and the point at which the hepatic veins enter the vena cava). The walls of large veins have some elastic tissue to resist the back pressure of right atrial systole.
   c. **Valves** in many veins limit flow proximally, toward the heart.
      (1) Valves occur primarily in veins of limbs and movable viscera, but not in the cerebral veins.
      (2) They are especially prevalent at junctions between tributaries and large veins.
2. **Parallel pattern** of arteries and veins
   a. Venous patterns are far more variable than arterial patterns.
      (1) Large veins are usually single.
      (2) Medium-sized veins frequently form venae comitantes, whereby the vein is doubled, tripled, or even forms a network.
      (3) In several regions, arterial patterns and venous patterns are separate and distinct (e.g., in the brain, liver, lungs, and penis).
   b. The angulation of entry of venous tributaries is less related to hemodynamic factors than it is in arteries, because venous pressure is lower.
   c. The frequency of venous cross-anastomoses reflects retention of embryonic conditions.
   d. The **counter-current pattern** of parallel arteries and veins is often related to transfer and conservation of water or heat, as in the kidney and the extremities, respectively.
3. **Hemodynamics.** Pressure gradients between the periphery and the right side of the heart control venous flow. Several factors establish these gradients.
   a. An arterial pressure of approximately 10 mm Hg transmitted through the capillary bed to the venous side

b. The sucking bulb-syringe action of the heart during right ventricular diastole (vis afronte)
c. The negative endothoracic pressure, relative to atmospheric pressure, produced by the thoracic cage during inspiration
d. The contractile activity of the muscles of the extremities, which "milks" the venous blood toward the heart
   (1) This action is aided in the limbs by the disposition of deep and superficial veins.
      (a) The muscles of the extremities, between which the deep veins lie, are surrounded by relatively inelastic fascial septa.
      (b) With shortening, muscles become wider, thereby increasing the pressure on the deep veins and moving the blood in a direction guided by the internal valves.
   (2) Movements of the limb segments facilitate venous flow in the superficial veins.
4. **Venipuncture.** The superficial veins provide convenient sites from which to draw blood samples. A snug tourniquet occludes superficial veins but not the underlying arteries, causing pooling of the blood and engorgement of the superficial veins distal to the tourniquet.

C. **Portal systems** begin in capillaries and subsequently form a large vein before breaking up into sinusoids.
1. **The hepatic portal system** drains most of the venous blood from the capillary beds of the intestinal tract to the hepatic portal vein, which branches to form the sinusoids of the liver. In this way, products of digestion and intestinal absorption are transported directly to the liver for metabolic or detoxification processes.
2. **The hypophyseal portal system** drains the base of the hypothalamic region toward the adenohypophysis (anterior pituitary gland). In this manner, neurohormonal releasing factors secreted by the hypothalamus are transported directly to the anterior lobe of the pituitary gland to induce secretion of various pituitary hormones.

D. **Arteriovenous (AV) anastomoses** permit direct transfer of blood from arterial to venous channels, bypassing the capillary bed.
1. **Distribution.** AV anastomoses are widely distributed.
   a. These vessels rise as a side branch of a terminal arteriole and join a venule.
   b. Each AV anastomosis has a thick muscular wall, abundantly supplied with vasomotor nerves (general visceral efferent, sympathetic), thereby forming a sphincter.
2. **Functions.** AV anastomoses usually occur in organs that function intermittently.
   a. In the gut, AV anastomoses are open except during periods of digestion.
   b. In the skin, these connections are especially numerous in apical parts (e.g., fingers, nose, lips, and ears), where they are involved in temperature regulation.

VI. LYMPHATIC SYSTEM

A. **Composition.** The lymphatic system is composed of an extensive network of extremely variable lymphatic vessels and nodes, which serve as filters and a source of lymphocytes and plasma cells.

B. **Function.** This system has evolved as a specialized mechanism to return to the bloodstream those fluids not taken up by the venous blood capillaries.
1. **The lymphatic vessels** contain lymph, which usually is clear and colorless. Milky lymph that contains triglycerides is called chyle.
   a. Triglycerides are absorbed from the gastrointestinal tract by mucosal lymphatics.

b. Chyle is channeled to the blood for distribution to appropriate organs, usually the liver.

2. **Lymph** contains proteins that enter the lymph stream from the tissue spaces.
   a. These proteins filter through lymph nodes on their way to the blood vascular system.
   b. This association is of particular importance if the protein is an antigen (i.e., "foreign"), because resident lymphocytes will initiate antibody production.
3. **Edema** results from blockage of lymphatic flow or from exceeding the capacity of the lymphatic drainage system.

C. Structure

1. **Lymph capillaries** begin as cul-de-sacs that drain the tissue spaces (see Figure 5-2).
   a. Lymph capillaries are wider than blood capillaries and are irregular in diameter (from a few micrometers to 1 mm).
   b. Lymphatic capillaries, found in most tissues of the body, are especially numerous in the mucous membranes, serous surfaces, and dermis of the skin. The brain, spinal cord, and eyeball lack these vessels; however, these structures have other channels for draining tissue fluids. Lymphatic capillaries are also absent in bone marrow and in parenchyma of the spleen.
2. **Lymphatic vessels** are formed by the convergence of lymph capillaries.
   a. Lymphatics have valves that give them a beaded appearance.
   b. They are more plentiful than veins, which they tend to accompany.
   c. The walls of the larger vessels become somewhat thicker as they acquire small amounts of smooth muscle and elastic tissue.
   d. Intercalated along the lymphatic vessels are the lymph nodes.
3. **Lymph nodes** are composed of lymphatic tissue.
   a. Nodes are usually grouped in specific regions and named accordingly.
   b. They are drained by lymph vessels, often terminating in another group of nodes. Lymph may filter through several nodes before reaching a major trunk.
   c. Bacteria or antigens are filtered by the phagocytes of the lymph nodes.
4. **Lymph ducts** are formed by the convergence of lymphatic vessels.
   a. Lymphatics from the lower extremities converge on lymph nodes located anteriorly and superficially in the uppermost part of the thigh.
   b. From the inguinal lymph nodes, the main duct from each lower extremity enters the abdomen.
   c. After passing through the lumbar lymph nodes, the main duct, referred to as the lumbar duct, enters the cisterna chyli.
5. **The cisterna chyli** is a dilated sac between the diaphragmatic crura, opposite the first lumbar vertebra, and behind the right side of the aorta.
   a. The cisterna chyli can vary in size and location.
   b. It contains some smooth muscle and is somewhat pulsatile.
   c. In addition to lumbar lymphatics and para-aortic lymphatics, the cisterna chyli also receives several lymph ducts, including the large common duct that drains lymph from most of the intestinal tract.
   d. It gives rise to the thoracic duct.
6. **The thoracic duct** originates from the upper end of the cisterna chyli.
   a. **Course.** The duct ascends on the anterior aspect of the vertebral column, inclining to the left and ending at the base of the neck, usually by entering the left brachiocephalic vein.
      (1) The thoracic duct thus receives lymph from both legs, the pelvis, the abdomen, and the left side of the thorax.
      (2) It may drain lymph from the left arm and from the left side of the head and neck, because the left jugular and left subclavian lymph ducts may join the thoracic duct in the base of the neck.
         (a) The **left jugular lymph duct** drains lymph from the left side of the head and neck, accompanying the left internal jugular vein in the neck.

(b) The **left subclavian lymph duct** drains the left axillary and subclavicular nodes of the left upper extremity and accompanies the left subclavian vein.
(c) The **left bronchomediastinal lymph duct,** which frequently joins the thoracic duct, drains the viscera on the left side of the thorax.
  b. **Flow rate.** In normal situations, the large volume of lymph flowing through the thoracic duct (12 ml/kg/hr) is derived mainly from the liver and alimentary tract.
7. **The right lymph duct** receives lymph from the right side of the head and neck, the right upper extremity, and the right side of the thorax, and returns lymph to the great veins at the base of the neck on the right side.
    a. **The right jugular lymph duct** drains the right side of the head and neck.
    b. **The right subclavian lymph duct** drains the right axillary and subclavicular nodes.
    c. **The right bronchomediastinal lymph duct** drains the right thoracic viscera.

D. Lymph flow.
The flow of lymph from a region is generally unidirectional toward the large veins at the base of the neck.
1. **Hydrostatic-osmotic pressure** from the volume of tissue fluids taken up by the lymphatic capillaries promotes lymph return.
2. **Mechanical forces**
    a. Pressure resulting from voluntary muscular activity on the valved lymph vessels is a factor. The flow of lymph in an immobile limb is almost negligible, but it becomes significant during muscular activity.
    b. The alternation of pressures within the thorax during ventilation propels lymph in the valved lymphatics of the mediastinum.
    c. Muscular contractions of the abdominal wall on expiration, coughing, and straining produce a positive abdominal pressure on the cisterna chyli that propels lymph.
    d. Pulsations of adjacent blood vessels have a massaging effect on lymph vessels to aid the flow of lymph.
    e. Rhythmic return flow is aided by the contraction of the smooth muscles in the walls of the cisterna chyli.
3. **Valves** occur in the thoracic duct and right lymph duct and prevent backflow.

E. Clinical considerations
1. **Lymphatics** are the major route by which carcinoma metastasizes.
    a. Malignant cells may be trapped in lymph nodes, where they proliferate.
    b. Malignant cells may be delivered to the venous system.
    c. Surgical removal of malignant tumors also necessitates resection of the major lymphatic vessels and nodes draining the involved region.
    d. Wound healing requires the regeneration of the lymphatics as well as the growth of blood capillaries.
2. **Lymph nodes**
    a. Bacteria or antigens filtered by the phagocytes of the lymph nodes may induce inflammation or a cell-mediated immune reaction, either of which can produce swelling of the node (**swollen glands**).
    b. When nodes are swollen by inflammation or blocked by metastatic cells, **edema** or drainage may occur along alternate lymphatic pathways.
3. **Thoracic duct**
    a. Injury to this structure may result in accumulation of fluid in the pleural cavities (**chylothorax**). The diagnosis is made by the high lymphocyte count in a pleural aspirate.
    b. Thoracic duct obstruction with backup of intestinal lymph to the kidney lymphatic capillaries may produce **chyluria.**
4. **Abnormal lymphatic drainage patterns**
    a. **Ascites** is the accumulation of extracellular fluid (usually from the liver) in the peritoneal cavity.

- (1) Lymphatic absorption of extracellular fluid in the peritoneal cavity occurs almost entirely through the peritoneal surface of the diaphragm. The rate of absorption is rapid, normally approaching 1 liter per day.
- (2) When transudation from dilated subcapsular and hepatic hilar lymphatics exceeds peritoneal uptake, lymph accumulates as ascites. It is commonly, but not exclusively, associated with conditions that raise intrahepatic blood pressure (e.g., right-ventricular heart failure or cirrhosis).
  b. **Pulmonary edema** is caused by either increased permeability or a hydrostatic-osmotic pressure imbalance in the pulmonary vascular bed. It results in fluid accumulation in the tissue spaces and transudation of fluid into the alveoli.
  c. **Pulmonary effusion** (hydrothorax) can be caused by similar processes, but with transudation into the pleural cavity.
  d. **Lymphedema** is the accumulation of tissue fluid, particularly in an extremity, as a result of lymphatic obstruction (e.g., **elephantiasis** caused by Filaria bancrofti).

PART I OVERVIEWS

STUDY QUESTIONS

DIRECTIONS: Each of the numbered items or incomplete statements in this section is followed by answers or by completions of the statement. Select the ONE lettered answer or completion that is BEST in each case.

1. An incision made through the skin parallel to the prevailing direction of the connective tissue fibers usually will

(A) tend to gape
(B) provide optimum surgical exposure
(C) heal with a prominent scar
(D) coincide with Langer's lines of cleavage

2. Which of the following statements about the lobes of the breast is true?

(A) They are defined by connective tissue septa that determine the posture of the breast
(B) They are located between the deep layer of the superficial fascia and the deep fascia
(C) They contain glandular tissue that is the source of breast size variation over the menstrual cycle
(D) They empty at the nipple via a common ampulla
(E) They number between 40 and 60

3. The principal lymphatic drainage from the right breast is toward the

(A) right axillary nodes
(B) right internal mammary nodes
(C) right internal thoracic nodes
(D) right supraclavicular nodes
(E) thoracic duct

Questions 4–7

A 14-year-old accomplished skier lost his balance getting off a ski lift. He grabbed the moving chair, wrenched his left shoulder, and was dragged to the ground. The following chair, which was empty, then struck the same shoulder. He was able to get up and cautiously begin his descent. One day later, the shoulder continued to be painful, and his parents sought medical attention.
The examining orthopedist first asked the patient to point to where it hurt, and then started gently palpating the shoulder, working toward the indicated areas of pain. Sometimes comparing his findings to the contralateral normal shoulder, the examiner noted that the left acromioclavicular joint (between the scapula and the clavicle) allowed more movement than the same joint on the uninjured side. Next, he passively moved the arm to check motion in all directions and the extent to which each movement was possible, ending with gentle circumduction. Because of the movement in the acromioclavicular joint and the reported pain over the clavicle, he ordered a radiographic examination of both shoulders.

4. Circumduction is movement that involves

(A) abduction/adduction and flexion/extension
(B) ball-and-socket joints only
(C) flexion/extension and medial rotation/lateral rotation
(D) medial rotation/lateral rotation and abduction/adduction
(E) one axis of rotation

5. The radiographs revealed a fine spiral fracture at about the midpoint of the clavicle; alignment was good. Also noted was a pronounced gap at the left acromioclavicular joint with some subluxation. A radiolucent band between the head and neck of the humerus had the appearance of another fracture. Longitudinal growth in a long bone may be interrupted prematurely by fracture through the

(A) anatomic neck
(B) diaphysis
(C) epiphyseal disk
(D) epiphyseal line
(E) metaphysis

6. Torn ligaments frequently are associated with bone fractures. In this case, subluxation of the left acromioclavicular joint is a diagnostic sign. Of the following statements about torn ligaments, which is correct?

(A) A torn ligament interrupts the connection between bone and muscle
(B) Torn ligaments can usually be visualized radiographically
(C) Torn ligaments, like fractured bone, repair rapidly and strongly
(D) Torn ligaments destabilize the joint

7. The healing of a bone fracture involves all of the following in sequence EXCEPT

(A) immediate deposition of mineral into the fracture site
(B) deposition of a connective tissue callus across the fracture
(C) bone resorption at the fracture
(D) calcification of the callus
(E) bone remodeling across the fracture
(end of group question)

8. The subdural space between the dura and arachnoid meningeal layers of the brain contains

(A) blood on hemorrhage of a cerebral artery
(B) blood on hemorrhage of a cerebral vein
(C) blood on hemorrhage of the middle meningeal artery
(D) cerebrospinal fluid

9. The muscles of the back receive motor innervation from

(A) dorsal primary rami
(B) dorsal roots
(C) posterior branches of lateral perforating nerves
(D) ventral primary rami
(E) white rami communicantes

10. A blood clot that develops in a peripheral vein and breaks free into the venous system will lodge in the vascular bed of the

(A) brain
(B) extremities
(C) heart
(D) kidneys
(E) lungs

11. Which of the following statements about end-arteries is correct?

(A) They are found around most highly movable joints
(B) They are frequently found in vital organs
(C) They form abundant anastomotic networks
(D) They provide a high margin of safety in the event of vascular blockage

12. Pain of visceral origin is not referred to dermatomes L3 through S1 because there is an absence of

(A) paravertebral sympathetic ganglia below L2
(B) sympathetic efferent (motor) supply to spinal nerves L3–S1
(C) visceral afferents in the lumbar splanchnic nerves
(D) visceral afferents in the sacral region of the spinal cord
(E) white rami communicantes to spinal nerves L3–S1

Study Questions   Part I Overviews   47

DIRECTIONS: Each of the numbered items or incomplete statements in this section is negatively phrased, as indicated by a capitalized word such as NOT, LEAST, or EXCEPT. Select the ONE lettered answer or completion that is BEST in each case.

13. The strength of a muscle is a function of all of the following EXCEPT the

(A) cross-sectional area
(B) degree to which the muscle is stretched in situ
(C) length of the lever arm
(D) length of the muscle in its fully stretched state in situ

14. A white ramus communicans that extends between the spinal nerve and a sympathetic trunk ganglion contains all of the following EXCEPT

(A) axons of presynaptic efferent neurons
(B) fibers that conduct general visceral afferent impulses
(C) fibers that directly activate smooth muscle in the skin
(D) myelinated fibers

15. Sympathetic postganglionic neurons, which join each spinal nerve via a gray ramus communicans, innervate all of the following EXCEPT

(A) arrector pili muscles
(B) skeletal muscle
(C) smooth muscle of the peripheral vascular bed
(D) sweat glands and modified sweat glands

16. All of the following regions usually drain via the thoracic duct EXCEPT the

(A) abdomen and pelvis
(B) left arm
(C) left side of the thoracic cage
(D) right leg
(E) right side of the face and neck

DIRECTIONS: Each set of matching questions in this section consists of a list of four to twenty-six lettered options (some of which may be in figures) followed by several numbered items. For each numbered item, select the ONE lettered option that is most closely associated with it. To avoid spending too much time on matching sets with large numbers of options, it is generally advisable to begin each set by reading the list of options. Then, for each item in the set, try to generate the correct answer and locate it in the option list, rather than evaluating each option individually. Each lettered option may be selected once, more than once, or not at all.

Questions 17–19

(A) Superficial layer of superficial fascia (Camper's f.)
(B) Deep layer of superficial fascia (Scarpa's f.)
(C) Deep fascia

For each of the following phrases, select the fascial layer that is most aptly described.

17. Membranous fascia that anchors sutures well

18. Connective tissue that overlies muscle and neurovascular structures, defining planes between them

19. Fatty layer that contains numerous blood vessels and nerves

Questions 20–22

(A) Amphiarthrosis
(B) Diarthrosis
(C) Synarthrosis
(D) Syndesmosis

For each of the following bone articulations, select the joint classification to which it is most appropriately associated.

20. Immovable cranial sutures

21. Slightly movable intervertebral disk

22. Highly movable glenohumeral (shoulder) joint

Questions 23–25

(A) Upper motor neuron
(B) Lower motor neuron
(C) Peripheral sensory neuron
(D) Central sensory neuron

For each of the following descriptions of the location of a cell body, select the most appropriate neuron.

23. Dorsal root ganglion

24. Gray matter of the cerebral cortex

25. Ventral horn of the spinal cord

ANSWERS AND EXPLANATIONS

1. The answer is D [Chapter 2 I D]. Langer's cleavage lines parallel the prevailing direction of the collagenous connective tissue fibers of the skin. An incision made parallel to these crease lines will sever fewer connective tissue bundles, will gape less, and will heal with a more cosmetic scar. Other factors must be considered, however, such as whether or not an incision along Langer's lines will provide adequate surgical exposure.

2. The answer is A [Chapter 2 III A 1 b]. The 15 to 20 lobes of the breast are separated by connective tissue septa (Cooper's ligaments), which run between the epidermis and the deep (Scarpa's) layer of superficial fascia. The suspensory ligaments determine the posture of the breast. The effect of estrogen on the fatty tissue accounts for the variation in breast size over the menstrual cycle.

3. The answer is A [Chapter 2 III B 2 a; Chapter 5 VI C 6 b (2) (b)] Approximately 75% of the drainage of the breast is through the axillary nodes, with a relatively small proportion draining through the internal thoracic (parasternal and internal mammary) nodes. Only the left axillary nodes drain into the thoracic duct.

4. The answer is A [Chapter 1 II D 8 c]. Circumduction is a combined movement involving both abduction/adduction and flexion/extension. It must, therefore, occur at joints with at least two degrees of freedom.

5. The answer is C [Chapter 3 I A 3 a (4) (c) (iii); 4 b (3)]. The epiphyseal plate (disk) in a child, composed of hyaline cartilage, is radiolucent and, thus, appears radiographically as would a fracture. Fracture through the unfused cartilaginous growth plate in a young person may impede or terminate longitudinal growth in a long bone. The cartilaginous epiphyseal disk is generally located toward one end of a long bone, between the diaphysis of compact bone and the metaphysis of trabecular bone. Fusion of the growth plate with cessation of growth leaves the epiphyseal line. A fracture of the disk, causing premature closure, results in asymmetric growth.

6. The answer is D [Chapter 3 I C 2 b 3]. Ligaments connect bones and stabilize joints. Because ligaments are radiolucent, radiographic evidence of a ligamentous tear or sprain occurs only if misalignment of bony structures accompanies such a tear. Although ligaments heal by scar formation, they do not undergo the extensive remodeling seen in bone, and thus a healed ligament usually is never as strong.

7. The answer is A [Chapter 3 I A 4 c]. The initial step in fracture healing is the formation of a cartilaginous callus about the fracture, which subsequently calcifies to provide a splint. After a few days, bone is resorbed from the fracture site, and an ingrowth of vessels and connective tissue spans the fracture. The unorganized connective tissue also calcifies, but the union is not particularly strong. Remodeling occurs along lines of force and compression so that new bone becomes as strong as the uninjured bone, often without trace of the original fracture.

8. The answer is B [Chapter 4 II A 2 b (3)]. The large distributing arteries lie in the subarachnoid space so that hemorrhage of one of these vessels will be evident by blood in the cerebrospinal fluid (CSF). A middle meningeal hemorrhage produces an epidural hematoma. Hemorrhage from a cerebral vein usually produces a subdural hematoma in the potential space between the dura and arachnoid. Neither the epidural nor subdural spaces have access to the subarachnoid space, which contains CSF.

9. The answer is A [Chapter 4 III C 2 a, b]. Dorsal roots (sensory only) and ventral roots (motor only) form the spinal nerves that bifurcate almost immediately into dorsal and ventral primary rami. The axial muscles and skin of the back are innervated by dorsal primary rami of the spinal nerves. Ventral primary rami, giving off lateral and ventral perforating branches, innervate the lateral and ventral regions of the body wall. White and gray rami communicantes are part of the autonomic (involuntary) nervous system.

10. The answer is E [Chapter 5 I A 1; II A 1; Figure 5-1]. A thromboembolus originating in the peripheral venous system will lodge in the vascular bed of the lung, producing a pulmonary embolus. Emboli that arise in the pulmonary veins or in the left side of the heart may lodge in the systemic vascular beds (i.e., of the brain, heart, kidneys, viscera, and extremities).

11. The answer is B [Chapter 5 III B 3]. End-arteries typically are found in vital organs, such as the brain, heart, kidneys, and organs of the gastrointestinal tract. These vessels supply distinct regions of tissue with no collateral blood supply. Given this low margin of safety, blockage of an end-artery results in ischemia and necrosis.

12. The answer is E [Chapter 4 V B 1 b, c]. The chain of paravertebral sympathetic ganglia extends along the length of the vertebral column. Every spinal nerve receives a sympathetic component from an associated gray ramus communicans; white rami communicantes occur only between T1 and L2. Visceral pain fibers traveling along the sympathetic pathways gain access to the spinal nerves via white rami. In those regions of the body in which cervical or lumbar splanchnic nerves bring afferent nerves to the sympathetic chain ganglia that have no white ramus, the sensory fibers follow the chain to reach the first ganglia that has a white ramus and its associated spinal nerve to reach the corresponding level of the spinal cord. The pain is perceived as originating from the dermatomes associated with that level of the spinal cord. Thus, pain of visceral origin is not referred to levels that lack a white ramus. In the pelvic region, visceral afferents may be carried by the nervi erigentes (pelvic splanchnics) to nerves S2–S4 and subsequently to the midsacral levels of the spinal cord. Visceral pain of pelvic origin is referred to dermatomes associated with those sacral levels of the spinal cord.

13. The answer is D [Chapter 3 II B 1 b; C 1, 2] The strength of a muscle is a function of the cross-sectional area (the number of contractile units in parallel), the length of the lever arm (the mechanical advantage), and the degree to which the muscle is stretched in situ (the amount of interaction between contractile filaments). The relative length of a muscle in its fully stretched state (the number of contractile units in series) has no bearing on the strength.

14. The answer is C [Chapter 4 V A 3 a (2) (b) (ii), (iii); B 1 b (1)]. Each white ramus communicans of the sympathetic chain is composed of myelinated presynaptic sympathetic neurons, which leave the spinal nerve to travel to a paravertebral or prevertebral ganglion. These fibers are presynaptic in the two-neuron autonomic motor pathway and therefore do not reach effectors. Instead, they synapse in ganglia on postsynaptic neurons that innervate viscera, sweat glands, smooth muscle associated with the hair follicles, and arterioles of the skin. Visceral afferent fibers reach the spinal nerve through white rami.

15. The answer is B [Chapter 4 V a 3 a (2); 3 (a) (i)]. Sympathetic preganglionic neurons leave the spinal cord in segments T1 through L2, gain access to the sympathetic chain via a white rami communicantes, course up and down the sympathetic chain, and synapse with a postsynaptic neuron. Those postsynaptic neurons that join a spinal nerve via gray rami communicantes innervate the arrector pili muscles, sweat glands, and the smooth muscle of the peripheral vascular bed.

16. The answer is E [Chapter 5 VI C 6]. The thoracic duct, originating as the cisterna chyli in the abdomen, receives lymph from the entire body below the diaphragm, as well as from the thoracic viscera, the left side of the thoracic cage, the left arm, and the left side of the head and neck. It enters the systemic venous system at the junction where the left subclavian and left internal jugular veins join to form the left brachiocephalic vein. The right arm, right side of the thoracic cage, and right side of the head and neck usually drain into the right lymph duct, which empties into the right brachiocephalic vein.

17–19. The answers are: 17-B, 18-C, 19-A [Chapter 2 II B]. The superficial fascia is divided into two layers. The superficial layer (Camper's fascia) is fatty and contains abundant blood vessels, lymphatics, and nerves. The deep layer (Scarpa's fascia) is membranous and, although thin, holds sutures well.

Deep (investing), and not superficial, fascia surrounds neurovascular bundles and muscles, providing fascial planes between these structures.

20–22. The answers are: 20-C, 21-A, 22-B [Chapter 3 I C 1 a–c]. Cranial sutures are synarthroses with no articular cartilage and in which little of no movement is possible. The fibrocartilaginous intervertebral disks are amphiarthroses in which the attachment of the cartilaginous connective tissue to the adjacent bones severely limits the amount of movement. The shoulder joint, like all joints with a capsule, synovial fluid, and articular cartilage capping the bones, is a diarthrosis, which allows extensive movement. A syndesmosis (a type of synarthrosis) is a broad fibrous joint that attaches two bones with little movement permitted, such as occurs between the tibia and fibula.

23–25. The answers are: 23-C, 24-A, 25-B [Chapter 4 IV A 2, B 2]. The somatic nervous system is divided into sensory and motor divisions, each of which has central and peripheral portions. The peripheral sensory neuron, which is stimulated by receptors, has its cell body in a dorsal root ganglion or homologous cranial nerve ganglion. The initial central sensory neuron, which receives input from the peripheral neuron, is located in the dorsal horn gray matter of the spinal cord or brain stem.

On the motor side, the upper motor neuron is located in the gray matter of the cerebral cortex. It synapses with a lower motor neuron in the ventral horn of the gray matter of the spinal cord or brain stem. The lower motor neuron travels in the peripheral nerves to a voluntary muscle. All central neuron cell bodies are located in gray matter; their myelinated processes form white matter.

SECTION II
EXTREMITIES AND BACK

PART IIA UPPER EXTREMITY

Chapter 6
Shoulder Region

I. INTRODUCTION TO THE UPPER EXTREMITY

A. Basic principles

1. **Primitive position.** The organization of the upper extremity can be understood by simulating the primitive position; that is, by abducting the arm until it is horizontal, with the palm facing forward, so that the whole upper extremity approximates the pectoral fin of a fish.
   a. **Primitive surfaces.** The **primitively ventral surface** generally corresponds to the anterior or flexor side. The **primitively dorsal surface** generally corresponds to the posterior or extensor side (see Figure 4-3).
   b. **Axial borders.** An imaginary plane drawn from side to side through the long axis of the extremity differentiates between a cephalad **preaxial border** and a caudal **postaxial border.** Although there generally is overlap between sequential dermatomes, there is no overlap across axial borders (see Figure 4-3).

2. **Early terrestrial adaptation** involves flexing the elbow 90° and extending the wrist 90° from the fin-like position to one that approximates the amphibian/reptilian upper extremity.

3. **Late terrestrial adaptation.** In mammals and humans, in particular, the amphibian/reptile brachium has rotated 90° caudally, bringing the upper extremity alongside the lateral body wall with the flexor surface facing anteriorly. The arm is then in the **anatomic position.**
   a. **Manipulation**
      (1) The upper extremity of the human, freed from its original role of support and propulsion, has developed high intrinsic mobility.
      (2) The **function of the human upper extremity** is to place its effector end, the hand, in a position to manipulate the environment. The hand can also be positioned so that it touches every part of the body.
   b. **Grasp.** The hand is developed for and is characterized by prehension. Only the human hand has a sophisticated two-point grasp between the thumb and index finger.
   c. **Sensation.** The hand is a sophisticated sensory organ.

B. Basic organization.
The free portion of the upper extremity (arm, forearm, wrist, and hand) is suspended from the axial skeleton by the embedded portion (the pectoral girdle).

1. **Bony support.** The only bony articulation between the pectoral girdle and the axial skeleton is through the clavicle at the sternoclavicular joint.

2. **Muscular support.** The attachment of the upper extremity to the thorax is primarily muscular.
   a. **Dynamic stability.** The muscles of the pectoral girdle provide support as well as stability and produce movement. They consist of extensors on the primitively dorsal side and flexors on the primitively ventral side.
   b. **Mobility.** The pectoral girdle is mobile at the expense of stability. Strenuous activity that involves extreme lateral rotation may result in dislocation at the glenohumeral joint.

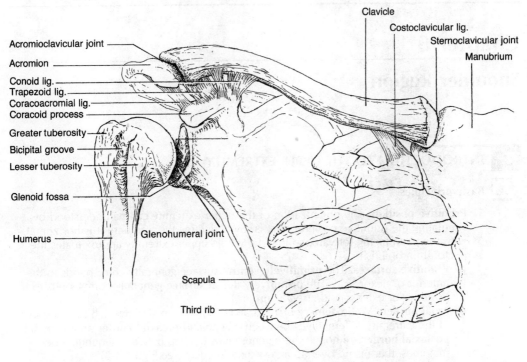

FIGURE 6-1. Right pectoral girdle. Anterior view of the right shoulder shows the sternoclavicular and acromioclavicular joints are strongly reinforced by ligaments, while the glenohumeral joint is stabilized primarily by muscles.

C. **Regions of the upper extremity**
   1. **The pectoral girdle or shoulder** consists of two bones, the clavicle and the scapula (Figure 6-1).
   2. **The arm or brachium** contains one bone, the humerus (see Figure 7-1).
   3. **The forearm or antebrachium** contains two bones, the radius and the ulna (see Figure 7-1).
   4. **The wrist or carpus** contains seven carpal bones in two rows and a sesamoid bone (see Figure 8-1, Chapter 9 I A).
   5. **The hand** contains five **metacarpals** and fourteen **phalanges** (see Chapter 10 I A 1, 2).

II. BONES OF THE PECTORAL GIRDLE

A. **Clavicle.** This S-shaped strut between the scapula and sternum is easily observed and readily palpable (see Figure 6-1).
   1. **Location.** The clavicle connects the scapula to the sternum. It articulates with the sternum at the **sternoclavicular joint** and with the scapula at the **acromioclavicular joint.**
   2. **Ossification.** It is the first bone to begin ossification, starting in the fifth week of fetal development, but it is the last to complete ossification, at about the twenty-first year. It is the only long bone formed by intermembranous ossification.

3. **Fracture.** The clavicle is one of the most commonly fractured bones in individuals younger than middle age.
   a. **Mechanism of injury.** During a fall onto an outstretched arm, forces transmitted through the clavicle from the arm to the sternum produce shear near the middle of the S-shaped clavicle.
   b. **Bone displacement.** Fracture in the middle one third of the bone results in upward displacement of the proximal fragment (because of the pull of the sternomastoid muscle) and downward displacement of the distal fragment (because of the pull of the deltoid muscle and the effect of gravity on the arm).
   c. **Complications.** Because major neurovascular structures lie between the clavicle and the first rib, the downward displaced distal clavicular fragment can produce nerve damage as well as severe, even fatal, internal bleeding.

B. **Scapula.** This structure, including its spine, acromion process, coracoid process, vertebral border, and inferior angle (apex), is palpable (see Figure 6-1).
   1. **Characteristics.** The scapula lies against the posterior aspect of the rib cage. This flat bone is shaped like an inverted triangle with superior, vertebral, and lateral borders and an inferior apex.
      a. The spine of the scapula (subcutaneous and horizontal) is a posterior reinforcing ridge. It divides the dorsal surface into a **supraspinous fossa** and an **infraspinous fossa.**
      b. The **acromion process** is the lateral expansion of the spine. It articulates with the clavicle and provides an attachment for the trapezius and deltoid muscles, as well as for one end of the coracoacromial ligament.
      c. The **coracoid process** projects anteriorly.
         (1) A remnant of the third bone of the pectoral girdle, it is palpable anteriorly, just medial to the head of the humerus.
         (2) It provides attachment for the pectoralis minor and coracobrachialis muscles, as well as the short head of the biceps brachii muscle. It also has several ligamentous attachments; for example, the **coracoacromial ligament** transmits tensile force from the coracoid process to the acromion process and spine of the scapula.
      d. The **glenoid fossa** is the site of articulation with the humerus. It faces laterally but slightly anteriorly and slightly superiorly.
         (1) This shallow fossa is deepened slightly by the **glenoid labrum** (lip of fibrocartilage).
         (2) The **supraglenoid tubercle** and **infraglenoid tubercle** provide attachments for the long heads of the biceps and triceps muscles, respectively.
   2. **Articulations.** The scapula articulates with the clavicle and the humerus.

C. **Humerus.** The proximal end of the long bone of the arm (humerus) consists of a head and two tuberosities separated by an intertubercular groove, all of which are palpable (see Figure 6-1).
   1. **The head** is covered by hyaline cartilage and articulates with the scapula at the **glenohumeral (scapulohumeral) joint.**
   2. **The anatomic neck** separates the head from the metaphysis and marks the location of an **epiphyseal plate** that normally fuses between the nineteenth and twenty-first years.
   3. **The lesser tuberosity** lies on the anterior surface of the humerus, just distal to the anatomic neck, and provides an attachment for the subscapularis muscle.
   4. **The greater tuberosity** lies on the lateral surface of the humerus, just lateral to the anatomic neck, and provides attachments for the supraspinatus, infraspinatus, and teres minor muscles.
   5. **The intertubercular (bicipital) groove** lies between the greater and lesser tuberosities and contains the tendon of the long head of the biceps brachii muscle.
      a. It is bridged by the **transverse humeral ligament.**
      b. The **lateral lip** of the groove provides muscular attachment for the pectoralis

major muscle, the **floor** for the latissimus dorsi muscle, and the **medial lip** for the teres major muscle.

III. ARTICULATIONS OF THE SHOULDER GIRDLE

A. Sternoclavicular joint

1. **Structure.** The **clavicle** articulates with the **manubrium** of the sternum, proximally. The joint has a fibrocartilaginous articular disk that divides the joint into two synovial capsules (see Figure 6-1).

2. **Movement.** This joint has two degrees of freedom.
   a. **Elevation/depression** of the shoulder occurs about an anterior–posterior axis.
   b. **Protraction/retraction** of the shoulder occurs about a vertical axis.
   c. **Circumduction** of the shoulder (the combined result of the two pairs of movements) occurs about both axes simultaneously.

3. **Support.** The fibrous joint capsule is reinforced by strong ligaments, so dislocation of this joint is uncommon.
   a. The **anterior and posterior sternoclavicular ligaments** run between the clavicle and the manubrium.
   b. The **costoclavicular ligament** runs between the clavicle and the first rib.

B. Acromioclavicular joint

1. **Structure.** The clavicle articulates with the acromion process of the scapula (see Figure 6-1).

2. **Movement.** This joint is a sliding articulation with two degrees of freedom. Rotation of the scapula occurs primarily at this joint about an anterior–posterior axis. Other movements at this joint are mainly accommodations to motions that occur at the sternoclavicular joint.

3. **Support.** The articular capsule is reinforced by strong ligaments.
   a. The **coracoclavicular ligament** runs from the **coracoid process** of the scapula to the **clavicle.** It is subdivided into a **conoid ligament** and a **trapezoid ligament,** both of which provide superior–inferior stability.
   b. The **acromioclavicular ligament** bridges the joint and provides anterior–posterior stability.
   c. **Acromioclavicular subluxation (shoulder separation)** is common despite these ligamentous reinforcements. Usually the result of traumatic downward displacement of the clavicle, this injury is relatively common among high school football players.

C. "Scapulothoracic joint"

1. **Structure.** No bony articulation exists between the scapula and the thoracic cage. This conceptualized pseudojoint, which lies between the subscapularis muscle and the serratus anterior muscle, appears to function as a joint, although its movement pairs actually occur at the sternoclavicular and acromioclavicular joints.

2. **Movement.** This "joint" has three degrees of freedom, which allow for considerable motion of the scapula on the posterolateral thoracic cage.
   a. **Protraction/retraction** occurs through a vertical axis at the sternoclavicular joint so that the scapula slides anterolaterally/posteromedially on the posterior thoracic wall.
   b. **Elevation/depression** occurs through an anterior–posterior axis at the sternoclavicular joint so that the scapula slides cranially/caudally on the posterior thoracic wall.
   c. **Rotation** occurs through an anterior–posterior axis at the acromioclavicular joint so that the scapula rotates on the posterior thoracic wall.

3. **Support.** This pseudojoint is supported by the clavicle and reinforced dynamically by the muscles that insert on the scapula from the axial skeleton, such as the rhomboids, serratus anterior, trapezius, and levator scapulae.

D. **Glenohumeral (scapulohumeral) joint**

1. **Structure.** The glenoid fossa is the site of articulation with the head of the humerus. This shallow fossa is deepened slightly by a lip of fibrocartilage, the **glenoid labrum** (see Figure 6-1).

2. **Movement.** As a ball-and-socket joint, it has three degrees of freedom.
   a. **Flexion/extension** occurs about the transverse axis.
   b. **Abduction/adduction** occurs about the anterior–posterior axis.
   c. **Internal and external rotation** occur about the vertical axis.
   d. **Circumduction** at this joint results from simultaneous movements about two or three axes.

3. **Ligamentous support.** Its lax fibrous capsule is reinforced by tough articular ligaments.
   a. **Superior, middle, and inferior glenohumeral ligaments** run from the glenoid lip to the anatomic neck of the humerus.
   b. **The coracohumeral ligament** between the coracoid process and the humerus supports the dead weight of the free portion of the upper extremity.
   c. **The coracoacromial ligament** runs between the coracoid process and acromion process.
      (1) Together with the acromion process, this ligament forms the **coracoacromial arch.**
      (2) This arch buttresses the superior aspect of the glenohumeral joint to prevent superior displacement of the humerus.
      (3) This ligament transmits tensile force from muscles that originate on the coracoid process to the acromion process and spine of the scapula.

4. **Dynamic stability.** This shallow, unstable joint is dynamically reinforced by muscle action.
   a. The **rotator (musculotendinous) cuff** (supraspinatus, infraspinatus, subscapularis, and teres minor muscles) acts as a dynamic "ligament," keeping the head of the humerus pressed into the glenoid fossa (see Figure 6-4). **Rotator cuff tears** may severely limit shoulder mobility.
   b. The **tendon of the long head of the biceps brachii muscle,** passing over the humeral head en route to the supraglenoid tubercle of the scapula, also forces the humeral head medially into the joint.

5. **Bursae.** Several bursae lie between the various musculoskeletal components.
   a. The **subacromial bursa** separates the acromion process from the underlying supraspinatus muscle.
      (1) This space is frequently the site of **subacromial bursitis** and **supraspinatus tendinitis** with calcium deposits.
      (2) Pain associated with **subacromial bursitis,** felt mainly during the initial stages of abduction and forward flexion, severely limits shoulder mobility.
   b. The **subdeltoid bursa** separates the deltoid muscle from the head of the humerus and the insertions of the rotator cuff muscles.
   c. **Communications.** The subacromial and subdeltoid bursae frequently communicate, but neither should communicate with the joint capsule. Such a finding is indicative of a **capsular tear.**

6. **Shoulder dislocation.** Despite the musculotendinous cuff and other stabilizing features, the extreme mobility of the glenohumeral joint and the shallowness of the glenoid fossa result in loss of stability, which predisposes to dislocation of the humerus.
   a. In **anterior dislocations,** the head of the humerus comes to lie inferior to the coracoid process. The axillary nerve is sometimes injured. Dislocation stretches the anterior capsule and may avulse the glenoid labrum, which increases the likelihood of recurrence.

b. In **posterior dislocations** (rare), the humeral head is displaced posteriorly. Because the dislocation is away from the brachial plexus, the incidence of associated neurovascular injury is low.

IV. MUSCLE FUNCTION AT THE SHOULDER JOINT

A. Movement of the pectoral girdle. Many movements are accomplished by combinations of primitively dorsal and primitively ventral muscles that run between the axial skeleton and the bones of the pectoral girdle and arm (see Table 6-1 and Figures 6-2 and 6-3).

1. **Organization of muscles acting on the pectoral girdle.** Shoulder movement occurs at the sternoclavicular, acromioclavicular, and "scapulothoracic" joints.
   a. **Major posterior muscles** are arranged in two layers (see Table 6-1).
      (1) The **superficial layer** includes the upper and lower portions of the **trapezius muscle** (Figure 6-2).
      (2) The **deep layer** includes the **levator scapulae** as well as the **major rhomboid** and the **minor rhomboid** (see Figure 6-2).
   b. **Major anterior muscles** are also arranged in two layers (see Table 6-1).
      (1) The **superficial layer** includes the **pectoralis minor** and **sternomastoid** (Figure 6-3).
      (2) The **deep layer** includes the **serratus anterior** and **subclavius** (see Figure 6-3).

2. **Group actions** (Table 6-1)
   a. **Elevation/depression** occurs through an anteroposterior axis through the sternoclavicular joint with a resultant vertical gliding at the scapulothoracic joint.
      (1) **Elevation.** Prime elevators, which pass superior to the anteroposterior axis, include the cervical portion of the **trapezius, levator scapulae, rhomboid major,** and **rhomboid minor.**
      (2) **Depression.** Depressors, which pass inferior to the anteroposterior axis, include the **pectoralis minor, subclavius,** and the **thoracic portion of the trapezius.**
   b. **Protraction/retraction** occurs about a vertical axis through the sternoclavicular joint with a resultant horizontal gliding at the scapulothoracic joint.
      (1) **Protraction (abduction).** Protractors, which pass anterior to the vertical axis, include the **pectoralis minor** (acting on the coracoid process) and the **serratus anterior** (acting on the vertebral border).
      (2) **Retraction (adduction).** Retractors, which pass posterior to the vertical axis, include the **rhomboids** and both heads of the **trapezius.**
   c. **Rotation** occurs about an anteroposterior axis through the acromioclavicular joint with resultant rotation at the scapulothoracic joint.
      (1) **Upward rotation.** Upward rotator muscles include the **cervical head of the trapezius** (acting on the acromion) and the **thoracic head of the trapezius** (acting on the spine of the scapula), as well as the **serratus anterior** (acting on the vertebral border of the scapula).
      (2) **Downward rotation.** Downward rotator muscles include the **pectoralis minor** and **rhomboids.**

3. **Group innervation.** Muscles acting on the scapula are innervated by the **spinal accessory nerve** (trapezius and sternomastoid), the **dorsal scapular nerve** (rhomboids), **long thoracic nerve** (serratus anterior), and twigs from the brachial plexus (see Table 6-1).

B. Movement of the arm. Many movements at the glenohumeral joint are accomplished by combinations of primitively dorsal and primitively ventral muscles that run between the axial skeleton and the humerus (see Table 6-1 and Figures 6-2 and 6-3), as well as by muscles that run between the pectoral girdle and the humerus (Figure 6-4; see Table 6-2).

Table 6-1. Muscles Acting on the Pectoral Girdle

Muscle	Origin	Insertion	Primary Action	Innervation
Posterior				
Trapezius:				
Upper portion	Superior nuchal line, ligamentum nuchae, and spine of C7	Lateral third of clavicle and acrominon process	Elevates and rotates scapula upward in elevation of arm	Spinal accessory (CN XI) and C3–C4
Lower portion	Spines of T1–T12	Spine of scapula	Depresses and rotates scapula upward	Spinal accessory n. (CN XI) and lower cervical nn.
Levator scapulae	Transverse processes C1–C4	Superior portion of vertebral border of scapula	Elevates scapula	Nn, to levator (C3–C4)
Rhomboids:				
Minor	Lower part of ligamentum nuchae and spines of C7–T1	Proximal portion of spine of scapula	Retracts and elevates scapula	Dorsal scapula n. (C5, posterior)
Major	Spines of T2–T5	Vertebral border of scapula inferior to the spine	Retracts and elevates scapula	Dorsal scapula n. (C5, posterior)
Anterior				
Sternomastoid	Mastoid process of cranium	Manubrium and proximal third of clavicle	Elevates sternum and clavicle; rotates the head	Spinal accessory n. (CN XI)
Subclavius	Costochondral junction of first rib	Middle third of clavicle	Depresses and protracts scapula	Nn. to subclavius (C5–C6, anterior)
Pectoralis minor	Outer surface of ribs 3–5	Coracoid process	Depression and protraction of scapula (elevates ribs if shoulder is fixed)	Medial pectoral n. (C8–T1, anterior)
Serratus anterior	Outer surface of ribs 1–9	Vertebral border of scapula	Protracts (abducts) and rotates scapula upward	Long thoracic n. (C7, anterior)

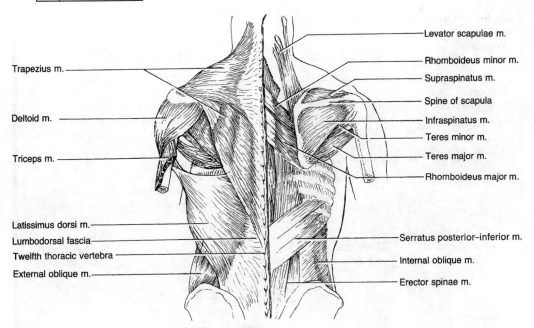

FIGURE 6-2. Posterior pectoral musculature. The left superficial layer contains the cervical and thoracic portions of the trapezius, the posterior and lateral portions of the deltoid, and the latissimus dorsi. The right deep layer contains the levator scapulae as well as the major and minor rhomboid muscles.

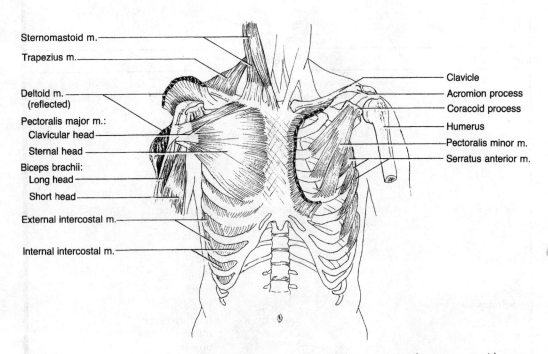

FIGURE 6-3. Anterior pectoral musculature. The right superficial layer comprises the sternomastoid, trapezius, clavicular and sternal heads of the pectoralis major, and anterior portion of the deltoid. The left deeper layer consists of the pectoralis minor and serratus anterior muscles.

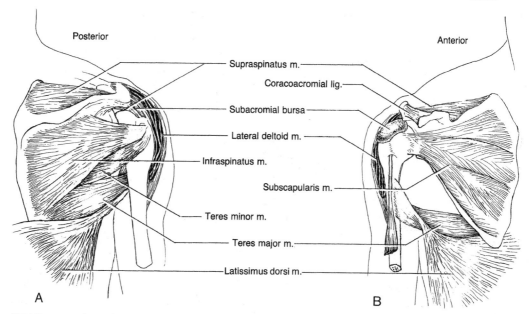

FIGURE 6-4. The right rotator cuff. *(A)* Posterior muscles are the supraspinatus, infraspinatus, and teres minor. *(B)* Anterior muscles, including the extensive subscapularis.

1. **Muscles acting on the arm**
   a. **Major posterior muscles** are arranged in two layers (see Table 6-2).
      (1) The **superficial layer** includes the posterior portion of the **deltoid and latissimus dorsi** and **triceps brachii** (see Figure 6-2).
      (2) The **deep layer** includes the muscles of the **rotator cuff.** From posterior–superior to anterior lie the **s**upraspinatus, **i**nfraspinatus, **t**eres minor, and **s**ubscapularis (see Figure 6-4). Mnemonic: "The rotator cuff **SITS** on the greater tuberosity."
   b. **Major anterior muscles** are also arranged in two layers (see Table 6-2).
      (1) The **superficial layer** includes the anterior portion of the **deltoid** and the **pectoralis major** (see Figure 6-3).
      (2) The **deep layer** includes the **coracobrachialis** and **biceps brachii** (see Figure 6-3).
2. **Group actions** (Table 6-2)
   a. **Abduction/adduction** occurs about an anteroposterior axis through the glenohumeral joint.
      (1) **Abduction.** Arm abductors, which pass superior to the anteroposterior axis, include the **supraspinatus** (initial 15°) and the **lateral part of the deltoid** (10°–100°).
      (2) **Adduction.** Arm adductors, which pass inferior to the anteroposterior axis, include the **pectoralis major, latissimus dorsi, coracobrachialis, teres major,** and the **anterior and posterior parts of the deltoid.**
   b. **Flexion/extension** occurs about a bilateral axis through the glenohumeral joint.
      (1) **Flexion.** Arm flexors, which pass anterior to the bilateral axis, include the anterior part of the **deltoid, coracobrachialis,** and the **long head of the biceps brachii.**
      (2) **Extension.** Arm extensors, which pass posterior to the bilateral axis, include the **posterior part of the deltoid, latissimus dorsi, teres major,** and **long head of the triceps.**
   c. **Rotation** occurs about a vertical axis through the glenohumeral joint.
      (1) **Medial rotation.** Medial rotators, which pass anterior to the vertical axis, in-

Table 6-2. Muscles Acting on the Arm

Muscle	Origin	Insertion	Primary Action	Innervation
Posterior				
Supraspinatus	Supraspinous fossa of scapula	Greater tubercle of humerus	Abducts arm first 15° and stabilizes shoulder joint	Suprascapular n. (C5–C6, posterior)
Deltoid:				
Lateral part	Acromion process of scapula	Deltoid tuberosity of humerus	Abducts arm between 10° and 100°	Axillary n. (C5–C6, posterior)
Posterior part	Spine of scapula	Deltoid tuberosity of humerus	Extends, adducts, and laterally rotates arm	Axillary n. (C5–C6, posterior)
Latissimus dorsi	Spinous processes of T6–L5, iliac crest and ribs 10–12	Floor of bicipital groove of humerus	Extends, adducts, and medially rotates arm	Thoracodorsal n. (C6–C8, posterior)
Teres major	Inferior angle of scapula	Medial lip of bicipital groove of humerus	Extends, adducts, and medially rotates arm	Lower subscapular n. (C5–C6, posterior)
Triceps brachii:				
Long head	Infraglenoid tubercle	Olecranon process of ulnar	Extends and adducts arm	Radial n. (C6–C8, posterior)
Infraspinatus	Infraspinous fossa of scapula	Greater tubercle of humerus	Laterally rotates arm and stabilizes shoulder joint	Suprascapular n. (C5–C6, posterior)
Teres minor	Inferior angle of scapula	Greater tubercle of humerus	Laterally rotates arm and stabilizes shoulder joint	Axillary n. (C5–C6, posterior)
Subscapularis	Subscapular fossa	Lesser tubercle of humerus	Medially rotates arm and stabilizes shoulder joint	Upper and lower subscapular nn. (C5–C6, posterior)
Anterior				
Deltoid:				
Anterior part	Lateral third of clavicle	Deltoid tuberosity of humerus	Flexes, adducts, and medially rotates arm	Axillary n. (C5–C6, posterior)
Pectoralis major:				
Sternal head	Sternum and costal cartilages 1–7	Lateral lip of bicipital groove of humerus	Adducts, flexes, and medially rotates arm	Medial pectoral n. (C8–T1, anterior)
Clavicular head	Proximal third of clavicle	Lateral lip of bicipital groove	Flexes, adducts, and medially rotates arm	Lateral pectoral n. (C5–C7, anterior)
Coracobrachialis	Coracoid process of scapula	Midhumerus	Flexes, adducts, and medially rotates arm	Musculocutaneous n. (C5–C7, anterior)
Biceps brachii:				
Long head	Supraglenoid tubercle	Radial tuberosity and antebrachial fascia	Flexes shoulder and stabilizes shoulder joint	Musculocutaneous n. (C5–C7, anterior)
Short head	Coracoid process	Radial tuberosity and antebrachial fascia	Flexes shoulder	Musculocutaneous n. (C5–C7, anterior)

       clude the **subscapularis, teres major, pectoralis major,** and the **anterior part of the deltoid.**
       (2) Lateral rotation. Lateral rotators, which pass posterior to the vertical axis, include the **infraspinatus, teres minor,** and the **posterior part of the deltoid.**
   d. Dynamic stability at the glenohumeral joint is provided in large part by the **rotator cuff muscles,** acting together.
   e. Displaced fractures
       (1) Fracture at the humeral neck results in abduction of the proximal fragment (from supraspinatus action) and medial displacement of the distal fragment (from traction of the pectoralis major and latissimus dorsi).
       (2) A **fracture in the proximal one third of the humerus** can lead to lateral and upward displacement of the distal fragment (from traction from the deltoid

muscles) and medial displacement of the proximal fragment (from traction of the pectoralis major and latissimus dorsi).

3. Group innervation
  a. The **anterior muscles** acting on the humerus are innervated largely by the **lateral pectoral nerve** (clavicular head of the pectoralis major), **medial pectoral nerve** (sternal head of the pectoralis major), and the **musculocutaneous nerve** (coracobrachialis and biceps brachii).
  b. The **posterior muscles** are innervated largely by the **axillary nerve** (deltoid and teres minor), **suprascapular nerve** (supraspinatus and infraspinatus), **thoracodorsal nerve** (latissimus dorsi), **upper subscapular nerve** (subscapularis), **lower subscapular nerve** (teres major), and the **radial nerve** (triceps brachii).

C. Combined movements of the shoulder and arm
  1. **Abduction of the arm.** Approximately the first 100° of abduction occurs at the glenohumeral joint; the remaining 80° occurs by elevation and upward rotation of the pectoral girdle.
  2. **Adduction of the arm.** In addition to adduction at the glenohumeral joint, the pectoral girdle is depressed and rotated downward.
  3. **Lateral rotation of the arm.** In addition to lateral rotation at the glenohumeral joint, the pectoral girdle is retracted.
  4. **Medial rotation of the arm.** In addition to medial rotation at the glenohumeral joint, the pectoral girdle is protracted.

V. AXILLARY VASCULATURE

A. The axilla is a pyramid-shaped space through which major neurovascular structures pass between the thorax and upper extremity.
  1. **The apex** is a triangular space limited by the first rib, the scapula, and the clavicle.
  2. **The anterior wall** is the **anterior axillary fold,** which overlies the pectoralis major and pectoralis minor muscles.
  3. **The posterior wall** is formed by the subscapularis muscle and the **posterior axillary fold,** which overlies the latissimus dorsi and teres major muscles.
  4. **The broad medial wall** is the serratus anterior muscle overlying the thoracic cage.
  5. **The narrow lateral wall** is the **intertubercular (bicipital) groove.**
  6. **The base** is formed by the skin and fascia of the axillary fossa (armpit).

B. Vascular supply (Figure 6-5)
  1. **The subclavian arteries** arise differently on the left and right sides of the body (see Figure 22-1).
    a. **Course.** The **right subclavian artery** arises from the brachiocephalic artery; the **left subclavian artery** arises directly from the aortic arch. Each subclavian artery is divided into three parts by the scalenus anterior muscle.
      (1) The subclavian arteries pass beneath the clavicles superior to the first rib between the insertions of the anterior and medial scalene muscles.
      (2) On each side, the subclavian artery continues as the axillary artery once it passes the distal edge of the first rib.
    b. **Branches.** Each subclavian artery gives off several major distributing arteries from its initial segment in the base of the neck (see Figure 6-5).
      (1) The **vertebral artery** provides a primary supply to the brain.

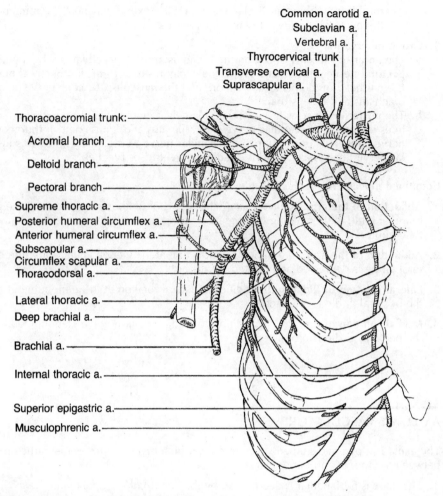

FIGURE 6-5. Arterial supply to the right shoulder and proximal arm. Note the extensive anastomoses between the various branches about the shoulder joint.

  (2) The **internal thoracic (mammary) artery** supplies the anterior thoracic and abdominal walls. It also can be a source of collateral supply to the heart when it is surgically diverted and anastomosed to a coronary artery.
  (3) The **thyrocervical artery (trunk)** supplies portions of the neck and shoulder. It usually gives off three significant branches.
    (a) The **inferior thyroid artery** supplies portions of the thyroid gland and larynx.
    (b) The **suprascapular artery** runs into the supraspinous fossa and passes through the great scapular notch superior to the transverse scapular ligament to reach the infraspinous fossa of the scapula.
      (i) It supplies the supraspinous and infraspinous muscles.
      (ii) It anastomoses with the circumflex scapular artery and the descending (dorsal) scapular artery.
    (c) The **transverse cervical artery,** when present, crosses the posterior triangle and bifurcates into two branches.
      (i) The **superficial cervical branch** supplies the trapezius. It anastomoses with the suprascapular, subscapular, and circumflex scapular arteries in the vicinity of the scapular apex. It may arise separately from the thyrocervical trunk.

- (ii) The **descending (dorsal) scapular branch** supplies the levator scapulae and rhomboids. It anastomoses with the suprascapular, subscapular, and circumflex scapular arteries in the vicinity of the scapular apex. It may arise separately from the third part of the subclavian artery.
- (iii) **Variations.** When the superficial cervical and descending (dorsal) scapular arteries arise separately (about 50% of the time), the transverse cervical artery does not exist.
- (4) The **costocervical artery** passes back over the dome of the pleura to the neck of the first rib where it divides into deep cervical and superior intercostal arteries.

2. **The axillary artery** is the continuation of the subclavian artery at the lateral edge of the first rib (see Figure 6-5).
   a. **Course.** This artery is divided into three parts by the tendon of the overlying pectoralis minor muscle.
      (1) In the axilla, it is surrounded by the cords of the brachial plexus.
      (2) Distally, the **axillary pulse** is palpable in the lateral wall of the axilla.
      (3) The axillary artery becomes the **brachial artery** at the distal margin of the teres major muscle.
   b. **Branches**
      (1) The **upper part** (proximal to the pectoralis minor) gives off one branch, the **supreme (highest) thoracic artery,** which supplies the more superficial subclavicular portion of the chest.
      (2) The **middle part** (deep to the pectoralis minor) gives off two branches.
         (a) The **thoracoacromial artery (trunk)** gives off four or five branches.
            (i) **Pectoral branches** descend to the pectoralis major and minor muscles.
            (ii) The **acromial branch** courses lateral to the deltoid muscle.
            (iii) A **clavicular branch** runs medial to the subclavius muscle.
            (iv) A **deltoid branch** supplies the deltoid and pectoralis major muscles and accompanies the cephalic vein in the deltopectoral groove.
         (b) The **lateral (long) thoracic artery** courses along the thoracic wall superficial to the serratus anterior muscle. It is a major supply to the mammary gland.
      (3) The **lower part** (distal to the pectoralis minor) gives off three branches. Here the anastomotic connections between the branches of the subclavian and axillary arteries about the shoulder joint are profuse and functional.
         (a) The **subscapular artery,** the largest branch, bifurcates.
            (i) The **circumflex scapular artery** curves around the lateral border of the scapula and passes through the **triangular space** (formed by the teres major, teres minor, and long head of the triceps brachii) to reach the dorsum of the scapula. Distally, it anastomoses with the suprascapular and dorsal (descending) scapular arteries.
            (ii) The **thoracodorsal artery** lies on the anterior surface of the latissimus dorsi muscle.
         (b) The **posterior humeral circumflex artery** accompanies the axillary nerve.
            (i) This artery passes posterior to the humerus, through the **quadrangular space** (formed by the teres major, teres minor, long head of the triceps brachii, and humerus).
            (ii) It anastomoses with the smaller anterior humeral circumflex artery.
         (c) The **anterior humeral circumflex artery** is a small branch
            (i) This vessel passes laterally beneath the coracobrachialis muscle as it courses anterior to the humerus.
            (ii) It anastomoses with the larger posterior humeral circumflex artery.
   c. **Collateral circulation.** Anastomoses about the shoulder are important because the axillary artery is often temporarily occluded by external pressure.
      (1) Branches of the axillary and subclavian arteries form anastomoses about the scapula (e.g., the branches of the thyrocervical trunk with the thoracoacromial trunk and subscapular artery).

(2) Branches of the axillary artery also anastomose about the glenohumeral joint (e.g., the branches of the thoracoacromial trunk with the subscapular artery and the humeral circumflex arteries).

3. **Veins of the axillary region**
   a. **The axillary vein** lies medial to the axillary artery.
      (1) This vein is formed by the joining of the **basilic vein** and the **brachial veins** near the distal margin of the teres major muscle.
      (2) It receives veins that correspond to the branches of the axillary artery with much variability.
      (3) It becomes the **subclavian vein** at the distal edge of the first rib.
   b. The **cephalic vein** lies in the deltopectoral groove (triangle), together with the deltoid branch of the thoracoacromial artery.
      (1) Although useful for the insertion of venous lines, it is absent in approximately 10% of the population.
      (2) It terminates in the axillary vein just distal to the clavicle.

4. **Lymphatic drainage**
   a. **Axillary lymph nodes** (10–30) are intercalated in the lymphatic drainage of the upper extremity and the thoracic walls, including the major portion of each breast.
      (1) **Apical (subclavian or level III) nodes** lie about the upper part of the axillary vein, proximal to the pectoralis muscle. All lower nodes drain through this group. The apical nodes on the right side drain to the right lymphatic trunk, and those on the left side drain to the thoracic duct. Some anastomotic drainage occurs with the deep cervical nodes. Thus, metastases may reach the lower cervical (supraclavicular or Virchow) nodes.
      (2) **Central (level II) nodes** lie about the axillary vein deep to the pectoralis muscle. They receive drainage from the pectoral, lateral thoracic, and subscapular nodes, corresponding to similarly named vasculature. Importantly, most of the drainage from the breast enters and passes through the central nodes.
      (3) **Brachial (level I) nodes** lie along the lower part of the axillary vein, receiving drainage from the arm, forearm, and hand.
   b. **Nodal swelling.** Metastatic tumor from the breast may cause swollen central and apical axillary nodes. Only infection within the upper extremity will result in changes involving the brachial nodes.

VI. BRACHIAL PLEXUS

A. Introduction. Innervation to and from the upper extremity runs through the brachial plexus, which arises from spinal nerves C4 through T1 and courses through the base of the neck and axilla before reaching the arm. Understanding the confusing yet consistent pattern of combinations and separations of nerves is the key to the diagnosis of injury.

B. Development

1. As embryonic somites migrate to form the extremities, they drag their nerve supply, so that each dermatome and myotome retains the original segmental innervation. With somite migration, some of the nerves come into close proximity and fuse in a particular fashion, forming a plexus.
2. The **brachial plexus** develops from the anterior primary rami of spinal nerves C4 through T1.
   a. The lateral branch of each anterior primary ramus contributes the posterior division of the plexus.
   b. The ventral continuation of each anterior primary ramus contributes the anterior division of the plexus.

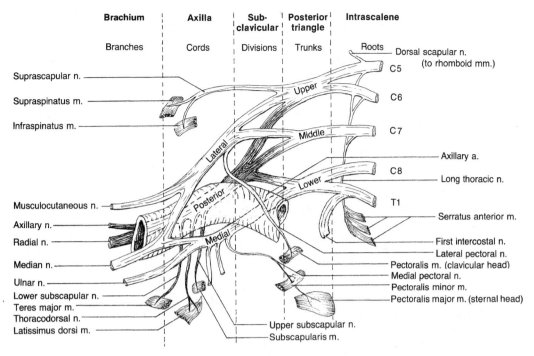

FIGURE 6-6. Formation and regions of the right brachial plexus. The medial, lateral, and posterior cords are shown in relation to the axillary artery. The posterior division is shaded, and the muscles innervated by branches of the roots and cords are indicated.

C. **Subdivisions.** The brachial plexus consists of roots, trunks, divisions, cords, and nerves (Figure 6-6).

1. **Roots** represent the anterior primary rami of spinal nerves C4, C5, C6, C7, C8, and T1. Each root innervates a specific **dermatome** and **myotome.** The roots lie between the anterior and middle scalene muscles (see Figure 6-6). Some roots give off branches.
    a. **Root C4** frequently (65%) contributes to the brachial plexus.
    b. **Roots C3–C5** contribute to the **phrenic nerve,** which innervates the diaphragm. (Mnemonic: "Roots C3, C4, and 5/Keep the diaphragm alive").
    c. **Roots C5–C7** give rise to two nerves.
       (1) The **dorsal scapular nerve** (C5) innervates the rhomboid and levator scapulae muscles.
       (2) The **long thoracic nerve** (of Bell, C5–C7) innervates the serratus anterior muscle. Injury to these roots is indicated by **"winged scapula"**—an ominous sign of severe brachial plexus injury.
    d. **Root T1** contributes the first intercostal nerve that innervates the thoracic wall.
    e. **Variations**
       (1) The plexus is **prefixed** when the C4 root is large and the T1 root is lacking.
       (2) The plexus is **postfixed** when the C4 root is lacking and the T2 contribution is significant.
    f. **Intrascalene (root) block.** The roots of the brachial plexus may be blocked by injection of anesthetic into the interval between the anterior and middle scalene muscles.

2. **Trunks** are formed by the joining of roots in the posterior cervical triangle (see Figure 6-6).
    a. The **upper trunk** is formed by the joining of roots C4, C5, C6. It gives rise to two nerves.

(1) The **suprascapular nerve** (C4–C6) runs posteriorly through the suprascapular notch and beneath the transverse scapular ligament to innervate the supraspinatus and infraspinatus muscles.

(2) The **subclavius nerve** (C5–C6) passes anteriorly to innervate the subclavius muscle (not shown).

b. The **middle trunk** is the continuation of root C7. It neither joins with adjacent roots nor gives off any branches.

c. The **lower trunk** is formed by the joining of roots C8 and T1. It does not give off branches.

3. **Divisions** are formed by bifurcation of trunks deep to the clavicle (see Figure 6-6).
   a. The **posterior divisions** of the trunks are equivalent to the lateral perforating branches that arise from the anterior primary rami of the spinal nerves (see Figure 4-2). The posterior division innervates the **primitively dorsal musculature** of the extremity (i.e., the **extensors**). The posterior divisions unite to form the **posterior cord**.
   b. The **anterior divisions** of the trunks represent the terminal portions of the anterior primary rami of the spinal nerves (see Figure 4-2). These divisions innervate the **primitively ventral musculature** (i.e., the **flexors**). The anterior divisions form the **lateral cord** and the **medial cord**.

4. **Cords** are formed by the joining of either anterior or posterior trunk divisions (see Figure 6-6). They lie in the axilla beneath the pectoralis minor muscle and adjacent to the axillary artery.
   a. The **lateral cord** (C4, C5–C7) is formed by the joining of the anterior divisions of the upper and middle trunks. Its name reflects its position relative to the axillary artery. It forms three major nerves.
      (1) The **lateral pectoral nerve** (C5–C7) innervates the clavicular head of the pectoralis major muscle.
      (2) The **musculocutaneous nerve** (C4–C6) innervates the flexor muscles of the brachium (coracobrachialis, biceps brachii, and brachialis muscles) before continuing as the **lateral antebrachial cutaneous nerve**.
      (3) The **median nerve** (C6–C8), to which both the medial cord and the lateral cord contribute, innervates numerous flexor and pronator muscles of the forearm, wrist, and hand.
   b. The **medial cord** (C8–T1) is the continuation of the anterior division of the lower trunk. It is named for its position relative to the axillary artery. It gives rise to five major nerves.
      (1) The **medial pectoral nerve** (C8–T1) innervates the pectoralis minor muscle and the sternal head of the pectoralis major muscle.
      (2) The **median nerve** (C6–C8), to which both the medial cord and the lateral cord contribute, innervates numerous flexor and pronator muscles of the forearm, wrist, and hand.
      (3) The **ulnar nerve** (C8–T1) innervates one and one-half flexor muscles in the forearm and wrist as well as numerous flexors in the hand.
      (4) The **medial antebrachial cutaneous nerve** (C8–T1) carries sensation from the forearm and a portion of the hand.
      (5) The **medial brachial cutaneous nerve** (T1) carries sensation from the medial arm.
   c. The **posterior cord** (C5–T1) is formed by the fusion of the posterior divisions of the upper, middle, and lower trunks. It is named for its position relative to the axillary artery. It gives off five major nerves.
      (1) The **upper subscapular nerve** (C5–C6) innervates the major portion of the subscapularis muscle.
      (2) The **thoracodorsal (middle subscapular) nerve** (C6–C8) innervates the latissimus dorsi muscle.
      (3) The **lower subscapular muscle** (C5–C6) innervates the teres major muscle and the inferolateral portion of the subscapularis muscle.
      (4) The **axillary nerve** (C5–C6) innervates the teres minor muscle and all of the

deltoid muscle. It passes through the quadrangular space with the posterior humeral circumflex artery.
- (5) The **radial nerve** (C5–T1) is the continuation of the posterior cord. It innervates all of the extensors and **supinators of the forearm and wrist.**

D. Lesions

1. **Injury to the roots**
   a. **Nerve root compression** is most commonly the result of **cervical spondylosis,** a condition in which the patient first complains of pain in the arm or shoulder in a dermatomal distribution, followed by progressive muscular weakness.
   b. **Trauma to the roots** produces not only complete paralysis of the upper extremity, but also winging of the scapula because of paralysis of the serratus anterior muscle innervated by the long thoracic nerve. In addition, patients may experience weakness of scapular retraction owing to involvement of the dorsal scapular nerve to the rhomboids.

2. **Injury to the trunks and cords** produces distinct syndromes.
   a. **Upper trunk injury** (Erb-Duchenne paralysis) is the most common problem. Violent downward displacement of the arm, such as results from being thrown from a horse or motorcycle, may tear the fifth and sixth roots or the upper trunk.
      (1) Patients lose function of all nerves that receive contributions from the anterior and posterior divisions of the upper trunk.
      (2) Loss of abduction, radial flexion, and external rotation causes the upper extremity to hang by the side in internal rotation, the "waiter's tip" position.
      (3) Involvement of the lateral pectoral nerve results in inability to touch the opposite shoulder.
   b. **Middle trunk injury,** by itself, is rare, but it may occur as the result of an attempted intrascalene anesthetic block.
   c. **Lower trunk injury** (Klumpke paralysis) is less common. It is caused by violent or prolonged upward displacement of the arm (as may occur in a difficult breach delivery), dislocation of the shoulder, apical tumors of the lung, a cervical rib, or scalene syndrome.
      (1) Function is lost in all nerves derived from the anterior and posterior divisions of the lower trunk.
      (2) Injury results in loss of ulnar flexion of the wrist and paralysis of many of the intrinsic muscle of the hand.
      (3) Involvement of the medial pectoral nerve produces inability to adduct the arm in the lowered position against resistance.
   d. **Cord injuries**
      (1) **Injury to the posterior cord** ("Saturday night palsy" or crutch palsy) or to the radial nerve results in loss of the extensor muscles of the arm, forearm, and wrist (wrist drop).
      (2) **Thoracodorsal nerve injury** produces paralysis of the latissimus dorsi muscle.
      (3) **Axillary nerve injury** results in deltoid and teres minor paralysis (loss of shoulder abduction and weak external rotation) with loss of sensation over the deltoid muscle.

3. **Scalene (thoracic outlet) syndrome** can be caused by spasm of the anterior and middle scalene muscles or by a cervical rib that restricts the size of the thoracic inlet.
   a. Spasm can compress portions of the brachial plexus, most commonly the lower trunk, causing pain along the medial border of the arm and atrophy of some of the small muscles of the hand.
   b. Spasm may also compress the subclavian artery, causing ischemia of the arm, which in turn can result in loss of nerve function and subsequent muscular paralysis.

Chapter 7
Elbow Region

I. BONES

A. Introduction

1. **The elbow region** consists of the distal part of the **arm (brachium)** and the proximal part of the **forearm (antebrachium).**
2. Bony landmarks
   a. The **humerus,** including its head, greater tuberosity, bicipital (intertubercular) groove, shaft, lateral epicondyle, and medial epicondyle, is palpable.
   b. The **ulna,** including its olecranon process and shaft, is palpable.
   c. The **radius,** including its head and shaft, is palpable.

B. Humerus. The **distal end of the humerus** consists of medial and lateral epicondyles separated by the capitulum and trochlea, which constitute the distal articular surfaces (Figure 7-1).

1. **Characteristics.** The distal humerus flattens anteroposteriorly and broadens transversely into **medial** and **lateral supracondylar ridges,** which terminate as the **medial** and **lateral epicondyles** (see Figure 7-1). The supracondylar ridges and epicondyles serve as muscle attachments.
2. **Articulations with the forearm**
   a. The **trochlea** is a medial, pulley-shaped articular surface that coapts the trochlear notch of the ulna to form the **humeroulnar joint** (see Figure 7-1).
      (1) **Anteriorly,** proximal to the trochlea, the **coronoid fossa** receives the coronoid process of the ulna when the forearm is flexed.
      (2) **Posteriorly,** proximal to the trochlea, the **olecranon fossa** receives the olecranon process when the forearm is extended.
   b. The **capitulum** is an anterolateral hemispherical articular surface that coapts the head of the radius to form the **humeroradial joint** (see Figure 7-1).

C. Radius and ulna. These bones constitute the forearm.

1. **The proximal ulna** articulates with the humerus medially in a strong joint that transmits principal forces from the forearm to the arm.
   a. **Characteristics.** The proximal end of the ulna terminates in the thickened **olecranon process** with a deep **trochlear notch.** The ulna narrows as it extends distally (see Figure 7-1).
      (1) The **olecranon process** provides attachment for the triceps muscle. It is separated from the subcutaneous tissue by the **olecranon bursa,** which is prone to **bursitis** when subjected to repeated and prolonged pressure **(student's elbow).**
      (2) The triangular **coronoid process,** just distal to the trochlear notch, provides the insertion for the brachialis muscle.
   b. **Articulations** with the humerus and radius (see Figure 7-1)
      (1) The deep **trochlear (semilunar) notch** on the anterior surface provides an articular surface for the **humeroulnar joint.**
      (2) The **radial notch,** lateral to the coronoid process, provides an articular surface for the **proximal radioulnar joint.**
2. **The proximal radius** articulates with the humerus laterally. It assumes a secondary role at the elbow because most of the force is transmitted to the humerus by the ulna.
   a. **Characteristics.** The proximal end of the radius terminates in a flattened head (see Figure 7-1). The radius broadens as it extends distally.

FIGURE 7-1. Elbow joint (right arm), consisting of the humeroulnar joint, humeroradial joint, and the proximal radioulnar joint. These articulations are reinforced by radial collateral and ulnar collateral ligaments as well as the annular ligament.

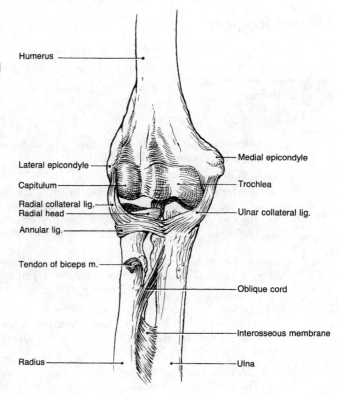

(1) The cylindric **head** is concave on the top at the proximal end of the radius.
(2) A short and narrow **neck** region lies distal to the head.
(3) The **radial (bicipital) tuberosity,** on the anteromedial aspect of the neck, receives the tendon of the biceps brachii muscle.
b. **Articulations** with the humerus and ulna (see Figure 7-1)
(1) The top of the head provides an articular surface for the **humeroradial joint**.
(2) The side of the head provides the articular surface for the **radioulnar joint**.

II. ARTICULATIONS

A. **Elbow joint.** The elbow region consists of three articulations between the humerus, radius, and ulna. Muscles that act across these joints produce movement in the forearm (see Figure 7-1). This combined joint has two degrees of freedom.
1. **The humeroulnar joint** is formed by the trochlea of the humerus and the trochlear notch of the ulna (see Figure 7-1).
   a. **Movement.** This joint has one degree of freedom, permitting flexion/extension about a transverse (bilateral) axis through the trochlea. The pulley-shaped trochlear articular surface that coapts the trochlear notch of the ulna permits movement only in the parasagittal plane.
   b. **Support** is by the **ulnar (medial) collateral ligament.**
      (1) This ligament is a strong, fan-shaped condensation of the fibrous joint capsule composed of anterior, intermediate, and posterior fiber bundles.
      (2) When this ligament is torn, there is abnormal abduction at the elbow joint.
   c. **Variation**
      (1) The shape of the **trochlea** is variable among individuals and determines the angle (3°–29°) that the extended forearm makes with the arm, the **"carrying angle."**

(2) The carrying angle is usually greater in women than men—the anatomic reason that most women tend to bowl with a natural curve.

2. The humeroradial (radiohumeral) joint is formed by the capitulum of the humerus and the head of the radius (see Figure 7-1).

  a. Movement. It has two degrees of freedom, permitting flexion/extension as well as pronation/supination of the forearm. Because the radial socket for the capitulum is so shallow, this joint offers little support.

  (1) Flexion/extension occurs about a bilateral (transverse) axis through the trochlea.

  (2) Pronation/supination (medial rotation/lateral rotation) occurs about a vertical axis that passes through both centers of curvature of the proximal and distal radioulnar joints. The center of curvature of the proximal radioulnar joint is the head of the radius, and that of the distal radioulnar joint is the head of the ulna.

  b. Support. This joint is reinforced by **the radial (lateral) collateral ligament.**

  (1) This strong, fan-shaped condensation of the fibrous joint capsule is composed of anterior, intermediate, and posterior fiber bundles.

  (2) Tearing of this ligament permits abnormal adduction at the elbow joint.

  (3) Tennis elbow seems to involve inflammation of this ligament, the periosteum about its insertion, or the small underlying bursa, or a strain of the common extensor tendon.

3. The proximal (superior) radioulnar joint is formed by the side of the head of the radius and the radial notch of the ulna (see Figure 7-1). It functions in concert with the distal (inferior) radioulnar joint (see Figure 8-1).

  a. Movement. It has one degree of freedom, permitting **pronation/supination** about a vertical axis that passes through both the proximal and distal radioulnar joints.

  b. Support. It is reinforced by two ligaments.

  (1) The annular ligament provides the major reinforcement.

    (a) Originating on the anterior lip of the radial notch of the ulna, it passes about the head and neck of the radius and inserts on the posterior lip of the radial notch of the ulna.

    (b) It permits rotation of the radius relative to the ulna.

  (2) Interosseous membrane (see Figures 7-1 and 8-1)

    (a) Structure. A broad sheet of strong connective tissue extends between the medial edge of the radius and the lateral edge of the ulna for nearly the entire length of the forearm. The **oblique cord** of this membrane runs inferolaterally between the ulna, just distal to the radial notch, and the radius, distal to the radial tuberosity.

    (b) Function

      (i) Because the hand is attached primarily to the radius, force is transmitted from the hand to the radius. Because the humeroulnar joint is the most stable joint of the elbow, force is transmitted from the radius through the **interosseous membrane** to the ulna and humerus.

      (ii) This membrane also serves as attachment for the deep extrinsic flexor and extensor muscles of the hand.

B. Joint stability

1. Support. In general, the elbow joint is strong and stable. The coronoid process and the coaptation of the trochlea and the trochlear notch at the humeroulnar joint preclude dislocation, except under extreme force or secondary to fracture.

2. Dislocation in children. The humeroradial joint cannot withstand excessive traction in children because the radial head is undeveloped and can escape from the annular ligament, producing a **pulled elbow.**

3. Fractures. The elbow joint is particularly susceptible to impacting fracture, such as:
  a. Radial head fracture with or without fracture of the capitulum
  b. Fracture of the coronoid process with posterior dislocation of the elbow joint

76 | Chapter 7 II B

      c. Fracture of the olecranon process or evulsion of the triceps brachii muscle insertion

III. MUSCLE FUNCTION AT THE ELBOW JOINT

A. Movement. Depending on the lines of action, muscles in the arm produce movement at the glenohumeral joint, at the elbow joint, or both (see Table 7-1 and Figures 7-2 and 7-3).

  1. The anterior (flexor) compartment corresponds to primitively ventral musculature that flexes the forearm at the elbow joint. It has two layers.
    a. The **superficial layer** consists of the **biceps brachii muscle** with its long and short heads (Figure 7-2A; see Table 7-1).
    b. The **deep layer** comprises the **brachialis** as well as the **coracobrachialis muscles** (see Table 7-1 and Figure 7-2B).
    c. The anterior compartment musculature is innervated by the **musculocutaneous nerve.**

  2. The posterior (extensor) compartment corresponds to primitively dorsal musculature that extends the forearm at the elbow joint.
    a. The **superficial layer** consists of the **long head** and the **lateral head of the triceps brachii muscle** (Figure 7-3A; see Table 7-1).
    b. The **deep layer** includes the **medial head of the triceps brachii muscle** and the **anconeus muscle** (see Table 7-1 and Figure 7-3B).
    c. The posterior compartment musculature is innervated by the **radial nerve.**

B. Group actions. Movement of the forearm at the elbow joint is accomplished by muscles that run from the pectoral girdle or the arm to the forearm or wrist (Table 7-1).

  1. Flexion/extension (150°) occurs about a bilateral axis through the trochlea.
    a. Flexion. Flexor muscles include the **biceps brachii** (acting on the radius), **brachialis** (acting on the ulna), and **brachioradialis** when the forearm is semipronated. In addition, several of the wrist and hand flexor and extensor muscles that origi-

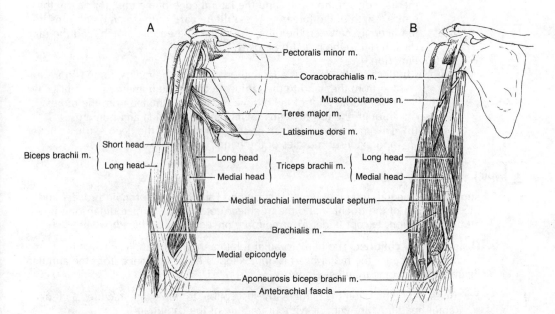

FIGURE 7-2. Flexor compartment of the right arm. *(A)* Superficial muscles; *(B)* deep muscles.

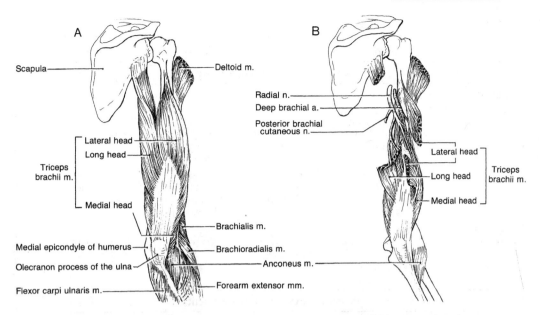

FIGURE 7-3. Extensor compartment of the right arm. *(A)* Superficial layer; *(B)* deep layer.

nate from the supracondylar ridges and epicondyles serve as weak forearm flexors (see Table 9-1).

    b. Extension. Extensor muscles include the **triceps brachii,** the **anconeus,** and the **brachioradialis** when the forearm is fully supinated and near full extension.

2. **Pronation/supination** (160°) occurs about an axis that passes through both the capitulum of the humerus and the distal ulna, as well as diagonally through the interosseous membrane.

    a. Pronation is accomplished by the **pronator teres** (long and short heads) and the **pronator quadratus,** both of which insert along the lateral edge of the radius and act by crossing the bones.

    b. Supination is accomplished by the **supinator** (a forearm muscle), which wraps about the radius and acts by unwinding, and the **biceps brachii,** which inserts on the radial tuberosity. The strength of the biceps as a supinator accounts for the fact that it is easier to drive a screw into wood with the right hand than with the left hand—even for left-handed persons.

C. Group innervations (see Table 7-1)

1. **Flexion** of the forearm is mediated principally by the **musculocutaneous nerve,** although the radial and median nerves are also involved in some secondary flexor function at the elbow. Thus, a lesion of any one of these nerves will greatly weaken, but not abolish, flexion.

2. **Extension** of the forearm is mediated by the **radial nerve.** Thus, injury to this nerve results in an inability to extend the elbow.

3. **Pronation** of the forearm is a function of the **median nerve,** although weak semi-pronation from a supinated position is possible through the brachioradialis muscle innervated by the radial nerve.

4. **Supination** of the forearm is a function of the **musculocutaneous and radial nerves.** The biceps brachii, more powerful than the supinator muscle, is innervated by the musculocutaneous nerve, the supinator by the radial nerve. A lesion of only one of these nerves will weaken, but not abolish, supination.

Table 7-1. Muscles Acting on the Forearm

Muscle	Origin	Insertion	Primary Action	Innervation
Anterior (Flexor Compartment)				
Superficial layer				
Biceps brachii:				
Long head	Supraglenoid tubercle of scapula	Radial tuberosity of the radius	Flex and supinate forearm at the elbow joint	Musculocutaneous n. (C5–C6, anterior)
Short head	Coracoid process of scapula	Radial tuberosity of the radius	Flex and supinate forearm at the elbow joint	Musculocutaneous n. (C5–C6, anterior)
Deep layer				
Brachialis	Distal half of anterior surface of humerus	Cornoid process of ulna	Flex forearm	Musculocutaneous n. (C5–C6, anterior)
Posterior (Extensor Compartment)				
Superficial layer				
Triceps brachii:				
Long head	Infraglenoid tubercle of scapula	Olecranon process of ulna	Extend forearm at elbow joint	Radial n. (C5–C6, posterior)
Lateral head	Posterior surface of humerus	Olecranon process of ulna	Extend forearm at elbow joint	Radial n. (C5–C6, posterior)
Deep layer				
Triceps brachii:				
Medial head	Posterior surface of humerus	Olecranon process of ulna	Extend forearm at elbow joint	Radial n. (C5–C6, posterior)
Anconeus	Posterior aspect of lateral epicondyle	Olecranon process	Extend forearm at elbow joint	Radial n. (C7–C8, posterior)

D. **Clinical considerations**

1. **Displaced midhumeral fractures** may damage the radial nerve and the deep radial artery as they wind about the posterior aspect of the humerus in the radial (musculospiral) groove.
2. **Supracondylar fractures** of the humerus may sever the brachial vessels and injure the median nerve. Traction of the triceps brachii muscle draws the proximal portion of the ulna posteriorly, and the brachialis muscle draws the distal humeral fragment anteriorly, jeopardizing the neurovascular bundle.
3. In **upper radial fractures,** the supinator muscles draw the proximal fragment laterally, and the pronator muscles draw the distal fragment medially.

IV. BRACHIAL VASCULATURE

A. **The brachial artery** supplies the brachium before bifurcating in the cubital fossa (Figure 7-4).

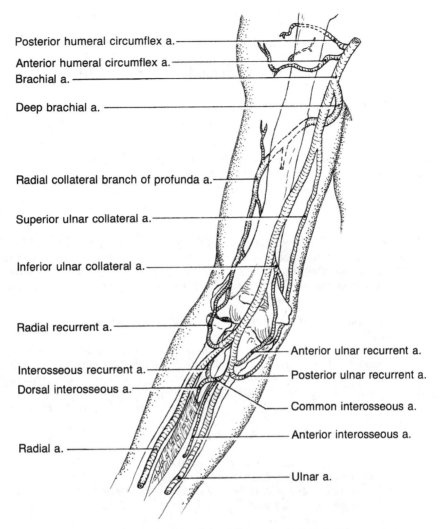

FIGURE 7-4. Arteries of the right arm and elbow region.

1. **Course.** This vessel, a continuation of the axillary artery, begins at the distal edge of the teres major muscle.
   a. It descends rather superficially along the medial border of the arm (see Figure 7-3B).
   b. In the cubital fossa, it lies deep to the bicipital aponeurosis, superficial to the brachialis muscle, and medial to the biceps brachii tendon. It is here that a **brachial pulse** is palpable.
2. **Major branches**
   a. The **deep brachial (profunda brachii) artery** supplies the posterior compartment of the arm.
      (1) **Course.** Together with the radial nerve, this deep branch passes posterior to and then lies lateral to the humerus in the **radial (musculospiral) groove.**
      (2) Three profunda branches anastomose with the posterior humeral circumflex, radial recurrent, and recurrent interosseous arteries.
   b. **Collateral circulation** about the elbow joint is rich. Temporary occlusion of the brachial artery during flexion, for instance, will not jeopardize circulation to the forearm.

(1) A **superior ulnar branch** passes posterior to the medial epicondyle to anastomose with the posterior recurrent ulnar artery.

(2) An **inferior ulnar branch** passes anterior to the medial epicondyle to anastomose with the anterior recurrent ulnar artery.

c. Distally in the cubital fossa, the brachial artery bifurcates into the **radial artery** and the **ulnar artery.**

3. **Variation.** The brachial artery may bifurcate anywhere in the brachium, resulting in a **superficial radial artery** (14% of the population) or a **superficial ulnar artery** (2% of the population).

4. **Volkmann's ischemic contracture.** Hemorrhage beneath the brachial or antebrachial fascia may compress uninjured collateral blood vessels, thereby producing ischemia of the forearm and hand musculature with paralysis and ultimate atrophy.

B. Venous return

1. **The deep veins of the brachium** parallel the brachial artery and its branches.
   a. The **radial** and **ulnar veins** of the forearm join to form two or three **brachial veins (venae comitantes brachiales),** which anastomose freely about the brachial artery.
   b. The **venae comitantes** join with the basilic vein in the region of the teres major muscle to form the axillary vein.

2. **The superficial veins of the brachium** lie in the subcutaneous tissue.
   a. **Cephalic vein**
      (1) This vein passes anterior to the lateral epicondyle along the anterior preaxial aspect of the brachium to the **deltopectoral triangle,** where it lies with the deltoid branch of the thoracoacromial artery. This site is commonly used for a surgical cut-down to insert an intravenous cannula.
      (2) It pierces the clavipectoral fascia and joins the axillary vein just distal to the first rib.
      (3) It is absent in about 10% of the population.
   b. **Basilic vein**
      (1) This vein passes anterior to the medial epicondyle and lies just medial to the biceps brachii along the brachium.
      (2) It penetrates the brachial fascia with the medial antebrachial cutaneous nerve, where it joins with the brachial veins near the teres major muscle to form the axillary vein.
   c. **Median cubital vein**
      (1) This connecting vein runs between the cephalic vein in the forearm through the cubital fossa to join the basilic vein in the arm. It is subject, however, to extreme variation. For example, the median cubital vein may be replaced by median cephalic and median basilic veins that join in the cubital fossa to form a median forearm vein.
      (2) It lies superficial to the bicipital aponeurosis, enabling it to be immobilized by the thumb of a phlebotomist for venipuncture.
      (3) It is a preferred site for **phlebotomy.**

C. Lymphatic drainage

1. **Deep lymphatics** accompany the brachial vein and drain through the brachial group of the **axillary lymph nodes.**

2. **Superficial lymphatics** drain along the superficial veins.
   a. Usually one or two **supratrochlear lymph nodes** are in the distal brachium, adjacent to the basilic vein just proximal to the medial epicondyle. These nodes frequently enlarge in association with superficial infections of the hand.
   b. The superficial drainage bypasses most of the axillary nodes, entering instead the **subclavian nodes.**

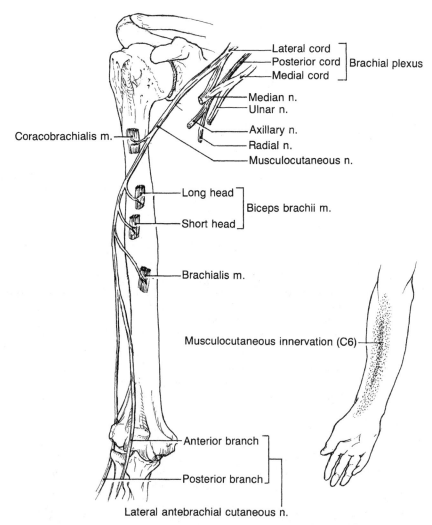

FIGURE 7-5. Course of the musculocutaneous nerve in the right arm and the muscles innervated by its branches. *(Inset)* the region of dermatomal innervation by the lateral antebrachial cutaneous nerve.

V. BRACHIAL INNERVATION

A. Musculocutaneous nerve (C5–C6, anterior)

1. **Course.** This nerve originates from the lateral cord of the brachial plexus and contains contributions from the anterior divisions of roots C5–C6 (Figure 7-5).
   a. It lies between the two heads of the coracobrachialis muscle as it passes toward the lateral side of the brachium.
   b. Immediately distal to the last muscular branch, it becomes the **lateral antebrachial cutaneous nerve,** which passes into the lateral aspect of the forearm.

2. **Distribution** (see Figure 7-5)
   a. **Motor.** This nerve innervates the **flexor muscles of the arm** (brachium), that is, the coracobrachialis, biceps brachii, and brachialis muscles.
   b. **Sensory.** The **lateral antebrachial cutaneous nerve** supplies the C6 dermatome along the preaxial (radial) side of the forearm.

3. Injury
   a. A lesion involving the musculocutaneous nerve produces the inability to flex and supinate the forearm strongly, **loss of the biceps tendon reflex,** and loss of sensation along the lateral aspect of the forearm.
   b. Some weak elbow flexion is possible despite this injury because of the secondary flexor action of the brachioradialis (radial innervation) and the forearm muscles that originate on the medial humeral epicondyle (median and ulnar innervation).

B. Median nerve (C6–C8, anterior)

1. **Course.** This nerve originates as a medial root from the lateral cord and a lateral root from the medial cord. The roots join anterior to the third portion of the axillary artery and contain contributions from the anterior divisions of roots C6–C8 (see Figure 9-6).
   a. It accompanies the brachial artery in the medial aspect of the brachium and does not give off branches in the brachium.
   b. It passes anterior to the medial epicondyle (see Figure 9-6A) and through the cubital fossa deep to the bicipital aponeurosis and usually (82%) medial to the brachial artery.

2. **Distribution**
   a. **Motor.** The median nerve has no motor function in the brachium. In the forearm, it supplies the pronator muscles and many of the flexors of the wrist and hand. Because some of these muscles originate from the medial epicondyle, they have secondary flexor function at the elbow joint.
   b. **Sensory.** This nerve has no sensory function in the brachium.

3. **Injury**
   a. A supracondylar fracture of the humerus may injure the median nerve and the brachial artery, resulting in deficits in the forearm and hand.
   b. The median nerve may be injured when drawing blood from the cubital fossa.

C. Ulnar nerve (C8–T1, anterior)

1. **Course.** This nerve is the direct continuation of the medial trunk, receiving contributions from the anterior divisions of roots C8–T1 (see Figure 9-6).
   a. It descends rather superficially along the medial side of the brachium, lying first medial to the brachial artery and then alongside the superior ulnar collateral artery.
   b. It passes posterior to the medial humeral epicondyle in the **ulnar groove** to enter the forearm.

2. **Distribution.** This nerve does not give off branches in the brachium. In the forearm, it supplies a few of the flexors of the wrist and hand, some of which originate from the medial epicondyle and, therefore, have a secondary flexor function at the elbow joint.

3. **Injury**
   a. Fracture of the medial epicondyle may produce ulnar nerve injury.
   b. Pressure on the ulnar nerve as it passes along the ulnar groove produces "funny bone" symptoms, with tingling along the hypothenar part of the hand and little finger.

D. Medial brachial cutaneous nerve (C8–T1, anterior).
This nerve originates from the medial cord of the brachial plexus and innervates the skin on the medial side of the arm.

E. Axillary (humeral circumflex) nerve (C5–C6, posterior)

1. **Course.** This nerve arises from the posterior cord of the brachial plexus, receiving contributions from the posterior divisions of roots C5–C6 (Figure 7-6).
   a. It runs lateral to the radial nerve and posterior to the axillary artery.
   b. It passes posterior to the humerus and anterior to the subscapularis muscle in company with the posterior humeral circumflex artery.
   c. It passes through the **quadrangular space** (teres minor, teres major, long head of

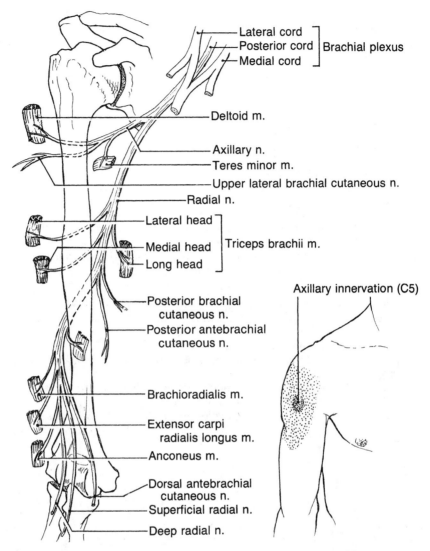

FIGURE 7-6. Course of the radial and axillary nerves in the right arm and the muscles innervated by their branches. *(Inset)* the region of dermatomal innervation by the axillary nerve in the arm.

the triceps, and surgical neck of the humerus) to reach the posterior aspect of the shoulder and brachium.

2. **Distribution** (see Figure 7-6)
   a. **Motor.** The axillary nerve supplies the teres minor and the deltoid muscles.
   b. **Sensory.** The nerve becomes superficial distal to the deltoid as the **superior lateral cutaneous nerve** of the arm, supplying the skin over the deltoid muscle.
3. **Injury.** Damage to this nerve results in deltoid paralysis, with total inability to abduct the arm and severe impairment of flexion and extension at the glenohumeral joint.

F. **Radial nerve** (C5–T1, posterior)

1. **Course.** This nerve is the direct continuation of the posterior cord of the brachial plexus once the axillary nerve diverges, thus receiving contributions from the posterior divisions of roots C5–T1 (see Figure 7-6).
   a. It runs posterior to the brachial artery.

b. It passes posterior to the medial head of the triceps in the **musculospiral (radial) groove** of the humerus in company with the deep radial artery (see Figure 7-3B).
   c. It passes anterior to the lateral epicondyle between the brachialis and brachioradialis muscles to enter the forearm.
2. **Distribution.** The radial nerve supplies the extensor and dorsal aspects of the arm and forearm (see Figure 7-6).
   a. Motor
      (1) Proximal to the musculospiral groove, medial muscular branches innervate the **extensor muscles of the arm,** that is, the long, medial, and lateral heads of the triceps brachii.
      (2) In the vicinity of the lateral epicondyle, lateral muscular branches innervate the anconeus, brachioradialis, and extensor carpi radialis longus muscles.
   b. Sensory. Branches arise from the radial nerve in the vicinity of the musculospiral groove deep to the long head of the triceps muscle.
      (1) The **lateral brachial cutaneous nerve** supplies the preaxial border of the arm; the **posterior (dorsal) brachial cutaneous** and **the posterior (dorsal) antebrachial cutaneous nerves** supply the dorsal arm and forearm.
      (2) In the vicinity of the lateral epicondyle, the **superficial branch of the radial nerve** arises and enters the forearm.
3. **Injury**
   a. Injury to the posterior cord **("Saturday night" or crutch palsy)** or to the radial nerve result in loss of the extensor muscles of the arm, forearm, and wrist **(wrist drop).**
   b. Midhumeral fracture may damage the radial nerve as it lies in the musculospiral groove, causing paralysis of the extensor muscles of the wrist and hand **(wrist drop).**
   c. Because the medial muscular branches of the radial nerve to the triceps arise proximal to this groove, extension of the forearm usually is not affected by midhumeral fracture, and some supination remains possible by biceps brachii action.

Chapter 8

Extensor Forearm and Posterior Wrist

I. BONES OF THE FOREARM AND PROXIMAL WRIST

A. Bony landmarks (Figure 8-1)

1. **The ulna,** including the shaft, head, and styloid process (medially), is palpable.
2. **The radius,** including the shaft, head, styloid process (laterally), and dorsal tubercle, is palpable.
3. **The carpal bones** are palpable.

B. Forearm bones

1. **The ulna** lies medially in the arm (see Figure 8-1).
   a. The proximal ulna is discussed in Chapter 7 I C 1.
   b. The **distal ulna** becomes secondary to the radius at the wrist with reference to force transmission (see Figure 8-1).
      (1) **Characteristics**
         (a) The ulna narrows as it extends distally, terminating in a slightly enlarged **distal head** and **styloid process.**
         (b) Anterolateral on the distal head is a surface that articulates with the **ulnar notch** of the radius, forming the **distal radioulnar joint.**
      (2) **Articulations**
         (a) The ulnar notch of the radius provides an articular surface for the **distal radioulnar joint.**
         (b) The **ulnocarpal joint** contains the **triangular (articular) disk.**
            (i) This disk fills the gap between the ulna and variable portions of the lunate and triquetral bones.
            (ii) It acts as a shock absorber, protecting the ulna and elbow joint from direct forces transmitted to the arm through the wrist.
2. **The radius** lies laterally in the arm.
   a. The proximal radius is discussed in Chapter 7 I C 2.
   b. The **distal radius** and its radiocarpal joint become primary at the wrist because it receives most of the force transmitted from the hand (see Figure 8-1).
      (1) **Characteristics.** The distal end of the radius widens transversely.
         (a) The posterior surface is grooved by the tendons of the extensor carpi radialis and the extensor pollicis longus muscles.
         (b) Laterally, the radius terminates in the **styloid process,** to which the brachioradialis muscle attaches.
         (c) The **ulnar notch,** on the medial surface of the distal head, coapts the head of the ulna and forms the **distal radioulnar joint.**
      (2) **Articulations** occur with the ulna and carpal bones at the wrist joint.
      (3) **Clinical considerations.** Fracture of the distal radius **(Colles' fracture)** is surpassed in frequency only by fracture of the clavicle, ribs, and digits.
         (a) The distal radius, composed of large mounts of cancellous bone within a thin rim of cortex, is the weakest point.
         (b) Anterior angulation of the proximal fragment may place the median nerve and radial artery in jeopardy.

C. Carpal bones. Two rows constitute the skeleton of the wrist.

1. **The proximal carpal row** consist of three true carpal bones and one sesamoid bone (see Figure 8-1).
   a. The **scaphoid (navicular) bone** is the most lateral of the proximal row.
      (1) It articulates with the radius proximally, with the lunate bone medially, and

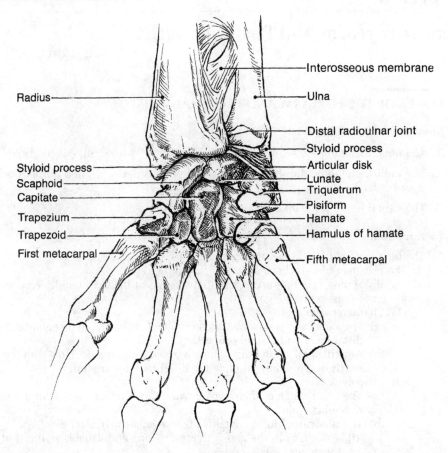

FIGURE 8-1. Bones and articulations of the distal forearm and wrist. Anterior view shows the radiocarpal, midcarpal, and carpometacarpal joints.

with the trapezium and trapezoid bones distally. It transmits forces from the abducted hand directly to the radius.
   (2) **Scaphoid fracture** is common because this carpal bone transmits forces from the abducted hand to the radius.
      (a) Because displacement usually is minimal and the bony cortex is thin, radiographic evidence of the fracture is subtle and often leads to misdiagnosis as a sprained ligament. If radiographic assessment is repeated 10 days after fracture (after the bone resorption stage of healing), the fracture will be apparent.
      (b) Because the nutrient artery enters the scaphoid distally, fracture may interrupt the blood supply to the proximal fragment and cause **avascular necrosis.**
   b. The **lunate bone** participates in both radiocarpal and ulnocarpal joints. It articulates with the radius and the triangular articular disk proximally, the scaphoid laterally, the triquetrum medially, and the capitate distally.
      (1) The degree of contact with the distal radial head and the articular disk varies with abduction/adduction.
      (2) **Lunate dislocation**
         (a) The lunate transmits forces from the adducted hand to the radius. In forced hyperextension of the wrist, such as occurs during a fall, the lunate may dislocate anteriorly into the carpal tunnel, compressing the median nerve and causing **carpal tunnel syndrome.**

(b) Because the lunate bone receives its major blood supply through vessels in the anterior and posterior radiocarpal ligaments, dislocation of the lunate can result in **avascular necrosis.**
  c. The **triquetral bone** participates in the ulnocarpal joint.
   (1) This bone articulates with the triangular articular disk proximally, the lunate bone laterally, the pisiform bone anteromedially, and the hamate bone distally.
   (2) The degree of triquetral contact with the articular disk is maximal in full adduction.
  d. The **pisiform bone,** which is not a true carpal bone, lies most medial in the proximal row of carpal bones.
   (1) It articulates only with the triquetral bone.
   (2) Developmentally, the pisiform is a **sesamoid bone,** embedded in the tendon of the flexor carpi ulnaris muscle. Thus, it is the only proximal carpal bone with a muscular attachment.
 2. **The distal carpal row** comprises four bones: the trapezium, trapezoid, capitate, and hamate (see Figure 8-1 and Chapter 9 I A 2 a–d).

D. **Wrist articulations**
 1. **Distal (inferior) radioulnar joint.** This articulation is formed by the ulnar notch of the distal radius and the distal head of the ulna. It functions in concert with the proximal (superior) radioulnar joint (see Figures 8-1 and 10-1).
  a. **Movement.** It has one degree of freedom, permitting **pronation/supination** about a vertical axis that passes through both the proximal and distal radioulnar joints.
  b. **Support**
   (1) The **articular (triangular) disk,** binding the distal ends of the radius and ulna firmly together, provides the major reinforcement.
   (2) The **dorsal** and **palmar radioulnar ligaments,** thickenings of the articular capsule, also support this joint.
   (3) **Interosseous membrane (septum)**
    (a) **Structure.** A broad sheet of strong connective tissue extends between the medial edge of the radius and the lateral edge of the ulna for nearly the entire length of the forearm.
    (b) **Functions**
     (i) **Force transmission.** Because the hand is attached primarily to the radius, force is transmitted from the hand to the radius. Because the humeroulnar joint is the most stable joint of the elbow, force is transmitted from the radius through the interosseous membrane to the ulna and humerus.
     (ii) **Muscle attachment.** This membrane also serves as attachment for the deep extrinsic flexor and extensor muscles of the hand.
 2. **Radiocarpal joint** (see Figures 8-1 and 10-1)
  a. **Structure.** This joint is located between the distal radial head and the scaphoid and lunate bones.
  b. **Movement.** The concave, ellipsoid articular surface of the distal end of the radius permits two degrees of freedom.
   (1) **Abduction/adduction** (radial deviation/ ulnar deviation) occurs about an anteroposterior axis through the head of the capitate.
    (a) In abduction (15°), the scaphoid bone makes maximal contact with the radius, and the lunate contacts the articular disk.
    (b) In adduction (45°), the lunate bone makes maximal contact with the radius, and the triquetrum contacts the articular disk.
   (2) **Flexion/extension** (170°) occurs about a transverse axis between the lunate and capitate bones. The radiocarpal joint is most stable in full flexion, when abduction/adduction is not possible.
   (3) **Circumduction** is possible because there are two degrees of freedom.
  c. **Support.** This joint is reinforced by several ligaments, but the high degree of mobility results in less stability and a predisposition to sprains.

(1) The **dorsal radiocarpal ligament** reinforces the dorsal side.
(2) The **palmar radiocarpal ligament** reinforces the ventral side.
(3) The **radial (lateral) collateral ligament** connects the radial styloid process and the scaphoid.

3. **Ulnocarpal joint** (see Figures 8-1 and 10-1)
   a. **Structure.** The end of the ulna is separated from the triquetral and lunate bones by an articular disk (the **triangular fibrocartilage**). This disk acts as a shock absorber, protecting the ulna and elbow joint from forces transmitted through the hand and wrist.
   b. **Movement.** The concave, ellipsoid articular surface of the triangular cartilage permits two degrees of freedom.
      (1) **Abduction/adduction** (radial deviation/ulnar deviation) occurs about an anteroposterior axis through the head of the capitate bone.
         (a) In abduction (15°), the lunate bone makes maximal contact with the articular disk.
         (b) In adduction (45°), the triquetrum makes maximal contact with the articular disk.
      (2) **Flexion/extension** (170°) occurs about a transverse axis between the lunate and capitate bones.
      (3) **Circumduction** is possible because there are two degrees of freedom.
   c. **Support.** This joint is reinforced by several ligaments, but the high degree of mobility results in less stability and a predisposition to sprains.
      (1) The **ulnar (medial) collateral ligament** connects the ulnar styloid process and the triquetrum.
      (2) An **ulnocarpal ligament** lies on the palmar side.

II. THE FOREARM EXTENSOR COMPARTMENT

A. **Antebrachial fascia** envelops the musculature of the forearm and divides the arm into an extensor and a flexor compartment (see Chapter 9 II).

1. **Organization**
   a. Septa lie between groups of muscles.
   b. The antebrachial fascia is the origin of some of the more superficial fascicles of the extensor muscles of the forearm.

2. **In the posterior compartment,** the antebrachial fascia condenses at the wrist to form the **extensor retinaculum** (Figure 8-2).
   a. **Compartments of the extensor retinaculum.** The extensor retinaculum is subdivided into six (occasionally seven) compartments by attachment in several places to bone. These compartments contain the tendons of the wrist and hand extensors (see Figure 8-2).
      (1) The first compartment (most medial) contains the extensor pollicis brevis and abductor pollicis longus. **De Quervain's syndrome** is the result of inflammation of the tendons of the first compartment (extensor pollicis brevis and abductor pollicis longus).
      (2) The second compartment contains the extensors carpi radialis longus and brevis. The third contains the extensor pollicis longus. The fourth contains the extensor digitorum and extensor indicis proprius. The fifth contains the extensor digiti minimi and the sixth the extensor carpi ulnaris.
   b. **Synovial sheaths** beneath the extensor retinaculum enclose the extensor tendons.

B. **Musculature.** The **primitively dorsal aspects** of the forearm, wrist, and hand can be considered a functional extensor unit. The extensor muscles of the forearm are mostly multijoint muscles, acting across the elbow, wrist, and carpal joints and, in some instances, the metacarpophalangeal joints as well.

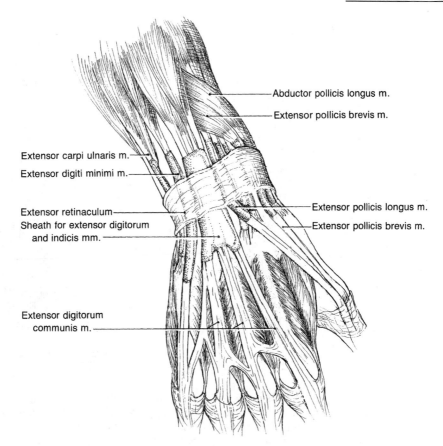

FIGURE 8-2. Extrinsic extensor muscles of the dorsum of the right hand.

1. **Organization.** The extensor muscles of the wrist and digits originate on the lateral (preaxial) aspect of the forearm.
   a. The **superficial group** of extensor muscles of the wrist and hand (Figure 8-3A) originates from the lateral supracondylar ridge and the lateral epicondyle of the humerus, as well as from the proximal radius. This group consists of the **brachioradialis, extensor carpi radialis longus, extensor carpi radialis brevis, extensor digitorum, extensor digiti minimi,** and **extensor carpi ulnaris.**
   b. The **deep group** of muscles (Figure 8-3B) originates from the midradius, the interosseous membrane, and the ulna. This group consists of the **supinator** (acting on the elbow), **abductor pollicis longus, extensor pollicis brevis, extensor pollicis longus,** and **extensor indicis.**

2. **Group actions at the wrist.** The action of each extrinsic extensor muscle at the radiocarpal joint is determined by the location of the tendon relative to the two axes of rotation. The primary actions are extension, abduction, and adduction of the wrist (Table 8-1).
   a. **Flexion/extension** occurs about a transverse axis through the radiocarpal and ulnocarpal joints (see L–M in Figure 9-4).
      (1) **Extensors of the wrist** are muscles that pass dorsal to the transverse axis of the radiocarpal joint. The prime extensors are the **extensor carpi radialis longus** and the **extensor carpi ulnaris,** acting together.
      (2) **Flexors of the wrist** are muscles that pass ventral to the transverse axis (see Chapter 9 II B 2).
   b. **Abduction/adduction** occurs about an anteroposterior axis through the radiocarpal and ulnocarpal joints (see A–P in Figure 9-4).

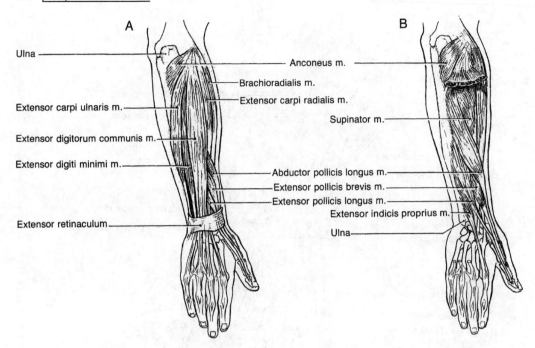

FIGURE 8-3. Extensor compartment muscles of the right forearm. *(A)* Superficial layer; *(B)* deep layer.

- **(1) Abductors of the wrist** are muscles that pass on the radial side of the anteroposterior axis of the radiocarpal joint. The prime extensor compartment abductor is the **extensor carpi radialis longus.** (The flexor carpi radialis muscle in the flexor compartment acts synergistically.)
- **(2) Adductors of the wrist** are muscles that pass on the ulnar (medial) side of the anteroposterior axis of the radiocarpal joint. The prime extensor compartment adductor is the **extensor carpi ulnaris.** (The flexor carpi ulnaris muscle in the flexor compartment acts synergistically.)
- **c. Dynamic stabilization of the wrist** is accomplished by the actions of the extensor carpi ulnaris longus and brevis muscles and the extensor carpi ulnaris muscle, together with the major flexors of the wrist.

3. **Innervation.** The extrinsic extensor muscles are innervated by branches of the **radial nerve** (see Table 8-1).

4. The dorsum of the hand has no intrinsic muscles.

III. INNERVATION OF THE EXTENSOR COMPARTMENT

A. Radial nerve (C5–T1, posterior). The extensor side of the forearm is innervated solely by the radial nerve and its branches (Figure 8-4).

1. **Course.** After leaving the musculospiral groove in the brachium, the radial nerve passes anterior to the lateral humeral epicondyle and between the brachialis and brachioradialis muscles, innervating those two muscles (see Figure 8-4).

2. **Distribution.** The radial nerve sends branches to three forearm muscles (anconeus, brachioradialis, and extensor carpi radialis longus) before dividing into the **superficial** and **deep branches** (see Table 8-1).
   a. **Motor.** The **deep branch** passes through and innervates the supinator muscle. Exiting the supinator muscle, it becomes the **posterior (dorsal) interosseous nerve,**

Table 8-1. Forearm Muscles Acting on the Distal Side of the Wrist and Hand

Muscle	Origin	Insertion	Primary Action	Innervation
Superficial layer				
Brachioradialis	Distal lateral surface of humerus	Styloid process of radius	Flexes, semipronates, and semisupinates forearm	Radial n. (C5–C6, posterior)
Extensor carpi radialis:				
Longus	Lateral epicondyle of humerus	Base of second metacarpal	Extends and abducts wrist	Radial n. (C6–C7, posterior)
Brevis	Lateral epicondyle of humerus	Base of third metacarpal	Extends wrist	Superficial or deep radial n. (C6–C7, posterior)
Extensor digitorum	Lateral epicondyle of humerus	Extensor expansion of index, middle, and ring fingers	Extends MP joint and, when fist is clenched, extends wrist	Post. interosseus branch of radial n. (C7–C8, posterior)
Extensor digiti minimi	Lateral epicondyle of humerus	All phalanges of fifth digit	Extends fifth digit	Post interosseus branch of radial n. (C7–C8, posterior)
Extensor carpi ulnaris	Lateral epicondyle of humerus	Base of fifth metacarpal	Extends and adducts wrist	Post. interosseus branch of radial n. (C7–C8, posterior)
Deep layer				
Supinator	Lateral epicondyle of humerus	Lateral aspect of midradius	Supinates forearm	Deep branch of radial n. (C5–C7, posterior)
Abductor pollicis longus	Post. surface of interosseous membrane	Base of first metacarpal, and ulna	Abducts thumb and wrist	Post. interosseous branch of radial n. (C8–T1, posterior)
Extensor pollicis:				
Brevis	Posterior midshaft of radius and interosseous membrane	Base of first phalanx of thumb	Extends thumb and abducts wrist	Post. interosseous branch of radial n. (C8–T1, posterior)
Longus	Post. surface of interosseous membrane and posterior ulna	Base of second phalanx of thumb	Extends thumb and abducts wrist	Post. interosseous branch of radial n. (C8–T1, posterior)
Extensor indicis	Interosseous membrane and ulna	Extensor expansion of index finger	Extends first digit and wrist	Post. interosseous branch of radial n. (C8–T1, posterior)

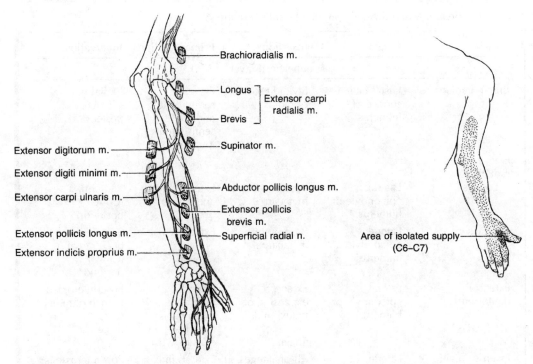

FIGURE 8-4. Course of the radial nerve in the right forearm and the muscles innervated by its branches. *Inset,* the region of dermatomal innervation by the superficial branch of the radial nerve.

which winds around the radius to lie immediately dorsal to the interosseous membrane. It innervates the remaining extensor muscles of the posterior (extensor) compartment of the forearm (see Figure 8-4).

  b. **Sensory.** The **superficial branch** usually arises at the level of the lateral epicondyle. It supplies the lateral aspect of the dorsum of the wrist and hand with a region of exclusivity in the dorsal web space between the thumb and index finger (see Figure 8-4, inset). Unless it supplies the extensor carpi radialis brevis, this branch has no motor contribution.

B. **Nerve injury.** Injuries involving the radial nerve produce signs that vary according to the level of the lesion.

  1. **Proximal to the epicondyles,** such as by a supracondylar or epicondylar humeral fracture, injury involves both superficial and deep branches.
     a. Deep branch injury results in pronation of the hand (loss of supination), **wrist drop** (inability to extend the wrist), and inability to extend the digits and thumb.
     b. Superficial branch injury results in loss of sensation to the dorsum of the hand and thumb.
  2. **Distal to the epicondyles,** such as by fracture of the proximal third of the radius, injury usually involves only the deep branch. **Wrist drop** is a common result, with inability to extend the digits and thumb but loss of neither supination nor sensation to the dorsum of the hand.

Chapter 9

Flexor Forearm and Anterior Wrist

I. BONES OF THE DISTAL WRIST

A. **Eight carpal bones,** lying in two rows, constitute the skeleton of the wrist (see Figure 8-1).
  1. **The proximal carpal row** consists of three true carpal bones (**scaphoid, lunate,** and **triquetral**) as well as one sesamoid bone (**pisiform**) (see Chapter 8 I C 1).
  2. **The distal carpal row** comprises four bones (see Figure 8-1).
     a. The **trapezium** or greater multangular bone is the most lateral of the distal carpal bones. It articulates with the scaphoid proximally, the trapezoid medially, and the first and second metacarpals distally.
     b. The **trapezoid bone** or lesser multangular bone articulates with the scaphoid proximally, the trapezium laterally, the capitate and the second metacarpal medially, and the first and second metacarpals distally.
     c. The **capitate bone** is the keystone of the carpal arch.
        (1) It articulates with the scaphoid and lunate bones proximally, the trapezoid laterally, the hamate medially, and the second, third, and fourth metacarpals distally.
        (2) It transmits forces from the second, third, and fourth fingers to the proximal row of carpal bones. **Lunate dislocation** is accompanied by proximal displacement of the capitate, as a result of which the middle digit extends not much further than the second and fourth digits.
     d. The **hamate bone** is the most medial of the distal row of carpal bones.
        (1) It articulates with the triquetrum proximally, the capitate laterally, and the fourth and fifth metacarpals distally.
        (2) The **hamulus,** a hook-like process that gives the hamate its name, projects toward the palm just distal to the pisiform bone.
           (a) Fracture of this structure is rare, and is seen most frequently in golfers. Radiographic diagnosis of this injury is difficult.
           (b) The superficial palmar branch of the ulnar nerve may become entrapped as it passes adjacent to the hamulus.
  3. **The carpal tunnel** lies in the medial aspect of the wrist (Figure 9-1).
     a. The **transverse carpal arch** is formed by the concave arrangement of the carpal bones, enhanced by the hamulus and pisiform on the ulnar side and the scaphoid on the radial side.
     b. The **transverse carpal ligament** bridges the transverse carpal arch to form the carpal tunnel.

B. **The midcarpal joint** lies between proximal and distal rows of carpal bones (see Figure 8-1).
  1. **Movement.** Small amounts of accommodative gliding movements occur in this joint during abduction/adduction and flexion/extension, and as the hand is flattened or hollowed.
  2. **Support.** The intercarpal joints are reinforced by numerous **dorsal intercarpal ligaments** and a palmar **carpal radiate ligament.**

II. THE FOREARM FLEXOR COMPARTMENT

A. **Antebrachial fascia.** The deep fascia of the forearm envelops the muscles of the forearm and divides the forearm into flexor and extensor (see Chapter 8 III) compartments.

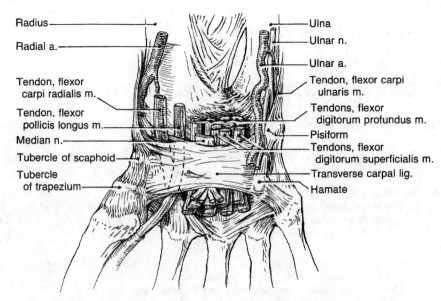

FIGURE 9-1. Carpal tunnel. Note the relations of the extrinsic flexor tendons of the hand as they pass beneath the transverse carpal ligament, as well as the relations of the median and ulnar nerves and the radial and ulnar arteries.

1. **Organization**
   a. Septa separate muscle groups.
   b. Proximally, on the anterior side, the **bicipital aponeurosis** of the biceps brachii muscle inserts into this fascial layer.
   c. The antebrachial fascia is the origin of some of the more superficial fascicles of the flexor muscles of the forearm.
2. **At the volar (palmar) wrist,** the antebrachial fascia condenses to form the **flexor retinaculum,** which consists of the more superficial **volar carpal ligament** as well as **the deeper transverse carpal ligament** (see Figure 9-1).
   a. The **volar carpal ligament** separates the tendon of the superficial palmaris longus muscle from the underlying ulnar artery and ulnar nerve.
   b. The **transverse carpal ligament** bridges the transverse carpal arch to form the **carpal tunnel.**
      (1) The **transverse carpal arch** is formed by the nearly semicircular arrangement of the carpal bones.
      (2) The **transverse carpal ligament** runs from the triquetrum, pisiform, and hamulus to the scaphoid and trapezium. It maintains the concavity of the carpal arch.
         (a) It prevents the tendons of the extrinsic flexor muscles of the hand from bowstringing on flexion.
         (b) Beneath it run the **median nerve,** as well as the **flexor digitorum superficialis, flexor digitorum profundus,** and **flexor pollicis longus muscles.**
3. **In the palmar hand,** the superficial layer of the antebrachial fascia continues as the thickened and fan-shaped **palmar aponeurosis.** Septa from this aponeurosis to metacarpal bones compartmentalizes the palm (Figure 9-2A).

B. Musculature. The flexor muscles of the wrist and hand originate from the postaxial (medial) aspect of the distal arm and proximal forearm. They correspond to the primitively ventral musculature. These mostly multijoint muscles act across the elbow joint and the wrist joint, as well as (in some instances) the carpal, metacarpal, and interphalangeal joints.

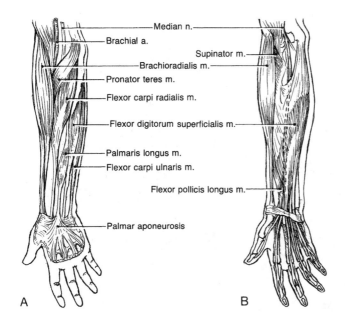

FIGURE 9-2. Muscles of the flexor compartment of the right forearm. *(A)* Superficial layer; *(B)* intermediate layers.

1. **Organization.** The flexor compartment muscles are arranged in three layers.
   a. The **superficial layer** includes the **pronator teres, flexor carpi radialis, palmaris longus,** and **flexor carpi ulnaris.** This layer comprises the primary flexors of the wrist, which originate from the medial supracondylar ridge and medial epicondyle of the humerus and the anterior aspect of the ulna (see Table 9-1 and Figure 9-2A).
   b. The **intermediate layer** consists of the **flexor digitorum superficialis.** In the forearm, it lies in the flexor compartment. In the wrist, this muscle divides into four tendons that lie in the **carpal tunnel** beneath the transverse carpal ligament (see Figure 9-2B). In the hand, the tendon to the index finger lies in the **thenar bursa** and the tendons for the third, fourth, and fifth digits lie in the **midpalmar bursa.** In the fingers, these tendons split near their insertions into the base of each side of the middle phalanx of the second through fifth digits (see Table 9-1 and Figure 10-4).
   c. The **deep layer** consists of the **flexor digitorum profundus, flexor pollicis longus,** and **pronator quadratus.** This group originates from the proximal ulna, the anterior surface of the interosseous membrane, and the anterior midradius (Figure 9-3; see Table 9-1).
      (1) The **flexor pollicis longus** muscle lies on the radial side of the forearm beneath the brachioradialis. In the wrist, the tendon of the flexor pollicis longus lies in the **radial bursa** within the **carpal tunnel** beneath the transverse carpal ligament (see Figure 9-3). In the hand, it turns laterally toward the thumb and passes between the two heads of the flexor pollicis brevis muscle to reach the distal phalanx of the thumb (see Figure 10-4).
      (2) The **flexor digitorum profundus** muscle lies in the flexor compartment of the forearm where it divides into four tendons (see Figure 9-3). In the wrist, the four tendons lie in the **carpal tunnel** beneath the transverse carpal ligament. In the hand, the tendon to the index finger lies in the **thenar bursa** and the tendons for the third, fourth, and fifth digits lie in the **midpalmar bursa.** In the fingers, these tendons pass through the split in the flexor superficialis tendon to insert into the base of the distal phalanx of the second through fifth digits (see Figure 10-4).

2. **Group actions.** The extrinsic flexor muscles flex the digits, thumb, and wrist (Table 9-1).

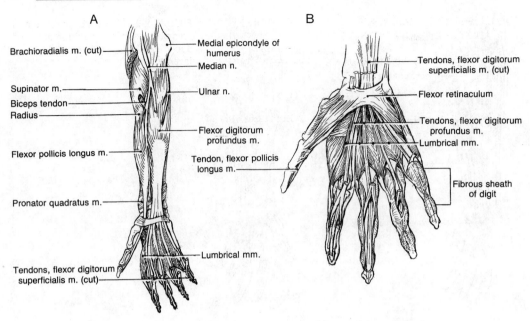

FIGURE 9-3. Deep muscles of the flexor compartment of the right forearm.

   a. Group actions at the wrist. The action of each muscle at the radiocarpal joint is determined by the location of the tendon relative to the two axes of rotation (Figure 9-4; see Table 9-1).
      (1) Flexion/extension occurs about a transverse axis (L–M in Figure 9-4).
         (a) Flexors of the wrist are the muscles that pass ventral to this axis (see Figure 9-4). The prime flexors are the **flexor carpi radialis** and **flexor carpi ulnaris**, with assistance from the **palmaris longus**.
         (b) Extensors of the wrist are the muscles that pass dorsal to the transverse axis of the radiocarpal joint (see Figure 9-4 and Chapter 8 II B 2).
      (2) Abduction/adduction occurs about an anteroposterior axis (A–P in Figure 9-4).
         (a) Abductors of the wrist are muscles that pass on the radial (lateral) side of the anterior–posterior axis (see Figure 9-4), primarily the **flexor carpi radialis**. The extensor carpi radialis longus in the extensor compartment acts synergistically.
         (b) Adductors of the wrist are muscles that pass on the ulnar (medial) side of this axis (see Figure 9-4), primarily the **flexor carpi ulnaris**. The extensor carpi ulnaris in the extensor compartment acts synergistically.
      (3) The wrist is **dynamically stabilized** by the actions of the flexor carpi radialis and flexor carpi ulnaris, together with the major extensors of the wrist.
   b. Group actions on the fingers (see Chapter 10 III B 2)
   c. Midradial fracture. This type of fracture is associated with less displacement than with proximal fractures because the action of the supinator is opposed by the pronator teres; however, the pronator quadratus still draws the distal fragment medially.

3. Group innervations. Except for one and one-half muscles, all flexor compartment muscles are innervated by the **median nerve** (see Table 9-1).
   a. The flexor carpi radialis and pronator quadratus, as well as all extrinsic flexor muscles of the hand, are innervated by the median nerve.
   b. The medial half of the **flexor digitorum profundus** and the flexor carpi ulnaris are innervated by the **ulnar nerve.**

Table 9-1. Forearm Muscles Acting on the Palmar Side of the Wrist and Hand

Muscle	Origin	Insertion	Action	Innervation
Superficial				
Pronator teres:				
Humeral head	Medial epicondyle of humerus	Lateral surface of midradius	Pronate and weakly flex forearm	Median n. (C6–C7, anterior)
Ulnar head	Proximal ulna	Lateral surface of midradius	Pronate and weakly flex forearm	Median n. (C6–C7, anterior)
Flexor carpi radialis	Medial epicondyle of humerus	Bases of second and third metacarpals	Flex wrist and weakly flex forearm	Median n. (C6–C7, anterior)
Palmaris longus	Medial epicondyle of humerus	Palmar aponeurosis	Flexes wrist	Median n. (C7–C8, anterior)
Flexor carpi ulnaris	Medial epicondyle of humerus and proximal ulna	Pisiform bone and base of fifth metacarpal	Flex wrist and weakly flex forearm	Ulnar n. (C8–T1, anterior)
Intermediate				
Flexor digitorum superficialis	Medial epicondyle of humerus, radius, and ulna	Base of second phalanx of second through fifth digits	Flexes proximal interphalangeal joint, metacarpal joint, and wrist	Median n. (C7–T1, anterior)
Deep				
Flexor digitorum profundus	Anterior proximal ulna and interosseous membrane	Base of third phalanx of second through fifth digits	Flexes distal and proximal interphalangeal joint, metacarpophalangeal joint, and wrist	Heads for second and third digits: median n. (C7-T1, anterior) Heads for fourth and fifth digits: ulnar n. (C8–T1, anterior)
Flexor pollicis longus	Anterior medradius and interosseous membrane	Lateral aspect base of second phalanx of thumb	Flexes thumb, meta-carpo-phalangeal joint, and wrist	Median n. (C7–T1, anterior)
Pronator quadratus	Anterior surface of ulna	Anterior surface of radius	Pronates forearm	Median n. (C6–C7, anterior)

III. ANTEBRACHIAL VASCULATURE

A. **Arterial supply.** The **radial artery** and a somewhat larger **ulnar artery** arise from the bifurcation of the **brachial artery** in the cubital fossa.

    1. The radial artery generally supplies the preaxial regions of the forearm and hand (Figure 9-5).

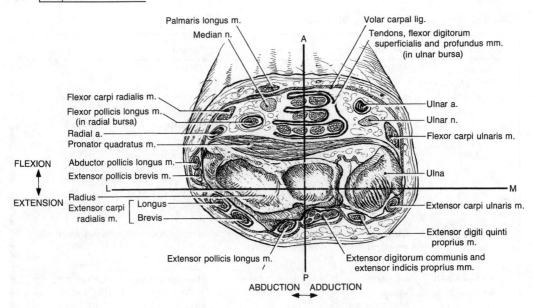

FIGURE 9-4. Flexor and extensor tendons of the wrist. The right wrist is supinated. The transverse axis (L–M) and the anteroposterior axis (A–P) of the radiocarpal joint are indicated. Muscles that lie to the lateral side of the A–P axis abduct; those to the medial side adduct. Muscles that lie to the palmar side of the L–M axis flex; those to the dorsal side extend.

FIGURE 9-5. Arterial supply to the forearm.

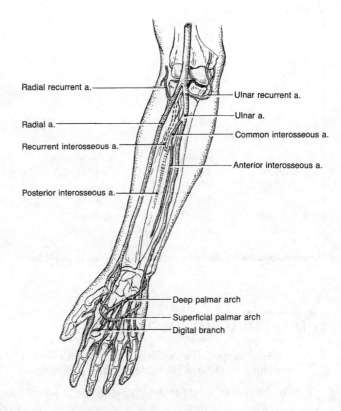

a. **Course.** This major vessel crosses the biceps brachii tendon deep to the bicipital aponeurosis, passes superficial to the pronator teres muscle, and descends along the anterior preaxial border of the forearm.
      (1) A **radial pulse** is palpable at the wrist between the tendons of the brachioradialis and the flexor carpi radialis muscles.
      (2) The **main stem** of this artery continues deep to the volar (palmar) carpal ligament. It then passes to the dorsal side of the wrist where it lies in the floor of the anatomic snuffbox. Here, a **radial pulse** is palpable. The artery continues toward the dorsal aspect of the hand.
   b. **Branches**
      (1) The **radial recurrent branch** returns anterior to the lateral epicondyle to anastomose with the radial (anterior) collateral of the deep brachial artery.
      (2) Numerous muscular branches are given off.
      (3) Distally, the **palmar carpal branch** contributes to the **anterior** and **posterior carpal networks (rete).**
      (4) The **superficial palmar branch** provides the smaller radial contribution to the **superficial palmar arch.**
      (5) It becomes the major contributor to the **deep palmar arch.**

2. **The ulnar artery** generally supplies the postaxial regions of the forearm and hand (see Figure 9-5).
   a. **Course.** This major vessel passes deep to the pronator teres muscle and deviates toward the anterior postaxial border of the forearm.
      (1) It continues deep to the palmar (volar) carpal ligament and superficial to the transverse carpal ligament to enter the hand.
      (2) The **ulnar pulse** is palpable just to the radial side of the pisiform bone. On occasion, the ulnar artery may be small or even absent.
   b. **Branches**
      (1) **Collateral branches** at the elbow
         (a) The **anterior ulnar recurrent branch** returns anterior to the medial epicondyle to anastomose with the inferior ulnar collateral of the brachial artery.
         (b) The **posterior ulnar recurrent branch** returns posterior to the medial epicondyle to anastomose with the superior ulnar collateral of the brachial artery.
      (2) The **common interosseous artery** bifurcates almost immediately.
         (a) The **anterior interosseous artery** descends with the anterior interosseous nerve and passes through the interosseous membrane distally to anastomose with the posterior interosseous artery.
         (b) The **posterior interosseous artery** passes through the interosseous membrane and descends in the forearm.
            (i) The recurrent interosseous branch passes behind the elbow to anastomose with the posterior branch of the deep brachial artery.
            (ii) Distally, the posterior interosseous artery returns through the interosseous membrane and anastomoses with the anterior interosseous artery at the wrist.
      (3) Near the pisiform bone, the ulnar artery gives off contributions to the **anterior** and **posterior carpal networks (rete).**
      (4) At the wrist, the **deep palmar artery** arises and contributes to the deep palmar arch.
      (5) It terminates as the major contributor to the **superficial palmar arch.**

B. Venous return

1. **Superficial veins** vary in course and size, but generally they drain more of the distal forearm and hand than do the deep veins.
   a. The **cephalic vein** originates from a venous plexus on the radial (preaxial) side of the dorsum of the hand.
      (1) At the wrist, it courses posterior to the styloid process of the radius. This site is commonly used for insertion of intravenous catheters.

(2) In the distal third of the forearm, it comes to lie anteriorly along the preaxial border.
(3) Just distal to the cubital fossa, it gives rise to the **median cubital vein,** which crosses the cubital fossa to join the basilic vein just superior to the medial epicondyle.

b. The **basilic vein** originates from a venous plexus on the ulnar (postaxial) side of the dorsum of the hand.
(1) This vein courses posterior to the ulna. In the middle third of the forearm, it comes to lie anteriorly along the postaxial border to reach the elbow region. Just below the medial epicondyle, it usually receives the median (anterior) antebrachial vein.
(2) Above the medial epicondyle, it receives the **median cubital vein.**

c. The **median cubital vein** crosses the cubital fossa between the cephalic and basilic veins just superior to the medial epicondyle. It is usually used for phlebotomy (venipuncture).

d. The **median (anterior) antebrachial vein** (when present) is variable, frequently consisting of a venous plexus.
(1) It drains the palm of the hand and anterior aspect of the forearm.
(2) It usually drains into the basilic vein, inferior to the medial epicondyle, but it may drain into the median cubital vein. If the latter is absent, it may drain into both basilic and cephalic veins.

2. **Deep veins** accompany the named arteries, often as **venae comitantes.**

3. **Phlebotomy (venipuncture)** is easily accomplished on the superficial veins of the cubital fossa.
   a. Care must be taken to keep the tip of the needle superficial to the bicipital aponeurosis to avoid the brachial artery.
   b. A snug, but not tight, tourniquet will occlude the superficial venous return, causing distention of the superficial veins. If a tourniquet is too tight, the arterial supply is occluded so that no venous return occurs.

C. Lymphatic drainage

1. **Superficial and deep lymphatics** accompany the superficial and deep veins.
2. A **supratrochlear node** lies just proximal to the medial epicondyle adjacent to the basilic vein. It receives drainage from the hypothenar aspect of the hand.

IV. INNERVATION OF THE FLEXOR COMPARTMENT

A. The median nerve (C6–C8, anterior) originates from the lateral and medial cords of the brachial plexus and innervates muscles of the flexor forearm and hand. It also has sensory function.

1. **Course** (Figure 9-6)
   a. **In the proximal forearm,** after passing anterior to the medial humeral epicondyle in the cubital fossa, the median nerve lies anterior to the brachialis muscles and medial to the brachial artery, which is immediately medial to the tendon of the biceps brachii muscle. It then passes between the humeral and ulnar heads of the pronator teres muscle.
   b. **In the middle of the forearm,** it passes between the humeroulnar and radial heads of the flexor digitorum superficialis muscle, briefly in company with the ulnar artery.
   c. **At the wrist,** it lies between the tendons of the palmaris longus and flexor carpi radialis muscles. It then passes through the carpal tunnel beneath the transverse carpal ligament to enter the hand (see Figure 9-1).

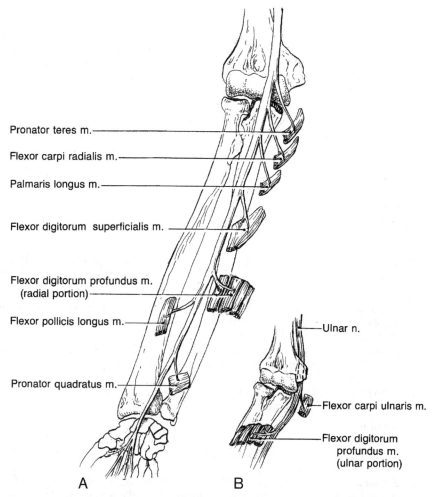

FIGURE 9-6. Course of the median *(A)* and ulnar *(B)* nerves in the forearm and the muscles innervated by their branches.

   2. Distribution
   a. Motor. The median nerve supplies all of the flexor muscles and pronator of the forearm, with the exception of the flexor carpi ulnaris and the ulnar half of the flexor digitorum profundus (see Figure 9-6). Innervation is by direct branches or by the anterior interosseous branch.
   (1) Muscular branches arise in the cubital fossa and innervate the superficial and intermediate groups of flexor muscles (pronator teres, flexor carpi radialis, palmaris longus, and flexor digitorum superficialis).
   (2) The **anterior interosseous branch** arises as the median nerve passes through the pronator muscle. It innervates the deep group (the radial half of the flexor digitorum profundus, flexor pollicis longus, and pronator quadratus muscles).
   b. Sensory. The median nerve supplies no sensation to the forearm. The **superficial palmar (palmar cutaneous) branch** arises just proximal to the carpal tunnel and passes superficially to supply the lateral proximal palmar surface of the hand.
   3. Injury. Injuries involving the median nerve produce signs that vary according to the level of the lesion.
   a. Proximal to the epicondyles, injury such as by a supracondylar or epicondylar humeral fracture produces a supinated forearm, weak flexion and abduction of the

wrist, paralysis of most of the muscles of the thenar side of the hand, and loss of sensation on the lateral side of the palm.
- b. **Superficial injury at the wrist,** such as by superficial lacerations, may sever the superficial branch of the median nerve with loss of sensation on the thenar side without loss of sensation along the major portions of the volar side of the first, second, and third fingers.
- c. **Carpal tunnel syndrome** results from compression of the median nerve within the carpal tunnel.
  - (1) It produces weakness of flexion and abduction of the thumb, inability to oppose the thumb, and inability to extend fully the first and second fingers, as well as loss of sensation along the volar aspects of the first, second, and third fingers.
  - (2) Because the superficial branch is not involved, no sensory loss is experienced along the radial side of the palm.
- d. **Deep injury at the wrist,** such as by a deep laceration, severs both the median nerve and its superficial branch, producing sensory and motor deficits in the hand.

B. **The ulnar nerve** (C8–T1, anterior) originates from the medial cord of the brachial plexus and innervates muscles of the flexor forearm and hand. It also has sensory function.

1. **Course** (see Figure 9-6)
   - a. **In the arm,** after passing posterior to the medial humeral epicondyle, this nerve enters the forearm by passing between the humeral and ulnar heads of the flexor carpi ulnaris muscle.
   - b. **In the forearm,** it is joined by the ulnar artery as it descends and passes lateral to the pisiform bone under the volar carpal ligament and superficial to the transverse carpal ligament.
   - c. **In the hand,** it divides into superficial and deep terminal branches at the base of the hypothenar eminence.
2. **Distribution**
   - a. **Motor.** The ulnar nerve supplies only the flexor carpi ulnaris and the ulnar half of the flexor digitorum profundus (see Figure 9-6). These motor branches arise in the vicinity of the elbow.
   - b. **Sensory.** The **superficial palmar (palmar cutaneous) branch** arises in the middle of the forearm to supply the medial proximal palmar surface of the hand.
3. **Injury** in the vicinity of the medial epicondyle, such as by a **supracondylar fracture** or **epicondylar fracture,** results in weakness of flexion and adduction of the wrist.

C. **The lateral antebrachial cutaneous nerve** (C6, anterior), a continuation of the musculocutaneous nerve, supplies the C6 dermatome along the preaxial (radial) side of the forearm.

D. **The medial antebrachial cutaneous nerve** (C8–T1, anterior) originates from the medial cord of the brachial plexus and innervates the skin on the ventromedial side of the forearm. It generally accompanies the basilic vein.

Chapter 10
Hand

I. BONES AND JOINTS

A. Bones. The bony skeleton consists of **metacarpals** and **phalanges,** all of which are palpable (Figure 10-1).

1. **Metacarpals.** Five metacarpal bones form the hand (see Figure 10-1).
   a. Each metacarpal bone has a proximal **base,** a **shaft,** and a distal **head.**
   b. The base of each metacarpal contacts the distal row of carpal bones to form a **carpometacarpal joint.**
   c. The head of each metacarpal contacts a proximal phalanx at a **metacarpophalangeal (MP) joint.**

2. **Phalanges.** Three rows of phalanges constitute the skeleton of the second through fifth digits (fingers); the first digit (thumb) has only two phalanges (see Figure 10-1).
   a. The proximal row articulates with the metacarpals at the **MP joints.**
   b. The middle row articulates with the proximal row of phalanges at the **proximal interphalangeal (PIP) joints.**
   c. The distal row articulates with the middle row of phalanges at the **distal interphalangeal (DIP) joints.**

3. **Fractures** are common.
   a. Fractures of the metacarpal shafts result in ventral displacement of the distal fragments because of the pull of the interossei muscles.
   b. Proximal phalangeal fractures usually displace toward the palm because of the pull of the intrinsic flexors.

B. Articulations. Four separate rows of joints make up the hand and digits.

1. **The carpometacarpal joints** lie within the palm of the hand (see Figure 10-1).
   a. **The first (thumb) carpometacarpal joint** is between the trapezium of the distal carpal row and the first metacarpal.
      (1) **Movement.** This saddle-shaped joint has 3 degrees of freedom.
         (a) **Flexion/extension** (75°) of the thumb occurs about a complex axis. The plane of flexion/extension is approximately 60° to that of the hand.
            (i) **Flexion** brings the thumb ventral to the plane of the hand toward the palm.
            (ii) **Extension** brings the thumb back into the plane of the hand.
         (b) **Abduction/adduction** (60°) of the thumb occurs about an axis perpendicular to the plane of the hand.
            (i) **Abduction** (15°) is movement of the extended thumb away from the index finger.
            (ii) **Adduction** (45°) brings the extended thumb against the index finger.
         (c) **Opposition/reposition** is a form of **circumduction,** which because of the sellar shape of the joint surfaces, involves rotational movement of the thumb.
            (i) **Opposition** is a medial rotation produced by flexion and adduction.
            (ii) **Reposition** is a lateral rotation produced by extension and abduction.
      (2) **Support.** Each joint is reinforced by **carpometacarpal ligaments.**
   b. **Second third, fourth, and fifth carpometacarpal joints** are condyloid, permitting two degrees of freedom.
      (1) **Movement** at this joint is slight, but important.
         (a) A slight amount of **flexion/extension** occurs. Little **abduction/adduction** is permitted by the deep transverse metacarpal ligaments.
         (b) **Circumduction** at these joints is essential for the fingers to meet. The com-

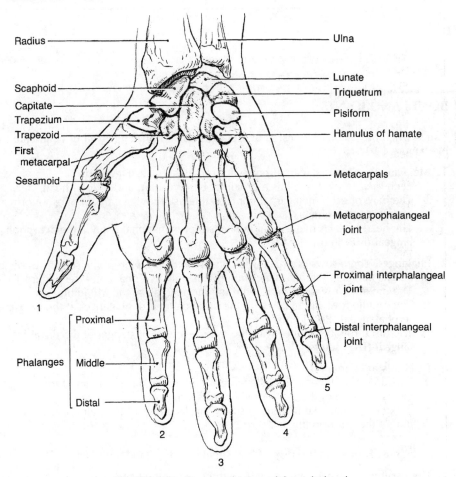

FIGURE 10-1. Bones and joints of the right hand.

bination of slight adduction and modest flexion accentuates the **transverse metacarpal arch** (the hollowing of the hand).
    (2) **Support.** Each joint is reinforced by carpometacarpal ligaments.
2. **The MP joints** represent the knuckles (see Figure 10-1).
    a. **Structure.** These joints are condyloid, with 2 degrees of freedom.
    b. **Movement** at these joints is considerable.
        (1) **Flexion/extension** (100°) occurs about transverse axes.
        (2) **Abduction/adduction** (as much as 30° for the index finger) occurs about anteroposterior axes through each MP joint. Such movement is with reference to the centerline of the hand, which passes along the middle finger.
        (3) **Circumduction** (the combination of movements) is possible because the MP joint has two degrees of freedom.
    c. **Support** is by MP collateral ligaments and the deep transverse metacarpal ligament.
        (1) The **collateral ligaments** that reinforce these joints are slack in the extended finger and taut in the flexed finger. They limit abduction.
        (2) The **deep transverse metacarpal ligaments** interconnect the heads of the metacarpals.
3. **Interphalangeal (IP) joints** consist of two rows: **the proximal interphalangeal (PIP) joints** and **distal interphalangeal (DIP) joints** (see Figure 10-1).
    a. **Structure.** The PIP and DIP are simple hinge joints.

(1) The **PIP joints** are between the proximal and middle phalanges.
(2) The **DIP joints** are between the middle and distal phalanges.
(3) The first digit (thumb) has only one IP joint.
  b. **Movement.** These joints have only one degree of freedom, with **flexion/extension** (90°) occurring about transverse axes.
  c. **Support.** These joints are strongly reinforced by **collateral ligaments,** which are slack when the fingers are extended.

II. DORSAL MUSCULATURE

A. **Fascia of the dorsal hand** is a continuation of the antebrachial fascia.
  1. **The antebrachial fascia** condenses at the dorsum of the wrist to form the **extensor retinaculum** (see Figure 8-2).
    a. This retinaculum is subdivided into six or seven compartments by attachment to bone.
    b. The tendons of the extrinsic extensor muscles of the hand pass through these compartments to gain access to the dorsum of the hand.
      (1) **Synovial sheaths** beneath the extensor retinaculum enclose the extensor tendons (see Figure 8-2). The sheaths terminate just distal to the extensor retinaculum where the tendons lie directly on the metacarpal bones.
      (2) The tendons proceed to the individual digits, between which are variable interconnections that tend to limit the individual movement of the fingers.
  2. **The deep dorsal fascia** is formed by the fusion of the antebrachial fascia on the dorsum of the hand with the extensor tendons (which are unsheathed on the dorsum of the hand).
  3. **The dorsal subcutaneous space** lies between the skin and the deep dorsal fascia or extensor tendons.
    a. The skin over the deep dorsal fascia is mobile. The loose connective tissue between these two layers represents the **dorsal subcutaneous space.** Cutaneous nerves course within this space as well as numerous small blood vessels and a rich lymphatic network.
    b. Lymphatic drainage of the palm is largely to the dorsum of the hand. Thus, digital and palmar infections may spread impressively to the dorsal subcutaneous space.
  4. **The dorsal subaponeurotic space** lies between the deep dorsal fascia and/or the extensor tendons and the deep fascia of the dorsal interossei and metacarpals. This space contains the **radial artery. Infection of the subaponeurotic space** may result from laceration of the knuckles with inoculation of pyogenic organisms (such as occurs with the violent meeting of fist and teeth). Otherwise, this space is not frequently involved in hand infections.

B. **Extrinsic muscles** of the hand lie in the extensor compartment of the forearm.
  1. **Organization.** The muscles on the dorsum of the hand include several of the extensor compartment muscles of the forearm (see Figure 10-2).
    a. **Composition.** This group includes the tendons of two layers within the posterior forearm (Figure 10-2; see Table 10-1). The superficial extensor group acts on the fingers (**extensor digitorum** and **extensor digiti minimi**). The deep extensor group acts on the index finger and thumb (**extensor indicis, abductor pollicis longus, extensor pollicis brevis, extensor pollicis longus**).
      (1) In the superficial layer, the tendons of the extensor digitorum lie in sheaths as they pass beneath the extensor retinaculum. In the hand, the extensor digitorum divides into four slips; each slip inserts into the **extensor aponeurosis** of the second through fifth digits (Figure 10-3A; see Figure 10-2).
      (2) In the deep layer, the **abductor pollicis longus and extensor pollicis brevis** lie immediately distal to the supinator. The tendons emerge from the extensor ret-

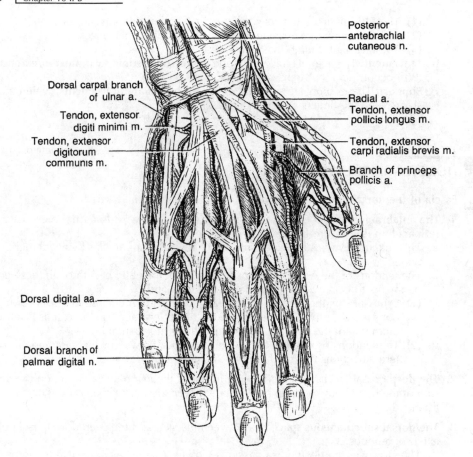

FIGURE 10-2. Extrinsic extensor tendons in the dorsum of the right hand.

inaculum on the lateral side of the radius to insert onto the base of the first metacarpal and the first phalanx, respectively. The **extensor pollicis longus** passes through a separate tunnel beneath the extensor retinaculum. Emerging, it veers toward the base of the thumb and continues to the distal phalanx. Its tendon is separated from that of the extensor pollicis brevis by the **anatomic snuffbox,** which contains the radial artery (see Figure 10-2).
  b. **Synovial sheaths** facilitate movement of the tendons as they pass beneath the extensor retinacula (see Figure 8-3). These sheaths may become inflamed secondary to the widespread synovial inflammation that accompanies rheumatoid arthritis, resulting in **subacute tenosynovitis.** Inflammation of the extensor tendons may lead to rupture.
  c. The tendinous **extensor aponeurosis (dorsal expansion** or **dorsal hood)** of each finger receives the extrinsic extensor muscle tendons, as well as the interosseous and lumbrical muscle tendons (see Figure 10-3A).
    (1) Each extensor aponeurosis passes over the MP joints and then trifurcates.
    (2) The **central portion** passes over the PIP joint to insert into the base of the middle phalanx.
    (3) The **two lateral bands** pass over the PIP joint as well as the DIP joint to insert into the base of the distal phalanx.
2. **Group actions.** Extension of the fingers at the MP joints (excluding the thumb) is accomplished by muscles of the superficial and deep antebrachial extensor groups. Flexion/extension of the MP joints occurs about a transverse axis.
  a. **Extensors of the digits** are the muscles that pass dorsal to the transverse axis of the MP joints (Table 10-1).

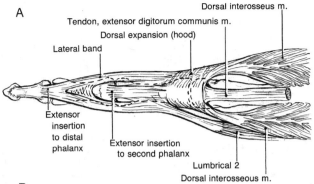
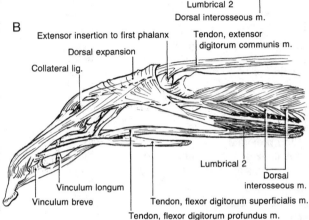

FIGURE 10-3. Extensor aponeurosis of the right third digit. *(A)* Dorsal view. The digital extensors, the extensor digitorum, interossei, and lumbricals, insert into the extensor aponeurosis (dorsal hood). The extensor aponeurosis inserts into the distal and middle phalanges. *(B)* Lateral view. The superficial flexor inserts into the middle phalanx, the deep flexor inserts into the distal phalanx, and the interossei and lumbricals insert into the extensor aponeurosis.

      (1) The first MP joint extensors are the **extensor pollicis brevis** and **extensor pollicis longus.**
      (2) The prime second to fifth MP joint extensors are the **extensor digitorum, extensor indicis proprius,** and the **extensor digiti minimi.**
      (3) The PIP and DIP joints are extended by intrinsic hand muscles.
  b. Flexors of the fingers are those extrinsic flexor compartment muscles and intrinsic hand muscles that pass ventral to the transverse axis of the MP joints (see section III B 2).
  3. Innervation. All of the extrinsic extensor muscles of the hand are innervated by the **posterior interosseous branch of the radial nerve.**

C. The dorsum of the hand has **no intrinsic muscles.**

III. PALMAR MUSCULATURE

A. **The fascia of the palm** is a continuation of the antebrachial fascia.
  1. Subcutaneous fascia lies superficial to the palmar aponeurosis.
    a. Fibrous fasciculi connect the palmar skin to the palmar aponeurosis, limiting movement of the palmar skin.
    b. The **palmaris brevis muscle** and the **recurrent (thenar) branch of the median nerve** lie in the subcutaneous fascia, as does the **palmar cutaneous branch** of the median nerve from the forearm.
  2. The antebrachial fascia condenses at the volar wrist to form the **flexor retinaculum,** which consists of the more superficial volar carpal ligament as well as the deeper transverse carpal ligament (see Figure 9-1 and Chapter 9 II A 2).

Table 10-1. Extrinsic Extensor Muscles

Muscle	Origin	Insertion	Action	Innervation
Acting on the Second Through Fifth Digits				
Superficial group:				
Extensor digitorum	Lateral epicondyle of humerus	Extensor hood and proximal phalanx of index, middle, and ring fingers	Extends MP joint and, when fist is clenched, extends wrist	Post. interosseous branch of radial n. (C7–C8, posterior)
Extensor digiti minimi	Lateral epicondyle of humerus and tendon of extensor digitorum	Extensor hood and proximal phalanx of little finger	Extends MP joint of little finger	Post. interosseous branch of radial n. (C7–C8, posterior)
Deep group:				
Extensor indicis	Interosseous membrane and ulna	Extensor hood and proximal phalanx of index finger	Extends MP joint of index finger and wrist	Post. interosseous branch of radial n. (C8–T1, posterior)
Acting on the Thumb				
Deep group:				
Abductor pollicis longus	Post. interosseous membrane	Base of first metacarpal and ulna laterally	Abducts thumb and wrist	Post. interosseous branch of radial n. (C8–T1, posterior)
Extensor pollicis brevis	Post. midshaft of radius and interosseous membrane	Base of first phalanx of thumb	Extends thumb and abducts wrist	Post. interosseous branch of radial n. (C8–T1, posterior)
Extensor pollicis longus	Post. surface of interosseous membrane and posterior ulna	Base of second phalanx of thumb	Extends thumb and abducts wrist	Post. interosseous branch of radial n. (C8–T1, posterior)

    a. The **volar carpal ligament** continues into the palm and joins the fan-shaped tendon of the palmaris longus to contribute to the **palmar aponeurosis.**
    b. The **transverse carpal ligament** continues into the palm as the **deep palmar fascia,** the outer surface of which fuses with the palmar aponeurosis.
  3. **Deep palmar fascia.** Septa from this dense layer form several spaces within the hand.
    a. The **palmar aponeurosis** spans the palm between the thenar and hypothenar eminences (see Figure 9-2).
      (1) Proximally, it receives the insertion of the **palmaris longus muscle,** enabling that muscle to assist flexion of the hand.

(2) Distally, it splits into four slips. Each slip has a superficial and a deep part.
  (a) The superficial parts insert into the skin at the base of the digits.
  (b) The deep parts divide and pass along the sides of the digits to join the fibrous flexor sheaths and the transverse metacarpal ligaments.
(3) Medially, it attaches along the length of the fifth metacarpal.
(4) Laterally, it attaches along the length of the first metacarpal.
(5) Centrally, in the palm, it attaches along the length of the third metacarpal, thus subdividing the central space.

b. The **thenar and hypothenar compartments** are formed as the deep palmar fascia envelops the thenar muscle group and the hypothenar muscle group.
  (1) Beneath the **thenar compartment,** the deep fascia is attached along the length of the first metacarpal.
  (2) Beneath the **hypothenar compartment,** it is attached along the length of the fifth metacarpal.

c. **The central (palmar) space** lies deep to the palmar aponeurosis between its medial and lateral attachments. The attachment of the palmar aponeurosis to the third metacarpal divides the space into a **thenar space** (not to be confused with the thenar compartment) and a **midpalmar space** (not to be confused with the palmar or central space).
  (1) The **thenar space** is lateral and contains the flexor pollicis longus tendons and the extrinsic flexor tendons of the index finger as well as the radial bursa. It overlies the adductor pollicis muscle.
  (2) The **midpalmar space** is medial and contains the extrinsic flexor tendons of the third, fourth, and fifth digits as well as the ulnar bursa.

4. **Dupuytren contracture** of the hand results from fibrosis and shortening of the palmar aponeurosis. It usually involves the left hand and is more severe toward the ulnar side.
  a. Because the palmar aponeurosis inserts into the proximal and middle phalanges, shortening of the aponeurosis results in progressive flexion of the fourth and fifth digits.
  b. A high correlation exists between Dupuytren contracture and coronary artery disease, possibly as a result of vasospasm produced by the effect of referred pain on the sympathetic innervation of the vasculature within the T1 component of the ulnar nerve distribution.

B. Extrinsic palmar muscles lie in the flexor compartment of the forearm.

1. **Organization.** The tendons of most of these forearm muscles lie in the carpal tunnel beneath the transverse carpal ligament.
  a. The **superficial layer** includes the **palmaris longus** (see Table 10-2 and Figure 9-2A). Lying in the middle of the forearm, its tendon passes superficial to the flexor retinaculum to insert into the palmar aponeurosis. The **palmaris longus** is absent in 13% of forearms.
  b. **The intermediate layer** consists of the **flexor digitorum superficialis,** which arises by the humeral, radial, and ulnar heads (see Table 10-2 and Figure 9-2B). In the forearm, this muscle lies in the flexor compartment. In the wrist, it divides into four tendons, which lie in the **carpal tunnel** beneath the transverse carpal ligament. In the hand, the tendon to the index finger lies in the **thenar space** and the tendons for the third, fourth, and fifth digits lie in the **midpalmar space.** In the fingers, these tendons split near their insertions into the base of each side of the middle phalanx of the second through fifth digits (Figure 10-4).
  c. The **deep layer** includes the **flexor digitorum profundus,** and the **flexor pollicis longus** (see Table 10-2, Figures 9-3, 10-3B, and 10-4). The deep group originates from the proximal ulna, the anterior surface of the interosseous membrane, and the anterior surface of the midradius.
    (1) **Flexor digitorum profundus.** In the forearm, this muscle lies in the flexor compartment where it divides into four tendons (see Figure 9-3). In the wrist, the four tendons lie in the **carpal tunnel** beneath the transverse carpal ligament. In the hand, the tendon to the index finger lies in the **thenar bursa** and the

FIGURE 10-4. Extrinsic flexor muscles of the right hand.

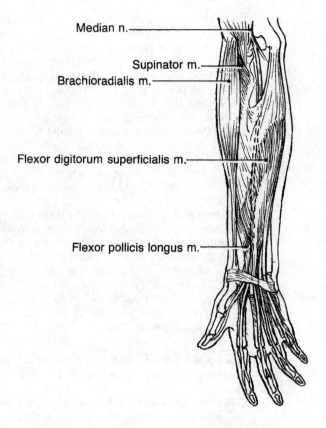

tendons for the third, fourth, and fifth digits lie in the **midpalmar bursa.** In the fingers, these tendons pass through the split in the flexor superficialis tendon to insert into the base of the distal phalanx of the second through fifth digits (see Figure 10-4).
  (2) **Flexor pollicis longus.** In the forearm, this muscle lies on the radial side beneath the brachioradialis (see Figure 9-3). In the wrist, the tendon of the flexor pollicis longus lies in the **radial bursa** within the **carpal tunnel** beneath the transverse carpal ligament. In the hand, it turns laterally toward the thumb and passes between the two heads of the flexor pollicis brevis muscle to reach the lateral aspect of the base of the second phalanx (see Figure 10-4).
 d. **Fibrous flexor sheaths** extend along each digit and attach the extrinsic flexor tendons to the adjacent bone.
  (1) The **flexor retinaculum** (**volar** and **transverse carpal ligaments**), which bridges the carpal arch to form the carpal tunnel, holds the flexor tendons in place at the wrist.
   (a) To facilitate movement as the tendons pass beneath the retinacula, the tendons are encased in synovial sheaths.
   (b) The **synovial sheaths** pass through the radial and ulnar bursae and enter the fibrous sheaths along the digits.
  (2) **Annular ligaments** are fibrous arches formed by condensations of the fibrous sheaths proximal and distal to each joint. Attached to the sides of the phalanges, they enclose the synovial sheaths.
  (3) **Cruciate ligaments** are formed across the flexor aspect of the joints by oblique fibers. They prevent the tendons from lifting away from a joint like a bowstring when the digit is flexed.
  (4) **Vincula** attach the superficial and deep flexors to the middle and distal phalanges, respectively, and provide routes for the tendinous vascular supply (see

Table 10-2. Extrinsic Flexor Muscles Acting on the Hand and Digits

Muscle	Origin	Insertion	Action	Innervation
Palmaris longus	Medial epicondyle of humerus	Palmar aponeurosis	Flexes wrist	Median n. (C7–C8, anterior)
Flexor digitorum superficialis	Medial epicondyle of humerus, radius, and ulna	Base of second phalanx of second through fifth digits	Flexes proximal interphalangeal (PIP) joint, metacarpophalangeal (MP) joints, and wrist	Median n. (C7–T1, anterior)
Flexor digitorum profundus	Anterior proximal ulna and interosseous membrane	Base of third phalanx of second through fifth digits	Flexes distal (DIP) and proximal (PIP) interphalangeal joints, metacarpophalangeal (MP) joint, and wrist	Heads for second and third digits: median n. (C7–T1, anterior) Heads for fourth and fifth digits: deep br. of ulnar n. (C8–T1, anterior)
Flexor pollicus longus	Anterior midradius and interosseous membrane	Lateral aspect base of second phalanx of thumb	Flexes thumb, metacarpophalangeal joint, and wrist	Median n. (C7–T1, anterior)

Figure 10-3). Swelling associated with **tenosynovitis** may compress the vascular supply, causing ischemic necrosis of the tendons.

   2. **Group actions.** The extrinsic flexor muscles flex the digits, thumb, and wrist (Table 10-2).
      a. **Flexion/extension** occurs at the MP and IP joints.
         (1) **Flexion at the MP joint** is accomplished primarily by the interossei. The flexors digitorum superficialis and profundus assist when the extensor digitorum is relaxed. Otherwise, these muscles act in concert to stabilize this joint.
         (2) **Flexion of the fingers at the PIP joints** is accomplished primarily by the flexor digitorum superficialis.
         (3) **Flexion of the fingers at the DIP joints** is accomplished by the flexor digitorum profundus.
         (4) **Flexion of the thumb at the DIP joints** is accomplished by the flexor pollicis longus.
      b. The **tension-generating capacity** of the long flexors of the digits is maximal when the wrist is extended, that is, in the stretched ("rest-length") position. This fact explains why a child can be made to release an object or an assailant can be made to release a knife (not recommended) by grasping and forcibly flexing the wrist.
   3. **Group innervation.** All of the extrinsic flexor muscles of the hand are innervated by the **median nerve,** except for the medial half of the flexor digitorum profundus, which is innervated by the deep branch of the **ulnar nerve** (see Table 10-2).
   4. **Palmar bursae** form by the fusion of synovial **tendon sheaths** that enclose the flexor tendons as they approach and traverse the carpal tunnel. Usually two **bursae** extend variably into the hand deep to the palmar aponeurosis.

a. The **ulnar bursa** surrounds the extrinsic digital flexor tendons as they pass through the **carpal tunnel** beneath the transverse carpal ligament.
   (1) This bursa is continuous with the synovial sheath of the tendons along the length of the fifth digit only.
   (2) The more distal portions of the sheaths of the second, third, and fourth digits usually are not continuous with the ulnar bursa; instead, they reform at the distal end of the metacarpals and continue to the proximal end of the distal phalanges.
b. The **radial bursa** surrounds the flexor pollicis longus tendon as it passes through the **carpal tunnel.**
   (1) This bursa is continuous with the synovial sheath of the flexor pollicis longus tendons along the length of the thumb.
   (2) The radial bursa may or may not be continuous with the ulnar bursa or an intermediate bursa.
c. An **intermediate bursa** occasionally (10 to 15%) invests the tendons of the flexor indicis, but usually will not extend to the digital sheath. The intermediate bursa may or may not be continuous with the radial bursa, the ulnar bursa, or both.
d. **Tenosynovitis.** The tendon sheaths may be involved in generalized synovial inflammatory processes or in the spread of infection from the fascial compartments of the palm and fingers.
   (1) Infection of the flexor sheaths of the thumb and fifth digit will track proximally into the radial and ulnar bursae, respectively.
   (2) If the radial and ulnar bursae communicate, infection may track to the opposite side of the hand, **"horseshoe abscess."**
   (3) Inflammatory swelling of the bursae beneath the transverse carpal ligament is one cause of **carpal tunnel syndrome,** wherein swelling entraps the extrinsic flexor tendons and compression of the median nerve results in thenar palsy.

C. **Intrinsic palmar musculature** may be divided into three groups.

1. **Palmar muscles** (Figure 10-5; see Table 10-3 and Figure 10-6)
   a. **Group composition.** The palmar muscles (along with the extensor digitorum) insert into the extensor expansion of each finger.
      (1) **Dorsal interosseous muscles** consist of four bipennate muscles that originate from the first through the fifth metacarpals (see Figure 10-5A). They insert on the lateral side of the proximal phalanges of the second and third digits and

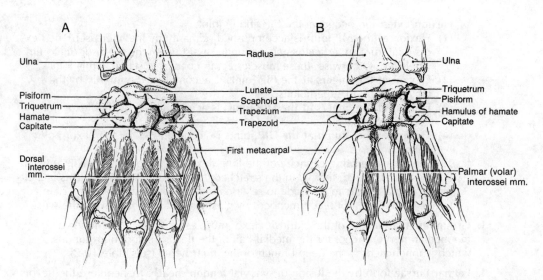

FIGURE 10-5. (A) The four dorsal interossei of the pronated right hand. (B) The three palmar interossei of the supinated right hand.

on the medial side of the proximal phalanges of the third and fourth digits (see Figure 10-3).

   (2) **Palmar interosseous muscles** consist of three unipennate muscles that originate along the medial side of second and lateral side of fourth and fifth metacarpals (see Figure 10-5). They insert into both the extensor aponeuroses and into the bases of the proximal phalanges medially on second digit, laterally on the fourth and fifth digits.

   (3) **Lumbrical muscles.** Each arises from a tendon of the flexor digitorum profundus in the deep palm (see Figure 10-6).

   (a) **Lumbricals I and II** are unipennate, taking origin from the flexor digitorum profundus tendon of the same digit. These muscles insert into the radial side of the extensor aponeurosis of the corresponding digits (see Figure 10-3).

   (b) **Lumbricals III and IV** are bipennate. Lumbrical III takes origin from the profundus tendon of the third and fourth digits, whereas lumbrical IV takes origin from the profundus tendons of the fourth and fifth digits. These muscles insert into the radial side of the extensor aponeurosis of the corresponding digits (see Figure 10-3).

b. **Group actions** (Table 10-3)

   (1) **Abduction at the MP joints.** The phalangeal insertions of the **dorsal interossei** abduct the second, fourth, and fifth digits away from the midline of the hand. The third finger is abducted toward both the ulnar and radial sides of the hand (see Figure 10-3A).

   (2) **Adduction at the MP joints.** The phalangeal insertions of the **palmar interossei** adduct the second, fourth, and fifth digits toward the midline of the hand. The middle finger has no palmar interosseous muscle because the dorsal interossei give it full mobility (see Figure 10-3A).

   (3) **Flexion of the MP joints** is by both dorsal and ventral interossei as well as by the lumbricals, because the line of action in the broad proximal portion of the extensor hood passes ventral to the transverse axes of the MP joints (see Figure 10-3B).

   (a) The insertions of the **dorsal** and **palmar interossei** into the opposite sides of the extensor aponeuroses produce flexion at this joint.

   (b) The insertions of the **lumbricals** into the sides of the extensor aponeuroses produce flexion at this joint. Because they have longer lever arms than the interossei, the lumbricals are stronger flexors of this articulation.

   (4) **Extension of the IP joints** is by both dorsal and ventral interossei and lumbricals, because the line of action in the narrow distal portion of the extensor hood is carried dorsal to the transverse axes of the PIP and DIP joints.

   (a) The insertions of dorsal and palmar interossei into the opposite sides of the extensor aponeuroses result in extension of the PIP and DIP joints.

   (b) By inserting into the extensor hood, the lumbricals extend the PIP and DIP joints (i.e., they are synergistic with the interossei).

c. **Group innervation.** The palmar muscles are innervated by the median or ulnar nerves.

   (1) The **median nerve** supplies lumbricals I and II.

   (2) The **ulnar nerve** supplies the dorsal interossei, the palmar interossei, and lumbricals III and IV.

2. **Thenar muscles** lie in the thenar compartment and move the thumb (Figure 10-6; see Table 10-3).

   a. **Group composition.** This muscle group comprises the adductor pollicis brevis, opponens pollicis, flexor pollicis brevis, and adductor pollicis.

   b. **Group actions.** The actions of these muscles are described by their names (see Table 10-3).

   (1) **Flexion** of the thumb is by the flexor pollicis longus muscle (at the IP joint) and flexor pollicis brevis muscle (at the MP joint).

Table 10-3. Intrinsic Muscles

Muscle	Origin	Insertion	Primary Action	Innervation
Acting on the Second Through Fifth Digits				
Dorsal interossei (4)	Medial side of first metacarpal; both sides of the second, third, and fourth metacarpal; and lateral side of fifth metacarpal	Tubercle of proximal phalanx and dorsal aponeurosis: laterally on second and third digits, medially on third and fourth digits	Abduct the second, third, and fourth digits from the midline of hand (third digit abducts to both sides); flex MP joint and extend PIP and DIP joints	Ulnar n. (C8–T1, anterior)
Palmar interossei (3)	Medial side of second and lateral side of fourth and fifth metacarpals	Tubercle of proximal phalanx and dorsal aponeurosis: medially on second digit, laterally on fourth and fifth digits	Adduct the second, fourth, and fifth digits to the midline of hand; flex MP joint and extend PIP and DIP joints	Ulnar n. (C8–T1, anterior)
Lumbricals 1 and 2	Tendons of the flexor digitorum profundus in the deep palm	Lateral side of dorsal expansion of second and third digits	Flex MP joint and extend PIP and DIP joints	Median n. (C8–T1, anterior)
Lumbricals 3 and 4	Tendons of the flexor digitorum profundus in the deep palm	Lateral side of dorsal expansion of fourth and fifth digits	Flex MP joint and extend PIP and DIP joints	Ulnar n. (C8–T1, anterior)
Acting on the Thumb				
Abductor pollicis brevis	Anterior surface of trapezium and scaphoid	Lateral aspect of first phalanx base	Abducts thumb	Median n. (C8–T1, anterior)
Opponens pollicis	Trapezium	Anteriolateral surface of first metacarpal	Medially rotates (opposes) thumb	Median n. (C8–T1, anterior)
Flexor pollicis brevis:				
Superficial head	Transverse carpal ligament and trapezium	Lateral side of first phalanx base	Flex thumb	Median n. (C8–T1, anterior)
Deep head	Lateral side of second metacarpal	Lateral side of first phalanx base		Ulnar n. (C8–T1, anterior)
Adductor pollicis:				
Oblique head	Anterior surface of capitate, second and third metacarpals	Medial side of base of first phalanx of thumb	Adduct thumb	Ulnar n. (C8–T1, anterior)
Transverse head	Distal half of third metacarpal	Medial side of base of first phalanx of thumb	Adduct thumb	Ulnar n. (C8–T1, anterior)

      (2) Extension of the thumb is by the extrinsic muscles—the extensor pollicis longus and brevis.
      (3) Abduction of the thumb is by the abductor pollicis longus muscle (at the carpometacarpal joint) and the abductor pollicis brevis muscle (at the MP joint).
      (4) Adduction of the thumb is by the adductor pollicis muscle (at the carpometacarpal joint).
      (5) Opposition of the thumb involves a medial rotation (at the first carpometacarpal joint), primarily by the opponens pollicis muscle and assisted by the flexor pollicis brevis muscle.
      (6) Reposition of the thumb involves a lateral rotation at the first metacarpal joint by the synergistic actions of the abductors and extrinsic extensors of the thumb.
    c. Group innervation (see Table 10-3)
      (1) The **median nerve** supplies the abductor pollicis brevis, opponens pollicis, and superficial head of the flexor pollicis brevis muscles.
      (2) The **ulnar nerve** supplies the deep head of the flexor pollicis brevis and the adductor pollicis muscles.

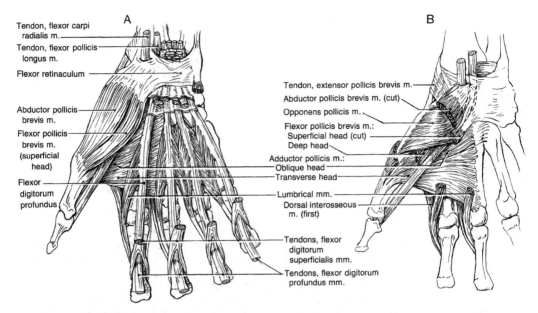

FIGURE 10-6. The supinated right hand. *(A)* Lumbrical muscles; *(B)* thenar muscles.

3. **Hypothenar muscles** lie in the hypothenar compartment and move the little finger (see Table 10-4 and Figure 10-6).
   a. **Group composition.** The hypothenar group consists of the **abductor digiti minimi, flexor digiti minimi brevis, opponens digits minimi, and palmaris brevis.**
   b. **Group actions.** The names of these muscles describe their actions (Table 10-4).
      (1) **Abduction** of the fifth digit. The abductor minimi abducts and flexes the proximal phalanx of the fifth digit, assisted by the opponens digiti minimi.
      (2) **Flexion** of the fifth digit. The flexor digiti minimi brevis flexes the fifth finger, assisted by the opponens digiti minimi.
      (3) **Opposition** of the fifth digit. The opponens digiti minimi abducts, flexes, and rotates the fifth metacarpal at the carpometacarpal joint.
      (4) **Corrugation** of the skin over the thenar eminence. The palmaris brevis draws the skin over the hypothenar eminence medially.

Table 10-4. Intrinsic Hypothenar Muscles

Muscle	Origin	Insertion	Primary Action	Innervation
Palmaris brevis	Medial border of palmar aponeurosis	Skin over hypothenar region	Corrugates palmar skin	Ulnar n. (C8–T1, anterior)
Abductor digiti minimi	Pisiform bone	Ulnar side of base of fifth proximal phalanx	Abduct the fifth digit	Ulnar n. (C8–T1, anterior)
Flexor digiti minimi	Flexor retinaculum and hamulus	Ulnar side of base of fifth proximal phalanx	Flex fifth metacarpo-phalangeal joint	Ulnar n. (C8–T1, anterior)
Opponents digiti minimi	Flexor retinaculum and hamulus	Ulnar side of fifth metacarpal	Flexion and opposition	Ulnar n. (C8–T1, anterior)

c. **Group innervation.** The hypothenar muscles are innervated by the **ulnar nerve** (see Table 10-4).

IV. VASCULATURE

A. Arterial supply

1. **The radial artery** generally supplies the lateral aspect of the hand (Figure 10-7).
   a. **Course.** It passes lateral to the scaphoid bone onto the dorsal side of the wrist between the tendons of the extensor pollicis longus and brevis muscles (the anatomic snuff box), where a **radial pulse** is palpable.
   b. **Branches.** It gives off a few minor branches before dividing into two major branches.
      (1) The **anterior (palmar) carpal branch** arises in the wrist and anastomoses with the carpal branches of the ulnar artery to supply the dorsal and volar aspects of the wrist.
      (2) The **superficial palmar branch** arises near the end of the radius.
         (a) It passes through the palmar aspect of the thenar compartment.
         (b) It provides a minor contribution to complete the **superficial palmar arch** by anastomosing with the superficial palmar branch of the ulnar artery.

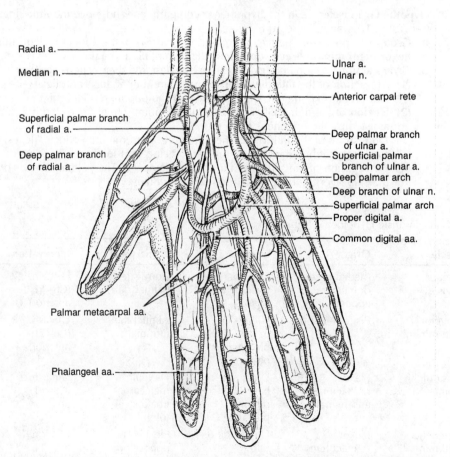

FIGURE 10-7. Blood supply to the right hand. The radial and ulnar arteries form or contribute to the superficial and palmar arches.

(3) The **deep palmar branch** begins at the end of the radius as the direct continuation of the radial artery.
    (a) In the wrist, it courses dorsally over the base of the thumb in the anatomic snuffbox, where a **radial pulse** is palpable.
    (b) In the hand, it passes dorsal to the first metacarpal and through the first interosseous space to enter the thenar space, where it bifurcates.
        (i) The **first metacarpal artery** supplies the medial side of the thumb and the lateral side of the index finger.
        (ii) The **deep palmar arch** anastomoses with the deep palmar branch of the ulnar artery. It gives off **palmar metacarpal arteries,** which join the common digital arteries just proximal to the bifurcation into phalangeal arteries.
2. **The ulnar artery** generally supplies the medial aspect of the hand (see Figure 10-7).
    a. **Course.** It continues into the hand on the medial side of the wrist, just lateral to the pisiform bone, where the **ulnar pulse** is palpable.
    b. **Branches.** It divides in the wrist to form three major branches.
        (1) **Carpal branches** arise in the wrist and anastomose with dorsal and volar carpal branches of the radial artery to supply the dorsal and volar aspects of the wrist.
        (2) The **deep palmar branch** arises just distal to the lateral side of the pisiform bone.
            (a) It passes through the palmar aspect of the hypothenar compartment.
            (b) This small branch makes the minor contribution to the **deep palmar arch** by anastomosing with the deep palmar branch of the radial artery.
        (3) The **superficial palmar branch** is the continuation of the ulnar artery.
            (a) It is the major contributor to the **superficial palmar arch.** This arch is larger than the deep palmar arch and anastomoses strongly with the superficial palmar branch of the radial artery.
            (b) Four branches arise from the superficial palmar arch.
                (i) The **proper digital artery** to the fifth digit arises from this arch.
                (ii) The three **common digital arteries** receive a collateral palmar metacarpal artery from the deep palmar arch before bifurcating into **phalangeal branches,** which run on the medial and lateral sides of the second, third, and fourth digits.

B. Venous return

1. **Superficial veins** provide a rich anastomotic network in the dorsal subcutaneous space.
    a. The **cephalic vein** is formed by a coalescence of dorsal veins on the radial side of the hand. It winds from the posterior preaxial border of the forearm to lie along the ventral preaxial border. On the dorsum of the hand, it courses posterior to the styloid process of the radius. This site is often used for insertion of intravenous catheters.
    b. The **basilic vein** is formed by a medial coalescence of dorsal veins on the ulnar side of the hand. It ascends nearly to the cubital fossa along the dorsal postaxial border.
2. **Deep veins (venae comitantes)** accompany the radial and ulnar arteries as well as other branches.

C. Lymphatic drainage

1. **Superficial lymphatics**
    a. Most lymphatic drainage from the thenar compartment, hypothenar compartment, and digits is toward the **dorsal subcutaneous space** of the hand, explaining the extreme swelling of this region that accompanies infections of the digits or volar surface.
    b. Lymphatics drain superficially along the cephalic and basilic veins, eventually draining through **clavipectoral nodes** and **supraclavicular nodes.** Thus, chronic

hand infections are usually accompanied by swollen axillary lymph nodes and red streaking up the arm along the course of the lymphatics.
2. **Deep lymphatics.** The thenar space, midpalmar space, and tendon sheaths drain through lymphatics that accompany the radial and ulnar vessels.

V. INNERVATION

A. **The radial nerve** provides sensation to the dorsum of the hand.
   1. **Course.** Only the **superficial branch** of the radial nerve reaches the hand (see Figure 8-4).
   2. **Distribution.** The superficial branch, containing contributions from the posterior roots of C6–C7, leaves the company of the radial artery at the level of the lateral epicondyle and passes dorsal to the wrist, where it divides into four **dorsal digital nerves** (see Figure 8-4).
      a. **Sensory**
         (1) The **dorsal digital nerves** supply the radial side of the dorsal hand from the level of the wrist to about the level of the PIP joints.
         (2) There is a region of exclusivity over the web space between the thumb and index finger.
      b. **Motor.** Extrinsic hand muscles supplied by the deep branch of this nerve include the extensor digitorum, extensor digiti minimi, extensor indicis, abductor pollicis longus, extensor pollicis brevis, and extensor pollicis longus. The radial nerve supplies none of the intrinsic hand muscles.
   3. **Nerve palsy.** Injuries of the radial nerve are rather common.
      a. **Arm and elbow region.** A lesion at the level of the elbow joint can produce **wrist-drop** and inability to extend the digits at the MP joints. Sensation over the radial side of the dorsum of the hand may be unaffected, however, if the lesion affects only the deep branch.
      b. **Wrist.** A lesion of the superficial radial nerve at the wrist results in loss of sensation on the radial side of the dorsum of the hand.

B. **The median nerve** innervates specific groups of intrinsic muscles and provides sensation to the lateral aspect of the palm, the thumb, and the index and middle fingers (Figure 10-8).
   1. **Course.** This nerve enters the hand through the carpal tunnel (see Figure 9-1).
   2. **Distribution.** On emerging from the transverse carpal ligament, the median nerve gives off the **recurrent branch** and then divides into four or five **volar (palmar) digital nerves,** which run along the sides of the fingers (see Figure 10-8).
      a. **Motor innervation** (see Tables 10-3 and 10-4 and Figure 10-8)
         (1) The median nerve supplies most of the extrinsic flexors of the hand, including the **flexor pollicis longus, flexor digitorum superficialis,** and the radial half of the **flexor digitorum profundus.**
         (2) The **recurrent branch** supplies most of the intrinsic thenar muscles, that is, the **opponens pollicis, abductor pollicis brevis, and flexor pollicis brevis** (superficial head only).
         (3) The **palmar (volar) digital branches** supply **lumbricals I and II** before becoming cutaneous.
      b. **Sensory distribution** (see Figure 10-8, insets)
         (1) The **palmar cutaneous branch** arises proximal to the flexor retinaculum, passes superficial to the flexor retinaculum, and supplies the radial side of the palm of the hand and volar aspect of the thumb.
         (2) The **volar (palmar) digital branches** supply the volar and dorsal surfaces distal to the PIP joints of the second and third digits with a variable portion of the fourth digit.

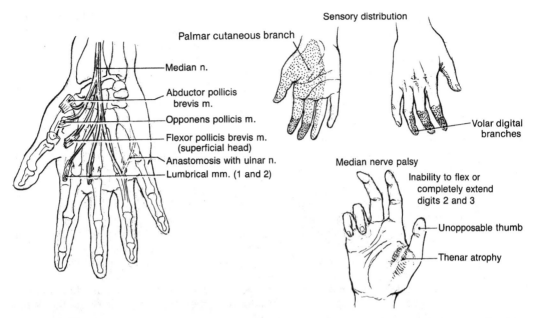

FIGURE 10-8. Median nerve innervation of muscles in the hand. *(Upper insets)* regions of sensory distribution of the palmar cutaneous branch and digital volar branches. *(Lower inset)* median nerve palsy produces the "hand of benediction."

3. **Median nerve injury.** Lesions can be classified according to the region or level of injury (see Figure 10-8, inset).
   a. **In the elbow region.** Damage to the median nerve, such as by medial epicondyle fracture, can produce paralysis of the flexor pollicis longus, the flexor digitorum superficialis, and the radial half of the flexor digitorum profundus, as well as the muscles of the thenar compartment and lumbricals I and II.
      (1) This injury produces the **sign of benediction,** in which the index and middle fingers cannot be flexed and the thumb cannot be opposed. (The ring and little fingers function normally.)
      (2) Patients experience a sensory loss over the radial side of the palm and digits lateral to the center line of the ring finger.
   b. **At the wrist**
      (1) **Carpal tunnel syndrome** is the result of compression of the median nerve within the carpal tunnel from **ulnar bursa inflammation** or **displacement of the lunate bone.**
         (a) Because of paralysis of the muscles of the thenar compartment (recurrent branch), flexion and abduction is weak, and opposition is not possible.
         (b) Paralysis of lumbricals I and II occurs, but the effect is negligible because the interossei are functional.
         (c) Sensation is lost in the second and third digits and a portion of the fourth digit (volar digital nerves).
         (d) No sensation is lost over the proximal palm because the palmar cutaneous branch passes superficial to the transverse carpal ligament.
      (2) **Laceration** deficits depend on whether the injury is superficial or deep.
         (a) **Superficial laceration** severs the palmar cutaneous branch of the median nerve, which produces loss of sensation over the proximal palm but no motor deficits.
         (b) **Deep laceration** severs both the palmar cutaneous branch and the median nerve. Median nerve injury produces paralysis of the muscles of the thenar compartment and lumbricals I and II, with loss of sensation in the second and third digits and a portion of the fourth digit. **Palmar cuta-**

neous branch injury associated with loss of sensation over the proximal palm.
  c. **In the palm.** Laceration of the **recurrent branch,** probably the most commonly injured nerve, produces loss of thenar musculature (weakness of flexion and loss of opposition) but no sensory deficit.

C. **The ulnar nerve** innervates specific groups of intrinsic muscles and provides sensation to the medial aspect of the palm as well as the fifth finger (Figure 10-9).

1. **Course.** This nerve enters the hand with the ulnar artery by passing deep to the volar carpal ligament, but superficial to the transverse carpal ligament, adjacent to the pisiform and hamate bones (see Figure 9-1).

2. **Distribution.** The ulnar nerve gives off a **superficial terminal branch** in the vicinity of the hamulus and continues as the **deep terminal branch.**
   a. **Motor innervation** (see Tables 10-3 and 10-4 and Figure 10-9)
      (1) The ulnar nerve supplies one and one-half extrinsic flexor muscles (the flexor carpi ulnaris and the ulnar half of the flexor digitorum profundus).
      (2) The **superficial terminal branch** supplies the palmaris brevis muscle before becoming cutaneous.
      (3) The **deep terminal branch** supplies the hypothenar musculature, lumbricals III and IV, the dorsal and palmar interossei, the adductor pollicis, and the deep head of the flexor pollicis brevis.
   b. **Sensory distribution** (see Figure 10-9, insets)
      (1) The **palmar cutaneous branch** arises in the forearm and innervates the ulnar side of both the dorsum and palm of the hand.
      (2) The **dorsal branch** arises proximal to the wrist and supplies the dorsal aspect of the fifth digit and a variable portion of the dorsum of the fourth digit.
      (3) The **superficial terminal branch** supplies the skin over the hypothenar eminence.
      (4) The **volar digital branches** supply the palmar side of the ulnar half of the fourth digit and the entire fifth digit.

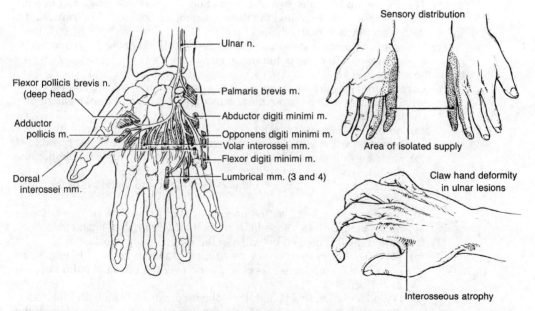

FIGURE 10-9. Ulnar nerve innervation of muscles in the hand. *(Upper insets)* regions of sensory distribution of the palmar cutaneous branch and dorsal digital branch and volar digital branches. *(Lower inset)* ulnar nerve palsy produces the "claw hand."

3. **Injury.** Ulnar nerve lesions can be characterized according to the region or level of injury (see Figure 10-9, inset).
   a. **Elbow region.** Ulnar lesions in this region produce **claw hand.**
      (1) The ulnar half of the flexor digitorum profundus, lumbricals III and IV, the dorsal and palmar interossei, and the hypothenar muscles are paralyzed. When the metacarpophalangeal joints are extended by the extensor digitorum, the PIP and DIP joints cannot be extended because the interossei and half of the lumbricals are nonfunctional. The result is a clawlike posture. Muscular atrophy over the affected portion of the hand occurs with time.
      (2) Sensory deficits on the ulnar side of the dorsal and palmar aspects of the hand result.
   b. **Fracture of the distal ulna.** This type of fracture, with its associated injury to the ulnar nerve, results in paralysis of the hypothenar muscles, most of the deep muscles of the hand, and some of the muscles of the thenar side of the hand. No loss of sensation occurs along the medial side of the palm because the superficial branch is not involved; however, there is sensory loss along the fifth finger.
   c. **Wrist region.** The ulnar nerve is superficial and susceptible to laceration or entrapment as it passes by the pisiform bone.
      (1) Laceration at the wrist leaves the innervation to the radial side of the digitorum profundus intact, but results in **claw-hand** and loss of sensation distally along the fifth finger.
      (2) Adduction of the thumb is lost, and therefore so is the ability to hold a piece of paper firmly between the thumb and the side of the index finger.
   d. **Referred pain.** Coronary artery disease can refer pain to the T1 dermatomic components of the ulnar nerve (along the ulnar side of the forearm, see Figure 4-3). This action can result in vasospasm that, over time, can produce **Dupuytren contracture.**

VI. FUNCTION

A. **The function of the upper extremity** is to position the hand to interact with the environment.

B. **The function of the hand** depends on extrinsic and intrinsic mobility, muscle strength, and sensation.
   1. **Stability and strength.** The hand has maximum stability and strength in the functional position, that is, when the wrist is partially extended and slightly abducted; when the digital joints are partially flexed and slightly abducted; and when the thumb is in partial abduction, flexion, and opposition.
   2. **Prehension.** The hand is characterized by several types of prehension.
      a. **Digital prehension by terminal opposition** (e.g., picking up a needle between the tips of the thumb and index finger) is the most precise type. This action is a test for median nerve integrity.
      b. **Digital prehension by subterminal three-point (chuck) opposition** (e.g., holding a pen between the thumb, index, and middle fingers) is the most powerful and stable type.
      c. **Digital prehension by subterminal lateral opposition** (e.g., holding a card between the thumb and side of the index finger by adduction of the thumb) is a test for ulnar nerve integrity.
      d. **Digital palmar prehension** is characterized by a grasp with opposition of the thumb, as in holding a bat.
      e. **Palmar prehension** involves a grasp with the thumb adducted into the plane of the hand, as in hanging onto a board fence.

PART IIA UPPER EXTREMITY

STUDY QUESTIONS

DIRECTIONS: Each of the numbered items or incomplete statements in this section is followed by answers or by completions of the statement. Select the ONE lettered answer or completion that is BEST in each case.

1. Which of the following statements concerning the teres major muscle, a contributor to the posterior axillary fold, is true?

(A) It is a major contributor to the stability of the posterior aspect of the shoulder joint
(B) It divides the axillary artery into three parts
(C) It inserts on the humerus just distal to the infraspinatus insertion
(D) It is active in adduction of the glenohumeral joint
(E) It is innervated by the same nerve that innervates the deltoid muscle

2. Which of the following statements describing the pectoralis minor muscle is correct?

(A) It attaches to the acromion process of the scapula
(B) It crosses the cords of the brachial plexus
(C) It is an adductor and medial rotator of the humerus
(D) It is innervated by the middle subscapular (thoracodorsal) nerve
(E) It originates deep to the axillary artery

3. Which of the following statements pertaining to the cephalic vein is correct?

(A) It accompanies the brachial artery
(B) It drains the ulnar side of the hand
(C) It is suitable for venipuncture anterior to the medial epicondyle
(D) It lies in the groove between the deltoid and pectoralis major muscles

4. Which of the following statements concerning the subacromial bursa is correct?

(A) It normally communicates with the synovial cavity of the glenohumeral joint
(B) It overlies the infraspinatus muscle
(C) It permits movement between the scapula and the thoracic wall
(D) It underlies the coracoacromial ligament

Questions 5–9

A new medical clerk is asked to place an intravenous line in the cubital region of the right arm of a patient. The student-physician knows that, despite variability, the major veins are relatively consistent in their anatomic relationship.

5. All of the following statements concerning the veins of the cubital region are true EXCEPT

(A) at the level of the midaxilla, the basilic vein is joined by the cephalic vein to form the axillary vein
(B) the basilic vein runs along the medial aspect of the forearm
(C) the cephalic vein originates on the radial side of the dorsum of the hand
(D) the median cubital vein links the cephalic and basilic veins in the cubital fossa
(E) the median cubital vein is separated from the brachial artery by the bicipital aponeurosis

6. The clerk applies a tight tourniquet approximately 2 inches proximal to the proposed site of venipuncture, but the veins do not distend when the patient is asked to vigorously flex the hand several times. The most probable reason for the failure of the veins to distend is

(A) that the arterial flow to the forearm is obstructed
(B) that the deep venous return from the forearm has not been obstructed
(C) that the superficial venous return has not been obstructed
(D) that the venae comitantes may be abnormally absent

7. After several attempts, a vessel in the cubital fossa is penetrated. A small amount of blood is aspirated into the line to ensure proper placement. The blood is bright red, which makes it likely that the needle is in the

(A) brachial artery
(B) deep brachial artery
(C) cephalic vein
(D) median cubital vein
(E) venae comitantes

8. The clerk reinserts the needle slightly more medially. The patient then experiences pain radiating down the ventral surface of the forearm and hand, including the thumb, index, and middle finger. The nerve pierced by the needle is most likely

(A) a branch of the musculocutaneous nerve
(B) the medial antebrachial cutaneous nerve
(C) the median nerve
(D) the radial nerve
(E) the ulnar nerve

9. Finally, placement of the intravenous line is successful, but the clerk fails to notice an approximately 1.5-ml air bubble in the line entering the vein. The region of the body at highest risk for air embolus in this instance is

(A) the brain
(B) the hand
(C) the heart
(D) the lungs

(end of group question)

10. Chronic traction on the ulnar nerve at the elbow may be relieved by medial condyle osteotomy, whereby the condyle is cut without disturbing the muscular attachments. The nerve is then relocated anterior to the humerus and the medial epicondyle is reattached. Which of the following muscles originates from the medial epicondyle and is, therefore, directly involved in medial condyle osteotomy?

(A) Brachioradialis
(B) Extensor carpi ulnaris
(C) Flexor carpi radialis
(D) Flexor pollicis longus
(E) Supinator

11. Which of the following statements describing the lunate bone is correct?

(A) It articulates maximally with the fibrocartilage disk during adduction
(B) It is a component of the carpometacarpal joint
(C) It can compress the median nerve if displaced anteriorly
(D) It provides an attachment for the transverse carpal ligament

12. The ulnar nerve innervates which of the following muscles of the thumb?

(A) Abductor pollicis brevis
(B) Abductor pollicis longus
(C) Deep head of the flexor pollicis brevis
(D) Opponens pollicis
(E) Superficial head of the flexor pollicis brevis

Questions 13–15

A tailor complains of difficulty picking up a needle and holding it while sewing. Physical examination reveals no sensory deficit. The patient is then tested for motor function of the second digit, as shown below.

13. The muscles being tested in the illustration include the

(A) flexor digitorum superficialis
(B) lumbrical 2
(C) palmar interossei
(D) radial portion of the flexor digitorum profundus

14. If paralysis or paresis (weakness) of the terminal phalanx of the second digit is detected in the patient described, one might also expect to find

(A) atrophy of the hypothenar eminence
(B) complete paralysis of the thumb
(C) paralysis or paresis of the fourth digit
(D) weakness of pronation

15. Nerves that supply the muscles that produce antagonistic actions to that illustrated include the

(A) median and radial
(B) median, ulnar, and radial
(C) median and ulnar
(D) radial and ulnar
(E) radial only

(end of group question)

DIRECTIONS: Each of the numbered items or incomplete statements in this section is negatively phrased, as indicated by a capitalized word such as NOT, LEAST, or EXCEPT. Select the ONE lettered answer or completion that is BEST in each case.

---

16. In the upper extremity, a pulse may be palpated at all of the following locations EXCEPT

(A) in the arm against the humerus just distal to the pectoralis minor
(B) in the cubital fossa medial to the biceps brachii tendon
(C) in the wrist at the radial side of the tendon of the flexor carpi radialis muscle
(D) at the wrist at the radial side of the pisiform bone
(E) in the hand between the tendons of the extensor pollicis longus and abductor pollicis longus muscles

17. With the upper limb adducted, hanging by the side and holding a heavy suitcase, support at the glenohumeral joint to prevent downward displacement is provided by all of the following EXCEPT the

(A) coracoacromial ligament
(B) coracohumeral ligament
(C) short head of the biceps brachii muscle
(D) supraspinatus muscle

18. Following complete severance of the musculocutaneous nerve, some weak flexion of the elbow is possible through contraction of muscles that are not innervated by that nerve. All of the following muscles also flex the elbow EXCEPT the

(A) brachioradialis
(B) flexor carpi radialis
(C) flexor carpi ulnaris
(D) ulnar head of the pronator teres

19. All of the following statements concerning the scaphoid bone are true EXCEPT

(A) it articulates maximally with the radius in adduction
(B) it is the most susceptible of the carpal bones to fracture
(C) it participates in the midcarpal joint
(D) it receives an attachment for the transverse carpal ligament

20. All of the following structures pass deep to the transverse carpal ligament EXCEPT the

(A) flexor digitorum superficialis tendon
(B) flexor digitorum profundus tendon
(C) flexor pollicis longus tendon
(D) median nerve
(E) ulnar artery

21. All of the following statements concerning the blood supply of the hand are true EXCEPT

(A) the arteries of the hand are accompanied by venae comitantes
(B) the blood supply reaches the long flexor tendons within the synovial sheaths through vincula
(C) the common digital arteries are branches of the deep palmar arch
(D) the deep and superficial palmar arches are supplied by the radial and ulnar arteries
(E) the dorsal carpal arch (rete) receives blood from the carpal branches of the radial, ulnar, and posterior interosseus arteries

22. Laceration of the recurrent branch of the median nerve results in paralysis of all of the following muscles of the thumb EXCEPT the

(A) abductor pollicis brevis
(B) deep head of the flexor pollicis brevis
(C) opponens pollicis
(D) superficial head of the flexor pollicis brevis

23. Reposition of the thumb is accomplished by all of the following muscles EXCEPT the

(A) abductor pollicis longus
(B) adductor pollicis
(C) extensor pollicis brevis
(D) extensor pollicis longus

Questions 24–25

A large splinter deeply embedded adjacent to the nail on the hypothenar side of the ring finger developed into an abscess.

24. All of the following statements describe possible sequelae of infection within the synovial sheath of the fifth digit (ulnar bursa) EXCEPT

(A) ischemic necrosis of the flexor tendons to the fifth finger
(B) distal spread into adjacent digits
(C) distal spread into the thumb via the radial bursa
(D) proximal spread through the carpal tunnel into the distal forearm

25. Before lancing to remove the fragment and drain the abscess, the surgeon blocked with anesthetic the appropriate

(A) dorsal digital branch of the radial nerve
(B) dorsal digital branch of the ulnar nerve
(C) palmar digital branch of the radial nerve
(D) palmar digital branch of the ulnar nerve
(E) superficial branch of the radial nerve
(end of group question)

DIRECTIONS: Each set of matching questions in this section consists of a list of four to twenty-six lettered options (some of which may be in figures) followed by several numbered items. For each numbered item, select the ONE lettered option that is most closely associated with it. To avoid spending too much time on matching sets with large numbers of options, it is generally advisable to begin each set by reading the list of options. Then, for each item in the set, try to generate the correct answer and locate it in the option list, rather than evaluating each option individually. Each lettered option may be selected once, more than once, or not at all.

Questions 26–28

(A) Infraspinatus muscle
(B) Subscapularis muscle
(C) Supraspinatus muscle
(D) Teres minor muscle

For each description, select the appropriate rotator cuff muscle.

26. A primary medial (internal) rotator of the arm

27. Initiates humeral abduction

28. Innervated by the axillary nerve

Questions 29–32

(A) Ulnar nerve
(B) Median nerve
(C) Median and ulnar nerves
(D) Radial nerve

For each muscle, select the most appropriate innervation.

29. Abductor pollicis longus muscle

30. Lumbrical muscles

31. Dorsal interosseous muscles

32. Adductor pollicis muscle

ANSWERS AND EXPLANATIONS

1. The answer is D [Chapter 6 IV B 2 a, b]. The teres major muscle, innervated by the lower subscapular nerve (C5–C6, posterior), inserts onto the medial lip of the bicipital groove of the humerus and, thus, is an adductor of the arm. The tendon of the pectoralis major muscle, not the teres major, divides the axillary artery into upper, middle and lower parts.

2. The answer is B [Chapter 6 VI C 4; Table 6–1]. The pectoralis minor muscle crosses the cords of the brachial plexus as well as the axillary artery as it attaches to the coracoid process of the scapula. It, therefore, has no direct action on the humerus. It is innervated by the medial pectoral nerve of the medial cord of the brachial plexus.

3. The answer is D [Chapter 7 IV B 2; Chapter 9 III B 1 a; Chapter 10 V B 1]. The cephalic vein drains the subcutaneous spaces of the radial (preaxial) side of the hand. It passes posterior to the styloid process, along the preaxial border of the forearm and anterior to the lateral epicondyle over which course it is readily accessible for venipuncture. Just distal to the cubital fossa, it gives rise to the median cubital vein, which joins the basilic vein in the vicinity of the medial epicondyle. The cephalic vein continues along the preaxial border of the brachium, passing in the deltopectoral groove and piercing the clavipectoral fascia to enter the axillary vein.

4. The answer is D [Chapter 6 III D 5 a, c; Figure 6–4]. The subacromial and subdeltoid bursae may be separate, may communicate, or may form one large bursa. These bursae lie deep to the deltoid muscle, extend inferior to the coracoacromial ligament, and lie superior to the supraspinatus muscle. Only in dislocations of the glenohumeral joint, when the joint capsule is torn, will the synovial cavity of the glenohumeral joint communicate with the subacromial or subdeltoid bursa.

5. The answer is A [Chapter 7 IV C; Chapter 9 III B 1 a, b]. At the level of the midaxilla, the basilic vein and the brachial venae comitantes join to form the axillary vein. The cephalic vein joins the axillary vein just distal to the clavicle.

6. The answer is A [Chapter 5 III D 2; Chapter 7 IV A 2, 4; Chapter 9 III B 3]. A tourniquet applied too tightly to the arm will compress the brachial artery and its collateral branches, preventing blood flow into the forearm. Consequently, there will be no venous return.

7. The answer is A [Chapter 7 IV A 2 a–c; Figure 7–4]. In the cubital fossa, the brachial artery, containing bright red oxygenated blood, lies beneath the bicipital aponeurosis. The median cubital vein, containing darker deoxygenated blood, lies superficial to the bicipital aponeurosis. The deep brachial artery with its accompanying venae comitantes does not pass through the cubital fossa.

8. The answer is C [Chapter 7 V B 1; Chapter 9 IV A 1,2; Figure 9-6A]. The median nerve passes through the cubital fossa deep to the bicipital aponeurosis and just medial to the brachial artery.

9. The answer is D [Chapter 5 I A]. An embolus in the systemic venous system will pass through the right side of the heart and enter the pulmonary circulation, where it will lodge in the small vessels of the lungs.

10. The answer is C [Chapter 8 Table 8–1; Chapter 9 Table 9–1]. The flexor carpi radialis and the other superficial forearm flexors (i.e., the pronator teres, palmaris longus, flexor carpi ulnaris, and a portion of the flexor digitorum superficialis) originate from the medial humeral epicondyle. The superficial group of extensors (i.e., brachioradialis, extensor carpi radialis longus and brevis, extensor digitorum communis, and extensor carpi ulnaris) originate from the lateral epicondyle as does the supinator of the deep extensor group. The flexor pollicis and the other deep flexor muscles originate from the shaft of the radius, interosseous membrane, and ulna.

11. The answer is C [Chapter 8 I C 1 b; D 3 b (1); 9 II A 2 b (2)]. The lunate bone articulates with the triangular fibrocartilage at the radiocarpal joint much more so in abduction than in adduction. When displaced anteriorly, the lunate bone can impinge on the long flexor tendons of the digits and compress the median nerve, producing the carpal tunnel syndrome. The transverse carpal ligament arches over the lunate bone.

12. The answer is C [Chapter 10 VI C 2 a (3)]. The thumb musculature is innervated by the ulnar, median, and radial nerves. The ulnar nerve innervates the deep head of the flexor pollicis brevis muscle as well as the adductor pollicis muscle. The median nerve innervates the opponens pollicis, abductor pollicis brevis, and superficial head of the flexor pollicis brevis. The radial nerve innervates the abductor pollicis longus as well as the extensor pollicis longus and extensor pollicis brevis.

13. The answer is D [Chapter 9 II B 2; Chapter 10 Table 10-1]. The radial half of the flexor digitorum profundus flexes the terminal phalanx of the second and third digits. The flexor digitorum superficialis flexes the middle phalanx of the second through fifth digits. The lumbricals and interossei are extensors of the middle and distal phalanges.

14. The answer is D [Chapter 9 IV A 2, B 2; Chapter 10 VI B 3 a]. The radial half of the flexor digitorum profundus, as well as the flexor digitorum superficialis and flexor pollicis longus, are innervated by branches of the median nerve in the forearm. A lesion that affected these muscles would most likely also affect the pronators teres and quadratus as well as some of the thenar muscles of the hand. Because the flexor digitorum profundus to the fourth and fifth digits is innervated by the ulnar rather than the median nerve paralysis of the fourth or fifth digit would not be expected to accompany paresis of the second digit. Also, because some thumb muscles are innervated by the radial and ulnar nerves, rather than the median nerve, injury of the median nerve would not cause complete paralysis of the thumb.

15. The answer is C [Chapter 10 III C 1 a (1); C 2 b; Table 10-1]. Extension of the terminal phalanx of the index finger at the distal interphalangeal joint is accomplished primarily by the dorsal and palmar interossei, which are innervated by the ulnar nerve, as well as the lumbrical 2, which is innervated by the median nerve. The extensors digitorum communis and indicis proprius, which are innervated by the radial nerve, primarily extend the metacarpophalangeal joint.

16. The answer is E [Chapter 9 III A 1 a, 2 a; Chapter 10 V A 1 a]. In the hand, a radial pulse is palpable in the "anatomic snuffbox" (i.e., between the tendons of the extensor pollicis longus and brevis). At the wrist, the radial pulse may be palpated between the flexor carpi radialis and brachioradialis muscles. The ulnar pulse is felt on the radial side of the insertion of the flexor carpi ulnaris into the pisiform bone. In the cubital fossa, the brachial pulse is palpable medial to the biceps brachii tendon. In the arm, the brachial pulse is palpable against the humerus just distal to the pectoralis minor.

17. The answer is A [Chapter 6 III D 3 b, c; Table 6-2]. The dead weight of the arm is supported primarily by the coracohumeral ligament. In addition, the supraspinatus muscle, coracobrachialis, the short head of the biceps brachii muscle, and the long head of the triceps brachii all provide dynamic support at the glenohumeral joint. The coracoacromial ligament prevents superior displacement of the humerus at the glenohumeral joint.

18. The answer is D [Chapter 7 III B 1 a, 3]. Aside from the brachialis and biceps brachii (both innervated by the musculocutaneous nerve), the brachioradialis (radial nerve), flexor carpi radialis (median nerve), and flexor carpi ulnaris (ulnar nerve) all take origin from the humerus and can weakly flex the elbow joint. The ulnar head of the pronator teres does not cross the humeroulnar joint.

19. The answer is A [Chapter 8 I C 1 a; D 2 c]. The scaphoid bone, articulating maximally with the radius in abduction, transmits the major forces from the hand to the forearm and is, therefore, the carpal bone most prone to fracture. The scaphoid provides a strong attachment for the transverse carpal ligament and participates in both the radiocarpal and midcarpal joints.

20. The answer is E [Chapter 9 II A 2 a, b]. The ulnar nerve and ulnar artery pass deep to the volar carpal ligament but superficial to the transverse carpal ligament. The median nerve is the only major neurovascular structure that passes through the carpal tunnel.

21. The answer is C [Chapter 9 III A 1 b (1), 2 b (4); Chapter 10 V A 1 b (2), (3), 2 b (2), (3)]. The deep palmar arch is supplied by the radial and ulnar arteries and gives off palmar metacarpal arteries. These arteries join the common digital arteries, which arise from the superficial palmar arch, before the latter bifurcate into phalangeal arteries.

22. The answer is B [Chapter 10 III C 2 c; VI B 3 c; Tables 10-1 and 10-2]. The recurrent branch of the median nerve in the palm innervates most of the muscles of the thenar compartment, including the abductor pollicis brevis, opponens pollicis, and the superficial head of the flexor pollicis brevis muscle. The deep head of the flexor pollicis brevis is innervated by the ulnar nerve.

23. The answer is B [Chapter 10 III C 2 b; Tables 10-1 and 10-2]. Reposition of the thumb, a combination of abduction and extension accompanied by lateral rotation, is accomplished primarily by the abductor pollicis longus and brevis muscles, as well as by the extensor pollicis longus and brevis muscles. Opposition of the thumb is accomplished by flexion accompanied by medial rotation.

24. The answer is B [Chapter 10 III B 4]. An infection within the synovial sheath of the tendons of the fifth digit may compress the vincula and cause ischemic necrosis of the tendons. It also can spread along the synovial sheath through the carpal tunnel and into the distal forearm. If the ulnar and radial bursae communicate, infection may spread distally into the thumb. Infection of the fifth digit cannot spread distally into the tendon sheaths of the second, third, and fourth digits.

25. The answer is B [Chapter 10 VI C 2 a, b]. The hypothenar (medial) side of the ring finger is generally innervated by the ulnar nerve. The dorsal digital branches innervate a region on the distal dorsum of the digits. The branches of the superficial radial nerve seldom extend beyond the proximal interphalangeal joints. The palmar digital branches innervate the volar surfaces of the digits.

26–28. The answers are: 26-B, 27-C, 28-D [Chapter 6 III D 4; IV B 1, 2]. The supraspinatus, infraspinatus, teres major, and subscapularis contitutee the rotator cuff that acts across the glenohumeral joint and dynamically stabilizes the shoulder. The subscapularis, forming the posterior wall of the axilla, is a strong medial rotator of the arm. It is innervated by the upper and lower subscapular branches of the posterior cord of the brachial plexus. The supraspinatus, innervated by the suprascapular nerve, initiates abduction of the arm through the first 15 degrees, at which point the deltoid muscle assumes abductive function. The infraspinatus and teres minor are both lateral rotators, the former innervated by the suprascapular nerve, the latter by the axillary nerve.

29–32. The answers are: 29-D, 30-C, 31-A, 32-A [Chapter 7 Table 7-1; Chapter 9 Table 9-1; Chapter 10 Tables 10-1 and 10-2]. Most of the muscles of the flexor compartment of the forearm are innervated by the median nerve. The two exceptions are the flexor digitorum profundus, which receives innervation from both median and ulnar nerves, and the flexor carpi ulnaris, which receives ulnar innervation. All of the muscles of the extensor compartment of the forearm, including the abductor pollicis longus, are innervated by the radial nerve. In the hand, the median nerve generally innervates the thenar muscles as well as lumbricals I and II. The ulnar nerve generally innervates the hypothenar muscles, both dorsal and ventral interossei muscles, the adductor pollicis muscle, and lumbricals III and IV. The flexor pollicis brevis usually receives innervation from both median and ulnar nerves.

PART IIB BACK

Chapter 11
Vertebral Column

I. GENERAL STRUCTURE

A. Composition. The vertebral column consists of 33 vertebrae connected by 23–24 intervertebral disks and supported by ligaments.

1. **Vertebrae.** The vertebral column is divided into cervical, thoracic, lumbar, sacral, and coccygeal regions (Figure 11-1). Vertebrae differ in appearance according to region.
2. **Intervertebral disks**
   a. **Location.** Intervertebral disks form **amphiarthroses (symphyses)** between the vertebral bodies.
      (1) In the cervical region, disks do not occur between the occiput and the atlas nor between the atlas and the axis.
      (2) Remnants of disks occur within the sacrum and coccyx.
   b. **Structure.** Disks vary in shape, corresponding closely to the associated vertebral bodies. They are composed of **fibrocartilage** and allow a limited amount of movement between adjacent vertebrae.
3. **Intervertebral ligaments.** Stabilizing ligaments run between the vertebral bodies, vertebral arches, transverse processes, and spinous processes.

B. **Spinal curvatures** are responses primarily to terrestrial adaptation and secondarily to the erect stance (see Figure 11-1).

1. **Normal curvatures** are attributable almost entirely to the shape of the intervertebral disks rather than the vertebral bodies.
   a. The **sacral curvature** (S1 to the coccyx) is concave anteriorly.
      (1) This curve is characteristic of the fetus and lower animal forms.
      (2) It is a **primary curvature.**
   b. The **lumbar curvature** (T12–L5) is concave posteriorly.
      (1) This curve appears between the ages of 12 and 18 months.
      (2) It is more pronounced in females than in males and presents a greater potential for instability than other regions.
      (3) It represents the major adaptation to upright posture and is thus a **secondary (adaptational) curvature.**
   c. The **thoracic curvature** (T2–T12) is concave anteriorly.
      (1) This curve is characteristic of the fetus and lower animal forms.
      (2) It is a **primary curvature.**
   d. The **cervical curvature** (C2–T2) is concave posteriorly.
      (1) This curve appears in the late intrauterine period and is established by 3–4 months after birth.
      (2) It represents an adaptation to quadrupedal posture and is a **secondary (adaptational) curvature.**

2. **Abnormal curvatures**
   a. **Kyphosis** ("hunchback") is an exaggerated thoracic curvature that usually results from congenital deformities, pathologic erosion, or traumatic collapse of one or more vertebral bodies.
   b. **Lordosis** ("swayback") is a greater than normal lumbar curvature that usually results from compression of the posterior portion of the lumbar intervertebral disks

FIGURE 11-1. The anterior view of the vertebral column (*left*) reveals the major differences between the vertebra of the cervical, thoracic, lumbar, sacral, and coccygeal regions. The lateral view (*right*) displays the primary sacral and thoracic curvatures as well as the secondary lumbar and cervical curvatures.

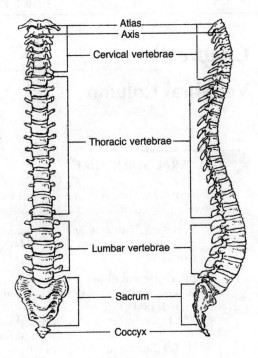

as compensation for a forward shift in the center of gravity, such as that accompanying the late stages of pregnancy or as occurs in obesity.
  c. **Scoliosis** is a lateral curvature.
    (1) The causes include congenital malformation, pathologic erosion, and unilateral weakness or paralysis of vertebral musculature.
      (a) Most commonly, this condition is idiopathic (of unknown causes).
      (b) It is usually accompanied by a compensatory curve in the opposite direction in another region of the spine.
    (2) Concomitant with its progression is rotation toward the convex side. This deformity progressively interferes with mechanical respiration.
    (3) Scoliosis is amenable to surgical correction, preferably during the late teenage years and definitely before the age of 30 years or anticipated childbearing.

II. STRUCTURE OF TYPICAL VERTEBRA

A. **The body** forms the major supportive portion of the vertebra (see Figure 11-3).

  1. **Structure.** The body is approximately cylindric. As an adaptation to upright posture, each subsequent body must support a greater proportion of the mass of an individual; therefore, bodies increase in bulk toward the sacrum.

  2. **Articulations**
    a. **Amphiarthroses.** The bodies of most vertebrae articulate with the superjacent and subjacent bodies through **intervertebral disks** that form **symphyses**.
    b. **Diarthroses.** The **superior and inferior articular processes** of most vertebrae articulate with the superjacent and subjacent vertebrae by **synovial joints**.

B. **The vertebral (neural) arch,** consisting of the pedicles and laminae, forms the posterior portion of the vertebra (see Figure 11-3).

1. **Pedicles** arise posterolaterally from the vertebral body.
2. **Laminae** are supported by the pedicles and fuse in the midline to form the roof of the vertebral arch.

C. Vertebral processes (see Figure 11-3)

1. **The spinous process** arises from the vertebral arch where the laminae fuse. It serves as a lever for the attachment of vertebral musculature and ligaments.
2. **Transverse processes** arise on each side of the vertebral arch where the pedicles and laminae fuse.
   a. They serve as levers for the attachment of vertebral musculature and ligaments.
   b. They articulate with the necks of ribs 1–10.
   c. Fused with the sacral costal processes, they form the sacral ala, which articulate with the ilia.
3. **Costal processes** arise on each side of the vertebral body anterior to the pedicle. They form the ribs in the thoracic region and contribute to the transverse processes of the cervical vertebrae as well as to the alae (wings) of the sacrum.
4. **Superior articular processes** (superior zygapophyses) are located on the superior surface of the pedicles.
   a. They form diarthrodial articulations with inferior articular processes of the vertebra immediately above.
   b. They generally face posteriorly in the cervical region, posteromedially in the thoracic region, medially in the lumbar region, and posteriorly in the sacrum. The orientation promotes certain movements and limits others.
5. **Inferior articular processes** (inferior zygapophyses) arise from the inferior surfaces of the pedicles.
   a. They form diarthrodial articulations with the superior articular processes of the vertebra immediately below.
   b. They generally face anteriorly in the cervical region, anterolaterally in the thoracic region, and laterally in the lumbar region, but become anterior by the fifth lumbar vertebra. The orientation promotes certain movements and limits others.

D. Foramina

1. **The vertebral foramen** is formed by the posterior surface of the vertebral body and the vertebral arch (see Figure 11-3).
   a. Successive vertebral foramina form the **vertebral (neural) canal.**
   b. These foramina contain the **spinal cord** and **nerve roots** with the associated meningeal coverings as well as vessels embedded in adipose tissue.
2. **Intervertebral foramina** are located bilaterally between adjacent vertebrae (see Figure 11-5).
   a. The intervertebral foramen on each side is formed by a deep notch in the inferior surface of each pedicle of one vertebra and a shallow notch in the superior surface of each pedicle of the vertebra immediately below.
   b. The intervertebral foramina transmit the **spinal nerves** as they exit the vertebral canal.
   c. These foramina also transmit the **intervertebral arteries,** which supply the nerve roots and spinal cord.

III. REGIONAL MODIFICATIONS OF VERTEBRAL CHARACTERISTICS

A. Cervical vertebrae (C1–C7)

1. **The atlas** is the first vertebra (Figure 11-2).
   a. **Structure.** Lacking a body and a spinous process, this vertebra consists of an anterior arch, the neural arch, and paired transverse processes.

FIGURE 11-2. Atlas and axis. The atlas articulates with the occipital condyles of the cranium, at which flexion and extension occur, and has no body or spinous process. The odontoid process (dens) of the axis, about which rotation occurs, represents the misplaced body of the atlas.

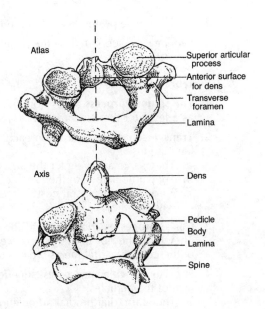

   b. **Articulations**
      (1) The atlas articulates superiorly with the basal portion of the occipital bone (**atlantooccipital joint**).
         (a) Two large concave facets receive the occipital condyles. The facets are nearly horizontal.
         (b) The atlantooccipital joint functions in **flexion/extension** ($\pm 15°$).
         (c) There is no intervertebral disk at the atlantooccipital joint, which accounts for the extensive range of flexion and extension.
      (2) The atlas articulates inferiorly with the axis through broad, flat inferior articular facets.
2. **The axis** is the second vertebra (see Figure 11-2).
   a. **Structure.** The axis differs from typical cervical vertebrae in that it has a superior projection, the dens, and large flat superior articular surfaces, which are nearly horizontal. The **dens (odontoid process)** represents the body of the atlas, which has become separated from the first vertebra and fused with the second.
   b. **Articulations**
      (1) The axis articulates superiorly with the atlas (**atlantoaxial joint**) both at the dens, which projects into the atlas to articulate with the anterior arch, and at the zygapophyses.
         (a) The **dens,** stabilized within the atlas by several ligaments, serves as a pivot about which rotation occurs.
            (i) The **cruciform ligament** of the atlas provides support for the atlantoaxial joint by running posterior to the dens and preventing subluxation of the joint.
            (ii) **Fracture of the dens** with posterior dislocation of the axis may crush the spinal cord at the level of the first cervical vertebra, with subsequent death from paralysis of all respiratory musculature.
         (b) The atlantoaxial joint functions primarily in **rotation** ($\pm 25°$).
         (c) There is no intervertebral disk at the atlantoaxial joint, which accounts for the extensive range of rotation.
      (2) The axis articulates inferiorly with C3 through an intervertebral disk and at the zygapophyses.
3. **Typical cervical vertebrae** are small but increase in size toward the thoracic region (Figure 11-3).
   a. **Structure.** They have a characteristic **transverse foramen** in each transverse process (see Figure 11-3).

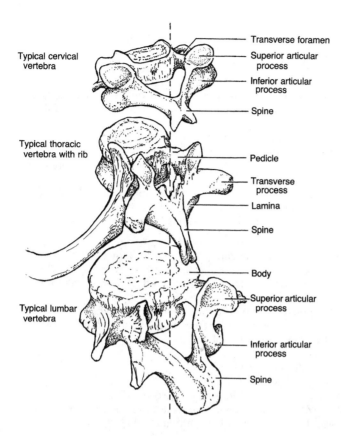

FIGURE 11-3. Typical cervical, thoracic, and lumbar vertebrae. *Dashed line,* the center of the spinal canal.

   (1) This foramen represents a hiatus that results from incomplete fusion of the costal and transverse processes.
   (2) Except for the seventh cervical vertebra (C7), the transverse foramen transmits the **vertebral artery,** which provides collateral circulation to the brain.
   (3) The cervical spinous processes angle sharply downward and often are bifid. The spinous process of C7, the **vertebra prominens,** is nearly horizontal and usually is not bifid. It is a most useful surface landmark.
  b. **Articulations.** The cervical articular processes have a characteristic orientation (see Figure 11-3).
   (1) The superior articular facets are slightly convex and more or less have a mutual radius.
   (2) These joints have **3 degrees of freedom,** allowing **rotation, flexion/extension,** and **lateral flexion.**
   (3) The superior articular surfaces generally face posteriorly, whereas the inferior articular surfaces generally face anteriorly.
   (4) The orientation of and the great mobility at the cervical zygapophyses predispose to dislocation.
 4. **Summary of movements** over the cervical region
  a. **Flexion:** 40°/**extension:** 90°, primarily at the atlantooccipital joint
  b. **Lateral flexion (abduction):** ~40°, primarily between C2 and C7
  c. **Rotation:** ±45°, half of which occurs at the atlantoaxial joint
  d. **Circumduction.** The combination of flexion/extension and lateral flexion permits circumduction of the head.

B. **Thoracic vertebrae** (T1–T12) are larger than the cervical vertebrae and increase in size toward the lumbar region. The typical thoracic vertebra has 10 articular facets in addition to the two amphiarthroses.

1. **Structure** (see Figure 11-3)
   a. The thoracic **spinous processes** are long and oriented caudally.
   b. The **costal processes** develop into ribs.
   c. The bodies and transverse processes have **articular surfaces** for the ribs (diarthrodial joints).
      (1) Vertebrae T1 through T10 have costal articular facets on the transverse processes and costal articular demifacets on the bodies near the origins of the pedicles.
         (a) The articular facets are for the necks of the ribs.
         (b) The articular demifacets are for the heads of the associated ribs.
      (2) The second through ninth ribs articulate with superior and inferior vertebrae at the level of the intervertebral disk.
         (a) The bodies of T1 and T10 through T12 have complete facets for the articulation of the associated ribs.
         (b) T2 through T9 have demifacets on the bodies for articulations with the associated ribs.

2. **Intervertebral articulations.** The thoracic zygapophyses have characteristic orientations (see Figure 11-3).
   a. The paired superior articular processes face posterolaterally, whereas the inferior articular processes face anteromedially.
   b. The articular surfaces lie approximately on an arc of a circle, the center of which is located in the region of the vertebral body, thereby allowing rotation. Noted over the twelve thoracic vertebrae are the following:
      (1) A total of ±35° of **rotation**
      (2) Less flexion (40°) and extension (20°)
      (3) A small amount of lateral flexion (20°)
   c. Movements of the thoracic vertebral column are limited by the ribs and their attachment to the sternum. Movement is possible because the anterior portion of the ribs is cartilaginous and the costal cartilages articulate with the sternum through diarthroses.

C. **Lumbar vertebrae** (L1–L5) are large and heavy with massive bodies, becoming progressively larger toward the sacral region (see Figure 11-3).

1. **Structure.** The spinous processes are stubby and project horizontally, allowing considerable extension and access to the intervertebral canal with a needle **(lumbar puncture).** The intervertebral notches on the inferior surface of the pedicles are deep.

2. **Articulations.** The articular processes have characteristic orientations.
   a. They lie on nearly parasagittal planes. The articular surfaces lie on an arc of a circle, the center of which is located near the tips of the spinous processes.
   b. The superior articular surfaces face medially, whereas the inferior articular surfaces face laterally.
      (1) This orientation allows considerable **flexion** (60°)/**extension** (35°).
      (2) Lateral flexion is moderate (20°).
      (3) Rotation is limited to ±5° over the lumbar region.
      (4) The combination of flexion/extension and lateral flexion allows considerable circumduction of the trunk.
   c. The articular facets for L5 are directed anteriorly and engage the posteriorly directed superior processes of the first sacral vertebra (S1).

D. **Sacral vertebrae** (S1–S5) fuse to form the sacrum (Figure 11-4).

1. **Structure** (see Figure 11-4)
   a. Four horizontal lines of fusion are evident on the anterior surface.
      (1) A rudimentary disk may be found between S1 and S2.
      (2) Rarely, **sacralization of L5** occurs whereby that vertebra becomes partially or completely fused with the sacrum.
   b. The fused spinous processes form the **median sacral crest.**
   c. The fused transverse processes, together with fused **costal processes,** form the

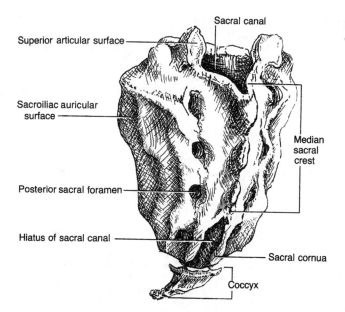

FIGURE 11-4. The sacrum and coccyx.

alae or sacroiliac articular processes. This fusion occurs around the spinal nerves to form **sacral foramina.**
  (1) The **intervertebral canal** remains between the fused vertebrae.
  (2) **Anterior and posterior sacral foramina** form canals that run anteroposteriorly, perpendicular to the intervertebral canal.
      (a) The **posterior sacral foramina** transmit the dorsal primary rami of the sacral spinal nerves.
      (b) The **anterior sacral foramina** transmit the ventral primary rami of the sacral spinal nerves.
  d. The lamina of S5 fail to form, resulting in the **sacral hiatus.** The pedicles of this vertebra form the **sacral cornua,** important landmarks for locating the sacral hiatus when administering **caudal anesthesia.**
2. **Articulations** include the lumbosacral, sacrococcygeal, and sacroiliac joints.
  a. The articular facets for S1 are directed posteriorly, thereby preventing the lumbar vertebrae from slipping forward on the anteriorly slanted body of the sacrum at the **lumbosacral joint.**
     (1) **Structure**
        (a) The large angle (about 140°) between the axes of L5 and the sacrum and the considerable angle (about 40°) of the body of S1 from the horizontal impose enormous forces on the superior articular processes of S1 (see Figure 11-1). Also, the lumbosacral intervertebral disk is wedge-shaped.
        (b) In addition to the articular processes of L5 and S1, the **anterior longitudinal ligament** also plays a role in resisting anterior displacement.
     (2) **Spondylolisthesis** is anterior displacement of the vertebral column, usually at the lumbosacral joint.
        (a) The **cause** is a loss of bony stability resulting from:
           (i) **Congenital defect** of the lamina of L5 **(spondylolysis),** found in 5% of the population
           (ii) **Fracture of the pedicles** or inferior articular processes of L5
           (iii) **Fracture of the lamina** or superior articular processes of the sacrum
        (b) **Subluxation** involves anterior dislocation of L5. Because the center of gravity normally passes through the lumbosacral joint, the vertebral column slides forward on the intervertebral disk at the lumbosacral joint. The anterior longitudinal ligament supports the entire vertebral structure.

(c) **Spondylolisthesis** may involve compression of spinal nerves S1 and S2, producing symptoms of **sciatica**. This syndrome presents a dermatomal pattern of pain and hypesthesia, sometimes including myotomal muscle weakness, associated with compression of the lower lumbar and upper sacral spinal nerves.

b. The sacrum articulates with the first coccygeal vertebrae to form the **sacrococcygeal joint**.

c. The **alae** articulate with the ilia at the strong **sacroiliac joint**.
  (1) **Structure**. Because of the synovial cavity within it, the sacroiliac joint is a diarthrosis.
  (2) **Ligamentous support**. With its extent of ligamentous reinforcement, this articulation is among the strongest diarthrodial joints in the body. These ligaments oppose the rotational effect of gravity at the sacroiliac joint.

E. Coccygeal vertebrae

1. **Structure**. The coccygeal vertebrae vary in number and appearance (see Figure 11-4).
   a. The four or five coccygeal vertebrae fuse to form the coccyx, but there may be a disk between the first and second coccygeal vertebrae.
   b. The first (and sometimes the second) coccygeal vertebra has rudimentary pedicles that form the **coccygeal cornua**.
   c. The second through fourth coccygeal vertebrae represent rudimentary vertebral bodies.

2. **Articulations**. The first coccygeal vertebra articulates with the sacrum. The **sacrococcygeal joint** usually has a small intervertebral disk, but fusion may occur in some individuals as they age.

F. Clinical considerations

1. **Compression fractures** of the vertebral bodies may result from disease or trauma.
   a. The internal collapse of the vertebral body may result in kyphosis or scoliosis.
   b. These fractures also may reduce the size of the intervertebral foramen and increase the possibility of spinal nerve compression.

2. **Fracture of the pedicles** dissociates a vertebral body from the stabilizing influences of the associated zygapophyses.
   a. Such fractures predispose to misalignment of the vertebral column with risk of compression of the spinal cord and damage to the exiting spinal nerves.
   b. Because the vertebral canal is relatively wide in the cervical region, a small displacement may not compress the spinal cord.
   c. Because the vertebral canal is relatively narrow in the thoracolumbar region, paraplegia is a common sequela to a thoracic fracture dislocation.

3. **Low back pain** is often misdiagnosed as sacroiliac sprain. Low back pain is usually noted in association with muscle spasm secondary to the pain of osteophytes; arthritis at the lumbar zygapophyses; or a herniated disk at the level of L4 or L5.

4. **Coccygodynia**. Fracture of the fused coccygeal joints or of a fused sacrococcygeal joint, as well as arthritis in these joints, can cause great pain.

IV. INTERVERTEBRAL DISKS

A. External structure.
The disks conform to surfaces of apposed vertebral bodies (Figure 11-5).

1. Disks represent about 25% of the total length of the vertebral column.
2. In the cervical region, the disks are small and thin. In the lumbar region, they are larger and thicker.

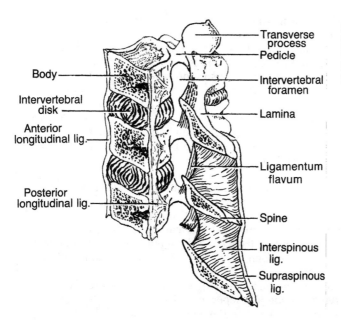

FIGURE 11-5. Principal ligaments of the vertebral column and intervertebral disks. Formation of intervertebral foramina from adjacent vertebrae is evident.

3. Differences in the thickness anteriorly and posteriorly define the **secondary vertebral curvatures** (i.e., they are thicker anteriorly in the lumbar and cervical regions).

B. Internal structure

1. **The annulus fibrosus** forms the outer portion of the disk.
   a. This structure is composed of lamelliform connective tissue bands. Each band is oriented in a different direction, providing considerable strength.
   b. It is firmly attached to and joins adjacent vertebrae, forming a symphysis type of amphiarthrosis.
   c. It supports the central nucleus pulposus.

2. **The nucleus pulposus** forms the central portion of the disk.
   a. This structure is a remnant of the embryonic **notochord.**
   b. It is surrounded and supported by the annulus fibrosus.
   c. It is composed of cartilaginous, mucinous tissue with fibrous and mucopolysaccharide complexes that contribute to its high osmotic pressure.
   d. The extremely high water content of this portion of the disk (70%–80%) accounts for the fact that a person may be as much as 1 cm taller on rising in the morning than at the end of a strenuous day.

C. Functions. As **amphiarthroses,** disks bind the vertebrae together and allow limited movement between adjacent vertebrae. The nucleus pulposus functions as a noncompressible but deformable pad, which distributes forces uniformly over the entire surface of the vertebra, regardless of the degree of flexion/extension, rotation, or lateral flexion.

D. Clinical considerations

1. **Disk degeneration** is associated with chronic dehydration of the nucleus pulposus.
   a. Loss of water and mucopolysaccharide leads to a narrowing of the intervertebral space and reduces the capacity of the disk to act as a cushion between vertebrae.
   b. Narrowing of disks results in diminished stature. It also decreases the size of the intervertebral foramina, increasing the possibility of spinal nerve compression.
   c. Occasionally, progressive calcification at the superior and inferior margins of the vertebral bodies forms **osteophytes.** Osteophytes in the cervical region may compress the spinal nerves, with resultant arm pain. In the lumbar region, they may cause soft tissue irritation, with resultant **chronic back pain.**

2. Disk herniation ("slipped disk") is the prolapse (extrusion) of the nucleus pulposus through the annulus fibrosus.
   a. Because of the lumbar curvature and the considerable mass of the body superior to this region, herniation usually occurs in the disk between L4 and L5 or between L5 and S1. Disk herniation also occurs in the cervical region (5%–10%).
   b. Rarely, prolapse occurs into the superjacent or subjacent vertebral bodies (Schmorl's node).
   c. Most commonly, herniation is directed posterolaterally, where the annulus fibrosus is not reinforced. The anterior and posterior longitudinal ligaments reinforce the underlying annulus fibrosus. These ligaments do not meet posterolaterally, forming a weak area predisposed to herniation.
   d. Nerve compression
      (1) Posterolateral disk prolapse impinges on the spinal nerve of the next lower vertebral level, causing symptoms associated with the dermatomic and myotomic distributions of that nerve.
         (a) In the cervical region, there are eight cervical nerves, but seven vertebrae; thus, herniation of the disk between C4 and C5 impinges on spinal nerve C5, which exits through the intervertebral foramen formed by these two cervical vertebrae.
         (b) In the lumbar region, the deeply notched pedicles allow the similarly numbered nerve to escape just superior to the disk. Herniation compresses the next spinal nerve as it descends across the disk to its intervertebral foramen.
         (c) A large herniation may involve several spinal nerves below the herniation and even contralateral nerves.
      (2) In addition to the pain referred to the distribution of the compressed spinal nerve, local pain produced from the stretched annulus fibrosus initiates painful spasm of the back muscles.

V. LIGAMENTS

A. The **supraspinous ligament** connects spinous processes at the tips and forms the **ligamentum nuchae** in the cervical region (see Figure 11-5).

B. **Interspinous ligaments** run between spinous processes (see Figure 11-5).
   1. They limit the range of motion of the vertebrae.
   2. Small tears in these ligaments, the result of hyperextension trauma, may be the basis for "whiplash" injury.

C. The **ligamentum flavum** (L. yellow) stretches between adjacent laminae. Except for a midline gap, the paired flaval ligaments, together with the vertebral arch, complete the posterior aspect of the vertebral canal (see Figure 11-5).
   1. Unlike other ligaments, the ligamentum flavum is composed of elastic tissue.
   2. Traumatic hyperextension of the neck may exceed the capacity of the ligamentum flavum to take up slack. The resultant buckling may injure the spinal cord.

D. The **anterior longitudinal ligament** connects the vertebral bodies and intervertebral disks anteriorly.
   1. This broad band runs from the sacrum to the occipital bone.
   2. It resists the gravitational tendency toward increased lordosis.
   3. It reinforces the annulus fibrosus anteriorly and directs herniation posteriorly.
   4. It can be used to splint fractured vertebrae when the trunk is cast in extension.

E. **The posterior longitudinal ligament** connects the vertebral bodies and the intervertebral disks posteriorly.

1. This denticulate ligament is wider posterior to the vertebral body and narrower posterior to the intervertebral disk.
2. It resists the gravitational tendency toward increased kyphosis.
3. It supports the annulus fibrosus posteriorly and directs disk herniation posterolaterally.

F. **The cruciform ligament** of the atlas supports the dens of the axis.

1. This small, but important, ligament runs transversely in the neural canal of the atlas posterior to the dens of the axis. It extends superiorly to the base of the occiput and inferiorly to the body of the axis.
2. Because the articular surfaces between the axis and atlas are nearly horizontal, this ligament is the principal structure preventing subluxation at the atlantoaxial joint.

Chapter 12
Soft Tissues of the Back

I. POSTERIOR VERTEBRAL MUSCLES

A. Overview. Three layers of posterior muscles extend, rotate, laterally flex, and stabilize the vertebral column (Figure 12-1).

B. Organization. The most superficial layer runs upward and obliquely outward, the middle layer runs parallel to the vertebral column, and the innermost layer runs upward and obliquely inward.

1. **The spinotransverse group** represents the superficial layer (Table 12-1; see Figure 12-1).
   a. **Attachments.** This group originates from spinous processes and the nuchal ligament, runs superiorly and obliquely laterally, and inserts onto upper cervical transverse processes or the base of the skull.
   b. **Divisions**
      (1) The **splenius capitis muscle** runs from the spines of the seventh cervical through the fourth thoracic vertebrae and the nuchal ligament to insert into the superior nuchal line and mastoid process of the skull.
      (2) The **splenius cervicis muscle** runs from the spines of the third through sixth thoracic vertebrae to the transverse processes of the second through fourth cervical vertebrae.
   c. **Group actions.** Acting unilaterally, these muscles rotate the head and neck toward the same side. Acting bilaterally, they elevate the head and extend the neck.

2. **The sacrospinalis group (erector spinae)** represents the middle layer (see Figure 12-1 and Table 12-1).
   a. **Attachments.** This group originates from the sacrum, iliac crests, and spinous processes of the lumbar and lower thoracic vertebrae; runs parallel to the vertebral column; and inserts into the ribs and transverse processes.
   b. **Divisions**
      (1) The **iliocostalis muscles,** including the lumbar, thoracic, and cervical portions, constitute the most lateral segment of the erector spinae.
      (2) The **longissimus muscle,** including the thoracic, cervical, and capitis portions, constitutes the middle segment.
      (3) The **spinalis muscle,** which is poorly developed, constitutes the medial segment.
   c. **Group actions.** Acting unilaterally, the sacrospinalis muscles flex the vertebral column to the same side. Acting bilaterally, they extend the vertebral column.

3. **The transversospinalis group** represents the innermost layer (see Figure 12-1 and Table 12-1).
   a. **Attachments.** This group originates from transverse processes, runs superiorly and obliquely medial, and inserts into spinous processes. The deeper muscles traverse shorter distances.
   b. **Divisions**
      (1) The **semispinalis muscle,** with its thoracic, cervical, and capitis portions, lies beneath the erector spinae muscle and generally passes over five or more vertebrae.
      (2) The **multifidus muscles (multifidi)** lie deep to the semispinalis muscle and pass over two to three vertebrae.
         (a) Except for their shorter bundles, they are nearly indistinguishable from the semispinalis muscle.
         (b) They are best developed in the lumbar and cervical regions.

FIGURE 12-1. Long muscles of the back. *(Left)* The spinotransverse and spinospinalis groups; *(right)* the transversospinalis group.

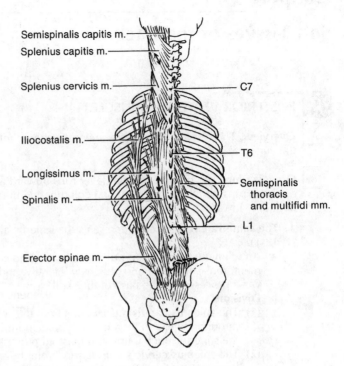

(3) **Long rotators** run from the transverse processes to the spinous process two vertebrae above (Figure 12-2).
(4) **Segmental muscles** run between adjacent vertebrae (see Figure 12-2).
  (a) **Short rotators** run between the transverse and spinous processes of adjacent vertebrae.
  (b) **Interspinales muscles** run between spinous processes and are best developed in the cervical region.
  (c) **Intertransversarii muscles** run between transverse processes and are especially well developed in the lumbar and cervical regions.
  c. **Group actions.** Acting unilaterally, the transversospinalis muscles rotate the neck and trunk to the opposite side. Acting bilaterally, they extend the vertebral column.
4. **The suboccipital muscles** (a special group of deep muscles) extend and rotate the head (Table 12-2; see Figure 30-6).
  a. The **inferior oblique capitis muscle** runs from the spine of the axis to the transverse process of the atlas.
  b. The **superior oblique capitis muscle** runs from the transverse process of the atlas to the occiput.
  c. The **posterior rectus capitis major muscle** runs from the spine of the axis to the occiput.
  d. The **posterior rectus capitis minor muscle** runs from the spine of the atlas to the occiput.

C. **Innervation.** The posterior muscles are innervated by **dorsal primary rami** of the spinal nerves.

II. ANTERIOR VERTEBRAL MUSCLES

A. **Overview.** An incomplete layer of anterior muscles primarily flexes, laterally rotates, and stabilizes the vertebral column.

TABLE 12-1. Posterior Vertebral Musculature-Posterior Group

Muscle	Origin	Insertion	Primary Action	Innervation
Spinotransverse group				
Splenicus capitis	Upper thoracic vertebrae and ligamentum nuchae	Mastoid process	Extend head	Dorsal primary rami of cervical plexus
Splenius cervicis	Spines of vertebrae C3–C6	Transverse process of vertebrae C1–C4	Extend neck	Dorsal primary rami of cervical plexus
Sacrospinalis (erector spinae) group				
Iliocostalis lumborum	Sacrum and iliac crest	Lower six ribs	Extend and laterally flex vertebral column	Dorsal primary rami of spinal nerves
Iliocostalis thoracis	Ribs T7–T12	Ribs T1–T6	Extend and laterally flex vertebral column	Dorsal primary rami of spinal nerves
Iliocostalis cervicis	Ribs T1–T6	Transverse processes of C4–C6	Extend and laterally flex neck	Dorsal primary rami of spinal nerves
Longissimus capitis	Transverse processes of vertebrae C2–C5	Mastoid process	Extend and rotate head to ipsilateral side	Dorsal primary rami of cervical plexus
Longissimus cervicis	Transverse processes of vertebrae C7–T5	Transverse processes or vertebrae C2–C6	Extend and laterally flex vertebral column	Dorsal primary rami of cervical plexus
Longissimus thoracis	Transverse processes of vertebra T6–L5	Ribs and transverse processes of vertebra T12–T1	Extend and laterally flex vertebral column	Dorsal primary rami of spinal nn. T1–L5
Spinalis thoracis	Spinous processes	Spinous processes	Extend vertebreal column	Dorsal primary rami
Spinalis cervicis	Spinous processes	Spinous processes	Extend vertebreal column	Dorsal primary rami
Spinalis capitis	Spinous processes	Spinous processes	Extend vertebreal column	Dorsal primary rami
Transverseospinalis group				
Semispinalis capitis	Transverse processes of vertebrae T1–T6	Occipital bone	Extend head and rotate head to opposite side	Dorsal primary rami of cervical plexus
Semispinalis cervicis	Transverse processes of vertebrae T1–T7	Spinous processes of vertebrae C1–C5	Extend vertebral column and rotate to opposite side	Dorsal primary rami of cervical plexus

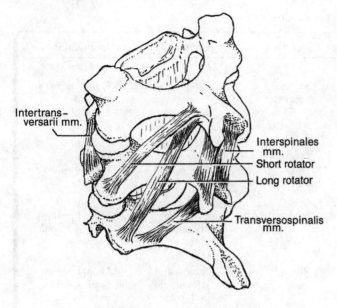

FIGURE 12-2. Short muscles of the back.

B. **Lateral group** (Table 12-3; see Figure 30-3)

1. **The scalene muscles** run from transverse processes of cervical vertebrae to the first and second ribs. Their primary action is elevation of the first and second ribs as accessory muscles of respiration. They are also involved in lateral flexion of the cervical vertebral column.

2. **The levator scapulae** runs from the transverse processes of the upper four cervical vertebrae to the scapular notch.

C. **Anterior group** (see Figure 30-4 and Table 12-3)

1. **The longus colli muscle** runs from the fourth through sixth cervical vertebrae to the occiput. It primarily flexes the cervical vertebral column.

2. **The lateral rectus capitis muscle** runs between the transverse process of the atlas and the jugular process of the occiput. It flexes and rotates the head.

3. **The anterior rectus capitis muscle** runs between the transverse process of the atlas

TABLE 12-2. Intrinsic Cervical Musculature—Suboccipital Group

Muscle	Origin	Insertion	Primary Action	Innervation
Rectus capitis post. maj.	Spine of axis	Occipital bone	Extend head	Spinal n. C1 (posterior)
Rectus capitis post. min.	Posterior tubercle of atlas	Occipital bone	Extend head	Spinal n. C1 (posterior)
Obliquus capitis inf.	Spine of axis	Transverse process of atlas	Rotate head to ipsilateral side	Spinal n. C1 (posterior)
Obliquus capitis sup.	Transverse process of atlas	Occipital bone	Extend head and rotate head to ipsilateral side	Spinal n. C1 (posterior)

TABLE 12-3. Anterior Vertebral Musculature

Muscle	Origin	Insertion	Primary Action	Innervation
Lateral group				
Scalenes:				
Anterior	Anterior tubercles of vertebrae C3–C6	First rib	Elevates first rib	Spinal nn. C4–C6 (anterior)
Middle	Posterior tubercles of vertebrae C3–C6	First rib	Elevates first rib	Spinal nn. C3–C8 (anterior)
Posterior	Posterior tubercles of vertebrae C3–C6	Second rib	Elevates second rib	Spinal nn. C3–C8 (anterior)
Levator scapulae	Posterior tubercles of vertebrae C2–C5	Superior angle of scapula	Elevates scapula	Spinal nn. C3–C4 (anterior)
Anterior group				
Longus cervicis (colli)				
Sup. oblique head	Anterior tubercles of vertebrae C3–C6	Spine of axis		
Vertical head	Upper thoracic vertebrae	Bodies cervical vertebrae C2–C4	Flexes neck	Spinal nn. C2–C7 (anterior)
Inf. oblique head	Upper thoracic vertebrae	Transverse processes of C4–C5		
Longus capitis	Anterior tubercles of vertebrae C3–C6	Occipital bone	Flexes head	Spinal nn. C1–C3 (anterior)
Rectus capitis anterior	Atlas	Occipital bone anterior to foramen magnum	Flexes head	Spinal nn. C1–C2 (anterior)
Rectus capitis lateralis	Atlas	Jugular process of occipital bone	Stabilizes atlantooccipital joint	Spinal nn. C1–C2 (anterior)
Quadratus lumborum	Iliac crest	Twelfth rib	Stabilizes and lowers 12th ribs during inspiration; abducts vertebral column	Twigs from nn. T12–L4 (ant)

and the basal part of the occipital bone. It primarily flexes the head at the atlanto-occipital joint.
4. **The quadratus lumborum muscle** runs from the iliac crests to the inferior borders of the twelfth ribs. It stabilizes the rib cage during ventilation.

III. FUNCTIONAL CONSIDERATIONS

A. **Normal function.** In addition to major movements of the vertebral column, the vertebral musculature keeps the center of gravity of the body over the first sacral vertebra.

B. **Adjustments**
   1. **Standing erect and holding a weight.** When a person stands erect (balanced over the center of gravity) with a 10-kg weight in each hand, 20 kg is distributed evenly over each vertebral disk.
   2. **Lifting a weight.** If the 20 kg is moved 20 cm anterior to the vertebral column by flexing the elbows, the forward shift in the center of gravity requires contraction of the back muscles to keep the individual erect.
      a. The muscles that insert into the tips of the spinous processes have a mechanical advantage because of the 2-cm lever arm. These muscles exert 200 kg of counterforce: 20 kg x (20 cm/2 cm) = 200 kg.
      b. This counterforce, pulling downward on one side of the vertebra, is additive to the weight of the object lifted. Thus, the total force distributed over each intervertebral disk is: 20 kg + 200 kg = 220 kg.
   3. **Bending forward to lift a weight.** If, in addition, the individual supports the 20-kg weight with the trunk flexed forward so that the weight extends 50 cm anterior to the fifth lumbar vertebra, the back muscles must generate additional force.
      a. The muscles that insert into the tips of the spinous processes have a mechanical advantage because of the 2-cm lever arm. These muscles exert 500 kg of counterforce: 20 kg x (50 cm/2 cm) = 500 kg.
      b. This muscular pull downward on the vertebral spine is additive to the weight of the object lifted: 500 kg + 20 kg = 520 kg. Thus, 520 kg must be distributed over each lumbar intervertebral disk.
      c. Because a disk may rupture when force exceeds 500–800 kg (1100–1760 lb), this individual is at risk for disk herniation.

Chapter 13
Spinal Cord

I. GENERAL ORGANIZATION

A. Characteristics. The spinal cord is an extension of the brain and a component of the central nervous system.

B. Vertebral (neural) canal. The spinal cord and its meningeal coverings lie within the osseofibrous vertebral canal (see Figure 11-5).

1. **Boundaries.** At the vertebral level, the canal is bounded by the **vertebral body** and the **neural arch,** which consists of the pedicles and laminae. At the intervertebral levels, it is bounded by the **intervertebral disks** and the **ligamenta flava,** which run between successive laminae on either side of the midline.

2. **The epidural space** between the boundaries of the vertebral canal and the dura mater is filled with loose fatty connective tissue. It contains the extensive **vertebral venous (Batson's) plexus** and the **spinal nerves,** which run through this space to reach the intervertebral foramina. Anesthetic agents are infused into this space for **epidural anesthesia.**

C. External structure. The spinal cord extends from the **medulla oblongata** at the level of the first cervical vertebra and terminates as the **conus medullaris** at the level of the intervertebral disk between the first and second lumbar vertebrae.

1. **Fetal location.** Until the third fetal month, the spinal cord is as long as the vertebral canal and extends to the level of the fourth sacral vertebra.

2. **The growth period and "law of descent."** After the fourth fetal month, the vertebral column outgrows the spinal cord. It appears that the cord regresses to the upper lumbar levels; however, the spinal roots descend through the dural sac to the appropriate intervertebral foramina—the law of descent. The resultant descending distribution of spinal roots forms the **cauda equina** (L. horse's tail; Figure 13-1).

3. **Postnatal location**
   a. In most individuals, the spinal cord terminates between the first and second lumbar vertebrae (see Figure 13-1). This statement is based, however, on a statistical distribution.
   b. Because the spinal cord extends below the second lumbar vertebra in about 1% of the population (especially in short-statured individuals), **neither lumbar puncture nor spinal (intrathecal) anesthesia should be attempted above the level of the third lumbar vertebra.**

D. Spinal meninges. Three layers envelop the brain and spinal cord (Figure 13-2). The dura is sometimes referred to as the **pachymeninx** (G. thick membrane). The combined arachnoid and pia can be referred to as the **leptomeninges** (G. delicate membranes).

1. **The dura mater** (L. tough mother) is the outermost meningeal layer (see Figure 13-2). It is a dense fibrous membrane that forms a sheath about the central nervous system.
   a. Beginning at the foramen magnum, where it is no longer fused with the periosteum of the cranial vault, the dura extends to the level of the second sacral vertebra. At this point, it forms the **coccygeal ligament,** extending to vertebra S4 or S5 and covering the pial **filum terminale.**
   b. The dura evaginates into each intervertebral foramen to surround the spinal nerve and fuse with the periosteum of the vertebrae. It becomes continuous with the epineurium, the connective tissue covering of each spinal nerve.
   c. It defines a true **epidural space** and the **subdural space.**

2. **The arachnoid** (L. spiderlike) is the intermediate meningeal layer (see Figure 13-2).

FIGURE 13-1. The spinal cord and roots. The spinal cord usually terminates just caudal to the L1 vertebral body. Below this point, the spinal roots form the cauda equina as each root courses to its respective intervertebral foramen.

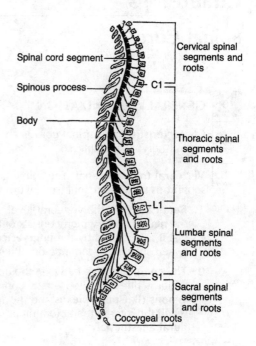

    a. Subdural space. Because the arachnoid is loosely adherent to the dura, a potential space exists between the two layers. This subdural space does not contain cerebrospinal fluid, although blood or pus may extravasate within it.
    b. Subarachnoid space. The arachnoid is attached to the underlying pia by numerous arachnoid trabeculae, creating a considerable cavity between the arachnoid and the pial layers. This subarachnoid space contains **cerebrospinal fluid,** which suspends the central nervous system and nerve roots. Large blood vessels pass within this space.
       (1) The **lumbar cistern** or **dural sac** (a misnomer) is that area between the conus medullaris (approximately vertebra L1) and the point at which the coccygeal ligament begins (about vertebra S2). The dural sac contains only spinal roots suspended in cerebrospinal fluid.
       (2) It is the lumbar cistern into which a needle is inserted for **lumbar puncture**

FIGURE 13-2. Spinal cord, spinal rootlets, and spinal meninges.

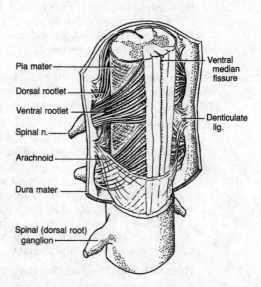

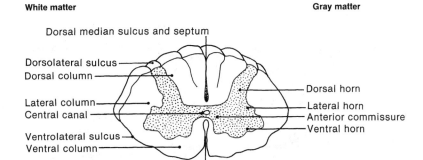

FIGURE 13-3. Cross section of the cervical spinal cord with characteristic regions of white and gray matter. (Reprinted with permission from DeMyer W: *Neuroanatomy.* New York, John Wiley & Sons, 1988, p 101.)

(spinal tap) and induction of **spinal anesthesia.** The suspended roots usually are displaced by the needle and, therefore, are not harmed.

3. **The pia mater** (L. tender mother) is the innermost meningeal layer (see Figure 13-2). It intimately invests the brain and continues through the foramen magnum to intimately invest the spinal cord. It contains the plexus of small blood vessels that supply the neural tissue.
   a. The **denticulate ligament** supports the entire spinal cord so that it is located in the center of the subarachnoid space. It consists of 21 pairs of lateral pial reflections, which pass through the arachnoid to attach to the dura. In this manner, the spinal cord is tethered in the cerebrospinal fluid.
   b. The **filum terminale** is an extension of the pia beyond the tip of the spinal cord (conus medullaris). At the distal end of the dural sac, it is covered by the dura, thereby contributing to the **coccygeal ligament.**

E. **Internal structure**

1. **Gray matter.** Groups of nerve cell bodies, organized into **nuclei** and **laminae,** are arranged into the dorsal horn, lateral horn, and ventral horn, as well as a central gray area (Figure 13-3).
   a. The **laminae of the dorsal horn** (lamina I–III) are sensory. Lamina II contains the second-order neurons of the spinothalamic sensory pathways and simple reflex arcs.
   b. The **nuclei of the lateral horn** (lamina VII) comprise the **preganglionic neurons for the sympathetic division** of the autonomic nervous system. The axons leave via the ventral roots between levels T1 and L2 and synapse with postganglionic neurons in a peripheral ganglion (see Figure 4-5).
   c. The **nuclei of the ventral horn** (lamina IX) comprise the **lower motor neurons** for voluntary activity **(alpha motoneurons)** and smaller neurons that regulate muscle tone **(gamma motoneurons).** The outflow tracts of these neurons leave via ventral roots (see Figure 4-4).
   d. The **central gray area** (lamina VIII) contains a multitude of small **interneurons,** many of which are associated with reflexes and some that are part of the spinoreticular tract.

2. **White matter.** Bundles of nerve axons are organized into **tracts.** Numerous descending (motor) and ascending (sensory) tracts in the spinal cord, some of which cross in the anterior commissure to the contralateral side, define three columns: the dorsal, lateral, and ventral funiculi (see Figure 13-3).
   a. **Motor (descending) tracts** are composed of **upper motor neurons** (Figure 13-4).
      (1) Two of these tracts originate primarily in the motor cortex (precentral gyrus) of the brain and descend in the brain stem and spinal cord. Because descend-

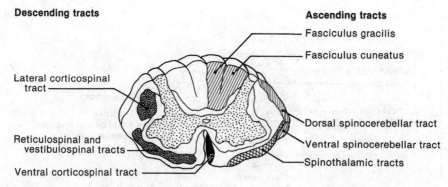

FIGURE 13-4. Cross section of the cervical spinal cord. *(Left)* The major descending tracts; *(right)* the major ascending tracts. (Reprinted with permission from DeMyer W: *Neuroanatomy*. New York, John Wiley & Sons, 1988, p 110.)

ing pathways exert their influence on **lower motor neurons,** they remain entirely within the central nervous system.
   (2) **Principal pathways**
      (a) The **corticobulbar tract** extends through the brain stem only and mediates voluntary activity in the head and neck. Its axons influence lower motor neuron activity in the **nuclei of the cranial nerves:** some ipsilateral, some contralateral, and some a mixture of both.
      (b) The **corticospinal (pyramidal) tract** mostly decussates in the lower brain stem so that each side of the brain controls the contralateral side of the body. The axons descend in the spinal cord as the lateral corticospinal tract and terminate on interneurons and lower motor neurons of the gray matter ventral horn. This pathway mediates voluntary movement below the neck.
      (c) **Extrapyramidal pathways,** such as the vestibulospinal and reticulospinal tracts, influence lower motor neuron activity and, thereby, affect muscular tone and visceral sympathetic function, respectively.
   b. **Sensory (ascending) tracts** convey different sensory modalities (i.e., light touch, pain, pressure, temperature) to the brain stem (see Figure 13-4).
      (1) Some are crossed, whereas others ascend on the ipsilateral side of the spinal cord to cross in the brain stem.
      (2) **Principal pathways**
         (a) The **dorsal columns (fasciculus cuneatus and fasciculus gracilis)** convey touch and proprioception; this is a three-neuron pathway. The **first-order neuron** resides in the **dorsal root ganglion,** and its central process enters the dorsal funiculus. These axons ascend ipsilaterally to the nuclei gracilis (for the lower extremity) or cuneatus (for the upper extremity) in the medulla. The **second-order neuron** lies in the **brain stem** and decussates (crosses) to reach the ventral posterolateral nucleus of the thalamus. The **third-order neuron** lies in the **thalamus** and projects to the sensory cortex (postcentral gyrus) of the brain.
         (b) The **lateral spinothalamic tract** conveys sharp pain and temperature; the **anterior spinothalamic tract** conveys light touch. Each tract is a three-neuron pathway. The **first-order neuron** resides in the **dorsal root ganglion,** and its central process terminates in the dorsal horn of the central gray matter. The **second-order neuron** lies in the **lamina of the dorsal horn** and decussates through the anterior commissure to the contralateral ventral column. It ascends to the midbrain, where synapse occurs with the third neuron in the thalamus. The **third-order neuron,** lying in the **thalamus,** projects from the ventral posterolateral nucleus of the thalamus to the sensory cortex (postcentral gyrus) of the brain.
         (c) The **spinoreticulothalamic tract** conveys dull and diffuse pain; this is a

multineuronal pathway. The **first-order neuron** resides in the **dorsal root ganglion,** and its central process terminates on the second neuron in the dorsal horn of the central gray matter. The **second-order neuron** lies in the **lamina of the dorsal horn** and ascends in the anterolateral white matter to the reticular formation of the medulla. **Higher order neurons** lie in the **reticular formation** and project to the autonomic nuclei of the hypothalamus, to the intralaminar nucleus of the thalamus, and to the limbic lobe of the cortex.
   (d) **Spinocerebellar pathways** convey subconscious proprioception and pressure to the cerebellum. The uncrossed dorsal spinocerebellar tract and the crossed ventral spinocerebellar tract run in the lateral column and enter the pons to be distributed to the cerebellar cortex for coordination.

F. Clinical considerations
1. **Paralysis**
   a. A lesion on one side that involves the corticospinal tract in the cortex or brain stem above the pyramidal decussation produces a contralateral hemiplegia; a lesion below the decussation produces ipsilateral hemiplegia.
   b. Bilateral involvement of the pyramidal tracts in the spinal cord produces paraplegia or quadriplegia, depending on the level of the lesion.
2. **Syringomyelia.** A cavitating lesion of the central canal involves the spinothalamic tracts for pain and temperature as the fibers decussate in the anterior commissure. The patient usually observes bilateral finger burns with little or no pain.
3. **Tabes dorsalis.** Involvement of the dorsal columns with loss of proprioception and sense of position produces a characteristic shuffling gait.
4. **Ipsilateral spinal cord crush injury (Brown-Sequard syndrome).** This injury results in paralysis on the ipsilateral side, loss of the sensation of pain and temperature on the contralateral side below the level of the injury, and bilateral anesthesia at the level of the injury.
5. **Spinal cord transection.** Complete paralysis and anesthesia below the injury result.

II. NERVE ROOTS

A. Spinal nerves. Thirty-one pairs of spinal nerves have their origin in the spinal cord: 8 cervical (spinal nerves C1 through C7 emerge above each respective cervical vertebra; spinal nerve C8 is below the seventh cervical vertebra), 12 thoracic, 5 lumbar, 5 sacral, and 1 coccygeal (each below the respective vertebra).

1. **Rootlets.** Each dorsal and ventral root is formed by six to eight dorsal (sensory) and ventral (motor) rootlets (see Figure 13-2).
   a. The **sensory rootlets** arise in linear fashion from the posterolateral sulci, and the **motor rootlets** arise from the ventrolateral sulci.
   b. The regions of the spinal cord that contribute rootlets to specific spinal nerves are termed spinal segments (levels).
2. **Roots.** Unequal growth of the neural canal and spinal cord results in a progressive descending obliquity of the nerve roots, forming the cauda equina (see Figure 13-1).
   a. **Cervical roots.** Only in the cervical region do the segments of the spinal cord correspond in position to the level of the corresponding vertebrae.
   b. **Thoracic, lumbar, and sacral roots.** Below the cervical region, each spinal nerve from a given cord segment travels inferiorly (sometimes several inches, e.g., the sacral rootlets) before exiting at the appropriate intervertebral foramen.
   c. **Cauda equina.** The nerve roots of the cauda equina float in cerebrospinal fluid, allowing a needle introduced into the dural sac to displace the roots without damage.

3. **Formation of spinal nerves.** Outside the dura, dorsal and ventral roots come together to form a spinal nerve (see Figure 4-4).

B. **Primary rami.** Each spinal nerve leaves the vertebral canal through an intervertebral foramen and bifurcates into dorsal and ventral primary rami (see Figure 4-2).

1. **Dorsal primary rami** supply the dermatomes and myotomes of the median region of the back.
2. **Ventral primary rami** supply the dermatomes and myotomes of the lateral and anterior parts of the body as well as the extremities in their entirety. The ventral primary ramus has two branches.
   a. The **lateral branch** supplies the lateral body wall and, in the regions of the brachial and lumbosacral plexuses, forms the posterior division of each.
   b. The **anterior branch** supplies the ventral body wall and, in the regions of the brachial and lumbosacral plexuses, forms the anterior division of each.

III. VASCULATURE

A. Arterial supply

1. **Vertebral artery branches. Spinal branches** of the vertebral artery give off ascending and descending branches, which run in the pia mater and fuse to form the **midline anterior spinal artery** and two **posterior spinal arteries.**
2. **Intervertebral arteries.** The **intervertebral branches** of the **deep cervical, intercostal, and lumbar arteries** accompany the spinal nerves through the intervertebral foramina. They supply parts of the spinal nerve before bifurcating into anterior and posterior branches.
   a. **Radicular arteries.** Most of the anterior and posterior intervertebral branches supply the dorsal and ventral roots and terminate in the pia associated with these roots as radicular arteries.
   b. **Medullary arteries.** In the lower cervical and upper lumbar regions, six to eight larger anterior and posterior intervertebral branches reach the spinal cord, where they course in the pia mater as medullary arteries.
      (1) **Anterior medullary arteries** reach the anterior median fissure of the spinal cord. Ascending and descending branches anastomose to continue the **anterior spinal artery,** which supplies the ventromedial aspects of the spinal cord.
      (2) **Posterior medullary arteries** reach the posterolateral sulcus. Their ascending and descending branches anastomose to continue the paired **posterior spinal arteries,** which supply the dorsal and lateral aspects of the spinal cord.

B. Venous return

1. Two **median longitudinal veins,** two **anterolateral longitudinal veins,** and two **posterolateral longitudinal veins** usually are distinguishable. They may, however, appear as a plexus in the pia. These veins drain into the intervertebral veins.
2. **Intervertebral veins** receive drainage from the spinal cord and rootlets as well as from the vertebral venous plexus. Intervertebral veins drain into the vertebral, intercostal, lumbar, and lateral sacral veins.
3. **The vertebral venous (Batson's) plexus** lies in the fatty tissue of the epidural space.
   a. The vertebral venous plexus communicates superiorly with the venous sinuses of the cranium and inferiorly with the venous sinuses of the deep pelvis.
   b. Batson's plexus offers not only a collateral pathway for venous drainage from the pelvis, but also a potential pathway for the metastatic spread of carcinoma of the pelvic viscera.

IV. CLINICAL CONSIDERATIONS

A. Lumbar puncture (spinal tap) is performed to obtain samples of cerebrospinal fluid for laboratory analysis.

1. This procedure is facilitated by the fact that the spinal cord rarely extends below the third lumbar vertebra. Also, the horizontal projection of the spinous processes of the lower lumbar vertebrae improves access to the vertebral canal. When a patient assumes a lateral decubitus position, thereby placing the vertebral column in maximal flexion, the space between the spinous processes of adjacent lumbar vertebrae is widened further.
2. **Landmarks.** A line tangential to the highest points of the iliac crests passes through the level of the fourth lumbar vertebra or the interspace between the fourth and fifth lumbar vertebrae.
3. **Procedure**
   a. To enter the subarachnoid space, local anesthetic is applied over the interspace between the fourth and fifth vertebrae, and a spinal needle is inserted.
   b. The needle passes successively through skin, superficial fascia, supraspinous ligament, interspinous ligament (between the paired flaval ligaments), epidural space, dura, and arachnoid to enter the subarachnoid space.

B. Spinal (intrathecal) anesthesia is used to block the roots of the spinal nerves within the dural sac (subarachnoid space).

1. The procedure is the same as that for lumbar puncture.
2. The number of segmental levels anesthetized is controlled by the amount of anesthetic injected and the position of the patient. The density of the anesthetic agent is greater than cerebrospinal fluid; therefore, the tilt of the operating table must never be such that the patient's head is below the horizontal.

C. Epidural anesthesia is used to block the spinal nerves in the epidural space in the lumbar region (i.e., external to the dural sac).

1. The procedure is similar to that for lumbar puncture, except that the dura is not penetrated.
2. The number of nerves anesthetized above or below the injection site is controlled by the amount of anesthetic injected into the connective tissue of the epidural space.

D. Caudal (sacral) anesthesia is used to block the spinal nerves in the epidural space (i.e., external to the dural sac).

1. The sacral cornua are palpated to locate the sacral hiatus.
2. A needle inserted cranially between sacral cornua penetrates the sacrococcygeal ligament to gain access to the sacral hiatus. The anesthetic is infiltrated into the **epidural space** around the dural sac.
3. The degree of ascent of anesthesia can be controlled (e.g., for parturition) so that afferent pain fibers from the perineum and cervix running in the sacral nerves can be blocked (saddle block) without affecting the lumbar fibers that control the abdominal musculature and those involved in reflex uterine contractions.

PART IIB BACK

STUDY QUESTIONS

DIRECTIONS: Each of the numbered items or incomplete statements in this section is followed by answers or by completions of the statement. Select the ONE lettered answer or completion that is BEST in each case.

1. Which of the following is a characteristic of the atlas?

(A) Bowl-shaped superior articular facets
(B) A characteristic body
(C) An odontoid process
(D) A prominent spinous process

2. Which of the following structures pass through the posterior sacral foramina?

(A) Anterior primary rami
(B) Dorsal primary rami
(C) Filum terminale
(D) Sensory rootlets
(E) Spinal nerves

3. The vertebral venous plexus is located in

(A) the dural sac
(B) the epidural space
(C) the subarachnoid space
(D) the subdural space

4. Which of the following characteristics best describes the anterior longitudinal ligament?

(A) It anchors the emerging spinal nerves in place
(B) It limits the direction of the nucleus pulposus extrusion during disk herniation
(C) It narrows anterior to the intervertebral disk
(D) It resists kyphosis

5. A small posterolateral herniation of the L4–L5 intervertebral disk will likely affect the

(A) fourth lumbar nerve
(B) fifth lumbar nerve
(C) fourth and fifth lumbar nerves only
(D) fourth and fifth lumbar as well as the first and second sacral nerves
(E) first sacral nerve

6. The posterior intrinsic muscles of the back receive motor innervation from the

(A) dorsal primary rami
(B) dorsal roots
(C) posterior branches of the lateral perforating nerves
(D) ventral primary rami

7. The dura mater terminates caudally as the

(A) coccygeal ligament
(B) conus medullaris
(C) denticulate ligament
(D) dural sac
(E) filum terminale

8. A spinal segment is defined as that region of the spinal cord that

(A) corresponds to a collection of nerves passing up or down within the white matter
(B) corresponds to the region of the vertebral column to which the spinal nerves are sent
(C) sends rootlets to a particular spinal nerve
(D) underlies the neural arch of a particular vertebra in the adult

9. In the adult vertebral column, the spinal cord usually terminates between

(A) T12 and L1
(B) L1 and L2
(C) L2 and L3
(D) L3 and L4
(E) L4 and L5

DIRECTIONS: Each of the numbered items or incomplete statements in this section is negatively phrased, as indicated by a capitalized word such as NOT, LEAST, or EXCEPT. Select the ONE lettered answer or completion that is BEST in each case.

---

10. A typical thoracic vertebra includes all of the following components EXCEPT

(A) a long spinous process
(B) inferior articular facets
(C) a neural canal
(D) superior costal facets
(E) transverse foramina

11. Support for the nucleus pulposus is provided by all of the following structures EXCEPT the

(A) annulus fibrosus
(B) anterior longitudinal ligament
(C) ligamentum flavum
(D) posterior longitudinal ligament

12. All of the following statements correctly pertain to the pia mater EXCEPT

(A) it forms one of the boundaries of the subarachnoid space
(B) it forms the filum terminale
(C) it is adherent to the spinal cord
(D) it supports the anterior and posterior spinal arteries
(E) it terminates at the foramen magnum

Questions 13–17

(A) Flexion/extension
(B) Rotation
(C) Abduction (lateral flexion)
(D) Little or no movement

For each joint or vertebral region, select the principal movements permitted.

13. Atlantoaxial joint

14. Atlanto-occipital joint

15. Thoracic region

16. Lumbar region

17. Sacral region

Questions 18–20

(A) Iliac crests
(B) Posterior–superior iliac spines
(C) Spinous process of T12
(D) Sacral cornua
(E) Coccyx

For each procedure, select the most appropriate landmark structure.

18. Induction of caudal anesthesia

19. Obtaining cerebrospinal fluid by means of a spinal tap

20. Induction of spinal (intrathecal) anesthesia

ANSWERS AND EXPLANATIONS

1. **The answer is A** [Chapter 11 III A 1, 2]. The atlas, the first cervical vertebra, has large concave superior facets that articulate with the occipital condyles. Considerable flexion/extension occurs at the atlanto-occipital joint. The inferior facets, which articulate with the axis, are flat, permitting considerable rotation at the atlantoaxial joint. The atlas has neither a body nor a prominent spinous process, the former being represented by the odontoid process of the axis.

2. **The answer is B** [Chapter 11 III D 1 c (2)]. The sacral spinal nerves leave the vertebral canal via intervertebral foramina, which bifurcate almost immediately into anterior and posterior sacral foramina. The dorsal primary rami pass through the posterior sacral foramina to innervate the back muscles and provide sensation to the dorsal portion of the dermatomes.

3. **The answer is B** [Chapter 13 I A 2, C 1 c; C 2]. The vertebral venous plexus (of Batson) lies in the loose fatty connective tissue of the epidural space. It drains through intervertebral veins. It provides a collateral route for venous drainage from the pelvis and, as such, is a pathway for the spread of pelvic malignancies.

4. **The answer is B** [Chapter 11 IV D 2 c; V D]. The anterior longitudinal ligament provides support for the vertebral column and reinforces the anterior and lateral aspects of the intervertebral disks. Consequently, herniation of an intervertebral disk tends to be in a posterolateral direction. Spinal nerves may become involved as they pass through the intervertebral foramina.

5. **The answer is B** [Chapter 11 IV D 2 d; V E]. Because the inferior intervertebral notch in the pedicle of the lumbar vertebrae is so deep, spinal nerve L4 escapes from the vertebral canal superior to the L4–L5 intervertebral disk. Posterolateral herniation of the L4–L5 intervertebral disk, however, places pressure on the subsequent nerve (L5 in this case) as it passes toward its intervertebral foramen. If the herniation is substantial, it may also involve subsequent roots; if very large, even roots across the midline are affected.

6. **The answer is A** [Chapter 12 I B; Table 12-2]. The intrinsic muscles of the back receive innervation from the dorsal primary rami of the spinal nerves. The ventral primary rami innervate the superficial muscles of the back as well as those of the anterior and lateral body wall and extremities.

7. **The answer is A** [Chapter 13 I C 1 a]. From the dural sac at the level of the second sacral vertebra, the dura mater extends as the coccygeal ligament to about the level of S4 or S5, covering the pial filum terminale.

8. **The answer is C** [Chapter 13 II A 1 b, B 1, 2; Figure 13-2]. A spinal segment is defined as that region of the spinal cord that sends rootlets to a particular spinal nerve. Because the vertebral column grows postnatally while the spinal cord does not, most of the vertebrae are displaced from the corresponding spinal segments in the adult, which accounts for the cauda equina.

9. **The answer is B** [Chapter 13 I B]. In most individuals, the spinal cord extends from the medulla oblongata at the level of the first cervical vertebra and terminates as the conus medullaris at the level of the first lumbar intervertebral disk (between vertebrae L1 and L2). In only 1% of the population does the conus medullaris extend beyond the L2 vertebra.

10. **The answer is E** [Chapter 11 III A, B]. Transverse foramina are characteristic of cervical vertebrae and are formed by the fusion of the transverse and costal processes. Except for the seventh cervical vertebra, the transverse foramina contain the vertebral artery as it passes toward the foramen magnum.

11. **The answer is C** [Chapter 11 IV B 1 c; V D 3, E 3]. The annulus fibrosus forms the outer portion of the intervertebral disk and provides principal support for the nucleus pulposus. Additional support is provided by the anterior and posterior longitudinal ligaments.

Disk herniation occurs most frequently in the posterolateral aspect of the disk, where these ligaments do not meet. The ligamentum flavum runs between adjacent laminae.

12. The answer is E [Chapter 13 I C 2 b, 3; III A 1]. The pia mater, the innermost layer of the meninges, tightly invests the spinal cord and supports its arterial supply. The space between the pia and arachnoid constitutes the subarachnoid space. The pia continues beyond the termination of the spinal cord as the filum terminale; it also continues through the foramen magnum as the pia of the brain.

13–17. The answers are: 13-B, 14-A, 15-B, 16-A, 17-D [Chapter 11 III A 1, 2, B 2, C 2, D 2]. In the cervical regions, most of the flexion/extension occurs at the atlanto-occipital joint; most of the rotation occurs in the atlantoaxial joint. In the midcervical region, principal movements are flexion/extension, rotation, and abduction. A moderate amount of both rotation and flexion–extension is seen in the thoracic region, in large part because of the elasticity of the costal cartilages. The principal movement in the lumbar region is flexion/extension because the zygapophyseal facets are oriented in anterior–posterior planes. The fused sacral vertebra permit no movement.

18–20. The answers are: 18-D, 19-A, 20-A [Chapter 11 III D 1 d; Chapter 13 IV A]. The sacral cornua (rudimentary pedicles of the fifth sacral vertebra) are important landmarks for the location of the sacral hiatus through which caudal (sacral epidural) anesthesia is induced. Because the spinal cord always ends by the L3 level and the spinous process of L4 is nearly horizontal, the interspace between L4 and L5 is the preferred site for a spinal tap to sample cerebrospinal fluid as well as for inducing spinal (intrathecal) anesthesia. The intertubercular plane through the iliac crests, which passes through vertebra L4 or through the L4–L5 intervertebral disk, provides the best landmark for these procedures. The spinous process of the twelfth thoracic vertebra is most difficult to find and is not a useful landmark. The posterior–superior iliac spines lie at the L5–S1 level.

PART IIC   LOWER EXTREMITY

Chapter 14
Gluteal Region

I. INTRODUCTION

A. Basic principles

1. **Primitive position.** The basic pattern of the human lower extremity may be approximated by simulating the primitive position, that is, by abducting the thigh and leg until horizontal with the sole facing forward so that the lower extremity approximates the pelvic fin of a fish (see Figure 4-3).
   a. **Primitive surfaces.** The **primitively ventral surface** generally corresponds to the flexor side. The **primitively dorsal surface** generally corresponds to the extensor side.
   b. **Axial borders.** An imaginary line drawn through the long axis of the extremity will differentiate between a cephalad **preaxial border** and a caudal **postaxial border.**

2. **Early terrestrial adaptation.** From the original finlike position of the lower extremity, flexing the knee 90° and extending the foot at the ankle 90° approximates the amphibian/reptilian lower extremity.

3. **Late terrestrial adaptation**
   a. In mammals, the amphibian/reptilian thigh rotated (abducted) 90° cephalad, bringing the lower extremity alongside the lateral body wall so that the primitively dorsal surface of the leg faces anteriorly, that is, into the **anatomic position.**
   b. Rotation of the lower extremity is opposite from that of the upper extremity.

4. **Assumption of an erect posture** effectively extends the hip 90° so that the primitively dorsal surface of the entire lower extremity faces anteriorly and the primitively ventral surface of the limb faces posteriorly.
   a. **Pattern of innervation.** The arrangement of nerves in a limb reflects its developmental origin as a horizontal bud.
      (1) Each spinal nerve innervates a discrete sensory area, the **dermatome.** In the pelvic region, the dermatomes are drawn distally and appear to march down the preaxial side and back along the postaxial side (see Figure 4-3). While there is generally overlap between sequential dermatomes, there is no overlap across the axial lines.
      (2) Each spinal nerve innervates a discrete muscle group, a **myotome.** In the lower extremity, there is an orderly preaxial to postaxial progression of muscles of myotomal origin and, therefore, sequential innervation (see Figure 4-3).
   b. **Functional adaptation.** Although the human upper and lower extremities are homologous, structural differences between them reflect different functions. The lower extremity retains its original role of **support** and **propulsion** at the expense of dexterity and intrinsic mobility.
      (1) At the **pelvis,** extensive fusion among the bones of the pelvic girdle as well as between the pelvic girdle and the axial skeleton are adaptations to weightbearing and stability.
      (2) At the **hip joint,** bony buttressing and strong ligamentous support permit sufficient movement while providing maximal support.
      (3) The **leg (calf)** becomes fixed in the pronated position and loses its ability to supinate.
      (4) The **foot,** while retaining some prehensile characteristics, is adapted to bipedal support and locomotion.

FIGURE 14-1. Hip joint. The femur articulates with the pelvic bone at the acetabulum. The ligamentum teres has been cut and the femur disarticulated to show the articular surface of the acetabular fossa.

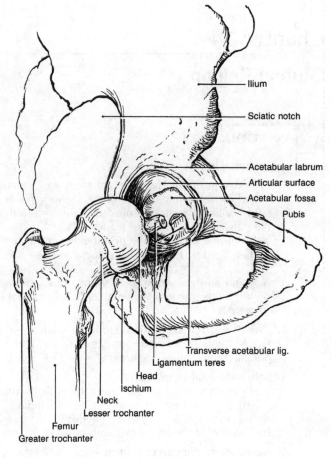

B. **Organization.** The free portion of the lower extremity (thigh, leg, ankle, and foot) is suspended from the axial skeleton by the embedded portion of the lower extremity (the pelvic girdle).

1. **The pelvic girdle** consists of three bones: the ilium, ischium, and pubis (see Figure 14-1).
2. **The thigh** contains one bone: the femur (see Figures 14-1 and 15-1).
3. **The leg** contains two bones: the **tibia** and **fibula** (see Figures 15-1 and 16-1).
4. **The ankle or tarsus** contains seven bones: a proximal row of three tarsal bones (talus, calcaneus, and navicular) and a distal row of four tarsal bones (the three cuneiform bones and the cuboid bone). Usually, the tarsal bones are included with the foot (see Figures 16-1 and 16-2).
5. **The foot** contains 19 bones: 5 metatarsals and 14 phalanges—2 in the great toe and 3 in each of the other toes (see Figure 16-1).

II. BONES OF THE PELVIS

A. **Bony landmarks**
   1. **Hip.** Portions of the ilium, ischium, and pubis are palpable (see Figure 14-1).
      a. Anteriorly, such landmarks include the **anterior–superior iliac spine, inguinal ligament, pubic tubercle, pubic symphysis,** and **ischiopubic ramus.**

b. Posteriorly, such landmarks include the **posterior–superior iliac spine, iliac crest,** and **ischial tuberosity.**
2. **Thigh.** Palpable landmarks of the proximal femur include the **femoral head, greater trochanter,** and **femoral shaft.**

B. **The pelvic (hip) girdle** consists of a hip (innominate or coxal) bone on either side of the sacrum. Each hip bone consists of three fused pelvic bones: ilium, ischium, and pubis.
   1. **Characteristics**
      a. The **ilium** lies superiorly (Figure 14-1).
         (1) The **medial auricular surface** forms the sacroiliac articulation and the expansive **alar plates** provide attachments for muscles of the abdomen and thigh.
         (2) The **iliac crest** defines the lower limit of the waist and terminates in the **anterior–superior iliac spine** and **posterior–superior iliac spine.**
         (3) The **posterior border** defines the upper border of the **sciatic notch** before fusing with the ischium.
      b. The **ischium** lies posteroinferiorly (see Figure 14-1).
         (1) The posterior surface defines the lower border of the **sciatic notch,** which is divided into greater and lesser portions by the **ischial spine.**
         (2) The lowermost **ischial tuberosity** provides an anchor for the strong sacrotuberous ligament and attachment for posterior thigh muscles.
         (3) The **ischial ramus** fuses with the inferior pubic ramus.
      c. The **pubis** lies anteriorly (see Figure 14-1).
         (1) The **body** articulates with its contralateral counterpart at the pubic symphysis.
         (2) The **superior ramus** fuses with the ilium; the **inferior ramus** fuses with the ischial ramus to form the **obturator foramen.**
   2. **Articulations** (see Figure 27-1)
      a. The **sacroiliac joint** is the diarthrosis between the pelvic girdle and the axial skeleton. It is formed by the articulation of the alar plates of the sacrum with the ilium.
      b. The **pubic symphysis,** anteriorly, is the amphiarthrosis between the left and right hip bones (see Figure 27-1).
      c. **The hip (coxal) joint** is the distal diarthrosis of each hip bone with the femur (see Figure 14-1).
         (1) The **acetabulum** is the fossa of the hip joint, which is formed by the fusion of the ilium, ischium, and pubis. The nearly hemispheric acetabulum (L. vinegar cruet) faces laterally but slightly posteriorly and slightly inferiorly.
            (a) The **acetabular articular surface** is horseshoe-shaped and lined with hyaline cartilage.
               (i) The **acetabular notch** is an inferior discontinuity in the articular rim of the acetabulum. It is bridged by the transverse acetabular ligament.
               (ii) The bone behind the articular surface is thick with bony trabeculae arranged along the lines of force.
            (b) The **acetabular labrum** increases the depth of the bony acetabulum.
               (i) This lip of fibrocartilage contributes to the stability of the hip joint.
               (ii) It continues across the acetabular notch as the transverse acetabular ligament.
            (c) The **acetabular fossa,** the rough, nonarticular medial wall of the acetabulum enclosed by the limbus of articular cartilage, is a thin sheet of bone.
         (2) The **ligamentum teres** of the femur originates from the transverse ligament over the acetabular notch and inserts into the **fovea (foveal notch)** of the femoral head.

C. **The femur** is the long bone of the thigh.
   1. **Characteristics.** The proximal end consists of a head, neck, greater and lesser trochanters, and shaft (see Figure 14-1).
      a. The **head** is approximately two thirds of a sphere.
         (1) It fits deeply into the acetabulum and is covered with hyaline cartilage, except at the **foveal notch,** where it receives the **ligamentum teres.**
         (2) An **epiphyseal plate,** which usually fuses between the nineteenth and twenty-first years, demarcates the head from the neck.

b. The **femoral neck** joins the head to the shaft.
   (1) The neck is normally angled upward at about 125° (slightly more in males and slightly less in females).
   (2) It is also angled forward (anteverted) by about 15°.
c. The **shaft** has a slightly anterior bow.
   (1) The **greater trochanter** on the lateral surface projects above the junction with the neck and provides attachment for the **gluteus medius and minimus muscles** as well as for the **piriformis muscle.**
   (2) The **lesser trochanter,** situated on the medial surface distal to the junction of the neck, receives the insertion of the **iliopsoas muscle.**
   (3) The **intertrochanteric crest** runs obliquely along the posterior surface of the shaft between the trochanters.

2. The internal structure of the femur enables the transmission of great forces between the pelvis and tibia.
   a. The **bony trabeculae** of the femoral head, neck, and trochanters are arranged along the lines of prevailing force, effectively transmitting these forces to the **compact bone** of the shaft.
   b. This trabecular arrangement attains the greatest development and strength during the active period of life. The trabeculae are reduced by the process of bone resorption during senescence, predisposing to fractures.

3. Clinical considerations
   a. Intracapsular fractures (within the joint capsule) may be subcapital or midcervical.
      (1) With resorption and weakening of the bony trabeculae (osteoporosis), these fractures may be the result of a fall or of a muscle spasm with fall subsequent to the fracture. Osteoporosis is so common in the elderly population that the initial treatment of choice may be joint replacement.
      (2) Fractures of this type endanger the blood supply to the proximal fragment; nonunion and avascular necrosis may result.
   b. Extracapsular fractures (intertrochanteric and subtrochanteric) occur in a region with abundant anastomotic blood supply. Healing occurs with few problems once reduction and internal fixation have been achieved.
   c. Fracture through the epiphyseal plate. The greater and lesser trochanters have separate ossification centers and may be avulsed, as may be the epiphyseal plate, before the age of 18 or 19 years. This injury is especially common in active, heavy, male teenagers.
   d. Fractures of the proximal femoral shaft result from violent trauma.
      (1) In fracture of the upper third of the femur, the proximal fragment is flexed by the iliopsoas, abducted by the gluteus medius and minimus, and externally rotated by the deep gluteal rotators.
      (2) Transverse fracture of the midfemur is associated with little displacement. Muscle spasm associated with an angled fracture produces overriding of the fragments, resulting in shortening of the thigh.

III. PELVIC ARTICULATIONS

A. **Sacroiliac joint** (see section II B 2 a)

   1. Structure. It is formed by the articulation of the alar plates of the sacrum with the ilium (see Figure 27-1).
   2. Movement. Unlike the articulation of the pectoral girdle with the axial skeleton, the sacroiliac joint is primarily ligamentous. Normally, no movement occurs between the axial skeleton and the pelvic girdle, endowing great stability.
      a. Changes with age. In young people, this joint is scarcely movable. By middle

age, it becomes completely immobile with ankylosis (fusion) typically occurring after age 50 years.
   b. **Changes with pregnancy.** The ligaments slacken somewhat under the influence of hormones to facilitate parturition.
3. **Numerous strong ligaments** make the sacroiliac joint the strongest and most stable in the body. No muscles act across this immovable joint (see Figure 27-1).
   a. **Dorsal sacroiliac ligaments** run between the posterior–superior iliac spine and the dorsum of the sacrum.
   b. **Ventral sacroiliac ligaments** cross the sacroiliac joint on the anterior surface of the pelvis.
   c. **Iliolumbar ligaments** run between the iliac crest on either side to the transverse processes of the fifth lumbar vertebra.
   d. The **sacrospinous ligament** runs from the ischial spine to the sacrum and divides the sciatic notch into a **greater sciatic foramen** and a **lesser sciatic notch**. Phylogenetically, it is a degenerated portion of the coccygeus muscle.
   e. The **sacrotuberous ligament** runs from the ischial tuberosity to the sacrum and provides the posterior boundary of the lesser sciatic notch, converting it into the **lesser sciatic foramen**. Phylogenetically, it represents the proximal portion of the tendons of the hamstring muscles.

B. **Pubic symphysis** (see section II B 2 b)
   1. An **amphiarthrosis,** the pubic symphysis permits little movement (see Figure 27-1).
   2. **Changes with pregnancy.** The fibrocartilage slackens somewhat under the influence of hormones to facilitate parturition.

C. **Major osseous and osseoligamentous foramina**
   1. **Obturator foramen** (see Figure 27-1)
      a. The **superior ramus** fuses with the ilium and the **inferior ramus** fuses with the ischial ramus to form the obturator foramen.
      b. The obturator nerve and obturator artery pass through the obturator foramen.
   2. **Greater sciatic foramen** (see Figure 27-1)
      a. The **sacrospinous ligament** divides the sciatic notch into a **greater sciatic foramen** and a **lesser sciatic notch.**
      b. The superior gluteal neurovascular bundle, the piriformis muscle, the inferior gluteal neurovascular bundle, the sciatic nerve, the internal pudendal vessels, and the pudendal nerve leave the deep pelvis through the greater sciatic notch and foramen.
   3. **Lesser sciatic foramen** (see Figure 27-1)
      a. The **sacrospinous ligament** divides the sciatic notch into a **greater sciatic foramen** and a **lesser sciatic notch.** The **sacrotuberous ligament** converts the lesser sciatic notch into the **lesser sciatic foramen.**
      b. The obturator internus muscle exits via the lesser sciatic foramen, but the pudendal neurovascular bundle reaches the perineum by passing into the lesser sciatic foramen.

D. **Hip (coxal) joint** (see section II B 2 c)
   1. **Structure.** The femur articulates with the hip bone at the **acetabulum** (see Figure 14-1). The hip joint transmits forces between the pelvis and thigh with great stability while allowing sufficient movement. Like the glenohumeral joint of the upper extremity, it is a ball-and-socket joint (see Figure 14-1).
      a. The **bony configuration of the acetabulum** has an overhanging slant, and the congruent configuration of the femoral head has an upward slant. The weight of the body is transmitted through the roof of the acetabulum to the femoral head and along the femoral neck to the shaft.
         (1) Maximum articular contact occurs when the hip is flexed, slightly everted, and slightly abducted—an approximation of the quadrupedal stance.

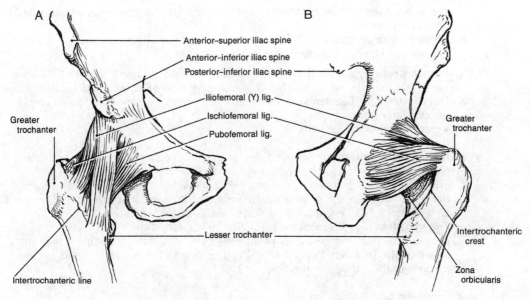

FIGURE 14-2. Fibrous capsule of the hip joint. *(A)* Anterior; *(B)* posterior.

      (2) Less articular contact is associated with the upright position; some stability is sacrificed for the bipedal stance.
   b. A **joint capsule** attaches to the margins of the acetabular lip as well as the transverse acetabular ligament and extends like a sleeve to the base of the femoral neck (see Figure 14-2).
      (1) The inner surface of the capsule is **synovial membrane** that secretes synovial fluid.
      (2) The **outer capsular layer** is fibrous.
         (a) Some fibers run circumferentially about the femoral neck, the **zona orbicularis.**
         (b) Other groups of fibers are reflected along the neck as the **retinacula,** carrying important blood vessels to the synovium and the femoral head.
         (c) The monolayer of **synovial fluid** between the extensive articular surfaces produces considerable surface tension, providing tremendous adhesion while permitting nearly frictionless gliding.
2. Movement. This ball-and-socket joint has **three degrees of freedom.** Range of motion is sacrificed for stability.
   a. Abduction/adduction. Abduction (30°) and adduction (20°) occur about the anterior-posterior axis.
   b. Flexion/extension. Flexion (120° with knee flexed) and extension (15°) occur about the transverse axis.
   c. Rotation. Medial rotation (60°) and lateral rotation (30°) occur about the vertical axis.
   d. Circumduction at the hip joint is not as great as at the shoulder, where muscles (not ligaments and bones) stabilize the joint.
3. Strong ligaments reinforce the hip joint. Also, the dynamic actions of the muscles that cross the joint contribute to its stability.
   a. Capsular ligaments are three fibrous condensations of the joint external to the fibrous capsule (Figure 14-2).
      (1) Iliofemoral ligament (**Y**-ligament of Bigelow)
         (a) It runs from the anterior–inferior iliac spine to the length of the intertrochanteric line. It is especially thick at the edges, giving the appearance of an inverted **Y**. The lateral limb, which runs to the greater trochanter, is the **iliotrochanteric band.**

(b) This strong ligament resists hyperextension and excess lateral rotation of the femur.
  (2) Ischiofemoral ligament
    (a) This flimsy ligament runs from the posterior surface of the acetabulum to the inner aspect of the greater trochanter and the lateral surface of the intertrochanteric line.
    (b) It is the thinnest of the capsular ligaments.
  (3) Pubofemoral ligament
    (a) It runs from the iliopubic ramus to the inferior surface of the intertrochanteric line. It is thin, and a tear may result in continuity between the joint capsule and the iliopectineal bursa.
    (b) This ligament resists excessive abduction.
  b. The **ligamentum teres** of the femur runs from the acetabular notch and transverse acetabular ligament to the femoral fovea (see Figure 14-1).
    (1) It plays little, if any, role in the stability of the hip.
    (2) It carries the **artery of the ligamentum teres,** which is inconsistent but may be an important blood supply to the femoral head in children.

4. Clinical considerations
  a. Dislocations. Because the hip is so stable, dislocation is usually the result of trauma severe enough to fracture the acetabulum.
    (1) Congenital dislocation. A stable hip is essential for abduction. In congenital dislocation of the hip, the femoral head lies outside an undeveloped acetabulum. Failure of abduction results in a waddling (Trendelenburg) gait.
    (2) Posterior dislocation, the most common type, occurs when the less stable flexed thigh comes into violent contact with an object, such as a dashboard.
      (a) The hip joint is least stable in the flexed position, which slackens the ligaments of the fibrous capsule.
      (b) The femoral head comes to lie posterior to the iliofemoral ligament. The sciatic nerve is especially vulnerable to damage.
    (3) Anterior dislocation, which is less common, results in the femoral head resting anterior to the iliofemoral ligament, either superiorly against the superior pubic ramus or inferiorly in the obturator foramen.
    (4) A tear in the articular capsule as a result of dislocation may jeopardize the blood supply to the femoral head.
  b. Osteoarthritis
    (1) The hip is a common site for degenerative arthritis. This condition produces pain and limited range of movement.
    (2) A cane used as support is held in the **opposite** hand. When the unaffected leg is raised to step, the arm and cane support the trunk. Strong abduction by the gluteus medius and minimus muscles to stabilize the pelvis is, therefore, unnecessary and the weightbearing joint is relieved of as much as 75% of the compressive loading and the accompanying pain.

IV. MUSCLE FUNCTION AT THE HIP JOINT

A. Organization. The musculature of the hip joint originates from the pelvic bones and is arranged into five compartments.

  1. The iliopsoas compartment consists of primitively ventral muscles that flex the thigh (Figure 14-3; see Table 14-1).
    a. Group composition. The **iliopsoas group** consists of the **iliacus** and **psoas muscles.** The iliacus (running from the iliac fossa) and the psoas (running from the transverse processes of the lumbar spine) unite as the **iliopsoas muscle,** with a single tendon that inserts into the lesser trochanter of the femur.
    b. Group action. In addition to flexing the hip, they flex the lumbar vertebral column if the leg is planted.

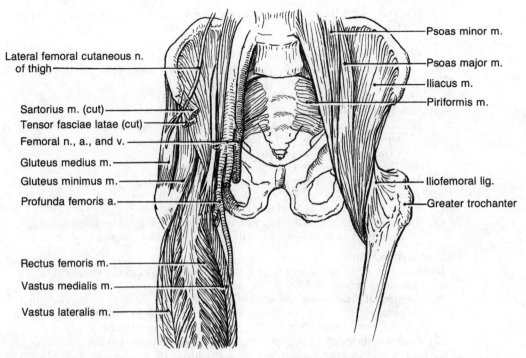

FIGURE 14-3. Principal flexor muscles of the thigh are identified in the left iliopsoas compartment and right anterior compartment. The right femoral neurovascular bundle lies in the femoral triangle.

2. **The anterior compartment** consists of primitively dorsal muscles that flex the thigh (see Figures 14-3 and 15-4A and Table 14-1).
   a. **Group composition.** This group includes the **rectus femoris** and **sartorius** muscles (see Figure 15-4A). The rectus femoris originates from the anterior inferior iliac spine and inserts into the proximal tibia through the patella. The sartorius originates from the anterior superior iliac spine and inserts into the medial tibial condyle.
   b. **Group action.** These muscles are involved in hip flexion.
   c. **Group innervation.** The anterior group is innervated by the **femoral nerve.**
3. **The posterior compartment** is composed of primitively ventral muscles that comprise the **hamstring extensor group** (Figure 14-4; see Table 14-2).
   a. **Group composition.** The hamstring group, which consists of the **semitendinosus muscle, semimembranosus muscle,** and the **long head of the biceps femoris muscle,** originate from the ischial tuberosity and insert in the vicinity of the knee.
   b. **Group action.** The hamstrings are extensors of the thigh at the hip joint.
   c. **Group innervation.** The hamstring muscles included here are innervated by the **tibial nerve.**
4. **The medial compartment** consists mostly of primitively ventral muscles and comprises the **adductor group** (see Table 14-4 and Figure 15-4).
   a. **Group composition.** The **adductor magnus** (anterior portion), **adductor longus, adductor brevis, pectineus,** and **gracilis** originate from the ischial rami and pubis and insert into the proximal femur.
   b. **Group action.** This muscle group adducts the thigh at the hip joint. They also flex and, to some extent, medially rotate the thigh.
   c. **Group innervation.** The adductor group is innervated primarily by the **obturator nerve.**
5. **The gluteal compartment** consists mostly of primitively dorsal muscles and is divided into superficial abductor and deep rotator groups (see Table 14-3 and Figure 14-4).

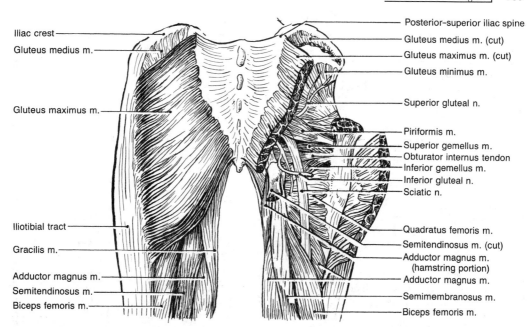

FIGURE 14-4. Gluteal musculature. *Left,* superficial (extensor) muscles; *right,* abductors and lateral rotators. The right sciatic nerve passes through the medial inferior quadrant of the buttock.

    a. The **superficial (abductor) gluteal group** consists of thigh abductors and extensors (see Table 14-3 and Figure 15-5).
       (1) **Group composition.** Muscles of the abductor group include the **gluteus medius, gluteus minimus,** and, to a lesser extent, the **tensor fasciae latae** and **gluteus maximus.** The principal abductors run from the ilium to the greater trochanter of the femur.
       (2) **Group action.** This group abducts the thigh.
       (3) **Innervation.** The gluteus medius and minimus are innervated by the **superior gluteal nerve.**
    b. The **deep gluteal group** consists of lateral rotators of the thigh (see Table 14-5 and Figure 15-5).
       (1) **Group composition.** The **piriformis, superior and inferior gemelli, obturators internus** and **externus,** and **quadratus femoris** originate largely from the sacrum or ischium and insert into or adjacent to the greater trochanter.
       (2) **Group action.** These muscles laterally rotate the thigh.
       (3) **Innervation.** The external rotators are usually innervated by twigs from the **sacral plexus;** the obturator externus is innervated by the obturator nerve.

B. **Group actions** occur about the transverse, vertical, and anterior-posterior axes. Movement is accomplished by muscles that run from the axial skeleton and the pelvic girdle across the hip joint to the femur, tibia, or fibula. Many movements of the hip joint are accomplished by combinations of primitively dorsal and primitively ventral musculature.
    1. Flexion/extension occurs about a transverse axis through the femoral head.
       a. **Flexion** (90° with knee extended, 120° with knee flexed) is accomplished by a muscle group with a line of action passing anterior to the transverse axis of the hip (Table 14-1).
          (1) The **iliopsoas** is the major flexor muscle of the hip, assisted by the **sartorius** and the **rectus femoris, pectineus,** and to a lesser extent by the tensor fasciae latae and the adductor longus. With the knee extended, flexion is limited primarily by tension in the hamstring muscles.
          (2) With the knee flexed, the hip may be flexed another 30° until it is stopped by apposition of the thigh to the abdominal wall.

TABLE 14-1. Flexor Muscles Acting at the Hip

Muscle	Origin	Insertion	Primary Action	Innervation
Iliopsoas group Iliopsoas				
Iliacus	Iliac fossa	Distal to lesser trochanter	Flexes thigh	Femoral n. (L2–L3, posterior)
Psoas	Transverse processes of lumbar vertebrae	Lesser trochanter	Flexes thigh	Lumbar nn. (L2–L3, posterior)
Anterior group Rectus femoris	Anterior-inferior iliac spine	Tibial tuberosity through patellar tendon	Flexes thigh	Femoral n. (L2–L4, posterior)
Sartorius	Anteriors-superior iliac spine	Medial tibial head	Flexes, abducts, and laterally rotates thigh	Femoral n. (L2–L3, posterior)
Pectineus	Iliopectineal line and pubis	Base of lesser trochanter	Flexes and adducts thigh	Femoral n. (L2–L3, posterior) Oburator n. (L2–L3, anterior)

    b. Extension (15°) is accomplished by muscles with lines of action that pass posterior to the transverse axis of the hip (Table 14-2).
       (1) The **gluteus maximus** muscle is the strongest extensor. It is used in climbing stairs and rising from the seated position.
       (2) The **posterior belly of the adductor magnus muscle** and the **hamstring group** (the long head of the biceps femoris, semitendinosus, and semimembranosus muscles) are also extensors.
  2. Abduction/adduction occurs about an anteroposterior axis through the femoral head (Table 14-3).
    a. Abduction (30°) is accomplished by muscles that pass superior to the anteroposterior axis, generally from the ilium to the greater trochanter (see Table 14-3).
       (1) The **abductors** overlap in three layers. These flat muscles include the **tensor fasciae latae, gluteus maximus** (superficially), and the **gluteus medius,** which overlies the **gluteus minimus** (the most anterior and deep).
          (a) The **gluteus medius** and **gluteus minimus** are strong abductors, important in maintaining pelvic stability while walking.
          (b) The **gluteus maximus** is primarily an extensor of the thigh.
          (c) The **tensor fascia lata** flexes the thigh as well as flexes and extends the leg at the knee joint.
       (2) The deep abductors are important in pelvic stability during walking because they keep the pelvis level when the opposite foot is raised from the ground.
          (a) This force of abductor contraction is approximately three times the body weight.
          (b) Within the joint, a force develops equal to four times the body weight (abductor action plus body weight).
          (c) For individuals with degenerative osteoarthritis of the hip, a cane held in the opposite hand relieves some of these forces.

TABLE 14-2. Extensor Muscles Acting at the Hip

Muscle	Origin	Insertion	Primary Action	Innervation
Gluteus maximus	Posterosuperior ilium and sacrum, associated ligaments and fascia	Gluteal tuberosity of femur and iliotibial tract	Extends, abducts, and laterally rotates thigh	Inferior gluteal n. (L5–S2, posterior)
Adductor magnus Posterior belly	Ischial tuberosity of femur	Adductor tubercle	Adducts, extends, and aids medial rotation of thigh	Tibial n. (L4–L5, anterior)
Semi-tendinosus	Ischial tuberosity	Medial surface of proximal tibia	Extends and aids medial rotation of thigh	Tibial n. (L5–S1, anterior)
Semi-membranous	Ischial tuberosity	Posterior side of medial tibial condyle	Extends and aids rotation of thigh	Tibial n. (L5–S1, anterior)
Biceps femoris Long head	Ischial tuberosity	Lateral side of fibular head	Extends and aids lateral rotation of thigh	Tibial n. (L5–S2, anterior)

TABLE 14-3. Abductor Muscles Acting at the Hip

Muscle	Origin	Insertion	Primary Action	Innervation
Gluteus medius	Superolateral surface of ilium	Greater trochanter	Abducts and aids medial rotation of thigh	Superior gluteal n. (L4–S2, posterior)
Gluteus minimus	Lateral surface of ilium	Greater trochanter	Abducts and aids medial rotation of thigh	Superior gluteal n. (L4–S1, posterior)
Tensor fasciae latae	Anterior-superior iliac spine and iliac crest	Iliotibial tract and lat. tibial condyl	Abducts, flexes, and medially rotates thigh	Superior gluteal n. (L4–S1, posterior)
Piriformis	Anterior sacrum	Greater trochanter	Laterally rotates and aids abduction of thigh	Nn. to piriformis (S1–S2, posterior)

TABLE 14-4. Adductor Muscles Acting at the Hip

Muscle	Origin	Insertion	Primary Action	Innervation
Adductor magnus				
Anterior belly	Ischial ramus and pubis	Distal portion of linea aspera	Adducts, flexes, and laterally rotates thigh	Obturator n. (L2–L4, anterior)
Posterior belly	Ischial tuberosity	Adductor tubercle	Adducts, extends, and medially rotates thigh	Tibial n. (L4–L5, anterior)
Adductor longus	Pubis between crest and symphysis	Linea aspera	Adducts and aids flexion and medial rotation of thigh	Obturator n. (L2–L3, anterior)
Adductor brevis	Body and inferior pubic ramus	Proximal portion of linea aspera	Adducts and flexes thigh	Obturator n. (L2–L4, anterior)
Pectineus	Iliopectineal line and pubis	Base of lesser trochanter	Adducts and flexes thigh	Femoral n. (L2–L3, posterior) Obturator n. (L2–L3, anterior)
Gracilis	Ischiopubic ramus	Medial side of medial tibial condyle	Adducts and flexes thigh	Obturator n. (L3–L4, anterior)

      (3) The tensor fasciae latae and the gluteus maximus, the superficial muscles of the buttock, insert into the **iliotibial tract.** As such, they are but weak abductors of the thigh.
    b. Adduction of the abducted limb (20°) is accomplished by muscles that pass inferior to the anteroposterior axis of the hip joint (Table 14-4).
      (1) The **adductors** run from the ischiopubic ramus to the medial aspect of the femur.
      (2) The major adductors of the thigh are included in the **adductor group,** including the **adductor magnus, adductor longus, adductor brevis, and gracilis muscles.**
  3. Rotation occurs about a vertical axis through the femoral head.
    a. Lateral rotation (eversion) is accomplished by muscles that act posterior to the vertical axis (Table 14-5).
      (1) The principal lateral rotators comprise the **pelvotrochanteric group** (the **piriformis, obturator internus, superior and inferior gemelli,** and the **obturator externus**) assisted by the **quadratus femoris.** The posterior portion of the gluteus maximus and the sartorius muscles also have external rotatory function.
      (2) These muscles pass nearly transversely, generally from the ischium or through its greater sciatic notch, to insertions in the vicinity of the greater trochanter and intertrochanteric crest.
    b. Medial rotation (inversion) is accomplished by muscles that act anterior to the vertical axis of the hip. There is no specific internal rotator group, but several muscles in the posterior, abductor, and adductor groups have secondary medial rotatory actions. Medial rotators include the **anterior portion of the gluteus min-**

TABLE 14-5. Lateral Rotators Acting at the Hip

Muscle	Origin	Insertion	Primary Action	Innervation
Gluteus maximus	Posterosuperior ilium and sacrum	Gluteal tuberosity of femur and iliotibial tract to tibia	Extends, abducts, and laterally rotates thigh	Inferior gluteal n. (L5–S2, posterior)
Piriformis	Anterior sacrum	Greater trochanter	Laterally rotates and aids abduction of thigh	Nn to piriformis. (S1–S2, posterior)
Obturator externus	Pubic and ischial rami, external side of obturator fascia	Greater trochanter	Laterally rotates and aids abduction of thigh	Obturator n. (L3–L4, anterior)
Obturator internus	Pubic and ischial rami, internal side of obturator fascia	Greater trochanter	Laterally rotates thigh	Nn. to obturator internus (L5–S2, anterior)
Superior gemellus	Ischial spine	Greater trochanter via obturator internus tendon	Laterally rotate thigh	Nn. to obturator internus (L5–S2, anterior)
Inferior gemellus	Ischial tuberosity	Greater trochanter via obturator internus tendon	Laterally rotate thigh	Nn. to obturator internus (L2–S2, anterior)
Quadratus femoris	Ischial tuberosity	Femur distal to greater trochanter	Laterally rotate thigh	Nn. to quadratus femoris (L4–S1, anterior)

imus, **tensor fasciae latae, semimembranosus, semitendinosus, posterior belly of the adductor magnus,** and **adductor longus** muscles.

4. **Circumduction** is produced by simultaneous movement about two or more axes.

V. VASCULATURE

A. **Gluteal region.** Blood is supplied from branches of the internal and external iliac arteries.

1. **The superior gluteal artery** is a branch of the posterior trunk of the internal iliac artery (Figure 14-5).
   a. It leaves the pelvic cavity via the suprapiriform portion of the greater sciatic foramen along with the superior gluteal nerve.
   b. The main trunk passes anteriorly, deep to the gluteus medius muscle.

FIGURE 14-5. Vasculature of the thigh and gluteal regions.

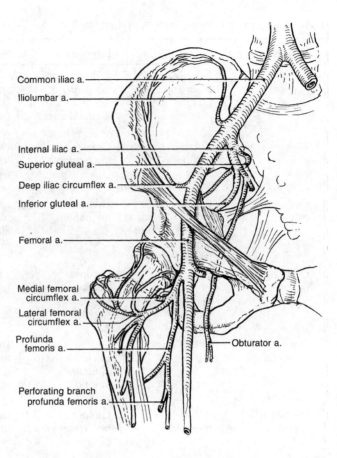

   c. It supplies the gluteus medius and minimus muscles, and a smaller portion of the gluteus maximus muscle.
2. **The inferior gluteal artery** is a branch of the anterior trunk of the internal iliac artery (see Figure 14-5).
   a. It leaves the pelvic cavity via the infrapiriform region of the greater sciatic foramen along with the inferior gluteal, pudendal, and sciatic nerves.
   b. It passes anteriorly on the deep surface of the gluteus maximus muscle.
   c. It supplies portions of the gluteus maximus muscle, the lateral rotators of the hip, and the most proximal portions of the hamstring group; it also sends twigs to the hip joint.
   d. It anastomoses with branches of the profundus femoris artery, forming part of the cruciate anastomosis.
3. **The obturator artery** is most often a branch of the anterior trunk of the internal iliac artery (see Figure 14-5).
   a. Occasionally (30%), this artery arises from the inferior epigastric artery and descends to the obturator foramen close to the femoral ring, where it may complicate surgical repair of a femoral hernia.
   b. It leaves the pelvic cavity via the obturator foramen.
   c. It supplies the proximal portions of the adductor group and sends twigs to the joint, including the artery of the ligamentum teres.
4. **The femoral artery,** the continuation of the external iliac artery, gives off the profunda femoris artery from which the medial and lateral femoral circumflex arteries arise (see Figure 14-5).
   a. The **medial and lateral femoral circumflex arteries** supply the proximal portions of the more lateral and anterior muscles of the thigh and are the major supply to the trochanters, neck, and head of the femur.

b. The two circumflex arteries pass transversely to anastomose with each other, with branches of the first perforating artery (inferiorly), and with the inferior gluteal artery (superiorly), forming the **cruciate anastomosis** posterolaterally.

B. Proximal femur

1. **Arterial supply**
   a. The **nutrient artery,** a branch of the profunda femoris artery, supplies the shaft.
   b. The **medial femoral circumflex, lateral femoral circumflex,** and **obturator arteries** supply the trochanters, the neck, and the head of the femur (see Figure 14-5). These vessels are clinically important.
      (1) The **retinacular (extracapsular) arteries** arise from the medial and lateral femoral circumflex arteries and give off branches that penetrate the joint capsule at the intertrochanteric line to supply the femoral neck and head.
         (a) Three groups of retinacular arteries with great variability run along the surface of the neck toward the femoral head.
         (b) These arteries supply the region distal to the epiphyseal plate in young people.
         (c) After fusion of the epiphyseal plate, the retinacular arteries anastomose with those of the ligamentum teres, and the latter may degenerate or become marginally significant in the adult.
         (d) Because of the course of the retinacular arteries, fracture of the neck jeopardizes the blood supply to the head.
      (2) The **artery of the ligamentum teres,** a branch of the obturator artery, is of principal importance in children, in that it supplies the femoral head proximal to the epiphyseal plate (see Figure 14-1). Once the growth plate fuses, this artery usually becomes atretic, the blood supply to the head coming from the retinacular arteries.
   c. **Clinical considerations**
      (1) **Septic arthritis.** In children, elevated intracapsular pressure associated with septic arthritis may compress the artery of the ligamentum teres and produce necrosis of the femoral head.
      (2) **Intracapsular fractures** endanger the blood supply to the proximal fragment. Nonunion and avascular necrosis may result, depending on the degree to which the retinacular arteries are torn and the effectiveness of the artery of the ligamentum teres.
2. **Venous return** consists of superficial and deep pathways. The deep veins parallel the arteries.

VI. INNERVATION

A. The lumbosacral plexus provides somatic innervation to the pelvis and lower extremities.

1. **Development.** As embryonic somites migrate to form the extremities, they drag their nerve supply, so that each dermatome and myotome retains the original segmental innervation.
   a. With somite migration, some of the nerves come into close proximity and fuse in a particular fashion, forming a plexus.
   b. The **lumbosacral plexus** develops from the anterior (ventral) primary rami of spinal nerves T12 through S3.
      (1) The lateral branch of each anterior primary ramus contributes the posterior division of the plexus (see Figure 4-2).
      (2) The ventral continuation of each anterior primary ramus contributes the anterior division of the plexus (see Figure 4-2).

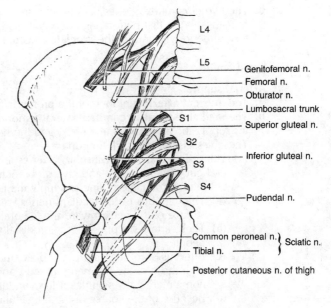

FIGURE 14-6. Lumbosacral plexus, anterior *(light)* and posterior *(shaded)* divisions.

2. **The lumbar plexus** (T12–L4) lies in the posterior abdominal wall and iliac fossa (Figure 14-6).
   a. The **anterior division** contributes the **genitofemoral** and **obturator nerves.** In general, these nerves provide sensation and motor innervation to the proximal anterior and medial aspects of the thigh, respectively.
   b. The **posterior division** contributes the **femoral nerve.** In general, these nerves provide sensation and motor innervation to the anterior thigh as well as sensory to the medial leg.
3. **The lumbosacral trunk** (L4–L5, anterior and posterior divisions) contributes to the sacral plexus and joins the lumbar and sacral plexuses (see Figure 14-6).
   a. The **anterior division** contributes the **tibial portion of the sciatic nerve.** In general, this nerve is sensory and motor to the posterior aspects of the thigh and leg as well as to the plantar surface of the foot.
   b. The **posterior division** contributes the **superior and inferior gluteal nerves** as well as the **common peroneal portion of the sciatic nerve.** In general, these nerves are sensory and motor to the posterior hip as well as to the anterior leg and dorsum of the foot.
4. **The sacral plexus** (L4–S3) lies in the deep pelvis and supplies the gluteal region, posterior thigh, leg, foot, and perineum (see Figure 14-6).
   a. The **anterior division** contributes to the **tibial portion of the sciatic nerve.** In general, this nerve is sensory and motor to the posterior aspects of the thigh and leg as well as to the plantar surface of the foot.
   b. The **posterior division** contributes to the **superior and inferior gluteal nerves** as well as to the **common peroneal portion of the sciatic nerve.** In general, these nerves are sensory and motor to the posterior hip as well as to the anterior leg and dorsum of the foot.

B. Major nerve branches
   1. **Superior gluteal nerve** (L4–S1, posterior)
      a. **Course.** This nerve exits the pelvis via the **suprapiriform portion** of the **greater sciatic foramen** along with the superior gluteal vessels. It lies in the superomedial quadrant of the buttock (see Figures 14-4 and 14-6).
      b. **Distribution.** It innervates the gluteus medius, gluteus minimus, and tensor fasciae latae muscles (see Figure 14-4).

c. **Clinical consideration.** Palsy results in "abductor lurch," a rolling (Trendelenburg) gait because of an inability to keep the pelvis level when the contralateral foot is raised off the ground. The lurch is a compensatory shift of the body weight directly over the ipsilateral femoral head.

2. **Inferior gluteal nerve** (L5–S2, posterior)
    a. **Course.** This nerve exits the pelvis via the **infrapiriform portion** of the **greater sciatic foramen,** along with the inferior gluteal vessels and the sciatic nerve. It lies in the inferolateral quadrant of the buttock (see Figures 14-4 and 14-6).
    b. **Distribution.** It innervates the gluteus maximus muscle.
    c. **Clinical consideration.** Palsy of this nerve results in difficulty in rising from a seated position and in climbing stairs because of weakness of hip extension.

3. **Sciatic nerve** (L4–S2, anterior and posterior)
    a. **Course.** It exits from the pelvis via the **infrapiriform recess** of the **greater sciatic foramen,** along with the inferior gluteal neurovascular bundle (see Figure 14-6).
        (1) It courses in an arc, starting halfway between the ischial tuberosity and the iliac spine.
        (2) It passes over the neck of the femur halfway between the ischial tuberosity and the greater trochanter and lies in the inferomedial quadrant of the buttock.
    b. **Distribution.** This nerve passes through the inferomedial gluteal region but does not contribute to gluteal innervation (see Figure 14-4).
    c. **Clinical consideration.** Intramuscular injection in the gluteal region should be in the upper lateral quadrant to avoid the sciatic nerve.

Chapter 15
Thigh and Knee Joint

I. BONES AT THE KNEE

A. Introduction. The thigh is that part of the lower extremity between the pelvic girdle and the knee joint.

1. **Basic organization.** The **primitively dorsal musculature** of the thigh lies anteriorly, flexes the hip joint, and extends the knee joint. Conversely, the **primitively ventral musculature** lies dorsally, extends the hip, and flexes the knee.
2. **Muscles acting at the knee.** Most of the major muscles that act at the knee joint are thigh muscles; however, muscles of the posterior compartment of the leg also act at this joint.

B. Distal femur (Figure 15-1)

1. **Characteristics**
   a. The **femoral shaft** broadens distally into epicondyles. The **linea aspera**, a ridge along the posterior aspect of the femur, provides attachment for septa that define the posterior compartment.
   b. The **lateral epicondyle**, a bony ridge lateral to the lateral condyle, serves as the attachment for the **lateral (fibular) collateral ligament** and the lateral head of the gastrocnemius muscle.
   c. The **medial epicondyle**, a bony ridge medial to the medial condyle, serves as a point of attachment for the **medial (tibial) collateral ligament** and the medial head of the gastrocnemius muscle. The **adductor tubercle,** superior to the medial epicondyle, receives the insertion of the posterior belly (hamstring portion) of the adductor magnus.
2. **Condyles** provide stable, weight-bearing articulations with the tibia (see Figure 15-1).
   a. A deep **intercondylar notch (fossa)** between the lateral and medial condyles divides the femorotibial articular surface inferiorly and posteriorly.
   b. The **patellar articular surface (trochlear groove)** bridges the condyles and gives the articular surface of the anterior aspect of the distal femur a U-shaped appearance. The patellar articular surface is separated from the tibial articular surface by a faint groove in each condyle.
   c. The **lateral condyle** is wider than the medial condyle. It lies along the principal line of force and is probably more important in weightbearing.
   d. The **medial condyle** has a longer tibial articular surface than the lateral condyle.
      (1) This length difference results in lateral rotation of the tibia with reference to the femur during the terminal phase of knee extension.
      (2) The medial condyle usually projects slightly further distally than the lateral condyle, producing an inward angle at the knee. Normal valgus is usually more pronounced in females (170°) than in males (175°).
         (a) **Genu valgum** (valgus or "knock-knees") is an exaggerated inward angling at the knee.
            (i) Valgus results from excessive prolongation of the medial femoral condyle or reduced growth in the lateral portion of the distal epiphyseal plate.
            (ii) If severe, the foot will also be somewhat everted (low arched) and medially rotated.
         (b) **Genu varum** (varus or "bow-legs") is an exaggerated outward angling at the knee. If varus is severe, the foot will also be somewhat inverted (high arched) and laterally rotated.
   e. The **transverse axis of the knee** is not fixed. The progressive radii of curvature of the condyles form a spiral evolute that defines the movement of the axis of rotation.

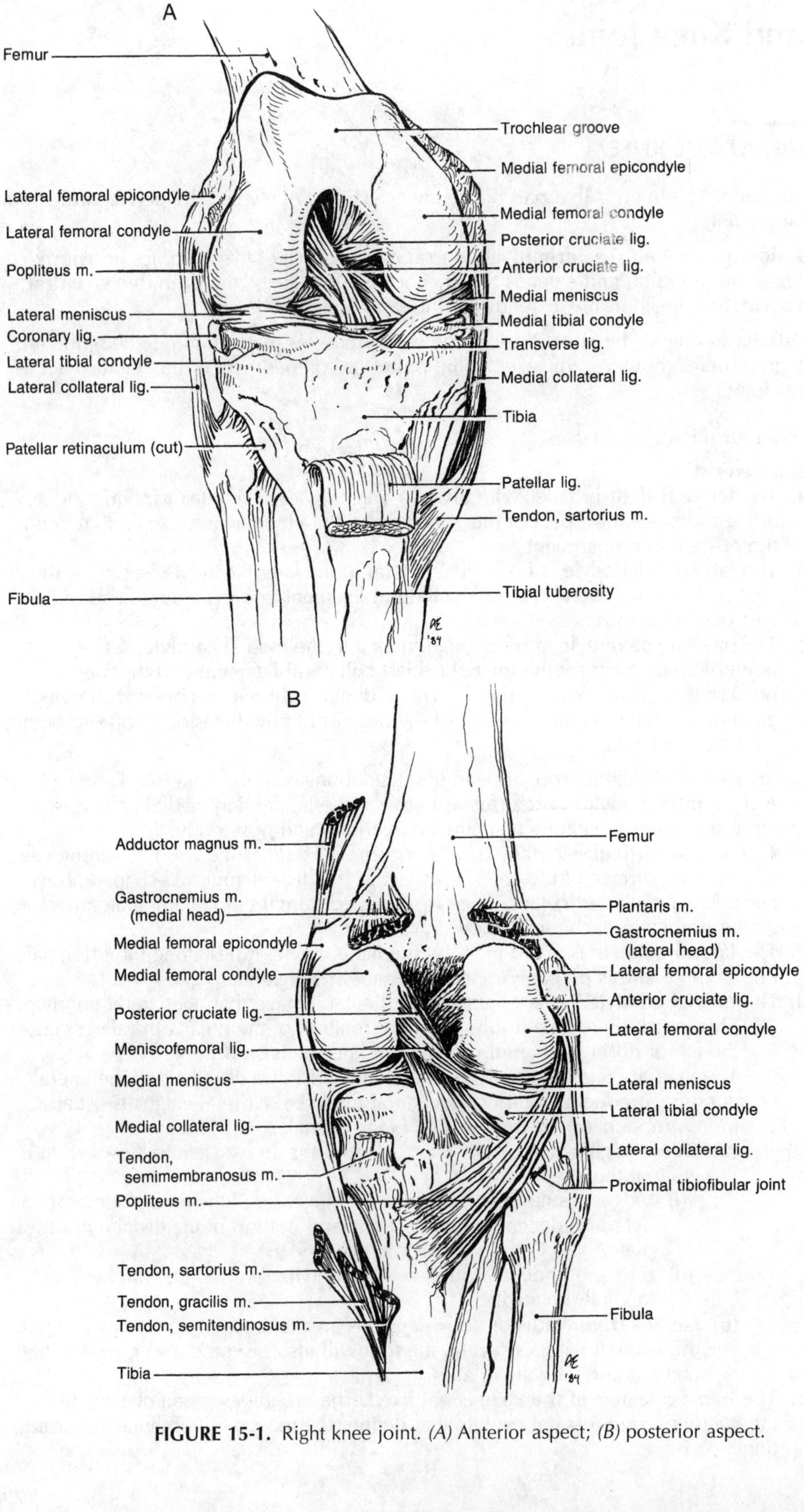

FIGURE 15-1. Right knee joint. *(A)* Anterior aspect; *(B)* posterior aspect.

(1) **Evolute of rotation.** As the knee is extended, the radii of curvature of both condyles more than double. The tibia is pushed distally, tightening the ligaments that connect the tibia and the femur.
   (a) **Posteriorly,** the radii are shorter (17 mm medial, 12 mm lateral).
   (b) **Anteriorly,** the radii are longer (38 mm medial, 60 mm lateral).
(2) In **flexion,** the transverse axis of rotation passes more posteriorly through the femoral condyles. Ligaments that attach to the femur anteriorly become taut.
(3) In **extension,** the transverse axis of rotation passes more anteriorly. The ligaments that attach to the femur posteriorly become tense. Because the tibia is pushed further distally in extension, however, all the ligaments tighten.

3. **Fractures**
   a. In a **supracondylar fracture,** the distal fragment is displaced posteriorly by gastrocnemius spasm. This positioning jeopardizes the large and important popliteal neurovascular structures, which, at this level, lie adjacent to the femur.
   b. An **intracondylar fracture** is T-shaped, combining a supracondylar fracture with a vertical fracture plane that separates the condyles. A valgus or varus knee deformity may result if this type of fracture is not reduced properly.

C. Proximal tibia

1. **Characteristics** (see Figure 15-1)
   a. The **tibial plateau** is formed by the **medial and lateral tibial condyles.** On the anterior tibial surface of the lateral condyle, a small elevation serves as insertion for the iliotibial tract.
   b. The **fibular head** articulates with the tibia just inferior to the lateral condyle.
   c. The **tibial tuberosity,** the insertion of the **patellar tendon,** lies inferior to the tibial head on the anterior surface of the shaft.
   d. The **tibial shaft** is triangular in cross section and tapers distally toward the **medial malleolus.**

2. **Condyles** provide stable, weightbearing articulations with the femur (see Figure 15-1).
   a. **Femorotibial (knee) joint**
      (1) The **medial tibial condyle** is concave in both frontal and sagittal planes. Like the medial femoral condyle, it is longer in the anteroposterior diameter.
      (2) The **lateral tibial condyle** is concave in the frontal plane but convex in the sagittal plane and more nearly round.
      (3) An **intercondylar eminence** between the tibial condyles lodges in the intercondylar notch of the femur and acts as a pivot for rotation around a vertical axis.
      (4) **Menisci** of fibrocartilage lie between the femoral and tibial condyles (Figure 15-2; see Figure 15-1). The menisci become slightly distorted as flexion/extension occurs, thereby compensating for the lack of congruity between the tibial and femoral articular surfaces and distributing weight evenly over the joint.
   b. **Proximal (superior) tibiofibular joint** (see Figure 15-1)
      (1) The fibula has lost contact with the femur. Instead, the fibular head articulates with the fibula laterally, just inferior to the lateral condyle. This gliding joint has extremely limited movement.
      (2) An **interosseous membrane** connects the fibula and tibia along their shafts. The connective tissue fibers run inferolaterally, functioning as shock absorbers and helping to stabilize the ankle joint.

D. Patella

1. **Characteristics** (see Figure 15-4)
   a. The **patella** (L. little pate) sits in the tendon of the quadriceps femoris muscle. This palpable sesamoid bone is triangular, with the apex directed inferiorly. Its posterior surface articulates with the femur at the **femoropatellar joint.**
   b. The **patellar tendon** (often termed ligament) runs between the apex and the tibial tuberosity.

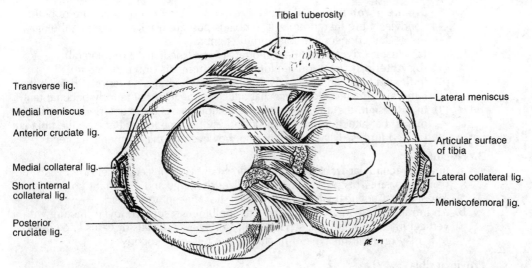

FIGURE 15-2. Articular cartilages and ligaments of the right knee from above. For orientation, note the medial and lateral collateral ligaments as well as the tibial tuberosity.

(1) The **patellar retinaculum** is an encapsulation of dense connective tissue derived from the quadriceps femoris tendon. It extends inferolaterally and inferomedially to insert into the tibial condyles, representing a broad tendinous insertion for the quadriceps femoris muscle.

(2) **Subcutaneous bursae** lie anterior to the knee.
  (a) The **prepatellar bursa** is superficial to the patella and deep to the epithelium. Inflammation associated with repeated abuse results in **prepatellar bursitis** or **"housemaid's knee."**
  (b) The **infrapatellar bursa** lies between the tibial tuberosity and the epithelium. Inflammation results in **infrapatellar bursitis** or **"vicar's knee."**
  (c) The **suprapatellar bursa** is deeper, extending from the joint capsule proximally between the quadriceps femoris tendon and the femur. Effusions at the knee joint distend this bursa (**suprapatellar bursitis** or **"water on the knee"**). On physical examination, the patella appears to float over the femur. Postinflammatory adhesions in the suprapatellar bursa reduce knee mobility (**"stiff knee"**).

2. **Functions**
   a. As a sesamoid bone, it obviates wear and attrition on the quadriceps tendon as it passes across the trochlear groove.
   b. It lengthens the lever arm of the quadriceps femoris muscle, thereby increasing the mechanical advantage.
   c. It changes the line of action of the quadriceps femoris group so that it passes anterior to the knee.

3. **Fracture** through the articular surface results in eventual **degenerative patellofemoral arthritis.** A congenitally bipartite patella may be misdiagnosed as a fracture.

II. KNEE JOINT

A. **Structure.** The knee joint is composed of two articulations.

1. **The femoropatellar joint** functions as a trochlea for the quadriceps femoris muscle group.

a. The **line of action** of the quadriceps femoris muscle passes inferomedially to reach the patella. From the patella, it passes inferiorly to reach the tibial tuberosity. The change in the line of action leads to instability.
   b. **Medial and lateral articular surfaces** are formed by a vertical ridge on the posterior surface of the patella. As a gliding joint, these articular surfaces slide in the trochlear groove of the femur (see Figure 15-1A).
      (1) In **flexion,** tension from the quadriceps femoris tendon and the patellar retinaculum tends to keep the patella deep in the trochlear groove between the femoral condyles.
      (2) In **extension,** the patella is drawn laterally.
         (a) **Stabilizing factors.** The prominent **lateral ridge of the trochlear groove, patellar retinaculum,** and especially the line of action of the **vastus medialis muscle** all stabilize the patella and prevent lateral dislocation.
         (b) **Destabilizing factors.** Because of the usual valgus at the knee, the line of action of the **rectus femoris, vastus lateralis,** and **vastus intermedius muscles** (through the quadriceps femoris tendon to the patella and thence through the patellar tendon to the tibial tuberosity) is somewhat lateral to the center of the patella. This **dog-legged line of action** with a **shallower trochlear groove** proximally (in extension), tends to draw the patella laterally.
         (c) **Recurrent patellar dislocation**
            (i) This condition results from underdevelopment of the lateral ridge of the trochlear groove or exaggerated genu valgum.
            (ii) It can be corrected surgically by transplanting the tibial tubercle slightly more medially and inferiorly, or by cutting the lateral edge of the patellar retinaculum.
   c. The **patella** lengthens the lever arm, thereby increasing the mechanical advantage of the quadriceps femoris muscle.
2. **The femorotibial joint** supports the body weight. Although adapted for weightbearing at any degree of flexion, stability is greatest in full extension.
   a. **Menisci (semilunar cartilages).** Composed of wedge-shaped fibrocartilage rings, the menisci lie between the femoral and tibial condyles and incompletely divide the femorotibial joint into a **suprameniscal compartment** and an **inframeniscal compartment** (see Figure 15-3).
      (1) The **lateral meniscus** approximates a closed circle, that is, it is O-shaped (see Figure 15-2).
         (a) **Insertions.** The **anterior** and **posterior horns** of the lateral meniscus insert close together into the lateral tubercle of the **intercondylar eminence.**
         (b) **Ligamentous attachments**
            (i) The **meniscofemoral ligament,** a posteromedial portion of the posterior cruciate ligament, attaches the lateral meniscus to the posterior aspect of the medial femoral condyle, posteriorly (see Figure 15-1B).
            (ii) The **coronary ligament** loosely attaches the periphery of the lateral meniscus to the joint capsule (see Figure 15-1A).
         (c) **Movement**
            (i) The close insertions, the loose attachment to the joint capsule, and little or no attachment to the lateral collateral ligament promote considerable movement of the lateral meniscus.
            (ii) During the terminal phase of extension with the foot planted firmly on the ground, the femur pivots and the lateral femoral condyle slides anteriorly with the lateral meniscus to lock the knee.
      (2) The **medial meniscus** is C-shaped (see Figure 15-2).
         (a) **Insertions.** The **anterior and posterior horns** insert external to those of the lateral meniscus (anterior and posterior to the intercondylar eminence, respectively), giving the medial meniscus a somewhat greater diameter than the lateral meniscus.
         (b) **Ligamentous attachments** (see Figure 15-2)
            (i) The **short internal collateral ligament** runs from the margin of the meniscus to the medial femoral epicondyle and tightly attaches the meniscus to the joint capsule.

(ii) The **medial collateral ligament** is attached to the medial meniscus medially.
(iii) The **coronary ligament** loosely attaches the periphery of the medial meniscus to the joint capsule (see Figure 15-1A).
   (c) Movement
      (i) The widely separated points of attachment, the medial attachment to the joint capsule, and the attachment to the medial collateral ligament restrict movement of the medial meniscus.
      (ii) This restriction makes the medial meniscus especially prone to **tearing.**
  b. Meniscal injury. During the rolling femoral movements of flexion and extension, the menisci are drawn posteriorly and anteriorly, respectively. The menisci can tear if they fail to follow the movements of the femoral condyles.
    (1) Violent hyperextension leads to transverse meniscal tears or detachment of the anterior horns.
    (2) Violent twisting movements, which combine lateral displacement with lateral rotation (especially in a semiflexed knee), pull the medial meniscus toward the center of the joint where it may be trapped and crushed by the medial femoral condyle. This event may lead to a longitudinal (bucket-handle) tear or partial detachment of the medial meniscus.
    (3) Torn menisci can become wedged between condylar surfaces, locking the knee in partial flexion and making full extension impossible.
  c. Osteoarthritis. The knee is a common site for this disease. Joggers and obese individuals are prone to its development.
    (1) Osteoarthritis appears radiographically as a narrowing of the joint cavity with degenerative changes.
    (2) Use of a cane, held on the either side of the body, relieves some of the weight borne by the joint surface and reduces the accompanying pain. Because the opposite arm goes forward with the forward swinging contralateral leg, using the cane on the **opposite side** from the affected knee is more natural and offers a wider triangle of support.

B. Ligamentous support

1. **The synovial capsule** of the knee is extensive.
   a. **Boundaries. Circumferentially,** the synovial capsule inserts about the margins of the articular surfaces. **Medially,** it attaches to the medial meniscus. **Posteriorly,** it invaginates into the intercondylar notch so that the cruciate ligaments are extracapsular. **Anterosuperiorly,** the **suprapatellar bursa** extends from the capsule between the quadriceps femoris tendon and the femur.
   b. **Effusion at the knee joint** expands the joint capsule, precluding a full range of motion. The knee assumes approximately 20° of flexion.

2. **The fibrous capsule** of the knee is weak.
   a. The **short internal collateral ligament,** a thickening of the fibrous capsule deep to the medial collateral ligament, extends from the medial femoral epicondyle to the medial meniscus (see Figure 15-2).
   b. The **coronary ligaments** are the thickened periphery of the capsule that loosely attaches each meniscus to the tibia (see Figure 15-1A).
      (1) Frequently (60%), it is thickened anteriorly to form the **transverse ligament,** which connects the anterior margins of the two menisci.
      (2) These ligaments are more likely than other knee ligaments to sustain injury (sprain) because, although lax, they are short.
   c. **Violent twisting movements,** which combine lateral displacement with lateral rotation (especially in a semiflexed knee), pull the medial meniscus toward the center of the joint and may sprain the coronary ligament.

3. **The collateral ligaments** and **patellar retinaculum** reinforce the fibrous capsule of the knee (see Figures 15-1 and 15-2).
   a. **The medial (tibial) collateral ligament** extends from the medial femoral epicon-

dyle and widens to insert into the shaft of the tibia below the level of the tibial tuberosity (see Figure 15-1).
- **(1) Structure.** This broad, fan-shaped ligament is effective at any degree of flexion.
  - **(a)** The posterior portion is fused to the joint capsule and is attached to the medial meniscus.
  - **(b)** Phylogenetically, this ligament is the degenerated tendon of insertion of the posterior portion of the adductor magnus muscle.
- **(2) Function.** The medial collateral ligament prevents abduction at the knee.
  - **(a)** Because it lies posterior to the axes of flexion/extension, it becomes taut on extension and, thus, also limits extension of the leg.
  - **(b)** A tear in this ligament is recognized by **abnormal passive abduction** of the extended leg.

b. The **lateral (fibular) collateral ligament** extends from the lateral femoral epicondyle to insert into the head of the fibula (see Figure 15-1).
- **(1) Structure.** It is a strong, narrow cordlike ligament.
  - **(a)** It is readily palpated when the legs are crossed and the ankle rests on the opposite knee. It is relatively free of the joint capsule and does not attach to the lateral meniscus.
  - **(b)** Phylogenetically, this ligament is the degenerated tendon of origin of the peroneus longus muscle.
- **(2) Function.** The lateral collateral ligament prevents adduction of the leg at the knee.
  - **(a)** Because it also lies posterior to the axes of flexion/extension, it becomes taut on extension and limits extension of the leg.
  - **(b)** A tear in this ligament is recognized by **abnormal passive adduction** of the extended leg.

c. The **posterior cruciate ligament** extends anteromedially from the tibia posterior to the intercondylar eminence and inserts into the medial femoral condyle anteriorly within the intercondylar notch (see Figures 15-1B and 15-2).
- **(1) Structure.** It is approximately parallel to the lateral collateral ligament and nearly perpendicular to the anterior cruciate ligament. The posterior cruciate ligament fans out toward its linear insertion along the medial femoral condyle.
  - **(a)** Because the fibers in different portions of the ligament are of unequal length, some are always under tension at any degree of knee flexion. Fibers inserting anteriorly on the femoral condyle tend to be taut in flexion. Fibers inserting posteriorly tend to be taut in extension.
  - **(b)** The **meniscofemoral ligament,** a subdivision of the posterior cruciate ligament, inserts into the posterior horn of the lateral meniscus and draws that meniscus posteriorly as the lateral femoral condyle slides posteriorly on the tibial plateau with flexion (see Figure 15-2).
- **(2) Function.** The posterior cruciate ligament is a key stabilizer of the knee joint, tightening maximally on extension.
  - **(a)** It checks anterior movement of the femur on the tibial plateau (posterior movement of the tibia with respect to the femur). In the flexed, weightbearing knee (descending stairs or walking down hill), this ligament prevents the femur from sliding forward.
  - **(b)** Because of perceived instability, an individual with a torn posterior cruciate ligament will lead with the opposite leg when descending stairs. A tear can be recognized by abnormal passive posterior displacement of the tibia **(posterior drawer sign).**

d. The **anterior cruciate ligament** extends posterolaterally from the tibia anterior to the intercondylar eminence and inserts onto the lateral femoral condyle posteriorly within the intercondylar notch (see Figures 15-1A and 15-2).
- **(1) Structure.** It is approximately perpendicular to the lateral collateral ligament and the posterior cruciate ligament.
  - **(a)** Slightly shorter than the posterior cruciate, it fans out toward its linear insertion along the lateral femoral condyle.

- **(b)** Because the fibers in different portions of the ligament are of unequal length, some are always under tension at any degree of knee flexion. Fibers inserting posteriorly on the femoral condyle tend to be taut in extension. Fibers inserting more anteriorly tend to be taut in flexion.
- **(2) Function.** The anterior cruciate ligament is a key stabilizer of the knee joint. Because it inserts posteriorly on the femur, it also tightens maximally on extension and provides a pivot around which rotation occurs during the terminal phase of extension.
  - **(a)** It checks posterior movement of the femur on the tibial plateau (anterior movement of the tibia with reference to the femur).
  - **(b)** A tear often is not noticed, except that occasionally the knee "suddenly gives way." This sensation occurs because the leg will not lock on full extension. A tear is recognized by abnormal passive anterior displacement of the tibia **(anterior drawer sign).**
- **e. Other factors.** The **oblique popliteal ligament** is a lateral extension of the semimembranosus tendon. It strengthens the joint capsule posteriorly, resisting hyperextension of the leg as well as lateral rotation during the terminal phase of extension. The **patellar tendon** reinforces the joint capsule anteriorly. The **patellar retinaculum** (into which the vasti muscles insert) reinforces the capsule medially and laterally [see section II A 1 b (1), (2)].

4. **Stability** is conferred by the collateral and cruciate ligaments.
   a. **Rotational stability**
      (1) **Medial rotation check.** Because the **collateral ligaments** are slightly coiled outward (i.e., nearly at right angles to each other), medial rotation of the femur tightens these coils, preventing further medial rotation.
      (2) **Lateral rotation check.** Because the **cruciate ligaments** are slightly coiled inward (i.e., nearly at right angles to each other), lateral rotation of the femur tightens both ligaments, preventing further lateral rotation.
   b. **Flexion/extension**
      (1) The **anterior cruciate ligament** tightens maximally on extension and checks posterior movement of the femur on the tibial plateau.
      (2) The **posterior cruciate ligament** tightens maximally on extension and checks anterior movement of the femur on the tibial plateau.
      (3) These ligaments represent the primary stabilizers of body weight at the knees in an individual in the erect posture with the knee extended and locked.
   c. The **dynamic stability** of the knee depends on the action of muscles that pass across this joint, especially when the knee is partially flexed.
      (1) The thigh muscles that insert below the knee and the leg muscles that originate above the knee contribute to this stability.
      (2) When the limb is immobilized in a cast, disuse atrophy of the muscles occurs. Physical therapy is necessary to restore full function and stability.

5. **Instability** results from torn ligaments.
   a. **Ligamentous sprains and tears.** Because the knee joint represents a compromise between stability and mobility, violent hyperextension, abduction, and adduction increases the potential for sprains (partial tears) and tears of one or more ligaments.
   b. **Combined knee injury,** such as the **"unhappy triad,"** is the result of fixation of a semiflexed leg with violent abduction and lateral rotation that occurs as the result of "clipping" in football or the "caught-edge" fall in skiing.
      (1) Three elements compose the unhappy triad.
         (a) **Medial collateral ligament injury** [see section II B 3 a (2)]. Because this ligament is attached to the joint capsule, the medial capsular (coronary) ligaments are torn as well.
         (b) **Anterior cruciate ligament injury** [see section II B 3 d (2) (b)] is associated with a bloody effusion in the joint capsule that may extend into any connecting bursa.
         (c) **Medial meniscus injury** is discussed in section II A 2 b.

(2) Treatment involves surgical reconstruction of the ligaments with every attempt to save as much of the meniscus as possible to prevent or limit future osteoarthritis. An arthroscope inserted into the knee joint through a small incision permits direct visualization of trauma or pathology. Corrective procedures can be accomplished through the instrument.

C. Dynamic action

1. **Movement.** Separate and distinct movements occur in the **suprameniscal compartment** and the **inframeniscal compartment** (Figure 15-3). The ligaments are arranged so that the femur and tibia are always held in close apposition with stability throughout an extensive range of movement. The proportion of each individual movement is dictated by the ligaments, the shape of the condyles, and the length of the articular surfaces.
   a. **Suprameniscal compartment movement** is rotational about a series of transverse axes (see Figure 15-3).
      (1) The femoral condyles rotate on the superior meniscal surfaces about the transverse axis, approximating a simple hinge.
      (2) The transverse axis of knee rotation changes through an evolute so that the knee does not function as a true hinge joint. The increasing radius with extension causes tightness of the collateral and cruciate ligaments.
   b. **Inframeniscal compartment movement** is anterior–posterior gliding and terminal rotation about a vertical axis (see Figure 15-3).
      (1) The inferior meniscal surfaces glide on the tibial plateau, the result of an anterior/posterior translation of the transverse axis.
      (2) In addition, at the end of extension, the femoral condyles pivot on the tibial plateau about a vertical axis.
   c. **Combined movements** in these two compartments are not independent but occur in conjunction with flexion/extension (see Figure 15-3).

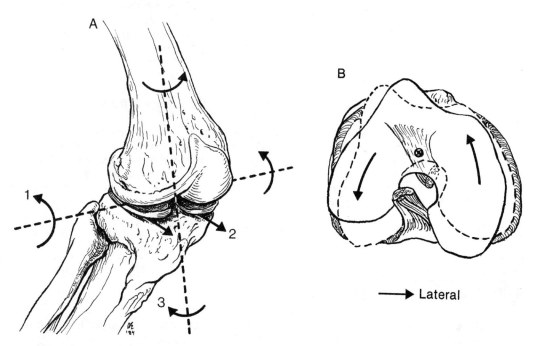

FIGURE 15-3. Dynamic action of the knee joint. *(A)* During extension, rotation occurs in the suprameniscal compartment about the transverse axis (1) and forward gliding occurs in the inframeniscal compartment (2). The medial rotation of the terminal phase of extension occurs about a vertical axis (3). *(B)* Looking down on the right tibial plateau in terminal extension, note the relative movement of the femur to the tibia *(from solid to dashed, with arrows)* about an axis of rotation (x) within the anterior cruciate ligament.

(1) **Flexion/extension** (140°) approximates a hinge joint.
(2) **Rotation** (20°) occurs at the terminal phase of extension or the initial phase of flexion.

2. **Phases of extension**
   a. **Initial phase** (see Figure 15-3A)
      (1) In the flexed knee, the anterior and posterior cruciate ligaments tether the femoral condyles on the tibial plateau. As the femorotibial joint begins to extend through the first 20°–30°, a concomitant anterior translation of the condyles occurs on the tibial plateau in the inframeniscal compartment.
      (2) This complex movement involves rotation in the suprameniscal compartment about a transverse axis through the femoral condyles and a concomitant anterior gliding movement in the inframeniscal compartment as that transverse axis moves anteriorly.
   b. **Intermediate phase** (see Figure 15-3A)
      (1) As extension continues, the increasing radii of curvature of the femoral condyles tighten the anterior cruciate ligament and the posterior fibers of the posterior cruciate ligament. This change prevents further forward gliding of the femur in the inframeniscal compartment.
      (2) The femoral condyles continue to rotate on the meniscal surfaces in the suprameniscal compartment.
   c. **Terminal phase** (see Figure 15-3A)
      (1) Toward full extension, the lateral condyle reaches the limit of its articular surface and ceases rotation.
         (a) The longer articular surface of the medial condyle continues to rotate in the suprameniscal compartment.
         (b) The lateral condyle is forced to slide forward and medially in the inframeniscal compartment about an axis formed by the taut anterior cruciate ligament.
      (2) The result is a **medial rotation of the femur** (or lateral rotation of the tibia) in the inframeniscal compartment about an axis through the intercondylar eminence at the insertion of the anterior cruciate ligament.
         (a) Medial rotation of the femur tightens the oblique popliteal, medial collateral, and lateral collateral ligaments and stretches the popliteal muscle so that the femorotibial joint becomes locked ("screwed home").
         (b) In terminal extension, the lateral femoral condyle glides about an axis formed by the anterior cruciate ligament.
      (3) At terminal extension, the angle between the femur and the tibia slightly exceeds 180°.
         (a) The weight of the body now passes anterior to the transverse axis of the femorotibial joint and tightens the knee ligaments.
         (b) Muscular activity is unnecessary to maintain the extended and locked position when standing. The line of action of the tensor fasciae latae and gluteus maximus muscles passes anterior to the transverse axis and ensures the locked position.

III. FASCIA AND MUSCLES OF THE THIGH

A. **The fascia lata** is the deep fascia of the thigh. This layer forms the iliotibial tract and is the source of septa that compartmentalize the thigh musculature.

1. **The iliotibial tract (band)** is a lateral thickening of the fascia lata. It runs from the iliac crest to the lateral tibial condyle. It functions as the tendon of insertion for the **tensor fasciae latae muscle** and the **gluteus maximus muscle** (see Figure 15-4).
   a. **At the hip.** Because the **tensor fasciae latae** and **gluteus maximus muscles** origi-

nate on the pelvis and pass over the hip joint, they are involved in extension of the thigh.
- b. **At the knee.** Because the iliotibial tract crosses the knee joint, the **tensor fasciae latae muscle** and the **gluteus maximus muscle** function both in extension and flexion of the leg.
  - (1) **When the knee is extended,** the line of action of the iliotibial tract passes anterior to the axis of rotation of the knee joint. Therefore, the tensor fasciae latae and gluteus maximus muscles keep the knee extended.
  - (2) **When the knee is slightly flexed,** the line of action of the iliotibial tract passes posterior to the axis of rotation of the knee joint, so these muscles now flex the knee joint. This configuration is evident when a person, standing erect and somewhat relaxed, has the knees knocked from behind. The line of action of the iliotibial tract shifts across the axis of rotation and the tensor fasciae latae and gluteus maximus muscles now flex the knee, producing the disconcerting sinking motion.

2. **Fascial septa** from the fascia lata divide the thigh musculature into **anterior** and **posterior compartments.**
   - a. The **lateral intermuscular septum** arises from the posterior border of the iliotibial tract and inserts into the lateral lip of the linea aspera on the posterior aspect of the femur. This septum provides an important insertion onto the femur for the tensor fasciae latae and the gluteus maximus muscles.
   - b. The **medial intermuscular septum** arises from the anteromedial border of the iliotibial tract and attaches to the medial lip of the linea aspera.

B. **Thigh musculature** acts on both the hip and knee joints.

1. **Organization.** The thigh is divided into an **anterior compartment** of the thigh contains primitively dorsal musculature and the **posterior compartment** contains primitively ventral musculature.
   - a. The **anterior compartment** is divided into the iliopsoas and anterior muscle groups.
     - (1) The **iliopsoas group** consists of the **iliacus** and **psoas muscles** (Figure 15-4A; see Table 15-1). In addition to flexing the thigh, if the leg is planted, they flex the lumbar vertebral column.
     - (2) The **anterior group** consists of the **sartorius** and **quadriceps femoris muscles (rectus femoris, vastus lateralis, vastus intermedius,** and **vastus medialis)** (see Figure 15-4 and Table 15-1).
       - (a) **Course.** The **sartorius** and **rectus femoris** arise from the ilium; the three **vasti** arise from the femoral shaft. The **sartorius** inserts onto the medial tibial condyle. The vasti insert into the patella via the quadriceps tendon.
       - (b) **Group actions.** All act principally as extensors of the leg at the knee.
         - (i) The rectus femoris and vastus intermedius muscles insert into the base (top) of the patella. The angle between the line of action of these muscles and the axis of the tibia tend to destabilize the patella laterally.
         - (ii) The vastus lateralis inserts onto the lateral aspect of the patella. This muscle also tends to destabilize the patella laterally.
         - (iii) The vastus medialis inserts onto the medial aspect of the patella. This muscle is occasionally divided visually and always divided functionally into a **vastus medialis longus,** which inserts into the quadriceps tendon, and a **vastus medialis obliquus,** which inserts into the medial aspect of the patella. The vastus medialis, especially its oblique portion, is a strong stabilizer of the patella.
       - (c) **Group innervation.** The adductor group is innervated by the **femoral nerve.**
   - b. The **posterior compartment** is divided into two muscle groups.
     - (1) In the **adductor group,** the **gracilis, pectineus,** and **adductor longus muscles** lie in an anterior layer; the **adductor brevis muscle** forms in an intermediate

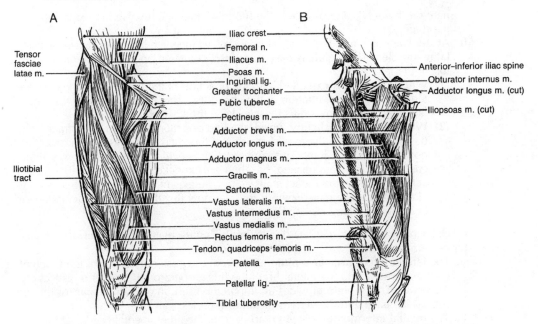

FIGURE 15-4. Anterior compartment of the thigh. *(A)* Superficial muscles; *(B)* deep muscles.

layer; and the anterior portion of the **adductor magnus muscle** constitutes the deep layer (see Table 15-2 and Figure 15-4B).
- **(a) Course.** Most of these muscles arise from the inferior pubic rami and insert into the femoral shaft medially; the gracilis inserts into the medial tibial condyle.
- **(c) Group actions.** Only the gracilis acts at the knee (flexion).
- **(d) Group innervation.** These muscles are innervated by the **obturator nerve.**
- **(2)** The **hamstring group** acts at both the hip and knee joints and consists of the **semimembranosus, semitendinosus,** the **long head of the biceps femoris,** and the **posterior portion of the adductor magnus muscle** (Figure 15-5).
  - **(a) Course.** This group arises from the ischial tuberosity and inserts into the tibial condyles, with the exception of the posterior portion of the adductor magnus, which inserts onto the adductor tubercle of the medial femoral epicondyle. The hamstring muscles insert about the knee—the adductor magnus, semimembranosus, and semitendinosus medially, the biceps femoris laterally.
  - **(b) Group actions.** The hamstrings that cross the knee flex that joint.
  - **(c) Group innervation.** The hamstring muscles are innervated by the **tibial nerve.**
  - **(d)** The **short head of the biceps femoris** technically is not considered part of the hamstring group because of its origin (femoral shaft) and innervation (common fibular nerve).

2. **Group actions at the knee**
   a. **Flexion/extension** (140°) occurs about a transverse axis through the femoral condyles.
   - **(1) Extension** is accomplished by the **anterior group** (Table 15-1), consisting of the quadriceps femoris, with the assistance of the tensor fasciae latae. Because the **rectus femoris** is a biarticular muscle (flexing the thigh at the hip joint and extending the leg at the knee joint), its efficiency at one joint depends on the position of the other joint. Thus, when the hip is extended, the rectus femoris is stretched and its force-generating capacity to extend the knee is greatest.

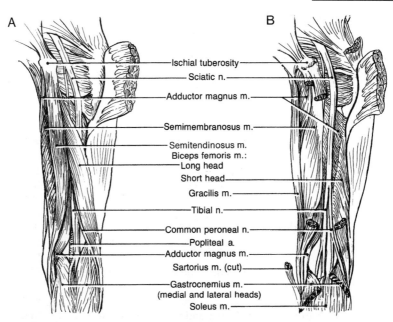

FIGURE 15-5. Posterior compartment of the thigh. *(A)* Superficial muscles; *(B)* deep muscles.

(2) **Flexion** is accomplished by the **hamstring group** and the sartorius (Table 15-2), which cross posterior to the transverse axis of the knee. Because the **hamstrings** are biarticular (extending the thigh and flexing the knee), their efficiency at the knee is greatest when the hip is flexed. The **sartorius muscle** of the anterior group as well as the popliteus and gastrocnemius muscles are also knee flexors.

TABLE 15-1. Anterior (Extensor) Compartment Musculature

Muscle	Origin	Insertion	Primary Action	Innervation
Rectus femoris	Anterior-inferior iliac spine	Tibial tuberosity through patellar tendon	Extends knee	Femoral n. (L2–L4, posterior)
Vastus lateralis	Proximal femur lateral to linea aspera	Tibial tuberosity through patellar tendon	Extends knee	Femoral n. (L2–L4 posterior)
Vastus intermedius	Anterior aspect of proximal femur	Tibial tuberosity through patellar tendon	Extends knee	Femoral n. (L2–L4 posterior)
Vastus medialis	Proximal femur medial to linea aspera	Medial patella and tibial tuberosity through patellar tendon	Stabilized patella and extends knee	Femoral n. (L2–L4 posterior)
Sartorius	Anterior-superior iliac spine	Medial side of tibial condyle	Flexes knee	Femoral n. (L2–L4, posterior)

TABLE 15-2. Posterior (Flexor) Compartment Musculature

Muscle	Origin	Insertion	Primary Action	Innervation
Semi-tendinosus	Ischial tuberosity	Medial surface of proximal tibia	Flexes knee	Tibial n. (L5–S1, anterior)
Semi-membranosus	Ischial tuberosity	Posterior side of medial tibial condyle	Flexes knee	Tibial n. (L5–S1, anterior)
Biceps femoris				
Long head	Ischial tuberosity	Lateral side of fibular head	Flexes knee	Tibial n. (L5–S2, anterior)
Short head	Posterior femoral shaft	Lateral side of fibular head	Flexes knee	Common fibular n. (L5–S2, posterior)
Gracilis	Ischiopubic ramus	Medial side of tibial head	Flexes knee	Obturator n. (L3–L4, anterior)
Gluteus maximus	Posterosuperior ilium	Femur and via iliotibial band to lateral tibial condyle	Flexes flexed knee, extends extended knee	Inferior gluteal n. (L5–S2, posterior)
Tensor fasciae latae	Anterior-superior iliac spine and iliac crest	Iliotibial band to lateral tibial condyle	Flexes flexed knee, extends extended knee	Superior gluteal n. (L4–S1, posterior)

    b. Rotation (20°) occurs about a vertical axis that passes through the intercondylar eminence of the tibia. Rotation occurs concomitantly with the terminal phase of extension. Although the knee cannot be actively rotated, the flexed knee may be passively rotated through 70°.

       (1) Lateral tibial rotation occurs during the terminal phase of extension, as a result of the differences in the radii of the lateral and medial femoral condyles as well as the limits on movement imposed by the cruciate and collateral ligaments. No muscles act directly to rotate the leg laterally.

       (2) Medial tibial rotation, unlocking of the knee, requires contraction of the **popliteus muscle** (see Figure 15-1B). Once the knee is unlocked, the flexors can initiate flexion.

C. **The femoral triangle** is located in the anterior proximal thigh.

    1. Boundaries. The femoral triangle is bounded by the inguinal ligament, the sartorius muscle, and the adductor longus muscle (see Figure 15-4A).

    2. Compartments. The area between the inguinal ligament and the superior pubic ramus is divided into two compartments.

      a. The **muscular compartment** is lateral. It contains the iliopsoas muscle and the femoral nerve.

        (1) The **iliopsoas muscle** flexes the thigh at the hip joint.

        (2) The **femoral nerve,** which is derived from the anterior division of the lumbar plexus, innervates the muscles of the anterior thigh.

      b. The **vascular compartment** is medial. It contains the femoral artery, femoral vein, and femoral ring, which contains lymphatics.

(1) Boundaries
   (a) Laterally, this compartment is bounded by the investing fascia of the muscular compartment.
   (b) Anterosuperiorly, it is bounded by the inguinal ligament.
   (c) Medially, it is bounded by the **lacunar ligament,** a reflection of the inguinal ligament.
   (d) Posteroinferiorly, the lacunar ligament becomes the **pectineal ligament** along the inferior pubic ramus.

(2) Contents
   (a) The **femoral artery** supplies the greater portion of the thigh and leg.
   (b) The **femoral vein** becomes the external iliac vein as it passes beneath the inguinal ligament.
   (c) The **femoral ring** contains lymphatics that drain the inguinal lymph nodes.

3. **Femoral ring.** The femoral artery, femoral vein, and **femoral canal** (the extension of the femoral ring) are wrapped in the **femoral sheath,** a continuation of the transversalis fascia leaving the abdominal cavity along with the femoral vessels.
   a. **Boundaries of the femoral ring**
      (1) Laterally, it is bounded by the **femoral vein.**
      (2) Anteriorly, it is bounded by the **inguinal** (Poupart's) **ligament,** which is the inferior border of the external oblique aponeurosis between the anterior–superior iliac spine and the pubic tubercle.
      (3) Medially, the boundary is the **lacunar** (Gimbernat's) **ligament,** which is a reflection of the inguinal ligament posteriorly to the pectinate line of the superior pubic ramus.
      (4) Posteriorly, the boundary is the **pectineal** (Cooper's) **ligament,** which is an extension of the lacunar ligament along the pectinate line of the superior pubic ramus.
   b. **Contents** consist of lymphatics from the legs, perineum, and lower abdominal wall. Within the peritoneal cavity, they drain to the lumbar (aortic) nodes.

4. **Femoral hernias** occur through the femoral ring.
   a. **Prevalence.** Fully 34% of all hernias in females are femoral, but this type of hernia is rare in males.
   b. **Course.** Femoral hernias enter the femoral ring and femoral canal inferior to the inguinal ligament. They usually lie in the thigh, presenting through the fossa ovalis of the great saphenous vein.
      (1) Because of the ligamentous boundaries, the potential for strangulation and necrosis of a herniated loop of bowel is high.
      (2) These hernias may run medially or laterally to, or may even bifurcate to run on both sides of, an **aberrant obturator artery,** which arises from the inferior epigastric artery. This situation can prove dangerous for the unsuspecting surgeon undertaking repair of a femoral hernia.
   c. **Characteristics.** Femoral (in contrast to inguinal) hernias lie inferior to the inguinal ligament and lateral to the pubic tubercle to which the inguinal ligament attaches.

IV. VASCULATURE OF THE THIGH

A. **Arterial supply** is primarily by the femoral and obturator arteries, with some contribution from the gluteal vessels.

1. **The obturator artery** supplies a small portion of the adductor compartment (see Figure 14-5).
   a. **Course**

FIGURE 15-6. Vasculature of the thigh, anterior perspective.

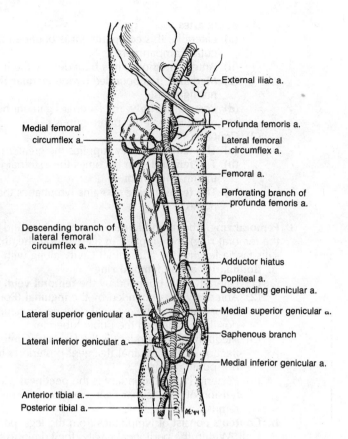

- (1) This artery usually arises from the **internal iliac artery** deep within the pelvis, but it may arise aberrantly from the inferior epigastric artery.
- (2) It exits the pelvic cavity with the obturator nerve through the **obturator foramen** and bifurcates.

b. **Distribution**
- (1) The **anterior branch** supplies the proximal portions of the muscles of the adductor group and anastomoses with the medial circumflex branch of the profunda femoris artery.
- (2) The **posterior branch** mainly supplies the proximal portions of the hamstring group and gives off the **acetabular branch** from which the **artery of the ligamentum teres** arises.

2. **The femoral artery** provides the principal supply of blood to the thigh (Figure 15-6).
   a. **Course.** This vessel is the continuation of the external iliac artery distal to the inguinal ligament.
   - (1) It enters the **femoral triangle** by passing between the inguinal ligament and iliopubic ramus.
     - (a) The triangle is bounded by the inguinal ligament, the sartorius muscle, and the adductor longus muscle.
     - (b) The **femoral pulse** is palpable high within the femoral triangle, at which point the femoral artery is a convenient source of arterial blood for blood-gas determinations and a preferred site for catheter insertion in angiography.
   - (2) At the inferior angle of the triangle, the femoral artery enters the **adductor (Hunter's) canal,** along with the femoral vein, the saphenous nerve, and the nerve to the vastus medialis.
     - (a) This canal passes between the adductor (anterior) and hamstring (poste-

rior) portions of the adductor magnus muscle and enters the popliteal fossa through the adductor hiatus.
- **(b)** The canal is a frequent site for femoral artery stenosis and popliteal artery (Hunter's) aneurysm.
- **(3)** Emerging from the adductor canal into the **popliteal fossa,** the femoral artery becomes the **popliteal artery.**

b. **Distribution.** The femoral artery gives off several branches in the femoral triangle (see Figure 15-6).
- **(1)** The **superficial epigastric artery** supplies the hypogastric region of the abdomen.
- **(2)** The **external pudendal artery** supplies the anterior–superior pudendal region.
- **(3)** The **deep femoral (profunda femoris) artery,** the largest branch, initially lies lateral to the femoral artery before passing posterior to it to lie on the psoas muscle, before passing into the posterior compartment. It gives off several branches.
  - **(a)** The **lateral femoral circumflex artery** arises from the lateral side of the deep femoral artery and gives off three branches.
    - **(i)** The **ascending branch** passes superiorly and anastomoses with the inferior gluteal artery.
    - **(ii)** The **transverse branch** passes around the femur to anastomose with the medial femoral circumflex artery at the **cruciate anastomosis.**
    - **(iii)** The descending branch anastomoses with both the descending genicular branch of the femoral artery and the superior lateral genicular branch of the popliteal artery.
  - **(b)** The **medial femoral circumflex artery** arises from the posterior aspect of the deep femoral artery and gives off three branches.
    - **(i)** The **ascending branch** passes into the gluteal musculature.
    - **(ii)** The **transverse branch** passes posterior to the femur to anastomose with the transverse branch of the lateral femoral circumflex artery at the **cruciate anastomosis.**
    - **(iii)** The **descending branch,** which is small, passes inferiorly.
  - **(c)** Three or four **perforating arteries** supply the major portion of the hamstring muscles.

3. **The popliteal artery** continues the femoral supply to the leg.
   a. **Course.** It emerges from the adductor hiatus, descends through the **popliteal fossa** (see Figure 15-6), and continues into the leg. It becomes the **posterior tibial artery** near the point at which the anterior tibial artery arises.
   - **(1)** The **popliteal fossa** is defined by several muscles.
     - **(a)** It is bounded superolaterally by the biceps femoris, supermedially by the semitendinosus, and inferiorly by the lateral and medial heads of the gastrocnemius.
     - **(b)** The popliteus muscle forms the floor.
   - **(2) The popliteal pulse** is best palpated by compressing the artery against the popliteus muscle with the leg flexed, because the artery lies deep within the popliteal fossa.
   - **(3) Aneurysm of the popliteal artery** (Hunter's aneurysm) may occur, often in association with stenosis of the femoral artery in the adductor canal.
   b. **Distribution.** The popliteal artery gives off several branches, which participate in the rich **geniculate anastomosis** that permits compressive occlusion of the popliteal artery for minutes at a time without consequence (see Figure 15-6).

B. Venous return of the thigh

1. **The great saphenous vein** is the principal superficial vein.
   a. **Course.** It lies on the superficial fascia in the medial aspect of the leg and thigh, passing on the flexor side of the ankle, knee, and hip joints. It contains about eight valves in the thigh. In the femoral triangle, it penetrates the fascia lata through the **saphenous hiatus,** an opening in the fascia lata that is incompletely covered by the **cribriform fascia.**

b. **Branches**
   (1) **Perforating veins** communicate between the saphenous vein and the deep veins of the leg and thigh.
   (2) The **superficial epigastric, superficial circumflex iliac,** and **external pudendal veins** join the great saphenous vein in the femoral triangle.
2. **The femoral vein** is the principal deep vein.
   a. As the popliteal vein passes through the adductor canal, it becomes the femoral vein. It accompanies the femoral artery, frequently as a subdivided **venae comitantes.**
   b. In the femoral triangle, it is a single vein that lies medial to the femoral artery, at which point it is a source of venous blood as well as a site for **venipuncture** and insertion of venous catheters.
   c. The femoral vein becomes the external iliac vein as it passes beneath the inguinal ligament.
3. **Venous hemodynamics and varicose veins**
   a. A **venous pump** results from the course of the deep veins between the deep muscles of the leg and thigh. This pump is essential for the return of blood to the heart against gravity.
   b. **Valves** in the saphenous vein direct the flow upward and protect that vein from the hydrostatic pressure produced by the standing column of blood. Valves in the **perforating veins** direct the flow inward to protect the saphenous vein from the pressure that results from the pumping action of the leg and thigh muscles on the deep veins. An important valve is that found where the saphenous vein joins the femoral vein.
      (1) Venous stasis associated with long periods of standing predispose the superficial vein to dilation, which reduces the competency of the valves.
      (2) Incompetent valves in the perforating veins allow blood to escape at relatively high venous pressure into the saphenous system, exacerbating the dilation and forming varicose veins.
      (3) Treatment for varicose veins varies from identification and ligation of the perforating veins with incompetent valves to removal of the entire saphenous vein.

C. **The major lymphatic pathway** from the lower extremity to the deep inguinal lymph nodes lies medial to the femoral vein.

V. INNERVATION OF THE THIGH

A. **Anterior compartment**
1. **The femoral nerve** innervates the muscles of this compartment as well as dermatomes of the anterior aspect of the thigh (Figure 15-7).
   a. **Course and composition.** The femoral nerve arises from the posterior divisions of lumbar plexus roots L2–L4. It leaves the greater pelvis through the muscular compartment inferior to the inguinal ligament.
   b. **Distribution**
      (1) **Motor.** Branches of the femoral nerve that innervate muscles in this compartment arise in the femoral triangle. The **knee-jerk (patellar) reflex** tests both afferent and efferent divisions of the L4 femoral nerve component.
      (2) **Sensory.** The **anterior femoral cutaneous nerve** innervates the dermatomes of the anterior aspect of the thigh (see Figure 15-7, inset). The cutaneous **saphenous nerve** does not pass completely through the adductor canal, leaving superficially to pass into the medial aspect of the leg.
   c. **Injury,** usually the result of trauma to the femoral triangle, produces weakness of hip flexion with an inability to extend the knee, as indicated by **loss of the patellar reflex** (L4). Also noted is anesthesia over the anterior thigh and medial leg.

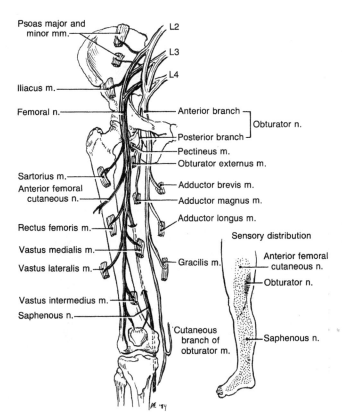

FIGURE 15-7. Innervation of the anterior and medial compartments of the thigh by the femoral nerve and the obturator nerve, respectively.

2. **The genitofemoral nerve** innervates dermatomes of the anterior aspect of the thigh.
    a. **Composition.** The **femoral branch of the genitofemoral nerve** arises from L1–L2 (anterior) and enters the thigh through the superficial inguinal ring.
    b. **Distribution.** It mediates sensation from the medial aspect of the proximal thigh and provides the afferent limb of the **cremaster reflex,** which is used to test the integrity of levels L1–L2 of the spinal cord.

B. Posterior compartment

1. **The obturator nerve** is associated with the medial thigh and medial compartment muscles.
    a. **Course and composition** (see Figure 15-7).
        (1) The obturator nerve arises from the anterior division of roots L2–L4 of the lumbar plexus.
        (2) It exits the pelvis via the obturator foramen in company with the obturator artery and bifurcates.
    b. **Distribution**
        (1) The **anterior (superficial) branch** is both motor and sensory.
            (a) **Motor.** This branch innervates the **adductor longus** and **adductor brevis** as well as the **gracilis muscles** (see Figure 15-7).
            (b) **Sensory.** It mediates sensation from a small area of the distal medial aspect of the thigh.
        (2) The **posterior (deep) branch** innervates the obturator externus and the anterior portion of the adductor magnus muscles as well as the medial portion of the pectineus muscle. It sends sensory twigs to the hip joint.
    c. **Injury** is uncommon. A penetrating groin wound may injure the obturator nerve with inability to adduct the thigh. Such trauma would also result in loss of sensation over a small area of the medial thigh.

FIGURE 15-8. Innervation of the posterior compartment of the thigh. Note the distribution of the common fibular (peroneal) and tibial divisions of the sciatic nerve.

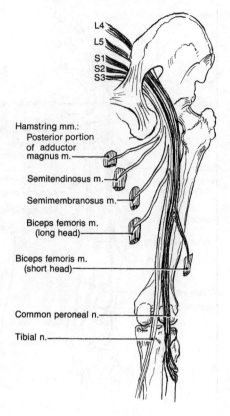

2. **The sciatic nerve** is associated with the posterior thigh, the leg, and the foot.
   a. **Course and composition.** This nerve arises from both anterior and posterior divisions of roots L4–S3 of both lumbar and sacral plexuses.
      (1) **Course** (Figure 15-8)
         (a) From the pelvis, it exits via the **infrapiriform recess of the greater sciatic foramen** along with the inferior gluteal neurovascular bundle.
         (b) In the gluteal region, it arches through the inferomedial quadrant of the buttock, starting halfway between the ischial tuberosity and the iliac spine. It passes over the neck of the femur halfway between the ischial tuberosity and the greater trochanter.
         (c) In the thigh, it passes through the posterior compartment and joins the popliteal artery in the popliteal fossa.
      (2) The **sciatic nerve** is formed by the fusion of the **tibial nerve** and the **common fibular nerve.** The extent of the fusion is variable.
   b. **Distribution**
      (1) The **tibial nerve** (L4–S3, anterior) innervates the semimembranosus and semitendinosus muscles and the long head of the biceps femoris muscle, as well as the posterior belly of the adductor magnus muscle (see Figure 15-8). It has no sensory component in the thigh.
      (2) The **common fibular nerve** (L4–S2, posterior) innervates only one muscle in the thigh, the short head of the biceps femoris muscle (see Figure 15-8). It has no sensory component in the thigh.
   c. **Injury**
      (1) **Sciatica**
         (a) The most frequent sciatic nerve injury results from compression of spinal roots that constitute the sciatic nerve by a herniated intervertebral disk.
         (b) The sciatic nerve may be injured by an intramuscular injection into the inferomedial quadrant of the buttock.

- (c) Occasionally, this nerve exits through the piriformis muscle, and spasm of that muscle may compress the nerve.
- (d) Injury may result from traction of the nerve as it passes posterior to the neck of the femur.
    - (2) In the thigh, tibial or sciatic nerve injury manifests as weakness of both thigh extension and knee flexion. The **hamstring reflex** tests the integrity of spinal nerve L5.
3. **The lateral femoral cutaneous nerve** (L2–L3, posterior) is sensory to the lateral thigh. It may become entrapped beneath the inguinal ligament, especially in an obese person, producing hyperesthesia or pain.
4. **The posterior femoral cutaneous nerve** (S1–S3, anterior and posterior) is sensory to most of the posterior thigh.

Chapter 16
Anterior Leg and Dorsal Foot

I. BONES OF THE DISTAL LEG AND TARSUS

A. **The leg** is that part of the lower extremity that lies between the knee and ankle and comprises the tibia and fibula.

1. **The tibia** has a palpable medial condyle, lateral condyle, anterior lip of the tibial plateau, tibial tubercle, medial surface of the shaft, and medial malleolus. The tibia corresponds to the radius of the forearm. (The upper and lower extremities have rotated in opposite directions, phylogenetically, in attainment of the bipedal stance.)
    a. **Characteristics**
        (1) Proximally, the **tibial tubercle (tuberosity)** receives the patellar tendon (see Figure 15-1).
        (2) The medial aspect or **shin** is subcutaneous.
        (3) Distally, the shaft narrows before expanding into the subcutaneous **medial malleolus,** which is notched posteriorly by the tendon of the tibialis posterior muscle (Figure 16-1).
    b. **Articular surfaces.** The distal end and medial malleolus articulate with the talus at the **talocrural (ankle) joint** (see Figure 16-1).
        (1) The articular surface for the talus is trapezoid (wider anteriorly) and concave. It continues medially as the articular surface of the medial malleolus.
        (2) Laterally, the **fibular notch** accommodates the fibula as the **distal tibiofibular joint.**
2. **The fibula,** with its palpable head, slender shaft, and lateral malleolus, corresponds to the ulna of the forearm. In the developmental process, it became fixed in the pronated position. It lost contact with the femur and, thus, takes no part in the knee joint.
    a. **Characteristics**
        (1) Proximally, a slightly bulbous head articulates with the lateral tibial condyle at the proximal tibiofibular joint (see Figure 15-1). The **styloid process** of the head receives the lateral collateral ligament.
        (2) The shaft is narrow.
        (3) Distally, the shaft expands slightly as the subcutaneous **lateral malleolus,** which extends further distally than the medial malleolus (see Figure 16-1). The posterior surface of the lateral malleolus is grooved by the tendons of the peroneus longus and brevis muscles.
    b. **Distal articular surfaces**
        (1) The proximal end articulates with the tibia at the **proximal tibiofibular joint.**
        (2) The distal end articulates with the tibia at the **distal tibiofibular joint** (see Figure 3-2B).
        (3) On the medial surface is a triangular facet for the talus at the **talocrural (ankle) joint.**
3. **The interosseous membrane** is a broad fibrous ligament connecting the adjacent borders of the tibia and fibula along their entire length. The most distal region is thickened as the **interosseous ligament.** Both of these structures stabilize the tibiofibular joints.

B. **The foot** is composed of **7 tarsal bones, 5 metatarsal bones,** and **14 phalanges,** all of which are palpable.

1. **The proximal tarsal group** consists of three bones, not in a row.
    a. The **talus** (astragalus) has a rounded head, a neck, and a cuboid body. Numerous ligaments, but no muscles, attach to the talus (see Figures 16-1 and 16-2).
        (1) The **head** articulates with the navicular bone anteriorly and with the calcaneus inferiorly.

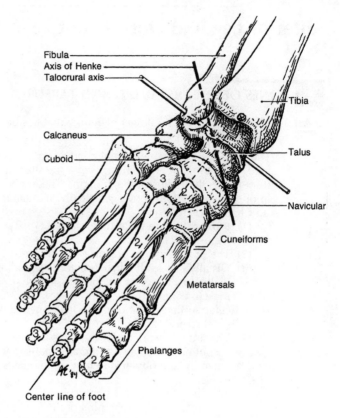

FIGURE 16-1. Bones of the distal leg, ankle, and foot. The talocrural axis and the axis of the subtarsal joint (of Henke) are shown.

  (2) The **neck** has a deep inferior groove, the **sulcus tali,** in which the interosseous ligament of the talocalcaneal joint attaches.
  (3) The **body** has a number of complex articular surfaces.
   (a) Superiorly, the **trochlea tali,** which is wider anteriorly, articulates with the tibia.
   (b) Laterally and medially, the body of the talus articulates with the tibial and fibular malleoli.
   (c) Inferiorly, a flat surface articulates with the calcaneus.
 b. The **calcaneus,** which forms the heel, is the largest tarsal bone (see Figure 16-2).
  (1) Posteriorly, the calcaneus receives the **calcanean (Achilles') tendon.** Posteroinferiorly, intrinsic plantar flexor muscles attach on both sides.
  (2) Superiorly, it has three articular surfaces for the talus, with the sulcus tali separating the middle and posterior surfaces.
  (3) Anteriorly, it articulates with the cuboid bone and has a tubercle for the attachment of the **long plantar ligament.**
  (4) Medially, the **sustentaculum tali,** a prominent shelf on the anteromedial edge, provides attachment for the **calcaneonavicular (spring) ligament.**
   (a) The **sustentaculum tali** and the calcaneonavicular (spring) ligament support the head of the talus.
   (b) The **calcaneonavicular (spring) ligament,** between the sustentaculum tali and the navicular bone, helps to maintain the longitudinal plantar arch. Stretching of this ligament results in **fallen arches** or **"flat feet."**
   (c) Inferiorly, the sustentaculum tali is grooved by the tendon of the flexor hallucis longus muscle.
 c. The **navicular (scaphoid) bone** articulates with all of the tarsal bones, except the calcaneus, to which it is strongly connected by the calcaneonavicular (spring) ligament (see Figures 16-1 and 16-2).
  (1) It curves posteriorly to receive the head of the talus at the apex of the arch of the foot.

(2) It has a prominent tuberosity, which is the principal attachment of the tibialis posterior muscle.
2. **The distal tarsal group** consists of a row of four bones.
   a. The **medial, intermediate, and lateral cuneiform bones** articulate with the navicular bone proximally (see Figures 16-1 and 16-2).
   b. The **cuboid bone** is lateral to the navicular bone and the cuneiform bones. It articulates posteriorly with the calcaneus at the **calcaneocuboid joint** (see Figures 16-1 and 16-2).
3. **Tarsal fractures**
   a. **Comminuted talar and calcaneal fractures** are seen in persons who fall or jump from a height.
   b. A painful **calcaneal spur** may develop where the intrinsic plantar muscles arise from the calcaneus.

II. ANKLE AND TARSAL JOINTS

A. Distal tibiofibular joint

1. **Structure.** The **fibular notch** of the tibia accommodates the medial aspect of the distal fibula to form the distal tibiofibular joint (see Figure 16-1).
2. **Movement.** The distal tibiofibular joint allows little movement. Because the superior articular surface of the talus is trapezoid, however, this joint can spread slightly to accommodate the wider portion of the talus during dorsiflexion.
3. **Ligamentous support** is provided by the interosseous membrane. The **interosseous ligament,** a distal thickening of this membrane, stabilizes the joint. The **anterior tibiofibular** and **posterior talofibular ligaments** strengthen this joint.
4. **Sprains.** Because this joint is inherently stable, tearing of the tibiofibular ligaments usually occurs as a result of violent adduction and internal rotation in association with fracture of one or both malleoli.

B. Talocrural (ankle) joint

1. **Structure.** The socket for this joint is formed by the tibia and its medial malleolus, the lateral malleolus of the fibula, and the inferior transverse ligament (see Figures 16-1 and 16-2). These form a mortise for the tenon of the trochlea tali.
   a. The **weight of the body** is transmitted from the tibia to the talus, which distributes the weight anteriorly and posteriorly within the foot.
   b. The **center of gravity** in the erect posture passes somewhat anterior to the ankle joint.
      (1) The wedge-shaped trochlea tali (wide edge forward) is forced between the malleoli on dorsiflexion, increasing the inherent stability of the joint. Conversely, this joint is loose on plantar flexion.
      (2) Strong tibiofibular ligaments prevent excessive separation of the malleoli.
      (3) Fan-shaped collateral ligaments keep the trochlea tali in the joint socket.
2. **Movement.** The ankle joint has one degree of freedom about a transverse axis that passes through the lateral malleolus and the trochlea tali but passes inferior to the medial malleolus (see Figure 16-1).
   a. The **axis of rotation** angles slightly anteromedially so that the long axis of the foot deviates by about 10° laterally in full dorsiflexion.
   b. **Flexion/extension** (25°/35°) occurs about this transverse axis (see Figures 16-1 and 16-4).
3. **Ligamentous support.** The capsule of the joint is attached to the edges of the articular cartilage and is supported by strong collateral ligaments.

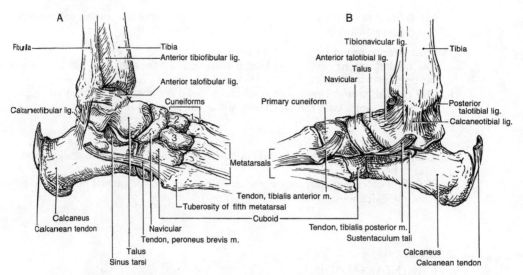

FIGURE 16-2. Ligaments of the ankle joint. *(A)* Lateral aspect; *(B)* medial aspect.

- a. **External (lateral) collateral ligaments** (Figure 16-2A)
  - (1) The **anterior talofibular ligament** passes from the tip of the lateral malleolus to the talus anteriorly. It tends to limit plantar flexion.
  - (2) The **calcaneofibular ligament** passes from the lateral malleolus to the calcaneus with the **talocalcaneal ligament** running at its base. It resists adduction.
  - (3) The **posterior talofibular ligament** passes from the tip of the lateral malleolus to the talus posteriorly. The **posterior talocalcaneal ligament** extends this band to the calcaneus. Both tend to limit dorsiflexion.
- b. **Medial collateral (deltoid) ligament** (see Figure 16-2B)
  - (1) The **superficial portion** is composed of the **tibionavicular ligament,** which runs anteriorly from the medial malleolus to the navicular bone, and the **calcaneotibial ligament,** which runs from the tip of the medial malleolus to the edge of the calcaneus. Both prevent abduction.
  - (2) The **deep portion** comprises the **anterior talotibial ligament** and **posterior talotibial ligament,** which run anteriorly and posteriorly between the medial malleolus and the talus. They tend to limit plantar flexion and dorsiflexion, respectively.

4. **Clinical considerations**
   - a. **Sprains** of the talocrural joint usually result from excessive movement at the subtalar joint.
     - (1) **Inversion sprains,** the most common ankle injury, involve the tearing of the lateral collateral ligaments. Usually the **anterior talofibular ligament** is torn first, then the **calcaneofibular ligament,** and, in severe cases, the posterior **talofibular ligament.**
     - (2) **Eversion sprains,** which are less common, usually involve tearing of the deltoid ligament.
   - b. **Midshaft and "boot-top" fractures** of the tibia and fibula are common. Usually the bones override somewhat because of muscular pull.
   - c. **Distal fractures of the tibia and fibula** are the result of severe external rotation. This type of fracture is common in skiers with high, plastic boots, because the bones rotate within the boot.

C. **The subtalar joint** is formed where the talus rests on the calcaneus and navicular.

1. **Structure.** Two plantar articular surfaces form the talocalcaneonavicular joint and a third lies between the calcaneus and talus (see Figure 16-2).
   - a. **Talocalcaneonavicular joint.** This ball-and-socket articulation is formed by the

ovoid head of the talus with a socket formed by the navicular bone, the calcaneonavicular (spring) ligament, and the anterior articular surface of the calcaneus.
- (1) **Anterior talocalcaneal joint.** On the medial side, the head of the talus articulates with the sustentaculum tali of the calcaneus and the spring ligament that spans between the sustentaculum tali and the navicular bone.
- (2) **Talonavicular joint.** On the lateral side, the head of the talus articulates with the navicular bone (anteriorly).
- (3) Articulation with the spring ligament completes the **talocalcaneonavicular joint.**
b. **Posterior talocalcaneal joint.** Posteriorly, a flat surface on the body of the talus articulates with the calcaneus.

2. **Movement**
   a. **Axis of rotation.** Because the anterior talocalcaneal articulation of the subtalar joint is also part of the transverse tarsal joint, the movements are conjoined. Thus, movement (inversion/eversion) occurs about a resultant axis—the axis of Henke (see Figures 16-1 and 16-4).
   b. **Inversion/eversion.** Movement of this joint can be described as abduction/adduction. Because the subtalar and transverse tarsal joints are conjoined, however, this movement does not occur in pure form. Abduction/adduction at the subtalar joint, in conjunction with pronation/supination at the transverse tarsal joint, results in inversion/eversion (see Figures 16-1 and 16-4).

3. **Ligamentous support.** The calcaneofibular and deltoid ligaments traverse the subtalar joint (see Figure 16-2), and three deep ligaments connect the talus and calcaneus (not shown).
   a. The **lateral talocalcaneal ligament** lies deep to the calcaneofibular ligament.
   b. The **medial talocalcaneal ligament** lies deep to the calcaneotibial ligament.
   c. The **interosseous talocalcaneal ligament** runs from deep within the sinus tarsi of the calcaneus to the sulcus calcanei of the talus. It provides an axis of rotation about which movement in the subtalar joint occurs.

4. **Sprains.** Excessive movements in this joint produce ligamentous sprains. Severe trauma causes more widespread injury and may produce fractures of the malleoli and ligamentous tears involving the talocrural joint.
   a. **Abduction** stresses the ligaments across the medial aspect of the joint.
   b. **Adduction** stresses the ligaments across the lateral aspect of the joint.

D. The **transverse (mid) tarsal joint** consists of conjoined portions of the subtalar and calcaneocuboid joints.

1. The **talonavicular joint** and the **anterior calcaneonavicular joint** (a portion of the subtalar joint) also participate in the transverse tarsal joint.
   a. **Structure.** The **talocalcaneonavicular joint** is a ball-and-socket articulation (see section II C 1 a).
   b. **Movement**
      (1) **Axis of rotation** (see section II C 2 a)
      (2) **Inversion/eversion.** Movement of this joint can be described as supination and pronation. Because the subtalar and transverse tarsal joints are conjoined, however, this movement does not occur in pure form, and instead, in conjunction with abduction/adduction in the subtalar joint, results in inversion/eversion.
   c. **Ligamentous support**
      (1) The **talonavicular** and **dorsal calcaneonavicular ligaments** support the joint dorsally.
      (2) The **plantar calcaneonavicular (spring) ligament** supports the head of the talus and maintains the longitudinal plantar arch. Laxity of this ligament results in fallen arches or "flat feet."
      (3) Numerous other small ligaments provide additional support.

2. **The calcaneocuboid joint** is part of the transverse tarsal joint.
   a. **Structure.** At this gliding joint, movement is an accommodation to the movements that occur conjointly at the subtalar and talocalcaneonavicular joints.
   b. **Ligamentous support** is provided by the **long plantar ligament.**

E. **Conjoined movements of the foot**
   1. **Subtalar and transverse tarsal joints** (posterior tarsal joints). Movements are mechanically linked by common joints so they function together with but 1 degree of freedom (see Figure 16-4).
      a. **Inversion of the foot** is produced by the conjoined movements of adduction in the subtalar joint and supination in the transverse tarsal joint (and to some extent dorsiflexion).
      b. **Eversion of the foot** is produced by the conjoined movements of abduction in the subtalar joint and pronation in the transverse tarsal joint (and to some extent plantar flexion).
      c. The **resultant axis of rotation** (of Henke), about which inversion/eversion occurs, passes inferomedially through the lateral side of the neck of the talus, through the sinus tarsi, and **along the interosseous ligament** to the sinus calcanei. It then emerges from the medial aspect of the posterior tubercles of the calcaneus.
   2. **Anterior tarsal joints.** Movements are conjoined because the articulations are mechanically linked (see Figure 16-4).
      a. The movements are primarily ones of accommodation to movements in the posterior tarsal joints.
      b. Because of the shape of the articular surfaces, these movements change the shape of the transverse plantar arch.
         (1) **Eversion and dorsiflexion** flatten the transverse plantar arch.
         (2) **Inversion and plantar flexion** heighten the transverse arch.

III. ANTERIOR AND LATERAL CRURAL COMPARTMENTS

A. **Fascia**
   1. **The superficial fascia** of the leg is continuous with that of the thigh. It supports the cutaneous nerves and superficial veins.
   2. **The deep (crural) fascia** is continuous with the **fasciae latae** of the thigh posteriorly. It also would be continuous anteriorly if both did not attach to the patella.
      a. **Anteromedially,** the crural fascia is attached to the patella and the medial surface of the tibia.
      b. **Laterally,** septa run to the anterior and posterior borders of the fibula.
         (1) The **anterior intermuscular septum** subdivides the extensor compartment into anterior and lateral compartments.
         (2) The **posterior intermuscular septum** and the fibula separate the lateral and posterior compartments.
         (3) The fibula, **interosseous membrane,** and tibia separate the anterior and posterior compartments.
      c. At the **ankle,** the crural fascia condenses to form retinacula for the tendons of the extensor, flexor, and peroneal muscles (see Figure 16-3).

B. **Musculature.** The muscles of the anterior and lateral compartments comprise the dorsiflexors (extensors) of the ankle and the extensors of the pedal digits.
   1. **Organization.** The primitively dorsal musculature of the leg, which lies anteriorly, is divided by the anterior intermuscular septum into two compartments. Because most of the muscles of the anterior and lateral compartments continue into the foot, the leg and foot may be considered a functional unit.

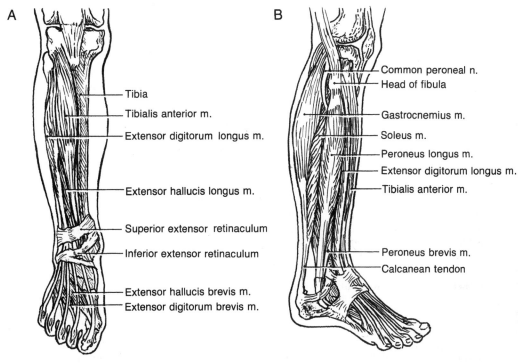

FIGURE 16-3. Musculature of the anterior *(A)* and lateral *(B)* crural compartments.

    a. The **anterior crural compartment** contains the dorsiflexors of the ankle as well as both invertors and evertor of the foot.
        (1) Composition. This group includes the **tibialis anterior, extensor hallucis longus, extensor digitorum longus,** and **peroneus tertius muscles** (Figure 16-3A; see Table 16-1).
        (2) Course. These muscles arise from the lateral tibial condyle as well as from the proximal two thirds of the tibial and fibular shafts and the interosseous membrane. They insert in the foot.
    b. The **lateral crural compartment** contains the peroneal muscles, which are plantar flexors of the ankle as well as the principal evertors of the foot.
        (1) Composition. This group includes the **peroneus longus** and **peroneus brevis muscles** (see Figure 16-3B and Table 16-1).
        (2) Course. The peroneal muscles arise from the anterior fibular shaft and pass posterior to the lateral malleolus. The peroneus longus inserts into the plantar surface of the medial cuneiform and the first metatarsal bones. The peroneus brevis inserts into the base of the fifth metatarsal bone ventrolaterally.
2. Group actions are determined by the location of the tendons relative to the axis of rotation of the ankle joint and that of the foot (Table 16-1).
    a. Flexion/extension occurs about a transverse axis through the ankle joint. Muscles that pass anterior to the transverse axis effect dorsiflexion; those that pass posterior to this axis effect plantar flexion (Figure 16-4).
        (1) Dorsiflexion (extension). All of the muscles of the anterior compartment pass anterior to the transverse axis and are dorsiflexors of the foot.
        (2) Plantar flexion. Both muscles of the lateral compartment pass behind the lateral malleolus. Because they are posterior to the transverse axis, they are plantar flexors of the foot.
    b. Inversion/eversion occurs about the complex axis of Henke. Muscles that pass medial to this axis are invertors; those that pass lateral to it are evertors (see Figure 16-4).

TABLE 16-1. Muscles of the Anterior Leg and Dorsal Foot

Muscle	Origin	Insertion	Primary Action	Innervation
Anterior crural compartment				
Tibialis anterior	Proximal half of anterior tibia	Medial cuneiform bone and base of first metatarsal on dorsal side	Dorsiflexes and inverts foot	Deep fibular n. (L4–L5, posterior)
Extensor hallucis longus	Front of fibula	Base of second phalanx of great toe	Extends terminal phalanx of great toe	Deep fibular n. (L4–S1, posterior)
Extensor digitorum longus	Lateral tibial condyle and interosseous membrane	Second and third phalanges of second to fifth toes	Extend metatarso-phalangeal joints	Deep fibular n. (L4–S1, posterior)
Peroneus tertius	Anterior fibula	Base of fifth metatarsal	Dorsiflexes and everts foot	Deep fibular n. (L4–S1, posterior)
Posterior crural compartment				
Peroneus longus	Proximal half of anterior tibia	Medial cuneiform bone and base of first metatarsal on plantar side	Plantar flexes and everts foot	Superficial fibular n. (L5–S1, posterior)
Peroneus brevis	Distal half of fibula	Base of fifth metatarsal	Plantar flexes and everts foot	Superficial fibular n. (L4–S1, posterior)
Dorsal foot				
Extensor digitorum brevis	Calcaneus	Extensor hoods (laterally) of second, third, and fourth toes	Dorsiflexes proximal phalanges at the metatarso-phalangeal joints	Deep fibular n. (L5–S1, posterior)
Extensor hallucis brevis	Calcaneus	Base of proximal phalanx of great toe	Dorsiflexes proximal phalanx of great toe	Deep fibular n. (L4–S1, posterior)

(1) **Inversion.** The medial muscles of the anterior compartment (tibialis anterior and extensor hallucis longus) are both invertors of the foot.

(2) **Eversion.** Both muscles of the lateral compartment are strong evertors of the foot. The lateral muscles of the anterior compartment (extensor digitorum longus and peroneus tertius) are also evertors.

3. **Group innervation**
   a. The **anterior compartment** is innervated by the deep (anterior tibial) branch of the common fibular (peroneal) nerve (see Table 16-1).
   b. The **lateral compartment** is innervated by the superficial branch of the common fibular (peroneal) nerve (see Table 16-1).

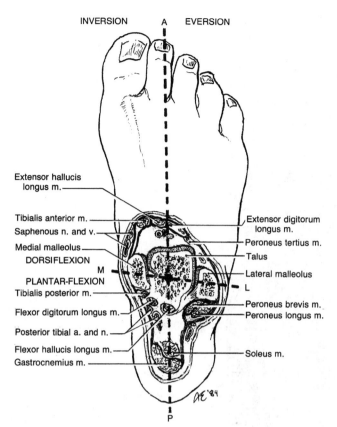

FIGURE 16-4. Action of the crural muscles at the ankle joint. Muscles passing anterior to the transverse (talocrural) axis of the ankle joint (M-L) dorsiflex the foot; those passing posterior to this axis plantar flex the foot. Muscles passing medial to the resultant axis (A-P) of the subtalar and midtarsal joints invert the foot; those passing lateral to that axis evert the foot.

IV. DORSUM OF THE FOOT

A. The **crural (deep) fascia** in this region is divided into two layers by the extrinsic extensor tendons. At the ankle, the crural fascia condenses to form retinacula for the tendons of the extensor, flexor, and peroneal muscles (see Figure 16-3).

1. **The superior extensor retinaculum** (transverse crural ligament) passes between the distal shafts of the tibia and fibula and contains the tibialis anterior and extensor hallucis longus muscles, the extensor digitorum longus tendon, and the peroneus tertius muscle (Figure 16-5).

2. **The inferior extensor retinaculum** is **Y-shaped** (see Figure 16-5). It diverges from a common lateral attachment on the calcaneus to both the medial malleolus and the deep fascia of the medial aspect of the foot. Unlike the superior extensor retinaculum, the inferior extensor retinaculum is subdivided into compartments for the individual tendons.

3. **The superior peroneal retinaculum** runs from the lateral malleolus to the calcaneus. The **inferior peroneal retinaculum** is a continuation of the inferior extensor retinaculum over the lateral surface of the calcaneus. It is attached to the peroneal trochlea between the peroneus longus and the peroneus brevis muscles.

B. Musculature

1. **Extrinsic muscle organization** (see Figures 16-3 and 16-5)
   a. The **tibialis anterior muscle** inserts into the medial aspect of the medial cuneiform bone and first metatarsal bone (see Figure 16-2B).
   b. The **extensor digitorum longus muscle** divides into four slips that insert into the

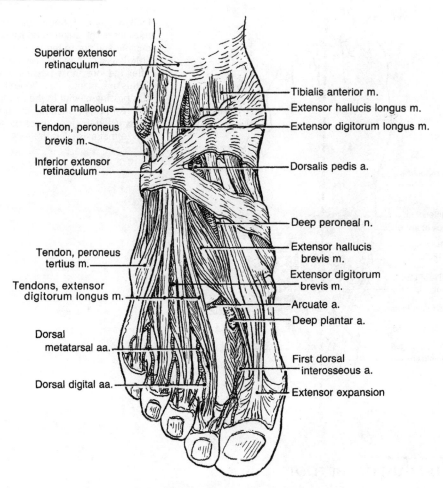

FIGURE 16-5. Muscles of the dorsum of the foot.

   extensor expansion (hood) of the second through fifth digits. The extensor expansion then attaches over the dorsal surfaces of the middle and distal phalanges (see Figure 16-5).
   c. The **extensor hallucis longus muscle** inserts into the distal phalanx of the great toe.

2. **Intrinsic muscle organization** (see Figure 16-5)
   a. The **extensor digitorum brevis muscle** arises from the calcaneus and divides into three slips that insert into the lateral margins of the extensor hoods of the second, third, and fourth digits.
   b. The **extensor hallucis brevis,** along with the extensor digitorum brevis, arises from the calcaneus and inserts into the base of the proximal phalanx of the great toe.

3. **Group actions.** The extensor digitorum brevis assists the action of the extensor digitorum longus in dorsiflexion and the metatarsophalangeal joints. The extensor hallucis brevis extends the proximal phalanx of the great toe, and the extensor hallucis longus extends the terminal phalanx of that toe. The tibialis anterior is a strong dorsiflexor of the foot (see Table 16-1).

4. **Group innervations.** The intrinsic extensor muscles of the foot are usually (78%) innervated by the deep fibular nerve, but they may be innervated by the superficial fibular nerve (22%).

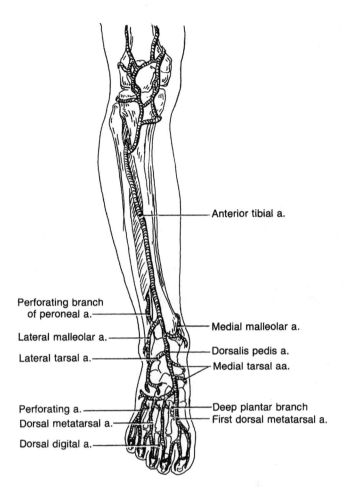

FIGURE 16-6. Vasculature of the anterior crural compartment and dorsum of the foot.

V. VASCULATURE

A. Arteries

1. **The popliteal artery** (see Chapter 15 IV A 3) gives rise to the **anterior tibial artery** in the popliteal fossa.

2. **The anterior tibial artery** supplies the anterior and lateral crural compartments (Figure 16-6).
   a. **Course.** Passing through a gap in the interosseous membrane, this artery enters the anterior crural compartment. It descends anterior to the interosseous membrane in company with the deep fibular (peroneal) branch of the common fibular (peroneal) nerve.
   b. **Distribution.** Below the knee, it gives off the anterior tibial recurrent artery. At the ankle, it gives off medial and lateral **malleolar branches,** which anastomose with branches of the posterior tibial and fibular (peroneal) arteries.

3. **The dorsal pedal (dorsalis pedis) artery** is the continuation of the anterior tibial artery onto the dorsum of the foot (see Figure 16-6).
   a. **Course.** This artery lies between the tendons of the extensor hallucis longus and extensor digitorum longus muscles, where the **dorsal pedal pulse** is palpable. It runs toward the great toe and terminates as the deep plantar artery.
   b. **Distribution**
      (1) **Lateral and medial tarsal arteries** supply the tarsus.

(2) The **arcuate artery** courses across the metatarsals, giving off dorsal metatarsal arteries, each of which bifurcate into **dorsal digital arteries**.
(3) The **first dorsal metatarsal artery** supplies adjacent sides of the first and second toes.
(4) The **deep plantar artery** dives between the heads of the first interosseous muscle to the plantar aspect of the foot and joins with the lateral plantar branch of the posterior tibial artery to form the **plantar arterial arch**.
   c. **Variation.** Occasionally (3.5%), the fibular artery is especially large and gives rise to the dorsal pedal artery via a perforating branch.

B. Veins

1. **Superficial veins**
   a. The **great saphenous vein** arises by the coalescence of a venous network into the medial marginal vein of the dorsal foot.
      (1) It passes anterior to the medial malleolus in company with the saphenous nerve, where it is accessible for venipuncture or insertion of intravenous lines.
      (2) It ascends along the medial aspect of the leg and passes posterior to the medial condyles of the knee. It contains about 12 valves in the leg.
   b. The **small saphenous vein** begins as the lateral marginal vein along the lateral side of the dorsum of the foot.
      (1) It passes posterior to the lateral malleolus and ascends along the lateral aspect of the leg, freely anastomosing with the great saphenous vein.
      (2) It penetrates the crural fascia to enter the popliteal fossa, where it joins the popliteal vein. It contains about 12 valves.
   c. The superficial veins are prone to varicosities (see Chapter 15 IV B 3).
2. **Deep veins** parallel the arteries (see Chapter 18 IV B 2).

VI. INNERVATION

A. Course and Composition.
The **common fibular (common peroneal) nerve** arises from the posterior portions of the lumbar and sacral plexuses (L4–S3) and forms the posterior division of the **sciatic nerve**.

1. **Above or within the popliteal fossa,** the sciatic nerve bifurcates into common fibular and tibial nerves (see Figure 15-8). The common fibular nerve diverges laterally in the popliteal fossa and gives rise to one or more articular branches.
2. **In the proximal leg,** the common fibular nerve passes posterior to the head of the fibula (the origin of its name: fibula, L. pin; perone, G. pin) and laterally across the neck of that bone. The **common fibular nerve gives off** the **sural communicating nerve** before dividing into the **superficial fibular (superficial peroneal)** and the **deep fibular (deep peroneal, anterior tibial) nerves** (Figure 16-7).

B. Distribution

1. **Motor distribution** (see Figure 16-7)
   a. The **superficial fibular peroneal nerve** innervates the muscles of the lateral compartment, that is, the peroneus longus and brevis.
   b. The **deep fibular peroneal, anterior tibial nerve** innervates the muscles of the anterior compartment (extensor hallucis longus, extensor digitorum longus, tibialis anterior, and peroneus tertius) and usually the intrinsic muscles of the dorsum of the foot (extensor hallucis brevis and extensor digitorum brevis).

2. **Sensory distribution** (see Figure 16-7, inset)
   a. The **sural communicating nerve** from the common fibular nerve contributes the lateral half of the **sural nerve**.

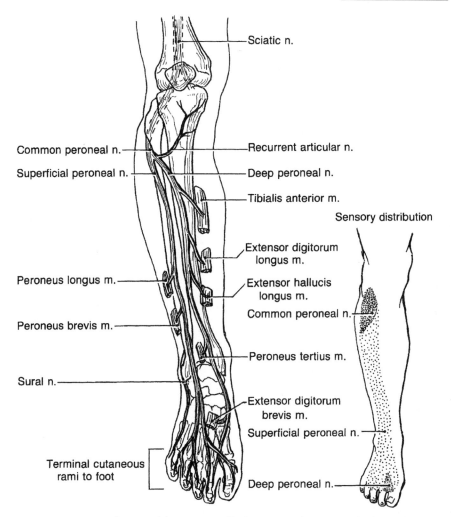

FIGURE 16-7. Distribution of the common fibular (peroneal) nerve in the leg and foot.

- **b.** The **common fibular (peroneal) nerve** gives off branches that innervate the region about the head of the fibula.
- **c.** The **superficial fibular (peroneal) nerve** supplies the anterolateral leg and sends a terminal cutaneous branch to the dorsum of the foot.
- **d.** The **deep fibular (peroneal, anterior tibial) nerve** sends a terminal cutaneous branch to an area between the first and second toes, which is specific for L5.

C. **Injury to the common fibular nerve** results in **foot drop**. Fibular nerve palsy produces a characteristic gait with "foot-slap," which results from inability to dorsiflex the foot.

1. **Proximal injury** may result from herniation of an intervertebral disk or intramuscular injection into the inferomedial buttock quadrant. Clinical evidence includes an inability to dorsiflex the foot or to stand back on the heels, as well as loss of sensation along the lateral aspect of the leg and dorsum of the foot.
2. **Distal injury** may result from pressure where the superficial or deep branch of this nerve passes across the fibula. This pressure can be as frank as blunt trauma or fracture of the proximal fibula. It may be as subtle as the constant pressure exerted by the edge of a plaster cast, which terminates just below the knee, or by sitting with the knees crossed for a prolonged period.

Chapter 17
Posterior Leg

I. MUSCULATURE

A. **The fascia** of the posterior compartment consists of two layers.

1. **The superficial fascia of the leg** is continuous with that of the thigh. It supports the cutaneous nerves and superficial veins.
2. **The deep (crural) fascia** is continuous with the fascia latae of the thigh posteriorly. It envelops the muscles of the leg.
   a. The **posterior intermuscular septum** (a septum of the crural fascia), tibia, interosseous membrane, and fibula form a barrier that separates the posterior compartment from the anterior and lateral compartments.
   b. The **transverse crural septum** (another septum of the crural fascia) subdivides the posterior compartment into superficial and deep compartments.
   c. The **flexor retinaculum (tarsal tunnel)** is a condensation of the crural fascia on the medial side of the ankle, extending between the medial malleolus and the calcaneus.
      (1) Posterior to the medial malleolus, it subdivides into four compartments, which contain and tether the **tibialis posterior muscle**, the **flexor digitorum longus muscle**, the **posterior tibial artery and nerve**, and **the flexor hallucis longus.**
      (2) A **posterior tibial pulse** normally is palpable in the vascular compartment immediately posterior to the medial malleolus.

B. **The muscles** of the posterior compartment lie in two subcompartments.

1. **Organization.** The flexor muscles of the posterior compartment (calf) are divided into superficial and deep groups by the transverse crural septum.
   a. The superficial group consists of the **triceps surae.**
      (1) **Composition.** This group comprises the **gastrocnemius, plantaris,** and **soleus muscles** (Figure 17-1; see Table 17-1).
      (2) **Course.** The two heads of the gastrocnemius muscle arise from the posterior aspects of the medial and lateral femoral epicondyles to define the lower limits of the popliteal fossa. The small plantaris muscle arises from the lateral supracondylar ridge of the distal femur and gives rise to a long thin tendon (humorously, the "nervus asinorum"). The soleus runs from the posterior surfaces of the fibula, interosseous membrane, and tibia. The plantaris and gastrocnemius muscles join the tendon of the triceps surae to form the calcanean or Achilles' tendon, which inserts onto the calcaneus.
   b. **The deep group**
      (1) **Composition.** These muscles include the **popliteus, flexor digitorum longus, flexor hallucis longus,** and **tibialis posterior** (Figure 17-2; see Table 17-1).
         (a) The **popliteus muscle** forms the floor of the popliteal fossa. It spans the knee joint and either **rotates the tibia medially** or the femur laterally, depending on which bone is fixed. It thereby **unlocks and initiates flexion of the knee.**
         (b) The **flexor digitorum longus,** a bipennate muscle, usually has two heads. The **medial head** arises from the posterior surface of the tibia. A **lateral head** (not acknowledged in most references) arises from the fibula in 52% of all individuals.
            (i) The fibular origin varies from a small proximal band to an extensive attachment along the fibular shaft, so compartmentalization of the tibialis posterior muscle is possible when the fibular attachment is extensive.

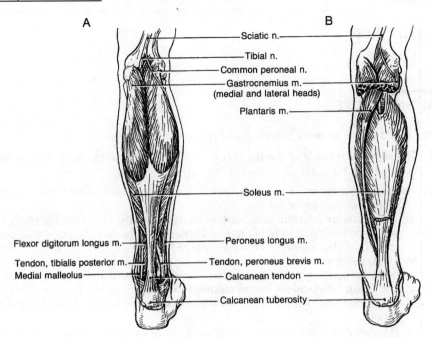

FIGURE 17-1. Superficial musculature of the posterior crural compartment. *(A)* Gastrocnemius muscle; *(B)* soleus and plantaris muscles.

(ii) When the flexor digitorum longus muscle originates from both tibia and fibula, use hypertrophy of the tibialis posterior may cause **deep posterior compartment syndrome.**

(c) The **flexor hallucis longus** flexes the great toe. It arises along the posterior aspect of the middle and distal fibula. The distal tendon of this muscle is associated with two sesamoid bones where it passes under the distal head of the first metatarsal before inserting into the distal phalanx of the great toe.

FIGURE 17-2. Deep musculature of the posterior crural compartment. *(A)* Flexor digitorum longus and flexor hallucis longus muscles; *(B)* tibialis posterior muscle.

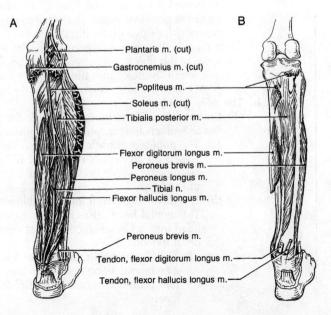

TABLE 17-1. Musculature of the Posterior Crural Compartment

Muscle	Origin	Insertion	Primary Action	Innervation
Superficial muscles				
Gastrocnemius	Posterior surfaces of medial and lateral femoral condyles	Calcanean tuberosity	Plantar flexes foot	Tibial n. (L5–S2, anterior)
Soleus	Proximal posterior tibia and fibula, interosseous membrane	Calcanean tuberosity	Plantar flexes foot	Tibial n. (L5–S2, anterior)
Plantaris	Lateral supracondylar ridge	Calcanean tuberosity	Plantar flexes foot	Tibial n. (L5–S2, anterior)
Deep muscles				
Popliteus	Lateral femoral condyle	Medially on proximal posterior tibia	Medially rotates and flexes knee	Tibial n. (L4–S1, anterior)
Tibialis posterior	Posterior shafts of tibia and fibular, interosseous membrane	Medial cuneiform bone and navicular on plantar side	Plantar flexes foot	Tibial n. (L5–S1, anterior)
Flexor hallucis longus	Distal third of posterior fibula	Phalanges of great toe	Flexes great toe	Tibial n. (L5–S2, anterior)
Flexor digitorum longus	Posterior surface of tibial and fibular shafts	Distal phalanges of second to fifth toes	Flexes distal phalanges of second through fifth toes	Tibial n. (L5–S2, anterior)

- (d) The **tibialis posterior muscle,** the deepest of the posterior group, arises from the tibia, interosseous membrane, and fibula to insert on the plantar surface of the navicular bone and medial (first) cuneiform bone, with slips to the second through fourth metatarsals.
- (e) **Shin splints** are most likely the result of tendinitis at the tibialis posterior origin on the tibia and interosseous membrane. Possible causes include an imbalance that may be corrected by using different running shoes or orthotic inserts.
  - (2) The **flexor retinaculum (tarsal tunnel)** tethers the tendons as they pass posterior to the medial malleolus. From medial to lateral lie the tendons of the **ti**bialis posterior muscle, the flexor **d**igitorum longus muscle, the posterior tibial artery **an**d nerve, and the tendon of the flexor **h**allucis longus. (Mnemonic: **T**om, **D**ick, **AN**d **H**arry.)
- 2. **Group actions at the ankle and tarsal joints**
  - a. **Flexion/extension** occurs about a transverse axis through the talocrural joint (Table 17-1; see Figure 16-4).
    - (1) **Plantar flexion.** All of the muscles of the posterior compartment pass poste-

rior to the transverse axis of the talocrural joint, that is, they are plantar flexors. In this action, they are synergistic with the muscles of the lateral compartment.
- (2) **Dorsiflexion** is a function of anterior compartment musculature.
- b. **Inversion/eversion** occurs about the axis of Henke, which is the resultant axis of the combined subtalar and midtarsal joints (see Figure 16-4).
  - (1) **Inversion.** All of the muscles of the posterior compartment pass medial to the resultant axis (of Henke), that is, they are invertors of the foot.
  - (2) **Eversion** is primarily a function of lateral compartment musculature.
3. **Group actions at the tarsometatarsal joints**
   a. **Flexion/extension** of the digits occurs about a transverse axes through the tarsometatarsal and interphalangeal joints.
      (1) **Plantar flexion.** The flexor hallucis longus and flexor digitorum longus flex the tarsometatarsal and interphalangeal joints. In this action, they are synergistic with the digital flexor muscles of the plantar foot.
      (2) **Dorsiflexion** is a function of musculature of the anterior crural compartment and the dorsal foot (see Table 17-1).
   b. **Abduction/adduction** of the digits occur about vertical axes through the metatarsophalangeal joints. This movement is a function of the intrinsic plantar musculature.
4. **Group innervation.** The muscles of the posterior compartment are innervated by the **tibial nerve.**

II. VASCULATURE

A. **Arteries** of the posterior compartment include the **posterior tibial artery** and its major branch, the **fibular (peroneal) artery** (Figure 17-3). In the foot, the posterior tibial artery gives rise to the **medial and lateral plantar arteries.**

1. **Popliteal artery** (see Chapter 16 V A 1)
2. **Posterior tibial artery.** This artery is the direct continuation of the popliteal artery (see Figure 17-3).
   a. **Course.** The posterior tibial artery descends medially in the superficial compartment. It gives rise to the **fibular artery** in the upper third of the leg. It continues posterior to the medial malleolus, where the **posterior tibial pulse** is palpable.
   b. **Distribution**
      (1) Proximally, it participates in the **geniculate anastomosis. Muscular branches** arise to supply the gastrocnemius and soleus muscles as the posterior tibial artery descends medially in the superficial posterior compartment.
      (2) The **fibular (peroneal) artery** arises from the posterior tibial artery in the upper third of the leg.
      (3) Distally, the posterior tibial artery gives off the **medial malleolar artery** in the vicinity of the medial malleolus as well as several communicating branches that anastomose with the fibular artery.
   c. **Variation.** Occasionally, the **anterior tibial artery** arises from the posterior tibial artery (instead of the popliteal artery) to supply the anterior compartment.
3. **Fibular (peroneal) artery** (see Figure 17-3)
   a. **Course.** It arises from the posterior tibial artery in the upper third of the leg and descends laterally in the deep posterior compartment between the tibialis posterior and flexor hallucis longus muscles.
   b. **Distribution**
      (1) Proximally, it participates in the **geniculate anastomosis** and gives off several branches to the muscles of the posterior and lateral crural compartments.

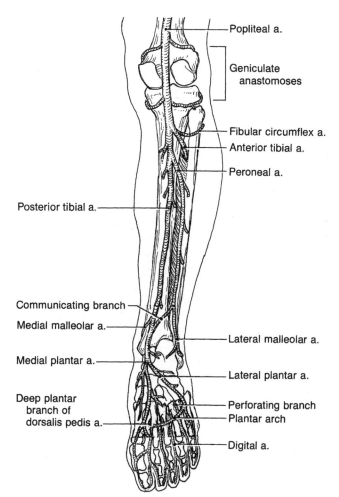

FIGURE 17-3. Vascular supply to the posterior and lateral crural compartments and the plantar foot, posterior perspective.

   (2) Distally, several communicating branches anastomose with the posterior tibial artery.
   - **(a)** The posterior lateral malleolar branch passes laterally about the ankle and anastomoses with the anterior lateral malleolar branch of the anterior tibial artery.
   - **(b)** The ankle joint has a profuse collateral circulation.

   (3) Occasionally (3.5%), the fibular (peroneal) artery is especially large and gives rise to the **dorsal pedal artery** via a perforating branch.

   4. The posterior tibial artery terminates by bifurcating into the **medial** and **lateral plantar arteries.**

B. Veins

   1. Superficial veins
   - **a.** The **great saphenous vein** (see Chapter 16 V B 1 a) and the **small saphenous vein** (see Chapter 16 V B 1 b) are the superficial veins of the posterior compartment.
   - **b. Varicosities.** These veins are especially susceptible to varicosities, for which there seems to be a genetic predisposition.

   2. Deep veins accompany the arterial supply.
   - **a.** The **posterior tibial vein** accompanies the posterior tibial artery, usually as venae comitantes.

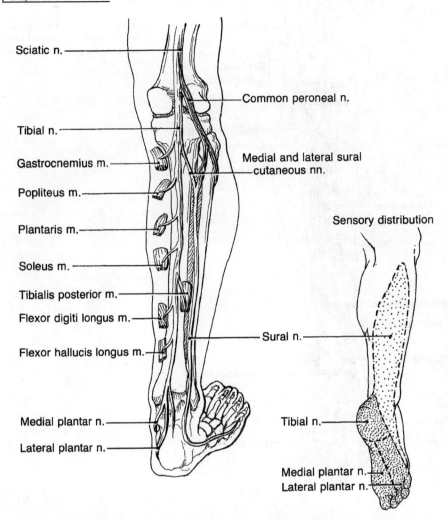

FIGURE 17-4. Distribution of the tibial nerve in the leg.

    b. The **popliteal vein** is formed by the joining of the posterior tibial vein and common peroneal vein in the popliteal fossa. The popliteal vein continues in the thigh as the **femoral vein**.

III. INNERVATION

A. Cutaneous sensory innervation

1. **The saphenous branch (L4) of the femoral nerve** innervates the medial aspect of the leg (see Figure 15-7).

2. **The superficial peroneal branch (L4–L5) of the common peroneal nerve** innervates the anterolateral aspect of the leg (see Figure 15-7).

3. **The sural nerve (S2),** formed by contributions from the tibial and common peroneal nerves, innervates the posterior aspect of the leg (Figure 17-4).

B. **Motor innervation**
   1. **Course and composition** (see Figure 17-4)
      a. **In the thigh,** the **tibial nerve** is derived from anterior roots of anterior portions of roots L4–S3 and travels with the common peroneal nerve as the sciatic nerve.
      b. **In the leg,** the tibial nerve separates from the sciatic nerve high in the popliteal fossa, passes between the heads of the gastrocnemius muscle, and descends in the superficial compartment.
      c. **At the ankle,** the tibial nerve passes posterior to the medial malleolus. On entering the foot, it bifurcates into **medial and lateral plantar nerves.**
   2. **Distribution**
      a. The **tibial nerve** (L4–S3) gives off branches to the musculature of the **superficial and deep posterior compartments** and accompanies the posterior tibial vessels behind the medial malleolus.
      b. The **medial plantar nerve** (L4–L5) generally supplies the intrinsic muscles on the medial side of the foot.
      c. The **lateral plantar nerve** (S1–S2) generally supplies the intrinsic muscles on the lateral side of the foot.
   3. **Injury. Herniated vertebral disks** in the lower lumbar region may compress the sacral roots, and **intramuscular injection** in the inferomedial quadrant of the buttock may injure the sciatic nerve. Injury to the tibial nerve is less frequent than injury to the common peroneal nerve because the course is deeper and it generally is more protected. Trauma at the popliteal fossa or at the medial malleolus, however, may injure the tibial nerve.
      a. A **high lesion** involves wasting of calf musculature and an inability to stand on tiptoe. Other signs include loss of the **ankle-jerk reflex** (S1) when the Achilles tendon is tapped with a rubber mallet, an inability to abduct or adduct the toes, and loss of plantar sensitivity.
      b. **Popliteal lesion**
         (1) **Superficial lacerations** may injure the sural nerve, associated with loss of sensation over the posterolateral calf and lateral foot.
         (2) **Deep laceration** may injure the tibial nerve, common fibular nerve, or both. Evidence of tibial nerve involvement is similar to that of a high lesion.
      c. **Deep posterior compartment syndrome,** elevated tissue pressure within the deep posterior compartment, may compress the fibular nerve and so paralyze the deep flexor muscles. Surgical fasciotomy is indicated.

Chapter 18
Plantar Foot

I. BONES

A. **The distal leg** is that part of the lower extremity that lies between the knee and ankle and consists of the tibia and fibula (see Chapter 16 I A).

B. **The foot** is composed of seven tarsal bones and five metatarsal bones. The tarsal bones (including the calcaneus and navicular bone), metatarsal bones, and phalanges are all palpable.

1. **The proximal tarsal group** consists of three bones, not in a row: the talus, calcaneus, and navicular bones (see Figure 16-1; Chapter 16 I B 1).
2. **The distal tarsal group** consists of a row of four bones (see Figure 16-1; Chapter 16 I B 2).
    a. The **medial, intermediate, and lateral cuneiform bones** articulate with the first three metatarsal bones distally.
    b. The **cuboid bone** is lateral to the navicular and lateral cuneiform bones. It articulates with the fourth and fifth metatarsals distally.
3. **Metatarsals.** Five metatarsal bones form the skeleton of the foot (see Figure 16-1).
    a. Each metatarsal bone has a proximal **base,** a **shaft,** and a distal **head.**
    b. The base of each metatarsal contacts the distal row of carpal bones to form the **tarsometatarsal joint.**
        (1) The first, second, and third metatarsals articulate, respectively, with the first, second, and third cuneiform bones.
        (2) The fourth and fifth metatarsals articulate with the cuboid bone.
        (3) Congenital shortening of the first metatarsal makes the second toe appear longer than the great toe (Morton's toe).
    c. The head of each metatarsal contacts a proximal phalanx at a **metatarsophalangeal (MP) joint.**
4. **Phalanges.** Three rows of phalanges constitute the skeletons of the second through fifth digits (toes); the first digit (great toe) has only two phalanges (see Figure 16-1).
    a. The proximal row articulates with the metacarpals at the **metatarsophalangeal (MP) joints.**
    b. The middle row articulates with the proximal row of phalanges at the **proximal interphalangeal (PIP) joints.**
    c. The distal row articulates with the middle row of phalanges at the **distal interphalangeal (DIP) joints.**
5. **Sesamoid bones,** usually two, lie within the flexor hallucis longus tendon in the vicinity of the first MP joint.
6. **Fractures** of the metacarpals and phalanges are common.
    a. **Fractures of the metatarsal shafts** result in ventral displacement of the distal fragments because of the pull of the interossei muscles.
    b. **Proximal phalangeal fractures** usually displace toward the sole because of the pull of the intrinsic flexors. Phalangeal fracture usually is not a problem. The injured toe may be splinted by taping it to an adjacent toe.

II. ARTICULATIONS

A. **Cuneonavicular, cuneocuboid, intercuneiform, and tarsometatarsal joints**

1. **Locations** of these joints are described by their names.
2. **Movement.** Ligaments across these joints permit small gliding movements of accommodation that change the shape of the transverse plantar arch during inversion/eversion.

3. **Support.** They are supported by a series of strong dorsal and plantar interosseous ligaments.

B. **Metatarsophalangeal (MP) joints** (see Figure 16-1)
   1. **Movement.** These joints have two degrees of freedom.
      a. **Flexion/extension.** Unlike the hand, and as an accommodation to walking, extension of the pedal digits (55°) is greater than flexion (35°).
      b. **Abduction/adduction.** Also unlike the hand, limited abduction/adduction takes place about vertical axes, but with reference to a long axis through the second toe.
      c. Unlike the thumb, the great toe is not capable of opposition.
   2. **Support.** They are supported by deep transverse metatarsal ligaments and collateral ligaments.

C. **Interphalangeal articulations** (see Figure 16-1)
   1. **Movement.** These basic hinge joints have one degree of freedom, permitting about 90° of **flexion** and **extension** about transverse axes.
   2. **Support.** They are supported by strong collateral ligaments on each side and a plantar ligament.

D. **Plantar arches and ligaments.** The bones of the foot and their supporting ligaments form two longitudinal and two transverse arches. These arches are maintained by the interlocking of the bony elements, strong supporting ligaments, and the dynamic action of extrinsic leg muscles.
   1. **Longitudinal plantar arches**
      a. The **medial arch** is pronounced and formed by the talus, calcaneus, spring ligament, navicular, cuneiforms, and medial three metatarsals.
      b. The **lateral arch** is composed of the calcaneus, cuboid, and lateral two metatarsals.
      c. **Functions**
         (1) These arches provide rigidity to support the body weight and act as a spring in ambulation.
         (2) Stretching of the plantar ligaments results in pronation of the foot (fallen or flattened arches). This condition affects ambulation and may produce leg cramps, as well as knee, hip, or lumbosacral disability.
   2. **Transverse plantar arches**
      a. The **posterior transverse arch** is formed by the navicular and cuboid bones.
      b. The **anterior transverse arch** is formed by the heads of the metatarsal bones. This arch is dynamically supported by the actions of the tibialis anterior and peroneus longus.
      c. **Function**
         (1) These arches are primarily involved in pronation and supination; they are flattened in the former and accentuated in the latter.
         (2) Flattening of the anterior transverse arch may produce **Morton's metatarsalgia,** in which body weight compresses the plantar nerves.
   3. **Plantar ligaments** arise from the calcaneus.
      a. The **plantar calcaneonavicular ("spring") ligament,** which runs from the sustentaculum tali of the calcaneus to the navicula, supports the talus.
      b. The **short plantar ligament,** which lies deep to the long plantar ligament, runs from the plantar surface of the calcaneus to the cuboid bone. It supports the longitudinal plantar arch.
      c. The **long plantar ligament,** which runs from the plantar surface of the calcaneus to the bases of the second, third, and fourth metatarsals, supports the longitudinal plantar arch.
      d. The **plantar aponeurosis,** spanning from the calcaneus to the bases of the terminal phalanges, also contributes to longitudinal arch support.

4. **Dynamic support.** The **tibialis posterior muscle** and the **peroneus longus muscle,** coming from opposite sides of the foot and inserting onto the medial cuneiform bone, act as a sling to provide dynamic support for the longitudinal plantar arch. The flexor hallucis longus passes under the sustentaculum tali and the first metatarsal to support the longitudinal plantar arch.

III. MUSCULATURE

A. Fascia

1. **The superficial fascia** of the leg continues onto the foot and supports the cutaneous nerves and superficial veins.
2. **The plantar (deep) fascia** of the foot is continuous with the crural fascia.
   a. **Septa** run between the skin on the sole of the foot and the plantar fascia, limiting the mobility of the plantar skin.
   b. The **plantar aponeurosis** is a central thickening of the plantar fascia (see Figure 18-1A).
      (1) It is attached to the calcaneus and divides into five slips, each of which bifurcates into superficial and deep fascicles. The superficial fascicle of each slip inserts into the skin of the toes; the deep fascicle divides to insert into the bases of the proximal phalanx of the second through fifth digits.
      (2) Strong septa arise from this layer to divide the foot into medial, intermediate, and lateral plantar compartments.
      (3) This layer is instrumental in maintaining the **longitudinal plantar arch,** along with the calcaneonavicular (spring) ligament and the deep plantar ligament.
   c. **Clinical considerations**
      (1) **Plantar fasciitis** in athletes, associated with pain, results from chronic stress inflammation of the plantar fascia.
      (2) **Infections.** If infection involves the **subaponeurotic space** in the central compartment between the plantar aponeurosis and the flexor digitorum brevis, this area may be accessed and drained by a medial incision away from the weightbearing surfaces.

B. Extrinsic muscles

1. **Group composition.** The **tibialis posterior, flexor digitorum longus, and flexor hallucis longus extrinsic muscles** arise in the deep posterior crural compartment. The tendons enter the foot via the **tarsal tunnel** posterior to the medial malleolus (see Figure 17-2 and Table 17-1).
   a. The **tibialis posterior muscle** passes deep to the intrinsic plantar musculature to insert onto the plantar surface of the cuneiform bones with slips to the second through fourth metatarsals (see Figure 18-2B).
   b. The **flexor digitorum longus muscle** divides into four slips that insert into the distal phalanges of the second through fifth toes (see Figure 18-1B).
   c. The **flexor hallucis longus muscle** inserts into the distal phalanx of the great toe (see Figure 18-1B). There are two sesamoid bones where its tendon passes under the head of the first metatarsal (see Figure 18-2B).
2. **Group actions.** All three muscles plantar flex and invert the foot (see Table 17-1). In addition, the tibialis posterior powerfully stabilizes the longitudinal plantar arch. The flexor digitorum longus flexes the distal phalanges at the DIP joint. The flexor hallucis flexes the distal phalanx of the great toe, playing an important role in the propulsive (toe-off) phase of gait. The flexor hallucis longus tendon, passing under the sustentaculum tali and the first metatarsal, also dynamically supports the longitudinal plantar arch.
3. **Group innervations.** All posterior crural compartment muscles are innervated by the **tibial nerve** (see Figure 17-4).

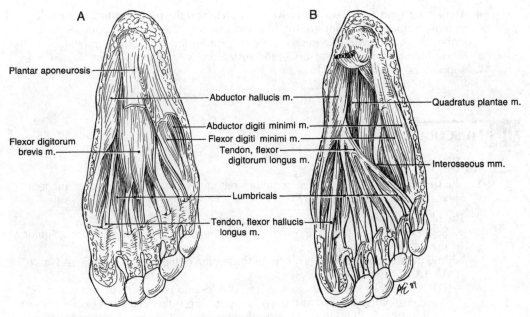

FIGURE 18-1. Superficial muscles of the plantar foot. *(A)* First layer; *(B)* second layer.

C. **Intrinsic muscles**

1. **Traditional organization.** The plantar muscles are commonly divided into four layers, a concept useful only for dissection (see Figures 18-1 and 18-2).
   a. The **first layer** (Figure 18-1A): abductor hallucis, flexor digitorum brevis, and abductor digiti minimi.
   b. The **second layer** (see Figure 18-1B): quadratus plantae and four lumbrical muscles, the tendon sheaths of the flexor hallucis longus and flexor digitorum longus.
   c. The **third layer** (Figure 18-2A): flexor hallucis brevis, adductor hallucis, flexor digiti minimi, the long plantar ligament, and the deeper short plantar ligament.

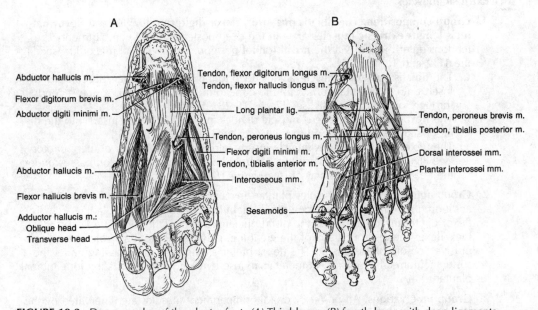

FIGURE 18-2. Deep muscles of the plantar foot. *(A)* Third layer; *(B)* fourth layer with deep ligaments.

TABLE 18-1. Plantar Musculature

Muscle	Origin	Insertion	Primary Action	Innervation
Medial group				
Abductor hallucis	Medial tubercle of calcaneus	Base of proximal phalanx of great toe	Abducts great toe	Medial plantar n. (L4–L5, anterior)
Flexor hallucis brevis	Cuneiform and cuboid bones	Base of proximal phalanx of great toe	Flexes first phalanx of great toe	Medial plantar n. (L5–S1, anterior)
Adductor hallucis Oblique head Transverse head	Calcaneus Metatarso-phalangeal ligaments	Base of first phalanx of great toe	Adducts great toe	Lateral plantar n. (S1–S2, anterior)
Intermediate group				
Flexor digitorum brevis	Medial tubercle of calcaneus	Splits to sides of middle phalanx of second to fifth toes	Flexes middle phalanx of second through fifth toes	Medial plantar n. (L5–S1, anterior)
Quadratus plantae	Calcaneus	Tendons of flexor digitorum longus	Changes line of action of flexor digitorum longus	Lateral plantar n. (S1–S2, anterior)
Lumbricals	Tendons of flexor digitorum longus	Medial sides of proximal phalanx of second to fifth toes	Flexes proximal phalanx and extends distal phalanges of second through fifth toes	Medial (I) and lateral (II, III, IV) plantar nn. (L5–S2, anterior)
Interossei Dorsal (4)	Between adjacent sides of metasal bones	First and second to both sides of second toe; third and fourth to lateral sides of corresponding toes	Abduct second, third, and fourth toes from the centerline of foot	Lateral plantar n. (S1–S2, anterior)
Plantar (3)	Medial sides of third to fifth metatarsals	Medial side of corresponding toes	Adducts third through fifth toes to the centerline of foot	Lateral plantar n. (S1–S2, anterior)
Lateral group				
Abductor digiti minimi	Calcaneus	Lateral side of first phalanx of fifth toe	Abducts fifth toe	Lateral plantar n. (S1–S2, anterior)
Flexor digiti minimi	Base of fifth metatarsal	Base of first phalanx of fifth toe	Flexes proximal phalanx of fifth toe	Lateral plantar n. (S1–S2, anterior)

  d. The **fourth layer** (see Figure 18-2B): the dorsal and plantar interossei, the tendons of the tibialis posterior and peroneus longus muscles.
 2. **Functional organization.** The plantar muscles may be organized functionally into three groups that lie in the **medial, intermediate,** and **lateral plantar compartments.**
   a. A **medial plantar group,** associated with the great toe, consists of the **abductor hallucis brevis, flexor hallucis brevis,** and **adductor hallucis,** as well as the tendon of the extrinsic **flexor hallucis longus** (Table 18-1; see Figure 18-2A).
   b. An **intermediate plantar group,** associated with the second through fifth digits, consists of the **flexor digitorum brevis, quadratus plantae, lumbrical muscles, dorsal interossei,** and **plantar interossei,** as well as the tendons of the extrinsic **flexor digitorum longus** muscle (see Figures 18-1 and 18-2 and Table 18-1).
     (1) The **flexor digitorum brevis** divides into four slips, which run to the second through fifth toes (see Figure 18-1A). Each slip splits to provide passage for the tendon of the extrinsic flexor digitorum longus before inserting into the sides of the middle phalanx of the second through fifth toes.
     (2) The **quadratus plantae** inserts into the tendon of the extrinsic flexor digitorum longus, changing the line of action of that muscle (see Figure 18-1B).

(3) The **lumbrical muscles** arise from the tendons of the flexor digitorum longus and insert into dorsal digital expansions over the proximal phalanges (see Figure 18-1B).
(4) The **dorsal interossei** arise between adjacent sides of the metatarsals and insert on both sides of the second toe and lateral sides of the third and fourth toes (see Figure 18-2B).
(5) The **plantar interossei** arise from medial sides of the third, fourth, and fifth metatarsals and insert into the medial sides of the same digits (see Figure 18-2B).

c. The **lateral plantar group,** associated with the fifth digit alone, consists of the **abductor digiti minimi** and the **flexor digiti minimi** (see Figure 18-1 and Table 18-1).

3. **Group actions**
   a. **Flexion/extension** of the pedal digits occurs about transverse axes through the MP and interphalangeal joints.
      (1) **Flexion.** The muscles of the plantar compartment pass inferior to the transverse axis of the MP and interphalangeal joints and, therefore, are flexors. In this action, they are synergistic with some of the long muscles of the posterior crural compartment.
         (a) **Metatarsophalangeal joints**
            (i) **Lumbricals I–IV** insert into the middle phalanges and flex the MP joints.
            (ii) The **flexor digitorum brevis** as well as the **flexor digitorum longus** with its associated **quadratus plantae** flex the MP joints, in addition to flexing the PIP and DIP joints.
            (iii) The **flexor digiti minimi** inserts into the proximal phalanx of the little toe and flexes the fifth MP joint.
            (iv) The **flexor hallucis brevis** inserts into the proximal phalanx of the great toe and flexes the first MP joint.
         (b) **Proximal interphalangeal (PIP) joints.** The **flexor digitorum brevis** tendons insert into the lateral aspects of the bases of digits two through five and flexes this joint.
         (c) **Distal interphalangeal (DIP) joints**
            (i) The **flexor digitorum longus** passes through the tendons of the flexor digitorum brevis to insert onto the plantar aspects the distal phalanges of digits two through five, and flexes this joint.
            (ii) The **quadratus plantae,** while changing the line of action of the flexor digitorum longus to the long axis of the foot by inserting into its tendons, also flexes the DIP joint.
            (iii) The **flexor hallucis longus** powerfully flexes the distal phalanx of the great toe. Passing under the sustentaculum tali and the first metatarsal, it supports the longitudinal plantar arch.
      (2) **Extension** of the interphalangeal joints is assisted by the lumbricals, although it is primarily a function of the muscles of the dorsal and anterior crural compartments.
   b. **Abduction/adduction** of the pedal digits occurs about vertical axes through the MP joints.
      (1) **Abduction** involves the four dorsal interosseous muscles as well as the abductor hallucis and abductor digiti minimi.
      (2) **Adduction** involves the three plantar interosseous muscles as well as the adductor hallucis.

4. **Group innervations.** The muscles of the plantar foot are innervated by the medial plantar and lateral plantar branches of the **tibial nerve.**
   a. The **lateral compartment musculature** is supplied by the lateral plantar nerve.
   b. The **intermediate compartment** musculature is supplied by both the medial and lateral plantar nerves.
   c. The **medial compartment** musculature is supplied by the medial plantar nerve.

IV. VASCULATURE

A. Arteries. In the foot, the posterior tibial artery gives rise to **medial and lateral plantar arteries** (see Figure 17-3).

1. **Lateral plantar artery** (see Figure 17-3)
   a. **Course.** This terminal branch of the posterior tibial artery deviates laterally across the long plantar arch and then turns medially across the bases of the metatarsal bones as the **plantar arch.**
   b. **Distribution.** It supplies the lateral side of the plantar portion of the foot, as well as the lateral and intermediate muscular compartments.
      (1) It joins with the **deep plantar branch of the dorsalis pedis artery** in the first interosseous space to complete the **plantar arch.**
      (2) It gives off the **proper digital artery** to the lateral side of the fifth toe and four **plantar metatarsal arteries.**
      (3) Each plantar metatarsal artery receives a **perforating branch** from the corresponding **dorsal metatarsal artery** and then bifurcates into **plantar digital arteries** to supply the sides of adjacent toes.

2. **Medial plantar artery.** This artery is smaller than the lateral plantar artery (see Figure 17-3).
   a. **Course.** This artery, which is smaller than the lateral plantar artery, passes between the abductor hallucis and the flexor digitorum brevis muscles. It anastomoses with the dorsal metatarsal branch of the dorsal pedal artery as well as with the lateral plantar artery through the plantar arch.
   b. **Distribution.** It supplies the medial side of the plantar portion of the foot and the medial muscular compartment and terminates along the lateral side of the great toe.

B. Veins

1. **Superficial veins** of the leg are the great saphenous vein and the small saphenous vein, which arise along the dorsal aspect of the foot.
2. **Deep veins** accompany the medial and lateral plantar arteries deep within the plantar aspect of the foot.

V. INNERVATION

A. Cutaneous sensory innervation

1. **The saphenous branch (L4) of the femoral nerve** innervates the medial aspect of the foot, including the great toe (see Figure 15-7).
2. **The superficial peroneal branch (L4–L5) of the common peroneal nerve** innervates the anterolateral aspect of the foot (see Figure 16-7).
3. **The deep peroneal branch (L5) of the common peroneal nerve** innervates the dorsal web space between the great and second toes. This location is used for testing the integrity of L5 (see Figure 16-7).
4. **The lateral and medial plantar branches (S1) of the tibial nerve** innervate the central and lateral aspects of the plantar foot, including the small toe (see Figure 17-4).

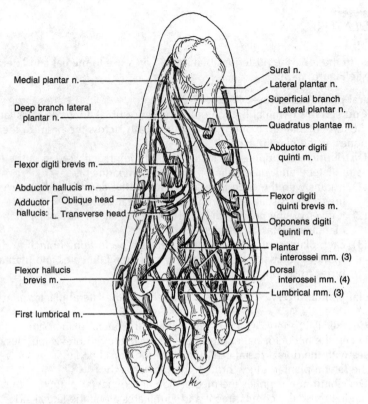

FIGURE 18-3. Distribution of the tibial nerve in the plantar foot.

B. Motor innervation

1. **Course and composition** (Figure 18-3)
   a. **At the ankle,** the **tibial nerve** (L4–S3), the anterior division of the sciatic nerve, passes posterior to the medial malleolus and accompanies the posterior tibial vessels (see Figure 17-4).
   b. On entering the foot, it bifurcates into **medial and lateral plantar nerves** (see Figure 18-3).

2. **Motor distribution** (see Figure 18-3)
   a. The **medial plantar nerve** (L4–L5) supplies the intrinsic muscles of the medial plantar compartment and only two muscles of the intermediate plantar compartment of the foot.
   b. The **lateral plantar nerve** (S1–S2) supplies the intrinsic muscles on the lateral plantar compartment and most of the muscles of the intermediate plantar compartment of the foot.

3. **Injury**
   a. An **ankle lesion,** such as medial malleolar fracture or tarsal tunnel syndrome (produced by entrapment of the nerve under the flexor retinaculum), results in an inability to abduct or adduct the toes (which many healthy people cannot accomplish) and loss of plantar sensitivity.
   b. **In the foot,** the first branch of the medial plantar nerve may become entrapped by the fascia of the quadratus plantae muscle (especially in long-distance runners), producing plantar pain that is more localized than that associated with plantar fasciitis.

Plantar Foot | **231**

VI. AMBULATION

A. Normal gait has been described as the translation of the body's center of gravity through space with a minimum expenditure of energy.

1. **Swing and stance phases.** In normal gait, each lower limb passes through a cycle consisting of alternating stance and swing phases.
   a. The **stance phase** accounts for two thirds of the gait period and consists of three parts.
      (1) **Heel-strike** or the restraining portion
         (a) The hip abductors and adductors stabilize the pelvic girdle over the newly weightbearing limb as the heel hits the ground.
         (b) The quadriceps femoris stabilizes the slightly flexed knee while the gluteus maximus and hamstrings extend the femur, propelling the center of gravity forward.
         (c) Relaxation of the dorsiflexors permits lowering of the foot to midstance, transferring the weight from the heel to the plantar arch.
      (2) **Midstance**
         (a) The center of gravity momentarily is directly over the limb and distributed evenly by the plantar arch to the heel and metatarsals.
         (b) The dorsiflexors pull the tibia forward, moving the center of gravity anteriorly.
         (c) The triceps surae begin to lift the heel off the ground.
      (3) **Toe-off** or the propulsive portion
         (a) The triceps surae maximally lift the heel and move the body forward by transferring the weight to metatarsals and phalanges.
         (b) The quadriceps femoris extends the slightly flexed knee.
         (c) The lateral compartment muscles evert the foot, transferring weight to the great toe.
         (d) The flexor hallucis longus gives a final strong plantar flexion, which pushes the body off from the great toe.
         (e) The momentum raises the limb free from the ground into the swing phase.
   b. **The swing phase** accounts for one third of the gait period and consists of two parts.
      (1) **Early swing portion**
         (a) The hip flexors accelerate the limb forward and the knee flexes passively, lifting the foot from the ground.
         (b) The tibialis anterior and synergistic dorsiflexors keep the foot clear of the ground.
      (2) **Late swing portion**
         (a) The dorsiflexors position the foot for heel-strike.
         (b) The hamstrings decelerate the limb just prior to heel-strike.
   c. **Summary.** The center of gravity is repeatedly moved forward, resulting in a momentary imbalance. The individual moves a lower limb in an attempt to regain stability. This results in an alternate shifting of dynamic equilibrium from one foot to the other.

2. **Ambulation** consists of reciprocally alternating cycles of gait.
   a. **Walking** involves overlapping stance phases.
   b. **Running** has no double-stance phase (i.e., left heel-strike with right toe-off).

B. Pathologic gait is an attempt to either minimize pain or reduce expenditure of energy by compensation.

1. **Hip disease** patients avoid heel-strike when walking because this part of the stance phase transmits force to the hip joint.

2. **Excessive valgus or varus** [see Chapter 15 I B 2 d (2)].
3. **Disease** in the first MP joint inhibits toe-off, thereby altering normal gait.
4. **Congenital anomalies** of the first metatarsal, such as Morton's toe, inhibit toe-off, thereby slightly altering normal gait.

PART IIC   LOWER EXTREMITY

STUDY QUESTIONS

DIRECTIONS: Each of the numbered items or incomplete statements in this section is followed by answers or by completions of the statement. Select the ONE lettered answer or completion that is BEST in each case.

1. An overweight, 57-year-old man has tingling, painful, or itching sensations in the lateral region of the thigh. The examining physician suspects nerve entrapment as a result of a bulging abdomen that compresses a nerve beneath the inguinal ligament. The nerve most likely affected is the

(A) anterior femoral cutaneous
(B) femoral branch of the genitofemoral
(C) genital branch of the genitofemoral
(D) ilioinguinal
(E) lateral femoral cutaneous

Questions 2–4

A person experiences weakness when climbing stairs. As part of the neurologic examination, the muscle action of the thigh is tested by asking the patient to extend the thigh against resistance. Extension of the right thigh is found to be weaker than that of the left thigh.

2. Muscles active in producing thigh extension include all of the following EXCEPT the

(A) adductor magnus
(B) gluteus maximus
(C) gluteus medius
(D) semimembranosus
(E) semitendinosus

3. While supine, with the hip and knee joints extended, the patient is asked to move the leg outward (abduct) against resistance. No weakness is observed. Abduction ability tests all of the following muscles EXCEPT the

(A) gluteus medius
(B) gluteus minimus
(C) piriformis
(D) quadratus femoris

4. Further neurologic examination reveals the knee-jerk reflex is normal and the patient can stand on the heels and toes. With this information, in addition to the knowledge that thigh extension is weak on the right side and thigh abduction and adduction are normal, it can be concluded that the innervation involved in this patient's problem includes which of the following nerves?

(A) Common peroneal
(B) Femoral
(C) Inferior gluteal
(D) Obturator
(E) Tibial

(end of group question)

5. A 7-year-old girl developed a left Trendelenburg lurch. She was found to have a tumor of the piriformis muscle that compresses a nerve as it passes through the suprapiriforme recess of the greater sciatic foramen. Injury of the nerve that passes through this recess would produce paralysis of which of the following muscles?

(A) Gluteus maximus
(B) Gluteus medius
(C) Obturator internus
(D) Piriformis
(E) Tensor fasciae latae

6. In the adult, avascular necrosis of the femoral head is a likely sequela to

(A) dislocation of the hip with tearing of the ligamentum teres of the femur
(B) intertrochanteric fracture of the femur
(C) intracapsular femoral neck fracture
(D) thrombosis of the obturator artery in the obturator canal
(E) all of the above

7. Which of the following actions takes place during the final phase of knee extension?

(A) The femur glides forward on the medial tibial condyle
(B) The medial femoral condyle pivots on the tibial plateau
(C) The leg undergoes medial rotation
(D) The popliteus muscle is stretched

8. Meniscal tears in the knee joint usually result from which of the following circumstances?

(A) Compression
(B) Hyperextension
(C) Hyperflexion
(D) Rotation in partial flexion
(E) Rotation in full extension

9. While stretching to the left for a backhand return in a game of tennis, a 35-year-old man feels a sharp pain in the calf of the weightbearing (left) leg. His doubles partner, a physician, examines the leg and notes a bulge in the distal part of the popliteal fossa. This finding, in addition to severe pain elicited on attempted plantar flexion against resistance, indicates a tendon has been torn. Which of the following muscles is likely to be injured when sudden plantar flexion coincides with strong knee extension?

(A) Peroneus brevis muscle
(B) Peroneus longus muscle
(C) Plantaris muscle
(D) Popliteus muscle

10. A marathon runner relates that she sprained her ankle when she stepped into a depression that bent her foot inward. Radiographs of the ankle reveal a small fragment of bone avulsed from the base of the fifth metatarsal, well away from the lateral collateral ligament. Which of the following muscles inserts into the tuberosity of the fifth metatarsal bone?

(A) Abductor digiti minimi
(B) Peroneus brevis
(C) Peroneus longus
(D) Tibialis anterior
(E) Tibialis posterior

11. Supracondylar fracture of the femur may injure which of the following structures?

(A) Anterior tibial artery
(B) Deep femoral artery
(C) Great saphenous vein
(D) Popliteal artery
(E) Sciatic nerve

12. Walking across a slope requires accommodative movements in the

(A) distal tibiofibular joint
(B) talocrural joint
(C) tarsometatarsal joint
(D) transverse tarsal joint

13. Muscles that attach to the plantar surface of the distal phalanx of the second through fifth toes include the

(A) flexor digitorum brevis
(B) flexor digitorum longus
(C) group of dorsal interossei
(D) group of lumbricals
(E) group of plantar interossei

14. Contraction of the plantar interosseous muscle between the fourth and fifth toes results primarily in

(A) abduction of the fourth toe
(B) adduction of the fourth toe
(C) adduction of the fifth toe
(D) flexion of the interphalangeal joints of the fourth toe
(E) flexion of the interphalangeal joints of the fifth toe

15. The dorsalis pedis artery is most commonly a continuation of which of the following arteries?

(A) Anterior tibial
(B) Fibular (peroneal)
(C) Lateral plantar
(D) Medial plantar
(E) Posterior tibial

Questions 16–18

The great saphenous vein of a patient scheduled for surgery needs to be cannulated for an intravenous line.

16. A convenient site for insertion of an intravenous line in the great saphenous vein is

(A) anterior to the medial malleolus
(B) inferior to the inguinal ligament
(C) posterior to the lateral malleolus
(D) medial to the popliteal fossa

17. During the cutdown and preparation of the vein for insertion of the cannula, the patient experiences pain radiating along the medial border of the dorsum of the foot. Which of the following nerves is most likely to be accidentally included in a ligature during the cannulation procedure?

(A) Lateral sural
(B) Medial femoral cutaneous
(C) Medial sural
(D) Saphenous
(E) Superficial fibular (peroneal)

18. Considering the pain distribution along the medial border of the dorsal foot, which spinal level is represented?

(A) L2
(B) L3
(C) L4
(D) L5
(E) S1

(end of group question)

DIRECTIONS: Each of the numbered items or incomplete statements in this section is negatively phrased, as indicated by a capitalized word such as NOT, LEAST, or EXCEPT. Select the ONE lettered answer or completion that is BEST in each case.

---

19. During full active extension of the knee joint, all of the ligaments are tightened EXCEPT the

(A) anterior cruciate
(B) coronary
(C) fibular collateral
(D) patellar
(E) tibial collateral

20. During an osteotomy to gain access to the femoral neck, the greater trochanter, with its muscles attached, is separated from the femur. All of the following muscles would be reflected with the greater trochanter EXCEPT the

(A) gluteus maximus
(B) gluteus medius
(C) gluteus minimus
(D) piriformis

21. A 53-year-old carpenter complains of soreness behind the knee and tingling sensations along the medial aspect of the leg. Examination reveals a pulsatile mass in the popliteal fossa, which is pathognomonic for aneurysm of the popliteal artery. During surgery, a replacement graft is run through the adductor (Hunter's) canal. This canal between the anterior and posterior portions of the adductor magnus muscle contains all of the following that must be identified EXCEPT the

(A) femoral artery
(B) great saphenous vein
(C) nerve to the vastus medialis
(D) saphenous nerve

22. Structures found within the tarsal tunnel (flexor retinaculum) include all of the following EXCEPT the

(A) flexor hallucis longus tendon
(B) peroneus brevis tendon
(C) posterior tibial artery
(D) tibial nerve
(E) tibialis posterior tendon

23. Factors that contribute to the stability of the ankle joint include all of the following EXCEPT the

(A) calcaneonavicular ligament
(B) deltoid ligament
(C) lateral ligament
(D) trapezoid shape of the trochlea tali

DIRECTIONS: Each set of matching questions in this section consists of a list of four to twenty-six lettered options (some of which may be in figures) followed by several numbered items. For each numbered item, select the ONE lettered option that is most closely associated with it. To avoid spending too much time on matching sets with large numbers of options, it is generally advisable to begin each set by reading the list of options. Then, for each item in the set, try to generate the correct answer and locate it in the option list, rather than evaluating each option individually. Each lettered option may be selected once, more than once, or not at all.

Questions 24–27

(A) Plantar flex the foot only
(B) Plantar flex and invert the foot
(C) Dorsiflex and evert the foot
(D) Dorsiflex and invert the foot

For each muscle group, select the appropriate action.

24. Anterior crural compartment musculature

25. Deep posterior crural compartment musculature

26. Lateral crural compartment musculature

27. Superficial posterior crural compartment musculature

Questions 28–32

(A) Common fibular (peroneal) nerve
(B) Femoral nerve
(C) Obturator nerve
(D) Tibial nerve

For each finding, select the appropriate nerve with which it is most apt to be associated.

28. Inability to stand on the toes by plantar flexion

29. Inability to stand on the heels by dorsiflexion

30. Inability to flex the knee

31. Loss of Achilles tendon reflex

32. Loss of knee-jerk reflex

ANSWERS AND EXPLANATIONS

1. **The answer is E** [Chapter 15 V B 3]. The lateral femoral cutaneous nerve passes between the inguinal ligament and the iliopubic ramus beneath the anterior superior iliac spine. Entrapment of the nerve in this location produces sensory symptoms along the lateral aspect of the thigh.

2. **The answer is C** [Chapter 14 IV A 3 a; Table 14–1]. The hamstring group, including the biceps femoris, semitendinosus, and semimembranosus, as well as the adductor magnus and gluteus maximus, are all extensors of the thigh. The gluteus medius is an abductor.

3. **The answer is D** [Chapter 14 IV A 4, 5; Tables 14-3 and 14-4]. Abduction of the thigh is primarily by the gluteus medius and minimus muscles, as well as the tensor fasciae latae. The piriformis laterally rotates and aids abduction of the thigh. The primary action of the quadratus femoris is lateral rotation of the thigh.

4. **The answer is C** [Chapter 14 VI B 2; Table 14–2]. The inferior gluteal nerve innervates the gluteus maximus muscle, any paralysis of which produces weakness in climbing stairs and in rising from a seated position. The normal knee-jerk reflex indicates the femoral nerve is not involved, and the ability to stand on the heals and toes indicates that the common peroneal and tibial nerves, respectively, are similarly unaffected. Normal thigh adduction rules out injury to the obturator nerve.

5. **The answer is B** [Chapter 14 VI B 1]. The superior gluteal nerve, which innervates the gluteus medius, minimus, and tensor fasciae latae muscles, passes through the suprapiriforme recess. The inferior gluteal nerve, which innervates the gluteus maximus muscle, as well as the sciatic and pudendal nerves pass through the infrapiriforme recess. In this case, paralysis of the hip abductors resulted in a compensatory lurch to overcome the inability to maintain a level position of the pelvis.

6. **The answer is C** [Chapter 14 V B 1 b (1), c (2)]. In the adult, blood flow through the artery of the ligamentum teres (when present) is usually insufficient to supply the head of the femur. The adult blood supply to this region of the femur is primarily via the retinacular branches of the medial and lateral circumflex arteries. Consequently, an intracapsular femoral neck fracture jeopardizes this vascular supply and may result in necrosis of the femoral head. Proximal thrombosis of the obturator artery, from which the artery of the ligamentum teres arises, could produce a similar problem, but only in a child. Before cessation of growth, the only supply to the femoral head is via the artery of the ligamentum teres.

7. **The answer is D** [Chapter 15 II C 2 c]. During this phase, the femur rotates medially about the anterior cruciate ligament, so the medial femoral condyle glides backward and the popliteus muscle is stretched. The medial rotation of the thigh is equivalent to a lateral rotation of the leg.

8. **The answer is D** [Chapter 15 II A 2 b]. Most meniscal tears are the result of lateral rotation of the partially flexed leg, when the knee joint is less stable. Because the knee joint is most stable in full extension, meniscal tears are unlikely in this position. Because full flexion is not compatible with mobility or support, this posture is rarely used and is unlikely to cause problems.

9. **The answer is C** [Chapter 17 I B 1 a (1); Table 17-1]. The plantaris muscle, a small but significant part of the triceps surae, acts with the gastrocnemius and soleus to plantar flex the foot. Originating from the lateral supracondylar ridge, it passes through the popliteal fossa and becomes tendinous between the soleus and gastrocnemius muscles to insert into the calcanean tendon. Because the plantaris muscle spans both the knee and ankle joints, its tendon may be avulsed when sudden plantar flexion coincides with strong knee extension. The popliteus muscle, a medial rotator of the knee, is not involved in plantar flexion. The peroneal muscles are in the lateral compartment.

10. **The answer is B** [Chapter 16, Table 16-1, Figure 16-2A]. The peroneus brevis muscle is

Answers and Explanations | Part IIC Lower Extremity | 239

a powerful evertor, and forceful inversion can avulse the tendinous insertion.

11. The answer is D [Chapter 15 IV A 2 b, 3 b; Chapter 16 V A 2; Chapter 17 II A]. This fracture jeopardizes the popliteal neurovascular structures, especially the popliteal artery. The sciatic nerve divides into common fibular and tibial nerves well above the knee region. The popliteal artery gives rise to the anterior tibial artery behind the tibia at the inferior border of the popliteal fossa before continuing as the posterior tibial artery. Only small anastomotic twigs of the deep femoral artery reach the vicinity of the knee to participate in the geniculate anastomosis.

12. The answer is D [Chapter 16 II D, E 1 a-c]. This movement necessitates inversion and eversion of the foot. It involves the subtalar and transverse tarsal joints, both of which share the talocalcaneonavicular and calcaneocuboid joints. Inversion and eversion occur about the resultant axis of rotation.

13. The answer is B [Chapter 17 I B 1 b (1) (b); Tables 17-1 and 18-1]. Only the flexor digitorum longus reaches the plantar surface of the distal phalanges of toes two through five. The flexor digitorum brevis splits at each of its insertions into the middle phalanges, so that the deeper flexor digitorum longus tendons may reach the distal phalanx. The lumbricals and interossei insert into the dorsal expansions of the pedal digits.

14. The answer is C [Chapter 18 III C 1 d, 2 b (5)]. The plantar interossei adduct the second through fifth toes toward the axis of the second toe. The dorsal interossei abduct the second, third, and fourth toes away from the second toe. The plantar and dorsal interossei also extend the proximal and distal interphalangeal joints and flex the metatarsophalangeal joint.

15. The answer is A [Chapter 16 V A 3; Chapter 17 II A 2 b, c]. The dorsal pedal artery usually is the direct continuation of the anterior tibial artery. Occasionally, it arises from the fibular artery, a branch of the posterior tibial artery.

16. The answer is A [Chapter 16 V B 1; Chapter 17 II B 1 a]. The lesser saphenous vein, which is not usually used for venipuncture, passes posterior to the lateral malleolus. The saphenous vein passes medial to the popliteal fossa and enters the femoral vein at the level of the inguinal ligament.

17. The answer is D [Chapter 17 III A 1]. The saphenous nerve comes into the proximity of the great saphenous vein posterior to the medial tibial condyle. It accompanies the vein along the anteromedial aspect of the leg and into the foot anterior to the medial malleolus.

18. The answer is C [Chapter 4, Figure 4-3; Chapter 15, Figure 15-7]. The saphenous nerve, the terminal cutaneous portion of the femoral nerve, contains the contribution from the L4 spinal root. The medial aspect of the foot is generally innervated by L4, and the lateral aspect is generally innervated by S1.

19. The answer is B [Chapter 15 I D 1 b; II B 3 a (2), b (2), d (2)]. During active extension, the quadriceps femoris pulls on the tibia through the patellar ligament. As the leg comes into full extension, the collateral and cruciate ligaments become taut because the radii of curvature of the transverse tibiofemoral joint increase. The anterior cruciate ligament provides an axis about which the femur rotates medially. The coronary ligaments that attach the menisci to the tibial plateau are usually lax, permitting meniscal movement.

20. The answer is A [Chapter 14 II C 1 c (2)]. The gluteus maximus inserts into the linea aspera of the femur.

21. The answer is B [Chapter 15 IV A 2 a (2)]. The great saphenous vein lies superficial to the fascia lata. The saphenous nerve enters the adductor canal, but leaves before the popliteal fossa to become superficial.

22. The answer is B [Chapter 17 I A 2 c; Chapter 18 V B 3 a]. The tendons of the deep muscles of the posterior crural compartment (i.e., tibialis posterior, flexor digitorum longus, and flexor hallucis longus) as well as the tibial artery and accompanying tibial nerve pass through the tarsal tunnel inferior to the medial malleolus. The peroneal muscles of the lateral compartment pass beneath the peroneal retinaculum inferior to the lateral malleolus. The sural nerve passes along the lateral side of the ankle.

23. The answer is A [Chapter 16 I B 1 b (3), (4); Chapter 18 III C 1 d]. The deltoid (from the medial malleolus to the navicula and calcaneus) and lateral (from the lateral malleolus to the talus and calcaneus) ligaments stabilize the ankle joint. The trapezoid shape of the trochlea tali increases stability in dorsiflexion. The calcaneonavicular (spring) ligament supports the talus and helps to maintain the longitudinal plantar arch.

24–27. The answers are 24-D, 25-B, 26-C, 27-A [Chapter 16 III B 2; Figure 16-4; Chapter 17 I B 2]. The muscles of the superficial crural compartment act only at the talocrural joint with pure plantar flexion. The muscles of the deep posterior crural compartment, passing posterior to the transverse axis of the ankle joint and medial to the resultant axis of the subtalar and transverse tarsal joints, plantar flex and invert the foot. The muscles of the lateral crural compartment, passing posterior to the transverse axis of the ankle joint and lateral to the resultant axis of the subtalar and transverse tarsal joints, plantar flex and evert the foot. The muscles of the anterior crural compartment, passing anterior to the transverse axis of the ankle joint and medial to the resultant axis of the subtalar and transverse tarsal joints, dorsiflex and tend to invert the foot.

28–32. The answers are 28-D, 29-A, 30-D, 31-D, 32-B [Chapter 15 V B 2 a (2), c; Tables 15-1 and 15-2; Chapter 16 VI B 1, C 1; Table 16-1; Chapter 17 III B 3; Table 17-1]. Injury to the tibial nerve affects the plantar flexors of the posterior crural compartment, as indicated by a loss of the Achilles tendon reflex and inability to stand on tiptoes. Injury to the common fibular nerve affects the lateral crural compartment, resulting in loss of dorsiflexion with foot-drop (foot-slap), an inability to rock back on the heels, as well as loss of eversion ability. A high sciatic nerve injury involving both the tibial and common fibular nerves, which innervate the hamstring muscles, results in an inability to extend the hip or flex the knee, in conjunction with paralysis of the leg and foot musculature. Femoral nerve injury is evident by loss of the knee-jerk reflex.

SECTION III
TRUNK

PART IIIA THORAX

Chapter 19
Thoracic Cage

I. INTRODUCTION

A. **Thorax.** The thorax comprises the upper bony portion of the trunk.

1. **Boundaries.** The thoracic region is delineated by the rib cage (consisting of ribs, costal cartilages, thoracic vertebrae, sternum, and intercostal musculature) and the diaphragm (see Figures 19-1 and 19-2).
   a. The **superior thoracic aperture (thoracic inlet)** defines the upper limit of the rib cage and communicates with the neck and upper extremities (see Figure 19-2).
   b. The **inferior thoracic aperture (thoracic outlet)** is closed by the **diaphragm.** The abdominal cavity and its contents extend cranially, under the thoracic cage, as far as the diaphragm (see Figure 19-2).

2. **Mammary glands** (see Chapter 2 III)
   a. In men, the nipple is a useful landmark lateral to the midclavicular plane at the fourth intercostal space.
   b. In women, the mammary glands are usually the most prominent superficial structures of the anterior thoracic wall (see Figure 19-2). The nipple is variably lower in women than in men.

3. **Contents.** The thoracic region includes the heart and lungs as well as neurovascular connections between the neck, upper extremities, and abdomen. The lungs and borders of the heart may be outlined during physical examination by **percussion** and **auscultation** (see Figure 20-1).

B. **Pectoral region.** Portions of the anterior chest wall provide attachments for muscles that insert onto the pectoral girdle and arms (see Figures 6-1 and 6-2).

1. **Landmarks**
   a. Each **clavicle** is palpable for the entire length (see Chapter 6 II A and Figure 6-1). A **midclavicular line** passes vertically through the center of each clavicle and approximately through the nipple in men and variably medial to the nipple in women (see Figure 23-1).
   b. The **scapula** is the most prominent dorsal landmark of the thorax. It has several palpable features: the acromion process, coracoid process, spine, vertebral border, and apex (see Chapter 6 II B and Figure 6-1).

2. **Primary function.** Although the pectoral region belongs to and primarily functions as part of the upper extremity, some pectoral muscles attach to the thoracic cage and contribute to inspiratory effort as **accessory muscles of inspiration.**

II. RIB CAGE

A. **Organization**

1. **Thoracic vertebrae** (see Chapter 11 II, III B; Figure 11-3). Proceeding caudally, the 12 thoracic vertebrae become slightly more massive to support the additional mass of the total body superior to it. An intervertebral disk lies between adjacent vertebrae (see Figure 11-3).
   a. The **body** of each vertebra (near the junctions of the pedicles) has demifacets for diarthrodial articulations with the heads of the immediately superior and inferior ribs (with some exceptions).

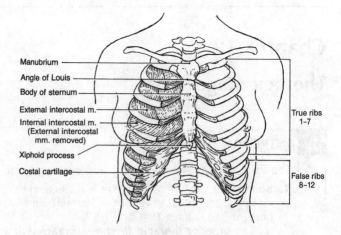

FIGURE 19-1. Musculoskeletal thoracic cage, anterior perspective. The intrinsic musculature lies between the 12 pairs of ribs. The external intercostal muscles terminate at the costochondral junction, but the layer continues to the sternum as the external intercostal membrane. The internal intercostal layer is shown in the right fourth and fifth intercostal spaces.

    b. Paired **transverse processes** arise at the junction between the pedicles and the laminae and have articular facets for the tubercles of ribs 1–10.

    c. The **spinous processes (spine)** of the 12 thoracic vertebrae are palpable and are used to define upper thoracic levels by number (see Figure 19-2).

        (1) The long spinous processes of upper thoracic vertebrae angle downward, with little to distinguish one from the other.

        (2) The spinous processes of lower thoracic vertebra begin to approach the horizontal, but become shorter, making them difficult to distinguish one from another.

        (3) The landmark **vertebra prominens** is the prominent spinous process of vertebra C7 where the counting of cervical and thoracic vertebral spines begins.

  2. Sternum. Having developed as a segmental structure, the adult sternum retains segmentation that divides it into the **manubrium, corpus (or body),** and **xiphoid process** (Figure 19-1). These readily palpable structures are separated by the **manubriosternal synchondrosis** and the **xiphisternal synchondrosis.** The articulation between the manubrium and corpus allows movement during respiration.

    a. Manubrium (from manus, L. handle)

        (1) Structure. The manubrium is shield-shaped. Its **jugular notch** is prominent and approximates an anterior boundary between the neck and thorax.

        (2) Articulations

            (a) Sternoclavicular joint. The manubrium is anchored on each side by articulation with the clavicle (see Figure 6-1). Motion in this joint is evident as the arm is circumducted.

            (b) Manubriocostal joints. It articulates by synchondroses (no joint capsule) with the costal cartilage of the first ribs. It also articulates by diarthroses (joint capsule present) with the superior demifacets of the second costal cartilages at the sternal angle (see Figure 19-1).

            (c) Manubriosternal synchondrosis. It articulates with the body of the sternum at the **manubriosternal synchondrosis** or **sternal angle** (of Louis).

                (i) Counting of ribs begins here because the second ribs articulate with the sternum at this readily palpable joint.

                (ii) This joint acts as a hinge, allowing for anteroposterior expansion of the rib cage during inspiration. Movement in this joint is evident over a full respiratory cycle.

    b. Corpus or body (see Figure 19-1)

        (1) Structure. This part of the sternum consists of four fused segments.

            (a) Because this bone has little overlying fat, regardless of body build, it is a preferred site for bone marrow aspiration. It may be split **(sternotomy)** for surgical access to the middle mediastinum and superior mediastinum.

            (b) It is sensitive to nociceptive stimulation; firm rubbing of the sternum with the knuckles can rouse a stuporous patient.

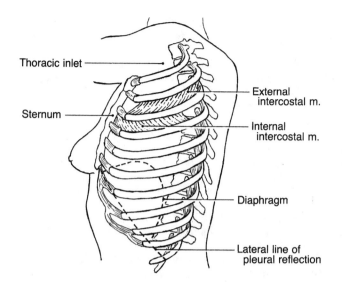

FIGURE 19-2. Musculoskeletal thoracic cage, lateral perspective. Differences in the directionality and extent of the external and internal intercostal muscles are shown in the second and third intercostal spaces, respectively. Also depicted is the normal end-expiratory position of the diaphragm.

   (2) **Articulations**
      (a) It articulates by synchondroses with the **manubrium** at the **sternal angle** (of Louis) as well as with the **xiphoid process.**
      (b) It articulates diarthrodially with the inferior demifacets of the second **costal cartilages** at the sternal angle as well as with the third through sixth costal cartilages.
   c. **Xiphoid process,** along with the adjacent costal margins, defines the **infrasternal angle** or **notch** (see Figure 19-1).
      (1) **Structure.** The xiphoid usually is pointed, but it may be bifid or project anteriorly. It can be palpated in the epigastrium in the infrasternal angle. Cartilaginous in the young, it calcifies during adolescence.
      (2) **Articulation** with the sternum occurs at the **xiphisternal synchondrosis,** which usually fuses in the elderly.
3. **The ribs,** the main portion of the thoracic cage, are usually visible and always palpable anteriorly, laterally, and posteriorly (Figure 19-2; see Figure 19-1).
   a. **Rib cage** (see Figures 19-1 and 19-2)
      (1) Seven pairs of **true ribs** (1–7) articulate with thoracic vertebrae and the sternum. The true ribs become progressively longer.
      (2) Five pairs of **false ribs** (8–12) articulate with thoracic vertebrae and the suprajacent costal cartilage to form the anterior costal margin. The false ribs become shorter. The last two pairs of false ribs (11 and 12) articulate only posteriorly with thoracic vertebrae and are sometimes called **floating ribs.**
   b. **Costal structure**
      (1) The **head** is expanded and articulates with two vertebrae.
      (2) The **neck,** a narrow portion between the head and tubercle, is directed posterolaterally.
      (3) The **tubercle,** where the neck joins the shaft, has a facet for articulation with the transverse process of the corresponding vertebra. Exceptions include the eleventh and twelfth ribs.
      (4) The **shaft** has several features.
         (a) The **costal angle** is where the rib turns anteriorly.
         (b) The **costal groove** indicates the location of the **intercostal neurovascular bundle.** Although the neurovascular structures are protected within this groove, rib fracture may result in injury of the contained structures.
      (5) The **costal cartilage** represents an unossified anterior portion of the rib. During inspiratory movements, the costal cartilages are bent upward and twisted.

This flexion and torsion stores energy, which is subsequently released and used to assist expiratory effort (**extrinsic elastic recoil**).

  c. **Costal articulations**
    (1) The **head** articulates with the costal facets of the corresponding vertebral body and the immediately suprajacent vertebral body via **inferior** and **superior articular facets** (see Figure 11-3). Exceptions include the first, eleventh, and twelfth ribs, which articulate only with their corresponding thoracic vertebrae.
    (2) The **tubercle** of ribs 1–10 articulates with the transverse process of the corresponding vertebra (see Figure 11-3).
    (3) The **costal cartilages** articulate with the manubrium and the body of the sternum (see Figure 19-1).
      (a) The first rib articulates with the manubrium below the sternoclavicular joint by a synchondrosis in which little or no movement is possible.
      (b) The second costal cartilage articulates with the sternum at the sternal angle with one synovial joint on the manubrium and another on the body of the sternum.
      (c) The third to the sixth or seventh costal cartilages of the true ribs articulate with the body of the sternum at the synovial joints.
      (d) The seventh or eighth to tenth costal cartilages of the false ribs articulate with the preceding costal cartilages at synovial joints.
      (e) The eleventh and twelfth ribs (and thirteenth, if present), as floating ribs, end in the musculature of the abdominal wall without anterior articulation.
  d. **Clinical considerations**
    (1) **Anomalous thirteenth ribs**
      (a) A **cervical rib** may arise from the transverse process of vertebra C7 and usually articulates with the anterior one third of the first rib. It may compress the lower trunk of the brachial plexus and the subclavian artery (misnamed the "**thoracic outlet**" **syndrome**).
      (b) A **lumbar ("gorilla") rib** may arise from vertebra L1.
    (2) **Enlargement of the intercostal arteries,** such as occurs with coarctation of the aorta, results in erosion of the costal grooves. This change appears radiographically as scalloping of the inferior edges of the ribs.
    (3) **Costochondritis** is painful inflammation of the synovial joints between the costal cartilages and sternum.
    (4) **Calcification of costal cartilages** may occur in old age. The associated loss of resiliency affects ventilation and predisposes to fracture.
    (5) **Deformities of or injuries to the thoracic cage** may impair ventilation of the lungs, which is the primary function of the thoracic cage.

B. **Movement** of the thoracic cage is necessary for ventilation as well as flexion, extension, abduction, and rotation of the vertebral column.

1. **Axis of costal rotation**
  a. **Definition of the axis.** On each side of the rib cage between ribs 1–10 and the corresponding vertebra are two function costovertebral joints. One occurs at the body (1 in Figure 19-3), the other at the transverse process (2 in Figure 19-3).
    (1) At the **costovertebral joints,** the axis of rotation passes between the head and vertebral bodies and between the tubercle and transverse process.
    (2) The **axis of rotation** extends across the midline anteriorly through the **contralateral costochondral junction.** The left and right axes of rotation cross anterior to the vertebral column.
    (3) Each axis of rotation is constrained by passing through two joints (A–B and C–D in Figure 19-3). Movement (rotation) in either direction about one axis of rotation is described as **one degree of freedom.**
  b. **Functional implications.** As the ribs on each side rotate upward, the crossing axes of rotation produce an upward and outward movement of the sternum—the **pump-handle effect** (see Figure 19-7B).

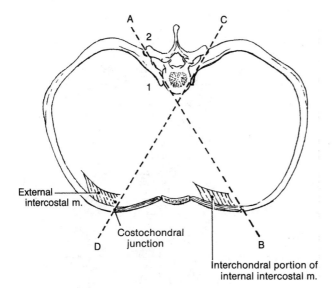

FIGURE 19-3. Rotation of the ribs occurs about an axis that passes through two joints, one between the body of the vertebra and head of the rib (1) and the other between the transverse process of the vertebra and neck of the rib (2). The defined axis of rotation (AB) passes through the contralateral costochondral junction. Because the axes of the left and right ribs cross, the resultant movement of the shared arcs elevates the sternum. Because both the external intercostal muscle and the interchondral portion of the internal intercostal muscle have parallel lines of action and lie on the same side of the axis of rotation (AB), they have the same action.

2. **Group actions.** Muscles either act with other muscles to produce the same action **(synergists)** or against each other to produce opposing actions **(antagonists)**.
   a. **Synergists** have one of the following:
      (1) A parallel fiber direction and pass on the same side of the axis of rotation
      (2) A perpendicular fiber direction and pass on opposite sides of the axis of rotation
   b. **Antagonists** have one of the following:
      (1) A parallel fiber direction and lie on opposite sides of an axis of rotation
      (2) A perpendicular lines of action but lie on the same side of the axis of rotation

III. MUSCULATURE

A. **Extrinsic inspiratory musculature** stabilizes and moves the pectoral girdle, upper limbs, and neck (Figure 19-4). The **pectoralis major** and **pectoralis minor** as well as the **sternomastoid** and **scalenes** (anterior, middle scalene, and posterior) attach to the thoracic cage and elevate the ribs. They are **accessory muscles of respiration** used during extreme inspiratory effort (Table 19-1).

B. **Intrinsic (body wall) musculature** of the thorax and the abdomen consists of an outer layer that runs inferomedially, an intermediate layer that runs superomedially, and a deep layer that tends to run transversely. This three-ply construction provides strength through bias bonding. All intrinsic muscles of the thorax and abdomen are used in respiration.

   1. **External intercostal muscles** constitute the outer layer (see Figures 19-1 and 19-2).
      a. **Attachments.** From the vertebrae, this layer runs between the ribs in a superolateral to inferomedial direction (Table 19-2; see Figure 19-6B). At the costochondral junction, the muscle fibers cease, but the layer of deep fascia continues to the sternum as the **external (anterior) intercostal membrane.**
      b. **Action.** These muscles elevate the ribs for inspiration (see Figure 19-3).
         (1) **Mechanical analysis.** Each fascicle of the external intercostal muscles has a longer lever arm on the lower rib than on the upper rib. Therefore, the lower rib is raised toward the upper when the muscles shorten.
            (a) **Definitions.** In Figure 19-5 I, points A and B are the **axes of rotation** of adjacent ribs; line CD represents the **line of action** of the external inter-

FIGURE 19-4. Extrinsic thoracic musculature. The pectoralis major muscle has been removed. In addition to their nonrespiratory actions, the pectoralis minor muscle elevates the ribs and the sternomastoid muscle elevates the sternum.

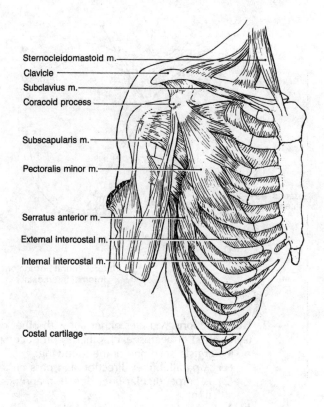

TABLE 19-1. Extrinsic Muscles of Respiration

Muscle	Origin	Insertion	Primary Action	Innervation
Inspiratory: Pectoral musculature				
Pectoralis major				
Sternal head	Sternum and costal cartilages 1–7	Lateral lip of bicipital groove of humerus	Elevates ribs 1–7 when scapula is fixed	Medial pectoral n. (C8–T1, anterior)
Pectoralis minor	Outer surface of ribs 3–5	Coracoid process	Elevates ribs 3–5 when scapula is fixed	Medial pectoral n. (C8–T1, anterior)
Sternomastoid	Mastoid process of cranium	Manubrium, proximal third of clavicle	Elevates sternum and clavicle	Spinal accessory n. (CN XI) and C2–C3
Scalenes	Vertebrae C1–C6	Ribs 1–2	Elevates sternum and clavicle	Twigs from spinal nn. C3–C8
Expiratory: Abdominal musculature				
Rectus abdominis	Pubic arch	Costal cartilages 5–8, lower sternum, xiphoid process	Stretches and raises diaphragm	Nn. T7–T12
External oblique	Ribs 10–12, iliac crest	Linea alba	Stretches and raises diaphragm	Nn. T7–T12
Internal oblique	Ribs 10–12, thoracodorsal fascia	Linea alba	Stretches and raises diaphragm	Nn. T7–T12, iliohypogastric and ilioinguinal nn. (L1)
Transverse abdominis	Ribs 5–12, thoracodorsal, fascia, iliac crest	Linea alba	Stretches and raises diaphragm	Nn. T7–T12, iliohypogastirc and ilioinguinal nn. (L1)
Quadratus lumborum	Iliac crest	Rib 12	Stabilizes and lowers 12th rib	Twigs from nn T12–L4

TABLE 19-2. Intrinsic Muscles of Respiration

Muscle	Origin	Insertion	Primary Action	Innervation
Inspiratory Muscles				
External intercostals	Inferior border of ribs 1–11	Superior border of ribs 2–12	Elevates ribs 2–12	Nn. T1–T12
Internal intercostals, interchondral portion	Inferior border of costal cartilages 1–9	Superior border of costal cartilages 2–10	Elevates ribs 2–10	Nn. T1–T9
Expiratory muscles				
Internal intercostals, intercostal portion	Superior border of ribs 2–12	Inferior border of ribs 1–11	Depresses ribs 1–11	Nn. T1–T12
Transverse thoracic	Lower sternum, xiphoid process	Ribs 2–6	Depresses ribs 2–6	Nn. T3–T7
Diaphragm	Costal margin	Central tendon	Lowers diaphragm	Phrenic n. (C3–C5), spinal nn. T12–L2

costals; line AC is the **lever arm** of the muscle on the upper rib; line BD is the **lever arm** of the lower rib.

  (b) **Movement.** Because lever arm BD is greater than lever arm AC and because muscle tension (T) is equal at attachment points C and D, then (T) × (BD) >> (T) × (AC). Thus, the lower rib will move toward the upper rib from point D to point D'.

 (2) **Summation effect.** Each subsequent thoracic level is raised not only by the action of its external intercostals but also by the sum of the amounts of contraction within the external intercostals of each superior thoracic level.

 2. **Internal intercostal muscles,** with a fiber direction opposite to that of the external intercostals, constitute the intermediate layer. Passage of the axes of rib rotation through the costochondral junctions divides these muscles into two functionally distinct groups (see Table 19-2).

  a. **Internal intercostal portion** or **internal intercostal muscles proper** (see Figure 19-1)

   (1) **Attachments.** Lateral to the costochondral junction, the intermediate layer runs between the ribs along an inferolateral to superomedial direction (see

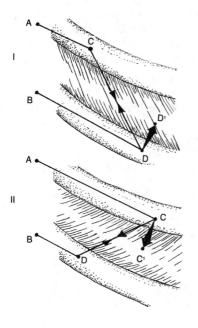

FIGURE 19-5. Vectorial analysis of intercostal muscle action on the right side. *(I)* On contraction of the right external intercostal muscles, lever arm BD is longer than lever arm AC, resulting in elevation of the lower rib toward the upper rib. *(II)* On contraction of the right internal intercostal muscles proper, the lever arm AC is longer than the lever arm BD, depressing the upper rib toward the lower rib.

Figure 19-6B and Table 19-2). At the angle of the rib, the muscle fibers cease, but the layer of deep fascia continues to the vertebral column as the **internal (posterior) intercostal membrane**.

  (2) **Action.** This portion of the intercostal layer depresses the ribs in forced expiration (see Figure 19-3).

  (a) **Mechanical analysis.** Each muscle fascicle has a longer lever arm on the upper rib than on the lower rib. Hence, the upper rib is drawn toward the lower rib.

  (i) **Definitions.** In Figure 19-5, A and B are the **axes of rotation** of adjacent ribs; line CD is the **line of action** of the internal intercostals; line AC is the **lever arm** of the muscle inserting onto the upper rib; and line BD is the **lever arm** of the muscles inserting onto the lower rib.

  (ii) **Movement.** Because lever arm BD is less than lever arm AC and because muscle tension (T) is equal at attachment points C and D, then $(T) \times (BD) << (T) \times (AC)$. Thus, the upper rib will move toward the lower rib with point C moving to C'.

  (b) **Summation effect.** The twelfth rib is stabilized by the quadratus lumborum. Each subsequent thoracic level is lowered, not only by the action of its internal intercostal muscle proper, but also by the sum of the amounts of contraction within each lower level.

  b. **Internal interchondral portion**

  (1) **Attachments.** Lying beneath the external intercostal membrane, these muscles run between the cartilaginous portions of ribs 1–10 in an inferolateral to superomedial direction (see Figure 19-1 and Table 19-2).

  (2) **Action.** This portion of the internal intercostal layer elevates the ribs for inspiration (see Figure 19-3).

  (a) **Mechanical analysis** is the same as for the external intercostal muscles in Figure 19-5.

  (b) **Synergists.** Because the internal interchondral portion of the internal intercostal lies on the same side of the axis of rib rotation and has the same fiber direction as the external intercostal muscle of the opposite side of the chest, but on the same side of the axis of rotation, these muscles are synergists.

  (c) **Antagonists.** Because the interchondral portion of the internal intercostal muscle and the internal intercostal proper have the same fiber direction but lie on opposite sides of the axis of rotation, these muscles are antagonists.

  c. **"Innermost intercostals"** are functionally part of the internal intercostal muscle group. As the intercostal neurovascular bundle passes obliquely through the intermediate muscle layer, it separates the innermost intercostals (see Figure 19-6B). As expected, their function is the same and the distinction is artificial.

3. **Transverse thoracic (sternocostalis) muscles,** the innermost layer, frequently are poorly developed (see Table 19-2).

  a. **Attachments** are from the xiphoid process and the lower body of the sternum to ribs 3–6.

  b. **Action** is to assist lowering the ribs in forced expiration.

C. **Diaphragm.** This musculofibrous structure separates the thoracic and abdominal cavities.

1. **Structure.** The diaphragm is dome-shaped, arching into the thoracic cavity (see Figures 19-2 and 26-4).

  a. **Regions.** Subdivision reflects the embryologic development.

  (1) The **costal portion** arises from the ribs that form the costal margin.

  (2) The **sternal portion** arises from the xiphoid process and lower sternum.

  (3) **Crura (legs) of the lumbar portion** arise from the bodies of vertebrae L1–L3 (see Figure 26-4).

  b. The **central tendon** is the common insertion of the diaphragm muscle. It is formed by the muscle arching upward and converging centrally. The right hemidi-

aphragm rises higher into the thoracic cavity than the left because the liver lies in the upper right quadrant of the abdomen.
  c. The **medial lumbocostal arch (medial arcuate ligament),** formed as the diaphragm arches from the vertebral bodies to the transverse process of L1, bridges the uppermost origins of the **psoas major muscle** and the **sympathetic chain** (see Figure 26-4).
  d. The **lateral lumbocostal arch (lateral arcuate ligament),** formed as the diaphragm arches from the transverse process of L1 to the tip of the twelfth rib, bridges the **quadratus lumborum muscle,** which stabilizes and depresses the twelfth rib during expiration (see Figure 26-4).

2. **Apertures** (see Figure 26-4)
  a. The **aortic hiatus,** between the diaphragmatic crura in the midline about the T12 vertebral level, contains the aorta, thoracic duct, and azygos vein.
  b. The **esophageal hiatus,** in the muscular portion of the left hemidiaphragm about the T10 vertebral level, contains the esophagus as well as the left and right vagus nerves. It is a frequent site of stomach herniation **(hiatus hernia).**
  c. The **caval hiatus,** in the tendinous portion about the T8 level, contains the inferior vena cava and branches of the right phrenic nerve.
  d. The **splanchnic nerves** and the **hemiazygos vein** usually pierce the crura.

3. **Development**
  a. The **source of diaphragmatic tissue** is six primordia (some of them paired) that fuse to form the definitive diaphragm.
    (1) A major portion of the diaphragm develops in the cervical region and descends into the coelomic cavity with the heart.
    (2) The diaphragm also includes contributions from the septum transversum, pleuroperitoneal membranes, and somatic wall.
  b. **Anomalies.** Failure of these primordia to fuse properly may result in anatomic defects through which herniation may occur.

4. **Action.** The diaphragm is involved in inspiration (see Table 19-2). The range of levels through which the diaphragm moves depends on body type, but maximal movement is between T5 and T11.
  a. **Contraction.** Because the muscle fibers of the diaphragm are radially arranged and insert into the central tendon, contraction pulls the dome inferiorly into the abdomen, thereby increasing the thoracic volume for inspiration.
  b. **Accommodations.** Because the pericardial cavity and posterior mediastinum rest on its central tendon and abdominal organs lie beneath it, the diaphragm exerts dynamic effects on related structures. The position of the thoracic and abdominal organs varies with the respiratory cycle. Conversely, postural changes alter the ventilatory effectiveness of the diaphragm.

5. **Innervation**
  a. The **phrenic nerves,** which arise from levels C3–C5, provide principal motor and sensory innervation to and from the diaphragm (see Figure 22-1).
    (1) A phrenic nerve lesion or compression may paralyze one hemidiaphragm.
    (2) Sensation, such as from pleurisy, traveling to the C3–C5 spinal levels refers to the dermatomes over the neck and shoulder.
  b. **Intercostal nerves** provide sensory innervation at the costal margin of the diaphragm (see Figure 29-6A). The **lower thoracic and upper lumbar nerves** provide innervation to the crura.
    (1) Peripheral pleural sensation, traveling to the T6–T11 spinal levels, is perceived at the costal margin.
    (2) A lumbar nerve lesion may paralyze one crus.
  c. Voluntary control of the diaphragm is subject to override by the respiratory center in the brain stem.
  d. **Hiccups** are recurrent spasms of the diaphragm. Phrenicectomy is sometimes undertaken as a last resort to relieve chronic hiccups.

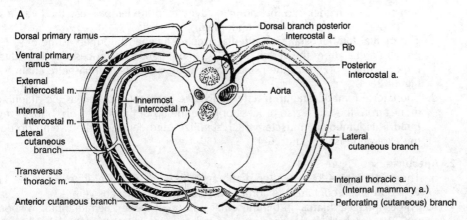
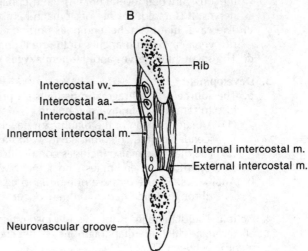

FIGURE 19-6. Thoracic innervation and blood supply. *(A)* Branches of the right spinal nerve in relation to the rib and the left posterior intercostal artery in relation to the intrinsic thoracic muscles. *(B)* Relations of the intercostal neurovascular bundle to the ribs and intrinsic thoracic musculature.

IV. VASCULATURE

A. **Organization.** The segmented skeleton of the thoracic cage is overlaid by segmentally innervated musculature (myotome) and skin (dermatome). On both sides, each thoracic segment has a rib, a **neurovascular bundle,** and three layers of intercostal musculature interposed between the ribs (Figure 19-6). Each **neurovascular bundle** is composed of a posterior intercostal vein, a posterior intercostal artery, and the anterior primary ramus of a spinal nerve.

1. **Posteriorly,** the neurovascular bundle lies immediately behind the inferior edge of each rib in the **neurovascular (intercostal) groove** (see Figure 19-6B).

2. **Laterally,** the neurovascular bundle gives off smaller collateral branches, which move inferiorly across the intercostal space to lie immediately above the next rib anteriorly.
   a. **Costal fractures,** posteriorly or laterally, may tear the associated vascular structures. If the parietal pleura that lines the pleural cavity is also torn by the sharp edges of a fractured rib, blood may enter the pleural cavity **(hemothorax).**
   b. **Pleural taps (thoracentesis)** usually are performed in the posterior or midaxillary line just above the superior edge of a rib. This location avoids interference with the main intercostal neurovascular bundle (see Figure 19-6B). This procedure usually is not performed below the ninth rib because of the proximity of the liver.

B. Arterial supply

1. **The aorta** supplies the third to eleventh posterior intercostal arteries and the subcostal artery (see Figure 19-6A).

2. **The internal thoracic (mammary) artery,** a branch of the subclavian artery, supplies the following (see Figure 19-6A):
   a. **First through sixth anterior intercostal arteries,** which anastomose with the highest and posterior intercostal arteries
   b. **Superior epigastric artery,** which anastomoses in the anterior abdominal wall with the inferior epigastric artery, a branch of the external iliac artery (see Figure 23-3)
   c. **Musculophrenic artery,** which anastomoses with the seventh to tenth posterior intercostal arteries

3. **The highest or superior intercostal artery** arises from the subclavian artery or the costocervical trunk. It supplies the first and second posterior intercostal arteries (see Figure 6-5).

4. **Pathways of collateral circulation**
   a. **Aortic stenosis** or **coarctation of the aorta** (a congenital or slow-onset narrowing to ultimate occlusion) promotes anastomotic connections between the **posterior** and **anterior intercostals.**
      (1) One important pathway is between the **posterior** and **anterior intercostals.** Hypertrophy of the intercostal arteries caused by increased flow through collateral pathways results in erosion of the inferior edges of the ribs along the neurovascular groove. This change appears radiographically as **notching of the ribs,** a scalloping of the inferior edges of the ribs.
      (2) Another important collateral pathway is that between the **superior** and **inferior epigastric arteries.**
   b. **Coronary bypass.** Because the collateral circulation within the chest wall is strong, the internal thoracic artery may be diverted into the coronary circulation to provide a reliable blood supply to the myocardium.

C. Venous return.
The veins of the chest wall parallel the arteries.

V. INNERVATION

A. Organization.
Each of the 12 pairs of **thoracic nerves** is numbered after the vertebra just craniad to the intervertebral foramen through which each nerve passes; for example, nerve T4 emerges below vertebra T4.

1. **Dorsal sensory roots,** which emerge from the dorsal aspect of the spinal cord, consist of afferent nerve fibers that mediate input from the sensory receptors of the body (see Figure 4-4).
   a. **Afferent nerve classification**
      (1) General somatic afferent (GSA) neurons carry sensation from the body wall.
      (2) General visceral afferent (GVA) neurons carry sensation from the viscera.
   b. **Dorsal root ganglia** within the intervertebral foramina contain cell bodies of the sensory neurons.
   c. A **dermatome** is the region of skin supplied by the dorsal root of each spinal nerve (see Chapter 4 III and Figure 4-3).

2. **Ventral (motor) roots,** which emerge from the ventral aspect of the spinal cord, consist of efferent nerve fibers that carry output from the spinal cord to the effectors (see Figure 4-4).
   a. **Efferent nerve classification**
      (1) General somatic efferent (GSE) neurons innervate somatic musculature.

(2) **General visceral efferent (GVE, autonomic)** neurons innervate involuntary effectors.
- b. The **ventral gray matter** of the spinal cord (lateral and ventral horns) contains cell bodies of ventral root neurons.
- c. A **myotome** is a muscle group supplied by the fibers of a ventral root (see Chapter 4 III and Figure 4-3).
3. **Spinal nerves** are formed by the merger of dorsal and ventral roots in the intervertebral foramen (see Figures 4-4 and 19-6A).

B. Distribution. Each spinal nerve divides almost immediately into a **dorsal primary ramus** and a **ventral primary ramus** (see Figure 19-6A).
1. **The posterior (dorsal) primary ramus** arches dorsally to innervate the posterior aspect of the associated dermatome and myotome.
2. **The anterior (ventral) primary ramus** continues ventrally as the intercostal nerve in company with the intercostal artery and vein. The intercostal nerve innervates the anterior aspect of the associated dermatome and myotome, for example, the parietal pleura, intercostal muscles, and skin, through two branches (see Figure 19-6A).
    a. The **lateral cutaneous (perforating) branch** penetrates the internal and external intercostal muscle layers at approximately the midaxillary line. It subsequently bifurcates into **anterior** and **posterior branches of the lateral cutaneous nerve** to supply the lateral aspect of the dermatome.
    b. The **anterior cutaneous (perforating) branch** penetrates the internal intercostal muscles and the external intercostal membrane just lateral to the sternum. It bifurcates immediately into **medial** and **lateral branches of the anterior cutaneous nerve** that supply the anterior aspect of the dermatome with some overlap across the midline.
    c. **Clinical considerations**
        (1) **Landmark.** Nerve T4 supplies the dermatome that contains the nipple; T10 supplies that containing the umbilicus.
        (2) **Intercostal nerve block** is accomplished by injecting an anesthetic immediately beneath the inferior edge of a rib, posteriorly.
        (3) **Injury** to an intercostal nerve, producing paralysis of intercostal musculature, is evident as a sucking-in of the affected intercostal space on inspiration and a bulging on expiration.

VI. RESPIRATORY MECHANICS

A. Mechanical respiration. Ventilation is divided into **costal respiration** and **diaphragmatic respiration.** In both, air is moved into and out of the lungs by muscular and mechanical forces.

1. **Definitions** (relative to atmospheric pressure)
    a. **Intrathoracic (endothoracic, intrapleural) pressure** is between lungs and thoracic wall.
    b. **Intrapulmonic (pulmonary) pressure** is within the lungs.
2. **Dynamics of air flow.** The second law of thermodynamics states that flow (of air) must always be from higher energy (pressure) to lower. Boyle's law relates the reciprocal relationship between pressure and volume: $P_1V_1 = P_2V_2$.
    a. **Inspiration** increases thoracic volume with a concomitant decrease in the intrapulmonic pressure. Therefore, air will flow into the lungs until the pressures equalize.
    b. **Expiration** decreases thoracic volume with a concomitant increase in intrapulmonic pressure. Therefore, air must flow out of the lungs (if the glottis is open) until the pressures equalize.

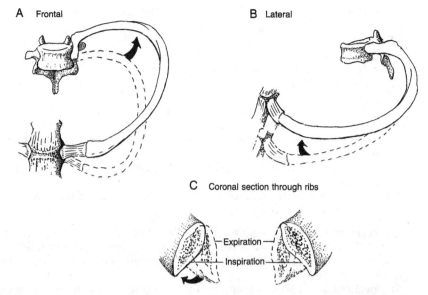

FIGURE 19-7. Respiratory movements of the ribs. *(A)* Bucket-handle motion of the left ribs. Rotation about the axes produces elevation of the ribs, which increases the transverse diameter of the thoracic cage. *(B)* Pump-handle motion of the ribs and sternum. Rotation about the crossed axes produces elevation of the sternum, which increases the anteroposterior diameter of the thoracic cage. *(C)* Because the cross-sectional shape of the ribs is ellipsoid, rib rotation with elevation also increases the diameter of the thoracic cage.

B. **Costal ventilation** changes thoracic volume by moving the ribs.
   1. **Mechanics** (see section II B 1)
      a. **"Bucket-handle" effect.** The external intercostals and the interchondral portion of the internal intercostals move the ribs upward and outward, thereby **increasing the transverse thoracic diameter** (Figure 19-7A).
      b. **"Pump-handle" effect.** Concomitantly, because the axes of costal rotation on each side cross in front of the vertebral column, elevation of the ribs moves the sternum forward and upward. This movement **increases the anteroposterior thoracic diameter** (see Figure 19-7B). The pump-handle action changes the sternal angle from about 160° in full inspiration to about 180° in full expiration.
      c. **Rotation effect.** Because the interior surface of the rib is concave and the rib rotates around its axis of rotation, **the transverse thoracic diameter increases** further when the ribs are elevated (see Figure 19-7C).
   2. **In normal (relaxed) costal inspiration,** the external intercostal muscles and the interchondral portion of the internal intercostals increase thoracic volume. The effect is to lower intrathoracic and intrapulmonary pressures from about -4 cm H_2O to between -5 and -10 cm H_2O, producing the **tidal volume.** With a tidal volume of 0.5 L/breath and a respiratory rate of 16–20 breaths/minute, the lungs are ventilated with approximately 7 L of air per minute.
      a. The **external intercostals** elevate the ribs, thereby increasing the transverse diameter of the rib cage.
      b. The contralateral **interchondral portion of the internal intercostals** is on the same side of the axis of rotation, has the same directionality, and, as such, acts synergistically with the external intercostal muscles.
   3. **Relaxed costal expiration** requires little, if any, muscular effort. **Tidal volume** is normally expelled by elastic recoil.
      a. **Extrinsic elastic recoil (thoracic compliance)**
         (1) Elevation of the ribs during inspiration deforms the costal cartilages. When the external intercostals relax, stored energy is released, rolling the ribs back

to their resting (end-expiratory) position. The opposite occurs when the rib cage is forcibly depressed in full expiration.
- (2) The importance of this mechanism diminishes with age because calcification of the costal cartilages decreases thoracic compliance.
- b. **Intrinsic elastic recoil (pulmonary compliance)**
  - (1) The elastic fiber component of the lung parenchyma causes the lung to collapse. Thus, the lung recoils after inspiratory stretching.
  - (2) A thin film of fluid between the visceral and parietal pleura produces a high adhesive effect **(surface tension).** Thus, as the lungs deflate, they tend to pull the chest wall inward. Although this negative force, termed **intrathoracic pressure,** cannot be measured accurately, manometry suggests that it may range from -5 cm H_2O (relative to atmospheric pressure) to -15 cm H_2O during relaxed tidal respiration.
  - (3) Diseases that reduce pulmonary compliance, such as **emphysema,** affect this aspect of ventilation.

4. **Forced costal inspiration** is required for ventilation more than tidal volume and contributes to the **inspiratory reserve volume** (about 3 L). This mechanism, using the accessory muscles of respiration (see Figure 19-4), comes into action when respiratory requirements exceed 7–10 L/minute.
   a. The **pectoral muscles,** which attach to the thoracic cage, assist maximal elevation of the ribs to increase the transverse diameter of the thorax.
   b. The **sternocleidomastoid and scalene muscles** elevate the manubrium and first rib to increase the anteroposterior diameter of the thorax. (Recall the posture of a runner's neck and shoulders at the finish line.)

5. **Forced costal expiration,** in addition to expelling the inspiratory reserve volume and the tidal volume, provides **expiratory reserve volume** (about 1.2 L).
   a. **Internal intercostal** and **transverse thoracic muscles** decrease the transverse and anteroposterior diameters of the thoracic cage so that intrathoracic pressure becomes positive (reaching as high as 300 mm Hg during a sneeze).
   b. The **quadratus lumborum,** running from the iliac crest to the twelfth rib, lowers and stabilizes the twelfth rib so the thoracic cage can be depressed effectively.

C. **Diaphragmatic ventilation** changes thoracic volume by lowering the diaphragm.
   1. **Diaphragmatic inspiration** contributes to both **tidal** and **inspiratory reserve volume.**
      a. **Diaphragmatic contraction** lowers the level of the dome, which increases the superior-inferior dimension of the thoracic cavity and thereby lowers intrathoracic pressure. Vigorous contraction may lower the dome as much as 10 cm.
      b. **Displacement of the abdominal contents** occurs during contraction of the diaphragm, permitted by relaxation of the abdominal wall muscles.
   2. **Abdominal expiration** normally contributes to **expiratory reserve volume,** but it may also contribute to tidal volume.
      a. The abdominal musculature is antagonistic to the diaphragm.
      b. Contraction of the abdominal muscles as a unit forces the abdominal viscera under the diaphragm (much like a fluid piston), thereby stretching the relaxed diaphragm into the thoracic cavity. This change decreases thoracic volume and thereby increases intrathoracic pressure.
   3. **Effects of gravity on ventilation** (Figure 19-8)
      a. When a patient is sitting or standing, the force of gravity on the abdominal and thoracic viscera tends to lower the diaphragm.
      b. In the horizontal position, gravity acts on abdominal viscera and pushes the diaphragm into the thoracic cavity, making contraction of the diaphragm more strenuous.
      c. The plus and minus effects of gravity on ventilation explain the rationale for propping up a bedridden person who has respiratory difficulty (especially if elderly or obese).

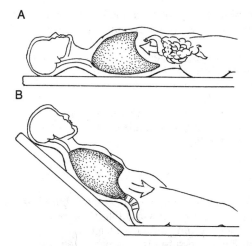

FIGURE 19-8. Effects of posture on respiratory effort. *(A)* In a horizontal position, the weight of the abdominal viscera tends to stretch the diaphragm into the thoracic cavity. The patient must overcome this additional gravitational force on inspiration. *(B)* A slight inclination redistributes the abdominal viscera and facilitates inspiratory effort.

D. **Approximate ventilation capacities**

1. **Tidal volume** (0.5–0.6 L): Limited to normal quiet inspiration and expiration. Inspiration is accomplished by the external intercostals, interchondral portions of the internal intercostal layer, and (to a variable extent) diaphragm. Return is by elastic recoil of the thoracic cage and the lungs (abdominal muscle contraction will occur if the diaphragm participated).

2. **Inspiratory reserve volume** (3 L): Begins at the end of normal inspiration. It is achieved by accessory muscles of inspiration along with the external intercostals, interchondral portion of the internal intercostal layer, and diaphragm. Return is by elastic recoil of the thoracic cage and lungs, assisted as necessary by expiratory muscles.

3. **Expiratory reserve volume** (1.2 L): Begins at end of normal expiration. It is accomplished by internal intercostals, the quadratus lumborum, and abdominal muscles. Return to the resting position is accomplished by an expansionary elastic recoil of the thorax.

4. **Vital capacity** (4.8 L): Sum of tidal volume plus inspiratory and expiratory reserve volumes

5. **Anatomic dead space** (1.2 L): Nonrespiratory volume that cannot be expelled from the lungs and airways

E. **Clinical considerations**

1. **Normal balance between costal and diaphragmatic ventilation** depends on body type, posture, age, sex, activity, state of health, and even clothing. Except in the following circumstances, ventilation tends to be an unequal mix of costal and diaphragmatic contributions.
   a. **Diaphragmatic ventilation.** Young children and elderly persons with reduced thoracic compliance tend to breathe diaphragmatically. Also, horn and wind instrument players and vocalists develop greater control over diaphragmatic and abdominal musculature.
   b. **Costal ventilation.** Obese people, persons wearing girdles or corsets, and women in advanced pregnancy cannot effectively contract the diaphragm and therefore favor costal ventilation.

2. **Chest fractures**
   a. **Fracture of several ribs** in two locations (usually anteriorly and posterolaterally) produces **"flail chest."**
      (1) The structural integrity of the thoracic cage is diminished with the loss of effective lever arms for the thoracic musculature. As the diaphragm is lowered,

the flail portion of the wall cannot resist decreased intrathoracic pressure and is drawn inward. Conversely, as the abdominal muscles contract in expiration, the flail portion is pushed outward.

(2) The result is reduced ventilation and a decreased ventilation-perfusion ratio.

b. **Pneumothorax and hemothorax.** Fractured ribs may penetrate the thoracic wall or tear visceral pleura. Either circumstance can cause **pneumothorax.** Tearing associated blood vessels can lead to **hemothorax.**

(1) The result is reduced ventilation and a decreased ventilation-perfusion ratio.

(2) A negative-pressure chest tube (-10 cm H_2O) will reinflate the lung and restore to normal the ventilation-perfusion ratio.

3. **A phrenic nerve lesion** may paralyze one hemidiaphragm.
    a. **Paradoxic respiratory movements** of a paralyzed hemidiaphragm result in reduced ventilation.
        (1) During inspiration, expansion of the thoracic cavity with decreased intrathoracic pressure draws a paralyzed hemidiaphragm upward into the thorax, resulting in little or no increase in thoracic volume on the affected side.
        (2) During expiration, increased intrathoracic pressure draws a paralyzed hemidiaphragm inferiorly toward the abdomen, resulting in little decrease in the thoracic volume on the affected side.
    b. **Paralysis of one hemidiaphragm** is evident on radiographs taken at full inspiration. A paralyzed hemidiaphragm does not lower and the lung appears more radiopaque.

4. **Respirators**
    a. **Negative-pressure devices,** such as the Drinker ("iron lung" or tank) respirator, cyclically lower the extrathoracic and, therefore, the intrapulmonary pressure below atmospheric pressure, mimicking natural negative-pressure breathing.
    b. **Positive-pressure devices** cyclically elevate the atmospheric pressure above normal and above intrapulmonary pressure so that air is forced into the lungs.

Chapter 20
Pleural Cavities and Lungs

I. INTRODUCTION

A. Basic principles

1. The **vertebrate body** consists of a **somatic wall** surrounding the **coelom,** within which the hollow viscera are suspended by mesenteries.
2. Division of the common coelom by the diaphragm produces the **thoracic cavity** and the **peritoneal (abdominopelvic) cavity.**

B. Subdivisions.
The thoracic cavity contains the pleural cavities and the mediastinum (see Figure 20-1).

1. **The pleural cavities** contain the normally inflated lungs, suspended by mesentery, as well as the pleural space. The **pleural space** separates the lungs and the walls of the pleural cavity.
   a. The **left and right pleural cavities** are independent of each other.
   b. The **pleural space** around the lung in each pleural cavity is a potential space, containing only a monolayer of fluid. It becomes apparent only when air or fluid displaces lung tissue.
2. **The mediastinum** is an intrapleural region that separates the pleural cavities. It is subdivided into four regions.
   a. The **anterior mediastinum** contains the inferior portion of the thymus gland and sternopericardial ligaments (see Figure 22-1).
   b. The **middle mediastinum** contains the pericardial cavity with its enclosed heart, phrenic nerves, and lung roots (see Figure 21-1).
   c. The **superior mediastinum** contains the superior portion of the thymus gland, aortic arch and associated great vessels, trachea and main stem bronchi, esophagus, superior vena cava and its branches, and autonomic nerves to the heart, lungs, trachea, and esophagus (see Figure 22-1).
   d. The **posterior mediastinum** contains the esophagus, thoracic aorta, thoracic duct, inferior vena cava, azygos and hemiazygos veins, sympathetic chains, and vagus nerves (see Figure 22-4).

II. PLEURAL CAVITIES

A. Parietal pleura.
This layer of serous mesothelium lines the pleural cavities. It lies beneath the endothoracic fascia, which underlies the ribs as well as the internal intercostal and transverse thoracic muscles.

1. **Divisions.** Topographic terms describe the relation of the parietal pleura to adjacent structures.
   a. **Costal pleura** lines the thoracic wall.
   b. **Diaphragmatic pleura** lines the thoracic surface of the diaphragm.
   c. **Mediastinal pleura** lines the mediastinum and sides of the pericardial cavity.
   d. **Cervical pleura** (cupula of the pleura) lies in the base of the neck at the apex of the pleural cavities.

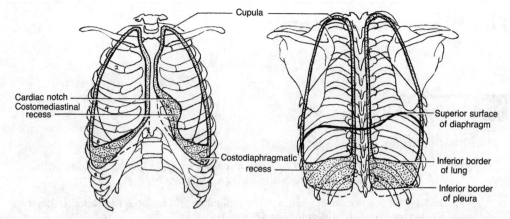

FIGURE 20-1. Pleural cavities and lungs. Because the lungs normally do not completely fill the pleural cavities, costodiaphragmatic and costomediastinal recesses occur. The pleural recesses provide potential space for lung expansion during inspiration. The lines of pleural reflection vary between the extremes (*dashed lines*).

   2. Lines of reflection define the pleural cavities (Figure 20-1).
      a. Anterior border of the **right pleural cavity**
         (1) This border runs from the cupula (bounded by the base of the neck, clavicle, and trapezius muscle) through the right sternoclavicular joint. It then passes through the sternal angle in the midline and drops in the midline to the level of the sixth costal cartilage or xiphoid process. It then swings laterally to run just above the anterior costal margin. The lower extent of this line of reflection is variable.
         (2) A right subcostal incision carried into the sternocostal angle can nick the right pleural cavity, producing pneumothorax.
      b. Anterior border of the **left pleural cavity**
         It is similar to that of the right pleural cavity. At the level of the fourth rib, it deviates laterally and then inferiorly to form the **cardiac notch.** It then diverges laterally at the level of the sixth or seventh costal cartilage to parallel the anterior costal margin.
      c. The **inferior line of reflection** in both pleural cavities moves slightly superior to the costal margin. It passes over the costochondral junction of the eighth rib and over the midshaft of the eleventh rib.
      d. The **posteroinferior border** of both pleural cavities runs horizontally across the twelfth rib at midshaft to reach the vertebral column at about the twelfth intervertebral level.
         (1) The posteroinferior borders are variable.
         (2) A posterior abdominal incision inadvertently carried along the twelfth rib runs a high probability of penetrating the pleural cavity and causing a pneumothorax.
      e. The **vertebral border** of both pleural cavities runs vertically between the twelfth intervertebral level and the cupula.
   3. Innervation
      a. The **costal pleura** is innervated by spinal nerves T1–T11.
      b. The **peripheral diaphragmatic pleura** is innervated by the lower **intercostal nerves.** Pleuritic pain occurs over the costal margin near the source of irritation.
      c. The **central diaphragmatic pleura** and **mediastinal pleura** are innervated by the **phrenic nerves,** which arise from C3–C5 (see Figure 22-1). Pain may be referred over the distribution of nerves C3–C5 to the neck and shoulder.
   4. Clinical considerations
      a. Inflammation of the pleura (**pleurisy**) is accompanied by pain (accentuated by respiratory movement). A rasping sound (**friction rub**) may be heard on auscultation. If an effusion of fluid occurs, the friction rub disappears.

    b. Mesothelioma, a highly malignant tumor originating in the pleura, results in considerable pain in the thoracic cage.

B. **Mesenteries (pulmonary ligaments)** run along the medial aspect of each lung from the hilus (where the vessels and air passages enter the lung) to the base (diaphragmatic surface). These ligaments suspend each lung from the mediastinal wall.

  1. **A mesentery** (or ligament) is formed by two layers of back-to-back mesothelium and supports a visceral structure from the body wall. The pulmonary ligament (a true mesentery) is formed where the parietal pleura of the anterior and posterior walls of the thoracic cavity is reflected off the mediastinal wall toward the lungs. Mesenteries also provide routes for neurovascular structures.

  2. **Surgical release** of the pulmonary ligament is necessary for mobilization of a lung and in order to visualize the diaphragmatic surface and the contents of the posterior mediastinum.

C. **Visceral pleura.** This layer of serous mesothelium covers the surface of the lungs.

  1. The mesothelial layers of the pulmonary ligament diverge at the edge of the lung and pass onto each side of it, becoming the visceral (pulmonary) pleura.

  2. The parietal and visceral pleurae, with the intervening pulmonary ligament, present a continuous surface.

D. **The pleural space** is the potential space between the parietal and visceral pleural layers, occupied by a thin layer of fluid.

  1. **Pleural recesses** are extensions of pleural spaces that are devoid of lung tissue (see Figure 20-1). They exist because the pleural cavities are larger than pulmonary volumes. Parietal (costal) pleura contacts parietal (diaphragmatic or mediastinal) pleura, except in deep inspiration, when the lung expands into, but never completely fills, the pleural recesses.

    a. The **costomediastinal recess** is on the left side between the mediastinal pleura and the costal pleura adjacent to the cardiac notch (see Figure 20-1). It has less significance on the right side.
      (1) The depth of this recess is variable.
      (2) Needle insertion into the pericardial cavity or heart adjacent to the costomediastinal recess avoids penetration of the pleural cavities and the associated risk of pneumothorax.

    b. The **left and right costodiaphragmatic recesses** lie between the costal pleura and the diaphragmatic pleura (see Figure 20-1).
      (1) With a patient standing or sitting, the lowest part of this recess is the point at which the pleura passes over the eleventh rib posterolaterally. In this position, pleural fluid accumulates in the costodiaphragmatic recess.
      (2) Because the lung floats, excessive fluid produces a level that can be determined **radiographically or by percussion.**

  2. **Pleural fluid,** normally a thin film or monolayer, lubricates the parietal and visceral pleura. It allows the visceral surface to slide easily over the parietal surface with respiratory movements. This thin fluid layer also provides **surface tension** between the parietal and visceral pleural layers that keeps the lung inflated close to the limits of the pleural cavities.

E. **Functional considerations**

  1. **Normal conditions.** The sizes of the pleural recesses vary with the respiratory cycle and with the degree of inspiratory and expiratory effort. As the thoracic cavities and lungs expand on inspiration, a volume of lung tissue equivalent to the inspired volume of air moves into the pleural recesses. The changes in pulmonary volume between inspiration and expiration can be delineated by percussion.

  2. **Air in the pleural space (pneumothorax)** is a life-threatening condition that reduces ventilation capacity.

    a. Etiology. Surface tension normally holds the lung surface against the thoracic wall. Thus, the lung remains inflated against its own intrinsic elasticity.
       (1) Air in the pleural space disrupts the surface tension so that the lung collapses. Conditions include penetrating thoracic wounds, spontaneous rupture of a pulmonary bulla, a tear of abnormally fused pulmonary and parietal pleura, and iatrogenic piercing of the pleural cavity.
       (2) Unbalanced intrinsic elasticity results in collapse of the lung—**pneumothorax.**
    b. Classification
       (1) Sucking pneumothorax occurs if air enters and leaves the pleural space through the wound. This condition is accompanied by hyperexpansion of the chest wall on the normal side and a **mediastinal flutter** (a slight shift of the mediastinal mass toward the normal side during inspiration and toward the injured side during expiration).
       (2) Tension pneumothorax occurs if the wound acts as a check valve so that air leaking into a pleural space during inspiration cannot be expelled from the pleural space during expiration. The resulting increase in intrathoracic pressure permanently forces the mediastinal contents toward the normal side **(mediastinal shift)**, thereby reducing the vital capacity of the normal lung.
    c. Reduced ventilation–perfusion ratio
       (1) Because blood continues to perfuse through a collapsed lung (little or no ventilation is occurring), a significant portion of the cardiac output is not oxygenated. The result is **dyspnea** and even **cyanosis.**
       (2) Mediastinal flutter, or worse, a mediastinal shift, further reduces gas exchange by interfering with ventilation of the normal lung. Clearly, pneumothorax is a surgical emergency.
    d. Diagnosis. In addition to dyspnea and cyanosis, pneumothorax produces a **hyper-resonant percussive tone.** Radiographically, the pleural cavity is radiolucent, and the free edge of the collapsed lung is visible.

3. Excess fluid in the pleural space reduces vital capacity of the lung and decreases the ventilation–perfusion ratio.
    a. Fluid from pathologic processes (**hydrothorax** or **pyothorax,** if infected) or traumatic events (**hemothorax** or **chylothorax**) pools in the costodiaphragmatic recesses when the person is upright.
       (1) Bleeding may occur from large or small thoracic vessels. Chyle may issue from a torn thoracic duct. Fluid may also result from direct or indirect pathologic processes, such as pulmonary inflammation or congestive heart failure.
       (2) Fluid levels may be ascertained radiographically or by percussion.
         (a) As little as 500 ml of fluid is seen on a posteroanterior thoracic radiograph as blunting of a costodiaphragmatic recess. A lateral view reveals even smaller accumulations of fluid.
         (b) The level at which the percussive tympanic tone characteristic of air-filled lung changes abruptly to a dull tone reveals the fluid level.
    b. Clinical considerations
       (1) Thoracentesis (pleural tap) is performed posterior to the midaxillary line while the patient is seated.
         (a) The fluid level is determined by percussion. The site of insertion is one or two intercostal spaces below the fluid level, but not below the ninth intercostal space because of the proximity of the liver.
         (b) A needle is inserted immediately above the superior margin of the rib to avoid injury to the main intercostal neurovascular bundle.
       (2) Chest tubes at -10 cm H_2O are used to reduce pneumothorax and to drain accumulating fluid from the pleural cavity.
         (a) For reduction of pneumothorax uncomplicated by fluid accumulation, a tube is inserted into the pleural cavity anteriorly through the second intercostal space immediately superior to the third rib.
         (b) If fluids are present, the chest tube is usually inserted in the **posterior axillary line through the fifth or sixth intercostal space.**

III. LUNGS

A. Overview

1. **Development.** The developing lung buds can be thought of as expanding into the pleural cavities, pushing the primitive mesothelium of the mediastinal pleura ahead of them. In this manner, the lung buds are covered by mesothelium (visceral pleura).

2. **Differences.** Although the right lung is shorter because the liver is on the right side, it is wider because the heart is offset to the left. The right lung has a slightly larger capacity than the left lung.

B. External structure.
The lung surfaces are named to correspond to the related thoracic structures, for example, the costal, diaphragmatic, and mediastinal surfaces (see Figure 20-2).

1. **The apex** lies in the **cupula,** which extends above the first rib into the base of the neck. Percussion and auscultation of the lung at the base of the neck and shoulder are performed routinely during physical examination.

2. **The base** is defined by the diaphragmatic surface.

3. **The radix** or **root of the lung** is on the mediastinal surface. Structures that enter or leave each lung at the hilus include the main stem (primary) bronchus, pulmonary artery, pulmonary veins, bronchial lymphatics, and hilar lymph nodes.

4. **The pulmonary ligament** extends along the mediastinal surface from the hilus to the base and supports each lung in the pleural cavity (see section II B).

5. **Lobes.** Fissures of the visceral pleura divide the lungs into lobes (Figure 20-2; Table 20-1).
   a. **Upper lobes** lie anteriorly; **lower lobes** are posterior.
   b. A **middle lobe** lies anteroinferiorly on the right side. The **lingula,** its equivalent on the left side, is formed by the pronounced **cardiac incisure.**
   c. A **lobar (secondary) bronchus** supplies each lobe, accompanied by the lobar branch of the pulmonary artery and lobar branches of pulmonary veins.

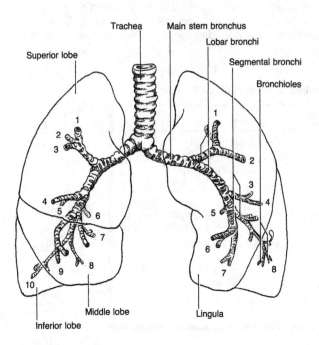

FIGURE 20-2. Bronchial tree and lobes of the lung. The right main stem bronchus divides into upper, middle, and lower lobar bronchi, which in turn give rise to 10 segmental bronchi (numbered). The left main stem bronchus is more horizontal, divides into upper and lower lobar bronchi, and gives off 8–10 segmental bronchi (numbered).

TABLE 20-1. Lobes and Fissures of the Lungs

Right Lung (3 lobes)		Left Lung (2 lobes)	
Anteriorly		Anteriorly and Posteriorly	
Upper lobe	} Horizontal fissure	Upper lobe	} Oblique fissure
Middle lobe	Oblique fissure	Lower lobe	
Lower lobe			
Posteriorly			
Upper lobe	} Oblique fissure		
Lower lobe			

    6. **Bronchopulmonary segments** are the functional units of the lungs (Table 20-2; see Figure 20-2).
        a. **Structure.** A bronchopulmonary segment has three characteristics.
            (1) **A segmental (tertiary) bronchus** lies centrally.
            (2) **A segmental artery** accompanies the segmental bronchus.
            (3) **Intersegmental veins** coalesce from a peripheral venous plexus in the intersegmental septa.
        b. **Distribution.** The right lung has 10 bronchopulmonary segments: 3 in the upper lobe, 2 in the middle lobe, and 5 in the lower lobe (see Figure 20-2 and Table 20-2). Depending on the text reference, however, 8, 9, or even 10 are described in the left lung (see Figure 20-2 and Table 20-2).

C. **Clinical considerations**

    1. **Mendelson's syndrome.** Because the **superior segmental bronchi** of the left and right lower lobes (6 in Figure 20-2) are the most posterior, they are most frequently involved in **aspiration pneumonia.** When the body is supine, inhaled material (such as aspirated vomitus) drains into the superior bronchopulmonary segments of the lower lobes, producing inflammation (pneumonia).
    2. **Segmental pulmonary resection.** Pathologic processes frequently are confined to a discrete bronchopulmonary segment.
        a. Surgical removal of a bronchopulmonary segment is now a common procedure.
        b. A bronchopulmonary segment may be defined surgically by dissecting along the airway until the desired segmental bronchus is reached. Clamping abolishes perfusion of the segment, the surface of which darkens as a result of the unoxygenated blood.

TABLE 20-2. Bronchopulmonary Segments of the Lungs

Right Lung	Left Lung
Right upper lobe (3)	Left upper lobe (4)
1. Apical	1–2. Apicoposterior
2. Posterior	3. Anterior
3. Anterior	
Right middle lobe (2)	
4. Medial	4. Superior lingular
5. Lateral	5. Inferior lingular
Right lower lobe (5)	Left lower lobe (4–5)
6. Superior	6. Superior
7. Basal medial	7–8. Basal anteromedial
8. Basal anterior	9. Basal lateral
9. Basal lateral	10. Basal posterior
10. Basal posterior	

3. **Pancoast tumor,** a lesion of the upper lobe of either lung, may compress several important structures.
   a. Subclavian or brachiocephalic vein compression results in venous engorgement and edema of the face and arm on one side.
   b. Subclavian artery compression results in diminished pulse in that extremity.
   c. Phrenic nerve compression results in paralysis of a hemidiaphragm.
   d. Recurrent laryngeal nerve compression results in vocal hoarseness.
   e. Sympathetic chain compression results in **Horner's syndrome** (miosis, pseudoptosis, anhydrosis).
4. **Primary carcinoma of the lung** is common. It may produce secondary lung abscess and metastasize to mediastinal structures.
5. **Secondary metastases** to the lungs by hematogenous routes are frequent.

IV. BRONCHIAL TREE

A. **The trachea** lies in the superior mediastinum. It is approximately 12 cm long and 2 cm wide (Figure 20-3; see Figure 20-2).

   1. **Cartilaginous C-shaped rings** reinforce and maintain patency of the trachea. The posterior discontinuity of the cartilage allows momentary esophageal expansion into the trachea for deglutition.

   2. **The carina** marks the bifurcation into left and right main stem bronchi behind the sternal angle at vertebral level T5.

B. **Main stem (primary) bronchi** are located posteriorly in the hilus behind the pulmonary artery (intermediate location) and the pulmonary veins (anterior). The right and left main stem bronchi supply the entire right and left lungs, respectively (see Figures 20-2 and 20-3).

   1. **Right main stem bronchus**
      a. It is wider, shorter, and more vertical than the left main stem bronchus.
      b. It is the probable resting place for large aspirated objects.
      c. It divides external to the lung into three lobar bronchi.

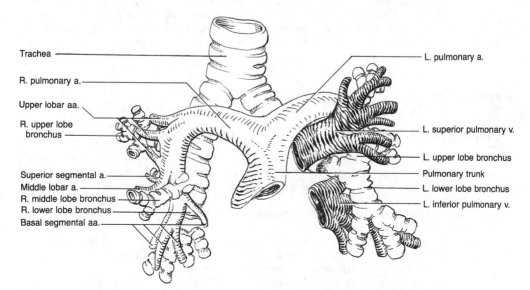

FIGURE 20-3. The bronchial tree, right pulmonary arteries, and left pulmonary veins.

2. **Left main stem bronchus**
   a. It is narrower (because of smaller left lung volume), longer, and more horizontal than the right main stem bronchus (because the heart is toward the left).
   b. It divides external to the lung into two lobar bronchi.

C. **Lobar (secondary) bronchi** (see Figure 20-2)
   1. **Right upper, right middle,** and **right lower lobar bronchi** supply the upper, middle, and lower lobes of the right lung, respectively (see Figure 20-3).
      a. The **right lower lobar bronchus** is most vertical, most nearly continues the direction of the trachea, and is larger in diameter than that on the left side. It is, therefore, the most common resting place for small aspirated objects.
      b. Such foreign bodies may block the airway and cause inflammation, **atelectasis** (local collapse), or even abscess.
   2. **Left upper and lower lobar bronchi** supply the upper and lower lobes of the left lung.

D. **Segmental (tertiary) bronchi** arise from lobar bronchi and supply bronchopulmonary segments (see Figure 20-2).
   1. **Structure.** A segmental bronchus enters the center of each bronchopulmonary segment accompanied by a segmental branch of the pulmonary artery. Air and unoxygenated blood enter the center of the bronchopulmonary segment, while the oxygenated blood returns by intersegmental veins about the periphery.
   2. **Distribution.** Within a bronchopulmonary segment (see section III B 6 b), the segmental bronchus may branch 6–18 times to produce 50–70 respiratory bronchioles, which give off alveolar sacs composed of alveoli. Alveoli constitute the lung parenchyma.

E. **Clinical considerations**
   1. **Bronchiectasis.** The segmental bronchi to the basal segments of the lower lobes are prone to this condition. When these lowermost airways become mucus filled, dilated, and nonrespiratory, the ventilation–perfusion ratio is reduced. Treatment usually involves resection of the lower lobes. Although such excision decreases reserve volume, it returns the ventilation–perfusion ratio to normal.
   2. **Atelectasis.** An entire lobe, a bronchopulmonary segment, or a small region of a lung may collapse, usually as a result of blockage of an airway with resorption of the contained air and subsequent adherence of the airway walls by surface tension. It is a common postsurgical complication because of accumulation of mucus in the airway. Postoperative respiratory therapy is a key preventative step.

V. VASCULATURE

A. **Arterial supply.** Two separate arterial systems supply the lungs.
   1. **The pulmonary trunk** arises from the **conus arteriosus** of the right ventricle. After approximately 5 cm, it bifurcates into left and right pulmonary arteries. **Pulmonary arteries** supply the respiratory (alveolar) parenchyma, carrying deoxygenated blood from the right ventricle.
      a. The **right pulmonary artery** passes posterior to the ascending aorta and superior vena cava (see Figure 20-3).
         (1) It crosses the right main stem bronchus to lie superior to that bronchus.
         (2) It lies posterior to the right superior pulmonary vein in the hilus.
         (3) It divides into three lobar, then segmental, arteries.
      b. The **left pulmonary artery** passes anterior to the arch of the descending aorta to which it is connected by the ligamentum arteriosum (see Figure 20-3).

(1) It crosses the left main stem bronchus to lie superior to that bronchus.
(2) It lies posterior to the left superior pulmonary vein in the hilus.
(3) It divides into two lobar, then eight to ten segmental, arteries.
c. **Segmental arteries** enter the center of each bronchopulmonary segment with a segmental bronchus (see Figure 20-3). These arteries continue to subdivide to supply the capillary plexuses about each alveolus.
d. **Clinical considerations**
  (1) **Pulmonary embolus** is a blocking body, such as a thrombus (thromboembolism), that is transported to the pulmonary arterial circulation, where it becomes impacted. Thrombi may develop within the systemic venous system (usually as a result of venous stasis or injury) or from the right side of the heart (usually related to valvular disease or myocardial infarction).
  (2) An embolus results in local **ischemia.** If the ischemia is severe, the tissues may undergo **necrosis.** The area of necrosis is an **infarct.**
    (a) A large embolus in a main stem or lobar artery may result in neurogenic shock and death within minutes.
    (b) A small embolus may produce only mild pain, pleurisy, and perhaps hemoptysis before it resolves with healing.
    (c) Pulmonary embolism is one of the greatest single causes of death in elderly injured or postoperative patients. It is estimated that fewer than one half of all pulmonary emboli are correctly diagnosed.
2. **Bronchial arteries** supply the conducting (nonrespiratory) portions of the lung with oxygenated blood.
  a. **Left and right bronchial arteries** arise from the descending aorta at the level of the tracheal bifurcation (T4–T5).
  b. During the course of pulmonary disease, bronchial arteries increasingly anastomose with the pulmonary circulation. These anastomoses may enlarge significantly to provide blood to the respiratory parenchyma.

B. Venous return. After transalveolar gaseous exchange, oxygenated blood returns to the left atrium through the pulmonary veins.
1. **Venous capillaries** arise from the alveolar capillary plexuses and coalesce peripherally in the connective tissue septa of the bronchopulmonary segments as the intersegmental veins.
2. **Intersegmental veins** run in the connective tissue between the bronchopulmonary segments. They combine to form the **lobar veins.**
3. **Pulmonary veins** empty into the left atrium (see Figure 20-3).
  a. **Superior lobar and middle lobar (or lingular) veins** usually unite to form the **superior pulmonary veins.**
  b. **Inferior lobar veins** continue as **inferior pulmonary veins.**
4. **Variations.** Left and right superior and inferior pulmonary veins generally enter the left atrium separately, but variations are common. The left atrium may have three, four, or five openings.
  a. The left superior and inferior veins may unite to enter the left atrium as one vein (about 25%).
  b. The right superior, middle, and inferior lobar veins may enter the left atrium separately (2%), or by a common pulmonary vein (3%).

C. Lymphatics
1. **Pulmonary lymph nodes,** associated with the lobar bronchi, drain from the alveolar regions.
2. **Hilar (bronchopulmonary) nodes** at the root of the lung receive lymph from the pulmonary nodes.
  a. These nodes are usually involved in bronchogenic carcinoma.
  b. Hilar nodes that are enlarged because of infection or carcinoma make surgical dissection and identification of hilar structures difficult and time-consuming.

3. **Tracheobronchial (bronchotracheal) nodes** are next along the main stem bronchi.
4. **Left and right bronchomediastinal (tracheal) nodes** unite with the **parasternal nodes** to form the mediastinal lymph trunks that drain into the left and right brachiocephalic veins, respectively.
5. **Clinical considerations.** Metastatic spread of lung carcinoma may occur along lymphatic pathways. When lymph nodes become blocked by metastatic growth, alternate drainage channels include lymphatic connections to the contralateral lung or to the celiac nodes in the epigastric region of the abdomen, bringing metastases to these regions.

VI. INNERVATION

A. **Course and composition.** General visceral afferent and general visceral efferent (autonomic) innervation to the lungs are carried by branches from the thoracic sympathetic chain and the vagus nerve (CN X).

1. **Sympathetic nerves** (Figure 20-4; see Chapter 4 V A 3)
    a. The left and right sympathetic chains provide thoracic splanchnic branches to the **pulmonary plexuses,** which are located about the main stem bronchi on each side.
    b. Sympathetic neurons mediate vasoconstrictive function in the pulmonary vascular bed and secretomotor activity to the bronchial glands.
2. **Parasympathetic nerves** (see Figure 20-4 and Chapter 4 V A 3)
    a. **Left and right vagi** pass anterior to the main stem bronchi and give off twigs to the **pulmonary plexuses.**

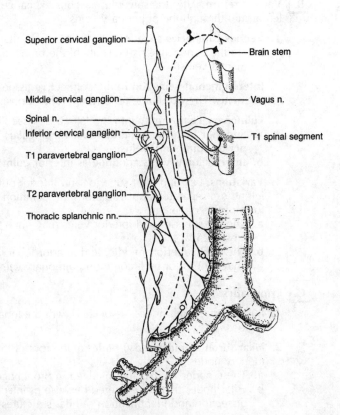

FIGURE 20-4. Autonomic innervation of the bronchial tree. The first (preganglionic) sympathetic neurons lie in the thoracic spinal cord, send axons to the spinal nerves through the ventral roots, and leave the spinal nerves via the white rami communicantes to gain access to the sympathetic chain. The second (postganglionic) sympathetic neurons lie in the sympathetic chain and send axons along splanchnic nerves to reach the bronchial tree. The preganglionic parasympathetic neurons lie in the brain stem and send axons along the vagus nerve (CN X) to the bronchial tree. The postganglionic parasympathetic neurons lie in the wall of the bronchial tree. Reflex sensation from the bronchial tree travels to the brain along afferent neurons (*dashed lines*) in the vagus.

b. Parasympathetic neurons from the vagus innervate bronchial smooth muscle. Excessive stimulation produces **asthma.**
c. The vagus nerves also carry respiratory reflex afferents.

B. Regulation of respiration

1. **Stretch reflexes.** The lungs contain stretch receptors innervated by afferent neurons that run in the **vagus nerve.**
   a. When the lung is stretched, these afferents inhibit the respiratory center in the brain stem (end-inspiratory reflex).
   b. The end-expiratory reflex is similar.

2. **Pressure and chemical reflexes.** The **glossopharyngeal nerve** (CN IX) and **vagus nerve** (CN X) innervate the **carotid body and sinus** and the **aortic body,** respectively. These small structures, located in conjunction with the branchial arches, monitor changes in the partial pressure of oxygen and hemodynamic pressure, respectively.
   a. A decrease in the partial pressure of oxygen increases inspiration through reflex mechanisms in the brain stem.
   b. An increase in blood pressure decreases respiration through similar reflex mechanisms.

VII. FUNCTIONAL ANATOMY OF THE RESPIRATORY SYSTEM

A. Basic principles.
The lungs are the organs through which ambient oxygen passes into the blood and ultimately reaches the intracellular mitochondria for oxidative metabolism. Similarly, the lungs are the organs through which the products of oxidative metabolism (carbon dioxide and some water vapor) are expelled to the atmosphere.

B. Pulmonary function

1. **Ventilation,** the mechanical displacement of gases, is effected by the muscles of the thoracic cage and abdominal wall, intrinsic elastic recoil of the thoracic cage, intrinsic elastic recoil of the lungs, and gravity.
   a. The airway. Ventilation moves gases through a series of airways with constantly decreasing diameters. Functionally, the airway is divided into two portions.
      (1) The **conducting portion** consists of the trachea and bronchi as far as the terminal bronchiole. The air in the conductive portion at the end of inspiration does not exchange gases with the blood. Similarly, the air in this portion at the end of expiration cannot be expelled (anatomic dead space).
      (2) The **respiratory portion** consists of respiratory bronchioles, alveolar ducts, alveolar sacs, and alveoli.
   b. Clinical correlation. Reduced vital capacity or airway compromise reduces ventilation, results in dyspnea, and may produce cyanosis.
      (1) Reduction of vital capacity can result from flail chest, diaphragmatic paralysis, pneumothorax, pleural effusion, and loss of intrinsic or extrinsic elasticity.
      (2) Obstruction of the airway can result from foreign material, mucus, asthma, or neoplasm.

2. **Respiration** is the exchange of gases between alveoli and the surrounding capillary beds. Gaseous exchange occurs primarily in alveoli, the total surface area of which is more than 70 m, or about the surface area of a tennis court.
   a. Alveolar respiration is the exchange of gases across the **blood–air barrier** brought about by the differences in partial pressure between venous blood and alveolar air.
      (1) The blood–air diffusion barrier consists of the alveolar epithelium and capillary epithelium with the intervening basal lamina.

- **(2)** Because gas exchange is efficient, freshly oxygenated blood in the pulmonary veins normally reaches the same partial pressures as alveolar air.
- **(3)** Ideally, alveolar ventilation is matched volume-for-volume by pulmonary vascular perfusion.
- **b. Gaseous diffusion** across the alveolar membrane is a function of membrane thickness, surface area, and partial pressure differentials. Several conditions reduce gaseous diffusion.
  - **(1) Alveolar–capillary block. Pulmonary fibrosis, pulmonary edema, bronchiectasis,** and **hyaline membrane disease** increase the effective thickness of the diffusion pathway.
  - **(2) Reduction of surface area. Emphysema** reduces the surface area available for gas exchange.
  - **(3) Reduced ventilation–perfusion ratio**
    - **(a)** Ineffective ventilation decreases pressure differentials so that gas exchange becomes inefficient, and the blood perfusing the alveoli becomes incompletely oxygenated—in effect, an **arteriovenous shunt.**
    - **(b)** Blockage of an airway (by tumor, foreign body, or mucus), collapse of the lung, and inability to expand the lung reduce this ratio.
3. **Perfusion.** The lung parenchyma must be perfused with blood. Any condition that interferes with perfusion, such as pulmonary embolus, lowers the ventilation–perfusion ratio.

Chapter 21

Pericardial Cavity and Heart

I. MIDDLE MEDIASTINUM

A. **Boundaries.** This region is bounded by the left and right pleural cavities, diaphragm, anterior and posterior leaves of the fibrous pericardium, and superior aspect of the left and right pulmonary arteries (see Figure 21-1A).

B. **Contents.** The middle mediastinum contains the **roots of the lungs,** the **ascending aorta,** the **pericardial cavity** and **heart,** the **pulmonary trunk** and left and right **pulmonary arteries,** the left and right superior and inferior **pulmonary veins,** the superior and inferior **venae cavae,** the arch of the **azygos vein,** the left and right **phrenic nerves,** and **bronchomediastinal lymph nodes** (see Figure 21-1A).

II. PERICARDIAL CAVITY

A. **Characteristics.** The pericardial cavity (sac) contains the heart and the roots of the great vessels (Figure 21-1A).

  1. **Location.** It lies behind the sternum and second through sixth costal cartilages on the right side. On the left side, it extends inferolaterally beyond the costochondral junction of the second through fourth ribs as far laterally as the midclavicular line (see Figure 21-11).
  2. **Surgical access.** Left and right **phrenic nerves** run along its lateral borders between the mediastinal pleura and the fibrous pericardium. For this reason, access to the pericardial cavity is usually by **midline sternotomy** followed by longitudinal incision of the pericardium.

B. **Structure.** Two layers of pericardium form the pericardial sac.

  1. **Fibrous pericardium**
     a. This outer layer of tough connective tissue adheres to the central tendon of the diaphragm and is attached to the sternal periosteum by numerous small **sternopericardial ligaments.**
     b. It prevents transient overdistention of the heart, but it will stretch to accommodate chronic heart enlargement resulting from disease, such as **congestive heart failure.**
     c. It is continuous with the adventitia of the aorta, upper pulmonary trunk, superior and inferior venae cavae, and pulmonary veins.
  2. **Serous pericardium**
     a. **Parietal serous pericardium** is serous mesothelium lining the pericardial cavity.
     b. **Visceral serous pericardium (epicardium)** is a reflection of the mesothelial layer over the great vessels and surface of the heart to form the outermost layer of the heart wall. This layer can be thought of as being formed by invagination of the pericardial cavity by the developing heart.
     c. **Pericardial fluid** exudes through the serous mesothelium.
        (1) It keeps the surface layers moist and slippery, allowing nearly frictionless beating of the heart within the pericardial cavity.
        (2) It develops **surface tension** between the pericardial layers. This effect results in slight expansion of the heart during inspiration, so in conjunction with lowered inspiratory intrathoracic pressure, blood return to the right side of the heart is increased.

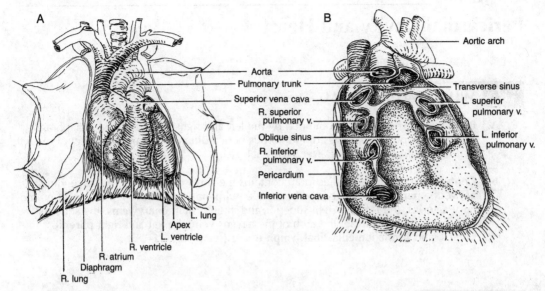

FIGURE 21-1. *(A)* Middle mediastinum. The medial portions of the lungs were retracted to show the location of the heart in the pericardial cavity. *(B)* Pericardial cavity. With the heart removed, the line of pericardial reflection about the vessels that enter and leave the heart forms the oblique sinus and transverse sinus.

- **C. Pericardial sinuses.** Irregularity of the line of reflection of the parietal pericardium results in the formation of sinuses.
    1. **The oblique sinus** represents a cul-de-sac in the pericardial cavity posterior to the heart. It is formed by an arch of reflected parietal pericardium (see Figure 21-1B).
        a. **Boundaries.** Inferiorly on the right, it encompasses the inferior vena cava, right inferior and superior pulmonary veins, and superior vena cava. It extends transversely between the right and left superior pulmonary veins and then inferiorly to encompass the left inferior pulmonary vein.
        b. **Mesenteric attachments.** This pericardial reflection is equivalent to a mesentery and could be called the "**cardiac ligament.**" It suspends the heart in the pericardial cavity.
    2. **The transverse sinus** is a passage behind the ascending aorta and pulmonary trunk (see Figure 21-1B).
        a. **Boundaries.** It is separated from the oblique sinus by the reflection of the pericardium ("cardiac ligament") between the left and right superior pulmonary veins.
        b. During cardiac surgery, ligatures passed into the transverse sinus and then around the aorta and pulmonary trunk can be used to control hemorrhage or to secure cannulas placed in the great vessels.

- **D. Clinical considerations**
    1. **Pericardial tamponade.** Accumulation of fluid (blood or serous) in the pericardial sac compromises cardiac expansion.
        a. This condition results in reduced cardiac output but increased heart rate (indicated by a weak but rapid pulse) and increased venous pressure with associated jugular vein distention, pulsating liver, and edema.
        b. Immediate pericardiocentesis is necessary to restore cardiac output.
    2. **Pericardiocentesis and intracardiac injection**
        a. **Subcostal route.** A needle may be safely introduced into the pericardial sac via the **sternocostal angle.**

(1) The needle is inserted on the left side adjacent to the xiphoid process, angling upward at about 45° and to the left. Fluid drains effectively if the patient is propped up slightly.

(2) The risk of iatrogenic pneumothorax is small because the pleural cavities are avoided. This route also obviates puncturing the anterior descending or the right marginal branches of the coronary arteries.

b. **Parasternal route.** A needle is introduced into the pericardial cavity or heart through the left fourth or fifth intercostal space immediately adjacent to the sternum. The cardiac notch in the pleura on the left side eliminates the chance of inducing pneumothorax.

(1) The internal thoracic artery, which passes about 1 cm or more laterally to the margin of the sternum, is avoided if the edge of the sternum is located with the tip of the needle before plunging into the pericardial sac.

(2) This route is commonly used for intracardiac injection.

3. **Pericarditis.** Inflammation of the parietal serous pericardium results in pain referred to the precordium and the epigastrium. A systolic **friction rub** may be heard on auscultation.

III. EXTERNAL CARDIAC STRUCTURE

A. General characteristics

1. **Location.** The heart is contained in the pericardial sac and covered by visceral pericardium. Approximately one third of the heart lies to the right and two thirds to the left of midline (see Figure 21-11).

2. **Size.** The transverse cardiac diameter varies with inspiration and expiration, but it normally measures about 89 cm in transverse diameter on a radiograph at maximum end-inspiration in the standing position. Generally, the transverse diameter should not be more than one half the diameter of the chest.

B. Boundaries (see Figure 21-1A)

1. **The right (acute) border** delineates the superior vena cava, right atrium, and inferior vena cava.

2. **The inferior border** delineates the right ventricle.

3. **The apex** is the tip of the left ventricle, where the inferior and obtuse borders meet. An **apical pulse** or **point of maximal impulse (PMI)** is palpable and sometimes visualized in thin males at the fifth intercostal space just beneath the nipple.

4. **The left (obtuse) border** is formed by the left ventricle.

5. An upper border is indistinct radiographically.

C. Surface characteristics

1. **The coronary (atrioventricular) sulcus** completely encircles the heart and separates the atria from the ventricles (see Figure 21-1A). It marks the location of the **annulus fibrosus.**
   a. It has a nearly vertical orientation behind the sternum.
   b. It contains the **circumflex branch of the left coronary artery, right coronary artery, coronary sinus,** and **small cardiac vein,** which are all embedded in fat (see Figure 21-2A).

2. **The anterior interventricular sulcus** marks the interventricular septum (see Figure 21-1A).
   a. It runs diagonally along the sternocostal surface close to the left border of the heart.

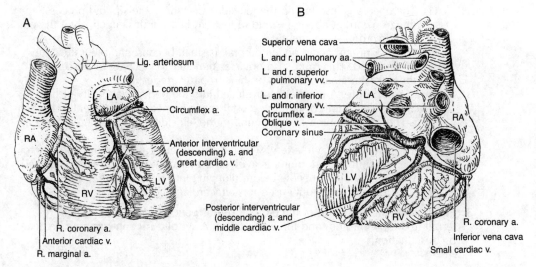

FIGURE 21-2. The heart and coronary circulation. *(A)* Anterior aspect; *(B)* posterior aspect.

    b. It contains the **anterior interventricular (descending) artery** (a branch of the left coronary artery) and the **great cardiac vein** (see Figure 21-2A).

  3. The posterior interventricular sulcus delineates the interventricular septum posteriorly (see Figure 21-2B).
    a. It is the continuation of the anterior interventricular sulcus around the apex onto the diaphragmatic surface of the heart.
    b. It contains the **posterior interventricular (descending) artery** (usually a branch of the right coronary artery) and the **middle cardiac vein**.

D. Divisions. The heart is divided into right and left sides, each of which is composed of two chambers.

  1. The right atrium forms the right border of the heart, part of the base, and part of the sternocostal surface (see Figure 21-1A).
    a. It gives off the **right auricular appendage,** which overlies the nodal artery to the right of the pulmonary trunk.
    b. It receives the **superior vena cava, inferior vena cava,** and **coronary sinus.**
    c. It discharges into the right ventricle through the tricuspid valve.

  2. The right ventricle forms the largest portion of the sternocostal surface and a small part of the diaphragmatic surface (see Figure 21-1A).
    a. It projects to the left of the right atrium toward the apex.
    b. It receives blood from the right atrium.
    c. It discharges through the pulmonary semilunar valve into the pulmonary trunk just to the left of the aorta.

  3. The left atrium forms the posterior surface (base) of the heart, lying entirely within the boundaries of the oblique sinus of the pericardial sac (see Figure 21-1A).
    a. It gives off the **left auricular appendage,** which overlies the left coronary artery to the left of the pulmonary trunk.
    b. It receives two **right pulmonary veins** (sometimes three) and two **left pulmonary veins** (sometimes one).
    c. It discharges into the left ventricle through the mitral valve.

  4. The left ventricle forms the left border, **the apex,** one third of the sternocostal surface, and two thirds of the diaphragmatic surface (see Figure 21-1A).
    a. It projects to the left of the left atrium toward the apex.
    b. It receives blood from the left atrium.

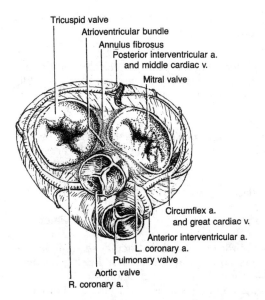

FIGURE 21-3. Dissection of the left and right atria reveals the fibrous ring (annulus fibrosus) that supports the atrioventricular valves and semilunar valves. Also shown are the origins and initial courses of the left and right coronary arteries.

   c. It ejects blood into the aortic vestibule and through the aortic semilunar valve into the **ascending aorta.**

IV. CORONARY CIRCULATION

A. **Arterial circulation.** The myocardium and epicardium are supplied by the **left** and **right coronary arteries** (see Figure 21-3).
   1. **The right coronary artery** (Figure 21-2)
      a. **Course.** It arises from the **right aortic sinus** (of Valsalva) and runs to the right in the **coronary sulcus.** It passes beneath the right auricular appendage and around the right cardiac margin to the posterior aspect of the heart (see Figure 21-3).
      b. **Major branches**
         (1) The **sinoatrial (S-A) nodal branch,** the first major branch in most (61%) individuals, courses toward the superior vena cava. It supplies the top of the right atrium and usually the S-A node.
         (2) The **right marginal branch** passes along the inferior border toward the apex to supply a portion of the right ventricle (see Figure 21-2A).
         (3) The **posterior interventricular (descending) branch** descends in the posterior interventricular sulcus. It supplies the posterior one third of the interventricular septum (see Figure 21-2B).
         (4) The **atrioventricular (A-V) nodal artery** arises from a deep loop of the right coronary artery immediately beyond the posterior interventricular branch. It supplies the A-V node and right branch of the atrioventricular bundle (of His).
         (5) The **terminal branches** of the right coronary artery usually continue a few centimeters beyond the posterior interventricular sulcus. They supply the posteromedial aspect of the left ventricle.
      c. **Distribution**
         (1) **Prevalence.** The right coronary artery generally (85–90%) supplies the right atrium (including the **S-A node**), the superior parts of the right ventricle, and the posterior one third of the interventricular septum, including the **A-V node** and the **right branch of the A-V bundle.**

(2) Exception. In 10–15% of the population, the posterior septum and A-V node are supplied by the left coronary artery.

2. **Left coronary artery** (see Figure 21-2)
   a. **Course.** It arises from the left (left posterior) aortic sinus (of Valsalva) and runs between the pulmonary trunk and left auricular appendage for a short, variable distance before bifurcating into the **anterior interventricular** and **circumflex arteries** (Figure 21-3).
   b. **Branches**
      (1) It occasionally (39%) gives off the **artery of Kugel,** which can give rise to the **S-A nodal artery.** When this branching occurs, the S-A nodal artery courses around the posterior aspect of the superior vena cava from left to right to reach the S-A node.
      (2) The **anterior interventricular (descending) artery** enters the anterior interventricular sulcus (see Figure 21-2A).
         (a) It supplies the anterior aspects of the left and right ventricles via **diagonal branches.**
         (b) It supplies the anterior two thirds of the interventricular septum via **septal branches,** including the **left branch of the A-V bundle.** It gives off twigs to the A-V node in about 40% of the population.
      (3) The **circumflex artery** runs in the coronary sulcus toward the left border and around to the base of the heart (see Figures 21-2 and 21-3).
         (a) The **left marginal artery,** a major branch, courses along the left border of the heart to supply the left ventricle.
         (b) It gives rise to the **posterior descending artery** in 10–15% of the population.
         (c) It generally supplies the posterior aspect of the left atrium and the superior portions of the left ventricle.
   c. **Distribution**
      (1) **Prevalence.** The left coronary artery generally supplies the left atrium, the anterior one third of the interventricular septum, the inferior parts of the right ventricle, and the apex as well as the superior aspects of the left ventricle.
      (2) **Exception.** In 10–15% of the population, the posterior septum and A-V node are supplied by the left coronary artery.

3. **Variations**
   a. **Balanced coronary circulation** is the coronary circulation just described.
      (1) It is present in only 60–65% of the population.
      (2) It is sometimes termed "right dominant" because the right coronary artery gives off the posterior descending branch, but this term does not distinguish balanced from right preponderant circulation.
   b. **Left preponderant coronary circulation** occurs when the circumflex branch of the left coronary artery gives rise to the posterior descending branch.
      (1) The entire interventricular septum, including the A-V node, is supplied by the left coronary artery.
      (2) Both arteries supplying the septum arise from the same stem. This common origin reduces or precludes the possible development of effective collateral circulation in the septal region. Left preponderance lowers the chances of survival after septal infarct. These individuals stand to benefit most from coronary bypass surgery.
      (3) It is present in about 10–15% of the population.
      (4) It is sometimes termed "left dominant" because the left coronary artery gives off the posterior descending branch.
   c. **Right preponderant coronary circulation** occurs when the right coronary artery, in addition to supplying the posterior descending artery, crosses the posterior interventricular septum to reach as far as the left marginal artery. The circumflex artery is diminutive, often terminating at the left margin.
      (1) A substantial portion of the diaphragmatic surface of the left ventricle is supplied by this extension of the right coronary artery.

(2) It is present in 20–25% of the population.
(3) It is sometimes called "right dominant" because the right coronary artery gives off the posterior descending branch, but this term does not distinguish this condition from balanced circulation.
  d. **Other variations** include the manner in which the coronary arteries arise from the aorta or even, lethally, from the pulmonary trunk.
4. **Functional considerations**
  a. **Anastomoses**
    (1) **Normal healthy heart.** In healthy young individuals, most of the arteries are true **end-arteries,** which supply discrete volumes of myocardium. If any anastomoses between left and right coronary arteries are present, they typically are inadequate to maintain effective circulation in the event of sudden-onset occlusion of a major branch of the coronary circulation.
    (2) With **slow-onset coronary occlusion** (such as with atherosclerotic coronary artery disease), anastomoses develop to an extent that collateral circulation is relatively effective.
    (3) **Coronary obstruction** can result from atherosclerotic blockage or vasospasm.
      (a) Obstruction in a coronary artery produces ischemia. Pain associated with ischemic cardiac tissue—**angina pectoris**—refers to the precordium, epigastrium, shoulder, and, frequently, the left arm.
      (b) Ischemia may result in necrosis and hemorrhage (an **infarct**) with severe and life-threatening damage to the myocardium.
      (c) If a patient survives the myocardial insult, the thrombus that forms over damaged endocardium may become dislodged and produce life-threatening emboli.
  b. **Coronary perfusion**
    (1) Because left ventricular systolic pressure and left ventricular transmural pressure are equal to aortic systolic pressure, blood flows through the coronary circulation of the left ventricle only during diastole when the diastolic aortic pressure is greater than the transmural pressure. (Recall that the second law of thermodynamics requires a pressure gradient for flow to occur.)
    (2) Because the heart is in systole about 50% of the time, it is perfused only 50% the time. Therefore, it has a **low margin of safety** with respect to the amount of tolerable perfusion loss before ischemia results.

B. Venous return

1. **The coronary sinus** receives most of the venous return from the epicardium and myocardium (see Figure 21-2B).
  a. **Course.** It is approximately 2.5 cm long and lies in the coronary sulcus, directed toward the diaphragmatic surface. It opens into the right atrium between the opening of the inferior vena cava and the right A-V valve.
  b. **Branches**
    (1) The **great cardiac vein** lies in the anterior interventricular groove (see Figure 21-2A).
      (a) It drains the anterior portion of the interventricular septum and the anterior aspects of both ventricles in the vicinity of the septum.
      (b) It accompanies the anterior descending (interventricular) branch of the left coronary artery.
    (2) The **middle cardiac vein** lies in the posterior interventricular groove (see Figure 21-2B).
      (a) It drains the posterior portion of the interventricular septum and the posterior aspects of both ventricles in the vicinity of the septum.
      (b) It accompanies the posterior descending (interventricular) branch of the right coronary artery.
    (3) The **small (lesser) cardiac vein** lies in the coronary sulcus to the right coronary sinus (see Figure 21-2B).
      (a) It drains the marginal aspect of the right ventricle.

(b) It accompanies the right coronary artery and its marginal branch.
(c) It usually terminates as the **right marginal vein,** although it may not reach the right marginal branch if the latter opens separately into the right atrium.
(4) The **oblique vein** (of Marshall) courses superolaterally across the posterior aspect of the left atrium (see Figure 21-2B). It is the embryologic remnant of the left common cardinal vein. It occasionally develops into a persistent left superior vena cava, whether or not the right vena cava is present.

2. **Anterior cardiac veins** consist of one to several small veins that drain the anterior surface of the right ventricle (see Figure 21-2A).
   a. They course across the coronary sulcus anterior to the right coronary artery.
   b. They open separately into the right atrium.
   c. Occasionally, the right marginal vein is an anterior cardiac vein, emptying directly into the right atrium.
3. **Least cardiac (thebesian) veins** or venae cordis minimae lie within the heart walls.
   a. They drain the endocardium and innermost layers of myocardium.
   b. They tend to empty directly into the left and right atria, with fewer draining into the left and right ventricles.

V. CARDIAC WALL

A. Endocardium. The inner lining of the cardiac chambers is a thin, smooth endothelial layer that is continuous with the epithelium of the great vessels.

B. Myocardium. This intermediate layer is composed of three layers of cardiac muscle, which are most substantial in the ventricles.
   1. **Organization.** The myocardium varies in thickness between chambers, depending on the functional requirements of that chamber.
      a. Each layer originates from the **annulus fibrosus** (fibrous ring, see Figure 21-3). The layers spiral nearly perpendicular to each other, producing a wringing maneuver that effectively reduces the volume of the chambers during systole.
      b. Myocardial fibers do not cross the coronary sulcus.
      c. In the atria, the myocardial fibers convey the electrical impulses. In the ventricles, specialized cells as well as myocardial fibers perform this function.
   2. **Excitable tissue** in the form of nodes and the cardiac conducting system are contained in the myocardium.
      a. **S-A node,** which initiates atrial systole, lies between the crista terminalis and the superior vena cava.
      b. **A-V node,** which is paced by the S-A node to initiate ventricular systole, lies in the right atrial floor near the interatrial septum, medial to the ostium of the coronary sinus and above the septal cusp of the tricuspid valve.
      c. **A-V bundle (of His),** which transmits electrical activity throughout the ventricles, originates at the A-V node, passes through the annulus fibrosus, and descends along the posterior border of the membranous part of the interventricular septum to enter the muscular portion of the interventricular septum.

C. Epicardium. The **serous visceral pericardium** and a variable amount of fatty tissue form the outermost layer of the heart.

D. Fibrous ring (annulus fibrosus)
   1. **Organization.** This layer of dense connective tissue is arranged in the A-V plane and demarcates the coronary sulcus (see Figure 21-3). The annulus fibrosus forms the "skeleton" of the heart, which provides the site of origin for muscle layers of the atria and ventricles.

2. **Function.** It surrounds and supports each valvular opening and electrically isolates the ventricles from the atria.
   a. **Support.** The connective tissue stroma of the left and right A-V valves, as well as the pulmonary and aortic valves, is continuous with and supported by the fibrous ring.
   b. **Electrical isolation.** It prevents the spread of spurious electrical activity from the atria to the ventricles. Only the atrioventricular bundle passes through the fibrous ring.

VI. CARDIAC CHAMBERS

A. **Right atrium.** The right atrium has the thinnest walls of the four chambers (Figure 21-4). The **crista terminalis (terminal crest)** divides this chamber into two regions.

1. **The sinus venarum** is the smooth-walled remnant of the embryologic sinus venosus.
   a. **Location.** It lies to the right of the crista terminalis.
   b. It receives numerous veins (see Figure 21-4).
      (1) **Superior vena cava** enters superiorly.
      (2) **Inferior vena cava** enters inferiorly and has an associated valve leaflet.
         (a) The **valve of the inferior vena cava** (eustachian valve) consists of one incompetent leaf.
         (b) In the fetus, the eustachian valve directs oxygenated inferior vena caval blood through the foramen ovale into the left atrium.
      (3) **Coronary sinus** enters the atrial floor between the inferior vena cava and the tricuspid valve. It also has an associated valve leaflet. The **valve of the coronary sinus** (thebesian valve) may reduce regurgitation of blood into the coronary sinus during atrial systole.
      (4) **Anterior cardiac veins** enter through the anterior inferior atrial wall.

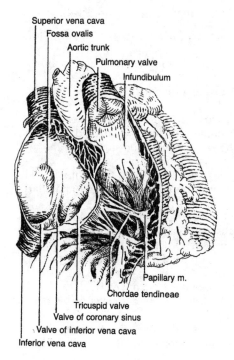

FIGURE 21-4. Chambers of the right side of the heart. The anterior heart wall has been dissected away, revealing the right atrium, tricuspid valve, right ventricle, and pulmonary valve.

2. **The atrium proper** is the anterior muscular portion of the atrium (see Figure 21-4).
   a. It lies to the left of the **crista terminalis.**
      (1) The **crista terminalis** is a muscular ridge running along the anterior atrial wall just medial to the openings of the superior and inferior vena cavae.
      (2) The crista terminalis gives rise to **pectinate muscles** that run across the wall of the atrium proper.
   b. The right atrium contains the **right auricular appendage,** which has muscular walls. It is a potential site for formation of thrombi that, if dislodged, produce pulmonary embolism.
   c. The **atrial septal wall** separates right and left atria.
      (1) The **fossa ovalis** is the embryologic remnant of the fetal **foramen ovale** (see Figure 21-4). It connects the atria in the fetus and represents the **septum primum** (floor) and **septum secundum** (limbus or upper lateral edge). It remains probe-patent in 10% of the population.
      (2) **Atrial septal defects (ASD)** are most common in the vicinity of the fossa ovalis.
         (a) **Septum primum defects** are rare and usually are accompanied by endocardial cushion defects involving valve structure.
         (b) **Septum secundum defects** (the typical patent foramen ovale) account for 10–15% of all cardiac anomalies.
         (c) Normal left atrial pressure is slightly greater than right atrial pressure, so either the patent valve leaflet is kept closed or a **left-to-right shunt** occurs through an open ASD.
            (i) Usually, ASD is not associated with cyanosis because oxygenated blood from the left side of the heart is shunted to the right side.
            (ii) An ASD is usually compatible with a normal life. Should extreme exercise, cardiac disease, or pulmonary disease alter chamber pressures, however, a right-to-left shunt will produce cyanosis.
   d. The **right A-V (tricuspid) valve** opens into the right ventricle.

B. **Right ventricle.** To overcome pulmonary vascular resistance, this chamber has a moderately thick wall with well-developed **trabeculae carneae.** These ridges and bridges of cardiac tissue have a characteristic radiographic appearance. The **supraventricular crest (crista supraventricularis),** which lies between the right A-V valve and the pulmonary orifice, separates the right ventricle into two regions (see Figure 21-4).

1. **The right ventricle proper** (the inflow region) has rough, muscular walls with trabeculae carneae and papillary muscles.
   a. The **right A-V (tricuspid) valve** transmits blood from the right atrium to the right ventricle.
      (1) **Structure.** It is composed of three cusps, chordae tendineae, and three or more papillary muscles (see Figure 21-4).
         (a) **Anterior, posterior, and septal cusps.** Each cusp has a connective tissue core that is attached to the annulus fibrosus and covered with endothelium (chordae tendineae). The septal cusp is the smallest of the three.
         (b) **Chordae tendineae.** Each valve cusp is directed into the ventricle with these connective tissue strands that run from the free edge to papillary muscles.
         (c) **Papillary muscles.** Chordae tendineae are associated with a large **anterior papillary muscle,** a small **posterior papillary muscle,** and usually several small **septal papillary muscles.**
      (2) **Function.** This valve prevents regurgitation from the right ventricle to the right atrium during ventricular systole.
         (a) During the isovolumic (pressure buildup) phase of ventricular systole, the papillary muscles tense to establish valvular competence by preventing eversion of the valve cusps.
         (b) Because the volume of the ventricle is reduced during the isotonic (blood ejection) phase of systole, shortening of the papillary muscles

takes up any slack in the chordae tendineae to maintain the competence of valve closure.

  b. The **interventricular septum** bulges into the right ventricle, causing that cavity to appear crescentic in cross section.
     (1) The large **muscular portion** gives origin to the septal papillary muscles.
     (2) A small, **upper membranous portion** is composed of connective tissue continuous with the annulus fibrosus.
        (a) This part of the intraventricular septum is the usual site of **ventricular septal defects (VSD).**
           (i) A small VSD may result in an inconsequential left-to-right shunt.
           (ii) In the presence of pulmonary stenosis, a VSD produces a **right-to-left shunt** with cyanosis and the "blue-baby" syndrome. A large VSD is a principal factor in **tetralogy of Fallot.**
        (b) The atrioventricular bundles passes into the interventricular septum just posterior to the membranous portion.

2. **The infundibulum,** the smooth-walled outflow region, leads into the pulmonary orifice, which contains the pulmonary semilunar valve (see Figure 21-4).
   a. The **pulmonary valve** consists of three semilunar leaflets (demilunes): left cusp, right cusp, and anterior cusp.
      (1) These cusps are attached peripherally to the annulus fibrosus, which supports the valves.
      (2) Their free edges, directed upward into the pulmonary trunk, are strengthened with a rim of connective tissue. A nodule (corpora Arantii) in the center of the free edges completely closes the lumen.
   b. The pulmonary semilunar valve prevents regurgitation of ejected blood (reverse flow from the pulmonary trunk into the ventricle during ventricular diastole).

3. **The pulmonary trunk** starts at the level of the pulmonary valves and lies to the left of the aortic outflow pathway for the first 3 cm.
   a. It bifurcates into the **right and left pulmonary arteries** at about 5 cm.
   b. **Pulmonary sinuses** (of Valsalva), slight depressions in the walls of the pulmonary trunk, lie behind each valve leaflet.

C. **Left atrium.** This chamber has slightly thicker walls than the right atrium because increased effort is required during atrial systole to overcome the elasticity of the extremely thick left ventricular walls.

  1. Right and left pulmonary veins empty into the left atrium—usually two from each side but occasionally three on the right or one on the left (Figure 21-5).
  2. The **fossa ovalis** on the septal wall is the remnant of the embryonic foramen ovale (see Figure 21-5).
  3. The **left auricular appendage,** which has muscular walls, arises anterosuperiorly from the left atrium (see Figure 21-5).

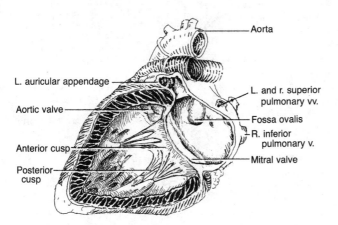

FIGURE 21-5. Chambers of the left side of the heart. The posterior heart wall has been dissected away, revealing the left atrium, mitral valve, left ventricle, and aortic valve.

a. It is a potential site for the formation of thrombi, which, if dislodged, can result in cerebral, systemic, or renal embolism.
b. It may be incised to provide access to the left atrium and mitral valve (for mitral commissurotomy).

4. The **left A-V valve** opens into the left ventricle.

D. **Left ventricle.** To overcome the vascular resistance of the systemic vascular bed, the wall of the left ventricle is three times as thick as the wall of the right ventricle. The left ventricle is divided into two regions.

1. The **left ventricle proper** has the most muscular wall of all the heart chambers and appears round in cross section (see Figure 21-5).
   a. The **left A-V (bicuspid, mitral) valve** transmits blood from the left atrium to the left ventricle.
      (1) **Structure.** It consists of two cusps, chordae tendineae, and two papillary muscles (Figure 21-6B; see Figure 21-5).
         (a) The annulus fibrosus supports a large **anterior cusp** and a small **posterior cusp.**
         (b) It has a greater number of **chordae tendineae** than the right A-V valve. The chordae run from the free edges of the cusps to a larger **anterior papillary muscle** and a smaller **posterior papillary muscle.**
      (2) **Function.** It prevents regurgitation of blood into the left atrium during ventricular systole.
         (a) During the isovolumic (pressure buildup) phase of ventricular systole, the papillary muscles tense to prevent eversion of the valve cusps.
         (b) Because the volume of the ventricle is reduced during the isotonic (blood ejection) phase of systole, shortening of the papillary muscles takes up any slack in the chordae tendineae to maintain the competence of the valvular closure.
      (3) **Mitral insufficiency** is heard as a low-pitched, late systolic blowing murmur. **Mitral stenosis** manifests on auscultation as a late diastolic murmur.
   b. The **interventricular septum** bulges into the right ventricle, giving the left ventricle a round cavity in cross section.

2. The **aortic vestibule** is superior and to the right of the mitral valve. It leads into the ascending aorta and contains the aortic valve (see Figure 21-5).

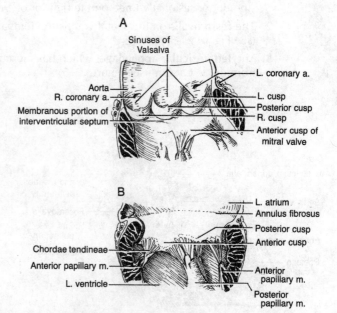

FIGURE 21-6. *(A)* Semilunar valve structure. The aorta has been opened between the left (posterior) and right semilunar valve cusps, behind which ostia open to the left and right coronary arteries. Except for the origins of the coronary arteries, the structure of the pulmonary valve is similar. *(B)* Atrioventricular valve structure. The left ventricle has been opened, showing the anterior and posterior valve cusps, chordae tendineae, as well as anterior and posterior papillary muscles of the mitral valve. Except for an additional septal cusp and septal papillary muscles, the structure of the tricuspid valve is similar.

a. **Structure.** The **aortic valve** consists of three semilunar leaflets (demilunes).
   (1) The **cusps** are named by association with the coronary arteries (see Figure 21-6A).
      (a) The **left cusp** or left posterior cusp is associated with the origin of the left coronary artery.
      (b) The **right cusp** or anterior cusp is associated with the ostium of the right coronary artery.
      (c) The **posterior (noncoronary) cusp** or right posterior cusp is not associated with a coronary artery.
   (2) These cusps are attached peripherally to the annulus fibrosus, which supports the valves.
   (3) Their free edges, directed upward into the pulmonary trunk, are strengthened with a rim of connective tissue. A nodule (corpora Arantii) in the center of the free edges completely closes the lumen.
b. **Function.** The aortic valve prevents regurgitation from the aorta into the ventricle during ventricular diastole.
   (1) A bicuspid aortic valve is a common anomaly that is especially prone to stenosis with age.
   (2) **Aortic insufficiency** manifests on auscultation as a diastolic murmur. **Aortic stenosis** is heard as a high-pitched systolic murmur.

3. **The ascending aorta** starts at the level of the aortic valve.
   a. The **aortic sinuses** (of Valsalva), slight depressions in the walls of the aorta, lie behind the valve cusps (see Figure 21-6A).
      (1) They accommodate the volume of the open valve leaflets to reduce turbulent flow during ejection.
      (2) They provide the origins of the **coronary arteries.**
         (a) The **right coronary artery** normally originates from the **right aortic sinus.**
         (b) The **left coronary artery** normally originates from the **left (left posterior) aortic sinus.**
   b. The **elastic properties of the aorta** accommodate the ejected blood volume and maintain the range of diastolic arterial pressure (see Chapter 5 III A).

VII. CONDUCTING SYSTEM

A. Composition. Specialized cardiac muscle cells in certain regions of the myocardium have highly developed sensitivity and autorhythmicity.

1. **The S-A (sinoatrial, sinuatrial) node** is the autorhythmic pacemaker and initiates the contraction cycle with approximately 72 depolarizations per minute, which spread over both atria and to the A-V node.
   a. **Location.** This node is about the size of a grain of rice (7 mm × 2 mm × 1 mm). It lies in the myocardium between the crista terminalis and the opening of the superior vena cava (Figure 21-7).
   b. **Blood supply.** It is supplied by the **nodal branch of the right coronary artery** (60%) and occasionally by a nodal branch that arises from the left coronary artery.
   c. **Function.** This node initiates contraction. The nodal electrical depolarization spreads throughout the musculature of the right and left atria in 0.09 second (see Figure 21-9). The atria have no pathways of modified cardiac tissue.
   d. **Innervation.** It is influenced principally by the parasympathetic division of the autonomic nervous system, which slows autorhythmicity.

2. **The (A-V) node** is autorhythmic, with approximately 40 depolarizations per minute. It is not connected to the S-A node by specialized conducting fibers, but it is stimulated by atrial depolarizations.
   a. **Location.** It lies in the right atrial floor near the interatrial septum, medial to the

FIGURE 21-7. Intrinsic electrical activity initiates within the sinoatrial (S-A) node and spreads throughout the atrial myocardium within 0.1 second, producing atrial systole and exciting the atrioventricular (A-V) node. From the A-V node, electrical activity enters the A-V bundle, which passes through the annulus fibrosus to reach the ventricles. Dividing into left and right bundle branches, activity spreads over the ventricular myocardium to produce ventricular systole within 0.22 second.

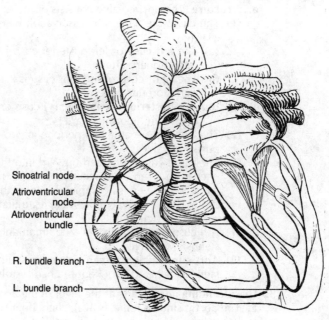

ostium of the coronary sinus and above the septal cusp of the tricuspid valve (see Figure 21-7).
- **b. Blood supply.** It usually (85–90%) is supplied by the right coronary artery.
- **c. Function.** This node synchronizes ventricular depolarization. Its slow conduction velocity (long latency) allows atrial depolarization to spread over the entire right and left atria (0.09 second), so the A-V bundle begins to depolarize at about 0.16 second after the S-A node (see Figure 21-9). Thus, the atria contract completely before the ventricles are stimulated.
- **d.** It gives rise to the A-V bundle.

3. **Atrioventricular (A-V) bundle** (of His)
   - **a. Location.** It arises from the A-V node, passes through the annulus fibrosus, and descends along the posterior border of the membranous part of the interventricular septum to enter the muscular portion of the septum (see Figure 21-7).
   - **b. Branches.** It divides into **left and right bundle branches (crura),** which descend in the interventricular septum and spread out into the walls of the ventricles. They eventually become indistinguishable from the cardiac muscle fibers, which continue to conduct depolarizations.
     - (1) The **right bundle branch** is compact and discrete. A significant branch passes to the anterior papillary muscle along a consistent trabecular bridge—the moderator band.
     - (2) The **left bundle branch** fans out within the septal wall, making this branch somewhat less vulnerable to a small ischemic lesion.
   - **c. Function.** The A-V bundle conducts electrical activity to the ventricles. It consists of modified muscle cells (Purkinje fibers), which have a rapid conduction rate (low latency). It begins to depolarize at about 0.16 second after the S-A node, so activity has spread throughout the ventricles by 0.22 second (see Figure 21-9).

B. Clinical considerations
1. **Electrocardiogram (ECG).** The contractile stimulus for any muscle fiber is an electrical depolarization of its surface membrane (see Figure 21-9). The heart has a large surface-to-volume ratio. Consequently, a large signal is generated that is detectable on the body surface by using an electrocardiogram (ECG or EKG from L. kardia).

2. **Heart block.** If a pathologic condition interrupts impulse propagation in the A-V bundle, **heart block** occurs, with asynchronous beating of the atria and ventricles. If depolarization is initiated outside of the nodal system, the ventricular muscle may then fibrillate, and no blood is pumped.
3. **Premature ventricular systole. Bundles of Kent** are occasional abnormal muscle bridges between the atria and the ventricles, which may cause ventricular pre-excitation by bypassing the A-V node (Wolff-Parkinson-White syndrome).

VIII. CARDIAC INNERVATION

A. **Motor control.** The heart rate and ejection volume are controlled by the autonomic nervous system (see Chapter 4 V and Chapter 22 V).

1. **Parasympathetic division**
   a. The **vagus nerve** sends fibers over the surface of the heart and to the nodal areas (Figure 21-8).
      (1) Long, myelinated preganglionic fibers from the brain stem leave the vagus nerves and recurrent laryngeal branches and pass through the cardiac plexus to reach the heart.
      (2) Synapse occurs with minute, unmyelinated postganglionic fibers in the myocardium.

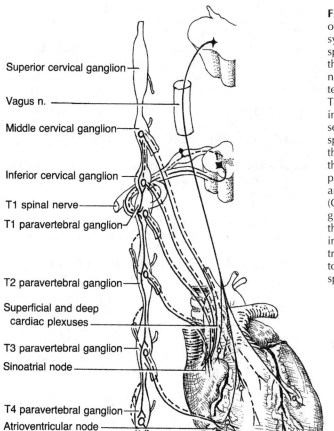

FIGURE 21-8. Autonomic innervation of the heart. The first (preganglionic) sympathetic neurons in the thoracic spinal cord (levels T1–T4) send axons through the ventral roots, spinal nerves, and white rami communicantes to reach the sympathetic chain. The second (postsynaptic) neurons lie in the sympathetic chain ganglia and send axons either through the cervical splanchnic (accelerator) nerves or through thoracic splanchnic nerves to the heart. The preganglionic parasympathetic neurons lie in the brain stem and send axons along the vagus nerve (CN X) to the heart. The short postganglionic parasympathetic neurons lie in the walls of the heart. Axons conveying painful sensation from the heart travel along the sympathetic pathways to the upper thoracic segments of the spinal cord.

b. Cholinergic vagal activity slows the heart rate and reduces the stroke volume.

2. **Sympathetic division.** Fibers from the sympathetic chain terminate in the vicinity of the S-A and A-V nodes.
   a. **Cervical splanchnic (cardiac accelerator) nerves** (see Figure 21-8)
      (1) Presynaptic fibers originate in the T1–T2 spinal levels and ascend the cervical chain to the cervical ganglia.
      (2) Postsynaptic neurons form cervical splanchnic (cardiac accelerator) nerves, which run in the neck from the superior, middle, and inferior ganglia of the cervical sympathetic chain and course through the cardiac plexus to the heart.
   b. **Thoracic splanchnic (cardiac) nerves** (see Figure 21-8)
      (1) The presynaptic fibers originate in the T1–T4 spinal levels.
      (2) The postsynaptic neurons form the **thoracic splanchnic nerves** from ganglia T1–T4 of the thoracic sympathetic chain. These nerves course through the cardiac plexus to the heart.
   c. **Adrenergic sympathetic activity** accelerates the heart rate and increases the stroke volume.

B. **Afferent sensation from the heart** runs along sympathetic pathways to reach levels T1–T4 of the spinal cord.

1. **Afferent nerve distribution** (see Figure 21-8). Fibers mediating painful sensation from the heart course along several nerve pathways to the sympathetic chain (see Chapter 22 V).
   a. **Cervical pathways.** Some afferent fibers course along the **middle** and **inferior cervical splanchnic (cardiac) nerves** to the middle and lower cervical ganglia of the sympathetic chain. From there, they descend to the **white rami communicantes** of the T1–T2 sympathetic ganglia to reach the respective spinal nerves and segmental levels of the cord (see Figure 21-8).
   b. **Thoracic pathways.** Other afferent fibers pass along the **thoracic splanchnic (cardiac) nerves** directly to ganglia T1–T4 of the sympathetic chain. From there, they travel along the T1–T4 white ramus communicantes to the respective spinal nerves and segmental levels of the spinal cord (see Figure 21-8).

2. **Referred cardiac pain.** Painful sensation from the heart originates from coronary artery insufficiency.
   a. Pain is referred to (perceived as coming from) the arm (usually left), shoulder, and precordium.
   b. The concept of pain referral is key to the proper evaluation of clinical signs and symptoms. Differentiating the pain of coronary artery insufficiency from that of esophageal or gastric origin is critical.
   c. **Mechanism**
      (1) Painful sensation mediated by visceral afferent fibers that enter the spinal cord at a particular level is referred to the somatic dermatome corresponding to that vertebral level.
      (2) This mechanism probably involves a final common pathway in the spinal cord with cerebral interpretation of a somatic rather than visceral location.

IX. CARDIAC DYNAMICS

A. **The cardiac cycle** is about 0.8 second in duration (Figures 21-9 and 21-10).

1. **Atrial systole** (contraction) begins the cardiac cycle (see Figures 21-9 and 21-10B).
   a. Contraction occurs for approximately 0.1 second and produces pressures of approximately 5 mm Hg in the right atrium and 11 mm Hg in the left atrium.
   b. Because the A-V valves are open during ventricular diastole, ventricular filling is primarily passive. The small amount of blood pumped into the ventricles dur-

Pericardial Cavity and Heart | **287**

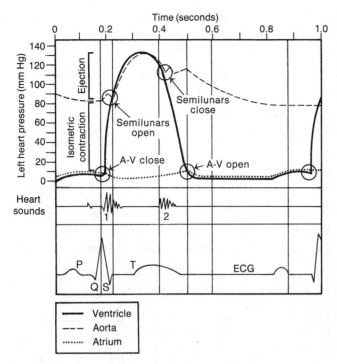

FIGURE 21-9. Pressures that develop during the cardiac cycle. The atrioventricular valves close (S1, "lub") early in ventricular systole when ventricular pressures exceed atrial pressures. The semilunar valves close (S2, "dup") early in ventricular diastole when arterial pressure exceeds ventricular pressures.

ing atrial systole stretches the ventricular walls, placing the heart muscle at the optimal point on the length-tension curve (see Chapter 5 II B).

2. **Atrial diastole** (relaxation) occupies most of the cardiac cycle (see Figure 21-9).
   a. Relaxation begins after 0.1 second into the cardiac cycle and is approximately 0.7 second long.
   b. It coincides with pressures of −5 mm Hg in the right atrium and 0 mm Hg in the left atrium.

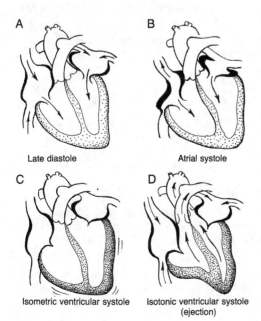

FIGURE 21-10. The cardiac cycle normally requires 0.8 second. *(A)* Ventricular diastole usually takes about 0.4 second, atrial diastole 0.7 second. *(B)* Atrial systole takes about 0.1 second. *(C)* Isometric ventricular systole begins with just before atrioventricular valve closure and takes about 0.1 seconds. *(D)* Isotonic ventricular systole begins with semilunar valve opening and ends just before semilunar valve closure; it takes about 0.3 second.

3. **Ventricular systole** follows atrial systole and occupies slightly less than one half of the cardiac cycle (see Figures 21-9 and 21-10C, D).
   a. Contraction begins approximately 0.1 second into the cardiac cycle and is approximately 0.3 second in duration.
   b. It produces pressures of approximately **23 mm Hg in the right ventricle** and approximately **120 mm Hg in the left ventricle.**
   c. Ventricular systole has two phases.
      (1) **Isovolumic phase.** The **A-V valves close** when ventricular pressure exceeds atrial pressure, thus producing the **S1** ("lub") **heart sound** and beginning a period of isometric contraction during which pressure builds (see Figure 21-10C).
      (2) **Ejection phase.** The semilunar valves open when ventricular pressure exceeds pulmonary diastolic pressure (10 mm Hg) and aortic diastolic pressure (80 mm Hg). This opening begins the period of **isotonic contraction,** during which blood is ejected into the outflow trunks and maximum systolic pressure is produced (see Figure 21-10D).
4. **Ventricular diastole** occupies slightly more than one half of the cardiac cycle (see Figures 21-9 and 21-10A, B).
   a. Relaxation begins at approximately 0.4 second and has a 0.5-second duration.
   b. It coincides with an approximate -7 mm Hg pressure in the right ventricle and 0 mm Hg pressure in the left ventricle.
   c. The **semilunar valves close** when pulmonary pressure (22 mm Hg) and aortic pressure (120 mm Hg) exceed the end-systolic ventricular pressures, producing the **S2** ("dup") **heart sound.** The momentary aortic pressure fluctuation (dicrotic notch) just after valve closure results from elastic rebound when valve closure stops the momentary reverse flow of blood.
   d. During ventricular diastole, rapid passive ventricular filling results from low intrathoracic pressure and the bulb-syringe effect (vis a fronte) of the ventricular walls (5 mm Hg right, 0 mm Hg left).

B. **Ejection volume, contraction rate, and blood pressure**
1. **Ejection volume.** The volume of blood ejected into the aorta during ventricular systole stretches that vessel and increases the elastic energy within its walls.
   a. This pressure moves the blood from the aorta during diastole and maintains the range of diastolic pressure.
   b. With an ejection volume of 60–70 ml/beat and a pulse rate of 80 beats/minute, the heart pumps 5 L/minute.
2. **Systolic pressure** can be a function of ejection volume (see Figure 21-9). The greater the volume of blood ejected, the more the elastic walls of the aorta are stretched and the higher the systolic pressure.
3. **Diastolic pressure** can be a function of heart rate (see Figure 21-9). The slower the rate, the lower the aortic diastolic pressure falls before the subsequent heart beat.
4. **Pulse pressure** normally is a combination of the factors just described, in addition to other variables, such as peripheral vascular resistance and blood volume. Pulse pressure produces the palpable pulse used to measure cardiac rate. Measured blood pressure is an indicator of cardiac and vascular function (see Chapter 5 III).

C. **Surface projection of heart sounds** permits valve auscultation.
1. **Valve locations.** The four heart valves lie in the plane of the annulus fibrosus, which is nearly vertical behind the sternum. Not only are the valves close together, but from an anteroposterior viewpoint, the semilunar valves as well as the A-V valves are nearly superimposed (Figure 21-11).
2. **Conduction velocity.** Because of the difference in conduction velocity through different tissues, the sound produced by each cycling valve is auscultated with maximal clarity and with minimal contribution from other valves over a distinct area of the thoracic wall.

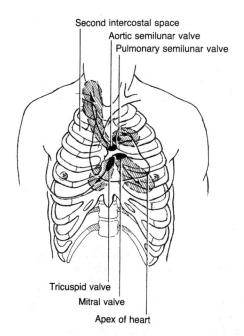

FIGURE 21-11. The atrioventricular and semilunar valves lie in the plane of the coronary sulcus, which is nearly vertical behind the sternum. The mitral valve sound projects to the fifth intercostal space in the left midclavicular line. The tricuspid valve sound is heard most distinctly just to the right of the sternum in the fourth or fifth intercostal space. The aortic valve sound projects to the second intercostal space at the right of the sternum and radiates to the neck. The pulmonic valve projects to the second intercostal space at the left of the sternum.

3. **Cardiac auscultation areas** (see Figure 21-11)
   a. **Tricuspid valve sounds** project to the midline and to the right side of the sternum at the fifth intercostal space.
   b. **Mitral valve sounds** project to the apex of the heart. The apex underlies the fifth intercostal space below the left nipple—the **point of maximum impulse (PMI)** or **apical pulse.**
   c. **Aortic semilunar valve sounds** project to the right of the sternum over the second intercostal space and also can be auscultated in the neck over the carotid artery. Inspiration lowers endothoracic pressure and increases venous return to the right side of the heart. This increased filling can result in closure of the pulmonary valve momentarily after the aortic valve, producing a "split S2."
   d. **Pulmonary semilunar valve sounds** project to the left of the sternum over the second intercostal space. The constant murmur produced by a patent ductus arteriosus is best auscultated just lateral to this point.

D. **Valvular defects and disease**
1. **Valvular insufficiency.** An incompetent or insufficient valve permits backflow. Valvular defects can be congenital or they can result from an infarct near a papillary muscle, rupture of the tendinous cords, or endocardial inflammatory processes.
   a. **Tricuspid valve insufficiency** leads to right-ventricular heart failure in which right ventricular pressure is transmitted back to the venous system to produce distended neck veins, a pulsating liver, abdominal ascites, and edema.
   b. **Mitral valve insufficiency** leads to left-ventricular heart failure in which left ventricular pressure is transmitted back to the lungs to produce pulmonary edema and pleural effusion (hydrothorax). Increased pulmonary pressure may induce **right-ventricular heart failure** (cor pulmonale).
2. **Valvular stenosis.** A stenotic valve restricts forward flow. Stenosis is either congenital or associated with endocardial inflammatory processes.
3. **Auscultation.** Both stenotic and incompetent valves produce **murmurs** that result from turbulent blood flow and can be localized by auscultation.
   a. **Atrioventricular insufficiency** is usually heard as a low-pitched, rushing, systolic murmur. **Atrioventricular stenosis** usually produces a low diastolic murmur before the first heart sound.

b. Semilunar insufficiency is associated with a diastolic murmur. **Semilunar stenosis** usually produces a high-pitched, musical systolic murmur.

4. **Congestive heart failure**
   a. As the volume of blood flow through the heart changes, the heart compensates by increasing or decreasing its work (pressure-volume product).
   b. Shunts, valvular stenosis, and valvular incompetence trigger cardiac compensation. A point is reached, however, at which overstretching of muscle fibers occurs with decompensation and resultant **cardiac failure.**
   c. Stenosis may be amenable to simple surgical correction, but insufficiency usually requires a prosthetic valve.

X. FETAL AND EARLY POSTNATAL CIRCULATION

A. **Fetal blood flow.** During fetal development, most of the blood passing through the heart must be shunted around the collapsed lungs (Figure 21-12). Three major blood channels present in the fetus are not functional postnatally.

1. **Ductus venosus.** Fetal blood is charged with oxygen and nutrients in the placenta and returns to the fetus via the umbilical vein.
   a. The **ductus venosus** diverts over one half of the blood around the liver into the inferior vena cava. The remaining blood from the umbilical vein drains through the portal vein and sinusoids of the liver before passing through the hepatic vein and joining the bypassed blood in the inferior vena cava.
   b. **Hepatic blood flow** is regulated by a muscular sphincter in the ductus venosus. When the sphincter relaxes, more blood passes through the ductus venosus; when the sphincter contracts, more blood is shunted to the portal vein and liver.
2. **The foramen ovale** in the interatrial septum functions as a pulmonary shunt (see Figure 21-12).

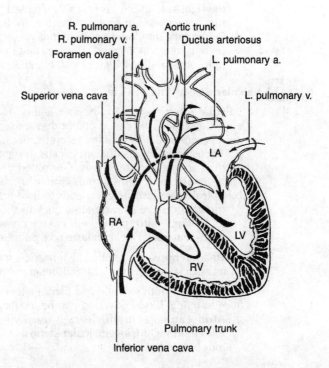

FIGURE 21-12. Fetal circulation. Oxygenated fetal blood, returning from the placenta through the inferior vena cava, is directed predominantly through the foramen ovale into the left atrium. It then passes through the left ventricle and supplies the important vessels of the heart and head. Deoxygenated fetal blood, returning from the head through the superior vena cava, passes predominantly into the right ventricle and then into the pulmonary trunk, most of it joining the aortic flow through the ductus arteriosus. Only a small amount of right ventricular blood passes through the lungs.

a. **Blood entering the right atrium** is from the inferior vena cava, which contains oxygenated blood from the umbilical vein as well as deoxygenated blood from the lower extremities, abdomen, and pelvis. Thus, blood from the inferior vena cava has a far higher oxygen saturation than blood in the superior vena cava.
      (1) The valve of the inferior vena cava (eustachian valve) guides well-oxygenated inferior caval blood through the **foramen ovale** into the left atrium.
      (2) The resistance to blood flow through the collapsed lungs makes the systolic pressure greater in the right atrium than in the left atrium. The valve of the foramen ovale is located on the left side of the septum and is therefore kept open by the interatrial pressure differential.
   b. In the **left atrium,** the well-oxygenated blood mixes with the small amount of deoxygenated blood returning from the lungs via the pulmonary veins. This blood then flows successively through the left ventricle and ascending aorta.
   c. The **coronary, carotid,** and **subclavian arteries** are the first major arteries to receive blood from the ascending aorta. Therefore, well-oxygenated and nutritional blood is directed to the heart, brain, head, neck, and upper extremities.
3. **Ductus arteriosus,** the remnant of the left sixth branchial arch, connects the left pulmonary artery and the aorta (see Figure 21-12).
   a. In the **right atrium,** deoxygenated blood returning from the head and upper extremities by way of the superior vena cava and from the heart via the coronary sinus is directed through the tricuspid valve into the right ventricle. It then flows into the pulmonary trunk.
   b. The **ductus arteriosus** shunts most of this blood into the descending aorta because of the high vascular resistance of the collapsed fetal lung. Here it mixes with well-oxygenated blood from the aortic arch. Because this admixture occurs after the coronary and carotid arteries have been given off, the head receives the blood with the highest possible oxygen content.
   c. Most of the mixed blood in the descending aorta passes to the placenta for reoxygenation. The remainder circulates through the lower part of the body.

B. **Changes during the postnatal period** are related to the cessation of placental flow and the commencement of pulmonary respiration.
   1. **Closure of the foramen ovale,** in two phases, occurs immediately on first inspiration.
      a. **Functional closure** is produced by **reversal of interatrial pressure.** On first expansion of the lungs, pulmonary vascular resistance drops significantly, making the systolic pressure lower in the right atrium than in the left atrium and causing the valve of the foramen ovale to close. During the first weeks of life, however, closure is reversible. Crying with hyperventilation creates a right-to-left shunt in newborns, accounting for cyanotic periods.
      b. **Anatomic closure** results from constant apposition of the valve leaflet with the septum, which leads to fusion after a few months, producing the **fossa ovalis.** In 10–15% of individuals, however, perfect anatomic closure is not obtained ("probe patency").
   2. **Closure of the ductus arteriosus.** This vessel begins to close soon after birth. During the first few postpartum days, a left-to-right shunt is common, and a murmur produced by a patent ductus may be auscultated.
      a. **Initial closure** is by temporary muscular contraction mediated by bradykinin, a substance released from the lungs during their initial inflation. During the postnatal period, the duct re-opens from time to time. Backflow during patency is less than might be expected from pressure differentials because the angle at which the ductus joins the aorta produces a Venturi effect.
      b. **Complete anatomic obliteration** by proliferation of the intima takes 1–3 months. In the adult, the obliterated **ductus arteriosus** is called the **ligamentum arteriosum.**
   3. **Closure of the umbilical arteries** is accomplished by contraction of the smooth muscles in the wall of the vessels, probably caused by thermal and mechanical stimuli and a change in oxygen tension.

a. Functionally, the arteries are closed a few minutes after birth, but the actual obliteration of the lumen by fibrous proliferation may take 2–3 months.
b. The distal parts of the umbilical arteries then form the medial umbilical ligaments; the proximal portions remain open as the superior vesical arteries.
4. **Closure of the umbilical vein and ductus venosus** occurs shortly after that of the umbilical arteries.
   a. The **umbilical vein** loses the lumen to form the **round ligament of the liver** (ligamentum teres hepatis) in the lower margin of the falciform ligament.
   b. The **ductus venosus,** which courses between the umbilical vein and the inferior vena cava, also undergoes fibrosis to form the **ligamentum venosus.**

C. **Congenital defects**
1. **Patent ductus arteriosus.** Although the ductus arteriosus may remain patent throughout life with no major problems, the patent ductus usually is ligated surgically once discovered.
2. **Septal defects** are only problematic when the shunt flows from right to left.
   a. An **atrial septal defect (ASD),** such as a patent foramen ovale, frequently is compatible with normal life and activity because the shunting is usually from left to right.
   b. A **ventricular septal defect (VSD)** may result in life-threatening **tetralogy of Fallot.**
      (1) **Characteristics of tetralogy of Fallot**
         (a) A **VSD** provides **interventricular communication.**
         (b) **Right ventricular hypertrophy** results from distribution of left systolic hydrostatic pressure through the VSD.
         (c) **Functional pulmonary stenosis** results from hypertrophy of the supraventricular crest so that myocardial contraction occludes the pulmonary outflow tract and forces a right-to-left shunt.
         (d) The opening of the **aorta appears to override** the VSD into the right ventricle.
      (2) **Sequelae.** The ensuing **right-to-left shunt** sends deoxygenated blood to the aorta, necessitating increased cardiac work to provide the same amount of oxygen to the body tissue with the possibility of cardiac decompensation and failure.
3. **Dextrocardia.** The heart is normal but in mirror image (0.01–0.02% of the population).
   a. A high incidence of situs inversus viscerum (reversed rotation of the gastrointestinal tract) in individuals with dextrocardia.
   b. Dextrocardia may be associated with either a right-sided or a left-sided aorta.

Chapter 22

Mediastinum

I. ANTERIOR MEDIASTINUM

A. Overview. The anterior mediastinum is the region between the thoracic cage and the pericardial cavity.

B. The thymus gland is the principal structure of this part of the mediastinum (Figure 22-1). It is an important lymphoid organ associated with immunologic recognition.

1. **Location.** Derived from the third embryonic branchial pouch, the thymus gland descends with the thyroid gland to its adult position in the superior mediastinum and the anterior mediastinum (see Figure 22-1).
   a. In the **anterior mediastinum,** the thymus gland lies beneath the manubrium and upper body of the sternum as well as anterior to the fibrous pericardium. It extends laterally beneath the upper four costal cartilages, sometimes as far as the hilum of the lung on either side.
   b. In the **superior mediastinum,** it lies under the sternohyoid and sternothyroid muscles, but anterior to the ascending aorta and brachiocephalic vein.

2. **Size.** It varies in size with age. At birth, it weighs 10–15 g. Before puberty, it grows to 30–40 g. During adult life, it regresses by involution and fatty atrophy to less than 10 g.

3. **Function.** The thymus is the source of T-lymphocytes, associated with immunologic recognition.
   a. The **normal aging process** may be driven by thymic atrophy.
   b. **Thymectomy** is undertaken as a palliative measure for **myasthenia gravis,** an autoimmune disorder in which thymocytes make antibodies to transmitter receptors at the neuromuscular junctions.

C. Sternopericardial ligaments run from the posterior surface of the sternum to the fibrous pericardium, tethering the anterior pericardium.

II. SUPERIOR MEDIASTINUM

A. Overview. The principal contents are the **thymus gland** and **vascular structures** running to and from the heart (see Figure 22-1).

1. **Thymus gland.** Located primarily in the anterior mediastinum, thymic tissue may also be found in the superior mediastinum.

2. **Vascular structures.** The veins course anteriorly and toward the right, whereas the arteries tend to course behind and to the left of the veins.

B. Neurovascular structures

1. **Brachiocephalic veins.** The left and right brachiocephalic veins receive drainage from the head, neck, and upper extremities.
   a. **Branches** (see Figure 22-1)
      (1) The **subclavian** and **internal jugular veins** on each side unite to form the left and right brachiocephalic veins.
      (2) The **right brachiocephalic vein** and the longer **left brachiocephalic vein** unite to form the **superior vena cava.**
      (3) The **superior vena cava** receives the **azygos vein** before emptying into the right atrium.

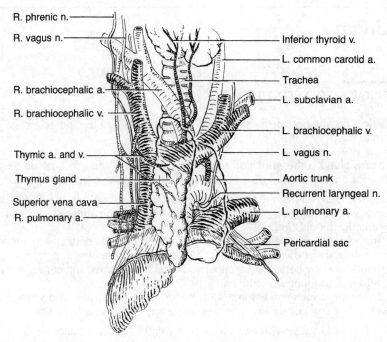

FIGURE 22-1. Anterior mediastinum and superior mediastinum. The lower part of the right half of the thymus gland is visible, and the remainder of this gland is ghosted to show the major mediastinal vessels.

   b. **Development** is from the complex system of embryonic cardinal veins that also form the oblique vein of the heart as well as the coronary sinus. Reversal or modification of the developmental sequence may result in a single left superior vena cava or bilateral venae cavae.
 2. **Ascending aorta and aortic arch.** The aorta emerges from the pericardial sac between the superior vena cava and the pulmonary trunk (see Figure 21-1).
   a. **Development.** The definitive aortic arch is derived from the fourth left branchial arch of the embryo. The right subclavian artery is derived from the fourth right branchial arch.
   b. **Course.** The **ascending aorta** runs superiorly and toward the right before arching posteriorly as the **aortic arch** and then inferiorly as the **descending aorta** (see Figure 22-2). The terminal portion of the aortic arch produces the "aortic knob," a landmark on the left side of the mediastinum in a posteroanterior chest radiograph.
   c. **Branches** (see Figure 22-1)
     (1) **Left and right coronary arteries** supply the heart (see Figure 21-3).
     (2) The **brachiocephalic artery** (on the right only) gives rise to the right common carotid and right subclavian arteries.
     (3) The **left common carotid artery** usually arises from the ascending aorta. In some individuals (25% whites, 50% blacks), it arises from the brachiocephalic artery.
     (4) The **left subclavian artery** supplies the left arm.
 3. **Nerves**
   a. **Phrenic nerves** innervate the diaphragm (see Figure 22-1).
     (1) Arising from cervical nerves C3–C5, the phrenic nerves descend through the superior mediastinum, passing between the subclavian arteries and veins.
     (2) They pass into the middle mediastinum, lateral to the pericardial cavity, to reach the diaphragm.
   b. **Vagus nerves** innervate mediastinal structures as they pass through the thorax (see Figures 22-1 and 22-2).

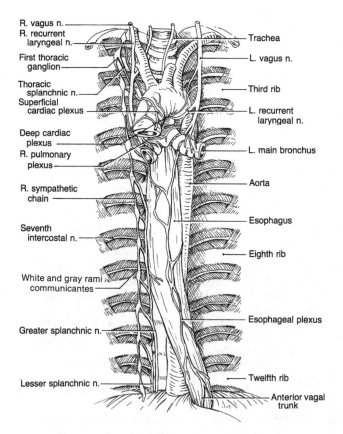

FIGURE 22-2. Posterior mediastinum and autonomic nerves of the thorax. The esophagus and aorta are the principal posterior mediastinal structures. The sympathetic trunk and vagus nerves bring sympathetic and parasympathetic innervation, respectively, to the thoracic viscera.

(1) The left and right vagus nerves leave the jugular foramen and descend through the neck into the superior mediastinum, passing anterior to the right subclavian artery and anterior to the aortic arch.

(2) They give rise to **left** and **right recurrent laryngeal nerves,** which innervate laryngeal derivatives of the sixth branchial arch and pouch.

   (a) The **right recurrent (inferior) laryngeal nerve** reverses course by passing inferior to the right subclavian artery, the remnant of the artery of the right fourth branchial arch. (The right fifth and sixth arches degenerate completely so the nerve slips up to the level of the fourth arch.)

   (b) The **left recurrent (inferior) laryngeal nerve** reverses inferior to the **ligamentum arteriosum,** the remnant of the artery of the sixth branchial arch.

(3) Vagus nerves give rise to numerous twigs to the cardiac and pulmonary plexuses.

III. POSTERIOR MEDIASTINUM

A. Overview. By definition, the posterior mediastinum lies behind the pericardial sac (Figure 22-2).

   1. Boundaries. It extends approximately between vertebral levels T1 and T4. It is bounded by the mediastinal pleura on either side and by the diaphragm below.

   2. Relations and contents. It is continuous with the superior mediastinum and lies behind the middle mediastinum. It contains the esophagus, trachea, descending aorta, thoracic duct, and numerous smaller nerves and vascular structures.

B. **The esophagus** extends from the lower laryngopharynx to the cardiac opening of the stomach (see Figures 22-1 and 25-1).

1. **Relationships.** It is approximately 25–30 cm long in men (in the upright position) and a few centimeters shorter in women.
   a. In the **neck,** the esophagus lies posterior to the trachea and anterior to the prevertebral fascia.
   b. In the **thorax,** its position changes slightly as it passes inferiorly toward the diaphragm (see Figure 22-2).
      (1) **High in the thorax,** it lies in the midline behind the trachea. Inferior to the tracheal bifurcation, it deviates to the right so that it lies slightly behind and to the right of the aorta.
      (2) **Low in the thorax,** it passes anterior to the aorta and through the esophageal hiatus, which is about 2 cm left of the midline in the muscular portion of the diaphragm.
   c. **In the abdomen,** below the diaphragm, it makes an abrupt turn to the left and enters the stomach (see Figure 25-1).

2. **Structure.** The esophagus is the most muscular segment of the alimentary tract.
   a. **Musculature.** Its upper quarter is composed primarily of striated muscle arranged in inner circular and outer longitudinal layers. Proceeding aborally, the amount of smooth muscle increases and the striated fibers disappear.
      (1) The **cricopharyngeal muscle** is the sphincter of the upper esophageal opening (see Figure 37-7). It remains closed except during deglutition (swallowing), eructation (belching), and emesis (vomiting).
      (2) The intra-abdominal portion of the esophageal musculature acts as a diffuse sphincter **(cardiac sphincter).**
   b. **The lumen** averages 1 cm in diameter orally and 2 cm aborally, except during deglutition.
      (1) The esophagus is normally flat and the shape of the lumen is irregular because tension within the inner (circular) layer of the muscularis externa causes formation of longitudinal folds (see Figure 30-1).
      (2) Because of this folding, the esophagus is distensible, accommodating anything that passes over or around the epiglottis.

3. **Vasculature**
   a. **Arterial supply** (see Figure 25-1)
      (1) **Inferior thyroid arteries** (branches of the subclavian arteries) supply the cervical portion of the esophagus.
      (2) **Esophageal branches of the aorta** as well as some **bronchial arteries** supply the middle portion.
      (3) **Esophageal branches of the left gastric artery** (a branch of the celiac trunk) supply the lower thoracic and abdominal portions.
   b. **Venous return** of the upper and lower esophagus is clinically important (see Figure 22-3).
      (1) Drainage from the upper portions is by the **inferior thyroid veins,** which usually drain into the left and right brachiocephalic veins.
      (2) Venous drainage of the middle portion is into the **azygos** and **hemiazygos veins.**
      (3) Drainage from the lower portions is by the **left gastric vein,** which drains into the **hepatic portal vein.**
   c. **Esophageal varices**
      (1) The veins of the middle part of the esophagus that drain into the systemic circulation anastomose with the veins of the lower esophagus that drain into the hepatic portal system.
      (2) These anastomoses occur in the **esophageal venous plexus** that lies in the submucosa.
      (3) In patients with **portal hypertension,** these anastomoses provide an important but potentially dangerous shunt that results in the development of **esophageal varices.**

(a) These tortuous and dilated veins immediately beneath the mucosa protrude into the esophageal lumen, usually in the distal one third of the esophagus. They are subject to mechanical trauma during deglutition, emesis, or the passage of diagnostic instrumentation.

(b) They are asymptomatic until they rupture, causing massive hematemesis. More than 50% of patients with advanced cirrhosis of the liver die as a result of rupture of an esophageal varix.

4. **Innervation**
   a. **Parasympathetic innervation** (see Figure 22-2) is by the vagus nerve (CN X). In the thorax, the **left and right vagus nerves** form the **esophageal plexus,** which reunites distally to form the vagal anterior (formerly left) and posterior (formerly right) **vagal trunks** (see Figure 22-2).
      (1) The long, myelinated **parasympathetic presynaptic neurons** leave the esophageal plexus and enter the esophageal wall.
      (2) The cell bodies of the **parasympathetic postsynaptic neurons** that lie in the **myenteric** and **submucosal plexuses** innervate esophageal musculature and glands.
      (3) Dysfunction of the parasympathetic pathway at the level of the myenteric plexus results in **achalasia (cardiospasm),** characterized by **dysphagia** and a progressively dilated and tortuous esophagus.
   b. **Sympathetic innervation** (see Figure 22-2)
      (1) **Sympathetic presynaptic neurons** to the esophagus emerge from the T1–T5 ventral roots and reach the sympathetic chain via associated **white ramus communicans.**
         (a) The thoracic portion receives presynaptic axons from chain ganglia T1–T4 via **thoracic splanchnic nerves.**
         (b) The abdominal portion receives splanchnic outflow from the T5 contribution to the **greater splanchnic nerve.**
      (2) **Sympathetic postsynaptic neurons**
         (a) These neurons contained in the thoracic splanchnic nerves (T1–T4) synapse either in the **sympathetic chain** or in small prevertebral ganglia.
         (b) The neurons to the terminal esophagus run in the T5 component of the greater splanchnic nerve and synapse in the **celiac ganglion.**
   c. **Afferent innervation** is along sympathetic pathways (see subsequent section V).
      (1) **Sensory pathways**
         (a) Afferent fibers mediating pain from the middle portion of the esophagus course along the **thoracic splanchnic nerves** to the sympathetic chain. Afferents then travel along the T1–T4 white ramus communicantes to the respective spinal nerves and segmental levels of the cord.
         (b) Afferent fibers mediating pain from the abdominal portion of the esophagus course along the **greater splanchnic nerve** to the sympathetic chain. Afferents then travel along the T5 white ramus communicans to the spinal nerve and segmental level.
      (2) **Referred esophageal pain.** Most painful sensation from the esophagus (such as from peptic ulceration) originates from the lower thoracic and abdominal segments.
         (a) Pain is usually mediated via splanchnic nerves associated with T4–T5 levels and so is referred to the **inferior thorax** and the epigastric **region** of the abdomen.
         (b) The differential diagnosis involving the pain of coronary artery insufficiency and that of esophageal origin is critical.

5. **Function.** The esophagus serves solely for the transport of food (deglutition).
   a. The **upper esophagus** is guarded at its opening by a sphincter, the **cricopharyngeus muscle** (see section III B 2 a).
   b. **The midesophagus** propels a bolus of food aborally by peristaltic activity.
      (1) A solid bolus entering the esophagus initiates an involuntary peristaltic wave, which reaches the cardiac sphincter in 5–6 seconds. The cardiac sphincter, normally closed, relaxes to allow the bolus to pass into the stomach.

(2) A fluid bolus outruns the peristaltic wave because of gravity, but the cardiac sphincter stops it until this sphincter is inhibited by the approach of the peristaltic wave. Thus, only one sound is heard when a solid bolus is swallowed, whereas two separate sounds are audible with a fluid bolus.
   c. **Lower esophagus.** The **cardiac sphincter** is located at the cardioesophageal junction.
      (1) **Structure.** The smooth muscle of the cardioesophageal junction is continuous with that of the stomach, and the muscle is not considered a structural sphincter. The sphincteric barrier is created by valvelike folds of mucosa maintained by the tonic activity of the muscularis mucosae.
      (2) **Function.** The cardiac sphincter maintains a zone of elevated luminal pressure at the aboral end of the esophagus.
         (a) When closed, the pressure within the stomach is always greater than that in the esophagus, preventing reflux of gastric contents.
         (b) This sphincter is relaxed by peristaltic inhibition, allowing the bolus to enter the stomach. In a series of rapid swallows, it remains inhibited so that fluid passes directly into the stomach unassisted by peristalsis (as in "chugging").
         (c) **Esophagitis.** Occasionally, gastric contents reflux into the esophagus. Acidic peptic chyme burns and inflames the unprotected stratified squamous epithelium of the esophageal mucosa, producing **regurgitative esophagitis,** or **"heartburn."** Repeated reflux can produce esophageal peptic ulcers, which can cause **dysphagia** and serious bleeding.

IV. VASCULATURE

A. **The descending aorta** is about 3 cm in diameter.
   1. **Course.** This vessel lies in the left side of the mediastinum and bulges into the left pleural cavity (see Figure 22-2).
      a. The **radiographic appearance** (posteroanterior view) of the aorta is a distinctive radiopaque border along the left side of the mediastinum, superior to the heart shadow.
      b. It enters the abdominal cavity through the midline **aortic hiatus** between the crura of the diaphragm.
   2. **Branches**
      a. The **ligamentum arteriosum** courses between the aortic arch and the left pulmonary artery. It is the remnant of the embryonic ductus arteriosus, a left sixth arch derivative.
      b. **Unpaired branches** (see Figure 25-1)
         (1) **Bronchial arteries** supply the bronchial tree and upper part of the esophagus.
         (2) **Esophageal branches** supply the middle portion of the esophagus.
      c. Ten pairs of **intercostal arteries** leave the thoracic part of the descending portion of the aorta (see Figure 19-6). They supply the thoracic wall and anastomose with the anterior intercostal branches of the internal thoracic arteries.

B. **Venous drainage**
   1. **The azygos vein** lies to the right of the esophagus on the vertebral column (Figure 22-3).
      a. It usually begins as a continuation of the right ascending lumbar vein.
      b. It receives the right posterior intercostal, right bronchial, esophageal, and hemiazygos veins.
      c. It enters the superior vena cava by passing to the superior mediastinum beneath the mediastinal pleura and arching anteriorly at vertebral level T4.
   2. **The hemiazygos vein** and the **superior (accessory) hemiazygos vein** may be separate or joined, and are subject to extensive variation (see Figure 22-3).

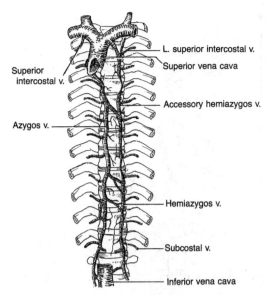

FIGURE 22-3. Azygos system of veins.

   a. The **hemiazygos vein** parallels the lower portion of the azygos vein.
      (1) It usually begins as a continuation of the left ascending lumbar vein. It is absent in 15% of the population.
      (2) It drains the lower eight or nine left posterior intercostal veins and esophageal veins.
      (3) It may anastomose with the inferior vena cava, left adrenal vein, or left renal vein in the abdomen.
      (4) In 54% of the population, it receives the accessory hemiazygos vein at the level of the eighth thoracic vertebra, where it turns right to join the azygos vein, sometimes (12%) by more than one connection.
   b. The **superior (accessory) hemiazygos vein** may drain downward, upward, or both.
      (1) It receives bronchial veins from the left lung and esophageal veins.
      (2) It receives the upper three to five left posterior intercostal veins.
      (3) It may join the inferior hemiazygos vein (54%), enter the azygos vein separately (15%), drain into the left superior intercostal vein (31%), or follow combinations of these courses.

C. **Thoracic duct.** Lymph from the body below the diaphragm, the left side of the thorax, the left arm, and the left side of the head and neck returns to the systemic circulation through the thoracic duct, which drains about 24 L of lymph daily.
   1. **Course.** It rises from the abdomen into the root of the neck on the left side (Figure 22-4).
      a. In the **abdomen,** the duct originates in the **cisterna chyli.**
         (1) The cisterna chyli is located at vertebral levels T12–L1 between the crura of the diaphragm and posterior to the aorta.
         (2) The cisterna has muscular walls that move lymph into the thoracic duct.
      b. In the **thorax,** the duct lies behind the esophagus and to the right of the aorta. It is delicate and may double or even triple.
      c. In the **superior mediastinum,** the duct arches over the cupula of the left pleura, posterior to the left subclavian vein. It enters the left brachiocephalic vein at the angle formed by the juncture of the subclavian vein with the left internal jugular vein.
   2. **Branches**
      a. The **intercostal nodes** drain the left thoracic wall and a small portion of the left breast.

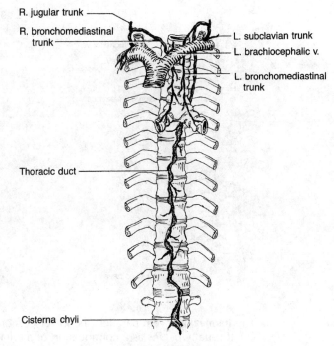

FIGURE 22-4. Lymph from the entire left side of the body, as well as from the lower half of the right side, returns to the left brachiocephalic vein by the thoracic duct.

- b. The **left subclavian trunk** drains the left arm, left breast, and left side of the neck.
- c. The **left jugular trunk** drains the left side of the head and neck.
- d. The **left bronchomediastinal trunk** drains the left bronchial tree.

3. **Chylothorax.** Chyle from a torn or severed thoracic duct may drain into the right or left pleural cavity.

D. Lymph nodes

1. **Preaortic nodes** lie anterior to the aorta. They drain the esophagus and empty into the thoracic duct.
2. **Para-aortic (intercostal) nodes** lie alongside the aorta.
   a. They receive lymph from the posterior intercostal spaces and drain directly into the thoracic duct.
   b. The anterior intercostal spaces are drained by **parasternal nodes** along the internal thoracic artery. Thus, the right parasternal nodes drain into the right lymphatic trunk; the left into the thoracic duct.
3. **Bronchomediastinal nodes** lie adjacent to the bronchi and trachea. They receive drainage from the hilar nodes. The **right bronchomediastinal nodes** drain into the right brachiocephalic vein via the **right bronchomediastinal trunk**. The **left bronchomediastinal nodes** usually drain into the thoracic duct or the brachiocephalic veins.
4. **Hilar nodes** are located around the root of the lung and within the lung parenchyma. They drain each lung and peripheral bronchial tree toward the bronchomediastinal nodes.

V. INNERVATION

A. Autonomic nervous system

1. **Overview.** The heart, lungs, trachea, and esophagus are innervated through this system.

a. **Splanchnic nerves** convey sympathetic stimuli.
b. The **vagus nerves** convey parasympathetic stimuli.
c. The **cardiac plexus** is a region of intermingling sympathetic and parasympathetic (vagal) fibers posterior to the arch of the aorta and anterior to the bifurcation of the pulmonary trunk (see Figure 22-2).
b. The **pulmonary plexus** is continuous with the cardiac plexus and lies about the roots of the lungs (see Figure 22-2).
c. The **esophageal plexus** is the continuation of the left and right vagus nerves (see Figure 22-2). It provides parasympathetic innervation to the esophagus and carries vagal fibers to the abdomen via anterior and posterior vagal trunks.

2. **Sympathetic pathways** consist of a myelinated preganglionic (presynaptic) neuron and an unmyelinated postganglionic (postsynaptic) neuron (see Figure 4-5 and Chapter 4 V A).
    a. **Preganglionic sympathetic fibers** run from the spinal cord to the sympathetic chain of ganglia (see Figures 20-4 and 21-8). The **white rami communicantes** provide pathways for these neurons between the spinal nerves and the corresponding thoracic sympathetic ganglion. Neurotransmission is cholinergic at the synapse.
    b. **Postganglionic sympathetic fibers** run from the chain of sympathetic ganglia to target organs (see Figures 20-4 and 21-8). Neurotransmission is adrenergic at the target receptors.
        (1) **Cervical splanchnic (cardiac) nerves,** containing predominantly unmyelinated postsynaptic neurons, arise from the superior, middle, and inferior cervical ganglia and run deep in the neck to the cardiac and pulmonary plexus to reach the target tissue.
        (2) **Thoracic splanchnic (cardiac) nerves,** containing predominantly unmyelinated postsynaptic neurons, arise from sympathetic ganglia T1 through T4 or T5 and run to the cardiac and pulmonary plexus to reach the target tissue.

3. **Parasympathetic pathways** consist of a myelinated preganglionic (presynaptic) neuron and an unmyelinated postganglionic (postsynaptic) neuron (see Chapter 4 V A). Neurotransmission is cholinergic at the synapse.
    a. **Preganglionic parasympathetic fibers** run in the vagus nerve from the brain to distal parasympathetic ganglia close to or within the target organ (see Figures 20-4 and 21-8).
    b. **Postganglionic parasympathetic fibers** arise from numerous small ganglia close to or within the target organ (see Figures 20-4 and 21-8). Neurotransmission is cholinergic at the target receptors.

4. **Visceral sensation.** Nerves mediating sensation from the thoracic viscera travel along autonomic pathways (Table 22-1).
    a. **Pain afferents** follow sympathetic pathways.
        (1) Afferents from the **heart** follow the middle and inferior cervical cardiac nerves as well as the upper four or five thoracic cardiac nerves to reach the sympathetic chain (see Figure 21-8).
        (2) **White rami communicantes** provide the pathway for pain afferents between the sympathetic chain and spinal nerves T1–T4.
            (a) **Cervical pathways.** Because the cervical region has no white rami, pain afferents arriving at the cervical ganglia descend to the T1 ganglion to reach white ramus to connect with a spinal nerve.
            (b) **Thoracic pathways.** Afferent nerves traveling with the thoracic splanchnic nerves enter the upper thoracic sympathetic ganglia and pass directly to the white rami associated with spinal nerves T1–T4.
        (3) The **cell bodies** of the visceral afferent fibers are in the **dorsal root ganglia** of the spinal nerves.
        (4) **Basis for referred pain**
            (a) Painful stimuli mediated by visceral afferent neurons, which enter the spinal cord at a specific spinal level, are interpreted by the brain as origi-

TABLE 22-1. Thoracic Visceral Afferent Pathways

Organ	Pathway	Spinal Levels	Referral Regions
Heart, Thoracic esophagus	a. Middle and inferior cervical splanchnic nn.	T1–T2 via cervical ganglia	Postaxial arm, high thorax
	b. Thoracic splanchnic nn.	T1–T4	High and middle thorax
Abdominal esophagus, Stomach, Gallbladder, Liver, Bile duct, Sup. duodenum	Celiac plexus and greater splanchnic n	T5–T9	Low thorax, epigastric region

nating from (referred to) the somatic dermatome corresponding to that particular spinal level (see Table 22-1).

   (b) As such, painful stimuli originating from the heart are referred to the T1–T4 dermatomes, which include the medial aspect of the upper arm and anterior thoracic wall as far as the level of the nipple.

   b. **Reflex afferents** follow parasympathetic pathways. Afferents for respiratory and cardiac reflexes travel exclusively along the vagus nerve.

B. **Sympathetic trunk.** Continuous along the length of the vertebral column and lying across the necks of the ribs, the sympathetic chain consists of interconnected **sympathetic (paravertebral) ganglia,** 12 of which are in the thorax.

   1. **Course and composition.** Each thoracic ganglion has five roots (see Figure 22-2).
      a. **White rami communicantes** bring presynaptic (myelinated) general visceral efferent neurons from the spinal nerve to or through the sympathetic chain. Those neurons passing through the ganglia without synapses may go up or down the chain to synapse at an adjacent level, travel in splanchnic nerves to synapse in prevertebral ganglia, or synapse in the ganglion at the level where they enter the chain.
      b. **Gray rami communicantes** return unmyelinated postsynaptic neurons to the spinal nerve. These nerves are involved in piloerection, sweating, and vasomotor functions in the associated dermatome.
      c. The **superior and inferior connectors** carry presynaptic and postsynaptic neurons to adjacent or remote levels of the chain. Because there are no white rami above T1 or below L2, the superior and inferior connectors bring sympathetic innervation to the cervical and lumbopelvic regions. They also convey visceral afferent neurons from the cervical and pelvic regions to the closest white rami.
      d. **Splanchnic nerves** carry presynaptic neurons to prevertebral ganglia, such as the small ganglia (of Wrisberg) within the cardiac plexus, the celiac ganglion, or the superior mesenteric ganglion, where they synapse.
         (1) From the prevertebral ganglia, postsynaptic neurons innervate visceral structures.
         (2) Other splanchnic nerves carry postsynaptic neurons from the sympathetic chain directly to visceral structures.

   2. **Distribution of splanchnic nerves** (see Table 22-1)
      a. **Ganglia T1–T4** contribute to the cervical chain because of the absence of white rami communicantes in the neck.
         (1) The **superior, middle, and inferior cervical ganglia** give off **cervical splanchnic (cardiac accelerator) nerves** to the heart.
         (2) All presynaptic sympathetic neurons to the head arise from upper thoracic levels.

b. Ganglia T1–T4 contribute **thoracic splanchnic nerves** to the heart, lungs, bronchial tree, and esophagus.
c. Ganglia T5–T9 contribute to the **greater splanchnic nerve,** which innervates the foregut.
d. Ganglia T10–T11 contribute to the **lesser splanchnic nerve,** which innervates the midgut.
e. Ganglion T12 contributes to the **least (small) splanchnic nerve,** which innervates the gonads, kidneys, and proximal ureters.

3. **Effects of sympathetic innervation** are those of **adrenergic neurotransmission:**
   a. Increases heart rate (tachycardia)
   b. Increases stroke volume (ejection fraction)
   c. Dilates coronary and pulmonary arteries
   d. Promotes bronchial secretion
   e. Produces piloerection, perspiration, and dilation or constriction of peripheral vessels in the dermatomes

C. **Vagus nerves (CN X)** provide **parasympathetic innervation** to the cervical and thoracic viscera as well as the foregut, and midgut.
1. **Right vagus nerve** (see Figure 22-2)
   a. This nerve gives off the **right recurrent (inferior) laryngeal nerve,** which passes around the right subclavian artery, a remnant of the right fourth branchial arch.
   b. It gives off twigs to the **cardiac plexus.**
   c. It runs behind the right main stem bronchus and contributes to the **anterior** and **posterior pulmonary plexuses.**
   d. It passes along the right surface of the esophagus to form an **esophageal plexus,** which re-forms as the **posterior vagal trunk** before passing through the esophageal hiatus of the diaphragm.
2. **Left vagus nerve** (see Figure 22-2)
   a. This nerve gives off the **left recurrent (inferior) laryngeal nerve,** which courses around the ligamentum arteriosum (remnant of the sixth left branchial arch).
   b. Together with the recurrent laryngeal nerve, it gives off twigs to the **cardiac plexus.**
   c. It passes posteriorly to the root of the lung, where it gives off major contributions to the **posterior and anterior pulmonary plexuses.**
   d. It passes along the left side of the esophagus to form an **esophageal plexus,** which re-forms as the **anterior vagal trunk** just above the diaphragm.
3. **Effects of parasympathetic innervation** are those of **cholinergic neurotransmission:**
   a. Decreases heart rate (bradycardia)
   b. Decreases stroke volume (ejection fraction)
   c. Produces bronchoconstriction
   d. Promotes peristalsis
4. **Clinical consideration.** The recurrent laryngeal nerves innervate the vocal musculature. Hilar or mediastinal tumors on the left side or an apical tumor on the right side, involving either of the respective nerves, cause hoarseness, often the first sign of these tumors.

PART IIIA THORAX

STUDY QUESTIONS

DIRECTIONS: Each of the numbered items or incomplete statements in this section is followed by answers or by completions of the statement. Select the ONE lettered answer or completion that is BEST in each case.

1. Two elderly persons, both diagnosed as having pleurisy, were comparing their illnesses. One had left lateral thoracic pain, whereas the other had left neck and shoulder pain. They were to the point of reviewing the academic qualifications of their respective physicians. Assuming the diagnoses to be correct, the explanation for these differences in pain distribution is that the pleura is innervated by

(A) cervical splanchnic and thoracic splanchnic nerves
(B) intercostal and phrenic nerves
(C) parasympathetic and sympathetic nerves
(D) phrenic and vagus nerves
(E) vagus and intercostal nerves

Questions 2–7

An 84-year-old man with a history of chronic congestive heart failure complains of being short of breath. His breathing is rapid and labored.

2. The movement of the ribs during thoracic inspiration involves all of the following EXCEPT

(A) elevation of the sternum
(B) increases in the transverse thoracic diameter
(C) mediastinal shift
(D) movement at the costovertebral joints
(E) movement at the manubriosternal joint

3. Physical examination of the patient in a sitting position reveals percussive dullness below the level of the fifth rib in the right midaxillary line. Fluid in the pulmonary cavity is suspected. Compromise of the perfusion-ventilation ratio explains the extreme respiratory effort. Elevation of the rib cage during inspiration may be aided by all of the following muscles EXCEPT the

(A) anterior, middle, and posterior scalene
(B) pectoralis major
(C) pectoralis minor
(D) sternocleidomastoid
(E) transverse thoracis

4. The presence of fluid in the thoracic cage reduces ventilation capacity. Thoracentesis to remove the fluid is best accomplished by inserting a needle

(A) adjacent to the sternum in the second intercostal space
(B) adjacent to the sternum in the fifth intercostal space
(C) in the midclavicular line in the fifth intercostal space
(D) in the midaxillary line in the seventh intercostal space
(E) in the twelfth intercostal space adjacent to the vertebral column

5. During the thoracentesis procedure, the physician is careful to avoid the intercostal neurovascular bundle, which is particularly susceptible to injury from a fractured rib because it lies

(A) behind the superior border of the rib
(B) beneath the inferior border of the rib
(C) between external and internal intercostal layers
(D) directly behind the midpoint of the rib
(E) halfway between two adjacent ribs

6. After withdrawal of 855 ml of fluid, the signs of respiratory distress are absent, yet the patient continues to breathe almost entirely with his diaphragm. Further clinical assessment identifies no noticeable bulging of the intercostal spaces on expiration. This type of diaphragmatic respiratory pattern in elderly persons is likely the result of

(A) injury to a phrenic nerve
(B) loss of extrinsic elastic recoil
(C) loss of intrinsic elastic recoil
(D) mediastinal shift
(E) paralysis of the thoracic musculature

7. A final respiratory function test is administered before hospital discharge and referral to a cardiologist for further cardiac management. While breathing into a spirometer, he is asked to inhale fully and then exhale fully. All of the following muscles contribute to forced diaphragmatic expiration EXCEPT the

(A) diaphragm
(B) external abdominal oblique
(C) internal abdominal oblique
(D) rectus abdominis
(E) transverse abdominis

(end of group question)

8. Which of the following structures transit the diaphragm via the esophageal hiatus?

(A) Azygos vein
(B) Hemiazygos vein
(C) Right vagus nerve
(D) Greater splanchnic nerves
(E) Thoracic duct

Questions 9 and 10

In a hospital emergency room, a 23-year-old man states that he "inhaled" a peanut a few days ago and has not been able to cough it up.

9. On bronchoscopy, a small aspirated object will most likely be located in the

(A) left lower lobar bronchus
(B) left main bronchus
(C) left superior segmental bronchus
(D) right lower lobar bronchus
(E) right superior segmental bronchus

10. Radiographic examination confirmed the foreign body in an airway and showed atelectasis (collapse) of a few bronchopulmonary segments. Which of the following statements best characterizes bronchopulmonary segments, the functional units of the lungs?

(A) Vessels containing oxygenated blood are located peripherally
(B) Invaginations of visceral pleura separate bronchopulmonary segments within a lobe
(C) Bronchopulmonary segments are supplied by a terminal bronchiole
(D) Oxygenated blood is received from bronchial arteries
(E) The bronchial tree and pulmonary vein are located centrally

(end of group question)

11. On a physical examination, a 37-year-old man with a history of alcohol abuse has a distinct region of percussive thoracic dullness. Radiographic evaluation reveals lightening of a particular bronchopulmonary segment. On the basis of these findings, a diagnosis of aspiration pneumonia is rendered. The bronchopulmonary segment usually involved in this disorder is the

(A) left or right apical
(B) left or right superior
(C) left or right basal anterior
(D) upper lingular
(E) lower lingular

12. The apex of the heart is normally located

(A) at the level of the fifth thoracic vertebra in the left midclavicular line
(B) at the level of the xiphoid process one finger breadth to the left of the midline
(C) deep to the third intercostal space in the left midclavicular line
(D) deep to the fifth intercostal space in the left midclavicular line

Questions 13-17

A 64-year-old man with severe angina pectoris is found by coronary angiography to have a 90% occlusion of the left coronary artery just proximal to its bifurcation.

13. Anginal pain of cardiac origin that radiates across the precordium and perhaps down the arm to the wrist is mediated by increased activity in afferent fibers contained in the

(A) cervical and thoracic splanchnic nerves
(B) phrenic nerves
(C) somatic nerves to the arm
(D) terminal portions of the intercostal nerves
(E) vagus nerves

14. After a midline sternotomy, the pericardium is incised by a longitudinal incision. Had a horizontal incision through the pericardium been used, which of the following structures might have been jeopardized?

(A) Azygos vein
(B) Internal thoracic arteries
(C) Phrenic nerves
(D) Vagus nerves

15. A venous homograph is not possible because of severe venous varicosities in the leg. Thus, the patient is treated surgically by diverting his left internal thoracic artery into the left coronary artery distal to the occlusion. Which of the following statements best describes the normal internal thoracic (mammary) artery?

(A) It bifurcates into the inferior phrenic and superficial epigastric arteries
(B) It descends posterior to the sternum
(C) It is a branch of the axillary artery
(D) It is accompanied by the azygos vein on the right and hemiazygos vein on the left
(E) It provides significant blood supply to the mammary gland

16. The surgeon decides to anastomose the internal thoracic artery side-to-side with the anterior interventricular artery and to continue to an end-to-side anastomosis with the circumflex artery. Careful technique is required to avoid compromise of adjacent vascular structures. The anterior interventricular branch of the left coronary artery is usually accompanied by the

(A) anterior cardiac vein
(B) coronary sinus
(C) great cardiac vein
(D) middle cardiac vein
(E) oblique vein (of Marshall)

17. On the basis of electrocardiographic findings, the coronary bypass appears to have been successful. Postoperatively, the chest wall region originally served by the internal thoracic artery receives blood flow from all of the following vessels EXCEPT the

(A) left costocervical trunk
(B) left inferior epigastric artery
(C) left pericardiacophrenic artery
(D) left posterior intercostal arteries
(E) right internal thoracic artery

(end of group question)

18. Which of the following statements correctly describes the papillary muscles in the heart?

(A) They are rudimentary and have no major function
(B) They compensate for decreased ventricular volume during systole
(C) They contract to close the atrioventricular (A-V) valve during ventricular systole (contraction)
(D) They contract to open the A-V valves during ventricular diastole (relaxation)

19. Blood returns to the left side of the heart via which of the following vessels?

(A) Anterior cardiac veins
(B) Coronary sinus
(C) Inferior vena cava
(D) Superior vena cava
(E) Thebesian veins

Questions 20–23

A 36-year-old male office worker complains of weakness and shortness of breath that are worse after moderate exercise. He describes a rapid, throbbing pulse when climbing two flights of stairs. Although the patient is not in pain at present, he admits to having occasional "heartburn" over the past year.

A physical examination reveals he has a weak and irregular pulse, a diastolic rumbling murmur attributed to the mitral valve, and rales (rattles) in the basal segments of the lungs.

20. The mitral valve is heard most distinctly on the

(A) left side, adjacent to the sternum in the second intercostal space
(B) left side, adjacent to the sternum in the fifth intercostal space
(C) left side, in the midclavicular line in the fifth intercostal space
(D) right side, adjacent to the sternum in the sixth intercostal space
(E) right side, in the midclavicular line in the sixth intercostal space

21. Percussion reveals that the area of cardiac dullness extends two finger breadths to the right of the sternum. A chest radiograph confirms that the heart is enlarged with specific right ventricular hypertrophy and right atrial hypertrophy. The preliminary diagnosis is mitral valve stenosis. A sequela of mitral valve stenosis is hypertrophy of the right ventricle, which is caused by all of the following EXCEPT

(A) impeded blood flow to the left ventricle
(B) rise in left atrial systolic pressure
(C) rise in pressure in the pulmonary vascular bed
(D) rise in right ventricular systolic and diastolic pressures
(E) fall in left ventricular pressure

22. A complication of mitral stenosis is the formation of thrombi on the walls of the enlarged left atrium with sequelae of thromboembolism. A thrombus dislodged from the left atrium may produce all of the following conditions EXCEPT

(A) a cerebral infarct
(B) a myocardial infarct
(C) renal necrosis and kidney failure
(D) gangrene in a lower extremity
(E) a pulmonary embolus

23. The patient is scheduled for open heart surgery to correct the valvular defect. The heart is approached by resection of the fifth rib on the left side. Commissurotomy, the surgical re-establishment of the divisions between the mitral valve cusps, occasionally is accomplished by insertion of a finger through the valve. The mitral valve is best approached through the wall of the

(A) left atrium, between the left and right superior pulmonary veins
(B) left atrium, between the left superior and inferior pulmonary veins
(C) left auricular appendage
(D) left ventricle on the posterior surface
(E) right atrium and foramen ovale

(end of group question)

DIRECTIONS: Each of the numbered items or incomplete statements in this section is negatively phrased, as indicated by a capitalized word such as NOT, LEAST, or EXCEPT. Select the ONE lettered answer or completion that is BEST in each case.

24. A Pancoast tumor at the apex of the right lung may produce all of the following symptoms EXCEPT

(A) edema of the lower right part of the body
(B) hoarseness of speech
(C) loss of sympathetic supply to the right side of the head
(D) paradoxic diaphragmatic movements
(E) venous engorgement of the face and arm on the right side

25. Circulatory changes that occur at birth normally include all of the following EXCEPT

(A) cessation of blood flow through the foramen ovale
(B) decreased right atrial pressure
(C) increased blood flow through the lungs
(D) increased left atrial pressure
(E) substantial backflow through the ductus venosus

26. All of the following neurons pass through the cardiac plexus EXCEPT

(A) those that innervate the diaphragm
(B) nonmyelinated postganglionic sympathetic fibers
(C) preganglionic parasympathetic fibers
(D) visceral afferent fibers mediating pain
(E) visceral afferent fibers mediating reflexes

DIRECTIONS: Each set of matching questions in this section consists of a list of four to twenty-six lettered options (some of which may be in figures) followed by several numbered items. For each numbered item, select the ONE lettered option that is most closely associated with it. To avoid spending too much time on matching sets with large numbers of options, it is generally advisable to begin each set by reading the list of options. Then, for each item in the set, try to generate the correct answer and locate it in the option list, rather than evaluating each option individually. Each lettered option may be selected once, more than once, or not at all.

Questions 27-30

(A) Anterior descending artery
(B) Circumflex branch of left coronary artery
(C) Right coronary artery

For each situation, select the artery that is most likely to make the correct association.

27. Usually supplies the S-A node

28. Usually supplies the A-V node

29. Usually supplies the posterior portion of the interventricular septum

30. Usually accompanies the coronary sinus

ANSWERS AND EXPLANATIONS

1. The answer is B [Chapter 20 II A 3 b, c]. The parietal pleura is innervated by branches of the intercostal nerves, and parietal pleuritic pain is localized to the region of irritation (lateral thoracic and anterior abdominal walls). The diaphragmatic pleura is innervated in large part by the phrenic nerves, arising from C3 to C5; thus, diaphragmatic pleuritic pain is referred to the midcervical dermatomes (neck and shoulder).

2. The answer is C [Chapter 19 II B 1, 2; VI B 1 a, b]. The inspiratory action of the external intercostal muscles and the interchondral portion of the internal intercostal muscles results in upward rotation of the ribs about an axis described by the costovertebral joints. Crossing of the axes of rotation anteriorly results in upward movement of the sternum and a flexion at the manubriosternal joint. The anteroposterior diameter of the thoracic cage is thereby increased. A mediastinal shift is abnormal, usually occurring as a result of pneumothorax.

3. The answer is E [Chapter 19 III A; Tables 19-1 and 19-2]. The pectoralis major and minor muscles, as well as the scalene muscles, attach to the ribs and can be used to elevate the ribs during exertional inspiration. The sternocleidomastoid muscle, attached to the manubrium, elevates the sternum, thereby increasing the anteroposterior thoracic diameter during exertional inspiration. The action of the transverse thoracis, an intrinsic muscle, is expiratory.

4. The answer is D [Chapter 20 II E 3 a (1)]. Thoracentesis involves inserting a needle to withdraw fluid on which the lung floats. The optimal site for insertion is in the midaxillary line one or two ribs below the fluid level determined by percussion, but not below the ninth intercostal space because of the proximity of the liver across the costodiaphragmatic recess.

5. The answer is B [Chapter 19 II A 3 a (4) (b); IV A 1, 2]. Each intercostal neurovascular bundle lies beneath the inferior border of a rib between the internal intercostal muscle and the innermost intercostal muscle, which is a division of the internal intercostal. Fracture of a rib may injure the nerve or blood vessels. If a fragment of rib also tears the pleura, hemothorax may result.

6. The answer is C [Chapter 19 II A 3 b (5), C (4)]. The elderly frequently experience progressive calcification of the costal cartilages, with resultant loss of thoracic cage elasticity that inhibits or even precludes thoracic respiratory movement. The person compensates by using the diaphragm for inspiration. Paralysis of the thoracic musculature results in bulging of the intercostal space during expiration. Paralysis of a hemidiaphragm produces paradoxic movements that are most evident radiographically. Thoracocentesis should have resolved any mediastinal shift.

7. The answer is A [Chapter 19 VI C 2; Tables 19-1 and 19-2; Chapter 19 VI C 2]. The diaphragm is inspiratory only. The muscles of the abdominal wall (including the external oblique, internal oblique, transverse abdominis, and rectus abdominis) are antagonistic to the diaphragm and participate in forced expiration.

8. The answer is C [Chapter 19 C 2 a, b]. The left and right vagus nerves, along with the esophagus, pass through the diaphragm via the esophageal hiatus. The azygos vein, the hemiazygos vein, and the thoracic duct pass through the aortic hiatus. The splanchnic nerves pass through the diaphragmatic crura.

9. The answer is D [Chapter 20 IV C 1 a]. Aspirated objects usually drop into the right main bronchus because it is more vertical than the left main bronchus and thus is more nearly in line with the trachea. Smaller objects tend to lodge in the right lower lobar bronchus, which more nearly continues the direction of the right main bronchus.

10. The answer is A [Chapter 20 III B 6 a]. The bronchopulmonary segment is characterized by the centrally located segmental bronchus and accompanying segmental branch of the pulmonary artery. The intersegmental veins, containing oxygenated blood, surround and define the bronchopulmonary segment. Lobes, but not bronchopulmonary segments, are separated by invaginations of visceral pleura.

11. The answer is B [Chapter 20 III C 1]. The left or right superior bronchopulmonary segments of the lower lobes are most dependent when a person is supine. Aspirated vomitus in the recumbent person drains into the nearly vertical superior segmental bronchi to produce a pneumonia. This condition results in a region of percussive dullness just medial to the vertebral border of the scapula (when the arm is elevated) as well as auscultatory rales.

12. The answer is D [Chapter 21 III B 3]. This location of the apex of the heart, formed by the tip of the left ventricle, is the point at which the apical pulse may be palpated and the mitral valve sound auscultated with maximum discrimination.

13. The answer is A [Chapter 21 VIII B; Chapter 22 V A 4 a]. Pain originating in the heart travels along visceral afferent nerve fibers, which take two pathways to reach the spinal cord. Some cardiac afferents run along the middle and inferior cardiac cervical (accelerator) nerves to reach the sympathetic chain and then travel inferiorly to the T1–T2 level. Other cardiac afferents travel along the upper thoracic splanchnic nerves to reach the sympathetic chain. The cardiac afferent fibers leave the sympathetic chain through the white rami communicantes and travel along the spinal nerves T1–T4 to reach the corresponding levels of the spinal cord. Pain of cardiac origin is referred to (appears as if originating from) dermatomes T1–T4. Reflex afferents travel along the vagus. The phrenic nerve, although motor to the diaphragm, also conveys sensation from the diaphragmatic pleura and peritoneum to spinal levels C3–C5.

14. The answer is C [Chapter 20 VI A 2; Chapter 21 I B; Chapter 22 IV B 1]. The phrenic nerves lie anteriorly along the lateral boundaries of the middle mediastinum between the fibrous pericardium and the mediastinal pleura. The left and right vagus nerves, which pass adjacent to the root of the each lung, as well as the azygos vein, are components of the posterior mediastinum.

15. The answer is E [Chapter 21 IV B 1 b (1), 2]. The internal thoracic artery is the second branch of the subclavian artery, arising opposite the origin of each vertebral artery. It descends posteriorly to the costal cartilages and supplies the chest wall, including portions of the mammary gland. It terminates by bifurcation into the musculophrenic and superior epigastric arteries.

16. The answer is C [Chapter 21 IV B 1 b (1), 2]. The great cardiac vein accompanies the anterior descending artery in the anterior interventricular sulcus. The anterior cardiac veins drain across the right coronary sulcus directly into the right atrium. The coronary sinus, into which the great cardiac vein drains, lies in the coronary sulcus along with the circumflex artery and also receives the middle cardiac vein from the posterior interventricular sulcus.

17. The answer is C [Chapter 19 IV B 1, 2]. The internal thoracic artery and its musculophrenic branch anastomose with the posterior intercostal arteries, which arise from the costocervical trunk (the superior intercostal artery) and the dorsal aorta. The superior epigastric branch anastomoses with the inferior epigastric artery. There are also anastomoses across the midline with the contralateral internal thoracic artery. The pericardiacophrenica artery supplies the pericardium and diaphragm and does not supply the chest wall significantly.

18. The answer is B [Chapter 21 VI B 1 a (1) (c), (2) (a)]. The papillary muscles attach the chordae tendineae of the atrioventricular valve cusps to the walls of the ventricular chambers. Contraction of the papillary muscles prevents eversion of the valve cusps as the ventricular chambers decrease in volume during the second (ejection) phase of systole.

19. The answer is E [Chapter 21 IV B 3]. Blood returns to the left side of the heart from the lungs via the pulmonary veins and from the innermost layers of the left atrial and ventricular walls via thebesian veins. The anterior cardiac veins, coronary sinus, superior vena cava, and inferior vena cava, as well as thebesian veins, return blood to the right side of the heart.

20. The answer is C [Chapter 21 IX C 3 b]. The sound of mitral valve closure projects to the apex of the heart and is heard most distinctly in the left fifth intercostal space in the midclavicular line. In men, this site usually is just below the left nipple.

21. The answer is E [Chapter 21 IX a 1-4]. Mitral valve stenosis, whereby blood flow to the left ventricle is impeded, results in elevated left atrial systolic pressure and consequent elevation of pulmonary pressure. This change necessitates that the right side of the heart work more to overcome the pulmonary pressure, with consequent hypertrophy. The elevated pulmonary pressure produces pulmonary effusion and results in rales.

22. The answer is E [Chapter 21 VI A 2 b, C 3]. Thrombi dislodged from the left atrium can result in cerebral, coronary, renal, or systemic embolism. Thrombi dislodged from the right atrium produce pulmonary embolism.

23. The answer is C [Chapter 21 VI C 3; Figure 21-5]. The best approach to mitral commissurotomy is to gain access to the left atrium through the left auricular appendage and thence direct the finger through the mitral valve into the ventricle.

24. The answer is A [Chapter 20 III C 3 a-e]. A tumor at the apex of the right lung (Pancoast tumor) may involve compression of the recurrent laryngeal nerve with hoarseness; of the sympathetic chain, with loss of sympathetic supply to the right side of the head (Horner's syndrome); of the phrenic nerve, with paradoxic hemidiaphragmatic movement related to paralysis; of the right subclavian vein, with venous engorgement of the face and arm on the right; and of the subclavian artery, with diminished pulses in the right upper extremity. Because the thoracic duct swings to the left side to enter the left brachiocephalic vein, lymph drainage should not be affected on the lower right side.

25. The answer is E [Chapter 21 X B 1, 2]. At birth, the lungs expand with a concomitant increase in the pulmonary blood flow, which both reduces the pressure in the right atrium and increases the pressure in the left atrium. This pressure reversal causes the foramen ovale to close, preventing a reverse flow of blood to the right atrium. Even though atrial pressure is greater than pulmonary pressure, the angle at which the ductus venosus joins the descending aorta establishes a Venturi effect that precludes significant backflow of blood.

26. The answer is A [Chapter 21, Figure 21-8; Chapter 22 V a 2-4]. The cardiac plexus contains postganglionic sympathetic neurons from the sympathetic chain, preganglionic parasympathetic neurons from the vagus nerve, and afferent neurons for both pain and reflexes. The phrenic nerves that innervate the diaphragm pass well lateral to the cardiac plexus.

27–30. The answers are: 27-C, 28-C, 29-C, 30-B [Chapter 21 IV A 1, 2]. The right coronary artery usually supplies both the sinoatrial node and the atrioventricular node and usually terminates as the posterior interventricular (descending) artery. The anterior portion of the interventricular septum is normally supplied by the anterior interventricular (descending) branch of the left coronary artery. The right coronary artery arises from the right aortic sinus; the left coronary artery arises from the left aortic sinus. Both the right coronary artery and the circumflex branch of the left coronary artery lie in the coronary sulcus.

PART IIIB ABDOMEN

Chapter 23
Anterior Abdominal Wall

I. INTRODUCTION

A. **Definition.** The abdominal region is below the **diaphragm** and above the bones and ligaments of the **pelvic brim.**

B. **Surface landmarks** (Figure 23-1A)
1. **Hard tissue landmarks** include thoracic and pelvic structures. Superiorly, they are the **xiphoid process** of the sternum and the **costal margin** of the thoracic cage. Inferiorly, they are the **iliac crest, anterior superior iliac spine, pubic tubercle,** and the **pubic crest** of the pelvis, as well as the **inguinal ligament** between the anterior superior iliac spine and the pubic tubercle.
2. **Soft tissue landmarks** include the **linea semilunaris** (Spieghel's line defined by the lateral border of each rectus abdominis muscle), the **linea alba** in the midline (defined by the medial borders of the rectus abdominis muscles), and the **umbilicus.**

C. **Abdominal subdivisions** are used to describe the location of structures as well as to describe symptoms and the results of physical examinations.
1. **Quadrants** (see Figure 23-1B). The abdomen is divided by **vertical** and **horizontal planes** through the umbilicus into the **right upper, left upper, right lower,** and **left lower quadrants.**
2. **Regions** (see Figure 23-1A). The abdomen is divided into nine regions by the horizontal **subcostal plane** (beneath the lowest extension of the thoracic cage) and the horizontal **intertubercular plane** (through the iliac tubercles), as well as two vertical **midclavicular planes** or **midinguinal planes.**
    a. **Left** and **right hypochondriac regions** underlie the rib cage on either side above the subcostal plane and lateral to the midclavicular planes.
    b. The **epigastric region** lies centrally between the midclavicular planes and above the subcostal plane.
    c. **Left** and **right lateral regions** (posterolateral lumbar and anterolateral flank) lie lateral to the midclavicular planes between the subcostal and intertubercular planes.
    d. The centrally located **umbilical region** is bounded by subcostal, intertubercular, and midclavicular planes.
    e. **Left** and **right inguinal (iliac) regions** lie inferior to the intertubercular plane and lateral to the midclavicular (midinguinal) planes.
    f. The **pubic (hypogastric) region** lies centrally, medial to the midclavicular (midinguinal) planes and inferior to the intertubercular plane.

D. **Internal structures**
1. **Organs and structures palpable through the abdominal wall** (see Figure 23-1B)
    a. The **aorta,** about 2.5 cm wide, can be outlined both above and below the umbilicus from its pulse. In thin individuals, the **aortic pulse** may be visible. An abdominal aortic aneurysm can be discerned on palpation.
    b. The **liver** usually is palpable in the right upper quadrant or right hypochondriac region, depending on a person's build. A palpable liver may or may not indicate the presence of a pathologic condition. In persons with **hepatomegaly,** this organ may extend to the iliac crest.
    c. The **gallbladder** is palpable if it is enlarged or contains choleliths (gallstones). It may be palpated in the right hypochondriac region at the junction of the linea semilunaris and the costal margin.
    d. The **stomach** is palpable only if engorged.

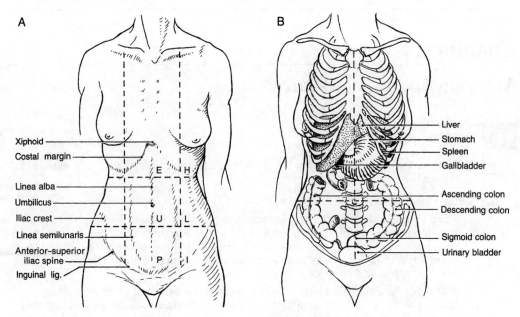

FIGURE 23-1. *(A)* The abdomen is divided into left and right hypochondriac (H), lateral (L), and inguinal (I) regions, as well as medial epigastric (E), umbilical (U), and pubic (P) regions. Prominent landmarks and surface features are identified. *(B)* Left and right upper and lower quadrants of the abdomen, with the principal underlying organs.

  e. The **spleen** normally is not palpable. When it is enlarged, however, it may be felt in the left hypochondriac region and sometimes in the lower quadrants. A grossly enlarged spleen must be palpated gently to avoid rupture and subsequent massive hemorrhage.
  f. The **kidneys** (only lower poles) usually can be palpated in the lateral regions.
  g. **Colon**
   (1) The **ascending colon** usually is palpable only if it is distended by gas or chyme.
   (2) The **descending colon and sigmoid colon,** containing solidifying feces, usually are palpable in the lower left quadrant.
  h. The **urinary bladder,** when distended, is palpable in the pubic (hypogastric) region.
  i. The **uterus** can be palpated bimanually in the pubic region with the cervix being supported by the fingers of one hand.
  j. The **ovaries,** if enlarged, may be palpated bimanually in the inguinal regions.
 2. **Organs percussed or auscultated through the abdominal wall**
  a. An **empty stomach** on percussion produces a tympanic tone in the left upper quadrant in an upright patient and in the epigastric region in a supine patient.
  b. The **bowel** normally produces tympanic tones on percussion. Hypertympanic tones are elicited when the bowel is distended by gas.
  c. **Bowel sounds** produced by peristalsis **(borborygmi)** normally are evident on auscultation. The frequency of borborygmi indicates the level of intestinal motility.
  d. The **abdominal aorta and common iliac arteries** may be auscultated for pulse sounds and pathologic bruits.

II. SKIN AND FASCIAE

 A. **Integument**
  1. **Principal layers** (see Chapter 2 I)
   a. The **epidermis** is a superficial cellular layer of stratified epithelium.
   b. **Dermis** is an underlying layer of loose, irregularly arranged connective tissue.

2. The **cleavage (Langer's) lines** of the skin are surgical considerations.
   a. **Orientation.** Cleavage lines are nearly horizontal on the anterior abdominal wall (see Figure 2-1).
      (1) The meshwork of collagen fibers provides overall mobility to the skin. Although these connective tissue fibers appear randomly oriented, each area of the body has a prevailing directionality.
      (2) The prevailing directionality of the connective tissue fibers of the dermis is evident as crease lines and defines Langer's lines.
   b. **Abdominal incisions.** The elasticity of the connective tissue fibers creates a slight tension. When these fibers are cut, the edges of the integument tend to retract.
      (1) An incision made **across** the cleavage lines cuts more connective tissue fibers and tends to gape, resulting in a prominent scar on healing.
      (2) An incision made **parallel to** the cleavage lines severs fewer connective tissue bundles, gapes less, and heals with a more cosmetic scar.
   c. **Elasticity.** The dermis resists tearing, shearing, and localized pressure but is flexible enough to permit some stretch.
      (1) Stretching for long periods, as in pregnancy or in obese persons, can cause rupture of the connective tissue fascicles with subsequent scar formation, called **striae** or "stretch marks."
      (2) **Striae** appear perpendicular to Langer's lines.

B. **Superficial fascia** (subdermis or hypodermis) contains the superficial blood vessels, lymphatics, and nerves. It is especially prominent in the abdominal region (see Figure 2-2), and has two distinct layers.
   1. **The superficial layer of the superficial fascia** (Camper's f.) underlies the epidermis and is predominantly an adipose layer, containing most of the fat of the subdermis.
      a. It continues over the chest as the superficial layer of the superficial thoracic fascia.
      b. It crosses the inguinal ligament to merge with the superficial fascia of the thigh.
   2. **The deep layer of the superficial fascia** (Scarpa's f.) is a membranous or fibrous layer that holds sutures.
      a. It continues over the chest as the deep layer of superficial thoracic fascia.
      b. It passes into the upper thigh, where it is attached to the fascia lata just below and parallel to the inguinal ligament.

C. **Deep (investing) fascia** underlies the superficial fascia (see Figure 2-2). By definition, it is the **investing fascia** of the musculature, aponeuroses, and large neurovascular structures. It cannot be separated easily from the underlying epimysium of muscle, perineurium of nerve, periosteum of bone, and perichondrium of cartilage.

D. **Transversalis fascia** is the inner layer of connective tissue that enwraps the abdominal cavity and supports the peritoneum. Loose extraperitoneal connective tissue contains varying amounts of fat and separates the transversalis fascia from the peritoneum.

III. MUSCULATURE

A. **Organization.** The **intrinsic abdominal musculature** is arranged in three layers, each having a different direction (Figure 23-2; see Table 23-1).
   1. **The external oblique muscle** lies most superficially (see Figure 23-2A).
      a. **Course.** From the costal margin, it runs diagonally, inferiorly and medially, to become aponeurotic at approximately the midclavicular line. At the **linea alba,** it decussates with fibers of the contralateral aponeurosis. It inserts into the **iliac crest** as well as the **pubic tubercle** and **pubic crest.**
      b. The **inferior lumbar trigone** (of Petit) is formed by the free posterior border of the external oblique muscle (between the rib cage and iliac crest), the iliac crest, and the lateral border of the latissimus dorsi muscle.

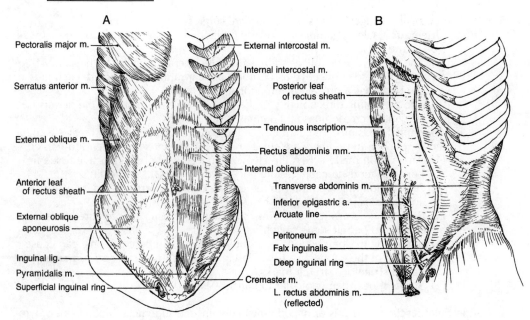

FIGURE 23-2. Abdominal musculature. *(A)* The right external oblique and left internal oblique muscles are depicted. On the right side, the anterior leaf of the rectus sheath is shown. On the left side, it is dissected away to show the left rectus abdominis muscle. *(B)* The left transverse abdominis and the right rectus abdominis muscles are shown. The left rectus abdominis muscle was removed to show the posterior leaf of the rectus sheath and arcuate line.

TABLE 23-1. Muscles of the Anterior Abdominal Wall

Muscle	Origin	Insertion	Primary Action	Innervation
Rectus abdominus	Pubic arch	Costal cartilages 5–8, lower sternum, xiphoid process	Flexes vertebral column and compresses abdomen	Nn. T7–T12
External oblique	Ribs 10–12 and iliac crest	Iliac crest, pubic crest, linea alba	Rotates vertebral column and compresses abdomen	Nn. T7–T12
Internal oblique	Thoracodorsal fascia, iliac crest, inguinal ligament	Ribs 10–12, linea alba, pubic crest	Rotates vertebral column	Nn. T8–T12, iliohypogastric and ilioinguinal nn. (L1)
Transverse abdominis	Ribs 10–12, thoracodorsal fascia, iliac crest	Linea alba, pubic crest	Compresses abdomen	Nn. T8–T12, iliohypogastric and ilioinguinal nn. (L1)

c. The **inguinal ligament** (of Poupart) is formed by the lower margin of the external oblique aponeurosis (between the **anterior superior iliac spine** and the **pubic tubercle**).
d. **The superficial (external) inguinal ring** is actually a triangular defect in the aponeurosis of the external oblique muscle.
   (1) **Structure.** Near the pubis, the aponeurosis of the external oblique muscle splits into two crura (see Figures 23-2A and 23-4A).
      (a) The **lateral crus** inserts onto the pubic tubercle with some fibers reflected to the **superior pubic ramus** as the **lacunar ligament** (of Gimbernat).
      (b) The **medial crus** inserts into the pubic symphysis.
      (c) **Intercrural fibers** strengthen the apex of the superficial inguinal ring.
   (2) **Contents.** This ring transmits the **spermatic cord** (in the male) or the **round ligament of the uterus** (in the female).

2. **The internal oblique muscle** is the intermediate layer of the abdominal wall musculature (see Figure 23-2A).
   a. **Course.** From the thoracolumbar (thoracodorsal) fascia and iliac crest, this layer runs superomedially to become aponeurotic near the linea semilunaris. At the linea alba, it decussates with fibers of the contralateral aponeurosis. It also inserts onto the lower costal margin and pubic symphysis.
   b. The **floor of the inferior lumbar trigone** is formed by the internal oblique muscle and lumbodorsal fascia. Because the abdominal wall is thinner at this location, this trigone may be a site of **lumbar herniation.** It is used in the surgical approach to the kidneys and ureters.
   c. The **cremaster muscle** of the spermatic cord is derived from the internal oblique layer. In the male, this muscle is derived from the internal oblique layer as the spermatic cord passes through the abdominal wall. Thus, the inguinal canal passes diagonally through and is formed in part by the internal oblique muscle.

3. **The transverse abdominal muscle** constitutes the innermost layer (see Figure 23-2B).
   a. **Course.** From the inferior borders of the lower costal cartilages, the lumbodorsal fascia, and the iliac crest, this muscle runs horizontally across the abdominal wall. At approximately the linea semilunaris, the muscle fibers cease and the layer becomes aponeurotic. It inserts into the linea alba by decussating with fibers of the contralateral aponeurosis. It also inserts on the pubic crest, thereby reinforcing the inferomedial portion of the inguinal region.
   b. The **falx inguinalis** is the arcing inferior free edge of the transverse abdominal muscle and aponeurosis (see Figure 23-2B).
      (1) It is usually (95%) craniad to the inguinal canal and takes no part in forming the spermatic cord.
      (2) The region below the falx inguinalis is weaker and is a potential site for inguinal herniation.
      (3) It is not synonymous with the conjoined tendon, but rather is a distinct and surgically important structure (see section VII B 3 b).

4. **The rectus abdominis muscle** represents the same layer as the occasional sternalis muscle in the thorax and the pubococcygeus muscle of the pelvis.
   a. **Course.** From the anterior costal margin and xiphoid, this muscle layer inserts into the pubic crest (see Figure 23-2A). Three or four **tendinous inscriptions** lie between the anterior leaf of the rectus sheath and the rectus abdominis muscle craniad to the umbilicus (see section III C).
   b. The **linea semilunaris** (Spieghel's line) is defined by the lateral margins of each rectus abdominis muscle.
   c. The **linea alba** in the midline is defined by the medial margins of the rectus abdominis muscles.

5. **The pyramidalis muscle** is usually undeveloped, absent, or occasionally unilateral. When present, it lies anterior to the lower rectus abdominis muscle between the linea alba and the pubic crest.

B. Actions. Depending on the type of motion, the muscles of the abdominal wall can be antagonistic to or synergistic with each other (Table 23-1).

1. **Flexion** of the vertebral column and pelvis is by contraction of the **external and internal oblique muscles,** bilaterally, as well as the **rectus abdominis muscle,** which is the most powerful abdominal flexor.
2. **Abduction** of the trunk occurs by ipsilateral contraction of the **external and internal obliques,** acting synergistically on one side.
3. **Rotation** of the trunk occurs by contraction of the **external oblique muscle** on one side with the **internal oblique muscle** of the contralateral side. The fibers of the external oblique on one side have the same prevailing direction as those of the internal oblique on the contralateral side.
4. **Respiration.** The abdominal musculature is antagonistic to the diaphragm. In **abdominal expiration,** all the abdominal muscles contract as a unit. This action forces the abdominal viscera (as a fluid piston) against the inferior surface of the diaphragm and causes the diaphragm to stretch and elevate into the thoracic cavity, thereby decreasing thoracic volume.
5. **Fixation of the abdominal wall**
   a. **Valsalva maneuver.** Fixation of the thoracoabdominal wall is by contraction of **abdominal musculature** in concert with the thoracic musculature with the glottis closed. The resultant increase in thoracoabdominal pressure is used in lifting heavy weights, coughing, sneezing, micturition, parturition, defecation, eructation, and emesis.
   b. **Abdominal herniation.** Excessive thoracoabdominal fixation, in combination with weakness of the abdominal wall, may result in herniation of mesenteric fat or viscera through the abdominal wall.

C. Rectus sheath invests the rectus abdominis muscle and is a surgical consideration. This sheath is formed by fusion of the aponeuroses of the internal oblique, external oblique, and transverse abdominis muscles (see Figure 23-2).

1. **Above the arcuate line** (the superior portion of the rectus sheath)
   a. The **anterior leaf** of this sheath is formed by fusion of the external oblique aponeurosis and the anterior portion of the split internal oblique aponeurosis. Two or three **tendinous inscriptions** (remnants of segmentation) attach the anterior leaf of the rectus sheath to the rectus abdominis muscle and frequently are visible surface features superior to the umbilicus.
   b. The **posterior leaf** of this sheath is formed by fusion of the posterior portion of the split internal oblique aponeurosis and the transverse abdominis aponeurosis. The posterior surface of the rectus sheath has no tendinous inscriptions.
2. **Arcuate line** (linea semicircularis of Douglas)
   a. About one or two inches below the umbilicus, all aponeurotic layers of the abdominal muscles pass anterior to the rectus abdominis muscle. The lower free edge of the posterior rectus sheath formed by this transition is the **arcuate line** (linea semicircularis).
   b. Two arcuate lines occur if the portion of the posterior rectus sheath contributed by the internal oblique aponeurosis passes anterior to the rectus abdominis superior to the portion contributed by the transverse abdominis aponeurosis.
   c. The junction of the arcuate line with linea semilunaris is the site of a **lateral ventral (spigelian) hernia** into the rectus sheath.
3. **Below the arcuate line**
   a. The **anterior leaf** is formed by fusion of the aponeuroses of all three lateral abdominal muscles.
   b. **Transversalis fascia** lies deep to the rectus abdominis muscle because there is no posterior leaf of the sheath below the arcuate line.

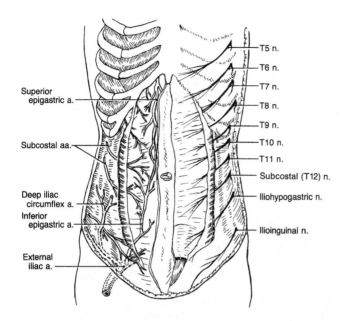

FIGURE 23-3. Right abdominal vasculature and left abdominal innervation. The vascular anastomoses between the superior epigastric, musculophrenic, intercostal, and inferior epigastric arteries are indicated. The tenth intercostal nerve supplies the umbilical region.

IV. VASCULATURE

A. Arterial supply (Figure 23-3)

1. **The epigastric arterial system** lies within the rectus sheath and provides a strong anastomotic connection between the thoracic aorta and the external iliac vessels (see Figure 23-2).
    a. **Superior epigastric artery:**
        (1) Arises from the bifurcation of the **internal thoracic artery,** where the musculophrenic artery diverges
        (2) Enters the rectus sheath beneath the costal margin
        (3) Anastomoses with the inferior epigastric artery within the rectus abdominis muscle
    b. **Inferior epigastric artery:**
        (1) Arises from the external iliac artery
        (2) Lies along the medial margin of the **deep inguinal ring** as it passes obliquely upward from its origin and enters the rectus sheath by passing superficially to the arcuate line
        (3) Anastomoses with the superior epigastric artery within the rectus abdominis muscle

2. **Other major anterior abdominal arteries** (see Figure 23-3)
    a. **Musculophrenic artery:**
        (1) Arises as the other bifurcating branch of the **internal thoracic (internal mammary) artery**
        (2) Courses along the costal margin
        (3) Supplies numerous branches to the anterior abdominal wall and to the diaphragm
    b. **Intercostal arteries 10–12:**
        (1) Arise from the **thoracic aorta** and extend ventromedially beyond the costal margin into the abdominal wall
        (2) Enter the rectus sheath laterally
        (3) Anastomose with the superior and inferior epigastric arteries
    c. **Deep circumflex iliac artery:**
        (1) Arises from the **external iliac artery**

(2) Courses deeply to the inguinal ligament toward the anterior superior iliac spine
(3) Supplies the deep inguinal region
(4) Anastomoses with the lower intercostal arteries
    d. **Superficial epigastric arteries:**
      (1) Arise from the **femoral artery**
      (2) Supply the superficial inguinal and pubic regions
      (3) Anastomose with lower intercostal arteries
  3. **Clinical considerations.** Anastomoses in the epigastric arterial system can provide important shunts to supply blood to the lower part of the body in conditions such as **coarctation of the aorta.**

B. Venous return
  1. **Superficial venous drainage** is toward the following veins:
    a. **Superior and inferior epigastric veins**
    b. **Axillary vein** via the thoracic and thoracoepigastric veins
    c. **Femoral vein** via the superficial branches of the great saphenous vein
    d. **Hepatic portal vein** via the paraumbilical veins that anastomose with the hepatic portal system
      (1) Pathologic obstruction of venous return from the gastrointestinal tract through the liver (portal hypertension) causes retrograde flow to the systemic venous return of the body wall via enlarged paraumbilical anastomoses **(caput medusae).**
      (2) Although caput medusae may not be readily apparent in cases of portal hypertension, it typically is demonstrable in a photograph of the abdominal wall taken with infrared film.
  2. **Deep venous drainage** is along veins that parallel the arterial supply.

C. Lymphatic drainage
  1. **Above the umbilicus**
    a. **Superficial lymphatics,** draining the subcutaneous tissues, drain primarily to the **axillary nodes,** with some drainage to the parasternal nodes.
    b. **Deep lymphatics,** draining muscles and investing fascia, drain primarily to the **parasternal nodes.**
  2. **Below the umbilicus**
    a. **Superficial lymphatics** drain toward the superficial **inguinal nodes.**
    b. **Deep lymphatics** drain through the **external iliac nodes** to the para-aortic nodes.

V. INNERVATION

A. **Intercostal nerves T7–T12.** The **ventral primary rami** of the spinal nerves supply the dermatomes and myotomes of the abdominal wall (see Figure 23-3).
  1. **Course.** They continue beyond the costal margin into the abdominal wall between the internal oblique and the transverse abdominis muscles. They penetrate the rectus sheath laterally.
  2. **Distribution.** They supply the skin, muscles, and parietal peritoneum.
    a. **Lateral cutaneous branches** arise approximately at the anterior axillary line. They pierce the abdominal musculature in the midaxillary line to reach the dermis, bifurcating into anterior and posterior branches.
    b. **Anterior cutaneous branches** arise as terminal branches of the anterior rami. Each enters the rectus sheath laterally and exits from it anteriorly to reach the dermis. They bifurcate into medial and lateral branches.

3. **Landmarks.** Nerve T10 supplies the dermatome that contains the umbilicus. Nerve T12 supplies the pubic region.

B. Lumbar nerves L1 and L2. The anterior primary rami of T12–L4 contribute to the **lumbar plexus** (see Figure 23-3).

1. **The lumbar plexus** has anterior and posterior divisions, both arising from the anterior primary rami of spinal nerves (see Figures 4-2 and 26-6).
2. **The anterior divisions** of T12 and L1 form the **iliohypogastric** and **ilioinguinal nerves.** The anterior divisions of L1 and L2 form the **genitofemoral nerve** (see Figure 2-6).
   a. The terminal portions of the **iliohypogastric** and **ilioinguinal nerves** course in the abdominal wall between the internal oblique and the external oblique muscles. They supply the pubic (hypogastric), inguinal, and anterior perineal regions (see Figure 23-3).
   b. Branches of the **genitofemoral nerve** run to the spermatic cord and anterior thigh.
3. **The posterior division** of L2 contributes to the **lateral femoral cutaneous nerve** and the **femoral nerve.**

VI. CLINICAL CONSIDERATIONS

A. Access to the peritoneal cavity takes into account the composition of the abdominal wall as well as the neurovascular structures within it.

1. **Via the rectus sheath** (see Figure 23-2B)
   a. **Paramedian incisions** cut through the skin and into the anterior leaf of the rectus sheath, medially. The rectus abdominis muscle may then be retracted laterally without compromising its vascular and nerve supply. A second incision through the posterior leaf of the rectus sheath provides extensive access to the peritoneal cavity. When the incisions in the posterior and anterior leaves are sutured, the undamaged rectus abdominis muscle slides medially and supports the incisions well.
   b. **Vertical incisions** through the rectus abdominis muscle sever the nerves supplying the muscle tissue medial to the incision, resulting in paralysis of the medial portion of the muscle and a weakened abdominal wall disposed to **ventral herniation.**
   c. **Horizontal incisions** through the rectus abdominis muscle transect the arterial supply. Moreover, re-anastomosis of a transected muscle (compared to sewing two paint brushes together) may come apart on Valsalva fixation (e.g., a cough or sneeze). Thus, the integrity of the abdominal wall is jeopardized until full healing has occurred.

2. **Via the linea alba** (see Figure 23-1A)
   a. A **midline incision** through this avascular region is relatively bloodless and provides extensive exposure. It may not heal well, however, without leaving a large, unsightly roll of tissue.
   b. Midline incisions predispose to **epigastric herniation,** a common postsurgical complication.

3. **Via the ventrolateral abdominal wall**
   a. **Transverse incisions,** although taking advantage of Langer's cleavage lines, may transect the nerve supply to the abdominal musculature unless precautions are taken. For example, the **muscle-splitting approach** (of McBurney) for appendectomy in the right lower quadrant uses Langer lines for the superficial incision.
      (1) The external oblique muscle is then split between its fascicles and retracted.
      (2) The internal oblique muscle is then split and retracted (locating and retracting the iliohypogastric nerve).

(3) Finally, the transverse abdominis muscle is then split and retracted.
(4) Each layer closes naturally with few, if any, sutures and provides a strong repair. Valsalva fixation (e.g., a cough or sneeze) tends to close a split muscle but can tear the sutures from a transected muscle.

b. **Vertical incisions** through the lateral abdominal musculature sever the nerves supplying the muscle tissue medial to the incision. The results are paralysis of abdominal musculature and a weakened abdominal wall disposed to **ventral herniation.**

B. Abdominal hernias

1. **Epigastric herniation** is a midline protrusion of fat or abdominal viscera through the linea alba (usually above the umbilicus). It is a frequent postoperative complication to a midline incision through the linea alba, but it is rare otherwise.

2. **Umbilical herniation** is a protrusion of abdominal viscera through the umbilical ring.
   a. **Congenital umbilical hernia,** which has an embryologic basis, occurs in about 1 in 50 births. It may be apparent at birth (omphalocele) or it may occur spontaneously in infants and children.
   b. **Acquired umbilical hernia** usually occurs in adult life.
      (1) It is frequently the result of repeated Valsalva fixation, pregnancy, ascites, or obesity along with a weakened abdominal wall. It also occurs as a postoperative complication.
      (2) Repair may be complicated by portal hypertension, which causes the paraumbilical anastomoses to enlarge and bleed profusely when divided.

3. **Semilunar (spigelian) herniation** is a ventrolateral herniation that usually occurs at the junction of the arcuate line and the linea semilunaris. This type of hernia may also occur along the linea semilunaris inferior to the arcuate line. In addition, a spigelian hernia may enter the rectus sheath at the arcuate line.

VII. INGUINAL REGION

A. Introduction.
The testes descend through the inguinal region of the abdominal wall to reach the scrotum, leaving a potential site for herniation.

1. **Descent of the testes** provides a reference to describe the inguinal region.
   a. The **gonads** develop retroperitoneally from the urogenital ridge in the region of the kidneys. They migrate retroperitoneally within the transversalis fascia toward the scrotum during embryonic development. They seem to be guided by the **gubernaculum,** a ligamentous structure that runs between the lower pole of each gonad to each labial-scrotal fold.
      (1) By the third embryonic month, the retroperitoneal gonads have moved into the false pelvis. In females, migration continues into the deep pelvis. In males, a peritoneal pouch, the processus vaginalis, evaginates into the developing scrotum.
      (2) By the seventh embryonic month, the testes migrate retroperitoneally to reach the abdominal end of the inguinal canal.
      (3) Between the seventh and eighth embryonic months, the testes descend through the inguinal canal behind the processus vaginalis (retroperitoneally) into the scrotum, thereby forming the spermatic cord.
      (4) Before birth, the testes usually reach their definitive position in the scrotum intimately associated with the remnants of the processus vaginalis.
      (5) Perinatally, the processus vaginalis usually closes off from the peritoneal cavity. It may remain patent as a **funicular process,** thereby predisposing to congenital indirect hernia during the early years of life.
   b. The **definitive testes** retain their original innervation, blood supply, and lym-

phatic drainage, which explains the long intra-abdominal course of the spermatic vessels and nerves.

   c. The **definitive ovaries** migrate into the deep pelvis, also trailing their vessels and nerves.

2. **Cryptorchidism** is a failure of testicular descent, whereby the testes remain at the deep ring or within the inguinal canal. This condition may correct spontaneously during the first year of life. Failure to surgically correct undescended testes before puberty results in sterility and, if bilateral, secondary sex characteristics fail to develop. Uncorrected cryptorchidism is associated with a higher risk of testicular malignancy.

B. **Inguinal canal.** The anterior abdominal wall is breached by the inguinal canal from the deep inguinal ring to the superficial inguinal ring.

1. **The superficial (external) inguinal ring** is a triangular interruption in the external oblique aponeurosis (Figure 23-4A).
   a. **Structure and contents** (see section III A 1 d)
   b. **Physical examination.** Both **direct and indirect inguinal hernias** emerge through the superficial inguinal ring.
      (1) The superficial inguinal ring may be palpated (but not normally entered) by inversion of the scrotum with the little finger directed along the vas deferens toward the pubic tubercle.
      (2) A small inguinal hernia may be induced to protrude through the superficial ring during Valsalva fixation, for example, a cough. A large hernia is palpable or even apparent visually.

2. **The deep (internal) inguinal ring** is the site of the embryonic funicular process.
   a. **Structure.** It is covered by peritoneum (unless there is a patent funicular process) and is located just lateral to the inferior epigastric artery.
   b. **Contents.** The **vas deferens** and the **testicular neurovascular bundle** pass through the deep inguinal ring.

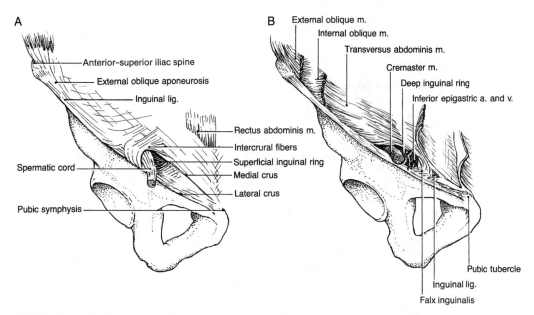

FIGURE 23-4. *(A)* The superficial inguinal ring is formed by the defect in the external oblique aponeurosis. The internal oblique muscle contributes the cremaster layer of the spermatic cord. *(B)* The deep inguinal ring is the location of the fused processus vaginalis. The inguinal canal passes beneath the inferior free edge of the transverse abdominis muscle and aponeurosis (falx inguinalis).

(1) Some of the **transversalis fascia** is drawn into the spermatic cord during descent of the testes, forming the **internal spermatic fascia.**
(2) An **indirect inguinal hernia** enters the spermatic cord via the deep inguinal ring.

3. **The walls of the inguinal canal** are formed by the muscular, aponeurotic, and fascial layers of the abdominal wall.
   a. The **anterior wall** is formed by **the external oblique aponeurosis** and the **internal oblique muscle** (see Figures 23-4A and 23-5).
      (1) Contraction of the internal oblique muscle occludes and strengthens the inguinal canal.
      (2) Because the free edge of the transverse abdominis muscle lies superior to the deep ring, it rarely (less than 5%) contributes to the inguinal canal or cremaster muscle.
   b. The **superior wall** (roof) is formed by the **falx inguinalis** (falx, L. sickle), which is the arcing inferior free edge of the **transverse abdominis aponeurosis** (see Figure 23-4B).
      (1) The **falx inguinalis** runs from the lateral portion of the inguinal ligament to the rectus sheath and pubic ramus, leaving a weak area inferolateral to this free edge.
      (2) The **falx inguinalis** and the **"conjoined tendon"** are neither synonymous nor homologous.
         (a) The **falx inguinalis** is the lower, curving portion of the transverse abdominis muscle and its aponeurosis.
         (b) The **conjoined tendon** results from fusion (conjoining) of the internal oblique and transverse abdominis aponeuroses as they form the anterior wall of the rectus sheath. The conjoined tendon rarely (less than 5%) extends more than 1 cm lateral to the rectus sheath.
   c. The **inferior wall** (floor) is formed by the **inguinal ligament** and the **lateral crus** of the superficial ring (see Figure 23-4B).
      (1) **Inguinal ligament** (see section III A 1 c)
      (2) The floor also receives a contribution from the lacunar ligament, which is formed by fibers reflected from the inguinal ligament posteriorly.
   d. The **posterior wall** is formed by fibers of the internal oblique muscle. The **transversalis fascia** and the **interfoveolar ligament** (a condensation of fascia around the inferior epigastric vessels) offer little support (see Figures 23-4B and 23-5).

C. **Spermatic cord**

1. **Embryologic development**
   a. **Processus vaginalis** develops as an evagination of the peritoneal cavity through the anterior abdominal wall. Normally, the processus vaginalis persists from the third to the ninth month.
   b. **Composition.** As the processus vaginalis and testis descend through the abdominal wall in the inguinal region, contributions from the abdominal wall form the spermatic cord (Figure 23-5).
      (1) A peritoneal remnant of the obliterated **processus vaginalis** or, if patent, the **funicular process** lies in the center of the cord.
      (2) The **internal spermatic fascia,** derived from the transversalis fascia, provides a bed for neurovascular structures.
      (3) The **cremaster muscle,** derived from the internal oblique muscle, forms the walls of the cord.
      (4) The **external spermatic fascia,** derived from the deep (investing) fascia of the external oblique muscle, invests the cord. Because the aponeurosis splits to form the superficial inguinal ring, there is no contribution from the external oblique muscle or aponeurosis.

2. **Contents in the male** (see Figure 23-5). The testis remains attached to the original excretory duct as well as its vascular and nerve supply.
   a. The internal spermatic fascia contains the **vas deferens, testicular artery, artery**

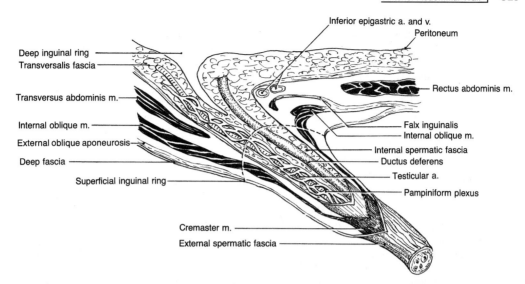

FIGURE 23-5. Spermatic cord. The deep fascia of the external oblique muscle contributes the external spermatic fascia. The internal oblique layer contributes the cremaster muscle. The transverse abdominis layer usually makes no contribution. The transversalis fascia continues into the cord about the neurovascular bundle as the internal spermatic fascia.

of the vas deferens, and **pampiniform plexus** of veins, as well as the **testicular nerves.**
  b. The external spermatic fascia contains the **genital branch of the genitofemoral nerve** (innervating the cremaster muscle) and the **ilioinguinal nerve** (sensory over the pubis).
3. **Contents in the female.** Only the **round ligament of the uterus,** which is a remnant of the **gubernaculum,** and a branch of the **ilioinguinal nerve** pass through the inguinal canal (of Nuck).
  a. The **round ligament of the uterus** usually loses its identity within the inguinal canal, but occasionally it may be followed to the labia majora.
  b. Because of the absence of a spermatic cord, the superficial inguinal ring in the female is small and usually does not significantly weaken the anterior abdominal wall. Thus, the canal of Nuck in the female rarely causes problems and, accordingly, receives less attention.

D. **Nerves of the inguinal region** arise from the lumbar plexus on each side. Posteriorly, they pass through the psoas major muscle and anterior to the quadratus lumborum muscle. Laterally, they course between the transverse abdominis and internal oblique muscles. Anteriorly, they become superficial (see Figures 23-3 and 26-6).

  1. **Iliohypogastric nerve (T12–L1, anterior)**
      a. This nerve is distributed to the skin of the hypogastric region just superior to the pubic crest and pubic symphysis.
      b. It penetrates the external oblique aponeurosis 1–2 cm superomedial to the superficial inguinal ring.
      c. It provides the afferent (sensory) limb and one of several efferent (motor) limbs of the **abdominal reflex.** Stroking or scratching the abdomen parallel to the inguinal ligament initiates a rippling of the rectus abdominis muscle or flank muscles.

  2. **Ilioinguinal nerve** (L1, anterior)
      a. This nerve accompanies the spermatic cord or round ligament of the uterus through the superficial inguinal ring.
      b. It is distributed to the medial aspect of the thigh, the base of the penis or mons, and the scrotum or labia majora.

3. **Genitofemoral nerve** (L1 and L2, anterior)
   a. **Course.** This nerve lies on the anterior surface of the psoas major muscle.
   b. The **femoral branch** passes beneath the inguinal ligament into the thigh. It is distributed to the anterosuperior aspect of the thigh and provides the afferent limb of the **cremaster reflex.**
   c. The **genital branch** lies laterally or posterolaterally to the spermatic cord as it passes through the superficial inguinal ring. It is distributed to the cremaster muscle and scrotum, and provides the efferent limb of the **cremaster reflex.**
   d. The **cremaster reflex** involves the elevation of a testis within the scrotum when the medial aspect of the thigh is stimulated.

E. **Inguinal triangle** (of Hesselbach)

1. **Boundaries.** The **inguinal (Hesselbach's) triangle** lies in the inferomedial inguinal region at the base of the **medial inguinal fossa** (Figure 23-6).
   a. **Linea semilunaris** (the lateral edge of the rectus sheath) forms the medial boundary.
   b. **Inferior epigastric artery** provides the lateral boundary as it leaves the external iliac artery and courses toward the arcuate line to enter the rectus sheath.
   c. **Inguinal ligament** between the inferior epigastric artery and the linea semilunaris marks the inferior boundary.

2. **Structure.** The triangle is an area of potential weakness and, thus, is the site of a **direct inguinal hernia.**
   a. The abdominal wall contributes little or no muscular support.
   b. Anteriorly, the triangle is covered by the external oblique aponeurosis. It is weakened by the superficial inguinal ring, a defect in the anterior abdominal wall (see Figure 23-6).
      (1) Behind the superficial ring, the **falx inguinalis** provides reinforcement of the triangle.
      (2) Support from the falx inguinalis varies among individuals. The more lateral its insertion on the superior pubic ramus, the greater its support.

3. **Inguinal hernias**
   a. **Prevalence.** Fully 97% of all hernias in males and 50% in females are inguinal.
   b. **Indirect inguinal hernias**

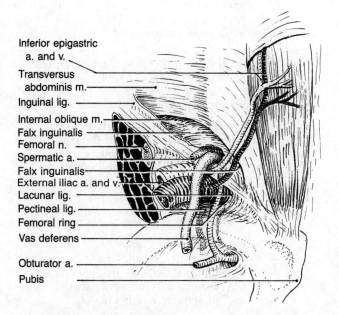

FIGURE 23-6. Inguinal triangle (viewed from behind). The boundaries are the border of the rectus abdominis muscle, the inguinal ligament, and the inferior epigastric artery. The falx inguinalis provides support to a variable extent.

(1) **Course.** Herniated viscus or fat lies within the inguinal canal and spermatic cord.
  (a) **Entrance.** Passing through the deep inguinal ring, the indirect hernia must pass lateral to the inferior epigastric artery to enter the inguinal canal, placing its origin in the **lateral inguinal fossa,** not the inguinal triangle.
  (b) **Exit.** This hernia exits the abdominal wall through the superficial inguinal ring. Because the hernial sac is within the spermatic cord, indirect hernias are directed toward the scrotum.
  (c) **Relations.** The risk of strangulation and infarct is high because the indirect hernia travels within the muscular inguinal canal.
(2) **Congenital (funicular) indirect inguinal hernias** occur through a patent **funicular process,** usually during infancy. They are common on the right side because that testicle normally descends after the left and the right funicular process remains open longer. This type of hernia is rare in females.
(3) **Acquired indirect inguinal hernias** occur through a weakened area behind the remnant of the fused processus vaginalis, usually in middle age or later.

c. **Direct inguinal hernia**
  (1) **Course.** The herniated viscus or fat lies adjacent to (not within) the spermatic cord.
    (a) **Entrance.** The direct hernia passes through the inguinal triangle at the base of the **medial inguinal fossa.** It passes medial to the inferior epigastric artery, avoiding the deep inguinal ring and inguinal canal.
    (b) **Exit.** This hernia exits the abdominal wall through the superficial inguinal ring. Because the hernial sac is adjacent to the spermatic cord, direct hernias seldom enter the scrotum.
    (c) **Relations.** The risk of strangulation is low because the hernia does not pass through muscular layers or strong tendinous confines.
  (2) Direct inguinal hernias are always **acquired.**
    (a) They are common in middle-aged men and half as common in women.
    (b) Genetic factors predispose to direct herniation. For example, the lateral extent of falx inguinalis attachment along the superior pubic ramus determines the strength of the abdominal wall in the inguinal triangle.

d. **Summary of characteristics**
  (1) **Both indirect and direct inguinal hernias** (in contrast to femoral hernias) lie **superior to the inguinal ligament** and **medial to the pubic tubercle** (to which the inguinal ligament and lateral crus of the superficial ring both attach), and **present through the superficial ring.**
  (2) **Indirect inguinal hernias** enter the inguinal canal through the deep ring (lateral to the inferior epigastric artery) and lie within the spermatic cord.
  (3) **Direct inguinal hernias** pass through the inguinal triangle (medial to the inferior epigastric artery) and lie medial to the spermatic cord.

F. Femoral ring

1. **Boundaries.** The area between the inguinal ligament and the superior pubic ramus contains the iliopsoas muscle and the femoral nerve, as well as the femoral artery, femoral vein, and the **femoral ring,** which contains lymphatics (Figure 23-7). The femoral ring is bounded by the **femoral vein** (laterally), **inguinal ligament** (anteriorly), **lacunar ligament** (medially), and the **pectineal ligament** (posteriorly).

2. **Structure.** The femoral artery, femoral vein, and **femoral canal** (the extension of the femoral ring) are wrapped in the **femoral sheath.** This continuation of the transversalis fascia leaves the abdominal cavity along with the femoral vessels.

3. **Femoral hernias**
  a. Femoral hernias (common in females, rare in males) **enter the femoral ring and femoral canal** inferior to the inguinal ligament. They usually lie in the thigh, presenting through the fossa ovalis of the great saphenous vein.
    (1) Ligamentous boundaries increase the potential for strangulation and necrosis of a herniated loop of bowel.

FIGURE 23-7. In the inguinal region, the lateral (muscular) compartment contains the iliopsoas muscle and the femoral nerve. The medial (vascular) compartment contains the femoral vein, femoral artery, and femoral ring, through which lymphatics pass.

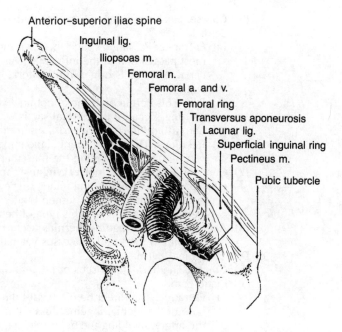

(2) An **aberrant obturator artery,** which arises from the inferior epigastric artery, presents a complication during surgical repair.

b. **Characteristics.** In contrast to inguinal hernias, these hernias lie inferior to the inguinal ligament and lateral to the pubic tubercle to which the inguinal ligament attaches.

Chapter 24
Peritoneal Cavity

I. INTRODUCTION

A. Basic principles. The peritoneal (abdominopelvic) cavity is the largest cavity of the body.

1. **The abdominal cavity** lies within the abdomen proper. It extends from the inferior surface of the musculotendinous diaphragm to the brim of the minor (deep) pelvis. It contains most of the gastrointestinal tract, accessory digestive glands, kidneys, and major vessels.

2. **The pelvic cavity** lies within the true pelvis and is about the size of a clenched fist. It extends from the minor (deep) pelvis to the pelvic diaphragm (levator ani muscle). It contains loops of small bowel, the sigmoid colon, the terminal portion of the gastrointestinal tract (i.e., the rectum and anal canal), and organs of the urogenital system with associated adnexa.

3. **Gender differences.** In males, the peritoneal cavity is completely enclosed by peritoneum. In females, it is perforated by the ostia of the uterine tubes. This pathway enables air and semen as well as pathogens to enter the peritoneal cavity from the external environment.

B. Peritoneum. Three divisions are defined.

1. **Parietal peritoneum,** the innermost layer of the abdominal wall, completely covers the inside wall of the abdominal cavity.
   a. **Composition.** It consists of mesothelial serous membrane, supported by transversalis fascia.
   b. **Innervation.** This layer is innervated by **somatic nerves** from lower intercostal nerves and lumbar nerves. Inflamed parietal peritoneum is exquisitely sensitive to palpation or stretching in a discretely localized area on the abdominal wall. Abdominal musculature tenses, producing a **rigid abdomen** (**guarding** or **splinting**) to minimize pain.

2. **Mesenteries** are formed by the reflection of mesothelium between parietal peritoneum and visceral peritoneum.
   a. As parietal peritoneum from one side leaves the body wall in the midline, it fuses with its contralateral counterpart. These back-to-back mesothelial layers form **mesentery,** or a **ligament.**
   b. As the mesentery reaches the viscera, the layers split and enclose the organ as **visceral peritoneum.**
   c. Mesenteries not only support the viscera, but also provide pathways for associated neurovascular structures and ducts.

3. **Visceral peritoneum** is the reflection of the delaminated mesentery over the abdominal and pelvic viscera.
   a. **Composition.** Histologically, it consists of a mesothelial serous membrane, the tunica serosa. The serosa constitutes the external surface of much of the gastrointestinal tract.
   b. **Innervation.** It is innervated by **splanchnic (visceral) nerves** that travel along autonomic pathways.
      (1) Relatively insensitive to pain, it is more sensitive to stretch.
      (2) Inflamed visceral peritoneum results in diffuse, crampy, or colicky abdominal pain, which is **referred** to specific dermatomes on the abdominal wall.

4. **Peritoneal fluid**
   a. **Serous fluid** that exudes through the peritoneal membranes provides a lubricating film for the parietal and visceral surfaces, facilitating free mobility with minimal friction.

b. Ascites. Inflammatory or other pathologic processes often result in large fluid collections in the abdominal cavity that require drainage **(abdominocentesis).**

II. DEVELOPMENT OF THE PERITONEAL CAVITY AND VISCERA

A. Early differentiation

1. **The embryonic disk** develops during the third week (Figure 24-1A). Rapid growth in the head, tail, and lateral regions forms infoldings of the embryonic disk (see Figure 24-1B).
2. **The primitive gut** is formed by incorporation of a portion of the upper pole of the **primitive yolk sac** into the developing embryo (see Figure 24-1C).
   a. The periphery of the embryonic disk delaminates, forming an outer **somatopleure** and an inner **splanchnopleure,** between which lies the **coelomic cavity.** The **left and right coelomic cavities** are continuous with the **extraembryonic coelom** (see Figure 24-3).
   b. Ventral fusion of the **head, lateral,** and **tail folds** forms the **enteric tube** and **body wall.**
      (1) The **splanchnopleure** (visceral primordium) of the two lateral folds, head fold, and tail fold meet and fuse at the umbilical stalk to form the **enteric tube.**
      (2) Similarly, ventral midline fusion of the **somatopleure** (body wall primordium) of these folds forms the **body wall** and separates the **left** and **right coelomic cavities** from the extraembryonic coelom (see Figure 24-3).
   c. The **enteric tube** is divided into three distinct regions.
      (1) The **foregut** is formed by the **head fold** (Figure 24-2).
         (a) It is closed by the **buccopharyngeal membrane** while a blind tube and eventually opens through the **stomodeum.**
         (b) It is supplied by the **celiac artery.**
      (2) The **midgut** is formed largely by **lateral folds** (see Figure 24-2).
         (a) It retains a connection with the yolk sac by the **omphalomesenteric (vitelline) duct,** which normally degenerates. It may persist as an **ileal diverticulum** (of Meckel).
         (b) It is supplied by the **superior mesenteric artery.**

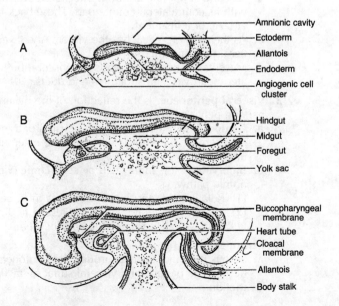

FIGURE 24-1. Early differentiation of the gastrointestinal tract. *(A)* At 19 days, the germ disk lies flat on the yolk mass. *(B)* By 21 days, head and tail folds form the early foregut and hindgut, respectively. *(C)* By 24 days, the head and tail folds have deepened, and lateral folds have come together to form the midgut.

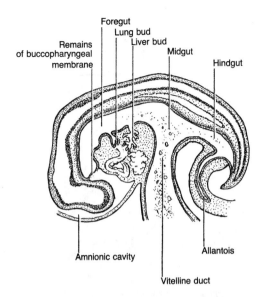

FIGURE 24-2. Later differentiation of the gastrointestinal tract. During the fourth week, flexion occurs with development of the body stalk. The midgut retains a connection with the yolk sac through the vitelline duct.

(3) The **hindgut** is formed by the **tail fold** (see Figure 24-2).
   (a) Although the hindgut is a blind tube closed by the **cloacal membrane,** it eventually opens through the **proctodeum.**
   (b) It is supplied by the **inferior mesenteric artery.**

3. **The coelomic cavity (coelom)** is the space enclosed between the splanchnopleure (the layer that forms the viscera) and the somatopleure (the layer that forms the body wall). It is lined by **pleura** in the thoracic region and by **peritoneum** in the abdominopelvic region (see Figure 24-3).
   a. **Visceral peritoneum,** composed of serous mesothelia, covers the outer surfaces of splanchnopleure.
   b. **Parietal peritoneum,** also serous mesothelia, covers the inner surfaces of somatopleure.
   c. **Dorsal and ventral mesenteries,** suspending the gastrointestinal tract within the coelom, divide the coelomic cavity into left and right halves (see Figure 24-3B).

4. **Embryonic mesenteries.** Back-to-back layers of parietal peritoneum reflect off the body wall to become mesenteries that, in turn, delaminate to become visceral peritoneum (Figure 24-3).
   a. The **ventral mesentery** degenerates, except for cephalad and caudal portions. The liver develops within the cephalad portion, dividing it into the **lesser omentum** and the **falciform ligament** (see Figure 24-4).
      (1) The **lesser omentum,** between the stomach and the liver, has two named portions.
         (a) The **gastrohepatic ligament** runs between the lesser curvature of the stomach and the liver (see Figure 24-5C).
         (b) The **hepatoduodenal ligament** runs between the liver and superior duodenum (see Figure 24-5C).
         (c) The **omental (epiploic) foramen** (of Winslow) is formed by the inferior free edge of the ventral mesentery.
      (2) **Coronary ligaments** attach the liver to the diaphragm and provide continuity between the falciform ligament and the lesser omentum (see Figure 25-5).
      (3) The **falciform ligament** attaches the liver to the ventral body wall (see Figure 25-4).
         (a) It extends between the ventral surface of the liver and the umbilicus.
         (b) It contains the **round ligament of the liver (ligamentum teres),** which is the remnant of the obliterated **umbilical vein.**
      (4) The **median umbilical fold** is a caudal remnant of the ventral mesentery.

FIGURE 24-3. Formation of dorsal and ventral mesenteries. *(A)* At about 21 days, the lateral folds delineate the intraembryonic coelom and midgut. *(B)* By 28 days, dorsal and ventral mesenteries are defined.

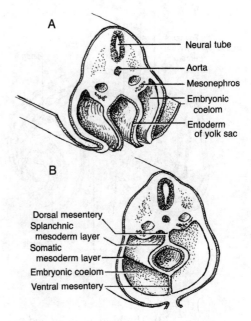

   (a) It runs from the apex of the urinary bladder to the umbilicus.
   (b) It contains the **urachus,** which is the remnant of the fused **allantoic stalk** (see Figure 24-2).
   (c) **Patent urachus** (rare) results in seepage of urine from the umbilicus. Isolated patency may produce a **urachal cyst.** Partial patency, extending from the apex of the bladder, may complicate surgery in the umbilical and pubic regions.
  b. The **dorsal mesentery** is the primary support of the gastrointestinal tract and provides passage for the blood vessels that supply the viscera (see Figure 24-4).
   (1) The **dorsal mesogastrium** has several ligaments.
      (a) The **gastrophrenic ligament** runs between the greater curvature of the stomach and the diaphragmatic peritoneum.
      (b) The **gastrosplenic ligament** runs between the stomach and the spleen. The **splenorenal (lienorenal, phrenicolienal) ligament** continues between the spleen and the dorsal body wall superior to the left kidney.
      (c) The **greater omentum** runs between the dorsal body wall and the greater curvature of the stomach (see Figure 24-6).
   (2) The **dorsal mesoduodenum** is an embryonic structure only.
      (a) Except for the superior segment, the duodenum becomes **secondarily retroperitoneal** by fusion of this mesentery to the peritoneum of the dorsal body wall. A secondarily retroperitoneal structure was supported by mesentery in the embryonic coelomic cavity, but it is fused with the peritoneum of the dorsal body wall in the adult.
      (b) The head and body of the pancreas become **secondarily retroperitoneal** by fusion of this mesentery, with the contained portions of pancreas, to the peritoneum of the dorsal body wall.
      (c) **Surgical mobilization.** Incision of the region of fusion (fascia of Toldt) mobilizes the duodenum and pancreas and their shared blood supplies.
   (3) The **dorsal mesointestine** is divided into regions, each of which supports a segment of bowel.
      (a) The **mesentery proper** supports most of the small intestine (see Figure 24-5C).
         (i) This ligament runs from the dorsal body wall to the jejunum and ileum. It is fan shaped, so a root about 9 inches long supports about 20 feet of small bowel.

(ii) The positions in which it supports the various loops of small bowel are variable.
  (b) The **ascending mesocolon** is an embryonic structure only.
    (i) The ascending colon becomes **secondarily retroperitoneal** (except for a small portion along the inferior cecum) as this mesentery fuses with the peritoneum of the dorsal body wall.
    (ii) **Surgical mobilization.** Incision of the line of fusion re-establishes the mesentery and "mobilizes" the ascending colon and its blood supply.
  (c) The **transverse mesocolon** supports the transverse colon.
    (i) It runs between the dorsal body wall and the transverse colon. It suspends the transverse colon in variable positions.
    (ii) It fuses with the **greater omentum** to form the **gastrocolic ligament** (see Figure 24-6).
  (d) The **descending mesocolon** is an embryonic structure only.
    (i) The descending colon becomes **secondarily retroperitoneal** as this mesentery fuses with the peritoneum of the dorsal body wall.
    (ii) **Surgical mobilization.** Incision of the line of fusion re-establishes the mesentery and "mobilizes" the descending colon with its blood supply.
  (e) The **sigmoid mesocolon** supports the sigmoid colon (see Figure 24-5C).
    (i) It runs between the dorsal body wall and the sigmoid colon.
    (ii) It suspends the sigmoid colon in variable positions within the deep pelvis.

B. **Rotation of the foregut**

1. **Early stage.** The definitive position of the gut in adults is attributable to rotation during development. The gastric anlage develops as a foregut dilatation during the fifth week (Figure 24-4; see Figure 24-5).
   a. A 90° rotation to the right occurs around the longitudinal axis of the gut.
   b. The left side of the stomach becomes anterior. As a result, the left vagus nerve innervates the anterior surface and the right vagus nerve innervates the posterior surface.

2. **Later stages**
   a. **Rotation of the stomach.** During rotation, the primitive dorsal surface of the stomach undergoes accelerated growth to produce the greater curvature.

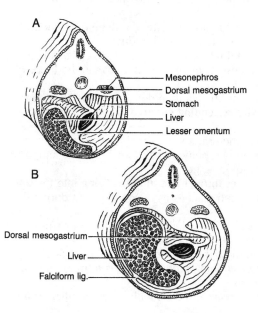

FIGURE 24-4. Rotation of the stomach. *(A)* During the fifth week, the stomach rotates so the primitive left side becomes the anterior surface. *(B)* By the sixth week, the greater omentum begins to form and the liver develops toward the right. The ventral mesentery caudal to the ligamentum teres begins to degenerate.

(1) Differential growth pulls the dorsal mesogastrium to the left.
    (a) This growth produces the **greater omentum** (see Figure 24-4).
    (b) Some of the right coelom becomes entrapped behind the stomach to form the **omental bursa** or **lesser sac** (see Figures 24-4 and 24-6). This space communicates with the remaining part of the coelomic cavity (greater sac) via the **omental (epiploic) foramen** (see Figure 24-7).
(2) The **pylorus** moves craniad and to the right by the concomitant growth of the intestines. The stomach assumes the definitive adult position.
(3) The **spleen** develops within the dorsal mesogastrium.
    (a) Its position divides this mesentery into **gastrosplenic** and **splenorenal ligaments.**
    (b) The **splenorenal ligament** is continuous with the peritoneum of the posterior abdominal wall and contains the tail of the developing pancreas.
(4) The dorsal mesogastrium continues to grow, doubling on itself.
    (a) Fusion of its mesothelial leaflets on the lesser sac side forms the **greater omentum** (see Figure 24-6).
    (b) The **omental bursa (lesser sac)** initially extends into the space between the folds of the dorsal mesogastrium.
    (c) Fat develops in the greater omentum.

b. **Rotation of the duodenum.** During rotation, the duodenum assumes a C-shape.
   (1) During the **sixth week,** this intestinal portion moves posteriorly and to the right because of gastric rotation to the left.
   (2) Late in the **second month,** it becomes **secondarily retroperitoneal** (except for the first segment) as the dorsal mesoduodenum fuses with the peritoneum of the dorsal body wall.
       (a) The pancreas (except for the terminal part of the tail) also becomes **secondarily retroperitoneal** with fusion of the dorsal mesoduodenum to the posterior body wall.
       (b) The duodenum passes first through a solid stage and then it reopens. Incomplete recanalization results in **pyloric stenosis** or **duodenal stenosis,** which must be corrected surgically.

C. Rotation of the midgut

1. **First stage** (Figure 24-5A)
   a. Rapid midgut growth during the fifth week results in the formation of the **primary intestinal loop,** which is connected to the yolk sac via the **omphalomesenteric (vitelline) duct.**
      (1) The **cranial limb** of the loop develops into the **distal duodenum, jejunum,** and **ileum** as far as the omphalomesenteric (vitelline) duct.
      (2) The **omphalomesenteric (vitelline) duct,** or its remnant (Meckel's diverticulum) in 3% of the population, represents the boundary between the limbs of the primary intestinal loop.
      (3) The **caudal limb** of the loop, beyond the vitelline duct, develops into the **terminal ileum, cecum** and **appendix, ascending colon,** and most of the **transverse colon.**
   b. A physiologic **umbilical herniation** of the primary intestinal loop normally occurs.
      (1) Because the liver and intestines grow rapidly, the abdominal cavity temporarily becomes too small for the developing gut.
      (2) The primary intestinal loop herniates through the **umbilical ring** into the **umbilical stalk** during the sixth week and reaches its maximal extent during the ninth week.
   c. The **primary intestinal loop** rotates 90° counterclockwise around an axis provided by the superior mesenteric artery, the artery of the midgut.
      (1) The cranial limb comes to lie on the right and the caudal limb on the left within the umbilical stalk. This positioning probably results from rapid growth of the cranial limb during this phase.
      (2) The cranial limb forms secondary loops and coils within the umbilical stalk.

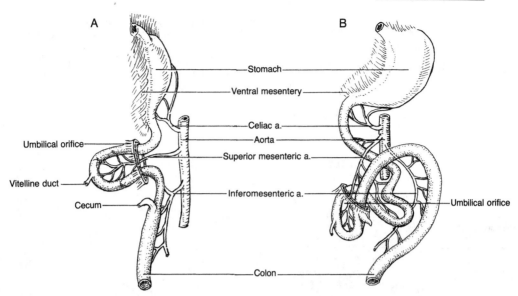

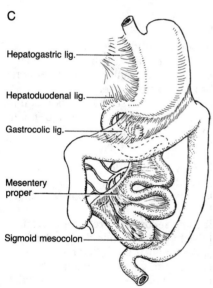

FIGURE 24-5. Rotation of the gut. *(A)* Stage I: Umbilical herniation. During the sixth week, the rapidly growing intestine herniates into the umbilical stalk. *(B)* Stage II: Reduction of the umbilical herniation. During the tenth week, the extraembryonic coils of intestine are rapidly withdrawn into the coelomic cavity. The cranial loop passes posterior to the caudal loop so an effective rotation results. *(C)* Stage III: Fixation of the gastrointestinal tract. Once the intestines have returned to the body cavity, their mesenteries fuse with the parietal peritoneum in several regions, making those portions of the tract appear retroperitoneal—thus the term secondarily retroperitoneal.

   (3) Abnormal (0.01%) clockwise rotation at this stage produces **situs inversus viscerum.**
   d. Rotation of the stomach and duodenal fixation to the dorsal body wall occur simultaneously with this stage of intestinal rotation.
2. **Second stage** (see Figure 24-5B)
   a. During the tenth week, the herniation is rapidly and completely withdrawn.
      (1) The abdominal cavity becomes larger as a result of slower hepatic growth.
      (2) The proximal jejunum is withdrawn first.
         (a) It passes posterior to the superior mesenteric artery and the caudal limb of the primary intestinal loop, coming to lie in the upper left quadrant. This change explains the position of the superior mesenteric artery anterior to the terminal duodenum.
         (b) With progressive reduction of the umbilical herniation, secondary intestinal loops lie progressively to the right as they enter the abdominal cavity.

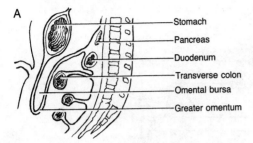
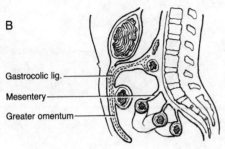

FIGURE 24-6. Formation of the gastrocolic ligament. *(A)* The omental bursa (lesser sac) extends between the redundant leaflets of the greater omentum. *(B)* Fusion of the redundant leaflets of the greater omentum and subsequent fusion with the transverse mesocolon form the gastrocolic ligament.

   (3) The caudal limb of the primary intestinal loop is withdrawn last.
      (a) The cecum, therefore, tends to lie to the right below the liver.
      (b) The caudal limb of the primary intestinal loop lies anterior to the former cranial limb and anterior to the superior mesenteric artery.
   b. Withdrawal of the physiologic herniation results in additional 180° rotation counterclockwise. Thus, the total rotation from the first and second stages is 270° counterclockwise about an axis formed by the superior mesenteric artery.

3. **Third stage** (see Figure 24-5C)
   a. The cecal region of the caudal limb of the primary intestinal loop moves inferiorly into the lower right quadrant, and the appendix develops.
   b. Fixation of portions of the gut occurs and the mesenteries assume their characteristic positions and appearance.
      (1) The ascending and descending colons become **secondarily retroperitoneal**.
      (2) The dorsal mesogastrium grows and fuses with itself, forming the **greater omentum** and obliterating that portion of the omental bursa that had been contained between the two redundant leaflets. The greater omentum, thus, consists of four layers of mesothelium, two of which are apposed and fused.
      (3) The inferior surface of the greater omentum also fuses with the superior surface of the transverse mesocolon, forming the **gastrocolic ligament** (Figure 24-6). This ligament, composed of six layers of mesothelium, divides the peritoneal cavity into **supracolic** and **infracolic compartments.**

4. **Clinical considerations.** Anomalies of rotation predispose to strangulation, intestinal obstruction, ischemic necrosis, and infarction. Although major developmental abnormalities of gastrointestinal tract rotation occasionally are incidental findings, they usually manifest clinically in neonates.
   a. **Nonrotation of the gut**
      (1) The primary intestinal loop may return to the abdominal cavity without rotation.
      (2) The jejunum and ileum may lie on the right, and the duodenum may remain peritoneal. The small intestine may twist around the superior mesenteric artery (volvulus), obstructing the intestine or compressing its blood supply. Volvulus results in ischemic necrosis and infarction of the small bowel.
      (3) The ascending colon and descending colon may lie on the left side so that there is, in effect, no transverse colon.
         (a) The descending colon alone may become fixed.

- (b) Peritoneal bands or adhesions may develop because of partial fixation and lead to intestinal obstruction.
- (c) The floating ascending colon may undergo volvulus.

  b. **Reversed rotation of the gut (situs inversus viscerum)**
  (1) Situs inversus is rare, occurring in about 1 in 10,000 births.
  (2) The external appearance is normal in females; in males, the right testis is lower. The visceral positions are a mirror image of normal.
  (3) This anomaly is caused by clockwise withdrawal of the primary intestinal loop into the abdominal cavity. Other than the reversed appearance, situs inversus predisposes to no particular problem.
  (4) Situs inversus can confuse the unaware.
  - (a) In one extensive study, 55% of situs inversus cases were recognized before surgery, 32% were recognized during surgery, and 13% were not recognized until after surgery.
  - (b) In the same study, only 14% of patients had a correct diagnosis and appropriate incision, 45% had an incorrect diagnosis, and 31% had an inappropriate incision. In 10% of cases, the procedure was aborted because the surgeon became confused.

  c. **Malrotation of the gut**
  (1) The stages or degrees of incomplete rotation vary. Depending on the degree of nonrotation and subsequent peritoneal fixation, volvulus or strangulation may occur.
  (2) Nonfixation occurs most frequently at the cecum, and always to a variable extent in the region of the sigmoid colon, so the normal length of the peritoneal portion of the sigmoid colon varies widely.
  (3) Malrotation may produce **paraduodenal hernias.** As the primary intestinal loop is withdrawn, the small intestine may pass between the ascending colon and the parietal peritoneum, producing a left paraduodenal hernia. If it passes between the ascending colon and parietal peritoneum, a right paraduodenal hernia develops. These hernias are readily detected radiographically.

  d. **Omphalocele, gastroschisis, and umbilical hernia**
  (1) **Omphalocele** is an anterior abdominal wall defect, covered only by peritoneum and perhaps a layer of amnion, through which viscera may herniate. The umbilical cord is inserted into the sac.
  (2) **Gastroschisis,** a defect in the abdominal wall related to delayed umbilical ring closure.
  (3) An **umbilical hernia** results from incomplete reduction of the physiologic umbilical hernia during the tenth week. Part of the primary intestinal loop is retained within the umbilical cord.
  - (a) This anomaly occurs in approximately 2% of infants.
  - (b) A small umbilical hernia may not be recognized immediately, and a loop of small intestine may be divided when the umbilical cord is sectioned and tied.

  e. **Meckel's diverticulum** is a persistent remnant of the vitelline duct that is present in 3% of adults.
  (1) It may remain attached to the umbilicus as either a fibrous cord or a vitelline fistula. Epithelium-lined cysts may persist along a fibrous connector.
  (2) Persistent vitelline vessels attach at the umbilicus and anastomose with somatic vessels of the abdominal wall.
  (3) Volvulus may occur around any connection between the ileum and umbilicus, with possible obstruction or strangulation.
  (4) Because Meckel's diverticula frequently contain gastric mucosa that secretes acid into the normally alkaline ileum, common complications include inflammation of the diverticulum and ulceration of the ileum.

III. PERITONEAL RELATIONSHIPS

A. The **definitive positions** and attachments of the gastrointestinal tract result from rotation of the gut (Table 24-1).

TABLE 24-1. Peritoneal Relationships Along the Gastrointestinal Tract

Organ	Classification	Mesenteric Support
Esophagus		
Thoracic	Retroperitoneal	Adventitia
Abdominal	Peritoneal	Lesser omentum and gastrophrenic lig.
Stomach	Peritoneal	Dorsal mesogastrium (gastrophrenic, gastrosplenic, phrenicolienal ligs.; greater omentum); ventral mesogastrium (gastrohepatic lig.)
Liver	Peritoneal	Lesser omentum (gastrohepatic, hepatoduodenal ligs.), falciform, coronary, triangular ligs.
Pancreas		
Head, body, most of tail	Secondarily retroperitoneal	(No mesentery)
Tip of tail	Peritoneal	Lienorenal lig.
Duodenum		
First part	Peritoneal	Hepatoduodenal lig., greater omentum
Second, third, and fourth parts	Secondarily retroperitoneal	(No mesentery)
Jejunum	Peritoneal	Mesentery proper
Ileum	Peritoneal	Mesentery proper
Cecum		
Terminal part	Peritoneal	Ileocecal fold of mesentery proper
Body	Secondarily retroperitoneal	(No mesentery)
Ascending colon	Secondarily retroperitoneal	(No mesentery)
Transverse colon	Peritoneal	Transverse mesocolon, gastrocolic lig.
Descending colon	Secondarily retroperitoneal	(No mesentery)
Sigmoid colon	Peritoneal	Sigmoid mesocolon
Rectum	Retroperitoneal	Adventitia

1. **Retroperitoneal (extraperitoneal) structures** develop outside the peritoneal cavity. They may be surrounded in part by adventitia or covered by peritoneum.
2. **Peritoneal structures** remain suspended in the peritoneal cavity by mesentery.
3. **Secondarily retroperitoneal structures** develop as peritoneal structures, but their associated mesentery secondarily becomes adherent to the peritoneum covering the dorsal body wall.

B. Peritoneal subdivisions. Peritoneal reflections as mesenteries divide the abdominal cavity into sacs, compartments, and several blind fossae or recesses (Figure 24-7), many of which are potential sites for infection.

1. **Sacs and compartments**
    a. The **greater sac** develops to the left of the dorsal and ventral mesenteries (see Fig-

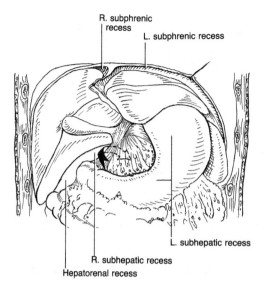

FIGURE 24-7. Named recesses of the supracolic compartment. The arrow passes through the omental (epiploic) foramen into the omental bursa (lesser sac).

ure 24-3B). The gastrocolic ligament (the fusion of the greater omentum and transverse mesocolon) divides the greater sac into **supracolic** and **infracolic compartments** (see Figure 24-8).
  b. The lesser sac (**omental bursa**) lies posterior to the stomach and lesser omentum (see Figure 24-7). This portion of the coelomic cavity, which develops to the right of the dorsal and ventral mesenteries, becomes isolated from the **greater sac** by the rotation of the stomach as well as by elevation and fixation of the duodenum. It communicates with the greater sac (the greater part of the coelomic cavity) via the **omental (epiploic, Winslow) foramen** (see Figure 24-7).
    (1) **Boundaries of the omental foramen** include the **caudate lobe** of the liver (superiorly), **inferior vena cava** (posteriorly), **superior duodenum** (inferiorly), and **hepatoduodenal ligament** (anteriorly) that contains the hepatic pedicle.
      (a) The **hepatic pedicle** contains the common bile duct, hepatic artery, and hepatic portal vein.
      (b) Intraoperative hepatic bleeding is controlled by compressing the hepatic pedicle between the thumb and the index finger inserted into the omental foramen.
    (2) **Omental herniation.** If a loop of bowel passes through the epiploic foramen and becomes incarcerated, none of the boundaries of the foramen can be safely incised. Instead, the bowel is deflated with a needle and the loop of gut is then withdrawn.
2. **Named recesses, spaces, and fossae** (Figure 24-8)
  a. The **subphrenic (suprahepatic) recess** is anterior and superior to the liver, beneath the diaphragm.
    (1) The falciform ligament divides it into **right** and **left subphrenic recesses.**
    (2) It is the second most frequently infected abdominal space. A pulmonary abscess may erode across the diaphragm, invading the abdominal cavity and involving the suprahepatic recess. This type of infection was a frequent cause of mortality before routine use of antibiotic therapy and remains a serious condition.
  b. The **infrahepatic recess** is inferior to the liver within the lesser sac.
  c. The **right subhepatic recess** and the **hepatorenal recess** form the **pouch of Morison.**
    (1) When a person is supine, the hepatorenal recess becomes the lowest portion of the abdominal cavity and is a site for fluid collection.
    (2) The pouch of Morison is probably the most frequently infected abdominal space.

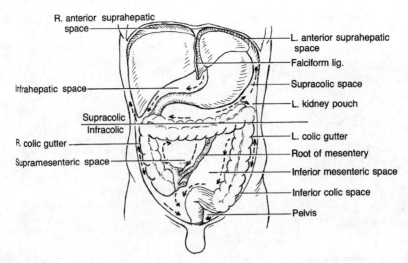

FIGURE 24-8. Peritoneal drainage pathways. Infection can spread from one part of the peritoneal cavity to another via recesses, spaces, and gutters.

        (a) If a patient has symptoms of abdominal infection without local abdominal signs, an abscess in the pouch of Morison is a likely diagnosis.
        (b) Primary causes of infection include appendicitis, perforated duodenal or gastric ulcer, liver abscess, perforated gallbladder or biliary tree, and surgical procedures.
    (3) The pouch of Morison communicates with the lesser sac via the omental foramen, right paracolic gutter, and the subphrenic recess. Infection may track between these spaces.
  d. Paraduodenal fossae occur in association with the terminal duodenum, inferior mesenteric vein, and left colic artery.
  e. Fossae of the cecal region
    (1) The **superior ileocecal fossa** is bounded by the mesentery proper and the superior ileocecal (vascular) fold.
    (2) The **inferior ileocecal fossa** is bounded by the mesoappendix and inferior ileocecal (bloodless) fold.
    (3) The **retrocecal fossa** lies beneath the cecum, is variable in extent, is formed by adhesion of the cecum to the posterior parietal peritoneum, and frequently contains the appendix.
  f. Paracolic recesses (gutters)
    (1) The **right colic recess (gutter)** lies lateral to the ascending colon. It communicates with the supracolic compartment, the pouch of Morison, and the pelvic cavity. It provides a route for the spread of infection between the pelvis and the upper abdominal region.
    (2) The **left colic recess (gutter)** lies lateral to the descending colon. It communicates with the supracolic and infracolic compartments and the pelvic cavity, but it seldom becomes infected.
  g. The **infrasigmoid (parasigmoid) fossa,** which lies beneath the sigmoid colon at the juncture of the sigmoid mesocolon and parietal peritoneum, is a potential site for intra-abdominal herniation of the sigmoid colon or small bowel.
  h. Abdominal wall fossae
    (1) **Lateral inguinal fossae** lie lateral to the lateral umbilical folds, which are formed by the underlying inferior epigastric arteries. These fossae contain the **deep inguinal rings,** where **indirect inguinal herniation** begins.
    (2) **Medial inguinal fossae** lie between the lateral umbilical folds (formed by the underlying inferior epigastric arteries) and the medial umbilical folds (formed by the underlying obliterated umbilical arteries). This fossa nearly coincides

with the **inguinal triangle** (of Hesselbach), where **direct inguinal herniation** begins.

(3) **Supravesical fossae** lie medial to the medial umbilical folds (formed by the underlying obliterated umbilical arteries) and lateral to the median umbilical fold (formed by the underlying urachus). These fossae are the sites of uncommon supravesical herniation.

C. Clinical considerations

1. **Paracentesis.** When a patient with a peritoneal infection is supine, inflammatory exudate collects in the subphrenic and hepatorenal recesses, as well as in the pelvic cavity.
   a. Diagnostic steps include percussion along the lateral abdominal walls and radiographic examination to determine a fluid level.
   b. Treatment includes puncture of the abdominal wall to withdraw fluid. A cannula inserted in the flank will pass through the skin, superficial fascia, deep fascia, aponeurosis of the external oblique muscle, internal oblique muscle, transverse abdominis muscle, transversalis fascia and extraperitoneal fat, and the parietal peritoneum.

2. **Peritonitis,** the "acute abdomen"
   a. **Inflammation** of the parietal and visceral peritoneum produces different symptoms.
      (1) The **parietal peritoneum** is innervated by somatic nerves of the anterior body wall. When inflamed, the parietal peritoneum is exquisitely sensitive to palpation or stretching in an area that is discretely localized on the abdominal wall. The abdominal musculature tenses (**guarding** or **splinting**) to produce a **rigid abdomen,** thereby minimizing the pain.
      (2) The **visceral peritoneum** is innervated by visceral nerves, which travel along autonomic pathways. When inflamed, the visceral peritoneum results in diffuse, crampy, or colicky abdominal pain, which is **referred** to specific dermatomes on the abdominal wall.
   b. Infection may gain access to the peritoneal cavity by several routes:
      (1) Perforation of the gastrointestinal tract
      (2) Traumatic or surgical penetration of the abdominal wall
      (3) The uterine tubes that normally open into the peritoneal cavity
      (4) Hematogenous or lymphatic spread

Chapter 25
Gastrointestinal Tract

I. INTRODUCTION

A. General principles

1. **Location.** The major portion of the gastrointestinal (GI) tract, as well as many of the associated digestive glands, is contained within the abdominopelvic cavity.

2. **Organization.** The abdominopelvic portion of the GI tract is divided into **foregut**, **midgut**, and **hindgut**.

3. **Vasculature.** The blood supply is derived from the **celiac artery** (to the foregut), **superior mesenteric artery** (to the midgut), and **inferior mesenteric artery** (to the hindgut). Venous return is through the **hepatic portal system**.

4. **Innervation.** Motor innervation is by the autonomic nervous system, the pathways of which are shared by visceral afferent nerves.

5. **Function.** The principal functions of the GI tract are specific to the region, but in general support digestion and absorption.

B. Developmental considerations.
The division, locations, and mesenteric attachments of the GI tract may be described best with reference to their embryonic development.

1. **The adult position** of the abdominal viscera is determined by rotation of the gut during early development (see Figure 24-5).

2. **Mesenteric attachments**
   a. Persistent mesenteries and localized peritoneal fusion subsequent to rotation of the gut anchor portions of the GI tract.
   b. Existing mesenteries limit movement of organs, support the viscera, and provide pathways for neurovascular structures.

C. Divisions

1. **The common organization** (but less useful) is by morphologically distinct organs, including the esophagus, stomach, small intestine, large intestine, and accessory glands.

2. **A functional organization** groups organs according to similarities in their **development, blood supply, innervation,** and **principal function.**
   a. The **foregut** consists of the esophagus, stomach, and duodenum, as well as the associated liver, pancreas, and spleen.
      (1) **Vasculature.** Blood is supplied primarily by the **celiac artery.** Venous return is by veins that drain primarily into the **hepatic portal vein.**
      (2) **Innervation.** It receives parasympathetic innervation from the **vagus nerve** (CN X) and sympathetic innervation from the **greater splanchnic nerve** (T5–T9).
      (3) **Functions** include a **conduit** (esophagus), a **reservoir** (stomach), **trituration** of the solid bolus into semifluid chyme (stomach), and **digestion** of chyme (stomach, duodenum, liver, and pancreas). Limited absorption also occurs (stomach and duodenum).
   b. The **midgut** consists of the jejunum, ileum, ascending colon (including the cecum and appendix), and transverse colon. Although the small intestine is arbitrarily divided into three parts, the morphologic and physiologic differences between duodenum and jejunum and between jejunum and ileum are slight and occur by transitions over considerable distances.

(1) **Vasculature.** Blood is supplied primarily by the **superior mesenteric artery**. Venous return is primarily to the hepatic portal system by the **superior mesenteric vein**.
(2) **Innervation.** It receives parasympathetic innervation from the **vagus nerve** (CN X) and sympathetic innervation from the **lesser splanchnic nerve** (T10–T11).
(3) **Functions** include **absorption of nutrients** (jejunum and ileum) and **conservation of fluid** (ascending and transverse colons), a process that converts chyme into semisolid feces.

c. The **hindgut** consists of the descending colon, sigmoid colon, and rectum.
(1) **Vasculature.** Blood is supplied primarily by the **inferior mesenteric artery**. Venous return is primarily into the hepatic portal system by the **inferior mesenteric vein**.
(2) **Innervation.** It receives parasympathetic innervation from the **pelvic splanchnic nerves** (S2–S3) and sympathetic innervation from the **lumbar splanchnic nerves** (L1–L2).
(3) Functions include **conservation of fluid** with solidification of the feces (descending colon), **storage** of feces (sigmoid colon), and **evacuation** of the feces (rectum and anal canal).

II. FOREGUT

A. Abdominal esophagus

1. **Position and relationships.** In the thorax, the esophagus passes through the superior mediastinum and posterior mediastinum (see Figure 22-4 and Chapter 22 III B). In the **abdomen**, it makes an abrupt turn to the left from the esophageal hiatus of the diaphragm and enters the cardiac opening of the stomach (Figure 25-1).

2. **Structure.** The esophagus is the most muscular segment of the alimentary tract (see Chapter 22 III B).

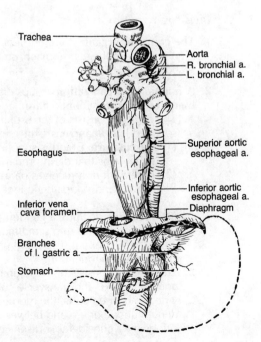

FIGURE 25-1. The esophagus passes through the esophageal hiatus of the diaphragm to gain access to the abdomen, where it joins the stomach. Blood is supplied by branches of the aorta and the left gastric artery.

3. **Vasculature**
   a. **Arterial supply** for the upper, middle, and lower portions is by respective branches from the **inferior thyroid arteries,** by the midline **esophageal arteries** and **bronchial arteries,** and by branches from the **left gastric artery.**
   b. **Venous return** parallels the arterial supply, except that the long middle portion drains into the **azygos vein** and the **hemiazygos vein.** The terminal portion drains into the hepatic portal system via the **left gastric vein.** The **esophageal venous plexus** is the site of important anastomoses between the azygous and gastric drainage pathways.
   c. **Lymphatic drainage** parallels the blood vessels, an important consideration in the spread of esophageal carcinoma to cervical, mediastinal, and celiac nodes.
4. **Innervation**
   a. Parasympathetic innervation is by the **vagus nerves** and esophageal plexus (see Chapter 22 III B 4 a).
      (1) In the thorax, the **left** and **right vagus nerves** form the **esophageal plexus,** which reunites distally to form the **vagal trunks.**
      (2) In the abdomen, the left and right vagus nerves change their relative positions to become **anterior** and **posterior vagal trunks,** respectively, because of the rotation of the stomach. Both trunks pass through the esophageal hiatus to enter the abdomen.
   b. **Sympathetic innervation** is by thoracic splanchnic nerves and by a portion of the greater splanchnic nerve (see Chapter 22 III B 4 b).
   c. **Afferent innervation** is along sympathetic pathways.
      (1) **Sensory pathways.** Afferent fibers mediating painful sensation course along the **thoracic splanchnic** and **greater splanchnic nerves** (see Chapter 22 III B 4 c).
      (2) **Referred esophageal pain.** Most painful sensation from the esophagus (such as from peptic ulceration) originates in the abdominal segment. Esophageal pain is, therefore, referred to the **lower thoracic and epigastric regions.**
5. **Function.** The esophagus serves solely for the transport of food deglutition (see Chapter 22 III B 5).
6. **Clinical considerations**
   a. **Esophagitis** refers pain to the precordium and epigastrium.
      (1) Occasionally, gastric contents reflux into the esophagus. The acidic peptic chyme burns and inflames the unprotected stratified squamous epithelium of the esophageal mucosa, producing **regurgitative esophagitis**—the uncomfortable sensation, or **"heartburn."**
      (2) Repeated reflux can result in peptic ulcers in the esophageal mucosa, which can cause **dysphagia** and serious bleeding.
   b. **Esophageal varices**
      (1) In individuals with **portal hypertension,** venous anastomoses between the systemic azygos drainage and the left gastric vein of the hepatic portal system provide an important but potentially dangerous shunt that results in the development of **esophageal varices.** These varices lie immediately beneath the mucosa where they are subject to mechanical trauma during deglutition, emesis, or the passage of diagnostic instrumentation.
      (2) They produce no symptoms until they rupture, causing massive hematemesis. More than 50% of patients with advanced cirrhosis of the liver die as a result of rupture of an esophageal varix.
   c. **Hiatus hernia** may produce discomfort and pain referred to the precordium and epigastrium.
      (1) Herniation of the stomach through the esophageal hiatus produces a saclike dilation above the diaphragm.
         (a) In 90% of cases, the esophagus ends above the diaphragm **(hourglass stomach** or **sliding hiatus hernia).**
         (b) In the remaining 10%, the cardiac region of the stomach dissects alongside the esophagus through a defect in the esophageal hiatus to produce an intrathoracic sac **(paraesophageal hernia).**

(2) Hiatus hernia may produce incompetence of the cardiac sphincter, resulting in **regurgitative esophagitis** and ultimately leading to **peptic ulceration** of the esophagus.

B. Stomach

1. **Position and relationships**
   a. **Location.** The stomach lies in the left hypochondriac and epigastric regions of the abdomen (see Figure 23-1).
   b. **Variations.** Because mesenteries suspend the stomach, it is mobile and easily displaced; that is, it has no fixed position. **Empty,** the stomach is almost tubular or J-shaped, except for the bulge of the fundus; it may lie almost entirely under the rib cage. **Full,** because it is distensible and can accommodate more than 2 liters, the stomach may pendulate as far as the pelvis.

2. **External structure and mesenteric attachments** (Figure 25-2)
   a. The **greater curvature** represents the primitive dorsal surface. It receives ligamentous support from the dorsal mesogastrium, a portion of the primitive dorsal mesentery.
      (1) The **greater omentum** is a redundant portion of the dorsal mesogastrium.
      (2) The **gastrophrenic ligament** extends between the fundic region and the dorsal body wall near the diaphragmatic crura.
      (3) The **gastrosplenic ligament** extends between the greater curvature of the stomach and the spleen.
      (4) The **splenorenal ligament** extends between the spleen and the dorsal body wall near the left kidney.
   b. The **lesser curvature** represents the primitive ventral surface. It receives ligamentous support from the derivatives of the primitive ventral mesentery, which becomes the **gastrohepatic (hepatogastric) ligament** of the **lesser omentum.**
      (1) The gastrohepatic ligament runs between the lesser curvature and the liver.
      (2) It contains the left and right gastric arteries.
   c. **Divisions** (see Figure 25-2)
      (1) The **cardia** is the region near the esophagus (see Figure 25-2).
         (a) The stomach receives the esophagus at the **cardiac opening,** which lies at the superior junction of the greater and lesser curvatures.

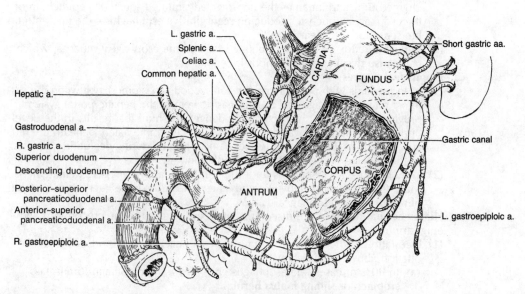

FIGURE 25-2. Regions of the stomach. The esophagus enters at the cardia. The duodenum leaves at the pyloric portion of the antrum. The window in the body shows the rugae and the location of the gastric canal. Blood is supplied to the stomach by three branches of the celiac artery.

(b) The **cardiac sphincter** at the oral end is a physiologic (functional) sphincter, even though it appears to lack the morphologic structure of a sphincter.
(c) The **cardiac notch**, the pronounced incisure between the esophagus and the fundus, is a radiographic landmark.
(2) The **fundus** is cephalad to the level of the esophageal juncture. It usually contains a variable amount of air (see Figure 25-2).
  (a) Air in the fundus is a useful radiographic landmark.
  (b) A tympanic note may be percussed from the fundic air bubble.
(3) The **body (corpus)** constitutes the major portion of the stomach (see Figure 25-2). This region is the most variable with respect to size and position.
(4) The **pylorus,** located distally, is divided into three regions (see Figure 25-3).
  (a) The **pyloric antrum** begins as a slight dilation, which produces the angular incisure in the lesser curvature.
  (b) The **pyloric canal** narrows abruptly over its 2- to 3-cm length.
  (c) The **pyloric sphincter** at the aboral end is not a functional sphincter (see Figure 25-3).
    (i) **Structure.** Unlike a true sphincter, the pylorus contracts synergistically with peristalsis and functions as a unit with the pyloric end of the stomach.
    (ii) **Function.** Although the pylorus does not control the rate of stomach emptying, it does control the size of the particles that enter the duodenum. A kernel of corn is about the limit.
    (iii) **Clinical correlation**. The pyloric sphincter may be abnormally thickened at birth. **Congenital hypertrophic pyloric stenosis** requires immediate surgical correction.

3. **Internal structure**
  a. **Rugae** are formed by the folding of the gastric mucosa.
    (1) **Permanent rugae** form the **gastric canal (gutter)** along the lesser curvature (see Figure 25-2).
      (a) Unlike rugae in other parts of the stomach, those of the gastric canal do not diminish as the stomach fills. They usually are evident on contrast-enhanced radiographic images of the upper GI tract.
      (b) These rugae direct fluids toward the pylorus. As such, the gastric canal is vulnerable to burns from accidental ingestion of caustic substances.
    (2) **Gastric rugae** in regions away from the gastric canal are temporary high folds of mucosa produced by muscular tension in the muscularis mucosae. They primarily run longitudinally and provide for stomach expansion, flattening as the stomach fills.
  b. The **gastric mucosa** contains small permanent furrows, **gastric sulci,** that divide the surface into **gastric (mamillated) areas** that do not flatten with stomach distention. **Gastric glands** (pits) open onto the surface of the gastric areas.
  c. **Muscularis.** Three muscle layers constitute the muscularis externa, but their predominant directions are difficult to characterize.

4. **Vasculature**
  a. **Arterial supply** (see Figure 25-2)
    (1) The **celiac artery** supplies the stomach through its three branches.
      (a) Two of three branches of the **common hepatic artery** supply the stomach (see Figure 25-2).
        (i) The **proper hepatic artery** usually gives rise to the **right gastric artery,** which supplies the inferior region of the lesser curvature (pylorus).
        (ii) The gastroduodenal artery gives rise to the **right gastro-omental artery,** which supplies the inferior portion of the greater curvature (pylorus and body).
      (b) The **left gastric artery** supplies the superior regions of the lesser curvature (body, cardia, and fundus) as well as the lower esophagus (see Figure 25-2).

(i) Its terminal branch courses along the lesser curvature of the stomach to anastomose with the right gastric artery.

(ii) An **esophageal branch** supplies the abdominal segment of that organ and anastomoses with the thoracic esophageal supply.

(iii) It occasionally (up to 35%) gives rise to an **aberrant left hepatic artery.**

(c) The **splenic artery** gives rise to two branches that supply the stomach and numerous branches that supply the spleen (see Figures 25-2 and 25-8).

(i) **Short gastric arteries** supply the left superior portions of the greater curvature (fundus and body).

(ii) The **left gastro-omental (gastroepiploic) artery** supplies the left portion of the greater curvature (body).

(2) The **collateral blood supply** to the body and fundus of the stomach is rich in extramural as well as intramural anastomoses. Any of the branches along the stomach may be ligated with little risk of ischemia and necrosis. The same is not true, however, at the pylorus.

(a) Anastomoses between the left gastric artery and the right gastric, esophageal, and splenic arteries as well as the gastro-omental arteries are constant and abundant.

(b) Anastomoses between gastric and duodenal arteries are scant, resulting in the so-called "**bloodless line**" at the pyloroduodenal junction.

(c) Because of the insecure blood supply, branches to this region cannot be ligated without risk of ischemia and necrosis. Therefore, the superior duodenum is removed when the distal portion of the stomach is resected.

b. Venous return (see Figure 25-15)

(1) **Course.** The veins of the stomach parallel the arterial supply and are similarly named, until they diverge from the arteries significantly to join the **hepatic portal system.**

(2) **Portal–systemic anastomoses**

(a) Anastomoses between the **left gastric (coronary) vein** and the **esophageal branches of the azygos or hemiazygos veins** are important shunts, but they may lead to life-threatening **esophageal varices.**

(b) Anastomoses between the esophageal veins and both the left gastroepiploic vein and short gastric veins may also become engorged if the splenic vein becomes occluded, such as by **pancreatic carcinoma.**

c. Lymphatic drainage pathways from the stomach follow the arteries and are so named (left and right gastric nodes, left and right gastroepiploic nodes), although alternative names are frequently used.

5. **Innervation** is by the autonomic nervous system.

a. **Parasympathetic innervation** is by the vagus nerve (see section VI). Parasympathetic presynaptic neurons that leave the vagus nerve synapse with the postsynaptic neurons in ganglia in the walls of the viscera.

b. **Sympathetic innervation** involves the greater splanchnic nerves (T5–T9) and celiac ganglia (see section VI).

(1) Sympathetic presynaptic neurons from spinal levels T5–T9 reach the sympathetic chain via an associated white ramus communicans and emerge from the chain to the **greater splanchnic nerve** that terminates in the **celiac ganglion.**

(2) The postsynaptic sympathetic axons leave the celiac ganglion and course along the perivascular adventitia of the celiac trunk as the **celiac plexus** along vessels to the organ to be innervated.

c. **Afferent innervation**

(1) **Pain pathways.** Painful sensation from the stomach is mediated via that portion of the greater splanchnic nerve associated with spinal segments T5 and T6. Stomach pain is referred to the **epigastric region.** Differentiating the pain of coronary artery insufficiency (T1–T4) from that of gastric origin (T5–T6) is critical.

(2) Reflex pathways. The vagus nerves carry reflex afferents to the brain that inhibit ingestion when the stomach expands excessively.
6. **Functions** of the stomach include storage, trituration, formation of chyme, acid enzymatic digestion, and some absorption.
    a. The **upper half of the stomach** serves as a reservoir for ingested food and expands passively as it is filled.
        (1) **Muscular activity.** Persistent tonic contracture around the cardia and fundus aids in the establishment of the cardiac sphincter and delivers food to the lower and more motile regions of the stomach.
        (2) **Secretion.** Boluses of food are bathed in gastric juice, which converts the surface into a liquid mixture, **chyme.**
            (a) **Hydrochloric acid** is secreted by gastric glands in both the fundus and the body of the stomach.
            (b) **Pepsin,** an enzyme secreted by gastric glands and activated at a low pH, converts proteins to polypeptides.
    b. In the **lower half of the stomach,** chyme is thoroughly mixed with the gastric secretions for trituration.
        (1) **Muscular activity.** Muscle layers increase in thickness from the cardia to the pylorus, and peristaltic activity increases in intensity along this gradient.
        (2) **Control of secretion.** Although the pylorus is devoid of acid-secreting cells, it contains most of the secretin-producing cells. This hormone influences acid secretion in the rest of the stomach.
    c. Factors controlling **motility** are intrinsic and extrinsic.
        (1) **Intrinsic rhythm of enteric neurons** regulates gastric peristalsis.
            (a) **Peristaltic waves,** which occur at about 20-second intervals, begin as ringlike contractions about midway in the body of the stomach. They propagate toward the pylorus with increasing vigor.
            (b) Once formed, chyme is moved by peristaltic activity toward and through the pylorus.
        (2) **Extrinsic factors.** After an initial adjustment period, the stomach empties its contents at a relatively constant rate, which is modified by nervous and hormonal influence.
            (a) Vagal activity accelerates peristalsis.
            (b) The **enterogastric reflex** releases the hormone **enterogastrone,** which inhibits gastric peristalsis. This reflex is initiated by introduction of acid chyme into the duodenum.
    d. **Absorption.** Some substances, such as alcohol, enter directly through the gastric mucosa.
7. **Clinical considerations**
    a. **Gastritis.** Excessive vagal activity may produce gastritis by excessively stimulating secretin production that activates acid-secreting glands. Substances that irritate the gastric mucosa, such as aspirin and steroids, also produce gastritis.
    b. **Peptic ulceration.** A variety of factors lead to the development of peptic ulcers. They occur in the nonacid-secreting regions of the upper GI tract, such as the duodenum (primarily), the antral region of the stomach along the lesser curvature (less frequently), and the lower esophagus (rarely).
        (1) **Complications.** Ulcers may produce severe bleeding, obstruction from edema or scarring, and peritonitis from perforation. Erosion into a subjacent blood vessel results in massive hematemesis or intra-abdominal bleeding with complicating peritonitis.
        (2) **Pain.** Pain from ulceration of the lower esophagus, stomach, or superior duodenum is referred to the fifth and sixth dermatomes, which include the epigastric region. Perforation of the stomach causes leakage of gastric contents, which leads to peritonitis. The peritoneum is innervated by twigs of the intercostal and lumbar nerves, so intense peritoneal pain is perceived at the location of irritation.
        (3) **Treatment.** Selective vagotomy (section of the gastric branches of the anterior and posterior vagal trunks) was used to reduce peptic secretion. Currently, drugs that block acid secretion usually are effective.

c. **Hemigastrectomy**
   (1) **Procedure.** Surgical treatment of duodenal and gastric ulceration frequently involves resection of approximately one third of the distal stomach and, because collateral blood supply is inconsistent, the superior duodenum and associated pancreas (Billroth II gastrectomy).
   (2) **Rationale.** Although the distal portion of the stomach produces little acid, it does produce most of the **secretin** that induces acid secretion in the fundus and body. Resection of the lower portion not only removes the ulcers, but also reduces acid secretion in the upper regions and preserves the storage capacity of the stomach.
d. **Malignant metastases.** Because of the venous and lymphatic drainage of the stomach, a malignant process at this site can spread to other organs and regions.

C. **Duodenum**
1. **Structure.** The duodenum is the first portion of the small intestine. It shares several characteristics with the jejunum and ileum. Differences occur by transitions over considerable distances.
   a. **Length and position.** The duodenum (L. twelve) is the shortest portion of the small intestine, about 12 finger breadths or 25 cm (10 inches) from the pyloric sphincter to the duodenojejunal flexure.
      (1) Although its position and size are variable, it loops in a C-shape to the right (see Figure 23-1B).
      (2) Except for the superior duodenum, which is supported by the **hepatoduodenal ligament,** the duodenum is **secondarily retroperitoneal.**
   b. **Divisions.** It is divided into four parts (Figure 25-3).
      (1) The **superior duodenum,** the first part, is 3–5 cm (1.5–2 inches) long. It receives chyme through the pylorus.
         (a) **Support.** This segment is **peritoneal,** supported by the **hepatoduodenal ligament,** which is part of the ventral mesentery and lesser omentum.
         (b) **Internal structure.** The mucosa of the superior duodenum lacks circular folds and, because of its characteristic smooth appearance on radiographs, it is termed the **duodenal cap.**

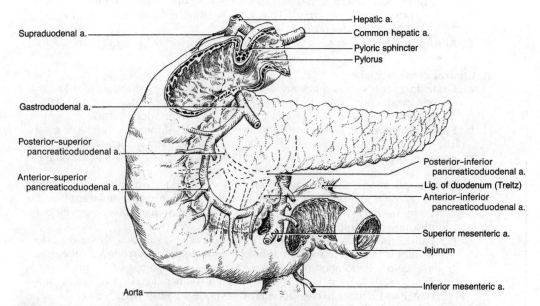

FIGURE 25-3. Duodenum and pancreas. The window in the superior portion of the duodenum shows the pyloric sphincter and beginnings of the plica circularis. The window at the end shows the well-developed plica circularis. The head of the pancreas is intimately associated with the duodenum, sharing a common blood supply that arises from both celiac and superior mesenteric arteries.

(2) The **descending duodenum,** the second part, is 8–10 cm (3.5–4 inches) long.
   (a) It is **secondarily retroperitoneal.**
   (b) Its **internal structure** promotes absorption. Circular folds of mucosa **(plicae circulares)** make their appearance, increasing in size and complexity toward the jejunum.
   (c) The **hepatopancreatic ampulla** and the **secondary pancreatic duct,** if present, enter this segment of the duodenum.
      (i) The common bile duct and the primary pancreatic duct usually join to form the **hepatopancreatic ampulla,** but variations are common.
      (ii) The **duodenal papilla** (of Vater) is a small protrusion where the hepatopancreatic ampulla enters the duodenum.
      (iii) The **hepatopancreatic sphincter** (of Oddi) surrounds the hepatopancreatic ampulla.
(3) **Inferior duodenum**
   (a) The **transverse portion,** the third part, is 2.5–5 cm (1–2 inches) long.
      (i) This horizontal segment is **secondarily retroperitoneal.**
      (ii) As a result of intestinal rotation during development, the **superior mesenteric vessels** are located anterior to the transverse portion. With wasting diseases or after severe dieting, the superior mesenteric vascular bundle may compress the underlying duodenum and produce intestinal obstruction, **superior mesenteric artery syndrome.**
   (b) The **ascending portion,** the fourth part, is 2.5–5 cm long.
      (i) This segment, also **secondarily retroperitoneal,** terminates at the **duodenojejunal flexure.**
      (ii) The **suspensory ligament (of Treitz)** secures the end of the ascending portion (see Figure 25-3). It runs to the right crus of the diaphragm and is composed of a tendon insinuated between slips of smooth muscle from the duodenal wall and fascicles of striated muscle from the diaphragmatic crus. This ligament is a palpable surgical landmark.
      (iii) At the **duodenojejunal flexure,** the small intestine becomes the peritoneal jejunum, supported by the **mesentery proper.**

2. **Function**
   a. As the terminal portion of the foregut, **digestion** is still the primary function of the duodenum.
      (1) Chyme is mixed with the secretory products of the liver and pancreas as well as with enzymes secreted by the duodenum.
      (2) The superior duodenum is relatively quiet. The activity of the descending duodenum is similar to a churning or milling action. True peristaltic activity becomes evident in the distal duodenum.
   b. The duodenum **regulates stomach and gallbladder emptying.** In response to receiving acid chyme, it secretes the hormone **enterogastrone,** which inhibits stomach peristalsis. In response to fatty chyme, it secretes **cholecystokinin,** which induces gallbladder contraction.

3. **Vasculature**
   a. **Arterial supply** to the duodenum is shared with the pancreas (see Figure 25-3).
      (1) The **celiac artery** supplies the proximal duodenum via the **right gastric** and **gastroduodenal arteries.**
         (a) **Blood supply to the first (superior) portion** is sparse, and almost precarious with few, if any, extramural anastomoses between the branches of the right gastric and gastroduodenal arteries.
            (i) The **supraduodenal artery** arises from the gastroduodenal artery, but it may arise from the right gastric artery or common hepatic artery. It is absent in 30% of individuals.
            (ii) **Retroduodenal arteries** vary in number and location. They may arise from the gastroduodenal artery, the right gastroepiploic artery, superior pancreatic artery, or any combination thereof.

(iii) Because of the insecure blood supply, the distal portion of the stomach is also removed when the superior duodenum is resected.
- (b) The **blood supply to the descending and inferior portions** is profuse. The **superior pancreaticoduodenal artery** (a branch of the gastroduodenal artery) divides into anterior and posterior branches.
- (2) The **superior mesenteric artery** supplies the distal duodenum via the **inferior pancreaticoduodenal artery.**
  - (a) The **inferior pancreaticoduodenal artery** (a branch of the superior mesenteric artery) also divides into anterior and posterior branches.
  - (b) Arcades from the anterior and posterior–superior pancreaticoduodenal arteries anastomose with the anterior and posterior branches of the inferior pancreaticoduodenal artery.
- b. **Venous return**
  - (1) The **proximal duodenum** drains into the **hepatic portal vein** via the right gastric, gastroduodenal, and superior pancreaticoduodenal veins.
  - (2) The **distal duodenum** drains into the **superior mesenteric vein** via the inferior pancreaticoduodenal vein.
  - (3) Because most of the duodenum is secondarily retroperitoneal, there are variable **transperitoneal anastomotic connections** (veins of Retzius) with the systemic veins of the body wall.
- c. **Lymphatic drainage** is to the **celiac** and the **superior mesenteric nodes,** as well as to retroperitoneal nodes of the dorsal body wall. These groups of nodes drain into the **cisterna chyli.**

4. **Innervation**
   a. **Parasympathetic innervation** is by the **vagus nerve** via the celiac plexus (see section VI).
   b. **Sympathetic innervation** is by the **greater splanchnic nerves, celiac ganglia,** and celiac plexus (see section VI).
   c. **Afferent innervation**
   - (1) **Sensory pathways.** Afferent fibers mediating painful sensation course along the greater splanchnic nerve to the sympathetic chain. The afferents then travel along a white ramus communicans to the respective spinal nerves and associated spinal segment.
   - (2) **Referred pain**
     - (a) Painful sensation from the superior and descending duodenum is mediated via that portion of the greater splanchnic nerve associated with spinal segments T7 and T9 **(epigastric region).**
     - (b) The differential diagnosis involving pain from the foregut and that from the midgut can be difficult. Often, the patient is observed over a few hours to see where peritonitis develops.

5. **Clinical considerations**
   a. **Mobilization of the duodenum.** Because the blood supply is from the medial side, incision of the peritoneum along the right edge of the descending duodenum mobilizes this viscus as well as the head of the pancreas. This action re-establishes the former dorsal mesentery with its contained blood supply.
   b. **Metastatic routes.** Because the venous and lymphatic channels may anastomose with those of the dorsal body wall, carcinoma of the duodenum and pancreas frequently is associated with a poor prognosis.
   c. **Duodenal ulcers.** Peptic ulceration in the duodenal cap is four times more frequent than gastric ulcers. Chronic duodenal ulcers may require Billroth I gastroduodenal resection.

D. Liver

1. **Developmental considerations**
   a. The **hepatic diverticulum** penetrates the ventral mesogastrium during the third week (see Figure 24-4).
   b. By the tenth week, the liver has assumed transient hematopoietic function.

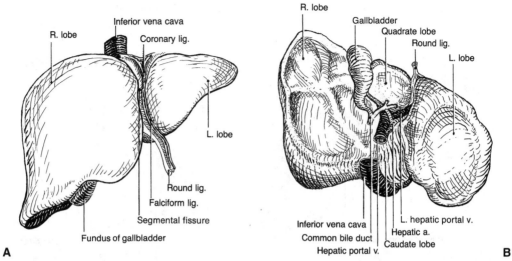

FIGURE 25-4. Liver. *(A)* Anterior surface shows the right and left lobes separated by the falciform ligament. *(B)* Inferior surface shows the right, quadrate, caudate, and left lobes, as well as the gallbladder located between the right and quadrate lobes.

      c. The developing liver is an important metabolic center in the umbilical venous channel and represents 10% of the embryonic weight.
- **2. Mesenteric attachments**
    - **a.** The **ventral mesogastrium,** within which the liver develops, is divided by the liver into two principal ligaments (see Figure 24-5C).
        - **(1)** The **lesser omentum** runs between the liver and the lesser curvature of the stomach (**gastrohepatic ligament**) as well as the first part of the duodenum (**hepatoduodenal ligament).**
        - **(2)** The **falciform ligament** is a continuation of the primitive ventral mesentery (see Figure 24-6). It runs from the anterior surface of the liver to the ventral body wall as far inferiorly as the umbilicus.
            - **(a)** The **round ligament (ligamentum teres) of the liver,** which is the remnant of the obliterated umbilical vein, is in the inferior free edge of this ligament (Figure 25-4A).
            - **(b)** Cannulation through the umbilicus can re-establish round ligament patency to measure portal venous pressure or to sample portal blood.
    - **b.** The **coronary ligaments** are formed by reflection of ventral mesentery between the superior pole of the liver and the inferior surface of the diaphragm (Figure 25-5).
        - **(1)** The **bare area** formed as the liver enlarged, forcing the two leaflets of the ventral mesogastrium apart superiorly (see Figure 25-5). Here, the tissues of the liver and diaphragm are in direct contact.
            - **(a)** Boundaries of the bare area on the left side are the **falciform ligament, the left anterior leaf of the coronary ligament,** the **left triangular ligament** (where the left anterior and posterior coronary leaflets unite), the **posterior left leaf of the coronary ligament,** and the gastrohepatic ligament (a portion of the lesser omentum).
            - **(b)** The boundaries are complementary on the right side.
        - **(2)** The **hepatic veins** enter the **inferior vena cava** as the latter passes through the bare area.
- **3. Position and relationships**
    - **a. Right hypochondriac and epigastric regions** are largely occupied by liver (see Figure 23-1).

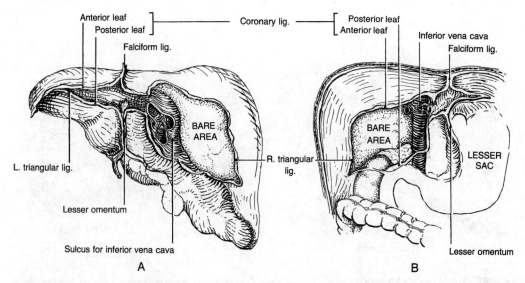

FIGURE 25-5. *(A)* The bare area of the liver, bounded by the coronary and triangular ligaments. *(B)* The corresponding bare area of the inferior diaphragmatic surface, where the ligaments become parietal peritoneum.

- (1) The right lobe extends from the diaphragm to just beneath the right costal margin, where it may be palpable, depending on its consistency.
- (2) With the fingers placed adjacent to the costal margin just lateral to the rectus abdominis muscle, an examiner may feel the liver as it slides inferiorly beneath the fingertips as the patient inhales.
- b. **Subphrenic (suprahepatic) recesses** (see Chapter 24 III B 2 a and Figure 24-7).
- c. **Infrahepatic recess** (see Chapter 24 II B 2 b)
- d. **Right subhepatic and hepatorenal recesses** (see Chapter 24 III B 2 c 1, 3 and Figure 24-7).

4. **Structure.** Morphologically, the liver appears to be one of the simplest organs. Functionally, it is one of the most complex.
   a. A **compound tubular gland,** the liver is the largest gland in the body.
      (1) It is enclosed in a thin, fibrous **hepatic capsule** (of Glisson) that lies just beneath the visceral peritoneum. From this capsule, septa project inward into the hepatic parenchyma.
      (2) Pain associated with transient venous congestion (**"runner's stitch"**) as well as acute hepatomegaly result from sudden stretching of Glisson's capsule.
   b. **Lobation**
      (1) **Anatomic lobation.** The liver has four lobes: a large right lobe and a smaller left lobe (see Figure 25-4), as well as two rudimentary lobes, the quadrate lobe (also called the medial segment of the right lobe) and the caudate lobe.
      (2) **Functional lobation.** Functionally, the liver is almost evenly divided.
         (a) **Right and left lobes** have separate biliary drainage and vascular supplies.
         (b) The **quadrate lobe** is functionally part of the left lobe because it receives blood from the left hepatic artery and drains the hepatic duct.
         (c) The **caudate lobe,** receiving blood from the left and right hepatic arteries and secreting bile into both left and right hepatic ducts, is functionally part of both the left and right lobes.

5. **Vasculature**
   a. **Arterial supply.** The liver is an extremely vascular organ, bleeding profusely when ruptured by blunt trauma or sectioned in surgery.
      (1) The **common hepatic artery,** one of the three branches of the celiac artery, becomes the proper hepatic artery distal to the gastroduodenal artery (see Figure 25-2).

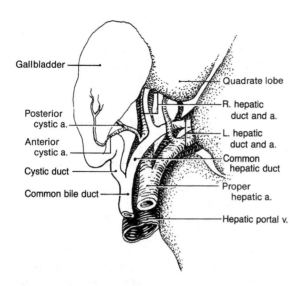

FIGURE 25-6. Hepatic pedicle. Most of the vascular supply to the liver and gallbladder passes through the cystohepatic triangle.

- **(2)** The **proper hepatic artery** ascends in the hepatoduodenal ligament and divides into the left and right hepatic arteries (Figure 25-6).
    - **(a)** The **right hepatic artery** supplies the right lobe and the right half of the caudate lobe as well as the gallbladder.
    - **(b)** The **left hepatic artery** supplies the left and quadrate lobes as well as the left half of the caudate lobe.
- **(3) Extrahepatic variation**
    - **(a)** The usual textbook representation of the hepatic blood supply is uncommon; 60–70% of individuals have some variation.
        - **(i)** The level of bifurcation into left and right hepatic arteries is variable.
        - **(ii)** The right hepatic artery may arise from the superior mesenteric artery via the inferior pancreaticoduodenal artery (15%).
        - **(iii)** The left hepatic artery may arise from the left gastric artery (25%).
        - **(iv)** The incidence of covariation with these two anomalies is high.
    - **(b)** Failure to appreciate variations in the right hepatic and cystic arteries may produce unexpected hemorrhage intraoperatively. Such hemorrhage may be controlled by compressing the hepatic pedicle between the thumb and a forefinger inserted into the omental foramen.
- **(4) Extrahepatic occlusion.** Hepatic and accessory hepatic arteries are not collateral arteries because they supply discrete regions of the liver parenchyma.
    - **(a)** Sudden occlusion of these vessels can produce ischemic necrosis of a region of the liver.
    - **(b)** Depending on location, ligation of the hepatic blood supply may be tolerated. For example, the common hepatic artery may be permanently ligated because ample anastomotic connections among vessels supply the stomach from the splenic artery and the terminal branches of the common hepatic artery. The proper hepatic artery, however, may be permanently ligated only proximal to the right gastric artery.
- **b. Venous "supply"** is by the **hepatic portal vein** (see Figures 25-4B, 25-6, and 25-15). A portal system is a venous system that begins as venous capillaries (such as in the stomach and intestine) and terminates in venous sinusoids.
- **c. Venous drainage** usually is by three **hepatic veins,** which drain directly into the inferior vena cava, usually in or about the bare area.

6. Function. The liver is primarily a metabolic center.
- **a. Intrahepatic perfusion.** The liver has a uniform structure, consisting of anastomosing sheets of cells separated by intervening venous sinusoids. All venous blood from the GI tract percolates through the liver sinusoids.

(1) Liver cells are supplied by blood from the hepatic portal vein (nutrient-rich, unoxygenated, sometimes toxic) and the hepatic artery (oxygenated).
(2) Because of this uncommon vascular arrangement, a lower than usual oxygen content bathes the liver cells, which may explain, in part, the high susceptibility of the liver to damage and disease.
(3) The sinusoids drain into the central veins of the lobules and ultimately into the **hepatic veins,** which join the inferior vena cava.

b. **Bile secretion.** The liver secretes as much as 1 liter of bile a day into the biliary tree. Among the constituents of bile are bile pigments and bile salts.
(1) **Bile pigments,** including bilirubin and biliverdin, are derived from the breakdown of hemoglobin in the spleen and liver. They produce the distinctive color of feces.
(2) **Bile salts,** formed from cholesterol, aid in digestion by emulsification of fats, thereby facilitating their absorption by the intestinal mucosa. Most bile salts are reabsorbed by the mucosa and returned to the liver by the portal veins for resection.

7. **Clinical considerations**
a. **Hepatic resection.** Although the liver is a vital organ, portions of it may be surgically excised without irreparable damage to the body. Such resections are based on functional lobation. The liver has tremendous regenerative power, which can be detrimental when undirected or impeded by scarring, such as in cirrhosis.
b. **Biliary overload and obstruction. Icterus (jaundice)** occurs either when the liver overloads beyond its capacity, such as by excessive hemolysis of blood or liver damage, or when the biliary tree is obstructed. In both instances, bile pigments accumulate in the blood, giving the skin and eyes a characteristic yellow hue, accompanied by itching.

E. Biliary tree and gallbladder

1. **The hepatic tree** is the excretory duct of the liver (see Figure 25-7).
a. The **intrahepatic portion** is formed by sheets of hepatocytes that secrete bile into bile canaliculi.
(1) **Bile canaliculi,** which traverse the sheets of hepatocytes, unite to form intrahepatic bile ducts. These ducts lie in portal triads along with branches of the hepatic portal veins and hepatic arteries.
(2) **Bile ductules** converge to form the left and right hepatic ducts. This pattern defines the functional segments of the liver.
b. **External structure** (Figure 25-7)
(1) **Right** and **left hepatic ducts** leave the respective lobes of the liver and join to form the common hepatic duct.
(a) The **left hepatic duct** drains the left and quadrate lobes of the liver as well as the left side of the caudate lobe.
(b) The **right hepatic duct** drains the right lobe and the right half of the caudate lobe.
(2) The **common hepatic duct** and the cystic duct converge to form the common bile duct. The patterns and location of this juncture are variable.
(3) The **cystic duct** to the gallbladder arises at the distal end of the common hepatic duct.
(a) Its length is variable, but usually it is 1–2 cm long.
(b) A spiral fold (valve of Heister) maintains its patency.
(c) Its course is variable.

(4) The **common bile duct** drains into the duodenum.
(a) It contributes to the **hepatic pedicle,** which lies in the free edge of the hepatoduodenal ligament and forms the anterior boundary of the omental foramen.
(i) The **common bile duct** lies anterior to the right within the hepatic pedicle.
(ii) The **hepatic artery** lies anterior to the left.

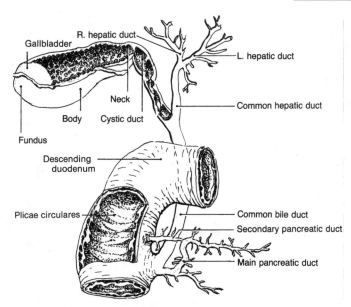

FIGURE 25-7. Biliary tree, gallbladder, and descending duodenum. Divisions of the biliary tree as well as the pancreatic drainage system are indicated.

        (iii) The **hepatic portal vein** lies posteriorly.
    (b) **Course**
        (i) The common bile duct passes posterior to the superior and descending duodenal segments. It is embedded in pancreatic parenchyma, but it is easily dissected free.
        (ii) It enters the descending duodenum about 3–8 cm beyond the pyloric sphincter. This intramural portion is its narrowest part.
    (c) It forms the **duodenal papilla** (of Vater), projecting into the descending duodenum (Figure 25-8A).
        (i) The duodenal papilla usually contains the **hepatopancreatic ampulla,** formed by the joining of the common bile duct and the pancreatic duct.
        (ii) The hepatopancreatic ampulla is variable (see Figure 25-8B and C). In about 65% of individuals, bile from the common duct and pancreatic juice from the pancreatic duct are mixed in a true hepatopancreatic ampulla. The two ducts may, however, open into the duodenal lumen separately, each with its papilla.
        (iii) **Autolytic pancreatitis.** Blockage of the hepatopancreatic ampulla by a bile calculus, muscle spasm, or edema may precipitate pancreatitis as a result of bile reflux along the pancreatic duct. Pancreatic enzymes are usually inactive until they reach the alkali duodenum. Refluxed bile can activate the proteolytic enzymes within the pancreas, however, producing autodigestion of the pancreas.
    (d) The **hepatopancreatic sphincter** (of Oddi) normally keeps the ostium of the hepatopancreatic ampulla closed. It is divided into three parts.
        (i) The **sphincter choledochus** is a well-developed portion around the termination of the common bile duct.
        (ii) The **papillary sphincter** surrounds the hepatopancreatic ampulla.
        (iii) The **pancreatic sphincter** is poorly developed around the termination of the pancreatic duct.
  c. Arterial supply
    (1) The common bile duct is supplied by a rich anastomotic network from the right hepatic, proper hepatic, retroduodenal, and posterior pancreaticoduodenal arteries.

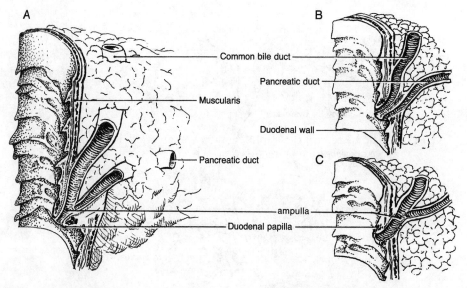

FIGURE 25-8. Duodenal papilla. *(A)* The usual configuration whereby the common bile duct and the primary pancreatic duct join to enter the duodenum through the hepatopancreatic duct. *(B)* The common duct and the primary pancreatic duct may enter the duodenum separately. *(C)* A more proximal joining of the biliary and pancreatic ducts produces an unusually long hepatopancreatic ampulla.

        (2) The twigs, variable and inconstant, course in the adventitia of the bile duct.
        (3) During surgical manipulation of the common duct, care is needed to avoid isolation of the blood supply, with resultant ischemia and infarction.
    d. Innervation
        (1) Parasympathetic innervation is by the vagus nerve via the celiac plexus. Cholinergic activity promotes gallbladder activity and relaxation of the hepatopancreatic sphincter.
        (2) The **afferent pathway** reaches levels T6–T8 of the spinal cord via visceral afferent fibers that travel with the greater splanchnic nerves.
        (3) **Biliary pain** is referred to the T6–T8 dermatomes within the upper quadrants of the abdomen.

2. Gallbladder
    a. Position and relationships. The gallbladder, which concentrates and stores bile, is 6–10 cm long with a capacity of about 45 ml. It is lies in a shallow fossa on the inferior hepatic surface between the right and quadrate lobes.
        (1) **Support.** The gallbladder may be embedded within the liver parenchyma (rare). It may be surrounded by peritoneum and have a short mesentery to the inferior surface of the liver (floating). Usually, it is adherent to the liver.
            (a) It lies adjacent to the superior duodenum and to the transverse colon.
            (b) It usually is not palpable. If it contains stones, it may be felt in the right upper quadrant at the angle between the costal margin and the rectus abdominis muscle.
        (2) The **cystohepatic triangle** (of Calot) lies between the liver, common hepatic duct, and cystic duct or gallbladder (see Figure 25-6). The boundaries include the **common hepatic duct** (left), **cystic duct and gallbladder** (right), and **liver** (superiorly).
    b. External structure. The gallbladder is divided into three parts: the **fundus,** the **body,** and the **neck** (see Figure 25-7).
        (1) The neck may contain a pendulous sacculation **(Hartmann's pouch).**
        (2) During surgery, gentle traction on Hartmann's pouch causes the cystic duct to stand out prominently. Hartmann's pouch may, however, be adherent to

the cystic duct, or even to the common bile duct, and may accidentally be included in a ligature passed around the cystic duct.
- c. **Vasculature** (see Figure 25-6)
  - (1) The **cystic artery** normally arises as a single stem from the right hepatic artery within the **cystohepatic triangle.**
    - (a) It subsequently bifurcates into **anterior** and **posterior cystic arteries.**
    - (b) Cystic arteries may arise from the right, left, or even proper hepatic arteries in any combination. The branches anastomose and even supply a portion of adjacent liver.
  - (2) **Cystic veins** may drain into the right branch of the hepatic portal vein or, more usually, drain directly into the liver.
- d. **Innervation**
  - (1) Parasympathetic supply is from the vagus nerve via the celiac plexus.
  - (2) Sympathetic supply is by the T6–T8 portion of the greater splanchnic nerves via the celiac ganglion and celiac plexus.
  - (3) **Visceral afferents** from the gallbladder reach levels T6–T8 of the spinal cord via visceral afferent fibers that travel with the greater splanchnic nerves. Biliary pain is referred to the T6–T8 dermatomes within the upper quadrants and epigastric region of the abdomen.
- e. **Function**
  - (1) **Storage of bile.** The sphincter choledochus and papillary sphincter normally are closed so that bile, secreted continuously by the liver, refluxes through the cystic duct into the gallbladder.
  - (2) **Bile concentration.** The gallbladder concentrates bile by mucosal absorption of electrolytes and water.
    - (a) The ability of the gallbladder to concentrate bile forms the basis for radiographic visualization of the biliary tree (cholangiography).
    - (b) When iodinated compounds are administered either orally or intravenously, they are excreted by the liver. As the bile is concentrated in the gallbladder, the amount of radiopaque dye increases so that the entire organ, and often the biliary tree, is visualized.
  - (3) **Control mechanism.** Biliary release is controlled by **cholecystokinin.**
    - (a) A spurt of fat through the pylorus into the duodenum causes release of cholecystokinin by intestinal glands in the duodenal mucosa.
    - (b) This hormone induces contraction of gallbladder musculature, causing the fundus to rise.
    - (c) The rise in biliary pressure overcomes the sphincter choledochus of the lower biliary tree, in which the muscular tone may have been decreased slightly by vagal activity, and bile spurts into the duodenum.
- f. **Clinical considerations**
  - (1) **Duct variation.** Numerous variations in the biliary tree make surgery challenging.
    - (a) Although procedures on the biliary tree are routine, complications in cholecystectomy (removal) and cholecystostomy (drainage) occur frequently.
    - (b) Inadvertent injury to the biliary tree with leakage of bile into the peritoneal cavity produces **bile peritonitis.** Subsequent edema or scar tissue may cause obstruction with pain and jaundice.
  - (2) **Arterial variation** within the **cystohepatic triangle** (of Calot) constitutes a major hazard in biliary surgery.
    - (a) The cystohepatic triangle contains most structures of importance in cholecystectomy (90% of all cystic arteries, 82% of all right hepatic arteries, 95% of all accessory right hepatic arteries, and 91% of all accessory bile ducts).
    - (b) Structures in the cystohepatic triangle and in the hepatoduodenal ligament should be neither clamped nor divided until dissected free and positively identified.
  - (3) **Gallstones (choleliths).** Overconcentration of bile, perhaps related to biliary stasis, results in the precipitation of bile salts and pigments, leading to concretions.

- (a) **Composition and size**
  - (i) They may be purely of cholesterol, which may calcify and become visible in an ordinary radiograph. They may be of purely bile pigments, which appear as filling defects in a cholangiogram. They may be mixed in concentric calcified and noncalcified layers.
  - (ii) In some women, it is possible to find two or more populations of similarly sized stones, each population representing a different pregnancy.
- (b) **Predisposition.** Gallstones are most common in overweight, multiparous women 40 years of age or older.
- (c) **Cholecystitis.** Whereas small stones may pass through the ducts without incident, those that are too large to enter the cystic duct may irritate the cystic mucosa (**cholecystitis**). Sequelae include possible ulceration and erosion of the gallbladder wall with subsequent peritonitis.
- (d) **Biliary obstruction.** Stones of appropriate size may enter the duct system and lodge at points of narrowing, such as where the duct penetrates the duodenal wall, the sphincter choledochus, and the ostium of the duodenal papilla.
  - (i) **Cholecystitis without jaundice** occurs if the stone lodges in the cystic duct, because bile produced by the liver drains into the duodenum.
  - (ii) **Cholecystitis with jaundice** occurs if the stone lodges in the common bile duct, preventing bile excretion.
- (e) **Biliary pain.** Biliary obstruction is especially painful when fatty chyme elicits the release of cholecystokinin, which elevates biliary pressure (**"gallbladder attack"**).
  - (i) **Referred pain.** Biliary pain is referred to the T6–T8 dermatomes within the upper quadrants of the abdomen. The afferent pathway reaches levels T6–T8 of the spinal cord via visceral afferent fibers that travel with the **greater splanchnic nerves.**
  - (ii) **Peritonitis.** When the somatically innervated parietal peritoneum adjacent to the gallbladder is involved, pain is localized in the right upper quadrant (right hypochondriac region).
- (4) **Anomalies** of the gallbladder include diverticula, duplication, and congenital absence, as well as ductal duplication, atresia, and stenosis.

F. Pancreas

1. **Developmental considerations.** The pancreas develops from two intestinal diverticula.
   a. The dual origin of the pancreas accounts for primary and secondary pancreatic ducts.
   b. Because of rotation of the gut, the head of the pancreas lies on the right side and becomes **secondarily retroperitoneal** as the duodenum fuses to the dorsal wall peritoneum.

2. **External structure**
   a. **Location.** The pancreas lies in the epigastric and left hypochondriac regions. The anterior surface is covered by the posterior wall of the lesser sac (omental bursa). The posterior surface is adherent to the parietal peritoneum of the dorsal body wall.
      (1) Incising the fascia of Toldt and re-establishing the mesentery mobilize the duodenum and pancreas while maintaining the body supply.
      (2) Because of its deep position, the pancreas is difficult to palpate.
   b. **Divisions** (see Figure 25-3)
      (1) The **head** (secondarily retroperitoneal) lies within the curve of the duodenum.
         (a) The intimate association between the head of the pancreas and the duodenum results from the shared mesentery and the shared **superior and inferior pancreaticoduodenal arteries.**

(b) The uncinate process is a portion of the head that projects from the left behind the superior mesenteric vessels.

(2) The **body** (secondarily retroperitoneal) lies posterior to the stomach. A perforated gastric ulcer may result in adhesions between the stomach and pancreas with subsequent erosion into the pancreatic parenchyma and resultant pancreatitis. **Hemorrhagic pancreatitis** may be rapidly fatal.

(3) The **tail** (peritoneal) projects into the **splenorenal ligament** toward the hilus of the spleen.

   (a) The splenic artery, usually embedded in the parenchyma of the tail, gives off a variable **great pancreatic artery** and often several **caudal pancreatic arteries.**

   (b) It contains the majority of the **pancreatic islets,** which is a major consideration in pancreatic resection.

3. **Internal structure.** Two pancreatic ducts drain the pancreas (Figure 25-9).
   a. The **primary pancreatic duct** is formed during development by the joining of the distal portion of the primitive dorsal duct to the proximal portion of the primitive ventral duct.
      (1) The primary duct courses the entire length of the pancreas.
      (2) It usually joins the common bile duct, which also develops in the ventral mesentery, to form the **hepatopancreatic ampulla (duodenal papilla)** [see section E 1 b (4) (c)].
   b. The **accessory (secondary) pancreatic duct** represents the proximal portion of the primitive dorsal duct.
      (1) It drains a small portion of the head and body.
      (2) It usually enters the duodenum separately at the **accessory duodenal papilla** about 2 cm above the primary duodenal papilla.
   c. **Ductal variations** (see Figure 25-8B,C)
      (1) The embryonic pattern persists in about 10% of individuals.
      (2) The distal end of the secondary duct may remain connected to the primary duct.

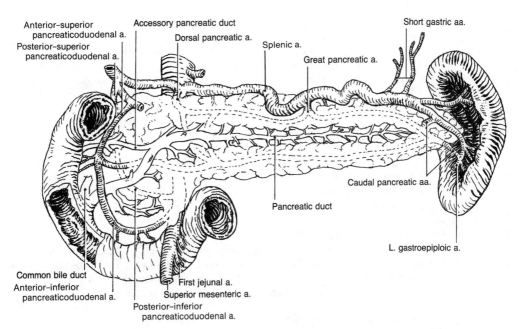

FIGURE 25-9. Pancreas and spleen. The primary and secondary (if present) pancreatic ducts empty into the descending portion of the duodenum. The pancreas receives blood supply from the branches of the celiac and superior mesenteric arteries. The spleen receives profuse supply from the splenic branch of the celiac artery.

(3) The primary pancreatic duct may not fuse with the common bile duct so that each opens separately into the duodenum.

4. **Vasculature**
   a. **Arterial supply** (see Figures 25-3 and 25-9)
      (1) The **superior pancreaticoduodenal artery** usually arises from the **gastroduodenal artery** and then bifurcates into anterior and posterior branches, which supply the head of the pancreas and the duodenum. These arcades anastomose with those from the inferior pancreaticoduodenal artery.
      (2) The **inferior pancreaticoduodenal artery** arises from the **superior mesenteric artery** and bifurcates into anterior and posterior branches, which supply the head of the pancreas and the duodenum. These arcades anastomose with those from the superior pancreaticoduodenal artery.
      (3) The **great pancreatic artery, caudal pancreatic arteries,** and an occasional **dorsal pancreatic artery** arise from the splenic artery and supply the body and tail of the pancreas with profuse anastomotic connections.
   b. **Pancreatic veins** drain into the hepatic portal system through the superior mesenteric vein as well as the common hepatic and splenic veins. Because the pancreas is mostly secondarily retroperitoneal, there are variable transperitoneal anastomotic connections (veins of Retzius) with the systemic veins of the posterior body wall.
   c. **Lymphatic drainage**
      (1) The head and body drain into the pyloric and splenic nodes of the celiac group as well as to the superior mesenteric nodes. Because of its secondary retroperitoneal position, it also drains by retroperitoneal channels.
      (2) The tail drains primarily via the splenic nodes.
      (3) The prognosis for recovery from pancreatic cancer is usually poor because metastases may spread widely by retroperitoneal channels.

5. **Innervation**
   a. The **afferent pathway** reaches levels T7–T11 of the spinal cord via visceral afferent fibers that travel with the greater as well as the lesser splanchnic nerves.
   b. **Pancreatic pain** is referred widely to the T7–T11 dermatomes.

6. **Function**
   a. **Endocrine function.** The **pancreatic islets** (of Langerhans) constitute the endocrine portion of the pancreas.
      (1) The islet cells secrete **insulin** and **glucagon** into the circulatory system.
      (2) The majority of islets are located in the pancreatic tail. As such, removal of the head of the pancreas along with the gastric antrum and duodenum (Billroth II procedure) usually does not significantly depress blood insulin levels.
   b. **Exocrine function.** As a compound tubuloacinar exocrine gland, the pancreas secretes 500–1200 ml of pancreatic juice a day into the duodenum.
      (1) This colorless, viscous fluid is alkaline because of a high bicarbonate content. It maintains the alkaline pH of the duodenum.
      (2) Pancreatic juice contains powerful digestive enzymes that are activated by an alkaline pH: trypsin, amylase, and lipase.
      (3) Secretion is regulated by local chemical stimuli as well as by parasympathetic innervation.

7. **Clinical considerations**
   a. **Obstruction and inflammation**
      (1) **Acute pancreatitis.** The pancreatic and common bile ducts join and enter the duodenum via a common ostium in 65% of individuals. Spasm of the hepatopancreatic sphincter or blockage of the duct at the duodenal papilla by a cholelith not only results in biliary stasis, but also may cause reflux of bile into the pancreas. This situation activates the pancreatic enzymes in situ with resultant **acute pancreatitis (autolytic pancreatic necrosis).**
      (2) **Pancreatic pain**
         (a) **Referred pain.** The afferent pathway reaches levels T7–T11 of the spinal cord via visceral afferent fibers that travel with the greater as well as the lesser splanchnic nerves.

(b) **Pain secondary to peritonitis.** When the parietal peritoneum to which the pancreas is fused is involved, sensation is mediated by the intercostal nerves. The intense, searing pain is localized in the middle of the back.
   b. **Carcinoma** of the pancreas is common.
      (1) The prognosis for recovery is usually poor because metastases may spread widely along retroperitoneal lymph channels.
      (2) Fully 80% of pancreatic carcinomas are located in the head.
         (a) Jaundice with no indication of biliary pathology or hepatic dysfunction may result from compression of the common bile duct by a pancreatic tumor. This signal often is lifesaving, as pancreatic tumors frequently are silent until it is too late for surgical intervention.
         (b) Pancreatic carcinoma frequently invades the splenic vein and even the portal vein, manifesting with symptoms of portal hypertension, such as esophageal varices, without indications of liver disease. By this time, metastases have usually spread widely by hematogenous routes.
   c. **Anomalies. Ectopic pancreatic tissue** (possible in the stomach, small intestine, gallbladder, or spleen) is a common finding. The pancreas may surround the descending duodenum (**annular pancreas**) with constriction or even obstruction of this portion of the small bowel.

G. Spleen

1. **Position and relationships.** The spleen lies in the left hypochondriac region (upper left quadrant) adjacent to the stomach (see Figure 23-1B).
   a. **Location.** It is superior to the left colic flexure and anterior to the left kidney.
      (1) The upper half is usually anterior to ribs 11 and 12; the lower pole may extend to the level of L2. This location makes it vulnerable to laceration by a broken rib.
      (2) The spleen is palpable only when enlarged.
   b. **Support.** Arising within the dorsal mesogastrium, its supports include the **gastrosplenic ligament** and the **splenorenal ligament.**
      (1) The tail of the pancreas may reach into the splenorenal ligament as far as the hilum of the spleen.
      (2) Because of its mesenteric support, its position is variable. It may prolapse into the false pelvis, especially in **splenomegaly.**
   c. **Variation.** One to ten accessory spleens in the dorsal mesogastrium are common.

2. **Structure.** The spleen (about the size and shape of half an orange) is composed of lymphoid tissue separated by blood-filled sinuses.

3. **Vasculature**
   a. **Arterial supply.** The spleen is supplied profusely by the **splenic artery,** which gives off a number of splenic branches that enter the hilus (see Figure 25-9). The splenic artery also supplies the fundus and greater curvature of the stomach, in addition to the body and tail of the pancreas.
   b. **Venous return** is by the **splenic vein,** a major branch of the hepatic portal vein (see Figure 25-15).
   c. **Lymphatic drainage** is not significant, even though the spleen is a major lymphatic organ situated in the circulatory system.

4. **Function.** Once thought to be the seat of "black humor" ("to vent one's spleen") and sadness, caused by an excess of "black bile" (melancholy), it is not a vital organ in the adult.
   a. **Immunogenic activity.** Splenic lymphocytes produce opsonins and antibodies to blood-borne antigens.
   b. **Phagocytic activity.** As a reticuloendothelial structure situated in the hepatic portal system, the spleen removes particulate matter and cellular residue from the blood. It also functions as the principal location of erythrocyte destruction.
   c. **Blood cell storage.** The spleen stores a variable amount of erythrocytes, which can be released into the circulatory system by contraction of intrinsic smooth muscle.

d. Hematopoiesis. It is hematopoietic in the fetus.

5. **Clinical considerations**
   a. **Splenomegaly.** An enlarged spleen may be a sign of chronic infection (e.g., mononucleosis), blood dyscrasias, lipid-storage disease, or lymphosarcoma.
   b. **Splenectomy.** When therapeutic splenectomy is indicated, all accessory splenic tissue must be located and excised.
   c. **Splenic laceration.** Lying against ribs 11 and 12, the spleen can be lacerated by a fractured rib. Laceration of this intensely vascular organ carries a real potential for fatal internal exsanguination. Immediate splenectomy is usually indicated.

III. MIDGUT

A. Jejunum and ileum

1. **Location and support.** The jejunum (jejunus, L. empty) and ileum (L. iliac, for its pelvic location) constitute the midgut portion of the small bowel. Supported by the **mesentery proper,** the jejunal loops lie in the left lateral region of the abdominal cavity and the ileal loops lie in the pelvic cavity.
   a. The **mesentery proper** originates from a root or radix (line of peritoneal attachment) that is 15–20 cm (6–9 inches) long and runs diagonally down the posterior coelomic wall from upper left to lower right. From the radix, the mesentery fans out to support more than 7 m (22 feet) of small bowel.
      (1) Except at their terminations, the jejunum and ileum are highly mobile. Their relationships, as seen radiographically, may change from day to day.
      (2) Excessive twisting (volvulus) of the small intestine may compress the blood supply that runs within the mesentery, possibly resulting in ischemia (intestinal angina) and infarction.
   b. **Regional differences.** The jejunal mesentery has less fat than that of the ileum.

2. **External structure.** The jejunum and ileum share a common blood supply, innervation, mesenteric support, and function.
   a. **Divisions.** Because the distinctive characteristics of the jejunum gradually become those of the ileum, there is no distinct demarcation between them.
   b. **Distinguishing characteristics** (Table 25-1). Although the differences between the jejunum and ileum are subtle, it is possible to distinguish between the proximal jejunum and the distal ileum.
      (1) **Visually.** The size, color, and number of arterial arcades and arteriae rectae,

TABLE 25-1. Distinguishing Characteristics of the Jejunum and Ileum

Characteristic	Jejunum	Ileum
Diameter	Greater (2 to 4 cm)	Less (2.5 to 3 cm)
Wall	Thick, heavy	Thin, light
Vascularity	Greater	Less
Color	Deeper red	Paler pink
Arteriae rectae	Tall	Short
Arcades	Fewer with large loops	Many with small loops
Mesentery	Less fat, translucent	Fatty, opaque
Plicae circularis	Tall, profuse	Short, rudimentary
Peyer patches	Few	Many

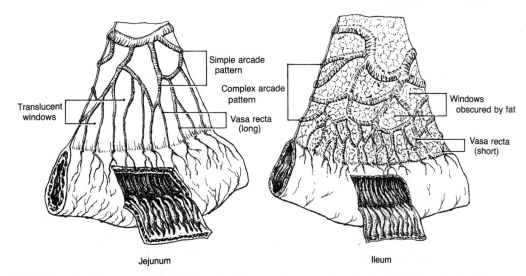

FIGURE 25-10. Jejunum and ileum. Several differences between the proximal jejunum and distal ileum and their mesenteries are indicated.

as well as the relative quantity of mesenteric fat between these vessels, are the more obvious distinguishing characteristics (Figure 25-10).
- (a) The jejunal chyme has greater fluid volume; thus, the diameter is larger.
- (b) The jejunal walls are thicker because the plicae circulares are well developed.

(2) **Radiographically.** With contrast enhancement, the differences are apparent by the diameter and the appearance of the plicae circulares.
- (a) The jejunum has a wider lumen than the ileum.
- (b) The jejunal lumen has a feathery appearance that reflects the abundant, tall, and branched plicae circulares. The ileum has a denser luminal content because of fewer and shorter plicae circulares and therefore more prominent peristaltic constrictions.

c. **Ileal (Meckel's) diverticulum** is a persistent remnant of the omphalomesenteric (vitelline) duct of the embryo found in about 3% of the population. When present, it occurs within 3 feet of the ileocecal junction (on the antimesenteric side of the ileum) and is usually less than 3 inches long. **Mnemonic:** Rule of 3's.

3. **Vasculature**
   a. **Arterial supply**
      (1) **Jejunal** and **ileal arteries** (up to 20) branch to form arterial arcades within the mesentery proper (see Figures 25-10 and 25-12).
      - (a) The arcades provide abundant collateral circulation, which is especially important during peristaltic movements when some of the arteries may be constricted.
      - (b) The arcades of the ileum are smaller and more numerous than those of the jejunum.

      (2) **Straight arteries** (arteriae rectae or vasa recta) run from arcades directly to the viscus wall, where they give off anterior and posterior branches to the sides of the viscus (see Figure 25-10).
      - (a) Adjacent vasa recta usually anastomose within the intestinal wall on the mesenteric side. On the antimesenteric side, the vessels are true **end-arteries,** which are, by definition, nonanastomotic and supply exclusive regions of tissue.
      - (b) The straight arteries of the ileum are shorter and more numerous than those of the jejunum.

(3) **Ileal twigs from the ileocolic branch** of the superior mesenteric artery supply the terminal ileum (see Figure 25-12).
   b. **Venous return** is through the **superior mesenteric vein,** which drains into the hepatic portal vein.
   c. **Lymphatic drainage.** Lymphatics of the small intestine absorb and transport triglycerides as well as lymph to the circulation by way of the thoracic duct. They are of minor clinical importance because of the low incidence of carcinoma of the jejunum and ileum.
4. **Innervation** (see section VI)
   a. **Parasympathetic innervation** is by the **vagus nerves** and superior mesenteric plexus. Activity promotes peristalsis.
   b. **Sympathetic innervation** is by the lesser splanchnic nerves (T10–T11), superior mesenteric ganglia, and the superior mesenteric plexus.
   c. **Afferent innervation**
      (1) **Sensory pathways.** Afferent fibers mediating painful sensation course along the **lesser splanchnic nerve** to the **sympathetic chain.** The afferents then travel along a white ramus communicans to the respective spinal nerves and associated spinal segment.
      (2) **Referred small bowel pain.** Painful sensation from the jejunum and ileum is mediated by the lesser splanchnic nerve and associated with spinal segments T10 and T11. Small bowel pain is, therefore, referred to the **umbilical region.**
5. **Function**
   a. **Intestinal motility** propels and mixes chyme.
      (1) **Segmentation** occurs at the rate of about 10 constrictions per minute and corresponds to the frequency of **borborygmi** (sounds auscultated with a stethoscope applied to the abdominal wall).
         (a) Segmentation is nonpropagating contraction of the circular muscle that divides columns of chyme into segments. These contractions are the result of reflexes initiated by enteric neurons on intestinal wall distention.
         (b) Subsequent contractions in adjacent regions both divide a segment of chyme and mix the chyme by reuniting previously divided adjacent segments.
      (2) **Pendular movements** are slight waves of contraction that pass bidirectionally and rapidly (12–13 per second) along short segments of the gut. These movements mix chyme with intestinal and pancreatic digestive juices and bile as well as assist absorption by continuously bringing chyme into contact with the surface mucosa.
      (3) **Peristalsis,** a unidirectional intestinal movement, begins in the terminal duodenum and progresses in the jejunum, ileum, and large intestine. Although not forceful in the jejunum, contrast-enhanced studies reveal that peristalsis is pronounced in the ileum.
         (a) **Peristaltic waves** are relatively slow (1–2 cm/second) advancing contractions that proceed aborally over relatively short distances (4–5 cm).
         (b) **Peristaltic rush** sweeps along the intestine, without pause, for great distances. It usually occurs after meals, especially the first of the day.
      (4) **Control of GI motility**
         (a) **Enteric reflexes.** Segmental and pendular movements are probably enteric in origin.
            (i) Many of the nerve cells of the myenteric plexuses are intrinsic and involve local reflexes.
            (ii) Enteric reflexes may explain the Bayliss-Starling law of the gut. A bolus of chyme, exerting transverse pressure on the intestinal wall, results in increased muscular tone immediately oral to the bolus and relaxation in the adjacent aboral region. This reflex mechanism ensures unidirectional oral to aboral flow of intestinal contents.
         (b) **Autonomic influences** affect peristaltic rate.
            (i) **Sympathetic (adrenergic) neurons** slow peristalsis by inhibiting the enteric reflexes. Sympathetic neurons from the celiac and superior

mesenteric ganglia synapse on the enteric neurons of the myenteric plexuses.
- (ii) **Parasympathetic (cholinergic) neurons** increase peristalsis by facilitating the enteric reflexes. Parasympathetic neurons of the myenteric plexuses are stimulated by the vagus nerve.

b. **Absorption** of carbohydrates, triglycerides and fatty acids, amino acids, vitamins, electrolytes, and water is maximized through structural modifications that increase the mucosal surface area.
   (1) **Length.** The 7 m of small intestine with a diameter of 2–4 cm provide substantial surface area for absorption.
   (2) **Plicae circulares,** circular folds, substantially increase the mucosal surface area. They increase in size, number, and complexity toward the ileum and decrease within the ileum.
   (3) **Intestinal villi,** surface projections on the plicae circulares, greatly increase the folded surface area.
   (4) **Microvilli,** cytologic modifications of the luminal surface of the mucosal cells, increase the absorptive surface area by magnitudes.

6. **Clinical considerations**
   a. **Ileojejunal resection.** Excision of as much as two thirds of the small intestine is compatible with normal life. Survival is possible with only 18 inches.
      (1) The stomach may be anastomosed to the jejunum (gastrojejunostomy) when the duodenum must be bypassed or resected (Whipple's procedure).
      (2) As a radical weight control measure, a segment of small intestine can be bypassed to reduce the absorptive area.
   b. **Ileal (Meckel's) diverticulum** may produce ileal peptic ulceration, ileal obstruction, ileal diverticulitis, or volvulus.
      (1) **Peptic ulceration** of the ileum. Meckel's diverticulum usually contains gastric mucosa with oxyntic (acid-secreting) cells. Peptic ulceration of adjacent ileal mucosa is a frequent complication. Associated edema may produce intestinal obstruction.
      (2) **Ileal diverticulitis.** An ileal diverticulum may become inflamed and generate symptoms that are similar to those produced by an inflamed appendix.
      (3) **Umbilical connections**
         (a) Rarely, the omphalomesenteric duct remains a patent connection with the umbilicus. If such a patency occurs in a newborn, chyme may be discharged to the exterior.
         (b) A connection to the umbilicus may persist with an obliterated lumen. The gut may revolve about this remnant **(volvulus),** with resultant obstruction of the intestinal lumen, occlusion of blood vessels, and intestinal infarction.
      (4) **Prophylactic resection.** During appendectomy and other surgical procedures, the terminal ileum is routinely explored for a possible ileal diverticulum. If found, it is resected as a prophylactic measure.
   c. **Obstruction.** Volvulus, edema, and intussusception can obstruct the small bowel. Initially, borborygmi increases as the gut works to move the chyme through the obstruction. With the onset of inflammation, borborygmi ceases with paralytic ileus, indicating the need for surgical intervention.
   d. **Paralytic ileus**
      (1) Inflammation or trauma (even surgical manipulation) to abdominal or related organs, such as the kidneys or gonads, usually leads to inhibition of the small bowel, **intestinal shutdown (paralytic ileus).**
      (2) The postoperative re-emergence of borborygmi signals re-established bowel function. Oral feeding may then resume.

B. **Ascending colon**
1. **Overview**
   a. **Divisions** of the **large bowel** are made according to its mesenteric attachments—or lack thereof—into **ascending, transverse, descending, sigmoid,** and **rectal regions.**

(1) The **ascending and transverse** portions are part of the midgut (see Figure 25-12).
(2) The **descending, sigmoid, and rectal** portions constitute the hindgut (see Figure 25-13).
b. **Structure** is related to formation, transport, storage, and evacuation of feces.
(1) The large bowel is approximately 2 m (5 feet) long.
(2) Fat-filled tags (**appendices epiploicae** or **epiploic appendages**) are scattered over its surface.
(3) The outer longitudinal layer of the muscularis externa is incomplete, being restricted to three longitudinal bands, the **teniae coli.**
  (a) The teniae coli, about 1 cm wide, are most obvious on the cecum and ascending colon.
  (b) The teniae converge and fuse to form a complete outer layer over both the appendix and rectum.
(4) Muscle tone in the teniae coli produces sacculations (**haustra**) along the large intestine.
c. The **principal function** is conservation of fluid by the converting liquid chyme into semisolid feces.
(1) The colon or large bowel is characterized by its capacity, distensibility, the amount of time it retains its contents, and the special arrangement of musculature within the tunica muscularis.
(2) The colonic mucosa is characterized by a mixture of absorptive cells and mucous cells that provide lubrication to facilitate movement of the forming stool toward the rectum.
d. **Innervation**
(1) **Parasympathetic innervation** is from the **vagus nerve** by way of the superior mesenteric plexus. Activity promotes peristalsis.
(2) **Sympathetic innervation** is by the **lesser splanchnic nerves** from spinal levels T10 and T11 via the **superior mesenteric ganglia** that contain the postsynaptic neurons. Activity inhibits peristalsis.
(3) **Afferent innervation** is along sympathetic pathways.
  (a) **Afferent pathways** are by way of the superior mesenteric plexus and the lesser splanchnic nerves to spinal segments T10 and T11 via the associated spinal nerves.
  (b) **Visceral pain** is referred to dermatomes T10 and T11, which correspond to the umbilical and hypogastric regions.

2. **The cecum,** the first part of the ascending colon, is a dilated pendulous sac inferior to the ileocecal juncture (Figure 25-11).
a. **Ileocecal juncture.** The ileum terminates by entering the cecum posteromedially with some variation.
(1) The **ileal papilla** is the conical projection at the point of entry. It is about 2–3 cm superior and medial to the lowest point of the cecum.
(2) The **ileocecal valve** consists of two folds that border the ileocecal ostium (see Figure 25-11).
  (a) Although this valve impedes regurgitation of cecal contents into the ileum, it is largely incompetent and of minor mechanical importance. A barium enema that fills the colon completely will penetrate the terminal ileum to a variable extent.
  (b) Backflow of chyme is prevented primarily by the direction of peristalsis.
(3) The terminal ileum is vulnerable to herniation through the ileocecal valve into the cecum (**intussusception**). The resultant functional obstruction necessitates immediate surgical intervention.
b. **Mesenteric support** (see Figure 25-11)
(1) The **superior ileocecal (vascular) fold** is an extension of the mesentery proper. It supports the pendulous position of the cecum, making it a peritoneal structure, and contains the **anterior cecal artery.**
(2) **Cecal fossae.** The more superior portions of the cecum, along with the rest of the ascending colon, are secondarily retroperitoneal. At the line of fusion be-

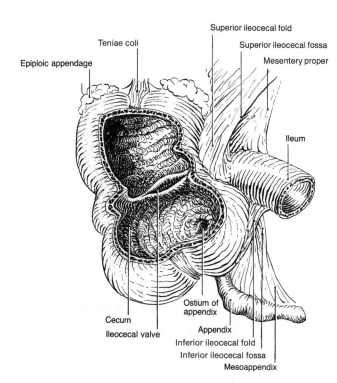

FIGURE 25-11. Mesenteries and mesenteric folds support the cecum and appendix. Also shown is the internal structure of the cecum.

tween the primitive ascending mesocolon and the parietal peritoneum, peritoneal folds define three cecal fossae.
- (a) The **superior ileocecal fossa** lies between the mesentery proper where it attaches to the terminal ileum and the superior ileocecal fold.
- (b) The **inferior ileocecal fossa** lies between the mesoappendix and the inferior ileocecal (bloodless) fold.
- (c) The **retrocecal fossa** is formed by the incomplete fusion of the visceral peritoneum of the cecum and the parietal peritoneum of the posterior abdominal wall. The appendix is frequently found within this fossa.

(3) **Mobilization.** The cecum can be mobilized along with its blood supply by incision along its lateral juncture with the parietal peritoneum.

c. The **three cecal teniae coli** converge toward the appendix. They provide useful landmarks for the often difficult task of locating the appendix during surgery.

d. **Arterial supply** to the cecum is by the **ileocolic artery.** This terminal branch of the superior mesenteric artery (see Figure 25-12) has four branches.
  (1) The **ascending branch** supplies the basal portion of the ascending colon.
  (2) **Anterior** and **posterior cecal arteries** supply the cecum. The anterior cecal artery runs in the superior ileocecal fold.
  (3) The **appendicular artery** supplies the appendix and lies in the mesoappendix.
  (4) **Ileal branches** of the ileocolic artery provide the major supply to the terminal ileum.
    - (a) Anastomoses between the last ileal arcade of the superior mesenteric artery and the ileal branches of the ileocolic artery are weak, if present at all.
    - (b) In cecal resection, with excision of the ileocolic artery, the terminal ileum up to the first substantial arcade is removed to avoid ischemic necrosis.

3. **Vermiform appendix.** A narrow, hollow, muscular viscus that arises from the posteromedial aspect of the cecum about 2–3 cm below the ileocecal orifice, the appendix represents the tip of the cecum that fails to enlarge (see Figure 25-11).

a. Structure. The length of the appendix varies between 0.5 cm and 25 cm (0.2 and 10 inches). Average length is 9 cm (3.5 inches).
  (1) A complete longitudinal muscular coat gives the appendix a smooth appearance.
  (2) Within the mucosa and submucosa are large accumulations of lymphatic tissue.
  (3) The region of the appendix next to the cecum has a thickened muscular coat and a slightly narrower lumen. The luminal diameter varies with age.
    (a) The lumen tends to be wide (6–8 mm) in infants and young children.
    (b) By middle age, it is often entirely obliterated.
    (c) The lumen is dangerously narrow in adolescents and young adults. During this period, it may be occluded by a fecalith or even by edema and swollen lymphatic tissue associated with mild inflammation, which explains the high frequency of **acute** and **chronic appendicitis** in this population.

b. Support and vasculature
  (1) The **mesoappendix** supports the appendix, making this organ peritoneal (see Figure 25-12). During surgery, the appendix may be found in the retrocecal fossa (65%), the iliac fossa (31%), the right paracolic recess (2%), the superior ileocolic fossa (1%), or the inferior ileocolic fossa (1%).
  (2) The **appendicular artery,** a terminal branch of the ileocolic artery, runs in the mesoappendix (see Figure 25-11).

c. Innervation is by the lesser splanchnic nerves and superior mesenteric ganglia. Afferent innervation travels along sympathetic pathways.

d. Appendicitis
  (1) The **pathogenesis** of this condition is related to the patency of the lumen.
    (a) If the lumen is blocked, local bacterial flora multiply rapidly and their toxins produce inflammation, nausea, and fever. The resultant distention of the appendix from gas or edema compromises the blood supply.
    (b) Local ischemia with breakdown of the ischemic blood vessels produces intramural hemorrhage, a surgical emergency. Delay results in gangrenous appendicitis with danger of rupture and complicating peritonitis.
  (2) **Appendicular pain**
    (a) **Initial colicky pain** is referred to dermatomes T10–T11, which include the **umbilical region,** because the appendix is innervated by neurons that travel along the **lesser splanchnic nerve.** This finding, however, is insufficient evidence to make a specific diagnosis because several structures refer pain to this region.
    (b) **Subsequent right inguinal peritonitis** develops as the adjacent parietal peritoneum becomes inflamed. This pain is exquisitely specific to the lower right quadrant. The overlying musculature exhibits the reflex spasm of the "acute abdomen" (guarding, splinting).
    (c) In situs inversus, peritoneal pain is noted on the left side.
  (3) The usual route for **surgical access** to the appendix is at **McBurney's point,** which is located about 5 cm (one third the distance) along a line from the anterior–superior iliac spine to the umbilicus. Surgeons frequently use **McBurney's muscle-splitting incision** for appendectomy.

4. The ascending colon proper lies along the right side of the abdominal cavity (see Figure 25-12). It is 15–20 cm long (about 8 inches).
  a. Structure and relationships (see Figure 23-1)
    (1) **Support.** The ascending colon proper is fused to the posterior abdominal wall, making it **secondarily retroperitoneal.**
      (a) Its relationships depend on the degree to which it adheres to the peritoneum.
      (b) It can be mobilized and its primitive mesentery with its contained blood supply re-established by incising along the line of peritoneal fusion.
      (c) Incomplete fixation may occur as a developmental anomaly, a situation that predisposes to volvulus.

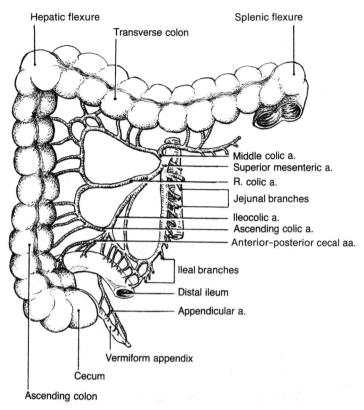

FIGURE 25-12. Distribution of the superior mesenteric artery.

(2) The **right paracolic gutter** is defined by the ascending colon. This recess communicates with the supracolic compartment (above) and the infracolic compartment (below). Infection may track between the pelvic cavity and the subhepatic recess along the right paracolic gutter.

(3) The **right colic (hepatic) flexure** of the ascending colon marks the transition between ascending colon and transverse colon.

b. **Vasculature**

   (1) **Arterial supply** is by the ileocolic, right colic, and middle colic branches of the **superior mesenteric artery** (Figure 25-12).

   (a) The **ascending colic branch of the ileocolic artery** supplies the most inferior portion of the ascending colon.

   (b) The **right colic artery,** by its ascending and descending branches, supplies most of the ascending colon. This artery is the most variable branch of the superior mesenteric artery and is absent in 18% of individuals.

   (c) The **middle colic artery** bifurcates into left and right branches. The right branch may supply a variable portion of the ascending colon near the hepatic flexure.

   (d) The **marginal artery** (of Drummond) is formed by the anastomoses of various branches of the superior mesenteric artery and inferior mesenteric artery. It lies in the mesentery adjacent to the colon and is an important vascular channel (see Figures 25-12 and 25-13).

   (2) **Venous return** is by way of the **superior mesenteric vein** that drains into the hepatic portal vein. The veins usually parallel the arteries until they reach the hepatic portal vein.

   (3) The **right colic nodes** drain into the **superior mesenteric nodes.** The **intestinal lymph trunk** collects the intestinal lymph and drains into the cisterna chyli.

c. Innervation is by the lesser splanchnic nerves to the superior mesenteric ganglia and plexus as well as by the vagus nerve. Afferent innervation is along the sympathetic pathways and is referred to dermatomes T10 and T11, which correspond to the umbilical and hypogastric regions.

C. Transverse colon

1. **Structure and relationships** (see Figure 25-12)
   a. This second and longest segment of the large bowel begins at the **hepatic flexure** and traverses the peritoneal cavity to the **splenic flexure** (see Figure 25-12). It is 45–50 cm (18–20 inches) long.
      (1) The **hepatic (right colic) flexure** is anterior to the right kidney and the inferior segment of the duodenum. It anchors the proximal end of the transverse colon to the secondarily retroperitoneal ascending colon.
      (2) The **splenic (left colic) flexure** is anterior to the pole of the left kidney and inferior to the spleen. It anchors the distal end to the secondarily retroperitoneal descending colon.
      (3) The position of the transverse colon between the hepatic and splenic flexures is variable and depends on its degree of fullness.
   b. **Support** is by the **transverse mesocolon** and **gastrocolic ligament.**
      (1) Mesenteric support between the hepatic and splenic flexures makes this bowel segment peritoneal.
      (2) The **gastrocolic ligament** is formed by the fusion of the transverse mesocolon with the inferior surface of the greater omentum. This ligament divides the peritoneal cavity into supracolic and infracolic compartments.

2. **Vasculature** (see Figure 25-12)
   a. **Arterial supply**
      (1) The **middle colic artery,** a branch of the superior mesenteric artery, lies within the transverse mesocolon.
      (2) Its left and right branches contribute to the **marginal artery;** however, anastomoses between the left branch of the middle colic artery and the ascending branch of the left colic artery (from the inferior mesenteric artery) may be poor or even absent near the splenic flexure—**a danger area.**
   b. **Venous return** is to the **superior mesenteric vein** and hepatic portal vein.
   c. The **middle colic nodes** drain into the **superior mesenteric nodes.** The **intestinal lymph trunk** collects the intestinal lymph and drains into the cisterna chyli.

3. **Innervation**
   a. **Parasympathetic innervation** is by the **vagus nerve** through the superior mesenteric plexus. This segment of bowel is the most caudal portion of the GI tract to be supplied by the cranial parasympathetic outflow.
   b. **Sympathetic innervation** is by way of the lesser splanchnic nerves, the superior mesenteric ganglia, and the superior mesenteric plexus.
   c. **Afferent innervation** follows sympathetic pathways. Visceral pain is referred to dermatomes T10 and T11, corresponding to the umbilical and hypogastric regions.

IV. HINDGUT

A. Descending (left) colon

1. **Structure and relationships**
   a. **Location.** The initial segment of hindgut lies along the left side of the abdominal cavity (see Figure 23-1). It is about 25 cm long (10 inches).
   b. **Support.** This segment of colon is fused to the posterior abdominal wall from the splenic flexure to the pelvic brim, making it **secondarily retroperitoneal**. It can be

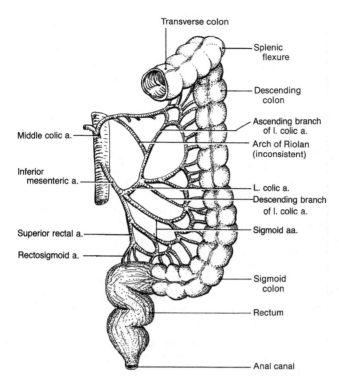

FIGURE 25-13. Distribution of the inferior mesenteric artery. Some individuals have few anastomoses between the middle colic and left colic arteries. An arch of Riolan is noted infrequently.

mobilized, along with its blood supply in the primitive descending mesocolon, by incising the line of peritoneal adhesion.
  c. The **left paracolic gutter** is defined by the descending colon. This recess provides communication between the pelvic fossa and the supracolic compartment.
  d. **External structure**
    (1) The descending colon begins at the splenic flexure, where the transverse mesocolon terminates (see Figure 25-13). It ends at the sigmoid colon, where the sigmoid mesocolon begins.
    (2) Because the bulk of feces decreases with the absorption of water, the diameter of the descending colon is smaller than either the ascending or transverse segments of large bowel.
2. **Vasculature**
  a. **Arterial supply** is by the **left colic branch** of the **inferior mesenteric artery** (Figure 25-13). The **ascending branch of the left colic artery** anastomoses inconsistently with the left branch of the middle colic artery—a danger area. The **descending branch** anastomoses with the arcades of the sigmoid arteries to contribute to the **marginal artery.**
  b. **Venous return** is by way of the **inferior mesenteric vein,** which may empty into the splenic vein or into the superior mesenteric vein.
  c. **Lymphatic drainage**
    (1) This segment of colon drains via the **left colic nodes,** which in turn drain into the **inferior mesenteric nodes,** the para-aortic nodes, and the thoracic duct.
    (2) Lymph from the descending colon may drain into the **celiac nodes** through lymphatic channels along the inferior mesenteric vein.
3. **Innervation** (see section VI)
  a. **Parasympathetic innervation** is from the **pelvic splanchnic nerves** or **nervi erigentes** (S2–S4). It flows retrograde through the pelvic plexus, inferior then superior hypogastric plexuses, aortic plexus, to the inferior mesenteric plexus.
  b. **Sympathetic innervation** is from the **lumbar splanchnic nerves** (L1–L2).

c. **Afferent innervation** from the descending colon reaches the L1–L2 levels of the spinal cord.
   (1) The afferents travel along the lumbar splanchnic pathways and through or up the sympathetic chain to reach the white ramus communicans of L2 and L1.
   (2) Pain originating within the descending colon is referred to the L1–L2 dermatomes of the inguinal region and thigh.

B. Sigmoid colon
1. **Structure and relationships**
   a. **Location.** This 25–40-cm (10–15-inch) long segment of the large bowel begins where the colon again becomes peritoneal, usually at the pelvic brim (see Figure 23-1). It terminates at the rectum, where the sigmoid mesocolon ends.
   b. **Support.** The sigmoid colon is **peritoneal.**
      (1) It is suspended from the posterior abdominopelvic wall by the **sigmoid mesocolon.** A large loop frequently occupies the deep pelvis.
      (2) The **intersigmoid recess,** an occasional site for intra-abdominal herniation, is located at the juncture of the inferior surface of the sigmoid mesocolon and the dorsal body wall.
   c. **Relationships.** Anteriorly, the sigmoid colon is related to the urinary bladder in males and to the posterior surface of the uterus and upper part of the vagina in females. Posteriorly, it is related to the external iliac vessels and the sacrum. The sigmoid colon continues into the rectum at the third sacral level.
2. **Function.** Its primary function is storage of feces. The internal structure is similar to the rest of the large bowel except that the mucosa is composed primarily of mucous cells.
3. **Vasculature**
   a. **Arterial supply** (see Figure 25-13)
      (1) **Sigmoid arteries** (usually four) anastomose profusely to form arcades that contribute to the **marginal artery.** The last sigmoid arcade may or may not anastomose with the rectosigmoid artery.
      (2) The **rectosigmoid artery,** the terminal branch of the **inferior mesenteric artery,** supplies the terminal sigmoid colon and the superior portion of the rectum.
         (a) Because the rectosigmoid artery may or may not anastomose with the last sigmoid arcade, this region is known as the **critical point** (of Sudeck). Adequate anastomoses between these two vessels occur in only 52% of individuals, a point of concern in colorectal resection.
         (b) The rectosigmoid artery may or may not anastomose inferiorly with the middle rectal (hemorrhoidal) artery, the terminal branch of the inferior mesenteric artery (see Figure 29-2).
   b. **Venous return** is by way of the **inferior mesenteric vein,** which drains into the splenic vein or superior mesenteric vein.
   c. **Lymphatic drainage** of the terminal portion of the large bowel is important because of the high frequency of colonic carcinoma.
      (1) Lymphatic drainage follows the veins.
         (a) The major drainage pathway is through the **left colic nodes** that drain, in turn, into the **inferior mesenteric nodes,** the para-aortic nodes, **lumbar nodes,** and the thoracic duct.
         (b) Lymph from the sigmoid colon may drain into the **celiac nodes** by way of the lymphatic channels along the inferior mesenteric vein if that vein enters the splenic vein.
         (c) Drainage may also involve the **deep pelvic nodes** through lymphatics associated with the middle hemorrhoidal vessels.
      (2) Because the drainage pathways are variable, carcinoma of the descending colon and sigmoid colon can spread to the liver hematogenously as well as metastasize widely to the lymph nodes of the abdomen and pelvis.
4. **Innervation** of the sigmoid colon is the same as for the descending colon.

5. **Clinical considerations**
   a. **Diverticulosis.** Diverticula are small saccular protrusions of intestinal mucosa through the colonic wall. They usually occur at the margins of the teniae coli next to penetrating blood vessels.
      (1) Diverticula result from a combination of an incomplete outer muscular layer, the penetration of the intestinal wall by arteriae rectae, and the more solid composition of the feces.
      (2) Prevalence of diverticula increases with age, associated with straining at stool with chronic constipation.
   b. **Diverticulitis.** Diverticula occluded by fecaliths may become inflamed (diverticulitis).
      (1) Ulcerative diverticulitis with bleeding into the intestinal lumen is a frequent cause of anemia in the elderly. An inflamed diverticulum may perforate, resulting in severe intra-abdominal bleeding and peritonitis.
      (2) Surgical resection of the involved portion of bowel is frequently the treatment of choice.
   c. **Hirschsprung's disease** (megacolon)
      (1) Congenital agenesis of the myenteric plexus (usually in the terminal portion of the sigmoid colon) precludes peristaltic activity and results in **Hirschsprung's disease.** This functional blockage with toxic megacolon manifests as chronic constipation in neonates or young children.
      (2) Bowel function may be restored surgically by an "abdominopelvic pull-through procedure," with resection of the aganglionic (distally narrow) portion of bowel and anastomoses of normal bowel to the region of the anal sphincter.

C. **Rectum.** This part of the GI tract lies retroperitoneally within the pelvic cavity (see Chapter 29 I).

D. **Anal canal.** This terminal part of the alimentary canal lies inferior to the pelvic diaphragm in the anal triangle of the perineum (see Chapter 28 III).

V. VASCULATURE

A. **Arterial supply to the GI tract**

1. **Celiac artery** (celiac trunk or celiac axis) (see Figure 25-2)
   a. **Origin and course.** This artery leaves the aorta in the midventral line just caudal to the aortic hiatus of the diaphragm.
   b. **Branches.** It trifurcates almost immediately (see Figure 25-2).
      (1) The **left gastric artery** supplies the right superior region of the stomach (cardia, fundus, and body) as well as the abdominal portion of the esophagus.
         (a) Its terminal branch lies along the lesser curvature.
         (b) It gives off the **esophageal branch,** which supplies the terminal esophagus. This artery provides the pathway for the posterior vagus nerve to reach the celiac plexus.
         (c) It occasionally (35%) gives off an **aberrant left hepatic artery.**
      (2) The **splenic artery** supplies the spleen and the entire left side of the stomach as well as portions of the pancreas and the greater omentum (see Figure 25-9).
         (a) **Short gastric arteries** supply the left superior portion of the stomach (fundus and body).
         (b) The **left gastro-omental (gastroepiploic) artery** supplies the left inferior region of the stomach and portions of the greater omentum.
         (c) The **great pancreatic artery** and several **caudal pancreatic arteries** supply the tail of the pancreas.
         (d) **Splenic branches** supply the spleen.

(3) The **common hepatic artery** supplies the liver and gallbladder as well as portions of the stomach, duodenum, and pancreas (see Figure 25-2).
  (a) The **right gastric artery** (which may also arise from the proper hepatic artery) supplies the right inferior region of the stomach.
  (b) The **gastroduodenal artery** gives off two major branches.
    (i) The **right gastro-omental (gastroepiploic) artery** supplies the right inferior region of the stomach and a portion of the greater omentum (see Figure 25-2).
    (ii) The **superior pancreaticoduodenal artery** supplies the upper portion of the duodenum and the head of the pancreas (see Figure 25-3). This artery divides into **anterior–superior** and **posterior–superior branches**.
  (c) The **proper hepatic artery** is the continuation of the common hepatic artery (see Figure 25-6).
    (i) The **left hepatic artery** supplies the left lobe, the quadrate lobe, and the left half of the caudate lobe.
    (ii) The **right hepatic artery** supplies the right lobe, the right half of the caudate lobe, and the gallbladder by the cystic artery.
    (iii) The **cystic artery,** usually a branch of the right hepatic artery, bifurcates into the anterior and posterior branches.

2. **Superior mesenteric artery** (see Figure 25-12)
   a. **Origin and course.** This artery leaves the aorta in the midventral line a few centimeters caudal to the celiac artery. It courses slightly to the left of the aorta in the mesentery proper.
   b. **Branches.** It gives off numerous branches that supply that portion of the GI tract derived from the primitive midgut.
     (1) The **inferior pancreaticoduodenal artery** divides into **anterior** and **posterior branches** to supply the descending and inferior portions of the duodenum as well as the head of the pancreas.
     (2) **Jejunal and ileal branches** supply the peritoneal portion of the small bowel. The plethora of arcades provides abundant collateral circulation.
     (3) The **ileocolic artery** is the terminal branch of the superior mesenteric artery. Branches include the **anterior and posterior cecal arteries,** the **appendicular artery,** the **ascending colic artery,** and **ileal branches.**
     (4) The **right colic artery** usually supplies the ascending colon. It frequently is missing. When present, it is the most variable of this group.
     (5) The **middle colic artery** supplies the transverse colon.
   c. **Clinical considerations**
     (1) **Intestinal angina.** Stenosis of the vascular supply may produce intestinal angina (colicky abdominal pain referred to the umbilical region).
     (2) **Superior mesenteric infarction.** Given the pattern of blood supply, occlusion of the superior mesenteric artery by a thrombus results in infarction of the small intestine and of the large intestine as far as the vicinity of the splenic flexure.

3. **Inferior mesenteric artery** (see Figure 25-13)
   a. **Origin and course.** This artery originates from the aorta distal to the gonadal arteries and supplies that portion of the GI tract derived from the primitive hindgut.
   b. **Branches**
     (1) The **left colic artery** supplies the descending colon. The ascending branch anastomoses variably with the left branch of the middle colic artery; the descending branch anastomoses with the sigmoid arteries.
     (2) **Sigmoidal arteries** supply the sigmoid colon and anastomose with the descending branch of the left colic artery.
     (3) The **rectosigmoid artery** supplies the terminal sigmoid colon and superior rectum, variably anastomosing with sigmoid and superior hemorrhoidal arteries.
     (4) The **superior rectal (hemorrhoidal) artery** supplies the upper rectum (see Figure 25-13). This artery has poor anastomotic connections with sigmoid arter-

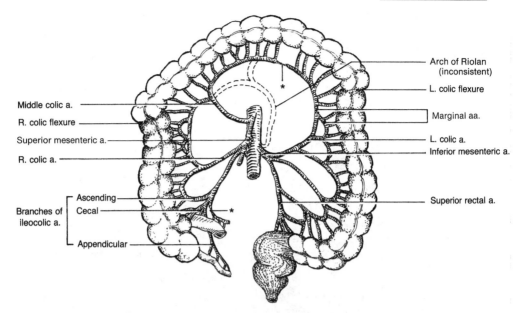

FIGURE 25-14. Marginal artery. Anastomoses occur between the branches of the superior mesenteric arteries as well as between the branches of the inferior mesenteric artery to form the marginal artery. Anastomoses are inconsistent in the regions of the ileocecal junction, the splenic flexure, and the rectosigmoid junction (*).

ies but strong anastomoses with the middle rectal arteries that arise from the internal iliac arteries.

c. **Clinical considerations. Stenosis** (narrowing) or **thrombosis** (occlusion) of the inferior mesenteric artery results in ischemia and potential infarction of the descending colon and sigmoid colon. The clinical presentation includes colicky abdominal pain referred to the inguinal regions and anterior thigh.

4. **Marginal artery** (Figure 25-14)
   a. **Origin and course.** Anastomoses between branches of the **superior mesenteric** and **inferior mesenteric arteries** result in the formation of a continuous arterial trunk, the **marginal artery** (of Drummond). The marginal artery lies in the mesentery, close to the border of the large intestine, and runs from the ileocecal valve to the rectosigmoid junction. It provides collateral circulation around the midgut and hindgut.
   b. **Clinical considerations.** At several "critical points," the marginal artery may not provide adequate collateral supply.
      (1) In the **ileocolic region,** anastomoses between the ileal branches of the superior mesenteric artery and the ileal branches of the ileocolic artery seldom provide adequate collateral circulation in the ileocecal region.
      (2) At the **left colic flexure,** there may be few anastomotic connections between the left branch of the middle colic artery and the ascending branch of the left colic artery.
         (a) Occlusion or ligature of one of these vessels may result in ischemic necrosis of a segment of large bowel.
         (b) An **arch of Riolan** occasionally provides a strong anastomotic connection between the inferior mesenteric artery and the left branch of the middle colic artery.
      (3) At the **rectosigmoid junction,** anastomotic connection between the rectosigmoid (lowest sigmoid) branch and the upper rectal branch of the superior rectal (hemorrhoidal) artery frequently is inadequate (Sudeck's critical point).

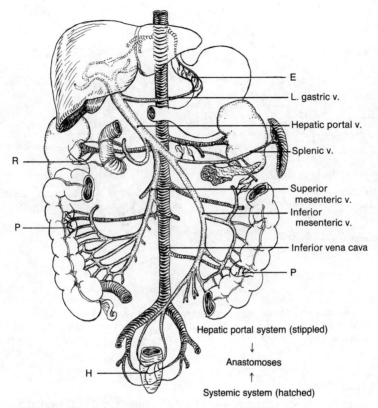

FIGURE 25-15. Hepatic portal system. Most of the blood from the gut returns to the liver. Anastomoses between the hepatic portal system and the systemic veins occur in several regions: E, esophageal veins; H, rectal (hemorrhoidal) veins; P, transperitoneal veins (of Retzius); R, renal veins; and with the paraumbilical veins (not shown). Varix development is possible in any of these anastomoses should liver obstruction occur.

B. **Venous return** from the GI tract (hepatic portal system). The hepatic portal system begins as the venous capillaries of the GI tract and ends as the venous sinusoids in the liver. This system delivers nutrient-rich (sometimes toxic) blood to the liver.

1. **The hepatic portal vein** brings venous return from the gastrointestinal tract to the liver (Figure 25-15).
   a. **Course.** This vein lies in the hepatoduodenal ligament just posterior and slightly to the left of the common bile duct and the proper hepatic artery.
   b. **Distribution** (see Figure 25-15)
      (1) The **superior mesenteric vein** drains the small intestine and large intestine as far as the splenic flexure. It is the largest contributor to the hepatic portal vein.
      (2) The **inferior mesenteric vein** drains the descending and sigmoid colons. It usually enters the superior mesenteric vein but may enter the splenic vein.
      (3) The **splenic vein** drains the spleen as well as portions of the stomach and pancreas. This vein joins the superior mesenteric vein to form the **hepatic portal vein.**
      (4) **Celiac veins.** Small veins corresponding to the named branches of the celiac artery drain directly into the hepatic portal vein.

2. **Hepatic veins**
   a. **Origin and course.** Hepatic sinusoids coalesce into interlobular veins, which come together to form the hepatic veins.
   b. **Variations.** Typically, three hepatic veins drain the left, right, and quadrate lobes

of the liver directly into the inferior vena cava, usually at the bare area. The hepatic veins may join before emptying into the inferior vena cava, or they may enter separately.

3. **Clinical considerations**
   a. **Portal hypertension.** Because the hepatic portal system has no valves, blood need not flow toward the liver. Liver disease (such as cirrhosis) or local compression of a vein results in blood shunting through the anastomotic connections in the systemic venous system.
      (1) **Esophageal varices** (see Figure 25-15E)
         (a) The branches of the **left gastric vein** anastomose profusely with the **azygos vein** and the **hemiazygos vein.**
         (b) With portal hypertension, increased shunting through the esophageal anastomoses results in **esophageal varices.**
         (c) **Hematemesis.** A varix may rupture when a rough bolus is swallowed, on emesis, or on the passage of diagnostic instrumentation. Uncontrollable and frequently fatal hemorrhage results from variceal rupture. Hematemesis associated with esophageal varices is the most common terminal event in alcohol-related liver disease.
      (2) **Caput medusae**
         (a) The **periumbilical veins** in the falciform ligament (even a recanalized ligamentum teres) anastomose abundantly with **superior** and **inferior epigastric veins** and other veins of the anterior abdominal wall. Periumbilical anastomoses may bleed profusely during midline surgical approaches.
         (b) Although the grossly dilated veins of a caput medusae appear to be rare, this condition is always present in association with portal hypertension, and can be demonstrated by infrared photography.
      (3) **Hemorrhoids** (see Figure 25-15H)
         (a) The **superior rectal vein** anastomoses profusely with the **middle rectal vein,** which is a branch of the internal iliac vein. In addition, the middle rectal vein anastomoses with the inferior rectal vein, which is a branch of the internal pudendal vein.
         (b) Increased shunting through these anastomotic channels results in varices within the submucosa of the terminal rectum **(internal hemorrhoids)** and of the anal canal **(external hemorrhoids).**
      (4) **Transperitoneal veins** (see Figure 25-15P)
         (a) The veins of a secondarily retroperitoneal structure may anastomose with the veins of the dorsal body wall, for example, lumbar veins.
         (b) These varices usually are silent and produce few problems. They may bleed profusely, however, when a secondarily retroperitoneal structure is mobilized.
   b. **Surgical correction.** The life-threatening sequelae of portal hypertension are amenable to surgical correction to provide routes for visceral blood to bypass the obstructed liver.
      (1) Surgical creation of a **portacaval fistula** produces a side-to-side anastomosis between the hepatic portal vein and the inferior vena cava, across the epiploic foramen.
      (2) A **splenorenal anastomosis** is also a common procedure. Splenectomy is followed by anastomosis of the splenic vein to the left renal vein.
      (3) Although shunts reduce the pressure-related dangers inherent in varices, toxic blood shunted around the obstructed liver has a long-term effect on the central nervous system. Thus, encephalopathy results from alcohol-related liver disease.

C. **Lymphatic drainage of the GI tract** is by way of regional nodes to the cisterna chyli.

1. **Lymphatic pathways** generally follow the vascular system. **Lymph nodes,** interposed in the lymphatic channels, are of great importance in the lymphatic spread of carcinoma of the GI tract.
   a. **Celiac nodes** are located around the celiac axis.

(1) **Four drainage areas** about the stomach gain significance from the high prevalence of gastric carcinoma.
    (a) **Left gastric nodes** lie along the lesser curvature of the stomach and receive drainage from the right superior side of the stomach and abdominal esophagus. The left gastric nodes provide a potential route for spread of carcinoma from the stomach to the esophagus and vice versa.
    (b) **Gastroepiploic nodes** lie along the inferior portion of the greater curvature, pyloric antrum, and pyloric canal. They receive lymph from the stomach, duodenum, and pancreatic head. This pathway gains significance from the high prevalence of gastric carcinoma (the most common carcinoma in the region) as well as the high prevalence of duodenal and pancreatic carcinomas.
    (c) **Splenic nodes** lie along the lateral portion of the greater curvature and the fundus. They receive lymph from the left side of the stomach, the spleen, and the pancreatic tail.
    (d) **Right gastric nodes** lie along the superior region of the pyloric antrum and the pyloric canal.
(2) **Celiac nodes** drain into either a separate celiac trunk or the intestinal lymph trunk, which, in turn, drains into the cisterna chyli.
b. **Superior mesenteric nodes** are located along the root of the superior mesenteric artery. These nodes and the celiac nodes give rise to the **intestinal trunk,** which drains into the **cisterna chyli.**
c. **Inferior mesenteric nodes**
    (1) **Location.** These nodes drain into the superior mesenteric nodes and celiac nodes. They gain significance from the high prevalence of colon and rectal carcinoma.
        (a) **Left colic nodes** are associated with the descending colon and the sigmoid colon.
        (b) **Superior rectal nodes** are associated with the rectum.
    (2) **Drainage course and anastomoses.** Carcinoma of the rectum may metastasize, resulting in wide dissemination of the tumor.
        (a) **Abdominal pathways.** Depending on the course of the inferior mesenteric vein, inferior mesenteric nodes drain into the superior mesenteric nodes or the celiac nodes.
        (b) **Pelvic pathways.** Profuse anastomotic connections occur among the superior, middle, and inferior rectal nodes. **Middle rectal nodes** drain along the internal iliac vessels to the hypogastric nodes. **Inferior rectal nodes** drain along the external iliac vessels to the inguinal and hypogastric nodes.
2. **The cisterna chyli** receives lymphatic drainage from the abdomen, pelvis, and lower extremities.
    a. **Location.** The cisterna chyli lies between the diaphragmatic crura behind and slightly to the right of the aorta. It receives the **celiac trunk** and **intestinal trunk,** thus draining the major portion of the gastrointestinal tract. It also receives drainage from the **hypogastric (lumbar) nodes** to which pass the systemic lymphatics of the lower body wall, lower extremities, and pelvis.
    b. **Drainage course.** It drains into the thoracic duct, which empties into the left subclavian vein at the base of the neck.

VI. INNERVATION

A. **The autonomic nervous system** (the visceral motor system) innervates the GI tract.
1. **Parasympathetic (craniosacral) division** (Figure 25-16)
    a. **The vagus nerve (CN X)** brings parasympathetic presynaptic neurons to the GI tract almost as far as the splenic flexure of the transverse colon.

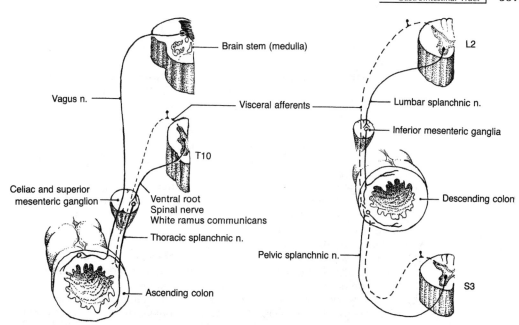

FIGURE 25-16. Innervation of the foregut and midgut *(left)* and the hindgut *(right)*. Preganglionic sympathetic neurons *(solid cell body)* in the thoracic spinal cord levels (T5–L2) send axons through the ventral roots, spinal nerves, white rami communicantes, the sympathetic chain, and splanchnic nerves to reach prevertebral ganglia. The postsynaptic sympathetic neurons *(open cell body)* send adrenergic axons to effectors in the gut walls. The cranial presynaptic parasympathetic neurons *(solid cell body)* in the brain send axons along the vagus nerve to enteric ganglia within the walls of the viscera as far as the splenic flexure. The sacral presynaptic parasympathetic neurons *(solid cell body)* in the spinal cord (levels S2–S4) send axons through the spinal nerves and pelvic splanchnic nerves to reach enteric ganglia in the colon distal to the splenic flexure. The short postsynaptic parasympathetic neurons *(open cell body)* of the enteric ganglia send cholinergic axons to the effectors. Afferent axons *(dashed)*, which convey painful sensation from the gut, travel along the autonomic pathways to both thoracolumbar and pelvic segmental levels of the spinal cord.

(1) **Course.** The vagus nerve reaches abdominal viscera by way of the esophageal plexus. Left and right vagal plexuses consolidate into anterior and posterior vagal trunks as they pass through the diaphragm at the esophageal hiatus.
  (a) The **anterior vagal trunk (left vagus nerve)** courses along the anterior surface of the abdominal esophagus.
    (i) **Hepatic branches** course through the lesser omentum to the **hepatic plexus** on the hepatic artery to innervate the liver, gallbladder, and bile duct. Some fibers course retrograde along the hepatic artery to join the **celiac plexus** and innervate portions of the duodenum and the sphincter of Oddi.
    (ii) **Gastric branches** continue along the anterior surface of the stomach as far as the pylorus to innervate the smooth muscle and glands of the stomach.
    (iii) The **celiac branch** courses along the left gastric artery, contributing to and coursing within the celiac plexus and the superior mesenteric plexus. It supplies the small and large intestines approximately as far as the splenic flexure by coursing along the arteries.
  (b) The **posterior vagal trunk (right vagus nerve)** courses along the posterior surface of the esophagus and stomach.
    (i) **Gastric branches** continue along the posterior surface of the stomach, not quite reaching the pylorus, to innervate smooth muscle and glands of the stomach.
    (ii) The **celiac branch** courses along the left gastric artery, contributing

to and coursing through the celiac plexus and superior mesenteric plexus. It supplies the small and large intestines approximately as far as the **splenic flexure** by coursing along the arteries.
- **(2) Functional considerations**
  - **(a)** The **vagus nerve** consists of preganglionic (presynaptic) neurons (see Figure 25-16). These neurons synapse with postganglionic (postsynaptic) neurons within the enteric plexuses, using acetylcholine as the neurotransmitter (**cholinergic transmission**).
  - **(b)** The short postganglionic neurons are located in the **enteric plexuses** (of Auerbach and Meissner) within the walls of the gut (see Figure 25-16). They synapse with effectors, such as smooth muscle cells and gland cells, and are also **cholinergic**.
  - **(c)** The parasympathetic division functions primarily to promote digestion. Vagal activity increases acid, alkaline, and enzymatic secretion from the various regions of the GI tract. It also enhances peristaltic activity and decreases sphincteric tone.
- **(3) Clinical considerations**
  - **(a) Selective vagotomy** (section of the gastric branches of the anterior and posterior vagal trunks) reduces peptic secretion. This procedure is used in the treatment of chronic gastric and duodenal ulcers.
  - **(b) Total vagotomy** (section of the anterior and posterior vagi on the surface of the abdominal esophagus) has other sequelae, such as a decreased rate of stomach emptying, gallbladder dilation, biliary stasis with predisposition to stone formation, and dumping syndrome.

b. **Nervi erigentes (sacral nerves, pelvic splanchnic nerves)** represent the sacral portion of the parasympathetic division of the autonomic nervous system (see Figure 25-16).
- **(1) Course.** The nervi erigentes arise from spinal levels **S2, S3, and S4** and innervate the lower gastrointestinal tract from the splenic flexure to the pectinate line of the anal canal. They contribute to the **lateral pelvic plexus** in the rectal adventitia.
  - **(a)** Some fibers course retrograde along the rectum, sigmoid colon, and descending colon.
  - **(b)** Other fibers course retrograde in the hypogastric plexus as far as the branches of the inferior mesenteric artery, which they follow to the descending bowel.
- **(2) Functional considerations**
  - **(a)** The long **preganglionic neurons** of the pelvic splanchnic nerves synapse with postganglionic fibers in enteric plexuses, using **cholinergic neurotransmission.**
  - **(b)** The short **postsynaptic neurons** have their cell bodies in **enteric plexuses**. These neurons synapse with effectors, such as smooth muscle cells and secretory cells, and are also **cholinergic**.
- **(3) Clinical considerations**
  - **(a) Aganglionosis** (Hirschsprung's disease) is a genetic defect wherein the parasympathetic ganglia do not develop in the walls of the descending colon.
  - **(b)** This condition is associated with a loss of initiation of propulsive function within the sigmoid colon—a functional blockage—with ensuing megacolon. Extreme stretching of the colon results in ischemia with abdominal signs. **Toxic megacolon** is usually treated surgically.

2. **Sympathetic (thoracolumbar) division** (see Figure 25-16)
   a. **Preganglionic sympathetic neurons** arise from levels T1–L2, or sometimes, L3 (see Figure 4-5). All sympathetic preganglionic neurons use acetylcholine as their neurotransmitter (**cholinergic transmission**).
   - **(1)** The **greater splanchnic nerve** arises from **T5–T9** and synapses with postganglionic fibers in the **celiac ganglion** (see Figure 25-16).
   - **(2)** The **lesser splanchnic nerve** arises from **T10 and T11** and generally synapses

with postganglionic fibers either in the sympathetic chain or in the **superior mesenteric ganglion.**
- (3) The **least splanchnic nerve** arises from **T12,** may be incorporated into the lesser splanchnic nerve, and synapses with postganglionic fibers in the **aorticorenal ganglion** (see Figure 25-16).
- (4) **Lumbar splanchnic nerves** arise from **L1–L2** (occasionally L3) and synapse with postganglionic fibers either in the sympathetic chain or in small ganglia located in the **aortic plexus.**

b. **Postganglionic sympathetic neurons** that innervate the GI tract generally lie in prevertebral ganglia (see Figure 4-5).
- (1) **Prevertebral ganglia** are paired bilaterally and vary in size—sometimes fused on either side to a variable extent.
  - (a) **Celiac ganglia** are located in the celiac plexus and supply sympathetic nerves to that portion of the viscera associated with the celiac artery. Presynaptic input is from spinal segments T5–T9 through the **greater splanchnic nerve.**
  - (b) **Superior mesenteric ganglia** are usually located in the superior mesenteric plexus and supply sympathetic nerves to that portion of the viscera associated with the superior mesenteric artery. Presynaptic input is from spinal segments T10–T11 through the **lesser splanchnic nerve.**
  - (c) **Aorticorenal ganglia** lie in the celiac or aortic plexuses and supply the kidneys. Presynaptic input is from spinal segment T12 by way of the **least splanchnic nerve.**
  - (d) The **ganglia of the aortic plexus** (small ganglia within the adventitia of the aorta) supply viscera associated with the inferior mesenteric artery. Presynaptic input is from spinal segments L1–L2 by **lumbar splanchnic nerves.**
- (2) Postganglionic sympathetic neurons contribute to the various named plexuses (celiac, superior mesenteric, inferior mesenteric, aortic, and hypogastric) and follow the arteries to the effectors, usually the musculature of blood vessels.
  - (a) These fibers are long relative to postganglionic parasympathetic fibers.
  - (b) Postsynaptic sympathetic neurons are **adrenergic,** using norepinephrine as the neurotransmitter.
- (3) **Adrenal medulla**
  - (a) The cells of the adrenal medulla are embryonic equivalents of sympathetic postganglionic neurons; both are derived from neural crest tissue.
  - (b) It secretes **epinephrine** and **norepinephrine** into the circulation, whereby they act on distant effector sites over a longer time.
  - (c) It receives preganglionic innervation from greater, lesser, and least splanchnic nerves.
- (4) **Functional aspects**
  - (a) Activity within the sympathetic division results in constriction of the blood vessels of the GI tract.
  - (b) **Sympathectomy** has no appreciable effect on the gut.
    - (i) It has been used to promote GI vasodilation in patients with extreme hypertension.
    - (ii) It has been used to relieve intractable visceral pain, because the afferents travel along the sympathetic pathways.

B. **Visceral afferent nerves**

1. These nerves are not considered part of the autonomic (visceral motor) nervous system, although they do share pathways.
2. **Reflex and pain pathways.** Except for the pelvis, parasympathetic pathways carry reflex afferents, whereas sympathetic pathways convey pain afferents.
   a. **Afferent reflex neurons** from the foregut (e.g., the gastric filling reflex) travel with the vagus fibers to the brain.
   b. **Pain and pressure afferents** travel along the thoracic splanchnic (sympathetic),

TABLE 25-2. Abdominal Visceral Afferent Pathways

Organ	Pathway	Spinal Levels	Referral Regions
Thoracic esophagus	Thoracic splanchnic nn.	T1–T4	Middle thorax
Abdominal esophagus			
Stomach			
Gallbladder	Celiac plexus and greater splanchnic n.	T5–T9	Low thorax, epigastric region
Liver			
Bile duct			
Sup. duodenum			
Inf. duodenum			
Jejunum			
Ileum	Superior mesenteric plexus and lesser splanchnic n.	T10–T11	Umbilical region
Appendix			
Ascending colon			
Transverse colon			
Kidneys	Aorticorenal plexus, least splanchnic n. and upper lumbar splanchnic nn.	T12–L1	Lumbar and inguinal regions (ipsilaterally)
High ureters			
Gonads			
Desc. colon	Aortic plexus and lower lumbar splanchnic nn.	L1–L2	Pubic and inguinal regions, ant. scrotum or labia, ant. thigh
Sigmoid colon			
Midureters			

lumbar splanchnic (sympathetic), and pelvic splanchnic (parasympathetic) pathways to the spinal cord (see Figure 4-5).
   (1) **Pain pathways** (Table 25-2)
      (a) **Foregut** pain pathways travel along the celiac plexus and greater splanchnic nerve to spinal levels T5–T9 (see Figure 25-16). Pain is referred to the low-central thoracic and epigastric regions.
      (b) **Midgut** pain pathways travel along the superior mesenteric plexus and lesser splanchnic nerve to levels T10 and T11 (see Figure 25-16). Pain is referred to the umbilical and hypogastric regions.
      (c) **Hindgut** (see Figure 25-16)
         (i) **From the descending and sigmoid colon,** pain pathways travel along the inferior mesenteric plexus and then along the least splanchnic nerve and lumbar nerves to spinal levels T12–L2. Pain is referred to the pubic and inguinal regions as well as to the anterior and lateral thigh.
         (ii) **From the rectum,** pain pathways travel along the pelvic plexus to the pelvic splanchnic nerves to reach spinal levels S2–S4. Pain is referred to the perineum and posterior thigh.
   (2) **Referred pain.** The particular afferent pathway explains the basis for referred pain (see Table 25-2).
      (a) Visceral nociceptive input at any thoracic or lumbar level of the spinal cord usually cannot be distinguished from that of somatic origin.
      (b) Pain is referred to the somatic dermatome associated with the particular splanchnic nerve.

Chapter 26
Kidneys and Posterior Abdominal Wall

I. KIDNEYS

A. Introduction

1. **Location.** The paired kidneys lie posteriorly in the abdominal cavity.
   a. The **kidneys are retroperitoneal,** embedded in a considerable amount of perinephric and paranephric fat (see Figure 26-5).
   b. About the size of a clenched fist, each kidney normally weighs about 145 g.

2. **Relationships.** Each kidney lies on the ventral surface of the quadratus lumborum muscle, just lateral to the psoas muscle and vertebral column. Both lie against the inferior surface of the diaphragm posterosuperiorly and are capped by an adrenal gland anteromedially (see Figure 26-2).
   a. The **right kidney** is lower than the left (2–8 cm, 1–2 inches) because of the presence of the liver on the right side. Posteriorly, it is related to the twelfth rib. Anteriorly, it is related to the liver, duodenum, and hepatic flexure of the large bowel.
   b. The **left kidney** is related posteriorly to the eleventh and twelfth ribs. Anteriorly, it is related to the pancreas, spleen, and splenic flexure of the large bowel.

3. **The renal fascia** (false capsule or Gerota's fascia) is a discrete fascial layer that surrounds each kidney (see Figure 26-5).
   a. **Structure.** This layer arises from the prevertebral fascia that splits to enclose each kidney with its associated adrenal gland. Fusion does not occur inferiorly or medially.
      (1) **Divisions.** The renal fascia divides the fat associated with the kidney into two distinct regions.
         (a) **Perirenal (perinephric) fat** is contained within the renal fascia and periureteral sheath. It is immediately external to the true renal capsule and extends inferiorly along the ureter.
         (b) **Pararenal (paranephric) fat** is external to the renal fascia. It is most abundant posterolaterally.
      (2) **Extensions.** Although the renal fascia fuses around the renal vessels, it is open inferiorly along each ureter, forming the **periureteral sheath.**
         (a) Infection or renal abscess may be limited by the renal fascia, but it can track within the periureteral sheath to the pelvis.
         (b) Air injected into the retroperitoneal tissue of the pelvis rises within the renal fascia to surround the kidney. This technique was used to enhance radiographic visualization of the kidneys and adrenal glands.
   b. **Support.** Because the kidneys are retroperitoneal, they are not supported by ligament or mesentery.
      (1) **Position.** Deposits of paranephric and perinephric fat maintain the position of each kidney.
         (a) This fat (similar to periorbital and ischiorectal fat) is slow to be reabsorbed in wasting disease or acute starvation.
         (b) Renal vessels provide some support, although it is uncertain how much.
      (2) **Mobilization.** Normally, kidneys do not exceed a 9-cm (3–4-inch) excursion when an individual changes position (supine to erect).
      (3) **Nephroptosis** (floating kidney) occurs frequently among truck drivers, horseback riders, and motorcyclists.
         (a) An aberrant inferior polar artery may compress the ureter with resultant hydronephrosis.
         (b) The pain of nephroptosis apparently results from traction on the renal vessels.

FIGURE 26-1. Right kidney. Note the division of the renal parenchyma into cortex and medulla. Renal pyramids empty into calyces, which coalesce into the renal pelvis—the dilated proximal portion of the ureter. The renal sinus contains the renal vasculature, renal pelvis, and perinephric fat.

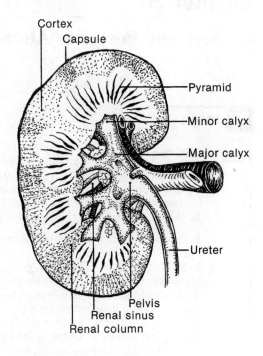

B. **External structure**

1. **The renal capsule,** the true fibrous capsule of the kidney, provides a barrier against the spread of infection.

2. **The renal sinus,** the cavity of the kidney, opens medially at the **hilus.** It contains several important structures that enter and leave the kidney (Figure 26-1).
   a. The **renal artery** branches extensively within the sinus to supply **renal segments.**
   b. The **renal vein** leaves the kidney at the renal hilus.
   c. The **renal pelvis,** the funnel-shaped proximal end of the ureter, is continuous with the ureter proper distally and with the major calyces within the kidney.
   d. **Perirenal fat** fills the spaces between the various structures within the renal sinus.

C. **Internal structure**

1. **The basic unit** of the kidney is the microscopic **nephron.**
   a. The **location and orientation** of the different parts of the nephron produce the gross characteristics of the kidney parenchyma, that is, cortex and medulla.
   b. **Characteristics of the nephron**
      (1) **Renal corpuscle.** Located in the cortex, an expanded renal corpuscle (Bowman's capsule) begins the nephron. The profuse capillary supply (glomerular capillary tuft) to the renal corpuscles gives the cortex a reddish hue.
      (2) **Proximal tubule.** The renal corpuscle empties into the proximal convoluted tubule, which lies in the cortex, straightens, and then descends into the medulla.
      (3) **Distal tubule.** In the medulla, the straight portion of the convoluted tubule turns (loop of Henle) toward the cortex, becoming the straight portion of the distal tubule. This segment returns to the cortex and becomes convoluted.
      (4) **Collecting ducts.** Numerous distal tubules join collecting ducts that descend through the medulla and empty into a minor calyx.
      (5) **Minor calyx.** About 500 collecting ducts, distributed within 9–14 pyramids in the normal adult kidney, terminate in 7–9 minor calyces.

2. **Divisions.** The renal parenchyma is divisible into two parts (see Figure 26-1).
   a. The **cortex** consists primarily of renal corpuscles, proximal convoluted tubules, and distal convoluted tubules.

b. The **medulla** is formed by 12–18 **renal pyramids,** each of which consists primarily of straight tubules and collecting ducts.
3. **Subdivisions.** The kidney is subdivided into lobes or pyramids (see Figure 26-1).
   a. **Each renal pyramid,** with the overlying cortex and adjacent regions of renal columns, forms the basic functional unit of the kidney (the lobe). Fusion of lobes (pyramids) at the apices forms 8–12 renal papillae.
   b. **Each renal papilla** opens into the cup-shaped end of a **minor calyx,** which is directed anteriorly or posteriorly.
4. **Collecting system** (see Figure 26-1)
   a. **Minor calyces** (7–9) receive one to three renal papillae (each formed from one or two pyramids) before fusing into major calyces.
   b. **Major calyces** (2–4) join in the renal sinus to form the funnel-shaped renal pelvis.
   c. The **renal pelvis** narrows in the renal hilus to form the ureter.

D. Vasculature

1. **Renal arteries.** The renal blood supply is profuse. The kidneys receive approximately one fifth of the cardiac output.
   a. **Origin and course.** The left and right renal arteries arise from lateral aspects of the aorta, usually between L1 and L2 just below the origin of the superior mesenteric artery. The **right renal artery** passes posterior to the inferior vena cava.
   b. **Extrarenal branches** that arise from the left and right renal arteries (Figure 26-2) include:
      (1) Inferior suprarenal arteries to the adrenal glands
      (2) Numerous ureteral twigs
      (3) Occasional gonadal arteries (15%)
      (4) Occasional inferior phrenic arteries (10%)
   c. **Intrarenal branches.** Renal arteries divide into **segmental arteries** within the renal sinus. They are end-arteries, with little or no anastomoses.
      (1) **Renal vascular segments.** Five renal segments are supplied by similarly named arteries (see Figure 26-2):
         (a) Superior (apical)
         (b) Anterior–superior (upper)
         (c) Anterior–inferior (middle)
         (d) Inferior (lower)
         (e) Posterior
      (2) **Renal divisions.** The distributions of the anterior–superior and anterior–inferior branches, as distinct from the posterior branch, divide the kidney into **anterior** and **posterior divisions.**
         (a) Few anastomoses occur between the anterior segments and the posterior segment, appropriately named the avascular line (Brodel's white line). A longitudinal incision along the **avascular line** produces minimal damage

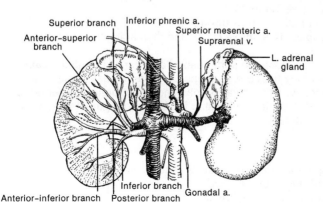

FIGURE 26-2. Blood supply and venous drainage of the kidneys and adrenal glands. *(Left)* The five segmental arteries of the right kidney, as well as the three arteries that supply the right adrenal gland. *(Right)* Venous drainage of the kidney and adrenal gland. Note the relationship between the renal vasculature and the great abdominal vessels.

to the blood supply with little bleeding. This approach is used for removal of renal (staghorn) calculi.
- (b) Ligation of a segmental artery results in necrosis of the entire segment.
- d. **Variations.** Aberrant or supernumerary segmental arteries are common (32%). They derive from the fetal lobation pattern with failure of the renal arterial segments to fuse into a single renal artery.
  - (1) About one half of all **aberrant segmental arteries** are **hilar** (arising from the renal artery); the other half are **polar** (arising directly from the aorta).
    - (a) **Polar arteries** tend to be larger than hilar arteries. Inferior polar arteries typically pass anterior to the ureter. Compression of the ureter by an **inferior polar artery** produces **hydronephrosis.**
    - (b) **Supernumerary renal arteries** are not considered collaterals because each supplies one segment. Ligation usually produces necrosis of a renal segment.
  - (2) **Angiographic studies** before renal surgery are advisable because lack of awareness of arterial variation can result in overwhelming hemorrhage. A torn polar artery usually separates where it arises from the aorta, leaving no stump to clamp for control of resulting hemorrhage. Control is especially difficult on the right side, behind the inferior vena cava.
2. **Renal veins** (see Figure 26-2)
   - a. The **right renal vein** enters the inferior vena cava at a point lower than the left renal vein. It usually has no significant tributaries.
   - b. The **left renal vein** is longer than the right renal vein and passes anterior to the aorta. It receives the left gonadal, left suprarenal, and left inferior phrenic veins, and communicates with the azygos vein.
   - c. **Variations**
     - (1) Multiple renal veins are less common (14%) than supernumerary arteries (32%).
     - (2) The most common single variation is doubling of a renal vein (6%).

E. Innervation

1. **Sympathetic pathways** principally supply the renal vasculature.
   - a. **Sympathetic preganglionic nerves** from T12–L2 pass through the sympathetic chain ganglia and run in the least splanchnic nerve and possibly in the lumbar splanchnic nerves.
   - b. **Sympathetic postganglionic cells** lie in aorticorenal ganglia and course within the renal plexus to the kidneys.
   - c. **Sympathetic function** is vasomotor.
     - (1) Excessive sympathetic stimulation reduces urine production.
     - (2) Sympathectomy, interrupting input to the aorticorenal ganglion, produces renal vasodilation and transient renal diuresis.
2. **Afferent pathways**
   - a. **Visceral afferent nerves** from the kidneys and abdominal ureters reach the spinal cord along two pathways (see Table 25-2).
     - (1) The principal afferent pathway runs through the renal plexus, aorticorenal ganglion, and **least splanchnic nerve** (T12), and then along the corresponding white ramus communicans.
     - (2) Some afferent fibers run along the aortic plexus and the first **lumbar splanchnic nerve,** and then along the corresponding L1 white ramus communicans.
   - b. **Pain** derived from the kidney, its vascular supply, and upper ureter is referred to the T12–L1 dermatomal distribution (subcostal, lumbar, and inguinal regions, as well as the anterosuperior thigh).

F. Functions

1. **Electrolyte balance.** Principal functions of the kidney are conservation of minerals and maintenance of ionic balance in body fluids.
   - a. Each kidney contains 106 or more **nephrons.**

(1) In the **renal corpuscle,** an ultrafiltrate of plasma (provisional urine) exudes through the glomerular capillaries into Bowman's capsule.
(2) Along the **proximal tubule,** select substances are actively absorbed from the provisional urine, resulting in osmotic uptake of water. In addition, some substances are secreted into the provisional urine by the tubular epithelium.
(3) Along the **distal tubule,** sodium and water may or may not be absorbed, depending on the circulating concentrations of antidiuretic hormone and aldosterone.
(4) From numerous nephrons, **collecting ducts** receive the fluid that remains as urine.
(5) Each **minor calyx** receives about 500 collecting ducts.
  b. Although approximately 150–200 L of provisional urine are produced each day, resorption leaves only 1–2 L of urine to be voided by micturition.
 2. **Blood pressure regulation.** The kidneys also produce vasoactive substances (angiotensin) or their precursors, which control blood pressure. One such precursor, **renin,** is produced by the juxtaglomerular apparatus where the distal convoluted tubule joins the afferent glomerular arteriole.
 3. **Clinical considerations**
    a. **Systemic hypertension.** A kidney with a stenotic or occluded renal artery, or one that is nonfunctional concerning production of urine, produces an overabundance of renin, with resultant systemic hypertension.
    b. **Intravenous pyelography (IVP).** This examination offers visualization of the renal outflow tract. Contrast material, injected intravenously, is excreted by the kidney and concentrated in the outflow tract, where it appears radiopaque. Images obtained at intervals demonstrate this process.
    c. **Kidney stones.** Urinary components (calcium compounds and urea) may precipitate in the calyces, and concretions (nephroliths) may develop. Nephroliths usually are small enough to pass through the ureter and are eliminated. They may be of sufficient size, however, to lodge within and obstruct a ureter. Large staghorn concretions may completely occlude a calyx.

II. URETERS

A. **Course and relations.** The ureters are the excretory ducts between the kidneys and the urinary bladder.
 1. **Within the kidney,** the **renal pelvis,** lying in the renal sinus, narrows at the ureteropelvic junction to form the ureter (see Figure 26-1). A staghorn calculus may lodge at this point.
 2. **In the abdomen,** each ureter descends retroperitoneally within the **periureteral sheath** (Figure 26-3).
    a. The **right ureter** (see Figure 26-3) passes posterior to the descending portion of the duodenum and posterior to the root of the mesentery proper. It then passes posterior to the right gonadal vessels, which contribute important vascular twigs.
    b. The **left ureter** (see Figure 26-3) passes posterior to the left colic vessels and posterior to the sigmoid mesocolon. It is next to the left gonadal vessels, which contribute important vascular twigs to the left ureter.
    c. **Each ureter** passes anterior to a psoas muscle and the common iliac vessels, before diving over the brim of the deep pelvis—a second point of narrowing.
      (1) Calculi may lodge at this point, one of the narrowest parts of the ureter.
      (2) At this location, the ureters may receive twigs from the common and internal iliac arteries. Also, a ureter may be affected by iliac artery aneurysm.
 3. **In the deep pelvis,** each ureter courses retroperitoneally anterior to the sacrum to reach the urinary bladder (see Figure 26-3).
    a. After crossing the common iliac vessels, each ureter receives vascular twigs from

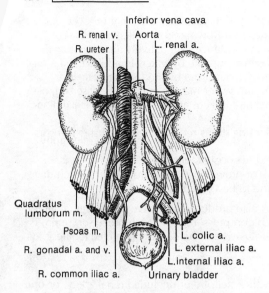

FIGURE 26-3. Ureters and urinary bladder. The course of the left and right ureters is shown. The blood supply to the ureters is from numerous sources.

the common and internal iliac arteries. As each ureter approaches the urinary bladder, it receives twigs from the internal iliac vessels.
  b. In **males,** the ureters pass inferior to the vas deferens and its associated deferential artery.
  c. In **females,** the ureters pass inferior to the cardinal (lateral cervical) ligaments and uterine vessels, where they may be inadvertently clamped or sectioned along with uterine vessels.
  d. The ureters converge to enter the bladder posteroinferiorly.
  e. The ureter narrows as it passes through the bladder wall, providing a third point where nephroliths may lodge.

B. Vasculature

1. **Arterial supply.** The blood supply to the ureters is diffuse and variable. Each ureter receives small arterial twigs from the **renal arteries, aorta,** small arteries of the **posterior abdominal wall, gonadal arteries,** and **common and internal iliac arteries,** as well as from the **inferior vesical arteries** (see Figure 26-3).

2. **Mobilization** of a ureter, or even traction on a ureter during surgery, should be avoided, because interruption of its delicate blood supply results in ischemic necrosis.

C. Innervation

1. **Motor pathways** are specific to each region of the ureter.
   a. **Upper abdominal ureter.** The **least splanchnic nerve** from T12, with synapses in the **aorticorenal ganglion,** supplies sympathetic innervation to the renal pelvis.
   b. **Lower abdominal ureter.** The **lumbar splanchnic nerves** from L1–L2 supply sympathetic innervation to the abdominal and pelvic portions of the ureter.
   c. **Pelvic ureter.** The **pelvic splanchnic nerves** from S2–S4 supply parasympathetic innervation to the entire ureter.

2. **Visceral afferents** also travel along the least splanchnic, lumbar splanchnic, and pelvic splanchnic nerves, with pain referred accordingly (see Table 25-2).
   a. **Upper ureter.** Obstruction and distention of the renal pelvis cause referral of pain to the lumbar region (T12 and L1).
   b. **Abdominal ureter.** Obstruction and distention of the abdominal portion cause pain referral to inguinal and pubic regions, the anterior scrotum or mons pubis, and the superoanterior thigh (L1 and L2).

c. Pelvic ureter. Obstruction and distention of the pelvic portion cause referral of pain to the perineum and sometimes to the posterior thigh and leg (S2–S4).

D. Clinical considerations

1. **Urination**
   a. The ureters contain a thick muscular wall, composed of circular and longitudinal layers, that moves the urine toward the bladder by waves of contraction (urination) that occur one to six times per minute.
   b. Evidence of normal **peristaltic activity** is seen during intravenous pyelography.

2. **Nephroliths** (concretions of calcium compounds and urea) may lodge in or pass slowly through regions of ureteral narrowing.
   a. The ureter narrows at the junction between the renal pelvis and the ureter proper; where the ureter crosses the pelvic brim; and at the bladder, where the ureter passes through the wall.
   b. Ureteral distention (**hydronephrosis**) proximal to the stone causes intense, excruciating pain (renal colic).
   c. If the stone fails to move despite copious imbibition, prolonged hydronephrosis may produce kidney damage.
      (1) Stone fragmentation is possible by passing instrumentation through the urethra and urinary bladder into the obstructed ureter.
      (2) A noninvasive treatment involves shattering the calculus ultrasonically (lithotripsy).

3. **Developmental anomalies**
   a. Renal and ureteral anomalies noted in 4% of individuals include **polycystic kidneys,** in which the collecting tubules fail to join a calyx; **horseshoe kidney,** in which the two kidneys are joined across the midline; **lobated kidneys,** which maintain their fetal pattern of segmentation; and **ectopic kidneys,** which may be located in the pelvis or both on the same side of the body.
   b. Numerous branching patterns are possible in the ureter.
   c. Dural ureters have a slower flow of urine than normal ureters. They are, therefore, more susceptible to infection, which can spread to the kidneys from the urinary bladder.

4. **Injury.** Careful identification and subsequent avoidance of the ureters are major objectives of abdominopelvic surgery. In female patients, ureters are frequently injured as a result of inadvertent clamping, ligation, or sectioning along with the ovarian vessels or the uterine vessels. Injuries must be recognized and repaired to prevent extravasation of urine and subsequent peritonitis.

III. ADRENAL GLANDS

A. Location. The adrenal glands are anteromedial to the kidneys, between the upper renal poles and the diaphragm (see Figure 26-2). They lie within the perirenal fat (adipose capsule) deep to the renal fascia.

1. **The right adrenal** is pyramidal and generally more horizontal, across the apical pole of the right kidney.

2. **The left adrenal** is semilunar and generally more vertical, adjacent to the apical pole of the left kidney.

B. Structure. A **hilus** is located on the anterior surface from which the adrenal vein emanates. The adrenal parenchyma is divided into two zones.

1. **The adrenal cortex,** the outer zone, is derived from the mesoderm of the embryonic urogenital ridge. It produces three classes of steroid hormones.
   a. **Mineralocorticoids** (such as **aldosterone**) are necessary for salt metabolism.

(1) Deficiency of mineralocorticoids results in **Addison's disease**.
(2) Adrenal hyperplasia with excessive mineralocorticoid secretion results in **Conn's syndrome**.
   b. **Glucocorticoids** (such as **cortisone**)
      (1) These hormones affect metabolism, especially that of the connective tissues.
      (2) Adrenal hyperplasia with excessive glucocorticoid secretion leads to **Cushing's syndrome**.
   c. **Male and female sex hormones** (**estrogens** and **androgens**) are produced in the adrenals of both sexes.
      (1) Hyperplasia with excessive secretion of these hormones produces the adrenogenital syndrome, in which adipose and hair patterns are affected. Depending on the hormone produced, the result is either **feminization of males** or **masculinization of females**.
      (2) Sex hormones produced by the adrenals are metabolized in the liver. Liver dysfunction in males (often associated with alcohol-related liver disease) results in high circulating levels of estrogen, with **gynecomastia** and the development of a feminine escutcheon.
2. **The adrenal medulla,** the inner zone, is derived from embryonic neural crest tissue, which also gives rise to sympathetic ganglia. It consists primarily of chromaffin (pheochrome) cells, which produce adrenergic hormones.
   a. **Innervation.** The adrenal medulla is innervated by presynaptic sympathetic neurons that travel variably in the greater, lesser, least, and lumbar splanchnic nerves. The chromaffin cells are homologous to postsynaptic sympathetic neurons.
   b. **Function.** It secretes **epinephrine (Adrenalin)** into the circulation in response to cholinergic stimulation. This action provides longer term sympathetic control than that mediated directly by postganglionic sympathetic nerves.

C. Vasculature

1. **Adrenal arteries** arise from three sources (see Figure 26-2).
   a. **Superior suprarenal arteries** (one or more) arise from each inferior phrenic artery.
   b. **Middle suprarenal arteries** arise from the aorta on each side. The artery to the right adrenal lies behind the inferior vena cava and is more difficult to ligate.
   c. **Inferior suprarenal arteries** (one or more) arise from each renal artery or even a superior polar artery.
2. **One suprarenal vein** exits at the hilus of each adrenal gland (see Figure 26-2).
   a. The **right adrenal vein** usually drains medially into the inferior vena cava.
   b. The **left adrenal vein** usually drains inferiorly into the left renal vein.

D. Clinical considerations

1. **Pheochromocytoma** is a tumor of the adrenal medulla. Excessive or life-threatening bursts of epinephrine and norepinephrine, which result in paroxysms of hypertension, may be released from these tumors on sympathetic activation or even abdominal palpation. Care is needed when palpating the abdomen of a patient in whom pheochromocytoma is suspected to avoid excess epinephrine release.
2. **Adrenalectomy**
   a. The suprarenal vein must be ligated before manipulation of the adrenal gland so that catecholamines do not escape into the circulation.
   b. The right adrenal is more difficult to approach surgically than the left adrenal because it is in part posterior to the inferior vena cava.
3. **Ectopic adrenal tissue**
   a. Ectopic adrenal tissue usually consists of both cortex and medulla.
   b. Pheochromocytomas may occur in this tissue anywhere along the embryonic urogenital ridge.

TABLE 26-1. Muscles of the Posterior Abdominal Wall

Muscle	Origin	Insertion	Primary Action	Innervation
Diaphragm	Costal margin, vertebrae L1–L4	Central tendon	Inspiratory: lowers diaphragm	Phrenic n. (C3–C5), twigs from spinal nn T12–L2
Iliopsoas	Iliac fossa and transverse processes T12–L4	Lesser trochanter of femur	Flexes vertebral column	Spinal nn. T12–L4
Quadratus lumborum	Iliac crest, transverse processes L4–L5	Rib 12	Stabilizes and lowers 12th rib; abducts vertebral column	Spinal nn. T12–L4

IV. POSTERIOR ABDOMINAL WALL

A. Structure. The posterior abdominal wall is formed by the diaphragm, the bilateral quadratus lumborum muscle, the iliopsoas muscle, and the thoracolumbar fascia (see Table 26-1).

1. **The diaphragm** is a dome-shaped muscular sheet separating the thoracic from the abdominal cavity (Figure 26-4). The upper surface is covered by **parietal pleura** and **pericardium.** The lower surface is covered by **parietal peritoneum.**
   a. Attachments. The **diaphragm** originates from the sternum, costal margin, and lumbar vertebrae. It inserts into itself at the central tendon.
      (1) The **sternal origin** is the posterior surface of the xiphoid process.
      (2) The **costal origin** comprises the inner surfaces of the costal margin (costal cartilages 7–9 as well as the tips of ribs 10–12).
      (3) The **lumbar origins** are by two **crura** that arise from the lateral aspects of the upper three or four lumbar vertebrae.
         (a) The **lateral arcuate ligament** is formed by the free edge of the diaphragm between the tip of the twelfth rib and the transverse process of L1. Beneath it runs the quadratus lumborum muscle.
         (b) The **medial arcuate ligament** is formed by the free edge of the diaphragm between the transverse process of L1 and vertebral bodies of L1 and L2. Beneath it runs the most cephalad portion of the psoas muscle.

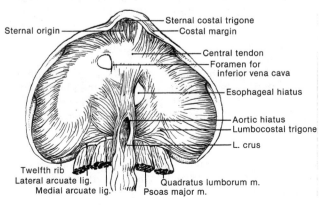

FIGURE 26-4. Diaphragm (inferior surface), with its three hiatuses.

b. Action. The diaphragm is an inspiratory muscle. Diaphragmatic contraction pulls the central tendon caudally, thereby increasing thoracic volume and decreasing intrathoracic pressure. The muscles of the abdominal wall are antagonistic to the diaphragm.
   c. Hiatuses provide access between the thoracic and abdominal cavities (see Figure 26-4).
      (1) The **aortic hiatus** is in approximately the midline about the T12 level between the diaphragmatic crura. It transmits the aorta, azygos vein, thoracic duct, and occasionally, splanchnic nerves.
      (2) The **caval hiatus** (foramen of the inferior vena cava) is at about the T8 level to the right of midline and transmits the inferior vena cava.
      (3) The **esophageal hiatus** sits at the T10 level and transmits the esophagus and vagal trunks.
         (a) This hiatus lies slightly left of midline in the muscular portion of the diaphragm.
         (b) It is the site of **hiatus hernia.**
            (i) Hiatus hernia is relatively common (1%), accounting for 98% of all diaphragmatic hernias.
            (ii) The cardiac end of the stomach may herniate through this hiatus into the thorax (sliding hiatus hernia).
      (4) The splanchnic nerves and the hemiazygos vein usually pierce the crura.
   d. Innervation
      (1) Motor innervation
         (a) The **phrenic nerves** (C3–C5) innervate the dome, which is logical because portions of the diaphragm develop in the cervical region along with the heart (see Figure 22-1).
         (b) Lumbar nerves (L1–L2) provide motor innervation to the crural regions.
      (2) Sensory innervation
         (a) Phrenic nerves. Sensation from the larger central region is carried by phrenic nerves to segments C3–C5 of the spinal cord. Pain associated with irritation of the diaphragmatic peritoneum (such as by free air or blood) is referred to the **shoulder region.**
         (b) Intercostal nerves. Sensation from the peripheral region is supplied by intercostal nerves. Pain from the periphery of the diaphragm is referred to the **thoracic and abdominal walls.**
   e. Weak areas are potential sites for hernias.
      (1) The **sternocostal triangle** lies between the sternal and costal portions. This area transmits superior epigastric vessels and is a potential site for herniation (usually acquired).
      (2) The **lumbocostal triangle** is situated between costal and lumbar portions.
         (a) This area may be nonmuscular, with only two layers of serous membrane separating the thoracic from the abdominal cavity, or it may be incomplete (foramen of Bochdalek).
         (b) Although the stomach is the most commonly herniated viscus, herniation of most peritoneal viscera (except the sigmoid colon) has been reported.
         (c) Diaphragmatic hernia is 8–10 times more common on the left side than on the right side.
2. **The quadratus lumborum muscle** is the most lateral muscle of the posterior abdominal wall (Table 26-1; see Figure 26-4).
   a. Attachments. This muscle runs between the iliac crest and the inferior border of the twelfth rib. It passes beneath the lateral arcuate ligament of the diaphragm.
   b. Actions. This muscle is involved in abduction (lateral flexion) of the vertebral column as well as fixation of the twelfth rib during inspiration, and depression of the twelfth rib during forced expiration.
3. **The iliopsoas muscle** has two separate muscular heads (see Table 26-1).
   a. Attachments. The **iliac head** (from the iliac fossa) and **psoas head** (from vertebrae T12–L4) fuse inferiorly to form the iliopsoas muscle, which inserts into the lesser trochanter of the femur (see Figure 14-3).

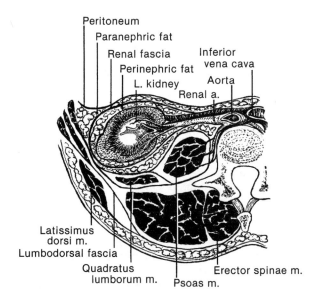

FIGURE 26-5. Posterior abdominal wall and renal fascia.

   b. Actions. This muscle is involved in flexion of the vertebral column as well as of the pelvis at the hip joint, when the leg is firmly planted. It also flexes the thigh at the hip joint when the leg is not planted.
   c. This muscle is related to the kidneys (anterolaterally), the ureters (crossing anteriorly), the sympathetic chain (anteromedially), and the common iliac vessels (medially). The lumbar plexus passes through the psoas muscle.

4. **Lumbodorsal (thoracodorsal) fascia** (Figure 26-5)
   a. **Attachments.** The posterior abdominal wall lateral to the quadratus lumborum muscle consists largely of lumbodorsal fascia. This layer is formed by fusion of the posterior aponeuroses of the internal oblique and transverse muscles.
   b. **Divisions.** This fascia splits into three leaves, which form two muscular compartments.
      (1) The **posterior leaf** inserts into the vertebral spinous processes and provides the site of origin for the latissimus dorsi muscle.
      (2) The **middle leaf** inserts into the tip of the vertebral transverse processes. With the posterior leaf, it encloses the erector spinae muscle.
      (3) The **anterior leaf** inserts midway along or at the base of the transverse processes. The psoas fascia arises about midway along the middle leaf.
   c. The **lumbar trigone** (of Petit) is bounded by the posterolateral free edge of the **external oblique muscle,** the anterolateral border of the **latissimus dorsi muscle,** and the superior aspect of the **iliac crest.**
      (1) The floor of the trigone (triangle) consists of the lumbodorsal fascia and, sometimes, fibers of the internal oblique muscle.
      (2) Because the abdominal wall is thin at this point, this trigone may be the site of a lumbar hernia. It is used in surgical approaches to the kidneys and ureters.

B. Posterior abdominal vessels (see Figure 26-3)

1. **The abdominal aorta** enters the abdomen through the **aortic hiatus** between the diaphragmatic crura. It descends on the anterior aspect of the vertebral column, deviating slightly to the left as it nears the aortic bifurcation.
   a. **Relations.** The aorta lies to the left of the inferior vena cava.
      (1) Passing immediately anterior to the aorta are the pyloric antrum, the body and a portion of the head of the pancreas, left renal vein, and inferior duodenum.
      (2) It is associated with the celiac and superior mesenteric plexuses and the in-

cluded ganglia as well as with the aortic, inferior mesenteric, and hypogastric plexuses.
- b. **Branches.** Before bifurcating into the **common iliac arteries,** the abdominal aorta gives off numerous paired and unpaired branches.
  - (1) **Paired branches** include the inferior phrenic, middle adrenal, renal, gonadal, lumbar, and intervertebral arteries.
  - (2) **Unpaired branches** include the celiac (to the foregut), superior mesenteric (to the midgut), and inferior mesenteric (to the hindgut) arteries, as well as numerous ureteral twigs. The aorta terminates as the middle sacral artery.
- c. **Clinical considerations**
  - (1) The **aortic pulse** is palpable through the anterior abdominal wall and often can be observed in lean individuals.
  - (2) **Aneurysms** of the aorta commonly involve the origins of the aortic branches. A large aneurysm may erode into an adjacent vertebra, with resultant back pain.
  - (3) **Coarctation** (occlusion) of the aorta
    - (a) Stenosis superior to the renal arteries is more likely to be fatal, because the kidneys require far more blood than can flow through collateral vessels.
    - (b) Slow occlusion below the renal arteries results in progressive development of collateral pathways.
      - (i) A major shunt between the internal thoracic artery and the inferior epigastric artery develops, with enlargement of the intercostal and abdominal arteries.
      - (ii) The marginal artery (of Drummond) may hypertrophy when occlusion occurs between the origins of the superior and inferior mesenteric arteries.

2. **The inferior vena cava** lies to the right of the aorta along the anterior aspect of the vertebral column (see Figure 26-3). It enters the thorax through the **caval hiatus** (foramen of the inferior vena cava).
   - a. **Branches.** From its formation by the coalescence of the common iliac veins, the vena cava receives several branches, including the lumbar, right gonadal, renal, right adrenal, right inferior phrenic, and hepatic veins.
   - b. **Variations.** The inferior vena cava develops from three distinct embryologic venous pathways that fuse and atrophy in various places. At one time, there are two parallel venae cavae, which explains why the adult vena cava does not receive the left adrenal and gonadal vessels directly.
     - (1) Numerous collateral pathways and variations form, depending on which embryonic sections are retained or atrophy.
     - (2) The vein may remain doubled or it may be absent, its drainage functions served by an enlarged azygos system entering the thorax through the aortic hiatus in the diaphragm.

C. Lymphatic channels

1. **Common iliac nodes** are adjacent to the common iliac vessels. They receive drainage from the lower limbs, the perineum, and the gluteal region by way of the **inguinal nodes.** Three nodes drain into the aortic nodes.

2. **Aortic (lumbar) nodes** lie next to the aorta and form the lumbar lymphatic trunk, often one on each side of the aorta. They receive drainage from the posterior abdominal wall and iliac nodes and drain into the cisterna chyli.

3. **The cisterna chyli** lies posterior and slightly to the right of the aorta, usually between the crura of the diaphragm at the aortic hiatus.
   - a. It receives drainage from the intestinal lymphatic trunk and from the celiac and superior mesenteric nodes, as well as from the aortic (lumbar) lymphatic trunks.
   - b. It represents the lower expanded portion of the thoracic duct.
     - (1) The **thoracic duct** courses through the posterior mediastinum between the aorta and the vertebral column (see Figure 22-2).

(2) It empties into the left subclavian vein where this vessel joins the internal jugular vein.

D. Nerves

1. **Sympathetic nerves** of the posterior abdomen arise from the sympathetic chain (see Chapter 25 VI A).
   a. Paired **sympathetic trunks** run in a groove between the psoas muscle and the vertebral column (see Figure 26-6).
   b. **Thoracic splanchnic nerves** (greater, lesser and least), arising from T5–T12, innervate portions of the gastrointestinal tract, kidneys, and ureters (see Figure 4-5). These nerves pierce the diaphragm to reach the celiac, superior mesenteric, and inferior mesenteric ganglia.
   c. **Lumbar splanchnic nerves,** arising from ganglia L1–L2, innervate portions of the gastrointestinal tract, kidneys, and ureters.

2. **Somatic nerves** of the posterior abdomen arise from the lumbar plexus (Figure 26-6).
   a. The **lumbar plexus,** the upper portion of the **lumbosacral plexus,** is formed by anterior primary rami of spinal nerves L1–L5 as the extremities develop from the segmental dermatomes and myotomes. The lumbar plexus divides anteriorly and posteriorly.
      (1) The **posterior division** of the lumbar plexus is equivalent to the lateral branch of a spinal nerve. It supplies the lateral region of the body wall and the primitively dorsal surface of the lower extremity.
      (2) The **anterior division** of the lumbar plexus is equivalent to the continuation of a spinal nerve distal to the lateral branch. It supplies the ventral region of the body wall and the primitively ventral surface of the lower extremity.
   b. **Major nerves** of the lumbar plexus innervate the abdominal musculature and convey sensation from the skin and parietal peritoneum (see Figure 26-6).
      (1) **Iliohypogastric nerve** (T12–L1, anterior)
         (a) It emerges from the lateral side of the psoas muscle to run within the musculature of the abdominal wall.
         (b) Its **iliac branch** provides sensation to the upper gluteal region.

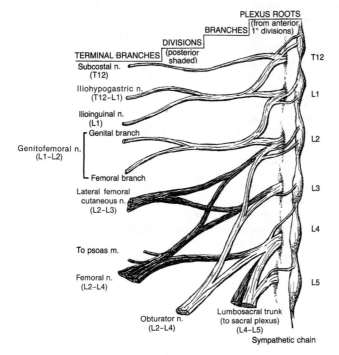

FIGURE 26-6. Lumbar plexus and major nerves from T12 to L5. The lumbar plexus (*light,* anterior divisions; *shaded,* posterior divisions) is formed by the anterior primary rami of the lumbar spinal nerves. The lumbosacral trunk from L4 to L5 joins the sacral plexus. Spinal nerves T12, L1, and L2 have both white and gray rami communicantes. Spinal nerves L3, L4, L5, and S1 have only gray rami communicantes.

(c) Its **hypogastric branch** provides motor innervation to the internal oblique and transverse muscles of the abdomen and conveys sensation from the pubic (hypogastric) region. As such, it provides both afferent and efferent limbs for the **abdominal reflex,** whereby stroking the lower abdominal wall produces a rippling of the underlying abdominal muscles.

(2) **Ilioinguinal nerve** (L1, anterior)
  (a) This nerve emerges from the lateral border of the psoas muscle and runs parallel to, and sometimes anastomoses with, the iliohypogastric nerve. Its terminal branch accompanies the spermatic cord through the superficial inguinal ring.
  (b) It provides motor innervation to the internal oblique and transverse muscles of the abdomen. It conveys sensation from the upper medial part of the thigh, the root of the penis or mons pubis, and the anterior scrotum or superior portion of the labia major.

(3) **Genitofemoral nerve** (L1–L2, anterior)
  (a) This nerve emerges anterior to the psoas muscle and runs along its anterior surface.
  (b) Its **genital branch** (the external spermatic nerve) passes through the inguinal canal. This branch provides motor innervation to the cremaster muscle (the **efferent limb of the cremaster reflex** in men) and conveys sensation from the anterior scrotum or superior portion of the labia major.
  (c) Its **femoral branch** (the lumboinguinal nerve) passes into the thigh beneath the inguinal ligament, medial to the psoas muscle. This branch conveys sensation from the anterosuperior aspect of the thigh and provides the **afferent limb of the cremaster reflex** in men, whereby stroking the anteromedial thigh produces elevation of the testes within the scrotum.

(4) **Lateral femoral cutaneous nerve** (L2–L3, posterior)
  (a) This nerve emerges lateral to the psoas muscle, runs across the iliac fossa, and passes beneath the inguinal ligament close to the anterior–superior iliac spine.
  (b) It conveys sensation from the lateral aspect of the thigh.

(5) **Femoral nerve** (L2–L4, posterior)
  (a) This nerve emerges from the lateral border of the psoas muscle just proximal to the inguinal ligament and passes beneath the inguinal ligament.
  (b) It provides motor innervation to the anterior aspect of the thigh, sensation from the anterior and medial aspects of the thigh, and, by its saphenous branch, sensation from the anterior and medial aspects of the leg. This nerve provides both afferent and efferent limbs of the **knee-jerk reflex,** whereby transient stretch of the patellar tendon produces brief contraction of the quadriceps femoris muscle.

(6) **Obturator nerve** (L2–L4, anterior)
  (a) This nerve emerges from the medial border of the psoas muscle and enters the thigh through the obturator foramen.
  (b) It provides motor innervation to, and occasionally sensation from, the medial aspect of the thigh.

(7) **Lumbosacral trunk** (L4–L5, anterior and posterior)
  (a) This large connector joins the lumbar plexus with the sacral plexus.
  (b) The posterior division contributes to the common peroneal portion of the sciatic nerve. In general, this nerve provides sensory and motor innerveration to the anterior leg and dorsum of the foot.
  (c) The anterior division contributes to the tibial portion of the sciatic nerve. In general, this nerve conveys both sensation from and motor innervation to the posterior thigh, posterior leg, and plantar aspect of the foot.

PART IIIB   ABDOMEN

STUDY QUESTIONS

DIRECTIONS: Each of the numbered items or incomplete statements in this section is followed by answers or by completions of the statement. Select the ONE lettered answer or completion that is BEST in each case.

1. Which of the following statements concerning the superficial inguinal ring is most accurate?

(A) It is formed in part by the falx inguinalis
(B) It is formed in part by the rectus sheath
(C) It is a perforation in the external oblique aponeurosis
(D) It is a perforation in the transversalis fascia

2. The cremasteric reflex draws the testes from the scrotum toward the superficial inguinal ring. The afferent limb of this reflex is carried centrally by the

(A) femoral branch of the genitofemoral nerve
(B) genital branch of the genitofemoral nerve
(C) iliohypogastric nerve
(C) subcostal nerve
(D) pelvic splanchnic nerves
(E) testicular nerve

3. The medial umbilical folds are created by peritoneum overlying the

(A) falciform ligament
(B) inferior epigastric arteries
(C) lateral borders of the rectus sheath
(D) obliterated umbilical arteries
(E) urachus

Questions 4–11

A 43-year-old advertising executive is brought to the emergency room with searing pain to the middle of the back and "coffee grounds" hematemesis. The present episode began shortly after eating a heavy lunch, which included two drinks. His history includes periodic attacks of heartburn, nausea, and pain in the epigastric region. These attacks are often relieved by antacids or eating. He smokes two packs of cigarettes daily and admits to moderate social use of alcohol.

4. All of the following structures refer pain via the greater splanchnic nerve to the epigastric region EXCEPT the

(A) abdominal esophagus
(B) descending duodenum
(C) gallbladder
(D) ileum
(E) stomach

5. Physical examination reveals abdominal rigidity, but no region that is particularly tender on palpation. Results of an EKG are normal. The patient continues to complain of pain in the middle of the back. The sharp pain felt in this location is explained by stimulation of neurons traveling along the

(A) greater splanchnic nerve
(B) lesser splanchnic nerve
(C) least splanchnic nerve
(D) lumbar splanchnic nerves
(E) spinal nerves (T12–L2)

6. Perforation of a peptic ulcer is suspected, and the patient is prepared for surgery. Under general anesthesia, the skin and anterior wall of the rectus sheath are incised in a paramedian incision from the xiphoid process to the umbilicus. The rectus abdominis muscle is retracted laterally, and the posterior wall of the rectus sheath is incised. As a result of its incision and manipulation, what adverse effect on the rectus abdominis muscle might be expected?

(A) Ischemic necrosis superior to the umbilicus
(B) Ischemic necrosis inferior to the umbilicus
(C) Paralysis in the region medial to the incision of the sheath
(D) Paralysis in the region lateral to the incision of the sheath
(E) No adverse effect is expected

7. On exploration of the abdominal cavity, a slight amount of blood and other fluid is observed in the pouch of Morison and subsequently aspirated. The pouch of Morison (subhepatic and hepatorenal recesses) is directly continuous with the

(A) inferior mesenteric space
(B) infracolic compartment
(C) left paracolic gutter
(D) left paraduodenal fossa
(E) lesser sac

8. No sign of perforation is seen on the anterior aspects of the stomach or duodenum. An incision is made in the membrane between the stomach and liver, avoiding the blood vessels of the lesser curvature as well as those of the portal triad. The membrane through which the incision passed is the

(A) falciform ligament
(B) gastrohepatic ligament
(C) greater omentum
(D) hepatoduodenal ligament
(E) left anterior coronary ligament

9. After aspiration of the lesser sac, a 1-cm perforation is seen on the superoposterior aspect of the pyloric antrum. The ulcer has eroded a small arterial branch, which is bleeding. The artery involved in this ulceration is most likely a branch of

(A) the esophageal artery
(B) the gastroduodenal artery
(C) the left gastric artery
(D) the left gastroepiploic artery
(E) a short gastric artery

10. Requiring more exposure to resect the ulcer, the surgeon extends the previous opening in the lesser omentum toward the esophagus. Suddenly, the field fills with bright red blood. The surgeon gains control of the bleeding by clamping and ligating both ends of the artery. Within a few moments, he notices that a portion of the left lobe of the liver blanched, indicating ischemia. The ligated artery was most likely the

(A) left gastric artery
(B) common hepatic artery
(C) right gastric artery
(D) right hepatic artery
(E) left gastroepiploic artery

11. The gastric ulcer is resected successfully and the surgical pathologic examination shows no evidence of malignancy. The postoperative course is stormy with spiking temperatures, but the patient gradually improves. The patient did not succumb to liver necrosis in this instance because the aberrant left hepatic artery was supplied by abundant anastomoses between the distal portion of the left gastric artery and all of the following EXCEPT the

(A) esophageal arteries
(B) right gastric artery
(C) short gastric arteries
(D) right hepatic artery
(E) splenic arteries

(end of group question)

Questions 12–20

A 34-year-old man comes into the emergency room with excruciating abdominal pain in the left flank and inguinal region. Evidence of blood in a urine sample suggests the patient is in the process of passing a renal calculus. He soon complains that the pain is worse and has moved into the anterior and medial aspects of the left thigh as well as to the pubic region. Plain film radiographs of the kidneys, ureters, and bladder suggest the presence of a calculus in the lower left quadrant.

12. The initial pain in the flank and inguinal region resulting from a nephrolith is attributable to afferent pain fibers, which travel via the

(A) greater splanchnic nerve
(B) iliohypogastric nerve
(C) least splanchnic nerve
(D) lumbar splanchnics
(E) vagus nerve

13. The subsequent renal colic referred to the medial and anterior aspects of the left thigh as well as to the pubic region is the result of afferent pain fibers, which travel via the

(A) genitofemoral nerve
(B) ilioinguinal nerve
(C) lesser splanchnic nerve
(D) lumbar splanchnic nerves
(E) vagus nerve

14. The pain in the lateral thigh and pubic regions continues to be intense and unremitting. The patient is admitted to the hospital for observation and possible surgical intervention. Given this particular pattern of renal colic and the anatomy of the renal system, it is suspected that a stone has stopped moving and is lodged

(A) at the junction of the renal pelvis and ureter
(B) midureter as it passes beneath the gonadal vessels
(C) at the pelvic brim
(D) in the intramural portion of the ureter where it penetrates the bladder
(E) in the urethra

15. The patient is scheduled for immediate surgery. In the operating room, the patient is draped and the surgeon indicates the location and extent of the anticipated incision. All of the following layers are severed in the anterior abdominal incision EXCEPT

(A) Camper's fascia
(B) the deep fascia
(C) the renal fascia
(D) Scarpa's fascia
(E) the transversalis fascia

16. The peritoneal cavity is opened by incision of the skin, fascia, and musculature of the abdominal wall from the iliac crest to the pubis above and parallel to the left inguinal ligament. Two large nerves are encountered and preserved in the process of making the abdominal incision. The location of these nerves is most apt to be

(A) between the internal oblique and external oblique muscles
(B) between the transverse abdominis and internal oblique muscles
(C) in the superficial fascia
(D) in the transversalis fascia
(E) within the rectus sheath

17. The surgeon mobilizes the descending bowel from the posterior abdominal wall, gently lifting the bowel and separating its mesentery from the peritoneum covering the dorsal body wall. This mobilization procedure reestablishes a mesentery that contains the

(A) left lumbar arteries
(B) left renal artery
(C) left testicular artery
(D) left colic artery
(E) superior mesenteric artery

18. With the descending colon and its mesentery drawn to the midline, the ureter is visible beneath the peritoneum. The stone is palpated and removed through a small longitudinal incision. The ureter is then probed both proximally and distally for any additional stones. Care is taken to preserve the blood supply to the ureter, which is obtained from all of the following vessels EXCEPT the

(A) common iliac artery
(B) gonadal artery
(C) inferior mesenteric artery
(D) inferior vesical arteries
(E) renal artery

19. With further inspection, the surgeon reports that the blood supply to the kidneys appears to be normal. All of the following statements concerning the blood supply in the vicinity of the kidneys are correct EXCEPT that the

(A) left gonadal artery usually originates from the left renal artery
(B) left gonadal vein usually drains into the left renal vein
(C) left renal vein passes anterior to the aorta
(D) left adrenal vein drains into the left renal vein

20. The abdominal incision (above and parallel to the left inguinal ligament from the iliac crest to the pubis) is closed in layers. The patient is ambulatory on postoperative day 1 and is released from the hospital on day 5. As a result of the location and direction of the incision, healing would likely be accompanied by

(A) paralysis of a portion of the rectus abdominis muscle
(B) minimal scarring
(C) ischemia to the rectus abdominis muscle
(D) significant weakness of a portion of the lateral abdominal wall

(end of group question)

21. A 57-year-old man is diagnosed with esophageal cancer. The consulting surgeon suggests a radical esophagectomy and a functional reanastomosis between the pharynx and stomach. She describes how a living segment of transverse colon can be brought up into the thorax and neck on a mesenteric pedicle containing the original blood vessels and nerves. Important considerations in performing the esophagectomy-colon interposition procedure would be to

(A) anastomose the hepatic flexure end of the transverse colon to the stomach and the splenic flexure end to the pharynx
(B) preserve the esophageal venous plexus
(C) preserve the middle colic vessels to the transverse colon
(D) avoid severing the pelvic splanchnic nerves

DIRECTIONS: Each of the numbered items or incomplete statements in this section is negatively phrased, as indicated by a capitalized word such as NOT, LEAST, or EXCEPT. Select the ONE lettered answer or completion that is BEST in each case.

---

22. A middle-aged man is examined for a large hernia in the left groin. Because the hernia is directed toward the scrotum, the examining physician suspects it is an indirect inguinal hernia. Which of the following descriptions does NOT pertain to an indirect inguinal hernia?

(A) Enters the deep inguinal ring
(B) Is found within the spermatic cord
(C) Lies medial to the inferior epigastric artery
(D) Presents through the superficial inguinal ring

23. Correct statements that characterize an ileal (Meckel's) diverticulum include all of the following EXCEPT

(A) it is about 3 inches long
(B) it is located within 3 feet of the ileocecal valve
(C) it is present in about 3% of the population
(D) it is a remnant of the urachus
(E) it is usually lined by gastric mucosa

24. Considering the incidence of lymphatic metastatic carcinoma from the rectum, knowledge of lymphatic drainage in this region is important. Near the middle rectal valve, lymphatic drainage is to all of the following EXCEPT the

(A) inferior mesenteric nodes
(B) internal iliac nodes
(C) superficial inguinal nodes
(D) superior mesenteric nodes

DIRECTIONS: Each set of matching questions in this section consists of a list of four to twenty-six lettered options (some of which may be in figures) followed by several numbered items. For each numbered item, select the ONE lettered option that is most closely associated with it. To avoid spending too much time on matching sets with large numbers of options, it is generally advisable to begin each set by reading the list of options. Then, for each item in the set, try to generate the correct answer and locate it in the option list, rather than evaluating each option individually. Each lettered option may be selected once, more than once, or not at all.

Questions 25–28

(A) C3–C5
(B) T1–T4
(C) T5–T9
(D) T10–T11
(E) L1–L2

For each inflammatory condition, select the dermatomes to which pain is initially referred.

25. Cholecystitis

26. Diverticulitis of the sigmoid colon

27. Free air under the diaphragm

28. Intussusception of the ileum at the ileocecal valve

Questions 29–32

(A) Celiac trunk
(B) Inferior mesenteric artery
(C) Inferior phrenic arteries
(D) Superior mesenteric artery

For each structure, select the artery from which the major blood supply is derived.

29. A portion of the left adrenal gland

30. The appendix

31. The tail of the pancreas

32. A portion of the rectum

ANSWERS AND EXPLANATIONS

1. The answer is C [Chapter 23 VII B 1]. The superficial inguinal ring is an opening in the external oblique aponeurosis that transmits the spermatic cord in males and the round ligament of the uterus in females.

2. The answer is A [Chapter 23 VII D 3 b, c]. The afferent limb of the cremaster reflex is carried by the femoral branch of the genitofemoral nerve to spinal levels L1–L2. The efferent limb to this muscle is carried by the genital branch of the genitofemoral nerve. The iliohypogastric nerve carries both limbs of the abdominal reflex.

3. The answer is D [Chapter 24 III B 2 h (2)]. The medial umbilical folds are formed by peritoneum draping over the umbilical ligaments, which are remnants of the umbilical arteries. The median umbilical fold, a remnant of the ventral mesentery, is formed by peritoneum overlying the urachus. The lateral umbilical folds are formed by the inferior epigastric arteries.

4. The answer is D [Chapter 25 II B 7 b (2); VI B; Table 25-2]. The visceral afferent neurons from the lower esophagus, stomach, duodenum, gallbladder, and bile duct travel through the celiac plexus. These neurons gain access to the greater splanchnic nerve, pass through the ganglia of the sympathetic chain (levels T5–T9), reach the associated spinal nerves via white rami communicantes, and enter the spinal cord via the dorsal root. Therefore, pain is referred to the epigastric region. The visceral afferents from the terminal duodenum, jejunum, ileum, ascending colon, and transverse colon travel through the superior mesenteric plexus, then along the lesser splanchnic nerve and refer pain to the umbilical region.

5. The answer is E [Chapter 24 III C 2 a (1)]. Sharp localized pain is an indication of peritonitis. The parietal peritoneum is innervated by twigs from the anterior primary rami of the spinal nerves.

6. The answer is E [Chapter 23 VI A 1 a-c]. Because the blood supply enters the rectus abdominis muscle from the superior and inferior ends and the innervation enters this muscle from the lateral aspect, the incision and manipulation described will compromise neither the vascular supply nor the innervation.

7. The answer is E [Chapter 24 III B 2 c (3)]. The hepatorenal and subhepatic recesses communicate directly with the lesser sac (via the omental foramen), the supracolic compartment, the right paracolic gutter, and the subphrenic space. It does not communicate directly with the infracolic compartment (divided into superior and inferior mesenteric spaces by the mesentery proper) or the left paracolic gutter. The paraduodenal fossae open into the infracolic compartment.

8. The answer is B [Chapter 24 II A 4 a (1)]. The lesser omentum runs between the foregut and the liver. The portion between the stomach and liver, the gastrohepatic ligament, is relatively avascular. The portion between the liver and duodenum, the hepatoduodenal ligament, contains the portal triad, which includes the major blood supply to the liver.

9. The answer is B [Chapter 25 II B 4 a (1), 7 b (1); V A 1 b (3) (a)]. The arteries most likely involved in pyloric peptic ulceration are the right gastric, common hepatic, gastroduodenal, and right gastroomental arteries. In addition, ulceration of the posterior wall of the body of the stomach may involve the splenic artery, with profuse bleeding.

10. The answer is A [Chapter 25 II D 5 a (3) (a) (iii), (4)]. An aberrant left hepatic artery frequently (35%) arises from the left gastric artery. Because the arteries supplying the liver are end-arteries, ligation of an aberrant artery (as in this case) usually produces ischemia of part of the liver.

11. The answer is D [Chapter 25 II B 4 a (2), 5 a (4)]. Few anastomoses occur between the left and right lobes of the liver, thus precluding anastomotic flow between the right and left hepatic arteries. Abundant anastomoses among the left gastric artery, the esophageal arteries, right gastric artery, and short gastric branches of the splenic artery hypertrophy to supply the aberrant left hepatic artery with sufficient blood to supply the left lobe of the liver.

12. The answer is C [Chapter 25 VI A 2 a (3), B 2; Table 25-2; Chapter 26 II C 2 a]. The renal calyces, renal pelvis, and most of the proximal ureters generally receive afferent innervation from spinal segment T12 via the least splanchnic nerve. Because somatic portions of spinal segment T12 supply the lumbar and inguinal regions of the abdominal wall, pain is referred accordingly.

13. The answer is D [Chapter 25 VI A 2 a (4), B 2; Table 25-2; Chapter 26 II C 2 b]. The middle section of each abdominal ureter generally receives innervation from spinal segments L1–L2. As such, the afferent pathway is via the lumbar splanchnic nerves. Because the somatic portions of spinal segments L1–L2 supply the pubic region and anterior aspects of the scrotum or labia majora as well as the anterior-superior aspect of the thigh, pain is referred accordingly.

14. The answer is C [Chapter 26 II A 2 c (1)]. The ureters narrow at the renal pelvis, at the pelvic brim, where they cross the iliac vessels and enter the deep pelvis, and at the urinary bladder. Thus, the probability of a nephrolith lodging in the ureter at the pelvic brim is high and is confirmed by the pattern of referred pain.

15. The answer is C [Chapter 23 II B 1–2, VI A 3 a; Chapter 24 III C 1 b; Chapter 26 II A 2 b]. An incision through the anterior abdominal wall passes through Camper's fascia (the superficial layer of superficial fascia), Scarpa's fascia (the deep layer of superficial fascia), the deep fascia investing the abdominal musculature, and the transversalis fascia before reaching the peritoneum. The renal fascia is retroperitoneal. After gaining access to the peritoneal cavity, the posterior peritoneum and the periureteral sheath (an extension of the renal fascia) are incised in the process of accessing the ureter.

16. The answer is B [Chapter 23 V A 1, B 2 a; Chapter 26 IV D 2 b (1) (a), (2) (a)]. The nerves of the anterior abdominal wall initially are located between the transverse abdominis and internal oblique muscles. More anteriorly, the nerves pass diagonally through the internal oblique muscle. The two large nerves encountered would be the iliohypogastric and ilioinguinal nerves.

17. The answer is D [Chapter 24 III A 3; Chapter 25 IV A 1 b, 2 a; Chapter 26 II A 2 b]. During development, the mesentery of the descending colon becomes adherent to the peritoneum of the dorsal coelomic wall anterior to the left ureter and the left gonadal neurovascular bundle. As such, the left colic artery lies in the descending mesocolon anterior to the left ureter.

18. The answer is C [Chapter 26 II B 1]. The ureter is supplied by twigs of numerous vessels, including renal, gonadal, common iliac, internal iliac, and inferior vesical arteries. Although the left colic artery comes into close proximity to the ureter, normally no anastomoses occur between this vessel and those of the ureter.

19. The answer is A [Chapter 26 I D 1, 2; III B 2]. The left gonadal vein usually drains into the left renal vein, but both gonadal arteries usually arise directly from the abdominal aorta just distal to the renal arteries.

20. The answer is B [Chapter 23 I D; Chapter 23 II A 2 b (2)]. In the lower abdominal wall, the cleavage lines run in a direction nearly parallel to the inguinal ligament. An incision approximating this direction gapes less and produces minimal scarring.

21. The answer is C [Chapter 25 III C 2 a; V A 2 b (5)]. The transverse colon receives its blood supply by the middle colic branch of the superior mesenteric artery and its parasympathetic innervation from the vagus nerve through the superior mesenteric plexus. Failure to preserve the vascular pedicle results in denervation and necrosis. To ensure proper peristaltic function, the hepatic end must be anastomosed to the pharynx and the splenic end to the stomach.

22. The answer is C [Chapter 23 VII E 3 b (1), c (1), F 3 b]. Although both direct and indirect hernias pass through the superficial ring, they reach it via different routes. The indirect inguinal hernia, by passing through the internal inguinal ring, enters the inguinal canal lateral to the inferior epigastric artery and passes within the spermatic cord. The direct hernia occurs within the inguinal triangle, medial to the inferior epigastric vessels. Direct inguinal hernias are adjacent to the spermatic cord. Because the direct inguinal hernia is not within the spermatic cord, it need not pass into the scrotum.

23. The answer is D [Chapter 24 II C 4 e; Chapter 25 III A 6 b]. An ileal diverticulum (of Meckel) is a remnant of the vitelline duct, which connects the midgut to the yolk sac during embryonic development. The urachus is a remnant of the duct that connects the bladder to the allantois. A rule of 3's applies to Meckel diverticulum: it is usually 3 inches long, within 3 feet of the ileocecal valve, and occurs in about 3% of the population. It is often lined with gastric mucosa, the secretions of which produce peptic ulceration of the ileum.

24. The answer is D [Chapter 25 IV B 3 c, V C 1 c (1) (b); Chapter 29 I C 3 a–c]. Lymphatic drainage from the middle of the rectum is along inferior mesenteric, internal ileal, and external iliac nodes. In general, the lymphatic drainage pathways follow the arteries and veins. Thus, because the rectum has three sources of blood supply, carcinoma of the rectum may disseminate widely within the abdomen and pelvis as well as to the inguinal nodes.

25–28. The answers are: 25-C, 26-E, 27-A, 28-D [Chapter 25 VI B; Table 25-2; Chapter 26 IV A 1 e (2)]. Afferents from the stomach, biliary tree, gallbladder, and upper duodenum travel along the greater splanchnic nerve to spinal levels T5–T9. The small intestine from the terminal portion of the duodenum to the ileocecal valve and the large bowel to the splenic flexure send afferent nerves along the lesser splanchnic nerve to spinal segments T10 and T11. Afferents from the descending colon and sigmoid colon travel along lumbar splanchnic nerves to the L1 and L2 spinal levels. The peritoneum of the diaphragm is innervated primarily by the phrenic nerves, which arise from spinal segments C3–C5 and refer pain to the neck and shoulder. Small peripheral regions of diaphragmatic peritoneum, however, are innervated by twigs from the somatic intercostal nerves.

29–32. The answers are: 29-C, 30-D, 31-A, 32-B [Chapter 25 I C 2 a (1), b (1), c (1); II F 2 b (3); III B 2 d (3); IV C; Chapter 26 II B 1, III B 1; Chapter 29 I C 1 a]. The adrenal glands obtain their blood supply from inferior phrenic arteries, the aorta directly, and the renal arteries. The celiac axis supplies foregut derivatives, including the esophagus, stomach, liver, and portions of the pancreas, including the pancreatic tail, which receives blood supply from the splenic artery. The superior mesenteric artery supplies midgut derivatives, such as the head of the pancreas and bowel from the middle of the duodenum to the splenic flexure. The inferior mesenteric artery supplies the large bowel from the splenic flexure to the middle of the rectum.

PART IIIC PELVIS

Chapter 27
Pelvis

I. BONES AND JOINTS

A. General structure. The pelvis is formed by bilateral coxal (hip, pelvic) bones, the sacrum, and the coccyx (see Figure 27-1).

1. **The coxal (hip or pelvic) bone** on each side of the pelvis is formed by fusion of the **ilium, ischium,** and **pubis.** These bones join and fuse at the **acetabulum,** which is the fossa of the hip joint. The two hip bones form the pelvic girdle and belong to the appendicular skeleton.
    a. The **ilium** is the most superior portion of the coxal bone (Figure 27-1).
       (1) A flared **alar plate** provides attachment on its outer surface for back and thigh musculature. The **iliac fossa,** the concave inner surface, provides attachment for the iliacus muscle.
       (2) The **auricular surface,** the posterior portion of the ala, articulates with the sacrum at the **sacroiliac joint.**
       (3) The **iliac crest** at the free superior edge of the ala runs between the anterior–superior and posterior–superior iliac spines. It provides attachment for back and abdominal musculature as well as several named ligaments.
       (4) The **greater sciatic notch** of the ilium provides an aperture through which neurovascular as well as muscular structures pass from the deep pelvis to the thigh (see Figure 27-1B).
           (a) The **piriformis muscle** originates from the anterior aspect of the sacrum, passes through the greater sciatic notch, and inserts at the base of the greater trochanter (see Figure 14-4).
           (b) The **suprapiriform recess** is that portion of the greater sciatic foramen superior to the piriformis muscle. It contains the **superior gluteal neurovascular bundle.**
           (c) The **infrapiriform recess** is that portion of the greater sciatic foramen inferior to the piriformis muscle. It contains the **inferior gluteal neurovascular bundle,** the sciatic nerve, and the **pudendal neurovascular bundle.** Compression of the sciatic nerve if it passes through the piriformis muscle produces **sciatica.**
       (5) The **linea terminalis** delineates the pelvic inlet and provides the boundary between the greater pelvis and the deep pelvis.
    b. The **ischium** is the most inferior portion of the coxal bone (see Figure 27-1).
       (1) The **ischial ramus** fuses with the inferior pubic ramus.
       (2) The **ischial tuberosity,** a palpable landmark, provides attachment for posterior thigh muscles and the **sacrotuberous ligament.**
           (a) In a standing person, the gluteus maximus muscles cover the ischial tuberosities.
           (b) In a sitting person, the gluteus maximus muscles move laterally, so one sits on the ischial tuberosities.
       (3) The **ischial spine** provides attachment for the **sacrospinous ligament,** which bridges the greater sciatic notch just superior to the sacrotuberous ligament, thereby forming the **greater sciatic foramen** (see Figure 27-1B).
       (4) The **lesser sciatic notch** of the ischium lies between the ischial tuberosity and the ischial spine (see Figure 27-1B). It provides an aperture through which neurovascular as well as musculotendinous structures pass from the deep pelvis to the thigh and perineum.

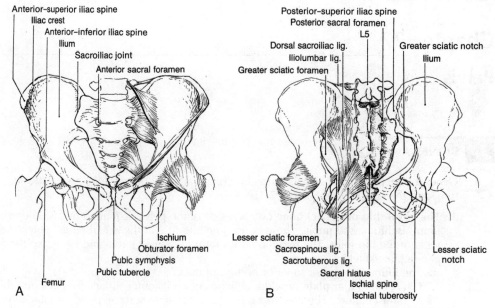

FIGURE 27-1. Female pelvis. *(A)* Anterior aspect. The anterior ligaments supporting the left lumbosacral joint are included. *(B)* Posterior aspect. The posterior ligaments supporting the left lumbosacral joint are included.

    (a) The **lesser sciatic foramen** is formed by the **lesser sciatic notch** of the ischium with the **sacrospinous** and **sacrotuberous ligaments.**
    (b) The **obturator internus muscle** originates on the internal aspect of the ilium and pubis, leaves the pelvis through the lesser sciatic foramen, and inserts at the base of the greater trochanter.
    (c) The **pudendal neurovascular bundle** arises in the deep pelvis, leaves through the greater sciatic foramen, and passes through the lesser sciatic foramen to reach the perineum.
  c. The **pubis** is the most anterior portion of the coxal bone (see Figure 27-1A).
    (1) The **superior pubic ramus** fuses with the ischium and ilium at the acetabulum.
    (2) The **inferior pubic ramus** fuses with the ischial ramus.
    (3) Anteriorly, left and right pubic bones articulate at the **pubic symphysis.**
      (a) The pubic symphysis, an amphiarthrosis, consists of hyaline cartilage at its articular surfaces separated by a fibrocartilaginous disk.
      (b) This joint relaxes under the influence of progesterone in the weeks before parturition, allowing the birth canal to widen.
 2. **The sacrum,** a portion of the axial skeleton, forms the posterior portion of the pelvis (see Figure 27-1B).
    a. **Structure.** Five sacral vertebrae fuse to form the sacrum, and four lines of fusion are evident on the anterior surface (see Figure 11-4).
      (1) Fused spinous processes form the **median sacral crest.**
      (2) Fused transverse processes together with fused costal processes form the **alae** (sacroiliac articular processes).
    b. **Sacral foramina** run anteroposteriorly through the sacrum in two parasagittal rows (see Figure 11-4).
      (1) **Posterior sacral foramina** transmit the **dorsal primary rami** of sacral spinal nerves.
      (2) **Anterior sacral foramina** transmit the **ventral primary rami** of sacral spinal nerves.
    c. The **sacral hiatus** is the caudal opening to the vertebral canal (see Figure 11-4).

(1) The laminae of the fifth sacral vertebra fail to form, resulting in the sacral hiatus.
(2) Pedicles form the **sacral cornua,** important landmarks used to locate the sacral hiatus for administration of caudal anesthesia.

3. **The sacroiliac joint** is formed by articulation of the alar plates of the sacrum with the ilium (see Figure 27-1).
   a. **Structure.** This joint is a diarthrosis with a synovial cavity.
      (1) Although mobility is slight, movement is greatest in the weeks before parturition under the influence of progesterone.
      (2) Mobility at this joint decreases with age. Ankylosis usually occurs by age 50 years.
   b. **Support.** As a result of ligamentous reinforcement, this joint is among the strongest diarthrodial joints. These ligaments oppose the rotational effect of gravity at the sacroiliac joint (see Figure 27-1).
      (1) The **sacrotuberous ligament** runs from the sacrum to the ischial tuberosity. It transmits to the sacrum the forces generated by the hamstring muscles and the gluteus maximus. It also provides a boundary for the lesser sciatic foramen.
      (2) The **sacrospinous ligament** runs from the sacrum to the ischial spine. It closes the greater sciatic notch and contributes a boundary to the lesser sciatic foramen.
      (3) The **iliolumbar ligament** runs between the iliac crest and the transverse processes of the fifth lumbar vertebra.
      (4) **Dorsal sacroiliac ligaments** run between the posterior–superior iliac spine and the dorsum of the sacrum.
      (5) The **ventral sacroiliac ligament** runs across the sacroiliac joint on the anterior surface of the pelvis.
      (6) The **interosseous ligaments** run between the sacroiliac articular surfaces. The tendency of the sacrum to rotate forward tightens these ligaments, locking the joint.
   c. **Low back pain** is discussed in Chapter 11 III F 3.

4. **The coccyx,** or tail bone, forms the inferoposterior portion of the pelvis (see Figure 11-4).
   a. **Structure** (see Chapter 11 III E)
   b. **Sex differences.** In the male, the coccyx angles anteriorly toward the pubis. In the female, the coccyx is more vertical, providing a larger pelvic outlet.
   c. **Coccygodynia,** a painful condition, may result from fracture of fused coccygeal joints or of a fused sacrococcygeal joint, or from arthritis in these joints.

B. **The birth canal** consists of the pelvic inlet, deep pelvis, and pelvic outlet.

1. **The major (greater or false) pelvis** is that portion of the bony pelvis between the iliac crests and the pelvic brim, superior to the pelvic inlet (see Figure 27-1A).
   a. Laterally, the **alar plates** (wings) of the ilia form the walls of the major pelvis.
   b. Posteriorly, it is bounded by the lumbar vertebrae L3, L4, and L5.
   c. Anteriorly, it is bounded by the anterior abdominal wall.
   d. With the exceptions of a gravid uterus and the fundus of the full urinary bladder, no pelvic organs lie in the major pelvis.

2. **The minor (lesser, deep, or true) pelvis** lies inferior to the pelvic inlet.
   a. The **pelvic inlet** demarcates the minor from the major pelvis.
      (1) **Boundaries.** The pelvic inlet is defined by the sacral promontory and the linea terminalis of the innominate bone. The **linea terminalis** includes the pubic crest, iliopectineal line, and arcuate line of the ilium.
      (2) **Principal measurements**
         (a) **Conjugate diameters**
            (i) The **diagonal conjugate** (about 10.5 cm) is measured from the sacral promontory to the inferior margin of the pubic symphysis. This measurement is noted along the index finger during pelvic examination.

(ii) The **obstetric conjugate** is the least anteroposterior diameter from the sacral promontory to a point a few millimeters below the superior margin of the pubic symphysis. Although this distance can only be measured radiographically (lateral projection), it may be approximated from the diagonal conjugate.
(b) The **transverse diameter** (13.5 cm) is the widest distance across the pelvic brim.

b. The **pelvic cavity** is defined by the minor pelvis. It forms the birth canal, and contains portions of the gastrointestinal tract (loops of ileum, sigmoid colon, rectum, and anal canal), the lower portion of the urinary tract, and certain reproductive organs.

c. **Pelvic outlet**
   (1) **Boundaries.** The pelvic outlet is defined by the coccyx, ischial tuberosities, inferior pubic ramus, and pubic symphysis (see Figure 27-2). It is closed by the pelvic diaphragm, covered by the perineum, and reinforced by the urogenital diaphragm.
   (2) **Principal measurements**
       (a) The **transverse (bituberous) diameter** is between the ischial tuberosities. It approximates a clenched fist in width (about 11.8 cm).
       (b) The **transverse midplane diameter** (about 10.5 cm) is measured between the ischial spines. If this diameter is less than 9.5 cm, there is nearly a 50% chance that surgical intervention will be necessary during labor.
   (3) **Pelvimetry,** although controversial, can be a valuable diagnostic tool in cases of cephalopelvic disproportion.

3. **Characterization of the pelvis**
   a. **Inlet shape** of the pelvic brim classifies the pelvis. In women, it plays a role in assessing the likelihood of a successful vaginal delivery.
       (1) The normal male pelvis is **android** and heart shaped, because of the prominence of the sacrum (32% of white women, 15% of black women).
       (2) The normal female inlet shape is **gynecoid** (41% of white women, 43% of black women).
       (3) An **anthropoid** pelvis has a lengthened anteroposterior diameter and a shortened transverse diameter (24% of white women, 40% of black women).
       (4) A **platypoid** pelvis is oval with a shortened anteroposterior diameter and a lengthened transverse diameter (3% of white women, 2% of black women).
       (5) An **asymmetric pelvis** may be associated with scoliosis or result from a congenital malformation.
   b. **Gender differences** relate primarily to the greater body mass of the average male and to childbearing function in the female (Table 27-1). Children of both sexes approximate the anthropoid pelvis.
       (1) The bones of the female pelvis are thinner than those of the male, the pelvic cavity is less funnel shaped, the surface for articulation of the sacrum with L5 is smaller, and the **subpubic angle** is greater (almost 90°).
       (2) Factors other than gender (such as race and nutrition) contribute to adult pelvic shape.

Table 27-1. Differences between the Male and Female Pelvis

Characteristic	Male	Female
Size	Smaller	Larger
Inferior pubic angle	Narrow	Wider
False pelvis	Taller	Flared
Pelvic inlet	Android	Gynecoid
Pelvic outlet	Smaller	Larger
Greater sciatic notch	Narrow	Wider
Ischial spine attitude	Medially	Posterior

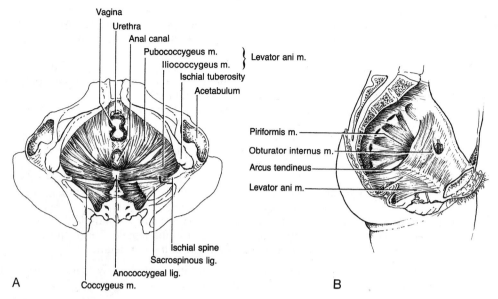

FIGURE 27-2. Muscles of the floor of the female pelvis (pelvic diaphragm). *(A)* Inferior view; *(B)* sagittal view.

II. MUSCULATURE

A. Organization

1. **The coccygeus (ischiococcygeus) muscle** forms an insignificant portion of the **pelvic floor** (pelvic diaphragm) (Figure 27-2; see Table 27-2).
   a. As a rudimentary "tail-wagger," this simple muscle originates from the ischial spine and inserts onto the lateral borders of the fourth sacral to the second coccygeal vertebrae as well as along the sacrospinous ligament.
   b. The **sacrospinous ligament** is a degenerated portion of the coccygeus muscle.
2. **The levator ani muscle** forms most of the pelvic floor (see Figure 27-2 and Table 27-2).
   a. **Attachments.** This muscle has a broad origin along the inner aspect of the pelvis, extending from the pubic symphysis, along the inferior pubic ramus, arcus tendineus (white line) of the obturator fascia, to the ischium and ischial spine.

Table 27-2. Muscles of the Pelvic Floor

Muscle	Origin	Insertion	Primary Action	Innervation
Coccygeus	Ischial spine	Vertebrae S4–Cx2	Supports pelvic viscera	Nn. S4–Cx2
Levator ani				
Iliococcygeus	Arcus tendineus	Coccyx	Supports pelvic viscera	Nn. S5–Cx2
Pubococcygeus	Pubic arch and arcus tendineus	Coccyx	Supports pelvic viscera	Nn. S4–Cx1
Puborectalis	Pubic arch	Midline and rectum	Rectal continence	Nn. S3–S5

b. **Subdivisions**
  (1) The **iliococcygeus muscle,** a rudimentary "tail-wagger," arises from the ischial spine and the arcus tendineus (white line) of the obturator fascia. It inserts onto the coccyx.
  (2) The **pubococcygeus muscle,** another rudimentary "tail-wagger," arises from the arcus tendineus laterally and the pubic arch anteriorly. It inserts primarily onto the coccyx and into the **anococcygeal raphe.**
  (3) The **puborectalis muscle** is the most medial and most massive portion of the levator ani muscle.
    (a) Arising from the pubic arch medial to the pubococcygeus muscle, the fibers loop around the rectum to form the **rectal sling,** the principal mechanism for fecal continence.
    (b) Some fibers of this muscle pass posterior to the vagina in the female **(pubovaginalis)** and posterior to the male prostate **(puboprostaticus)** to insert onto the central tendon of the perineum.
c. The **urogenital hiatus** is a midline defect in the levator ani muscle through which pass the urethra in both sexes and the vagina in the female. This hiatus is reinforced inferiorly by the **urogenital diaphragm** of the perineum.

3. **Innervation.** The muscles of the pelvic floor are innervated by twigs from the sacral plexus.

4. **Group actions** (Table 27-2)
  a. **Support** of pelvic viscera
    (1) With assumption of the erect posture and loss of a functional tail, the pelvic floor assumed a major role in visceral support and retention of viscera during the Valsalva maneuver.
    (2) In the female, the pelvic floor stabilizes the vagina and indirectly participates in uterine support.
  b. **Urinary continence.** The pelvic floor supports the urinary bladder and participates in urinary continence.
  c. **Fecal continence.** The **rectal sling,** by drawing the rectum anteriorly, is the principal component of fecal continence.

B. **Other muscles.** The iliacus, obturator internus, and piriformis originate within the pelvis but are associated with the lower extremity and act across the hip joint (see Figures 14-3, 14-4, and 27-2B).

III. FASCIA AND PERITONEUM

A. **Fascia of the pelvis** consists of transversalis and endopelvic divisions.

1. **Parietal fascia** is a continuation of the transversalis fascia into the pelvis.
  a. **Location.** As investing fascia, it covers the piriformis muscle and obturator internus muscle.
  b. **Subdivisions.** This fascial layer, which attaches to the arcuate line of the pubis and ilium, thickens over the obturator internus muscle to form the **arcus tendineus,** the origin of portions of the levator ani muscle (see Figure 27-2B).
    (1) At the arcus tendineus, the parietal fascial layer splits to cover both superior and inferior surfaces of the levator ani muscle as the **superior and inferior fasciae of the pelvic diaphragm.**
      (a) The **superior fascia** reflects onto the pelvic viscera as visceral fascia.
      (b) The **inferior fascia** contributes to the superior fascia of the urogenital diaphragm.
    (2) Inferior to the arcus tendineus, the parietal fascia splits around the pudendal neurovascular bundle to form the **pudendal canal** (of Alcock), which runs along the medial surface of the obturator internus muscle.

2. **Visceral (endopelvic) fascia** lies between the peritoneum and the pelvic viscera. It is a continuation of the extraperitoneal connective tissue. In the pelvis, this layer ensheathes retroperitoneal viscera and forms septa between retroperitoneal organs.

3. **Pelvic ligaments** are thickened fascial continuities between the parietal and visceral fasciae.
   a. Condensations of endopelvic fascia support pelvic viscera and provide pathways for the associated neurovascular bundles.
   b. These ligaments are especially important in the support of the uterus and urinary bladder.

B. **Peritoneum** lines the abdominal peritoneal cavity and pelvic viscera.

1. **Parietal peritoneum** overlies portions of the pelvic wall and endopelvic fasciae.

2. **Visceral peritoneum** lines the uterus, ovaries, and uterine tubes. The rectum and bladder are covered in part by peritoneum, remaining retroperitoneal.

3. **Mesenteries** are continuities between parietal and visceral peritoneum.

IV. VASCULATURE

A. **Arterial supply**

1. **Common iliac arteries** arise at the aortic bifurcation and in turn bifurcate into external and internal iliac arteries (Figure 27-3).
   a. **Origin and course.** Left and right common iliac arteries arise with the bifurcation of the aorta at the fourth lumbar vertebral level.

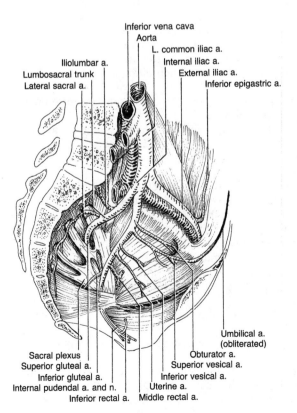

FIGURE 27-3. Vasculature of the pelvis.

(1) They lie anterior to the fourth and fifth lumbar vertebrae and medial to the psoas muscles.
(2) The common iliac vessels bifurcate to form **internal** and **external iliac arteries.**
  b. **Distribution.** The common iliac arteries give off no major branches, with the exception of small branches to the psoas muscle, twigs to the ureters, an occasional iliolumbar artery, or an accessory inferior renal artery.

2. **The external iliac artery** supplies somatic structures (see Figures 15-6 and 27-3).
   a. **Course.** The external iliac artery courses in the greater pelvis just above the pelvic brim. The external iliac artery exits the pelvis beneath the inguinal ligament and becomes the **femoral artery.**
   b. **Distribution.** Near the inguinal ligament, the external iliac artery gives off the **inferior epigastric** and **deep iliac circumflex arteries** before continuing as the principal supply to the leg.

3. **The internal iliac artery** supplies both somatic and visceral pelvic structures (see Figure 27-3).
   a. **Course.** This artery dives into the deep pelvis where it divides into anterior and posterior trunks.
   b. **Distribution.** Each internal iliac artery supplies the walls of the pelvis, the pelvic viscera, and the perineum. The branches of the internal iliac vessels are extremely variable in origin, arrangement, number, and position.
      (1) The **posterior trunk** provides only somatic branches (see Figure 27-3).
         (a) The **iliolumbar artery** courses upward behind the external iliac vessels as far as the medial border of the psoas muscle, where it divides into a lumbar branch and an iliac branch.
            (i) The **lumbar branch** supplies the psoas and quadratus lumborum muscles as well as the lower lumbar vertebra and dural sac. The **iliac branch** supplies the iliacus muscle. Its anastomoses with the distributions of the superior gluteal artery, iliac circumflex artery, and lateral femoral circumflex artery provide collateral circulation about the hip.
            (ii) The iliolumbar artery is especially vulnerable to tearing in association with posterior pelvic fractures. Severe hemorrhage results.
         (b) The **lateral sacral artery** supplies the region of the sacrum.
         (c) The **superior gluteal artery** is the largest branch of the internal iliac artery.
            (i) This artery leaves the pelvic cavity by way of the **suprapiriform portion of the greater sciatic foramen.**
            (ii) It supplies portions of the gluteus maximus, gluteus medius, and gluteus minimus muscles and sends twigs to the hip joint.
      (2) The **anterior trunk** supplies both somatic and visceral structures (see Figure 27-3).
         (a) The **inferior gluteal artery** supplies somatic structures of the buttock.
            (i) Inside the pelvic cavity, this vessel supplies portions of the coccygeus, piriformis, and levator ani muscles.
            (ii) This artery leaves the pelvic cavity through the **infrapiriform portion of the greater sciatic foramen** to supply portions of the gluteus maximus muscle, the lateral rotator muscles, and the hip joint.
         (b) The **internal pudendal artery** supplies somatic structures in the perineum.
            (i) This vessel exits the pelvis through the **infrapiriform portion of the greater sciatic foramen.**
            (ii) It enters the ischiorectal fossa of the perineum by way of the **lesser sciatic foramen.**
         (c) The **obturator artery** supplies somatic structures in the anteromedial thigh.
            (i) This vessel runs ventrally along the pelvic wall, medially to the obturator fascia, and leaves the pelvic cavity through the **obturator canal** in the obturator foramen to supply the proximal portions of the ad-

ductor muscles. It gives rise to the artery of the ligamentum teres of the femoral head.

      (ii) An aberrant obturator artery may arise from the inferior epigastric artery (30%). Such an anomalous vessel may complicate surgical intervention about the femoral ring.

(d) The **umbilical artery** retains a lumen for a short distance beyond the internal iliac artery and gives off one or two **vesical branches** to the urinary bladder.

      (i) The **superior vesical artery** supplies small branches to the cranial portion of the urinary bladder. The **middle vesical arteries** are variable and, when present, supply the fundus of the urinary bladder.

      (ii) The medial umbilical ligaments are the obliterated remnants of the **umbilical artery.**

(e) The **uterine artery** is homologous to the **deferential artery** in the male.

      (i) This artery courses medially on the superior surface of the levator ani toward the cervix. It passes superior to the ureters in the transverse cervical (cardinal) ligament and ascends within the broad ligament to reach the uterus. This artery hypertrophies greatly during pregnancy.

      (ii) **Tubal branches** supply the oviducts and anastomose with the distribution of the ovarian arteries. **Vaginal branches** supply the inner portions of the vagina and anastomose with the vaginal branches of the internal pudendal artery.

(f) The **deferential artery** is homologous to the **uterine artery** in the female. It supplies the vas deferens and epididymis. This vessel passes superior to the ureter and passes through the inguinal canal with the vas deferens to anastomose with the spermatic artery.

(g) The **middle rectal (hemorrhoidal) artery** supplies the rectum. It anastomoses with the inferior mesenteric artery by way of the superior rectal artery and with the internal pudendal artery through the inferior rectal artery.

(h) **Inferior vesical arteries** (one or two) supply the neck of the urinary bladder and, in the male, portions of the prostate and the seminal vesicles. They may give rise to the deferential artery.

B. Venous return

1. **Course.** Small veins of the pelvic cavity generally follow the arteries but with even greater variation. They drain into the **internal iliac vein,** which reaches the inferior vena cava by way of the **common iliac vein** (see Figure 27-3).

2. **Venous anastomoses**

    a. The **rectal venous plexus** more or less surrounds the rectum.

        (1) Abundant anastomoses exist among superior rectal, middle rectal, and inferior rectal (hemorrhoidal) veins.

        (2) These anastomoses connect the hepatic portal system with systemic drainage.

    b. The **vesical venous plexus** lies about the base of the bladder and, in the male, the prostate.

        (1) This venous plexus receives drainage from the penis or clitoris by way of the deep dorsal vein and from the urinary bladder, as well as from the vas deferens, seminal vesicle, and prostate in the male.

        (2) Because drainage from this plexus is diffuse and without valves, prostate cancer may spread by several hematogenous routes with potential for metastatic rests along the courses.

            (a) Principal drainage is into the **internal iliac vein.**

            (b) Collateral drainage is into the **vertebral venous (Batson's) plexus** within the neural canal, **obturator veins** (which frequently drain into the external iliac veins), and the **rectal plexus.**

C. Lymphatic drainage

1. **Course.** Lymphatic drainage of the pelvis is primarily to two groups of nodes, but with important alternative routes.

a. **External nodes** lie near the pelvic brim along external and common iliac vessels. They receive lymph from superficial and deep inguinal nodes, draining the lower extremity and portions of the external genitalia.

b. **Internal (pelvic) nodes** lie within the deep pelvis along the internal iliac vessels. They drain the pelvic floor and pelvic viscera.

2. **Lymphatic anastomoses** occur between rectal lymphatics and those of the abdomen as well as with the superficial lymphatic drainage of the perineum. Lymphatic drainage is a major consideration in procedures involving carcinoma of the pelvic viscera.

V. LUMBOSACRAL PLEXUS

A. **Organization.** The lumbosacral plexus provides somatic innervation to the pelvis and lower extremities.

1. **The lumbar plexus** (T12–L4) lies in the posterior abdominal wall and iliac fossa (see Figure 26-6).

2. **The lumbosacral trunk** (L4–L5, anterior and posterior divisions) contributes to the sacral plexus and joins the lumbar plexus with the sacral plexus (Figure 27-4).
   a. The **anterior division** contributes the **tibial portion of the sciatic nerve.**
   b. The **posterior division** contributes the **superior and inferior gluteal nerves** as well as the **common peroneal portion of the sciatic nerve.**

3. **The sacral plexus** (L4–S3) lies in the deep pelvis and supplies the gluteal region, posterior thigh, leg, and foot, as well as the perineum (see Figure 27-4).

4. **Pelvic splanchnic nerves (nervi erigentes)** arise from spinal nerves S2–S4 (see Figures 29-4 and 29-14).

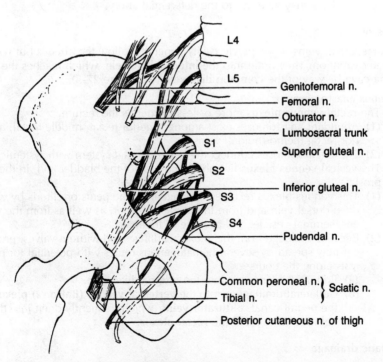

FIGURE 27-4. Anterior *(light)* and posterior *(shaded)* divisions of the sacral plexus, along with the major nerves that innervate the pelvis and lower extremities.

a. They consist of parasympathetic preganglionic fibers.
b. They also contain visceral afferent fibers from specific portions of the pelvic viscera.

5. **The coccygeal plexus** (S4–Cx) supplies the terminal portion of the vertebral column.

B. Major branches of the lumbar plexus (see Chapter 26 IV D and Figure 26-6)

C. Major branches of the sacral plexus (see Figure 27-4)

1. **Superior gluteal nerve** (L4–S1, posterior)
   a. This nerve exits the pelvis through the suprapiriform portion of the greater sciatic foramen.
   b. It supplies the gluteus medius and gluteus minimus muscles.
   c. Severance results in **abductor lurch,** a rolling (Trendelenburg) gait resulting from loss of abductive power in the hip.

2. **Inferior gluteal nerve** (L5–S2, posterior)
   a. This nerve exits by way of the infrapiriform region of the greater sciatic foramen.
   b. It supplies the gluteus maximus muscle.
   c. Paralysis results in difficulty climbing stairs or rising from a chair.

3. **Common peroneal nerve** (L4–S2, posterior)
   a. This nerve forms the posterior component of the sciatic nerve.
   b. It exits by way of the infrapiriform region of the greater sciatic foramen.
   c. It innervates the anterior aspect of the leg and the dorsum of the foot.
   d. Paralysis results in the inability to dorsiflex **(foot drop)** and evert the foot.

4. **Tibial nerve** (L4–S3, anterior)
   a. This nerve forms the anterior component of the sciatic nerve.
   b. It exits from the infrapiriform region of the greater sciatic foramen.
   c. It innervates the posterior region of the thigh and leg and the plantar surface of the foot.
   d. Paralysis is indicated by an inability to stand on the toes and loss of the Achilles' tendon reflex.

5. **Pudendal nerve** (S2–S4, anterior)
   a. **Course.** This nerve exits the pelvis through the infrapiriform region of the greater sciatic foramen. It then enters the ischiorectal fossa of the perineum by way of the lesser sciatic foramen. It comes to lie in the **pudendal (Alcock's) canal,** a condensation of the obturator fascia on the medial surface of the obturator muscle.
   b. **Branches** of this nerve supply the perineum.
      (1) The **inferior rectal nerve** provides motor innervation to the external anal sphincter and sensation to the inferior portion of the anal canal and the anal triangle of the perineum.
      (2) The **perineal nerve** conveys motor innervation to the muscles of the urogenital diaphragm and external genitalia and sensation to the urogenital triangle of the perineum, including the external genitalia.
   c. **Pudendal block** is used to anesthetize the perineum.
      (1) With the patient in the lithotomy position, the ischial tuberosities are palpated. After cutaneous anesthesia, the needle is passed just medial to the tuberosity to a depth of about 2 cm.
      (2) An alternative route involves intravaginal palpation of the ischial spine followed by needle insertion through the lateral vaginal wall to a point approximately 1 cm medial to the ischial spine and 1 cm below the sacrospinous ligament.

Chapter 28
Perineum

I. INTRODUCTION

A. **General structure.** The perineum is the area between the **ischial tuberosities,** extending from the **pubis** to the **coccyx.** The perineum covers the **pelvic outlet** and is pierced by anal and urogenital canals, which incorporate sphincteric mechanisms.

B. **Divisions.** The transverse diameter (a line connecting the ischial tuberosities) divides the perineum into the anal triangle posteriorly and the urogenital triangle anteriorly (Figure 28-1). These triangles share the levator ani muscle and have the same innervation and blood supply. Each triangle, however, has special musculature.
   1. **Anal triangle.** The principal structures are the anal canal and the anus guarded by the external anal sphincter (see Figures 28-2 and 28-10).
   2. **Urogenital triangle.** The principal structures are the external genitalia with their associated erectile tissue and musculature, as well as the gonads in the male (see Figures 28-2 and 28-10). Although the differences between the superficial structures of the urogenital triangle of the male and female are obvious, all structures have homologous counterparts in both genders (Table 28-1).

II. DEVELOPMENT OF EXTERNAL GENITALIA

A. **Indifferent stage** (to week 36)
   1. **The cloacal membrane** stretches across the perineum, closing the hindgut to the exterior (see Figure 24-1C). **Cloacal folds** form on either side of the **cloacal membrane.**
   2. **The genital tubercle** is formed anteriorly by fusion of the cloacal folds.
   3. **The urorectal septum** divides the cloaca transversely into the **urogenital sinus** (anteriorly) and the **anal canal** (posteriorly) by the sixth week.
      a. Fusion of the septum with the cloacal membrane forms the **central tendon** of the perineum **(perineal body).**
      b. The urogenital septum divides the cloacal membrane into **urogenital** and **anal membranes.** The cloacal sphincter is similarly divided, becoming the **urogenital diaphragm** and the **external anal sphincter.** Also, the cloacal folds are divided into **urethral** and **anal folds.**
   4. **Genital swellings (labioscrotal folds)** develop lateral to the urethral folds.
   5. **The urogenital and anal membranes** degenerate shortly afterward, opening the alimentary and urogenital canals to the perineum. In the female, a portion of the urogenital membrane persists as the **hymen.**

B. **Male external genitalia**
   1. **The phallus** is produced by rapid elongation of the genital tubercle.
      a. During this growth process, the urethral folds are drawn anteriorly, producing the **urogenital (urethral) groove.**
      b. By the twelfth week, the urethral folds fuse over the urethral groove ventrally, forming the **penile urethra.**
      c. During the sixteenth week, the **glans penis** differentiates and the **external urethral meatus** connects with the penile urethra at the **fossa navicularis.**
      d. The skin of the body of the penis grows over the glans penis as the **prepuce.** Ini-

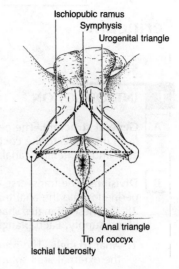

FIGURE 28-1. Boundaries and landmarks of the male perineum. The perineum is divided into anal and urogenital triangles by the transverse diameter of the pelvic outlet.

tially, the prepuce fuses to the glans, but it becomes separated by birth or shortly thereafter.
2. **The labioscrotal folds** migrate somewhat posteriorly and fuse, forming the **scrotal septum.** Each fold forms one half of the definitive **scrotum.**
3. **The testes** develop extraperitoneally, high in the posterior abdominal cavity, and descend into the scrotum (see Figure 29-5).
   a. Toward the end of the ninth week, the testes begin a retroperitoneal migration to the iliac fossa.
      (1) A fascial condensation, the **gubernaculum testis,** runs from the inferior pole of each testis, through the inguinal canals, to the respective scrotal fold.
      (2) The gubernaculum fails to lengthen during the subsequent period of rapid body growth, resulting in a shift of the gonads into the major pelvis.
   b. By the thirteenth week, the testes lie in the inguinal region but retain the original neurovascular connections.
   c. At about this time, the **inguinal canal** develops.
      (1) An evagination of the coelom, the **processus vaginalis (vaginal process),** penetrates the abdominal wall of the inguinal region and enters the scrotal fold.
      (2) The processus vaginalis draws with it some of the layers of the abdominal wall to form the **inguinal canal.**
   d. During the seventh month, the testes descend through the inguinal canal retroperitoneally (i.e., posterior to the vaginal process). In the scrotum, the vaginal process nearly surrounds the testes, forming the visceral and parietal layers of the **tunica vaginalis.**

Table 28-1. Homologous Perineal Structures

Male	Female
Scrotum	Labia major
Penis	Clitoris
Corpora spongiosum	Vestibular bulbs
Glans penis	Glans clitoridis
Penile urethra	Urogenital sinus
Urethral glands	Lesser vestibular glands
Bulbourethral glands	Greater vestibular glands
Prostate gland	Paraurethral glands

e. The **neck of the vaginal process** normally fuses by birth or shortly thereafter, isolating the **cavity of the tunica vaginalis** from the peritoneal cavity.
4. **Congenital anomalies of the penis** are common given the complex fusions that occur early in development. Anomalies include **hypospadias** (incomplete fusion of the penile urethra with the opening on the ventral surface), **epispadias** (with a urethral opening on the dorsal surface), **micropenis**, and doubling of the glans.

C. Female external genitalia
1. **The clitoris** is formed by a slight elongation of the genital tubercle.
   a. The urethral folds do not fuse ventral to the clitoris, but develop into the **labia minor**.
   b. The urogenital groove remains as the **vestibule (urogenital sinus)**.
2. **Labia major** develop from the genital swellings (labioscrotal folds).
3. **The ovaries** develop extraperitoneally, high in the posterior abdominal cavity. Subsequent to the ninth week, the ovaries migrate into the pelvis (see Figure 29-5).
   a. A fascial condensation (homologous to the gubernaculum) runs from the inferior pole of each ovary, through the inguinal canals, to the respective labial fold.
      (1) This ligament lengthens but slightly during subsequent rapid body growth, which results in a shift of the gonads into the major pelvis.
      (2) A small portion of this ligament fuses to the paramesonephric duct that develops into the uterus. This ligament is thus divided into a cranial **ovarian ligament** and a caudal **round ligament of the uterus.** The attachment of this ligament to the developing uterus redirects ovarian descent into the deep pelvis.
   b. By the thirteenth week, the ovaries lie in the minor pelvis but retain the original neurovascular connections that form the **suspensory ligament of the ovary.**
   c. At about this time, an evagination of the coelom, the **processus vaginalis (canal of Nuck),** penetrates the abdominal wall of the inguinal region and enters the labial fold.
   d. The neck of the processus vaginalis normally fuses by birth or shortly thereafter.

III. ANAL TRIANGLE

A. **The ischioanal (ischiorectal) fossa** is the subcutaneous region about the anal canal.
1. **Boundaries** are the transverse diameter of the pelvic outlet (between the ischial tuberosities) and lines from the tip of the coccyx to the ischial tuberosities (Figure 28-2).

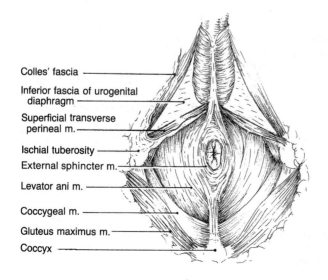

FIGURE 28-2. Anal triangle. Removal of fat from the ischioanal fossa reveals the underlying pelvic floor and external anal sphincter.

Colles' fascia
Inferior fascia of urogenital diaphragm
Superficial transverse perineal m.
Ischial tuberosity
External sphincter m.
Levator ani m.
Coccygeal m.
Gluteus maximus m.
Coccyx

a. The floor is formed by the **levator ani muscle** (superiorly). The walls are formed by the **obturator internus muscle** (anterolaterally) and the **sacrotuberous ligament** with the overlying gluteus maximus muscle (posterolaterally).
   b. An **anterior horn** extends forward into the urogenital triangle on each side superior (deep) to the urogenital diaphragm.
   c. The ischioanal fossa on each side communicates posterior to the anal canal.
2. **Contents**
   a. Considerable **ischioanal fat** is separated by stringy connective tissue fibers and septa.
      (1) This fatty tissue is a counterpart of the superficial layer of superficial fascia (Camper's fascia of the abdominal wall and Cruveilhier's fascia of the urogenital triangle). Ischioanal fat has different metabolic characteristics that make it somewhat resistant to resorption with starvation.
      (2) Tension in the gluteus maximus muscles (e.g., when standing) compresses the fat of the ischioanal fossae around the anal canal, contributing to **fecal continence.** The fat allows dilation of the anal canal during defecation, however, when the gluteal muscles are relaxed.
   b. **Inferior rectal (hemorrhoidal) vessels** and **nerves** pass through the ischioanal fossae, supported by the ischioanal fat (see Figures 28-10 and 28-15).
      (1) These neurovascular structures are branches of the **internal pudendal vessels** and the **pudendal nerve** that supply and innervate the **external anal sphincter.**
      (2) Inadvertent severing of nerves in this fossa results in **sphincteric incontinence** during brief peristaltic waves.
3. **Abscesses** in the ischioanal fossa may become large or may extend to the contralateral side around the anus ("horseshoe" abscess). When making incisions to allow drainage of such abscesses, the surgeon must avoid the inferior rectal neurovascular bundle or risk paralyzing the external anal sphincter.

B. Anal canal
1. **External structure**
   a. This part of the alimentary canal inferior to the pelvic diaphragm is 3–4 cm long and opens to the exterior through the **anus.**
   b. The most medial fibers of the **pubococcygeus muscle** (the **puborectalis** or **"rectal sling"**) form the **perineal flexure** ("carrying angle") of the rectum. Thus, the anal canal angles downward and backward from the rectum to the anal verge (see Figure 29-9). The puborectalis muscle is the principal mechanism of fecal continence.
2. **Internal structure.** The anal canal is divided between visceral and somatic portions (Figure 28-3).
   a. The **upper two thirds** of the canal belong to the intestine with respect to the mucosa, blood supply, and autonomic innervation.
      (1) The mucosa of this segment is thrown into longitudinal folds about 1 cm long, the **anal columns** (of Morgagni). The bases of these columns are joined by small semilunar folds of tissue, the **anal valves.** The bases of the anal columns and intervening anal valves define the **pectinate (dentate) line,** which approximates the division between visceral and somatic portions of the anal canal (see Figure 28-3).
         (a) Superior (deep) to the pectinate line, the epithelium is innervated by visceral afferents and is insensitive to touch.
         (b) The pectinate line also divides areas of venous drainage. Venous anastomoses between the portal system (draining the rectum) and the systemic system (draining the anal canal) are potential sites for varicosities (hemorrhoids).
      (2) Between the bases of the anal columns, behind the anal valves, are small blind sacs, the **anal sinuses (crypts),** into which **anal glands** open.
         (a) The **anal glands** are rudimentary circumanal glands, the secretions of which contain pheromones.

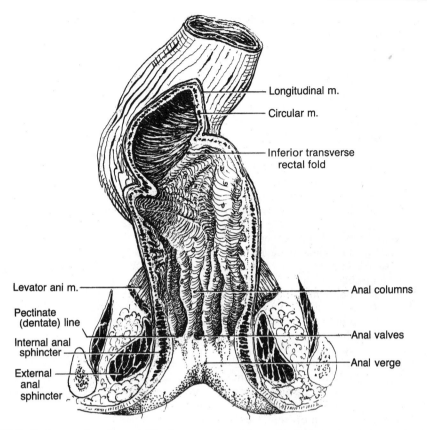

FIGURE 28-3. Rectum and anal canal. The levator ani muscle marks the boundary between the rectum and the anal canal.

- **(b) Anal fistulas.** These glands are prone to infection and may cause fistulous tracts that require surgical correction.
- b. The **lower one third** of the canal belongs to the perineum with respect to mucosa, blood supply, and somatic innervation. The mucosal lining of this segment is squamous epithelium with abundant somatic nerve endings that are extremely sensitive.
3. **Anal sphincters** keep the anal canal closed, except during passage of feces or flatus.
   a. The **sphincteric mechanism,** consisting of internal and external sphincters, can distinguish between and separate flatus from feces.
      (1) The **internal anal sphincter** is a tube of involuntary muscle that encloses the entire canal. It is a continuation of the circular muscular layer of the intestine.
         (a) Its intrinsic tone maintains this closure.
         (b) This sphincter is inhibited by the approach of a peristaltic wave and, thus, is not a major factor in fecal continence.
      (2) The **external anal sphincter** is striated muscle under voluntary control through the rectal branches of the pudendal nerve (see Figure 28-2).
         (a) Structure. A **superficial portion** arises from the coccyx, passes to either side of the anus, and inserts onto the perineal body. A **deep portion** surrounds the anal canal.
         (b) Function. The external sphincter can maintain a voluntary tonic contracture for about 10–20 seconds, which is sufficient to provide the voluntary contraction necessary to counter the passage of a peristaltic wave when the internal anal sphincter relaxes.

b. Continued anal continence is primarily a function of the **rectal sling,** part of the levator ani muscle group.

C. Vasculature

1. **Arterial supply**
   a. The **internal pudendal artery** (see Figures 28-10 and 28-15) and its **inferior rectal (hemorrhoidal) branches** supply the anal triangle.
   b. These vessels form anastomotic connections with the middle hemorrhoidal (rectal) arteries, direct branches of the internal iliac arteries (see Figure 29-2).

2. **Venous return**
   a. The **inferior hemorrhoidal (rectal) veins** parallel the inferior hemorrhoidal arteries and drain into the **internal pudendal vein** on each side (see Figure 29-2).
   b. The veins of the submucosal plexus form extensive anastomotic connections with the **middle rectal (hemorrhoidal) veins,** which drain directly into the internal iliac veins. They also anastomose with branches of the **superior rectal (hemorrhoidal) veins,** which drain into the hepatic portal system.
   c. **Anal hemorrhoids.** Because of the rich anastomoses between the hepatic portal system and the systemic venous drainage of the perineum, portal hypertension often results in hemorrhoidal varices.
      (1) **Internal hemorrhoids** occur above the pectinate line.
         (a) Innervation above this line is similar to that of the rest of the gut, with pressure receptors but no definitive pain receptors.
         (b) This type of hemorrhoid is **painless** because the epithelium is innervated by visceral afferents.
         (c) These hemorrhoids are likely to be large, dangerous, and silent, noticed only when they become large enough to bleed profusely or prolapse through the anus. Hemorrhoidal bleeding may be sufficient to produce **anemia,** especially common in the elderly.
      (2) **External hemorrhoids** occur below the pectinate line.
         (a) The area below the pectinate line is innervated somatically by the rectal (hemorrhoidal) branches of the pudendal nerve.
         (b) These hemorrhoids are painful; even small varices in this region demand attention.

3. **Lymphatic drainage**
   a. **Below the pectinate line,** lymphatic drainage of the anal canal is primarily toward the **superficial and deep inguinal nodes.**
   b. **Above the pectinate line,** lymphatic drainage is primarily toward the **internal iliac nodes.**
   c. Anastomoses between the lymphatics in the perianal region with those of the upper portion of the anal canal allow secondary drainage to occur toward the deep pelvic nodes.

D. Innervation

1. **The inferior rectal branch of the pudendal nerve** provides motor innervation to the external anal sphincter and sensation to the inferior portion of the anal canal as well as the integument of the anal triangle (see Figures 28-10 and 28-15).
   a. This nerve arises from the sacral plexus (levels S2–S4) and leaves the pelvis through the infrapiriform region of the **greater sciatic foramen.** It enters the ischioanal fossa of the perineum through the **lesser sciatic foramen.**
   b. It runs along the **pudendal (Alcock's) canal,** a condensation of the obturator fascia on the medial surface of the obturator muscle.

2. **The perineal branch of the posterior femoral cutaneous nerve** supplies cutaneous innervation to the lateral aspects of the anal and urogenital triangles. The small region between the anus and coccyx receives twigs from S5 and the coccygeal nerve.

3. **Anal incontinence.** Severance of the inferior rectal branch results in paralysis of the external anal sphincter and may result in momentary embarrassments when peristaltic waves reach the anus.

E. Functional considerations

1. **Fecal continence** involves the rectal sling, external anal sphincter, and ischioanal fat.
   a. When feces stored within the sigmoid colon suddenly move into the rectum, the rectal ampulla dilates and the urge to evacuate the rectum is perceived.
   b. Defecation is prevented primarily by the **puborectalis muscle (rectal sling),** which cants the lower part of the rectum forward **(carrying angle),** effectively kinking the lumen. Additionally, when a person is standing erect, the mass of **ischioanal fat** is compressed against the anal canal anteriorly and laterally by the location and tonic activity of the gluteus maximus muscle.
   c. To prevent defecation when peristaltic waves occur, occasional voluntary contraction of the **external anal sphincters** is required until the peristaltic waves subside (5–10 seconds).

2. **Defecation**
   a. When the time and place are propitious, compression of the anal canal by the ischioanal fat is released by suitable anatomic positioning. The puborectalis muscle relaxes, allowing the rectum to straighten and descend slightly. The anal sphincters relax: the internal, according to the Bayliss-Starling law, and the external, voluntarily.
   b. The muscular movements of the terminal portion of the alimentary canal, assisted by gravity, evacuate the rectum. The Valsalva maneuver, which increases intra-abdominal pressure, facilitates expulsion of feces.

3. **Restoration of fecal continence.** After passage of each fecal mass, the puborectalis muscle re-establishes the carrying angle, and the anal sphincters contract, thereby restoring anal continence.

IV. MALE UROGENITAL TRIANGLE

A. External genitalia

1. **The penis** consists of three bodies of vascular **erectile tissue.**
   a. **External structure** (Figure 28-4)
      (1) **Orientation.** Because the penis is suspended along the abdominal wall in lower mammals, the anterior surface in humans is termed the dorsal surface.

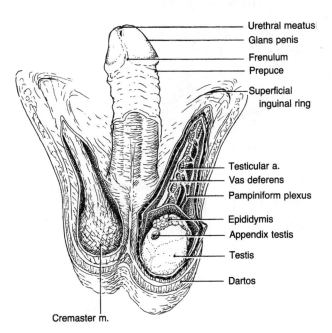

FIGURE 28-4. Male external genitalia. Contents of the left scrotum and spermatic cord are shown.

Fusion of the urogenital sinus forms the penile urethra and produces a **midline raphe** along the ventral surface. The penis and clitoris are homologous.

(2) **Composition.** Three vascular erectile bodies begin as the **radix (root) of the penis** in the superficial perineal pouch. The **body** of the penis is formed by fusion of these structures as they leave the perineum (see Figure 28-6A).

(a) The **central corpus spongiosum** (corpus cavernosum urethrae) is ventral (see Figures 18-5 and 28-6A).
    (i) It begins as the **bulb** of the penis in the superficial perineal pouch.
    (ii) It terminates as the **glans penis,** the base of which flares to define the **coronary sulcus** of the penis.
    (iii) It contains the **penile urethra.**
    (iv) Proximally, it is covered by the **bulbospongiosus muscle** (see Figure 28-8).

(b) The bilateral **corpora cavernosa** are dorsal (see Figure 28-6A).
    (i) Each corpus cavernosum begins as a **crus,** which attaches along an ischiopubic ramus in the superficial perineal pouch.
    (ii) These erectile bodies terminate proximal to the glans penis.
    (iii) Proximally, each is covered by an **ischiocavernosus muscle** (see Figure 28-8).

(3) **Epithelium and fascia.** The penis is covered by skin that is relatively hairless and has more pigment than other parts of the body.

(a) The **prepuce** or **foreskin** is an extension of the integument over the glans penis (see Figures 28-4 and 28-5). It usually is excessive over the tip of the glans. Often, it does not even cover the glans.
    (i) The foreskin attaches to the ventral raphe of the glans by the **frenulum.**
    (ii) **Circumcision** involves removal of redundant foreskin.

(b) The **superficial fascia of the penis** is an extension of the superficial fascia of the abdominal wall over the body of the penis (Figure 28-5).

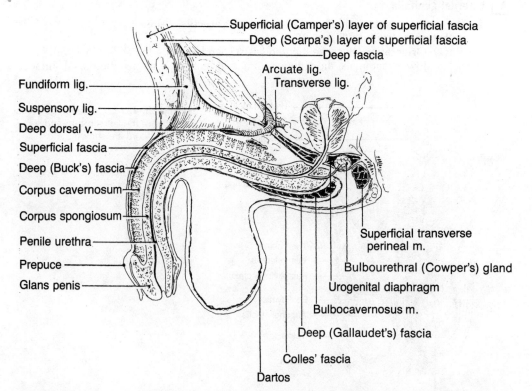

FIGURE 28-5. Perineal fascia in the male, midsagittal section.

(i) The superficial layer of the superficial fascia loses its fat and fuses with the deep layer of the superficial fascia to form the superficial fascia of the penis. It is loosely attached to the underlying tunica albuginea and is mobile.

(ii) This fascial layer does not extend over the glans penis, which is covered by mucous epithelium tightly bound to the underlying tunica albuginea.

(c) The **deep (Buck's) fascia of the penis** is an extension of the deep (Gallaudet's) fascia of the perineum (see Figure 28-5).
  (i) Proximally, this layer invests the muscles at the root of the penis.
  (ii) Distally, this layer extends as far as the glans, even though the muscle layer terminates where the penis becomes pendulous. It invests the corpus spongiosum and corpora cavernosa.

(4) **Ligamentous support** (see Figure 28-5)
  (a) The **suspensory ligament** arises from the linea alba and inserts onto the deep (Buck's) fascia of the penis.
  (b) The **fundiform ligament** is a continuation of the deep fascia of the abdominal wall as it reflects onto the penis.

b. **Internal structure**
  (1) **Erectile tissue** (Figure 28-6A)
    (a) The two lateral **corpora cavernosa** and the central **corpus spongiosum** consist of vascular sinusoids. The filling of these sinusoids with arterial blood results in penile erection.
    (b) Each erectile structure is surrounded by a distinctive connective layer, the **tunica albuginea.**
      (i) The paired **corpora cavernosa** are surrounded by tunica albuginea that is dense and not particularly elastic. When the penis becomes tumescent, this layer impedes venous return, which results in the extreme turgidity characteristic of the erect penis.

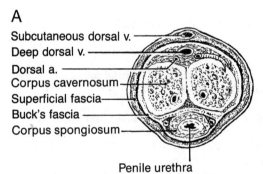

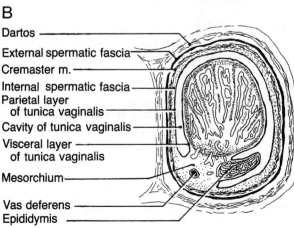

FIGURE 28-6. *(A)* Cross-section of the penis shows fascial layers and erectile bodies. *(B)* Cross-section through the scrotum and testis shows layers of peritoneum.

- **(ii)** The central **corpus spongiosum** is surrounded by tunica albuginea that is less dense, so this erectile tissue and its terminal glans penis do not become excessively turgid on erection. Therefore, the penile urethra is never occluded, permitting the passage of the ejaculate during erection.
- **(2)** The **penile urethra** extends along the entire penis within the corpus spongiosum (see Figure 28-6). The penile urethra is homologous to the urogenital sinus (vestibule) in the female.
  - **(a)** This portion of the urethra begins in the superficial perineal pouch just as the membranous urethra penetrates the urogenital diaphragm.
  - **(b)** After about 2.5 cm, it receives the ducts of the **bulbourethral (Cowper's) glands** (see Figure 28-5). More distally, mucous **urethral glands** (of Littré) line the penile urethra. Infection of these glands produces nonspecific urethritis.
  - **(c)** Terminally, the penile urethra widens within the glans penis, forming the **fossa navicularis,** which terminates as the slit-like **urethral meatus.**
    - **(i)** The **external urethral meatus** is the narrowest part of the urethra. An instrument that passes this point can be advanced into the urinary bladder.
    - **(ii)** Introduction of instruments into the male urethra must be accomplished carefully, with the penis straightened into an approximately erect posture.
- **c. Vasculature**
  - **(1) Arterial supply** (see Figure 28-10)
    - **(a)** The **superficial skin** of the penis is supplied by the perineal branches of the **internal pudendal arteries** as well as the **external pudendal arteries,** which arise from each femoral artery.
    - **(b) Erectile tissue** is supplied by the **internal pudendal arteries,** through the **deep dorsal, central,** and **bulbar arteries.**
  - **(2) Venous return** is by way of the **deep dorsal vein of the penis** to the prostatic venous plexus (see Figure 28-6A).
- **d. Innervation** is principally by branches of the pudendal nerve (see Figure 28-10).

2. The scrotum is suspended from the male perineum and contains the testes (see Figures 28-4 and 28-6B).
  - **a. Orientation.** The scrotum develops from the **labioscrotal folds,** which fuse in the male. The **midline raphe** marks this line of fusion.
  - **b. Epithelium and fascia**
    - **(1)** The **skin of the scrotum** is thin and has little fat, an important factor in maintaining lower testicular temperature.
    - **(2)** The **dartos layer** or **tunic** is formed by the fusion of the superficial and deep layers of the superficial fascia (see Figure 28-6B).
      - **(a)** This layer contains smooth muscle, which also functions in temperature regulation.
      - **(b)** Although core body temperature is 37°C, scrotal temperature is 33.9°C. A decrease in temperature causes the dartos to contract to bring the testes into close contact with the body to conserve heat. The dartos relaxes and becomes pendulous to dissipate heat.
    - **(3) Transillumination of the scrotum** defines the testicular outline to distinguish between fluid (hydrocele) and tissue (hernia or tumor).
  - **c. Innervation.** The scrotum is innervated by the **ilioinguinal, genitofemoral,** and **pudendal nerves,** as well as by the perineal branch of the **posterior femoral cutaneous nerve.**

3. The spermatic cord provides the route of neurovascular and ductal communication between the scrotum and the abdominopelvic cavity (see Figures 23-5 and 28-4).
  - **a. Orientation.** The spermatic cord forms during the descent of the testes.
  - **b. Structure.** Several layers correspond to those of the abdominal wall (see Figure 23-5).

- **(1) External spermatic fascia**
  - **(a)** This outermost layer is derived from the deep (investing) fascia of the external oblique muscle.
  - **(b)** Because the superficial inguinal ring is a defect in the external oblique aponeurosis, there is no direct contribution from that muscle.
- **(2) Cremaster muscle**
  - **(a)** This muscle, which forms the intermediate layer, is derived from the **internal oblique muscle** and its investing fascia.
  - **(b)** The **genital branch of the genitofemoral nerve** innervates this layer and provides the efferent limb for the **cremaster reflex** (L1–L2), which involves elevation of the testes within the scrotum when the inner thigh is stroked.
- **(3) Internal spermatic fascia**
  - **(a)** This innermost layer is derived from the **transversalis fascia.**
  - **(b)** It contains the testicular neurovascular bundle and the vas deferens with its accompanying deferential artery.
    - **(i)** The **testicular arteries** arise from the abdominal aorta just distal to the renal arteries.
    - **(ii)** Venous drainage from the testes forms the anastomotic **pampiniform plexus** of veins, which surrounds each testicular artery.
    - **(iii)** The **testicular nerves** arise from the sympathetic aorticorenal plexus. Testicular afferents follow this pathway back to the spinal cord.
    - **(iv)** The **artery of the vas deferens (deferential artery)** anastomoses with the testicular artery to provide collateral circulation to the testes and scrotum.
    - **(v)** The **vas deferens** rises from the inferior pole of the testes and passes into the spermatic cord.
- **(4)** The **tunica vaginalis** is a remnant of the processus vaginalis (see Figure 28-6B).
  - **(a)** The fetal **processus vaginalis** is an evagination of the peritoneal cavity into the scrotum. It usually occludes postnatally.
    - **(i)** A patent processus vaginalis predisposes to **indirect (congenital) inguinal hernia.**
    - **(ii)** Partial occlusion of a processus vaginalis can result in fluid accumulation **(hydrocele processus vaginalis),** which may not be distinguishable from a hernia or incompletely descended testes until surgery is performed.
  - **(b)** The tunica vaginalis consists of two layers of peritoneum with an intervening potential space **(cavity of tunica vaginalis)** that represents the detached portion of the peritoneal cavity (processus vaginalis) within the scrotum. It surrounds all but the posterior portion of the testes (see Figure 28-6B).
    - **(i)** The **parietal layer** is equivalent to the parietal peritoneum. It lines the outer wall of the potential space.
    - **(ii)** The **visceral layer** is equivalent to visceral peritoneum. It overlies the testes.
    - **(iii)** The **mesorchium** (mesentery of the testes) lies posteriorly. It is formed by reflection of the parietal layer to the visceral layer. Although the testes develop retroperitoneally and descend into the scrotal sacs retroperitoneally, the presence of a mesorchium supporting the testes justifies the classification of the postnatal descended testes as peritoneal structures.

B. Fascia. Underlying muscular and fascial structures support the external genitalia and reinforce the abdominal cavity.

1. **The superficial layer of superficial perineal fascia (Cruveilhier's f.)** is a fatty layer just beneath the dermis. It is continuous with Camper's fascia of the abdominal wall (Figure 28-7A).

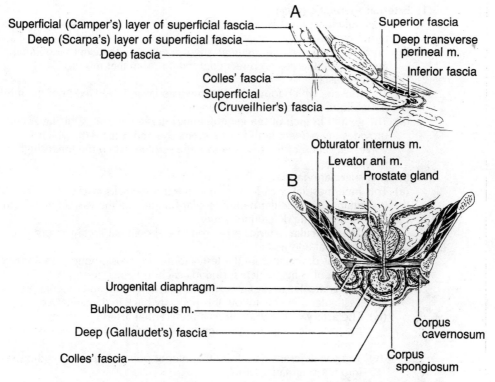

FIGURE 28-7. Perineal fascia in the male, parasagittal *(A)* and coronal *(B)* sections.

    a. It passes across the perineum to become the fat of the ischioanal fossa.
    b. It contributes to the **superficial fascia of the penis** by loosing its fat and fusing with the deep layer of the superficial fascia.
    c. It contributes to the **dartos layer** by loosing its fat and fusing with the deep layer of the superficial fascia.
  2. **The deep layer of superficial perineal fascia** (Colles' f.) is a membranous layer that attaches laterally and posteriorly to delimit the **superficial perineal pouch** (see Figure 28-7A).
    a. Anteriorly, it is continuous with Scarpa's fascia of the abdominal wall and fuses with Camper's fascia to pass over the penis as the **superficial fascia of the penis.**
    b. Laterally, it attaches to the ischiopubic rami.
    c. Medially, it inserts into the urethral adventitia.
    d. Inferiorly, it fuses with the superficial layer of the superficial fascia to pass over the scrotum as the **dartos layer.**
    e. Posteriorly, it passes superficial to the **superficial transverse perineal muscle** and inserts onto the posterior edge of the deep transverse perineal muscle (see Figure 28-7A).
      (1) Midline fibers join the **central tendon (perineal body).**
      (2) More lateral fibers merge with the inferior and superior leaflets of the fascia of the urogenital diaphragm, which enclose the **deep transverse perineal muscle.** These two leaflets define the **deep perineal pouch.**
        (a) The **superficial (inferior) layer** is termed the **inferior fascia of the urogenital diaphragm,** or the **perineal membrane.**
        (b) The deep (superior) layer is termed the **superior fascia of the urogenital diaphragm.**
  3. **The deep (investing) perineal fascia** (Gallaudet's f.) covers the muscles of the superficial pouch. This layer is continuous with the deep (Buck's) fascia of the penis (see Figures 28-5 and 28-7B).

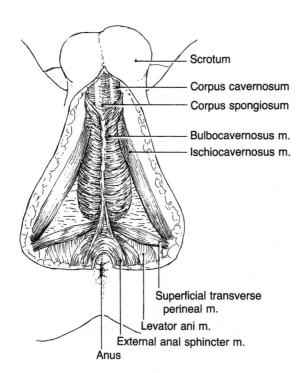

FIGURE 28-8. Superficial pouch in the male. Contents include the crura and bulb of the penis with associated muscles, as well as the superficial transverse perineal muscle.

C. **Superficial perineal pouch (space)**
  1. **Boundaries.** This space is bounded by Colles' fascia and lies between this fascial layer and the inferior fascia of the urogenital diaphragm (perineal membrane).
  2. **Contents** (see Figure 28-7B)
     a. **Corpora cavernosa** are paired bodies of vascular erectile tissue lying along the ischiopubic rami, each forming a **radix of the penis.**
        (1) These erectile bodies fuse near the pubic symphysis to form the body of the penis.
        (2) They are covered by the **ischiocavernosus muscles.** The male and female corpora cavernosa are homologous.
     b. The **bulb of the penis** is vascular erectile tissue.
        (1) It is covered by the **bulbospongiosus muscle.** The fused halves of the bulb of the penis in the male and the vestibular bulbs in the female are homologous.
        (2) It forms the corpus spongiosum and contains the penile urethra.
     c. **Ischiocavernosus muscles** cover each crus (Figure 28-8; see Table 28-2).
        (1) These muscles arise from the ischial rami and insert onto the penis just distal to the point at which the crura join to form the body of the penis.
        (2) They are covered by the deep investing fascia (Gallaudet's f.) in the superficial pouch, which becomes the deep fascia (Buck's) on the penile shaft.
        (3) They are innervated by the perineal branches of the pudendal nerve.
     d. **Bulbocavernosus (bulbospongiosus) muscles** overlie the bulb of the penis on each side (see Figure 28-8 and Table 28-2).
        (1) These muscles arise from the central tendon, posteriorly, and fuse at the median raphe. Each inserts into the deep fascia of the penis just after the crura fuse.
        (2) They are covered by the deep investing fascia in the superficial pouch, which becomes the deep (Buck's) fascia on the penile shaft.
        (3) They are innervated by the perineal branches of the pudendal nerve.
        (4) **Functions**
           (a) Once termed the compressor urethrae, the bulbospongiosus muscles expel residual urine from the penile urethra. They are rather ineffectual,

Table 28-2. Muscles of the Perineum

Muscle	Origin	Insertion	Primary Action	Innervation
Superficial transverse perineal	Ischial tuberosity	Central body of perineum	Stabilizes perineum	Pudendal n. (S2–S4)
Deep transverse perineal	Ischiopubic ramus	Midline raphe and central perineal body	Reinforces pelvic floor and stabilizes vagina	Pudendal n. (S2–S4)
External urethral sphincter	Circumurethral	Circumurethral	Urinary continence	Pudendal n. (S2–S4)
Ischiocavernosus	Ischiopubic rami	Body of penis or clitoris		Pudendal n. (S2–S4)
Bulbocavernosus	Central perineal body and midline raphe	Body of penis or clitoris		Pudendal n. (S2–S4)
External anal sphincter	Circumanal	Circumanal	Fecal continence	Pudendal n. (S2–S4)

however, because they do not extend beyond the base of the penile shaft.
- (b) Their role in the mechanism of erection is dubious because they can contract for only 5–10 seconds at a time. Spasm during orgasm may contribute to ejaculation.
- e. The **superficial transverse perineal muscle** (see Figure 28-8 and Table 28-2) arises on each side from the ischial tuberosity and inserts into the perineal body (central tendon).
  - (1) Although this small muscle may be absent unilaterally or bilaterally, it stabilizes the perineum.
  - (2) When present, it is a palpable surgical landmark that delineates the posterior extent of the superficial pouch.
- 3. **Perforations of the penile urethra** result in extravasation of urine into the superficial pouch.
  - a. From the superficial space, urine may extravasate between superficial and deep fascial layers along the abdominal wall (beneath Scarpa's fascia), into the scrotum (under the dartos tunic), and onto the penis (superficial to the deep penile or Buck's fascia).
  - b. Urine does not extravasate into the anal triangle because Colles' fascia defines the posterior boundary of the superficial pouch. It cannot extravasate into the thigh because Scarpa's fascia fuses with the fascia lata.

D. Deep perineal pouch (space)
1. **Boundaries.** This space is between the superior and inferior fascial planes of the deep transverse perineal muscle (Figure 28-9).
2. **Contents**
   a. The **deep transverse perineal muscle** originates along the ischiopubic rami and inserts into the central raphe of the urogenital diaphragm and the central tendon of the perineum (see Figure 28-9).
   - (1) This muscle and its associated superior and inferior fascial layers form the urogenital diaphragm.
   - (2) The **urogenital diaphragm** lies inferior to the **urogenital hiatus** of the **levator ani muscle** and supports this potentially weak region of the pelvic floor.

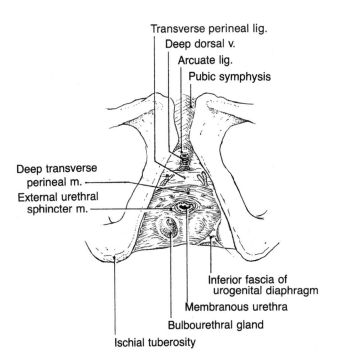

FIGURE 28-9. Deep perineal pouch in the male. Contents include the deep transverse perineal muscle with the external urethral sphincter as well as the bulbourethral glands.

    b. The **external urethral sphincter** is a modified portion of the deep transverse perineal muscle (see Figure 28-9).
       (1) Where the **membranous urethra** passes through the urogenital diaphragm, muscle fascicles of the deep transverse perineal muscle encircle the urethra.
       (2) The external urethral sphincter is voluntary, innervated by twigs from the perineal branch of the pudendal nerve.
       (3) This muscle is a prime factor in urinary continence, once the desire to void is perceived.
    c. The **paired bulbourethral (Cowper's) glands** are embedded in the deep transverse perineal muscle (see Figure 28-5). The ducts drain into the base of the penile urethra.
    d. The **inferior fascia of the urogenital diaphragm** (perineal membrane) bounds the deep pouch inferiorly. The **superior fascia of the urogenital diaphragm** bounds the deep pouch superiorly (see Figure 28-7A).
       (1) Anterior boundary
          (a) Merging of the inferior and superior fascial layers of the urogenital diaphragm anterior to the free edge of the deep transverse perineal muscle forms the **transverse perineal ligament** (see Figure 28-9).
          (b) The **arcuate ligament** lies behind the pubic symphysis and is the most anterior component of this layer (see Figure 28-9). The **deep dorsal vein of the penis** runs through the hiatus between the transverse and arcuate ligaments to reach the prostatic venous plexus.
       (2) Laterally, the attachments of the superior and inferior fascial layers of the urogenital diaphragm to the ischiopubic rami limit the deep pouch.
       (3) Posteriorly, the superior and inferior fascial layers of the urogenital diaphragm also fuse at the free edge of the deep transverse perineal muscle. Along this line of fusion, the deep layer of perineal (Colles') fascia attaches to define the posterior limit of the superficial perineal pouch.

3. Perforation of the urethra superior to the urogenital diaphragm from trauma or disease may produce intrapelvic, intraperitoneal, or extraperitoneal urinary extravasation.

FIGURE 28-10. Vasculature and innervation of the male perineum.

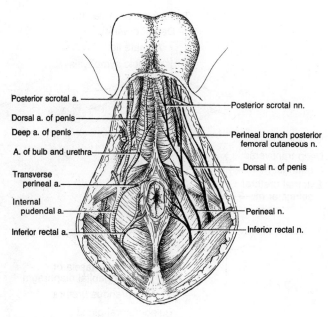

E. **Vasculature**
   1. **Arterial supply** (Figure 28-10; see Figure 27-3)
      a. The **internal pudendal artery** supplies most of the perineum.
         (1) This artery leaves the pelvic cavity through the infrapiriform portion of the **greater sciatic foramen.**
         (2) It then passes through the **lesser sciatic foramen** to enter the ischiorectal fossa of the perineum, where it courses in the **pudendal** (Alcock's) **canal** on the medial side of the obturator internus muscle.
         (3) Branches
            (a) **Inferior rectal (hemorrhoidal) arteries** supply the anal region.
            (b) **Superficial perineal arteries**
               (i) The **transverse perineal branch** supplies the superficial and deep perineal pouches.
               (ii) The **posterior scrotal branch** supplies the posterior scrotum.
            (c) **Deep perineal arteries** supply the erectile tissues of the penis.
               (i) **Dorsal arteries** lie beneath the deep fascia (of Buck) and external to the tunica albuginea.
               (ii) **Deep (central) arteries,** located within the corpora cavernosa, supply those erectile tissues.
               (iii) The **bulbar arteries** of the bulb of the penis supply erectile tissue of the corpus spongiosum and glans penis.
      b. The **external pudendal arteries,** which arise from the femoral arteries, supply the anterior superficial portion of the perineum and the subcutaneous tissue of the penis.
   2. **Venous return**
      a. **Perineal veins** generally parallel the branches of the internal pudendal artery and drain into the **internal pudendal vein.**
      b. **Superficial (subcutaneous) dorsal veins** of the penis and the veins of the anterior scrotum drain into the **external pudendal veins,** which enter the femoral vein.
      c. The **deep dorsal vein** of the penis drains the erectile tissue into the **prostatic venous plexus** (see section IV D 2 d (1) (b) and Figure 28-9).
   3. **Lymphatic drainage**
      a. **Penile lymphatics**
         (1) **Superficial lymphatics** drain toward the superficial inguinal nodes.

(2) **Deep lymphatics** drain into the internal iliac nodes.
  b. The **superficial pouch** drains largely into the superficial and deep inguinal nodes.
  c. The **testes** drain to the para-aortic nodes high in the abdomen.

F. Innervation

1. **Pudendal nerve** (see Figure 28-10)
   a. The **perineal branch** of this nerve is a mixed nerve, containing the general somatic afferent (sensory), general somatic efferent (motor), and sympathetic (vasomotor) nerves as components.
      (1) Perineal branches of the pudendal nerve are named according to the arteries they accompany. The superficial branches convey sensation to the skin. The deep branches provide motor innervation to the perineal musculature.
      (2) On the dorsum of the penis, the dorsal nerve of the penis (the terminal branch of the pudendal nerve) lies between the deep penile (Buck's) fascia and the tunica albuginea. It supplies the dorsum and glans of the penis.
   b. **Saddle block.** The perineum may be blocked by anesthesia of the pudendal nerve as it courses along the pudendal canal (see section V F 1 b).
2. **Ilioinguinal** and **genitofemoral nerves,** which arise from L1 and L2, supply the anterior part of the scrotum and the base of the penis.
3. **Perineal branches** of the **posterior femoral cutaneous nerve** arise from spinal segments S2–S4 and supply the posterolateral region of the urogenital triangle.

G. Functional considerations

1. **Erection** is a vascular condition principally controlled by the parasympathetic division of the autonomic nervous system.
   a. Parasympathetics from the **nervi erigentes** (S2–S4) reach the penis by way of the lateral pelvic plexus and prostatic plexus, crossing the urogenital diaphragm with the deep dorsal vein (see Figure 29-4). Stimulation relaxes the arterioles supplying the erectile tissue, increasing blood flow.
   b. The resultant tumescence compresses the veins that drain the erectile tissue as they pass through the tunica albuginea, impeding venous return and producing full erection.
2. **Ejaculation** is a primarily a sympathetic event.
   a. Sympathetics from the lumbar chain (L1–L2) follow the hypogastric plexus to reach the lateral pelvic plexus and prostatic plexus (see Figure 29-4).
   b. Stimulation produces vigorous contraction of the muscular vas deferens, seminal vesicles, and prostate gland, with expulsion of the seminal ejaculate.

V. FEMALE UROGENITAL TRIANGLE

A. External genitalia

1. **Vulva**
   a. It is the principal feature of the female urogenital triangle.
   b. Structure (Figure 28-11)
      (1) The **labia major** are two folds of hirsute skin, each supported by underlying fat pads. The labia major with the anterior and posterior commissures define the boundaries of the **pudendal cleft** (see Figure 28-11).
         (a) The **anterior commissure** is formed by the blending of the labia major over the pubic symphysis.
         (b) The **posterior commissure,** a fold of tissue anterior to the anus, connects the labia major posteriorly.
         (c) The labia major develop from the **labioscrotal folds.**
      (2) The **labia minor** are two small, hairless, sometimes pendulous, folds located

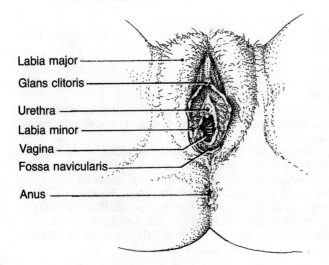

FIGURE 28-11. Female external genitalia.

medial to the labia major (see Figure 28-11). They enclose the **vestibule (urogenital sinus).**
   (a) Anteriorly, each labium minus divides into two parts.
      (i) Superior to the clitoris, the lateral portions fuse to form the **prepuce** of the clitoris, which variably covers the glans clitoridis.
      (ii) Inferior to the clitoris, the medial portions fuse to form the **frenulum** of the clitoris.
   (b) Posteriorly, even though the labia minor appear to blend with the labia major, a subtle fold, the **fourchette (frenulum of the labia),** connects the labia minor posterior to the vaginal introitus.
   (c) The labia minor develop from the **urethral folds.**
(3) The **vestibule,** or **urogenital sinus,** is bounded by the labia minor, frenulum of the clitoris, and fourchette.
   (a) The **external urethral ostium** is anterosuperior to the vaginal opening, approximately 2 cm inferior to the clitoris.
   (b) The **vaginal introitus** is the predominant feature of the vestibule.
      (i) The **hymen,** a remnant of the urogenital membrane, incompletely blocks the vaginal introitus.
      (ii) An imperforate hymen, which is rare, requires surgical correction prior to or at the time of first menses.
      (iii) The hymen usually stretches (sometimes tears) during first coitus. It often survives partially intact until the first vaginal childbirth.
   (c) The **fossa navicularis** (vestibular fossa of the vagina) lies between the vaginal introitus and the fourchette.
   (d) The **greater vestibular (Bartholin's) glands** lie on either side of the introitus.
      (i) These mucous glands lie in the superficial perineal pouch.
      (ii) They are homologous to the bulbourethral glands in the male.
      (iii) Infection of these glands results in Bartholin's cyst.
   (e) **Lesser vestibular glands** are numerous, small glands that secrete mucus directly into the vestibule.
   (f) **Paraurethral (Skene's) glands** open beside the external urethral ostium.
      (i) They also are prone to infection.
      (ii) They are homologous to the prostate gland in the male.
2. **Clitoris** (see Figure 28-11)
   a. **Orientation.** The clitoris and penis are homologous in most respects.
   b. **Structure.** Three bodies of vascular erectile tissue begin as the **radix** or **root of the clitoris** in the superficial perineal pouch. The filling of these sinusoids with arterial blood results in clitoral erection (see Figure 28-12).

(1) The bilateral **corpora cavernosa** fuse to form the body of the clitoris.
  (a) Each corpus cavernosum begins in the superficial perineal pouch as a **crus**, which attaches along an ischiopubic ramus and is covered by an **ischiocavernosus muscle.**
  (b) These erectile bodies terminate proximal to the glans clitoridis.
(2) The **vestibular bulbs** are also bilateral, but they do not contribute to the body of the clitoris.
  (a) They lie in the superficial perineal pouch, where they are deep to the labia minor and flank the vaginal introitus. Each is covered by a **bulbocavernosus muscle.**
  (b) Anteriorly, they continue as a slender thread of erectile tissue (pars intermedia), which fuses anterior to the urethra to form the thread-like **commissure of the vestibular bulbs.** The commissure extends along the ventral aspect of the clitoral body and contributes to the **glans clitoridis.**
  (c) When turgid, the vestibular bulbs tend to spread the labia minor and open the vaginal introitus.
(3) The **prepuce** or **foreskin** is an extension of the labia minor over the glans clitoridis. The amount is variable, but it usually amply covers the glans.
(4) The **suspensory ligament** arises from the pubic symphysis and inserts into the deep fascia of the clitoris. Distal to this attachment, the clitoris turns inferiorly, where it is mostly embedded in the fatty tissue of the mons pubis.
 c. **Vasculature** (see Figure 28-15)
   (1) **Arterial supply** to the clitoris is by the **internal pudendal arteries.** The superficial skin of the labia and vulva is supplied by both the internal pudendal arteries, which arise from each internal iliac artery, and the external pudendal arteries, which arise from each femoral artery.
   (2) **Venous return** from the clitoris is by the **deep dorsal vein** to the **vesical venous plexus** at the base of the urinary bladder. Return from the vulva is largely through the **internal pudendal vein.**
   (3) **Lymphatic drainage**
     (a) **Superficial lymphatics** of the clitoris drain toward the superficial inguinal nodes, whereas **deep clitoral lymphatics** drain into the internal iliac nodes.
     (b) **Lymphatics of the vulva and inferior vagina** drain largely into the superficial and deep inguinal nodes.
 d. **Innervation.** The vulva and clitoris are innervated by terminal branches of the **pudendal nerve** (see Figure 28-15). Numerous sensory nerve endings converge on the glans clitoridis.

B. **Fascia.** Underlying muscular and fascial structures support and contribute to the function of the external genitalia as well as reinforce the abdominal cavity.

1. **Superficial layer of the superficial fascia (Cruveilhier's f.)**
   a. This fatty layer is continuous with Camper's layer of the abdominal wall as well as the fat of the ischioanal fossa. It gives form to the mons pubis and labia major (see Figure 28-7A).
   b. The fat disappears as this layer extends into the labia minor.

2. **Deep layer of the superficial fascia (Colles' f.)**
   a. Anteriorly, this membranous layer is continuous with Scarpa's fascia (see Figure 28-7A).
   b. Laterally, it attaches to the ischiopubic rami.
   c. Medially, it inserts into the adventitia of the urethra and vagina.
   d. Inferiorly, it passes deep to the labia major.
   e. Posteriorly, it delimits the superficial pouch by passing superficial to the **superficial transverse perineal muscle.**
     (1) The midline fibers insert into the adventitia of the vagina and the **central tendon of the perineum.**
     (2) The more lateral fibers reach the posterior edge of the **deep transverse perineal muscle** before merging with the fascia of the urogenital diaphragm.

(a) The **inferior** or **external fascia of the urogenital diaphragm** overlies the inferior surface of the deep transverse perineal muscle.
(b) The **superior or internal fascia of the urogenital diaphragm** overlies the superior surface of the deep transverse perineal muscle.
(c) The transverse perineal ligament is formed by the fusion of these two layers anterior to the deep pouch.
   f. **Clinical considerations.** Because Colles' fascia in females does not fuse in the midline and the urethra is not enclosed, extravasation of urine beneath this fascial layer is all but impossible. Hematoma may occur in the superficial space, however, especially in association with parturition or trauma.
 3. **The deep, or investing, fascia** (Gallaudet's f.) covers the muscles of the superficial pouch.

C. **Superficial perineal space (pouch)**
 1. **Boundaries.** This space is bounded by Colles' fascia and lies between this fascial layer and the inferior fascia of the urogenital diaphragm. It is divided by the pudendal cleft (Figure 28-12).
    a. Anteriorly, it is continuous with the potential space in the anterior abdominal wall between the deep layer of the superficial fascia (of Scarpa) and the deep investing fascia.
    b. Posteriorly, the fusion of Colles' fascia to the peroneal body and the deep transverse peroneal muscle defines the limit of this space.
 2. **Contents** include the roots of the external genitalia, the crura of the clitoris, and the vestibular bulbs with the associated muscles, nerves, and vessels.
    a. **Corpora cavernosa** are paired bodies of erectile tissue that lie along the ischiopubic rami [see Figure 28-12 and section V A 2 b (1)].
    b. **Vestibular bulbs** are paired masses of vascular erectile tissue on either side of the vaginal introitus under the labia minor [see Figure 28-12 and section V A 2 b (2)].
    c. **Ischiocavernosus muscles** overlie the clitoral crura (Table 28-2; see Figure 28-12).
       (1) Their origin is the medial side of the ischial tuberosities and ischial rami.
       (2) They insert into the margin of the pubic arch and into each crus of the clitoris.

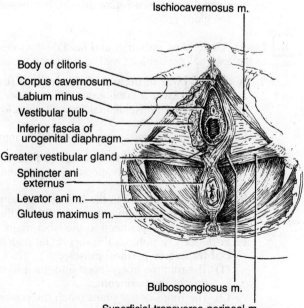

FIGURE 28-12. Superficial pouch in the female. Contents include the clitoris, crura of the clitoris, and vestibular bulbs with the associated muscles, as well as the greater vestibular glands.

d. **Bulbospongiosus (bulbocavernosus) muscles** overlie the vestibular bulbs (see Figure 28-12 and Table 28-2).
   (1) Their origin is the anterior portion of the central tendon.
   (2) The medial fascicles attach to the deep fascia of the dorsum of the clitoris.
   (3) The lateral fascicles attach to the inferior fascia of the urogenital diaphragm.
e. **Superficial transverse perineal muscle** (see Figure 28-12 and Table 28-2)
   (1) It originates along the anterior portions of the ischial tuberosities and inserts into the central tendon of the perineum (perineal body).
   (2) Although it is small and may be absent unilaterally or bilaterally, it stabilizes the perineum.
   (3) When present, it is a palpable surgical landmark that delineates the posterior extent of the superficial pouch.
f. **Greater vestibular (vulvovaginal) glands (of Bartholin)**
   (1) These glands lie at the posterior ends of the vestibular bulb (see Figure 28-12).
   (2) The ducts of these mucous glands open into the vestibule.
g. The **central tendon (perineal body)** is a common point of attachment of the fascia and muscles of the anal and urogenital triangles (see Figure 28-14). As such, it is an important structure of the perineum that must be sutured if torn or incised.

3. **Clinical considerations.** Tearing of the vascular vestibular bulbs during difficult parturition produces considerable bleeding into the superficial pouch, with extravasation of blood into the labia major and along the mons pubis and abdominal wall beneath Scarpa's fascia.

D. **Deep perineal space (pouch)**

1. **Boundaries.** This space is delineated by the superior and inferior fascial layers of the deep transverse perineal muscle (Figure 28-13).
   a. The **inferior fascia of the urogenital diaphragm** (perineal membrane) bounds the deep pouch inferiorly (see Figure 28-7B).
   b. The **superior fascia of the urogenital diaphragm** bounds the deep pouch superiorly.
   c. **Anterior, lateral, and posterior boundaries** (see Figure 28-13)
      (1) Arcuate and transverse ligaments lie anteriorly.

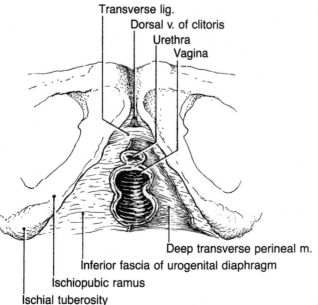

FIGURE 28-13. Deep perineal pouch in the female. Contents include the deep transverse perineal muscle with the external urethral sphincter.

FIGURE 28-14. Female reproductive tract, midsagittal section. The urethra is embedded in the anterior vaginal wall. The perineal body lies in the midline between the vagina and anal canal.

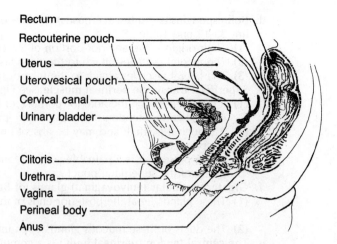

- (a) The merging of the inferior and superior fascial layers of the urogenital diaphragm anterior to the free edge of the deep transverse perineal muscle forms the **transverse perineal ligament.**
- (b) The **arcuate ligament** lies behind the pubic symphysis and is the most anterior component of this layer. The **deep dorsal vein** of the clitoris runs through the hiatus between the transverse and arcuate ligaments to reach the vesicular venous plexus.
- (2) Attachments of the superior and inferior fascial layers of the urogenital diaphragm to the ischiopubic rami limit the pouch laterally.
- (3) Superior and inferior fascial layers of the urogenital diaphragm fuse at the posterior free edge of the deep transverse perineal muscle. Along this line of fusion, the deep layer of perineal fascia (of Colles') attaches to define the posterior limit of the superficial perineal pouch.

2. **Contents** (see Figure 28-13)
   a. The **deep transverse perineal muscle** forms the bulk of the urogenital diaphragm (see section IV D 2 a).
   b. The **sphincter urethrae** (external urethral sphincter) is a modified portion of the deep transverse perineal muscle.
      (1) It is associated with the portion of the female urethra that passes through the deep perineal space.
         (a) The sphincter urethrae is incomplete in the female. Where the urethra passes through the urogenital diaphragm, muscle fascicles of the deep transverse perineal muscle arch anterior to the urethra. They do not, however, pass posterior to the urethra (as in the male) because the urethra is embedded in the adventitia of the anterior wall of the vagina at this level (Figure 28-14).
         (b) The particular arrangement predisposes to **urinary stress incontinence,** which may result from difficult parturition.
      (2) The external urethral sphincter is voluntary, innervated by twigs from the perineal branch of the pudendal nerve.
      (3) This muscle is a prime factor in urinary continence once the desire to void is perceived.
   c. The **dorsal artery** and **dorsal nerve** of the clitoris, the terminal branches of the internal pudendal artery and pudendal nerve, course through the deep perineal pouch.

E. **Vasculature**

1. **Arterial supply** (Figure 28-15)
   a. The **internal pudendal artery** supplies most of the external genitalia.

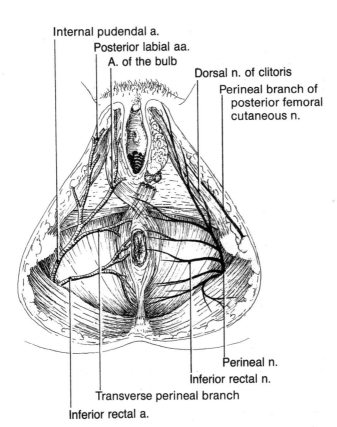

FIGURE 28-15. Vasculature and innervation of the female perineum.

(1) **Pelvic course.** This artery leaves the pelvic cavity through the infrapiriform portion of the **greater sciatic foramen.**
(2) **Perineal course.** It then passes through the **lesser sciatic foramen** to enter the ischiorectal fossa of the perineum, where it courses in the **pudendal (Alcock's) canal** on the medial side of the obturator internus muscle.
(3) **Branches**
  (a) **Inferior rectal (hemorrhoidal) arteries** supply the anal region.
  (b) The **superficial perineal artery** runs parallel to the superficial transverse perineal muscle.
   (i) The **transverse perineal branch** supplies the superficial perineal pouch and its musculature.
   (ii) The **posterior labial branches** supply the posterior vulval region.
  (c) **Deep perineal arteries**
   (i) The **bulbar branch** supplies the vestibule and erectile tissues of the vestibular bulbs.
   (ii) A **urethral branch** runs to the urethra.
   (iii) **Deep (central) arteries of the clitoris,** located in each corpus cavernosum, supply the erectile tissues of those bodies.
   (iv) The **dorsal artery of the clitoris,** which lies between the deep fascia of the clitoris and the tunica albuginea, supplies the glans clitoridis and superficial aspects of the clitoris.
 b. **External pudendal branches** of each femoral artery supply the mons pubis, anterior portions of the labia major, and subcutaneous tissue of the clitoris.
2. **Venous return**
 a. The **pudendal veins** generally parallel the arteries.
 b. The **deep dorsal vein** of the clitoris passes between the arcuate and transverse ligaments of the urogenital diaphragm to drain into the vaginal and vesicular venous plexuses (see Figure 28-13).

3. **Lymphatic drainage pathways** of the perineum are important clinically because of the high prevalence of cervical carcinoma.
   a. **Superficial lymphatics** of the clitoris drain toward the superficial inguinal nodes, whereas deep clitoral vessels drain into the internal iliac nodes.
   b. **Lymphatics of the vulva and inferior vagina** drain principally into the superficial and deep inguinal nodes.
   c. The **middle vagina** generally drains toward the internal iliac nodes.
   d. The **upper vagina** generally drains toward the internal iliac and common iliac nodes.

F. **Innervation** (see Figure 28-15)
   1. **The pudendal nerve** supplies most of the urogenital triangle by way of perineal branches.
      a. **Course.** The pudendal nerve arises from the sacral plexus (spinal segments S2–S4).
         (1) **Inferior rectal branches** convey sensation to most of the anal triangle and supply the external anal sphincter.
         (2) The **perineal branch** is a mixed nerve, containing the general somatic afferent (sensory), general somatic efferent (motor), and sympathetic (vasomotor) nerves as components. This branch gives off several branches that are named according to the arteries they accompany.
            (a) The **posterior labial branches** supply the skin of the labia major and minor.
            (b) The **deep (transverse) perineal branches** supply the musculature of the superficial and deep pouches, including the external urethral sphincter.
            (c) The **dorsal nerve of the clitoris** mediates sensation from that structure.
      b. **Saddle block**
         (1) Pain fibers from the perineum may be blocked by injecting an anesthetic in the vicinity of the pudendal nerve.
         (2) The needle is inserted between the perineal body and the ischial tuberosity, and then angled toward the ischial spine. Within a centimeter of the tuberosity, the needle is near the pudendal canal.
         (3) Temporary urinary incontinence results. Because the inferior rectal nerves arise near the sacrospinous ligament, rectal incontinence usually does not occur.
   2. **Ilioinguinal** and **genitofemoral nerves**, which arise from L1 and L2, supply the anterior part of each labium major as well as the corresponding half of the mons pubis.
   3. **Perineal branches** of the **posterior femoral cutaneous nerve** arise from spinal segments S2–S4 and supply the posterolateral region of the urogenital triangle.

G. **Clinical considerations**
   1. **Bartholin's cysts** result from infection of the greater vestibular glands.
   2. **Episiotomy** (section of the perineum) may be performed before parturition to prevent uncontrolled tearing. Suturing an incision is preferable to repairing a ragged tear. Moreover, episiotomy appears to reduce the incidence of stress incontinence.
      a. **Midline episiotomy**
         (1) Incision into the posterior vaginal wall is carried posteriorly in the midline through the fossa navicularis to divide the central tendon (perineal body).
         (2) The incision does not extend into the deep fibers of the external anal sphincter.
         (3) This type of episiotomy is relatively bloodless and painless, because no major vessels or nerves are transected. It does, however, provide limited expansion of the birth canal, with some possibility of tearing the anal sphincters.
      b. **Mediolateral episiotomy**
         (1) Incision into the posterior vaginal wall is carried posterolaterally into the ischioanal fossa.
         (2) This incision cuts skin over the ischioanal fossa, the bulbospongiosus muscle,

superficial transverse perineal muscle, fascia and muscle of the urogenital diaphragm, and transverse perineal branches of the internal pudendal artery and pudendal nerve.
(3) Mediolateral episiotomy allows greater expansion of the birth canal into the ischioanal fossa, but the risk of infection is increased because of contamination of the ischioanal fossa. Also, this incision is more difficult to close layer by layer.

Chapter 29
Pelvic Viscera

I. RECTUM

A. External structure.
The rectum is the pelvic portion of the gastrointestinal tract. Located within the deep pelvis, it is about 13 cm (5 in.) long and connects the sigmoid colon to the anal canal (Figure 29-1).

1. The upper limit of this **retroperitoneal structure** is where the sigmoid mesocolon disappears. The lower limit is the pelvic diaphragm.
   a. The **ampulla,** just above the pelvic floor, is the widest part of the rectum and is capable of considerable distension (see Figure 28-3).
      (1) Because feces are stored in the sigmoid colon, the ampulla is usually empty.
      (2) Dilation of the ampulla by movement of feces from the sigmoid colon to the rectum generates the sensation of rectal fullness and the urge to defecate.
   b. As the rectum pierces the levator ani portion of the pelvic diaphragm, it becomes the **anal canal** (see Figure 28-3).

2. **The rectal sling** is formed by the **puborectalis muscle** (see Chapter 28 III B 1 b and Figures 27-2 and 29-1).

B. Internal structure.
Three transverse **rectal folds** (plicae transversae, valves of Houston) are formed by the inner three layers of the intestinal wall (see Figure 28-3).

1. **The inferior rectal fold** projects from the left side about 2 cm (1 in.) above the anal canal and is palpable digitally.

2. **The middle rectal fold** projects from the right about 2 cm (1 in.) above the inferior fold and usually is not palpable.

3. **The superior rectal fold** projects from the left about 2 cm (1 in.) above the middle fold.

C. Vasculature

1. **Arterial supply** (Figure 29-2)
   a. The rectum receives branches from the inferior mesenteric artery by the **superior rectal (hemorrhoidal) artery,** from the internal iliac artery by the **middle rectal (hemorrhoidal) arteries,** and from the internal pudendal arteries via the **inferior rectal (hemorrhoidal) arteries.**
   b. Anastomotic connections are abundant in the submucosa between tributaries of these arteries.

2. **Venous return** (see Figure 29-2)
   a. The **superior rectal (hemorrhoidal) vein** drains into the inferior mesenteric vein. The **middle rectal (hemorrhoidal) vein** drains into the internal iliac vein. The **inferior rectal (hemorrhoidal) arteries** also drain into the internal iliac vein by way of the internal pudendal veins.
   b. Anastomotic connections among these three veins are especially rich. Enlargement of these submucosal anastomoses as a result of portal blood flowing to the systemic venous circulation produces **hemorrhoids** (see Chapter 28 III C 2 c).

3. **Lymphatic drainage** of the terminal portion of the large bowel is important given the high frequency of colorectal carcinoma. Lymphatic drainage follows the blood vessels.
   a. The **upper rectum** drains along the **superior rectal vessels,** which drain, in turn, into the **inferior mesenteric nodes,** the lumbar para-aortic nodes, and the thoracic duct.

446 | Chapter 29 I C

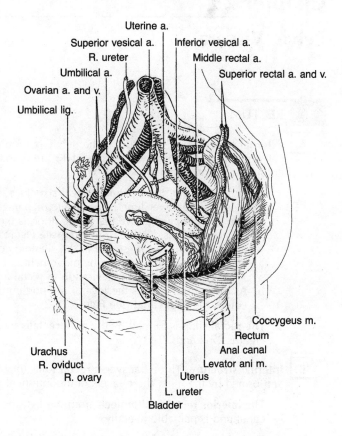

FIGURE 29-1. Female pelvic viscera. Parasagittal section indicates the relationships among the urinary bladder, uterus, and rectum, as well as the major pelvic vasculature.

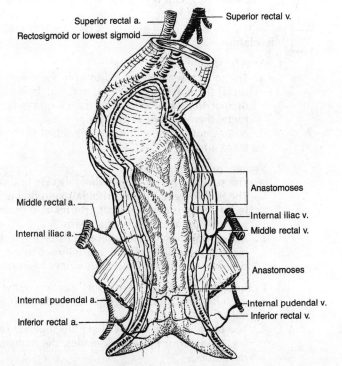

FIGURE 29-2. Vasculature of the rectum and anal canal, posterior view. The arterial supply from the inferior mesenteric artery (superior rectal artery) and the internal iliac arteries (middle and inferior rectal arteries) *(left side)* as well as the parallel venous drainage *(right side)* are shown. The venous anastomotic connections are especially rich.

b. The **midrectum** drains along the **middle rectal vessels** to the **internal iliac (pelvic) nodes,** which drain, in turn, into the para-aortic nodes, and thoracic duct.
c. The **lower rectum** drains along the **inferior rectal** and **pudendal vessels** to the **deep pelvic nodes,** as well as to the **inguinal nodes,** which drain to the **external iliac nodes** before draining to the para-aortic nodes.

D. Innervation (see Figure 25-16B)

1. **Parasympathetic innervation** is from the **pelvic splanchnic nerves** or **nervi erigentes** (S2–S4) to the inferior mesenteric plexus.

2. **Sympathetic innervation** is from the **lumbar splanchnic nerves** (L1–L2). It flows through the aortic plexus, the superior then inferior hypogastric plexuses, and then the pelvic plexus.

3. **Afferent innervation** reaches the S2–S4 levels of the spinal cord by the pelvic splanchnic nerves. Pain is referred to the posterior thigh and posterior leg.

E. Clinical considerations

1. **Hemorrhoids.** Portal system blood may shunt through the venous plexuses of the rectum and anal canal to the systemic venous system (see Figure 29-2) as a result of **portal hypertension** or local compression of the inferior mesenteric vein (because of chronic constipation with dilation of the sigmoid colon or the presence of a growing fetus). **Hemorrhoids** are variceal dilatations of the submucosal anal and perianal venous plexuses (see Chapter 28 III C 2 c).
   a. Hemorrhoidal varices may rupture during the evacuation of feces, with considerable blood loss. Chronically bleeding hemorrhoids may lead to anemia, particularly in the elderly.
   b. Hemorrhoids are amenable to treatment by ligature or cautery.

2. **Digital examination.** The high incidence of rectal, colonic, and prostatic carcinoma makes digital and sigmoidoscopic examination important in middle-aged and older men. Most rectal neoplasms occur distally and can be detected by digital palpation.

3. **Carcinoma of the rectum** may spread hematogenously and along lymphatic pathways as well as by direct invasion of adjacent tissues. Posterior spread may involve the sacral plexus with pain distribution along the posterior leg. Anterior spread may involve adjacent pelvic viscera, such as prostate and bladder or uterus and vagina.

4. **Prolapse of the rectum** through the anus, often related to damage of the levator ani muscle, is usually the result of obstetric trauma.

II. PELVIC PORTION OF THE URINARY SYSTEM

A. Ureters are discussed in detail in Chapter 26 II.

B. Urinary bladder

1. **External structure.** This extraperitoneal organ lies behind the pubic symphysis and the superior pubic rami (see Figures 29-1 and 29-3).
   a. The **fundus** is the expansive portion of the bladder.
      (1) With free expansion, this region rises above the pubic crest.
      (2) It normally accommodates 250–300 ml, but it may hold 400–500 ml.
   b. The **posterior surface** is covered by peritoneum.
      (1) In the male, the posterior surface is related to the **rectovesical pouch** (the lowest portion of the peritoneal cavity), **rectovesical septum** (Denonvilliers'

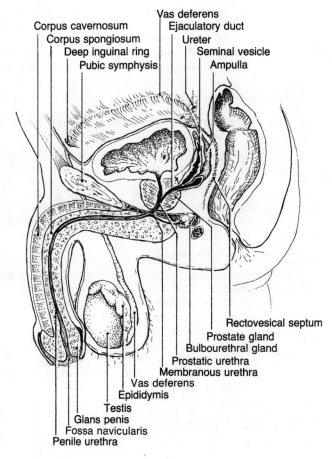

FIGURE 29-3. Male pelvic viscera. Midsagittal section depicts the lower portion of the male genitourinary tract.

fascia), and **seminal vesicles** (Figure 29-3). The rectovesical septum is formed in part by embryonic fusion of the inferior portion of the rectovesical pouch.
- (2) In the female, the posterior surface is related to the **vesicouterine pouch** and the **vesicovaginal septum** (see Figure 29-9).
- c. The **anterior surface** defines the **retropubic space,** a potential space filled with endopelvic fascia.
  - (1) The retropubic surgical approach to the prostate is through this space.
  - (2) Urethral trauma superior to the urogenital diaphragm or rupture of the urinary bladder results in extravasation of urine into the retropubic space.
- d. The **base** rests on the pelvic floor.
  - (1) The base has a relatively unexpandable portion that corresponds to the **trigone,** internally, where the ureters enter and the urethra leaves.
  - (2) The base is next to the prostate gland in the male.
- e. **Support**
  - (1) It receives support from the **urogenital diaphragm.**
  - (2) A condensation of **endopelvic fascia** at the base provides support.
  - (3) Paired **medial umbilical ligaments,** remnants of the **umbilical arteries,** support the bladder anterolaterally.
  - (4) The **median umbilical ligament** or **urachus** supports the bladder superiorly (see Figure 29-7).
    - (a) The **urachus** is a remnant of the allantoic duct (between the cloaca and the placental allantois).
    - (b) The urinary bladder forms in the base of the allantoic duct, which then obliterates between the bladder and umbilicus.

2. **Internal structure**
   a. The **walls** of the urinary bladder consist of a **substantial layer of smooth muscle,** the **detrusor muscle.**
   b. The **mucosal lining** consists of **transitional epithelium** that is thick and thrown into folds when the bladder is empty. As the bladder fills, the mucosa becomes smooth and attenuated.
   c. The **trigone** is a triangular area at the base of the bladder (see Figures 29-8 and 29-11).
      (1) It is defined posterosuperiorly by the ostia of the ureters (about 5 cm apart) and inferiorly by the urethra.
      (2) It is relatively indispensable and has few mucosal folds when the bladder is empty.
      (3) The ostia of the ureters are guarded by **ureterovesical valves.**
         (a) The angular penetration of the ureters through the walls of the detrusor muscle within the trigone results in a valve-like action.
         (b) Urine in the filling bladder exerts pressure against the bladder walls, compressing the anterior aspect of the intramural portion of the ureters to prevent reflux of urine from the bladder into the ureters.
      (4) The **uvula** of the bladder guards the urethra. It consists of longitudinal mucosal folding at the trigone apex, superoposterior to the urethra.
3. **Vasculature**
   a. **Arterial supply**
      (1) The **superior vesical arteries** arise from the patent portion of the **umbilical arteries** and supply the cranial portion of the bladder (see Figure 29-1).
      (2) The **middle vesical arteries are inconsistent.** When present, they may arise from a variety of sources to supply the fundus.
      (3) The **inferior vesical arteries** arise from the internal iliac arteries or, less frequently, the deferential artery. They supply the base and trigonal region (see Figure 29-11).
   b. **Venous return.** Veins tend to drain into the **prostatic venous plexus** at the base of the bladder. This plexus drains primarily into the internal iliac vein.
4. **Innervation** is by sympathetic and parasympathetic pathways, with afferent neurons using both pathways (Figure 29-4).
   a. **Motor innervation** to the bladder is by sacral parasympathetics that run along the **pelvic splanchnic nerves** (nervi erigentes).
   b. **Afferent pathways**
      (1) **Conscious perception** of bladder fullness occurs as the bladder wall stretches.
         (a) Sensory awareness of bladder fullness travels largely by way of afferent neurons that lie along sympathetic pathways **(hypogastric nerve)** to spinal segments T12–L2 (see Table 29-1).
         (b) As the bladder fills to the extreme, urine backs up and dilates the ureters. This transient hydronephrosis causes pain referred to upper lumbar dermatomes and then along the lowest thoracic dermatome (see Table 29-1).
         (c) Bladder content is first noted when it reaches 100–150 ml in the average male, somewhat less in the average female.
      (2) The **detrusor (micturition) reflex** initiates involuntary contraction of the bladder musculature as the wall stretches.
         (a) The **afferent limb** of this reflex travels along the parasympathetic pathways **(pelvic splanchnic nerves)** to spinal segments S2–S4.
         (b) The **efferent limb** of this reflex is mediated by sacral parasympathetics that run along the pelvic splanchnic nerves.
         (c) The detrusor reflex becomes active as the bladder fills beyond 100–150 ml, accentuating awareness of a filling bladder. Without this reflex, it is difficult to express urine from a nearly empty bladder.
5. **Functional considerations**
   a. **Urinary continence. Urination** is movement of urine down the ureters by peristaltic activity. On filling, the urinary bladder rises in the pelvic cavity.

Table 29-1. Pelvic Visceral Afferent Pathways

Organ	Pathway	Spinal Levels	Referral Regions
Urinary bladder	Hypogastric plexus and aortic plexus to lower lumbar splanchnic nn.	L1–L2	Pubic and inguinal regions, anterior scrotum or labia, anterior thigh
Abdominal ureters			
Gonads			
Uterine body			
Uterine tubes			
Rectum	Pelvic plexus and pelvic splanchnic nerves	S2–S4	Perineum and posterior thigh
Superior anal canal			
Pelvic ureters			
Cervix			
Epididymis			
Vas deferens			
Seminal vesicles			
Prostate gland			

- (1) **Initial urinary continence.** The internal urethral sphincter plays no role in urinary continence. Initial urinary retention in the empty to partially empty bladder (100–150 ml) is attributable to the **uvula** of the bladder.
  - (a) When the bladder is empty, uveal folds are pronounced at the apex of the trigone and occlude the urethral outlet.
  - (b) The sole function of the internal urethral sphincter in males is contraction during ejaculation to prevent retrograde ejaculation into the bladder.
- (2) **Voluntary urinary continence.** As the bladder rises with filling, the uveal folds become stretched and attenuated so the urethral outlet is unblocked. Urinary continence in the partially full to full urinary bladder is maintained by voluntary contraction of the **external urethral sphincter.**
  - (a) This sphincter is formed about the **membranous urethra** by circularly arranged fibers of the **deep transverse perineal muscle.** It is under voluntary control by the deep perineal branches of the **pudendal nerve** (see Figures 28-9 and 28-13).
  - (b) The most medial fibers of the **puborectalis muscle** of the pelvic floor contribute to bladder support and may play a role in urinary continence by raising the bladder and attenuating the urethral lumen (see Figure 29-1).
  - (c) **Urinary bladder capacity**
    - (i) Awareness begins with about 100–150 ml.
    - (ii) Distention becomes uncomfortable at a volume of 300–400 ml, beyond which painful sensations are experienced.
    - (iii) If unrelieved, involuntary micturition occurs by 400–500 ml.
    - (iv) In the female, capacity is lower than that in the male because the bladder is smaller and the external urinary sphincter does not completely surround the urethra.
- b. **Micturition** is the process of **emptying the urinary bladder** through the urethra, initiated by the detrusor muscle.
  - (1) **Detrusor reflex.** Stretch in the detrusor muscle elicits a reflex contraction of that muscle, enhancing then compelling the urge to void (see Figure 29-4).
    - (a) **Afferent limb.** The pelvic splanchnic nerve conveys impulses from the stretch receptors of the urinary bladder to segments S2–S4 of the spinal cord.

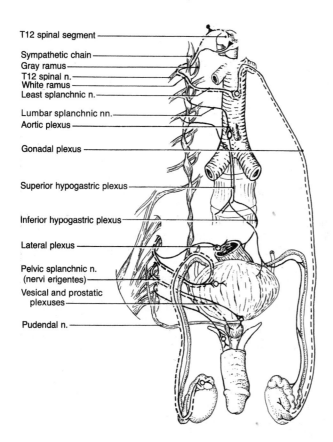

FIGURE 29-4. Innervation of the male bladder, reproductive tract, and genitalia. Sympathetic pathways arise from upper lumbar spinal levels (there are no white rami below L2) and reach the pelvic viscera by way of lumbar splanchnic nerves and hypogastric plexus. Parasympathetic pathways arise from the midsacral levels and reach the pelvic viscera through the pelvic splanchnic nerves. Visceral afferent fibers *(dashed)* from the pelvic viscera travel specifically along either one of the two autonomic pathways, producing specific patterns of referred pain. The pudendal nerve provides somatic innervation to and from the perineum.

- **(b) Efferent limb.** Pelvic nerves convey impulses from segments S2–S4 of the spinal cord to the detrusor muscle.
- **(2) Voluntary relaxation of the puborectalis muscle,** in anticipation of micturition, permits the bladder to descend slightly. This process relieves attenuation of the uvula and widens the initial segment of the urethra.
- **(3) Voluntary relaxation of the external urethral sphincter** allows the bladder to void. This process may be assisted by a Valsalva maneuver, which raises intra-abdominal pressure.
- **(4) Restoration**
    - **(a)** At the termination of micturition, strong contractions of the external urethral sphincter, pelvic floor, and the bulbocavernosus muscles in the male expel (incompletely) the residual urine from the urethra, and continence is restored.
    - **(b)** The detrusor muscle relaxes with the absence of stretch.

6. Clinical considerations
- **a. Transurethral cystoscopy** permits inspection of the urinary bladder and (in the male) the prostatic urethra. It may be combined with procedures to remove ureteral calculi and bladder stones. It is also an integral part of the transurethral approach to prostatectomy.
- **b. Patent urachus**
    - **(1)** A completely patent urachus (very rare) allows reflux of urine through the umbilicus.
    - **(2)** Approximately 33% of all individuals have patency of the lumen of the urachus to some extent. Severance of a patent urachus by a transverse suprapubic incision can result in extravasation of urine into the peritoneal cavity.
    - **(3)** An isolated patency may produce a **urachal cyst.**

c. **Pelvic fractures** may damage the urinary bladder, urethra, rectum, uterus, vagina, nerves, and, especially blood vessels, with a potential for severe internal hemorrhage.
   (1) The urinary bladder is most vulnerable in pelvic fracture, particularly when distended. Intrapelvic and subperitoneal extravasation of urine may result.
   (2) Following severe pelvic fractures, immediate surgical repairs to soft tissues may be essential.

C. **Urethra**
1. **The male urethra** is divided into three segments (see Figure 29-3).
   a. The **prostatic urethra** begins at the vesical neck at the apex of the trigone, extends through the prostate gland, and terminates at the superior fascia of the urogenital diaphragm (see Figure 29-8). Compared to the other two segments, this part of the urethra has the thickest walls.
      (1) The **urethral crest (crista urethralis)** lies on its posterior wall.
         (a) The numerous **prostatic ducts** open individually into the prostatic urethra on either side of the urethral crest.
         (b) The **colliculus seminalis (verumontanum)** is the widest portion of the urethral crest.
            (i) The **prostatic utricle (uterus masculinus)** is a 5-mm deep invagination that lies in the midline of the verumontanum. It may represent the terminal remnant of the fused terminal portion of the paramesonephric (mullerian) ducts, which form the uterus and oviducts in the female.
            (ii) This structure is prone to cyst formation.
         (c) **Ejaculatory ducts** enter the prostatic urethra through the urethral crest immediately lateral to the prostatic utricle.
      (2) Injury to the prostatic urethra above the urogenital diaphragm may result in intrapelvic, extraperitoneal extravasation of urine.
   b. The **membranous urethra** begins at the superior fascia of the urogenital diaphragm, extends through the deep transverse perineal muscle, and terminates at the inferior fascia of the urogenital diaphragm. This segment is short (10–12 mm long) and has thin walls. It is contained within the urogenital diaphragm (see Figures 29-3 and 29-8).
      (1) Except for the external urethral meatus, this segment is the narrowest and least distensible portion of the urethra, probably because of the tone of the surrounding **external urethral sphincter.**
      (2) Because of its narrowness and delicate walls, as well as the sharp angle between the penile and membranous portions, this segment is easily ruptured by trauma or passage of instrumentation.
         (a) If damage occurs above the urogenital diaphragm, extravasation of urine is intrapelvic and extraperitoneal.
         (b) With damage below this diaphragm, extravasation is within the superficial perineal space (see Chapter 28 IV C 3 a).
   c. The **penile (cavernous) urethra** begins at the inferior fascia of the urogenital diaphragm. It extends into the bulb of the penis, where it is surrounded by **corpus spongiosum (corpus cavernosum urethrae),** and terminates at the **external urethral meatus** of the glans penis [see Figure 29-3 and Chapter 28 IV A 1 b (2)].
      (1) It varies in length among individuals and with the state of erection.
      (2) **Damage** may result in extravasation of urine (see Chapter 28 IV C 3 a).

2. **The female urethra** is approximately 3.7 cm in length. It corresponds to the prostatic and membranous portions of the male urethra.
   a. **Relations.** Because the urethra fuses with the adventitia of the anterior vaginal wall, the fibers of the deep transverse perineal muscle do not pass posterior to it (see Figure 28-13).
      (1) The external urethral sphincter in the female is incomplete, which explains the high incidence of stress incontinence among women. It is especially

prevalent if the urogenital diaphragm is damaged during parturition or if **urethrocele** (herniation of the urethra into the vagina) develops.
(2) The intimate relationship of the urethra with the vagina predisposes the urethra to injuries and subsequent **cystocele** or **urethrocele** associated with difficult parturition.
   b. The **urethral glands** open into the urethra. **Paraurethral glands** (of Skene) open into the vestibule adjacent to the urethral orifice.
   c. The **urethral orifice** lies in the **vestibule** between the labia minor.
      (1) Its location is variable, which can make cannulation difficult.
      (2) Unlike in the male, once the orifice is located, insertion of the cannula is not problematic.

III. DEVELOPMENT OF UROGENITAL DUCTS

A. **Early stages.** The development of the kidneys proceeds through three sequential pairs of excretory organs that develop from the urogenital ridge. The reproductive system incorporates some of the early excretory ducts.

1. **The pronephros** develops by the fourth week but is never completely functional.

2. **The mesonephros** develops late in the fourth week and functions until development of the metanephric kidney.
   a. By the fourth week, rows of parallel nephrons drain into the **mesonephric (wolffian) duct,** which empties into the cloaca.
   b. By the sixth week, the mesonephros consists of elongated organs projecting into the coelomic cavity on either side of the midline of the posterior abdominal wall.
   c. By the eighth week, most of the mesonephros has degenerated, but the mesonephric duct remains.

3. **The metanephros,** the definitive kidney, develops during the fifth week and is fully functional by the eighth week. It develops from two primordia.
   a. A **metanephric diverticulum** arises from the posterior wall of the caudal region of the mesonephric duct.
   b. The metanephric mass arises from the posterior region of the urogenital ridge. The growth of the **metanephric duct** into the metanephric mass forms several kidney lobes.

B. **The indifferent stage (weeks 6–8)** is characterized by no apparent differences between the developing male and female (Figure 29-5).

1. **Genital ducts.** By the sixth week, two pairs of genital ducts have formed.
   a. **Mesonephric (wolffian)** ducts are urinary tracts exiting from the primitive mesonephros and draining into the unpartitioned cloaca. In males, these ducts form the genital tract. In females, they largely degenerate.
   b. **Paramesonephric (mullerian)** ducts arise parallel to the mesonephric ducts. In females, these ducts form the uterine tubes and uterus. In males, they largely degenerate.

2. **The urinary bladder** is formed by partition of the cloaca by the urorectal septum into an anterior urogenital sinus and a posterior anorectal canal.
   a. The **most cranial portion** of the urogenital sinus forms the urinary bladder.
      (1) This part of the urogenital sinus is continuous with the **allantois,** a diverticulum of the hindgut into the placenta.
      (2) When the lumen of the allantoic duct solidifies, it becomes the cord-like **urachus,** connecting the superior pole of the urinary bladder to the umbilicus and forming the **median umbilical ligament.**
   b. The **middle portion** of the urogenital sinus forms the **urethra** in the female,

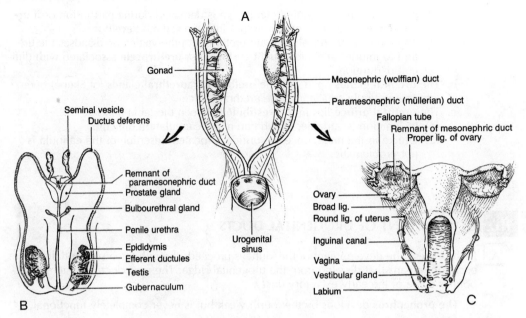

FIGURE 29-5. Development of the gonads and genital ducts. *(A)* In the indifferent stage, both mesonephric and paramesonephric ducts lead to the urogenital sinus. The gonads are high in the abdomen. *(B)* In the developing male, the mesonephric ducts develop into the vas deferens and epididymis. The testes descend through the abdominal wall into the scrotum. *(C)* In the female, the paramesonephric ducts partially fuse to form the uterus, and the unfused portions form the uterine tubes. The ovaries descend into the deep pelvis.

which is equivalent to the membranous and prostatic portions of the urethra in the male.

   c. The **inferior portion** of the urogenital sinus remains as the definitive urogenital sinus, which persists as the **vestibule** in the female and forms the **penile urethra** in the male.

3. **Metanephric ducts**
   a. The **mesonephric ducts** enter the urogenital sinus once the urorectal septum partitions the cloaca.
   b. As the walls of the terminal region of the mesonephric ducts become incorporated into the developing urinary bladder, **metanephric diverticula** separate from the mesonephric ducts so that the metanephric ducts enter the urinary bladder, whereas the mesonephric ducts enter the developing urethra.

C. **Development of the male reproductive tract** (see Figure 29-5B)
   1. **The testes** develop from the urogenital ridge.
      a. **Seminiferous cords** develop into seminiferous tubules, straight tubules, and the rete testis.
      b. Between the seminiferous cords, interstitial mesenchyme differentiates into testosterone-producing **interstitial cells** (of Leydig).
   2. **The mesonephric ducts** develop under the influence of the fetal testicular hormones to form the **vas deferens.**
      a. A diverticulum toward the caudal end of each mesonephric duct forms each **seminal vesicle.**
      b. Distal to the seminal vesicles, each mesonephric duct becomes the **ejaculatory duct,** which joins the prostatic urethra.
      c. As the mesonephros degenerates, some mesonephric tubules are incorporated into the developing genital duct system.

(1) Some of these tubules become the **efferent ductules (tubules),** which open into the portion of the mesonephric duct that develops into the epididymis.
(2) Some of the mesonephric tubules that fail to connect with the epididymis yet fail to degenerate become the vestigial **paradidymis** and the **appendix epididymis.**
(3) Most of the more cranial portions of the mesonephric duct degenerate, except for a small cranial portion that remains at the **appendix testis.**
d. The paramesonephric ducts degenerate, except for the **prostatic utricle,** which may be a site of cyst formation (mullerian cyst).
3. **The urethra** is the terminal portion of the male genital tract.
a. By the twelfth week, the **urethral folds** fuse across the urogenital sinus to form the **penile urethra.**
b. During the sixteenth week, the **glans penis** differentiates and the **external urethral meatus** connects with the penile urethra at the **fossa navicularis.**
c. The **prostate** develops as five groups of diverticula from the cranial portion of the urethra.
d. The **bulbourethral glands** develop as diverticula of the membranous urethra.

D. **Development of the female reproductive tract** (see Figure 29-5C)
1. **The ovaries** develop from the mesonephric (urogenital) ridge.
a. Cortical cords break up into the **primary follicles.** Two million primary follicles form during the early embryonic period. This number is reduced by attrition to about 500,000 by puberty.
b. The developing ovaries descend toward the major pelvis.
2. Paired **paramesonephric ducts** develop into the female genital tract in the absence of testicular hormone.
a. The cranial terminus of each duct opens directly into the peritoneal cavity as the **ostium** of the uterine tube.
b. These ducts fuse caudally to form the uterovaginal primordium.
(1) As paramesonephric fusion progresses cranially, the ovaries are drawn into the minor pelvis.
(a) This fusion brings the developing uterine tubes toward the midline, lifting them away from the pelvic walls.
(b) This elevation forms the **broad ligament,** which divides the inferior aspect of the pelvic cavity into an anterior **uterovesical pouch** and a posterior **uterorectal pouch.**
(2) The fused paramesonephric ducts form the corpus of the uterus and the cervix. Thus, the original paired sinovaginal bulbs and paired uteri usually become a single midline tube.
c. Two groups of vestigial mesonephric tubules remain near the ovary, the **epoöphoron** and the **paroöphoron,** which can be sites of cyst formation.
d. The **mesonephric ducts** degenerate except for the most cranial and most caudal portions, which persist as the appendix vesiculosa and Gartner's ducts, respectively. These structures also can be sites of cyst formation **(Gartner's cyst).**
3. **The vagina** develops along the posterior wall of the urogenital sinus from two solid **sinovaginal bulbs** that fuse in the midline.
a. The termination of the fused paramesonephric ducts (uterovaginal primordium) joins the sinovaginal bulbs.
b. The vaginal lumen is formed by cavitation of the sinovaginal bulbs, except for the most inferior portion (the urogenital membrane), which persists as the **hymen.** The hymen separates the vaginal cavity from the vestibule until the perinatal period.

IV. MALE REPRODUCTIVE TRACT

A. Testes
1. External structure
a. Each testis is approximately 4.5 cm × 3 cm × 2.7 cm.

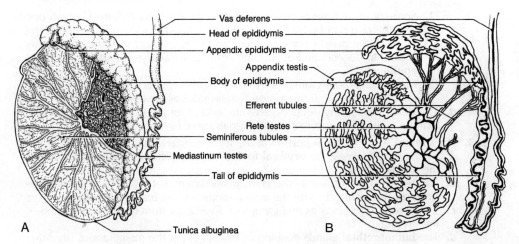

FIGURE 29-6. Testis. *(A)* Parasagittal section depicts the testicular parenchyma and the undissected epididymis. *(B)* Schematic depiction of the duct system.

   b. They are suspended in the cavity of the tunica vaginalis by the **mesorchium** (see Figure 28-6B).
   c. Within the scrotum, the left testis usually lies lower (see Figure 28-4).
   d. The **tunica albuginea,** a tough connective tissue layer, surrounds each testis.
   e. The **visceral layer** of the **tunica vaginalis** (visceral peritoneum) covers the testis on the medial, anterior, lateral, and inferior aspects (see Figure 28-6B).
      (1) This layer reflects off the testes at the mediastinum to form the mesentery of the testis (mesorchium).
      (2) The mesorchium reflects onto the parietal wall as the parietal layer of the tunica vaginalis.
      (3) Between these layers lies the cavity of the tunica vaginalis.

2. **Internal structure**
   a. Each testis is divided into approximately 300 lobules, each of which contains one to four **seminiferous tubules** (Figure 29-6A).
      (1) Each tubule is approximately 75 cm long. With about 750 m of tubules per testis, each male has over 1.5 km of tubules.
      (2) Sixty-one day cycles of spermatogenesis (development of spermatids) and spermiogenesis (maturation of spermatids into spermatozoa) occur in waves along the seminiferous tubules.
   b. The seminiferous tubules lead to about 25 **straight tubules,** which coalesce in the mediastinum of the testis as the **rete testis.** From the rete, 15–18 **efferent ductules** pierce the tunica albuginea to enter the head of the **epididymis** (see Figure 29-6B).

3. **Vasculature**
   a. Arterial supply
      (1) Each **testicular artery** arises from the aorta below the respective renal arteries and courses anterior to the psoas muscle (see Figure 26-3). It passes through the deep inguinal ring to enter the spermatic cord and enters the testis through the mesorchium (see Figure 28-4).
      (2) The testes have a rich collateral blood supply.
         (a) The **artery of the ductus deferens** arises from the internal iliac artery and is homologous to the uterine artery.
         (b) The **cremaster artery** arises as a branch of the inferior epigastric artery.
         (c) The **external pudendal** arteries arise as branches of the femoral arteries.
   b. **Venous return** is mainly by the **testicular (internal spermatic) veins.**
      (1) The **pampiniform plexus** in the spermatic cord coalesces to form each testicular vein (see Figure 28-4).

(a) This plexus functions as a **countercurrent heat exchanger** that maintains testicular temperature a few degrees below body temperature.
(b) It is frequently (80%) affected by **varicocele.**
  (i) Fully 90% of all varices are on the left side because of local venous hypertension resulting from compression of the left testicular vein by the feces-filled sigmoid colon.
  (ii) Varicocele thrombosis is often painful, affects temperature regulation, and may result in decreased sperm counts.
  (iii) Surgical ligation of the testicular vein superior to the deep inguinal ring alleviates the back-pressure resulting from the hydrostatic columns of venous blood within the abdomen and encourages collateral venous return.
(2) The **right testicular vein** drains into the inferior vena cava. The **left testicular vein** drains into the left renal vein (see Figure 26-3). Contrary to wide acceptance, this drainage pattern is not the cause of left varicocele.
(3) **Collateral venous drainage** is through the external pudendal, posterior scrotal, and cremaster veins, and the veins of the vas deferens.

c. **Lymphatic drainage**
  (1) The lymphatics from the testes parallel the testicular veins to drain directly into the **para-aortic lymph nodes.**
  (2) This pathway explains in part why testicular seminoma disseminates widely and rapidly in the midabdominal retroperitoneum.

4. **Innervation** (see Figure 29-4)
   a. The motor role of the autonomic nerves to the testes is uncertain.
   b. **Afferent nerves** from the testis travel to the spinal cord along the sympathetic pathways, parallel to the testicular vessels.
      (1) These afferents pass first through the **aorticorenal plexus** and then along the lesser and least splanchnic nerves to spinal segments T10–T12.
      (2) Pain originating in the testes is referred to the middle and lower abdominal wall (see Table 29-1).

5. **Testicular torsion**
   a. Testes may rotate (90–360°) about the spermatic cord.
   b. Precipitating factors seem to be strong contractions of the cremaster muscle and a congenitally long **mesorchium** between the parietal and visceral layers of the tunica vaginalis. It most commonly occurs in the adolescent.
   c. Torsion is usually external rotation and represents a serious urologic emergency.
      (1) Compression of testicular vessels leads to ischemic necrosis of the testes within 6 hours.
      (2) Failure to recognize and intervene immediately has resulted in loss of the testis in about 80% of cases, subsequent atrophy of the testis in 10% more, and fertile resolution in only 10%.
      (3) In the emergency room, the twisted testis can be manipulated medially by gentle external pressure, restoring some blood flow. At subsequent operation, **both sides are sewn** into position (orchiopexy) because the other testis (predisposed to torsion) is valuable.

B. Epididymis

1. **Structure.** The highly convoluted **efferent ductules** open into the portion of the **mesonephric (wolffian) duct** that develops into the epididymis (see Figure 29-6).
   a. About 6 m long, the epididymis is compactly coiled to form a **head** and a **body** over a few centimeters.
   b. In the **tail** of the epididymis, the duct becomes thicker and forms the **vas deferens.**
   c. The epididymis is partially covered by the visceral layer of the tunica vaginalis and is adherent to the superior pole of the testis (see Figure 28-6B).

2. **Function.** Storage of spermatozoa with continued maturation occurs in the epididymis.

3. **Blood supply** is predominantly by the **deferential artery** (artery of the vas deferens).
4. **Innervation** (see Figure 29-4)
   a. Parasympathetic and visceral innervation is by the nervi erigentes. Parasympathetic stimuli initiate peristaltic waves that move semen into and along the vas before emission.
   b. **Afferent nerves** from the epididymis and vas deferens travel along the vas deferens first to the prostatic plexus and then to the lateral pelvic plexuses, nervi erigentes, and sacral nerves to reach spinal segments S2–S4 (see Table 29-1). Pain originating in the epididymis, such as that accompanying epididymitis, is referred to the distribution of S2–S4 (perineum and posterior thigh).

C. **Vas deferens**
1. **Structure.** The vas deferens (a direct continuation of the duct of the epididymis) represents the **mesonephric (wolffian) duct** (see Figure 29-6). It has a muscular wall, which accounts for the cord-like composition.
   a. **External portion** (see Figure 28-4)
      (1) From the inferior pole of the testis, the vas deferens ascends in the center of the **spermatic cord** within the internal spermatic fascia.
         (a) It is palpable in the superolateral aspect of the scrotum, which is the usual site for **vasectomy.**
         (b) It may be double, an important factor if vasectomy is to be successful.
      (2) It passes through the inguinal canal within the internal spermatic fascia and enters the abdominal cavity.
   b. **Internal course** (see Figure 29-15)
      (1) **In the major pelvis,** the vas deferens lies within the **transversalis fascia.** At the deep inguinal ring, it immediately diverges from the testicular neurovascular bundle and passes extraperitoneally over the external iliac vessels and obturator neurovascular bundle to enter the deep pelvis.
      (2) **In the pelvic cavity,** the vas deferens lies within **endopelvic fascia.** It passes anterosuperior to the ureters and converges toward the prostatic urethra at the base of the urinary bladder.
   c. The **ampulla** is an enlarged, sacculated portion of the vas near the base of the **prostate gland** (Figure 29-7).

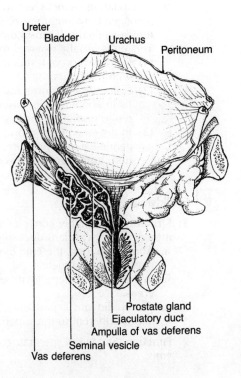

FIGURE 29-7. Posterior aspect of the urinary bladder and accessory glands of the male. The terminal portion of the left vas deferens, left seminal vesicle, and posterior portions of the prostrate gland are opened.

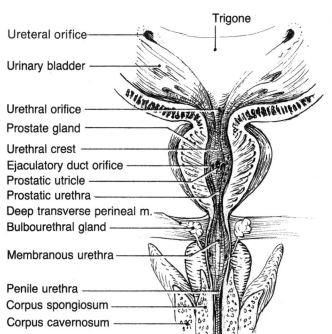

FIGURE 29-8. Male urethra.

  (1) Sperm is stored briefly in this portion of the vas during emission and before ejaculation.
  (2) The **seminal vesicles** join the vas deferens where the ampulla narrows as the **ejaculatory duct.**
 d. The **ejaculatory duct** is the narrow (0. 5 mm), thin-walled terminal intraprostatic portion of the vas deferens (see Figure 29-3).
  (1) It passes through the parenchyma of the prostate gland (see Figure 29-7).
  (2) It joins the **prostatic urethra** by an opening on the **colliculus seminalis** (verumontanum) on either side of the **prostatic utricle** (Figure 29-8).
2. **Blood supply** is by the **artery of the vas deferens (deferential artery),** which is homologous to the uterine artery in the female.
3. **Innervation** is primarily by parasympathetic fibers from the lateral pelvic plexus and sympathetic fibers from the hypogastric plexus (see Figure 29-4).
 a. **Parasympathetic innervation** produces **emission.** Slow peristaltic waves along the vas deferens move spermatozoa from the epididymis to the ampulla, where the sperm is stored before ejaculation.
 b. **Sympathetic innervation** produces strong contraction of the smooth muscle of the vas deferens, producing **ejaculation.**
 c. **Sensory afferents** travel through the pelvic plexus and along the pelvic splanchnic nerves to sacral levels S2–S4. As a result, pain of **deferentitis** or from prolonged engorgement of the ampulla is referred to the perineum (see Table 29-1).

D. **Seminal vesicles**
 1. **Structure.** Each seminal vesicle (about 3 cm long) is a dilated convoluted tube over 9 cm long (see Figure 29-7).
  a. They join the vas deferens where the ampulla becomes the ejaculatory duct.
  b. Located at the posterior base of the urinary bladder, they are palpable per rectum when swollen and inflamed.
 2. **Function.** They **contribute seminal fluid,** which contains fructose and choline among other substances.

- **a.** Spermatozoa require an outside source of sugar to produce energy for motility. Without fructose, sperm are incapable of reaching and fertilizing the ovum.
  - **(1)** Fructose, which is not produced anywhere else in the body, provides a forensic determination for the occurrence of rape.
  - **(2)** Choline crystals, however, form the preferred basis for the determination of the presence of semen (the Florence test).
- **b.** These glands do not store sperm.

3. **Innervation** is primarily by parasympathetic fibers from the lateral pelvic plexus and sympathetic fibers from the hypogastric plexus.
   - **a. Sympathetic innervation** produces strong contraction of the smooth muscle of the seminal vesicles, contributing seminal fluid to the ejaculate.
   - **b. Sensory afferents** travel through the pelvic plexus and along the pelvic splanchnic nerves to sacral levels S2–S4. As a result, pain of **vesiculitis** tends to be referred to the perineum (see Table 29-1). This pain, however, may not be particularly discrete.

E. Prostate gland

1. **Structure** (see Figures 29-7 and 29-8)
   - **a. Location.** The prostate gland lies between the base of the urinary bladder and the deep transverse perineal muscles (see Figure 29-3).
     - **(1)** The anterior surface is adjacent to the **retropubic space** (of Retzius).
     - **(2)** The posterior surface is adjacent to the seminal vesicles and the **rectovesical septum** (prostatoperitoneal membrane, Denonvilliers' fascia). It is palpable per rectum.
   - **b. Capsules.** The prostate has a true capsule and a false capsule (prostatic fascia) formed by endopelvic fascia. This composition enables the prostate to be "shelled out" surgically.
   - **c. Lobes.** The prostate gland forms as five diverticula of the prostatic urethra, and the prostatic parenchyma is subdivided into five lobes.
     - **(1)** The **left and right lateral lobes** are extensive and include what formerly were termed anterior lobes.
     - **(2)** The **left and right posterior lobes** also include the apex of the gland.
       - **(a)** The posterior lobes are most predisposed to malignant transformation.
       - **(b)** Malignant cells are found in the prostate gland of nearly 50% of all men 50 years of age or older. Carcinoma of the prostate is the most common malignancy in the adult male. Hard malignant nodes may be palpated rectally.
       - **(c)** Metastatic prostatic carcinoma represents the third most common cause of death from neoplastic diseases.
     - **(3)** The **medial lobe** surrounds the prostatic urethra.
       - **(a)** This lobe is predisposed to **benign prostatic "hypertrophy"** (BPH, middle lobe hypertrophy, Albarrán's lobe). This condition does not involve hypertrophy, but rather proliferative hyperplasia of the periurethral glands associated with the median lobe.
       - **(b)** Middle lobe hypertrophy results in obstruction of the urethra and visceral neck of the urinary bladder.
       - **(c)** This condition begins about age 45 years and occurs in 80% of all men by 80 years of age. In only 10% of these individuals, however, is surgical treatment required.
   - **d.** Numerous **prostatic ducts** from each lobe open individually into the prostatic urethra on either side of the urethral crest (see Figure 29-8).

2. **Function**
   - **a.** This compound tubuloalveolar gland secretes 0.52 ml of fluid daily. It contributes 10–20% of the ejaculate.
   - **b.** This fluid contains citric acid, acid phosphatase, prostaglandins, and fibrinogen.

3. **Vasculature**
   - **a.** The **arterial supply** is diverse, arising from the inferior vesical, differential, and inferior rectal arteries. Prostatectomy may result in extensive bleeding.

b. The **prostatic venous plexus** lies between the dense prostatic capsule and a layer of visceral pelvic fascia (prostatic fascia). It drains into the internal iliac veins as well as into the vertebral venous (Batson's) plexus, which may explain metastasis of prostatic cancer to the vertebral column and brain.
4. **Innervation** is primarily by parasympathetic fibers from the lateral pelvic plexus and sympathetic fibers from the hypogastric plexus.
   a. **Sympathetic innervation** produces strong contraction of the smooth muscle of the prostate gland, contributing seminal fluid to the ejaculate.
   b. **Sensory afferents** travel through the pelvic plexus and along the pelvic splanchnic nerves to sacral levels S2–S4. As a result, pain of **prostatitis** tends to be referred to the perineum (see Table 29-1). Prostatic pain, however, is not particularly discrete.

F. Clinical considerations
1. **Digital examination per anus** provides valuable clinical information. Palpable structures in the male are the membranous part of a catheterized urethra, posterior and lateral lobes of the prostate gland, rectovesical fossa, seminal vesicles if enlarged, bladder when distended, bulbourethral glands if enlarged, and ductus deferens when displaced or enlarged.
2. **Prostatectomy**
   a. **Suprapubic approach.** Incision through abdominal and vesical walls avoids the major portion of the retropubic space.
   b. **Retropubic approach.** The abdominal wall is incised to gain access to the retropubic space, which is then enlarged by pushing the bladder posteriorly to expose the prostate.
   c. **Perineal approach.** The incision is through the central tendon and the medial portion of the levator ani muscle. The dissection is anterior to the rectovesical (Denonvilliers') fascia, which "separates air (in the rectum) from water (in the bladder)." The rectum is pushed posteriorly and the urethra is pulled forward to visualize the prostate.
   d. **Transurethral resection.** Hypertrophic tissue is removed in conjunction with cystoscopy.
3. **Pain is referred** along two pathways.
   a. **Orchitis.** Because the afferent nerve pathways from the testis accompany the testicular arteries and veins, testicular pain is referred to the lower thoracic dermatomes (T10–T12).
   b. **Epididymitis, deferentitis, vesiculitis, and prostatitis.** Because the afferent nerve pathways from the epididymis, vas deferens, seminal vesicles, and prostate gland accompany the nervi erigentes, pain associated with these structures is referred to the perineum and posterior thigh (S2–S4).
4. **Stricture or blockage of the vas** reduces the ejaculate volume and may result in infertility.
   a. **Urinary tract infection** may spread along the vas to produce edema and stricture from scarring.
   b. A **cyst** involving the prostatic utricle may compress one or both ejaculatory ducts as they pass this mullerian remnant to reach the verumontanum.
   c. **Prostatic concretions** (corpora amylacea) within the prostate may become sufficiently large and calcified so that a nearby ejaculatory duct is compressed and blocked.

V. FEMALE REPRODUCTIVE TRACT

A. Peritoneum and pelvic mesenteries
1. **Pelvic peritoneum**
   a. **Parietal peritoneum** lines the pelvic cavity.

FIGURE 29-9. Female pelvic viscera. Midsagittal section depicts the lower portion of the female urinary tract and the genital tract.

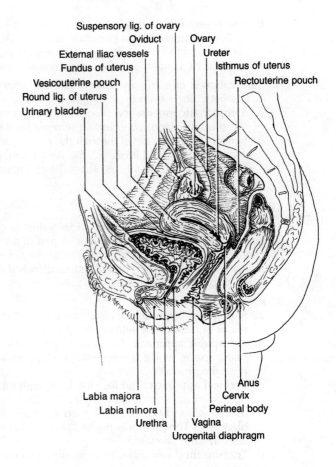

   b. **Visceral peritoneum** covers those pelvic viscera that protrude into the pelvic cavity.
   c. **Mesenteries** are formed by the reflections of serous mesothelium between parietal peritoneum and visceral peritoneum.

2. **Mesometrium (broad ligament)**
   a. **Structure.** This mesentery forms as the parietal peritoneum reflects from the walls of the pelvic cavity to cover the uterus and uterine tubes. It connects lateral margins of the uterus with the side wall of the pelvis (see Figure 29-10).
   b. **Relations.** The broad ligament and uterus partition the pelvic cavity transversely (Figure 29-9).
      (1) The **uterovesical pouch** is an extension of the pelvic peritoneal cavity between the posterior surfaces of the urinary bladder and the anterior surface of the uterus (with the broad ligament).
      (2) The **rectouterine or rectovaginal pouch** (of Douglas) is an extension of the pelvic peritoneal cavity between the posterior surfaces of the uterine cervix (with the broad ligament) and the anterior surface of the rectum. The floor of this pouch is apposed to the posterior fornix of the vagina.
   c. **Divisions** (Figure 29-10)
      (1) The **mesosalpinx** supports each uterine tube.
         (a) Beginning at the juncture of the mesovarium, this mesentery represents the superior limit of the broad ligament.
         (b) It contains the tubal branches of the uterine vessels as well as the vestigial epoöphoron and paroöphoron.
      (2) The **mesovarium** supports the ovary.
         (a) It is a posterior extension of the broad ligament.

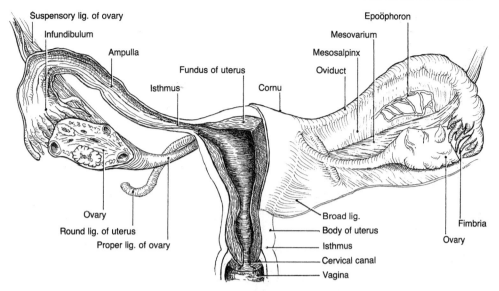

FIGURE 29-10. Female reproductive tract, posterior aspect. The left ovary, left uterine tube, and left side of the uterus are opened. The ligamentous support of the uterus, right uterine tube, and right ovary is intact.

- **(b)** It becomes continuous with the serosa or "germinal epithelium" of the ovary.
- **(3)** The **mesometrium** (the major portion of the broad ligament) supports the uterus on either side.
- **(4)** The **infundibulopelvic ligament (suspensory ligament of the ovary)** is an elevation of peritoneum between the pelvic brim and the mesovarium.
  - **(a)** It forms by a peritoneal elevation as the ovary descends into the deep pelvis.
  - **(b)** It contains the **ovarian artery** and **ovarian vein** as well as lymphatics and nerves.
- **d. Contents** (see Figure 29-10)
  - **(1)** A **uterine tube (oviduct)** lies in the superior free edge of this ligament on each side of the uterus.
  - **(2)** The **ovarian ligament (proper ligament of the ovary)** runs from the medial pole of the ovary to the uterine cornu, just below the origin of the uterine tube.
  - **(3)** The **round ligament of the uterus** runs from the lateral uterine border near the point of attachment of the ovarian ligament. It runs between the leaflets of the broad ligament and through the inguinal canal (of Nuck) to fuse with the dermis of the labium major.
  - **(4)** The **uterine neurovascular bundle** lies at the base of the broad ligament within the transverse cervical (cardinal) ligament.
  - **(5)** **Ovarian vessels** lie within the suspensory ligament of the ovary with branches in the mesosalpinx.
  - **(6) Vestigial structures**
    - **(a)** The **ovarian ligament** together with the **round ligament** of the uterus are homologous to the gubernaculum testis.
    - **(b)** Although the mesonephric (wolffian) ducts degenerate in the female, remnants may be sites of cyst formation.
      - **(i)** The **epoöphoron** constantly lies in the lateral portion of the mesosalpinx and mesovarium, represents residual mesonephric tubules, and is homologous to the epididymis.
      - **(ii)** The **paroöphoron** variably lies more medially in the mesosalpinx,

FIGURE 29-11. Anterior, lateral, and posterior condensations of endopelvic fascia provide ligamentous support of the cervix. Note the close relationship between the ureters and the uterine arteries in the cardinal ligament.

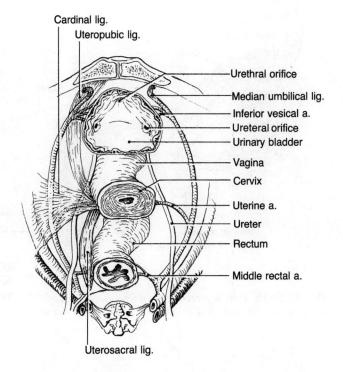

also represents residual mesonephric tubules, and is homologous to the paradidymis.
- (iii) **Gartner's ducts** lie in the mesometrium adjacent to the uterus, represent the mesonephric ducts, and are homologous to the vas deferens.
- (iv) A rete ovarii may persist next to the hilus of the ovary and is homologous to the rete testis.
- (v) The hydatid of Morgagni (appendix ovarii) represents the most cranial remnant of the mesonephric duct and is homologous to the appendix of the epididymis.

3. **Endopelvic fascia**
   a. **Location.** The endopelvic fascia, which lies between the musculoskeletal pelvic wall and the pelvic viscera, is the extraperitoneal tissue of the uterus **(parametrium),** vagina, urinary bladder, and rectum. It is continuous with the transversalis fascia of the abdomen.
   b. **Divisions.** Endopelvic fascia forms ligaments that support the uterus (Figure 29-11).
   (1) **Transverse cervical (cardinal) ligaments** (of Mackenrodt) provide lateral support.
      (a) These condensations of endopelvic fascia arise from the osteomuscular walls of the deep pelvis along the lines of attachment of the levator ani muscle and insert into the cervix.
      (b) Composed of white fibrous connective tissue with some smooth muscle, they are thickest about the uterine vascular bundles. They support the cervix laterally.
   (2) **Uterosacral ligaments** provide anterior stabilization (see Figure 29-11).
      (a) Fascial condensations run between the cervix and the sacrum, diverging to either side of the rectum to form the **rectouterine folds** of peritoneum.
      (b) These ligaments contain smooth muscle fibers (rectouterine muscles) and

maintain posterior tension on the cervix, drawing the cervix toward the rectum.
- (3) **Pubocervical (vesicovaginal) ligaments** provide anterior stabilization (see Figure 29-11).
  - (a) They arise from the cervix and cardinal ligament, and pass between the vagina and bladder to insert onto the pubis.
  - (b) This layer reinforces the anterior vaginal wall and helps to prevent cystocele.

B. Ovaries

1. **External structure.** Each almond-shaped ovary is approximately 3 cm × 1.5 cm × 1 cm.
   a. **Position.** The ovaries lie on the posterior side of the broad ligament and project into the **ovarian fossae,** triangular depressions on each side of the minor pelvic cavity bounded by the external iliac vessels and the internal iliac vessels (see Figure 29-9).
   b. **Ligamentous support** (Figure 29-12; see Figure 29-10)
      (1) The **mesovarium** attaches each ovary to the broad ligament (see Figure 29-12).
      (2) The **infundibulopelvic (suspensory) ligament** attaches each ovary to the lateral pelvic wall and contains the ovarian neurovascular bundle.
      (3) The **ovarian ligament (proper)** runs within the broad ligament from the inferior medial pole of each ovary to the uterus near the cornu. It is homologous to the cranial portion of the **gubernaculum** in the male.
      (4) Generally, one fimbria of the uterine tube infundibulum (the **ovarian fimbria**) attaches to the ovary.
2. **Internal structure**
   a. The **cortex** is the functional portion of the ovary (see Figure 29-10).

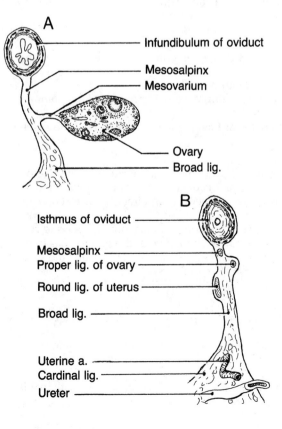

FIGURE 29-12. Ligamentous support of the ovary and uterine tubes. *(A)* Lateral parasagittal section through the infundibulum of the uterine tube shows the mesosalpinx and the mesovarium as extensions of the broad ligament. *(B)* Medial parasagittal section through the isthmus of the uterine tube shows the proper ligament of the ovary as well as the round ligament of the uterus within the broad ligament. The relationship of the uterine artery to the ureter in the cardinal ligament at the base of the broad ligament is also indicated.

(1) The ovarian surface is covered by modified peritoneum ("**germinal epithelium**") or serosa, which continues from the ovarian hilum as the mesovarium.
(2) In the adult, it contains several notable structures.
    (a) It has a multitude of **immature follicles** and developing follicles.
    (b) It may have one or more **mature (graafian) follicles,** each surrounding a maturing ovum.
        (i) These structures secrete **estrogen.**
        (ii) At peak maturity, the ovum bursts free of the mature follicle into the peritoneal cavity.
    (c) **Corpora lutea** comprise the remains of follicles immediately after ovulation.
        (i) These structures secrete **estrogen** and **progesterone.**
        (ii) Numerous follicles luteinize (become corpora lutea) each month without reaching ovulation.
    (d) **Corpora albicantia** are the end-stage scars of regressed corpora lutea.
(3) Primordial follicles (somewhat less than 500,000) develop in early fetal life.
(4) Ovulation occurs into the peritoneal cavity. The nearby or overlying uterine tube draws the ovum into the infundibulum by the combined movements of fimbria and cilia.
(5) After menopause, the ovary becomes small and atretic, and the follicles disappear.

  b. The **medulla** contains connective tissue and vasculature.

3. Vasculature
  a. Arterial supply (see Figure 29-13)
    (1) Each **ovarian artery** arises from the abdominal aorta near the renal arteries and runs inferiorly in the suspensory (infundibulopelvic) ligament and mesovarium to reach the ovary.
    (2) **Ovarian and tubal branches** of the **uterine arteries** anastomose with the ovarian artery in the mesosalpinx and can supply the ovary if the suspensory ligament is transected or the ovarian artery is occluded.
  b. Venous return
    (1) **Ovarian veins** drain asymmetrically (see Figure 26-3).
        (a) The **right ovarian vein** drains into the inferior vena cava near the renal vein.
        (b) The **left ovarian vein** drains into the left renal vein.
    (2) Twigs from the ovaries also drain along the tubal branches of the uterine veins.
  c. Lymphatic drainage is principally toward the para-aortic nodes near the renal arteries.

4. Innervation (see Figure 29-14)
  a. Autonomic innervation of the gonads is from the aortic plexus, but the role of the autonomic nerves regarding the ovary is unclear. It is noteworthy that in some mammals and perhaps in humans under certain circumstances, copulation induces ovulation.
  b. Visceral afferent fibers from the ovary run along sympathetic pathways to spinal segments T10–T12 and refer pain to lower abdominal and pubic regions (see Table 29-1). Intractable ovarian pain may be alleviated by transecting the suspensory ligaments, which contain visceral afferent fibers.

5. Clinical considerations
  a. Ovarian torsion
    (1) An abnormally long mesovarium and suspensory ligament predispose to this condition.
    (2) Torsion compresses the blood supply, producing ischemia and abdominal symptoms with pain referred to the lower thoracic and upper lumbar dermatomes.
    (3) This condition occurs as a complication in 10–20% of women with ovarian tumors.

b. Ectopic ovaries
(1) Migration of the ovaries may continue to the deep inguinal ring, into the inguinal canal, or, rarely, the labia major.
(2) Ovaries that fail to descend are termed retroperitoneal and lie in the posterior abdominal wall superior to the kidneys.
(3) Ovaries occasionally prolapse into the rectouterine pouch.

c. Congenital absence of one or both gonads is usually associated with a unilateral defect of the genital system (rare).

d. Ovarian tumors of numerous types are common.

C. Uterine tubes and uterus. Because of their common development, these structures constitute a functional unit.

1. **Uterine tubes** (fallopian tubes or oviducts) represent the unfused portions of the paramesonephric (mullerian) ducts.
   a. **External structure.** Approximately 10 cm long, each tube begins at an ostium that opens directly to the peritoneal cavity (see Figure 29-9).
   b. **Support.** They are located at the upper free edge of the broad ligament and are supported by **mesosalpinx** (see Figures 29-10 and 29-12).
   c. **Divisions** (see Figure 29-10)
      (1) The **infundibulum**, 2 mm wide, opens into the peritoneal cavity through the **internal ostium.**
         (a) The **internal ostium** is bordered by motile **fimbriae**.
         (b) The fimbriae generally sweep over the ovarian surface so that an expelled ovum is directed into the infundibular ostium.
         (c) One fimbria, the **ovarian fimbria,** frequently adheres to the ovary.
      (2) The **ampulla**, 2.5 mm wide, is thin walled and tortuous. It constitutes the major portion of the uterine tube. Fertilization of the ova usually occurs in the ampullary portion of the uterine tube.
      (3) The **isthmus**, 1 mm wide, is short, thick walled, and straight. It lies between the ampulla and the uterine wall.
      (4) The **intramural segment,** less than 1 mm wide, is the narrowest segment. It lies within the uterine wall and communicates with the uterine cavity.
   d. **Internal structure**
      (1) The serosa is visceral peritoneum. The muscularis consists of two layers of smooth muscle. The mucosa is ciliated columnar epithelium interspersed with mucous secretory cells, the number of which vary with the menstrual cycle.
      (2) The ova are swept along the uterine tube by peristaltic movement and ciliary action.
   e. **Clinical considerations**
      (1) **Pelvic inflammatory disease (PID).** The female genital canal is a direct communicating pathway from the urogenital sinus of the perineum into the peritoneal cavity. It is a ready pathway for infection.
      (2) **Salpingitis.** Inflammation of the uterine tube is a common cause of female sterility. Healing occurs by formation of scar tissue that may block the tube or result in areas that lack cilia and mucous cells. If fertilization does occur, the zygote may fail to traverse a subsequently scarred or dysfunctional uterine tube, resulting in ectopic tubal implantation.
      (3) **Ectopic implantation**
         (a) **Peritoneal implantation.** An ovum may be fertilized in the peritoneal cavity or even within an ovary. If the zygote fails to enter one or the other uterine tubes, ectopic implantation occurs.
         (b) **Tubal implantation.** This type of ectopic implantation (tubal pregnancy) results in rupture of the tube with hemorrhage and expulsion of the embryo into the peritoneal cavity, usually between the fourth week (if in the isthmus) and the tenth week (if in the ampulla). This occurrence is a gynecologic emergency, because the patient may die of internal hemor-

rhage. Ectopic implantation is most common in the ampullary portion of the uterine tube.
- (4) **Uterosalpingography,** in which radiopaque dye is injected into the uterus and uterine tubes, is used to determine the patency of the uterine tube.

2. **The uterus** represents the fused portions of the paramesonephric (mullerian) ducts (see Figure 29-5).
   a. **External structure.** Although variable in size, the uterus typically measures 8 cm × 2.5 cm × 5 cm (about two thumbs side by side) in the nulliparous woman. It is somewhat larger in the multiparous woman.
   b. **Divisions** (see Figure 29-10)
      (1) The **fundus,** the region superior to the uterine cornua, contributes most of the upper uterine segment during pregnancy.
      (2) The **cornu** on each side defines the entrance of a uterine tube.
      (3) The **body** (corpus) is the region inferior to the cornu and superior to the cervix.
         (a) The **uterine cavity** is shaped like a flattened, inverted triangle. At the superolateral angles, the uterine cavity extends into each uterine tube.
         (b) The **isthmus** is the dividing line between the cervix and the uterus. It corresponds with the **internal os.**
            (i) The isthmus is incorporated into the uterus during pregnancy as the lower uterine segment.
            (ii) The isthmus is the preferred site for cesarean section.
      (4) The **cervix** is the narrow inferior portion of the uterine body that projects into the vagina.
         (a) It is subdivided into three regions.
            (i) The **internal os** marks the junction between the cervical canal and the uterine body.
            (ii) The **cervical canal,** about 2.5 cm long, lies between the internal and external ostia.
            (iii) The **external os** is the opening of the cervical canal into the vagina. The nulliparous external os is round; the parous os is transverse.
         (b) Changes occur during pregnancy.
            (i) The cervix becomes softer, changing from feeling like a "nose" to feeling like "lips."
            (ii) The glands of the cervical canal secrete a protective mucous plug during pregnancy.
   c. **Support**
      (1) The **levator ani muscle** and **urogenital diaphragm** provide the principal support for the urinary bladder and the vagina. The **urinary bladder,** in conjunction with gravity, directly supports the anteverted and anteflexed uterus (see Figure 29-1).
      (2) The **mesometrium,** that portion of the **broad ligament** that attaches to the lateral aspects of the uterus, supports the uterus on either side (see Figure 29-10).
      (3) The **round ligament of the uterus** maintains the anteverted position. It stretches considerably during pregnancy (see Figure 29-10).
      (4) The **transverse cervical (cardinal) ligament,** with the contained uterine vasculature, contributes variable support in the cervical region (see Figure 29-11). When the uterus does not rest on the bladder (in retroversion), the cardinal ligaments provide greater support.
      (5) The **uterosacral and pubocervical ligaments** provide variable posterior and anterior support (see Figure 29-11).
   d. **Positions**
      (1) Normal uterine position is **anteflexed** (uterus bent forward on itself at the level of the internal os) and **anteverted** (angled forward about 90° to the vagina). Thus, the long axis of the uterus is nearly horizontal as the body of the uterus lies on the bladder.
      (2) The uterus often assumes other positions [see section V D 7 a (2) (d)].

(a) **Retroflexion.** The axis of the uterine body passes upward or even backward, but the angle at which the cervix enters the vagina is unchanged. The result is a bending back of the uterus, narrowing the cervical canal. This configuration may produce especially painful menstruation and reduce the likelihood of conception.

(b) **Retroversion.** The axis of the cervix with the vagina is upward or even backward, so the uterus may rest against the rectum, and either the external os or the posterior lip of the cervix presents first on digital examination. Retroversion predisposes to uterine prolapse because the weight of the uterus is no longer supported by the bladder.

e. **Relations** (see Figure 29-1)
   (1) Anteriorly, the **vesicouterine pouch** lies between and separates the uterus from the urinary bladder.
      (a) The supravaginal cervix is separated from the urinary bladder by the pubocervical (vesicovaginal) fascia.
      (b) The vaginal portion of the cervix lies immediately posterior to the **anterior fornix** of the vagina.
   (2) Posteriorly, the uterine body and cervical canal are separated from the rectum by the **rectouterine pouch** (of Douglas) and the **rectovaginal septum** (of Denonvilliers).
      (a) During development, the parietal peritoneum forming the most caudal portion of the rectouterine pouch fuses to form the **rectouterine septum.**
      (b) The rectouterine pouch is palpable across the **posterior fornix** of the vagina.
   (3) The **broad ligament** and its contents lie lateral to the uterus. The ureter is about 1 cm lateral to the lateral fornices.

f. **Internal structure**
   (1) The **endometrium** (mucosa) undergoes cyclic changes with the menstrual cycle.
      (a) The **proliferative (follicular)** phase occurs under the influence of estrogen from the maturing follicle. The endometrium is reformed.
      (b) The **secretory (luteal) phase** occurs under the influence of progesterone and estrogen from the new corpus luteum. It is marked by glandular proliferation and secretion as well as proliferation of the endometrial blood vessels so that the endometrial mucosa facilitates implantation.
      (c) The **menstrual phase** begins when the corpus luteum is regressing and the progesterone titers are decreasing. This phase involves sloughing of endometrium.
   (2) The **myometrium,** forming the bulk of the uterine wall, is smooth muscle.
   (3) The **mesometrium** is **visceral peritoneum** that reflects off the uterus as mesentery.

g. **Vasculature**
   (1) **Arterial supply** (Figure 29-13; see Figure 29-11)
      (a) **Uterine arteries** arise from each internal iliac artery and run in the transverse cervical ligaments at the base of the broad ligament.
      (b) They cross immediately superior to the ureters in the transverse cervical ligament, an important surgical consideration.
      (c) **Branches of the uterine artery** (see Figure 29-13)
         (i) The **ascending branch** gives off the **tubal branch** and anastomoses with the **ovarian artery.**
         (ii) The **descending (vaginal) branch** anastomoses with other vaginal branches and with the perineal branches of the **internal pudendal artery.**
      (d) **During pregnancy,** the uterine artery becomes hypertrophic, with an enormous volume of blood flowing to the uterus and placenta.
   (2) **Venous return** is along **uterine veins,** which generally parallel the arterial supply.
   (3) **Lymphatic drainage** parallels the blood supply.

FIGURE 29-13. Vasculature of the female reproductive tract and genitalia. Anastomoses occur between the ovarian and uterine arteries as well as between the uterine and deep perineal branches of the pudendal artery.

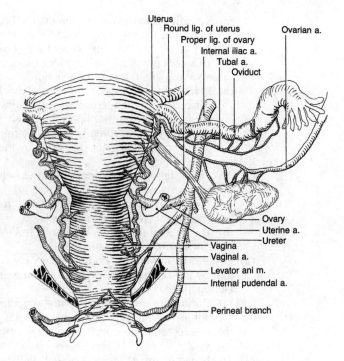

- (a) The uterine fundus drains mainly along the ovarian vessels to the **para-aortic nodes.**
- (b) The uterine body generally drains along the internal iliac vessels through the **internal iliac nodes,** but some lymphatics follow the round ligament to the **superficial inguinal nodes.**
- (c) The cervix drains in three directions.
    - (i) Principal drainage is along the uterine vessels to the **internal iliac nodes.**
    - (ii) Lymph pathways also follow along the internal pudendal vessels to the **internal iliac nodes.**
    - (iii) Other pathways course through the deep pelvis toward the **sacral nodes.**
- (d) Because of widespread lymphatic drainage, invasive carcinoma of the cervix may require extensive lymphatic dissection.
  h. **Innervation** is derived as the **uterovaginal plexus** from the lateral pelvic plexus (Figure 29-14).
    (1) **Sympathetic neurons** from T12–L2 generally synapse in the sympathetic chain. The **postsynaptic neuron** runs sequentially through the **lumbar splanchnic nerves, aortic plexus, superior hypogastric plexus, inferior hypogastric plexus, lateral pelvic plexus,** and **uterovaginal plexus** to reach the uterus.
    (2) **Parasympathetic neurons** from S2–S4 segments run in the **pelvic splanchnic nerves (nervi erigentes),** lateral pelvic plexus, and uterovaginal plexus to reach the uterus, where they synapse.
    (3) **Visceral afferents** follow two distinct pathways (see Table 29-1).
        - (a) **From the body of the uterus and uterine tubes,** visceral afferents run along sympathetic pathways. Pain associated with uterine spasm and salpingitis is referred to dermatomes L1–L2 (the middle of the back, the inguinal and pubic regions, and the anterior thigh).
        - (b) **From the cervical region,** visceral afferents run along the nervi erigentes. Pain associated with cervical dilation is referred to dermatomes S2–S4 (the perineum, gluteal region, posterior thigh, and leg).

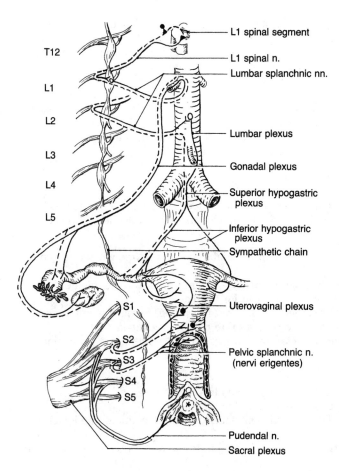

FIGURE 29-14. Innervation of the female reproductive tract and genitalia. Sympathetic pathways arise from the upper lumbar spinal levels *(black triangle)*. No white rami are found below L2. They reach the aortic plexus by way of lumbar splanchnic nerves. Synapse occurs in the aortic plexus *(white circles)*. The postsynaptic neurons reach the pelvic viscera by the hypogastric plexuses. Parasympathetic pathways arise from the midsacral levels and reach the pelvic viscera by the pelvic splanchnic nerves. Synapse occurs in the walls of the viscera *(black circles)*. Visceral afferent fibers *(dashed line)* from the pelvic viscera travel specifically along either the sympathetic or the parasympathetic pathway and produce specific patterns of referred pain. The pudendal nerve provides somatic sensation to and from the perineum.

i. **Clinical considerations**
   (1) **Pelvic examination.** Palpable per vagina are the cervix and ostium of the uterus, vagina, body of the uterus if retroverted, rectouterine fossa, and, under pathologic conditions, the ovary, uterine tubes, and broad ligament.
   (2) **Ureteral risk during surgery**
      (a) Because of the location of the ureters in the transverse cervical ligament, division of the ureter is possible when clamping and ligating the uterine arteries during hysterectomy.
      (b) Compression of the ureters by a growth in the pelvis may result in hydronephrosis and uremia, a potential cause of death.
   (3) **Uterine prolapse**
      (a) Retroversion of the uterus predisposes to prolapse because the weight of the uterus is no longer supported by the bladder.
      (b) Trauma to the uterine support as a result of difficult childbirth, as well as general relaxation of connective tissues in old age, may result in uterine prolapse.
   (4) **Anesthesia.** During parturition, pain from dilation of the cervix and stretch of the perineum may be blocked, leaving thoracic and lumbar nerves to control abdominal musculature.
      (a) **Caudal anesthesia** is most effective for the sacral nerves, including the pathways conveying pain from cervical dilation. The needle is inserted between the sacral cornua, penetrating the sacrococcygeal ligament across the sacral hiatus to gain access to the vertebral canal. The anesthetic is infiltrated into the epidural space external to the dural sac, and the spinal nerves are anesthetized as they leave the dura.

- **(b) Epidural anesthesia** is most effective for the lower lumbar nerves, including the pathways conveying pain from uterine spasm or contraction. The needle is inserted into the L4–L5 interspace, and the anesthetic is infiltrated into the epidural space about the dural sac.
- **(c) Spinal (intrathecal) anesthesia** is effective for the lumbar and sacral nerves, but it is more difficult to control and subject to more complications and side effects. The needle is inserted into the L4–L5 interspace and through the dural sac. The anesthetic is infiltrated directly into the subarachnoid space in which the nerve roots of the cauda equina (the lumbar and sacral nerves) are anesthetized.

D. Vagina

1. **Location.** From its opening at the vestibule of the vulva, the vagina passes superiorly and posteriorly through the pelvic diaphragm and surrounds the uterine cervix (see Figure 29-9).

2. **Structure.** The size of the vaginal canal is variable. Normally, the vaginal lumen is H-shaped.
   a. The **anterior wall**, which is approximately 7.5 cm (3 in.) long, terminates anterior to the projecting cervix as the **anterior fornix.**
   b. The **posterior wall**, approximately 10.5 cm (4 in.) long, terminates in the **posterior fornix** and is related to the **rectouterine pouch** (of Douglas).
   c. The **lateral walls** terminate in **lateral fornices** on either side of the projecting cervix.

3. **Relations** (see Figure 29-9)
   a. **Anterior** to the upper portion of the vagina lies the cervix.
      (1) The midvagina apposes the base of the bladder.
      (2) The lower half of the urethra is embedded in the adventitia of the lower vaginal wall.
   b. **Posteriorly,** the posterior fornix is separated from the rectum by the **rectouterine pouch.**
      (1) The upper quarter of the posterior vaginal wall is separated from the rectum by the **rectovaginal septum.**
      (2) The lower vagina is separated from the anal canal by the **perineal body.**
   c. **Laterally,** the vagina is flanked by the levator ani muscle and endopelvic fascia. The ureters and uterine arteries lie close to the lateral fornices.
   d. The **upper two thirds** of the vagina are within the pelvic cavity superior to the pelvic floor [see Chapter 27 II A 2 b (3) (b)].
   e. The **lower one third** of the vagina is within the perineum (inferior to the pelvic floor). It communicates with the exterior via the vaginal introitus [see Chapter 28 V A 1 b (3)].

4. **Internal structure.** The **vaginal epithelium** is stratified squamous, lubricated by cervical mucus and desquamated vaginal cells. The walls are rugose in the nulliparous female, becoming smoother after vaginal childbirth.

5. **Vasculature** (see Figure 29-13)
   a. **Arterial supply** is by the **vaginal branch of the uterine artery,** occasional vaginal branches of the internal iliac arteries, possible twigs from the **middle rectal arteries,** and branches from the **internal pudendal arteries.** The vaginal arteries course along the lateral vaginal walls.
   b. **Venous return.** The **vaginal venous plexus** drains into the internal iliac vein.
   c. **Lymphatic drainage**
      (1) The **upper one third** of the vagina drains into **external and internal iliac nodes.**
      (2) The **middle one third** drains into the **internal iliac nodes.**
      (3) The **lower one third** drains into the **internal iliac nodes** as well as into the **superficial inguinal nodes.**

6. **Innervation** (see Figure 29-14)
   a. The **upper two thirds of the vagina** is supplied by the **uterovaginal plexus** of nerves, a spray from the lateral pelvic plexus, that contains sympathetic, parasympathetic, and afferent fibers. The visceral afferents from this region travel along the nervi erigentes, so pain may be referred to the dermatomal distribution of spinal segments S2–S4 (see Table 29-1).
   b. The **lower (perineal) part of the vagina** is supplied by the **pudendal nerve,** which mediates all cutaneous sensations.
7. **Clinical considerations**
   a. **Vaginal examinations**
      (1) **Inspection with a speculum** reveals the vaginal walls and cervix.
         (a) Biopsy can be performed, cytologic smears taken, and first-order prolapse detected.
         (b) The posterior fornix of the vagina is the site for **culdocentesis,** whereby fluid in the rectouterine pouch may be tapped.
      (2) **Digital examination**
         (a) Through the anterior fornix, the urethra and bladder are palpable.
         (b) Through the posterior fornix, the rectum, coccyx, and sacral promontory are palpable across the pouch of Douglas.
         (c) Through the lateral fornices, the ovaries, uterine tubes, side wall of the pelvis, and occasionally the ureters are palpable.
         (d) At the apex of the vagina, the anterior and posterior cervical lips are palpable. In anteversion of the uterus, the anterior cervical lip presents first. In retroversion, either the external cervical os or the posterior lip of the cervix presents first.
      (3) **Bimanual examination** (two fingers in the vagina and the other hand palpating or exerting pressure on the lower abdomen) enables the examiner to determine the size and position of the uterus, palpate the ovaries and uterine tubes, and detect pelvic inflammation or neoplasms.
   b. **Herniation** is often the result of lax pelvic support structures weakened during parturition.
      (1) **Cystocele** is herniation of the urinary bladder through the anterior vaginal wall into the vaginal lumen.
      (2) **Rectocele** is herniation of the rectum through the posterior vaginal wall into the vaginal lumen.
   c. **Anomalies and variations**
      (1) **Uterine and vaginal duplication**
         (a) **Didelphia** (complete double uterus and double vagina) is rare.
         (b) **Uterus duplex** (double uterus with a single vagina) is uncommon.
         (c) **Bicornuate** or **septate uterus** (double or partitioned fundus) is noted more frequently.
         (d) **Unicornuate uterus** with complete absence of the uterine tube on one side is rare.
      (2) **Vaginal anomalies** are mainly associated with failure of the vaginal plate to canalize.
         (a) **Rectovaginal fistulas** commonly are coincident with an imperforate anus or perineal trauma.
         (b) **Vesicovaginal fistulas** (rare) are associated with continuous urinary discharge from the vagina.
         (c) **Urethrovaginal fistulas** are associated with urinary discharge from the vagina only on micturition.

VI. INNERVATION OF PELVIC VISCERA

A. **Pelvic plexus (lateral pelvic plexus)**
   1. **Location.** The **lateral pelvic plexus** lies along the inferior walls of the rectum. It contains postsynaptic sympathetic neurons from the hypogastric plexus, presynaptic

parasympathetic neurons from the nervi erigentes, small autonomic ganglia, and visceral afferent neurons.

   2. **Divisions** (see Figures 29-4 and 29-14)
      a. The **rectal plexus** provides autonomic innervation to the rectum and anal canal as well as parasympathetic innervation in a retrograde direction to the large bowel as far as the splenic flexure.
      b. The **uterovaginal or prostatic plexus** gives rise to the **cavernous plexuses,** which convey parasympathetic nerves to the erectile tissues as well as sympathetic nerves to smooth muscle associated with the reproductive tract.
      c. The **vesical plexus** is associated with bladder function [see section II B 4 b (2)].
   3. **Surgical interference** with the pelvic plexus, as for example in rectal resection, results in varying degrees of dysfunction.

B. **Sympathetic innervation** arises from spinal segments T12–L2 (see Figures 29-4 and 29-14).
   1. **Presynaptic pathways.** White rami communicantes T12–L2 carry preganglionic sympathetic motor neurons from the lowest thoracic and upper lumbar spinal nerves to the sympathetic chain (see Figures 4-5 and 25-16).
   2. **Sympathetic ganglia** include those of the sympathetic chain as well as small ganglia in the aortic plexus.
      a. Some neurons synapse within the paravertebral ganglia of the sympathetic chain before leaving via the **lumbar splanchnic nerves** to join the **aortic plexus.**
      b. Those neurons that do not synapse in the sympathetic chain do so in small ganglia in the aortic plexus.
      c. Presynaptic sympathetic neurons use **cholinergic transmission.**
   3. **Postsynaptic pathway.** After synapse, the axons run inferiorly in the aortic plexus (see Figure 25-16).
      a. The **superior hypogastric plexus** is the continuation of the **aortic plexus** over the bifurcation of the aorta. This pathway brings sympathetic nerves into the pelvis.
      b. The **inferior hypogastric plexuses (left and right hypogastric nerves)** are formed by the bifurcation of the superior hypogastric plexus at the pelvic brim.
         (1) These sympathetic pathways run along the anterior surface of the sacrum toward the rectum.
         (2) They provide the sympathetic contribution to the **pelvic plexus (lateral pelvic plexus)** on the walls of the rectum.
      c. Postsynaptic sympathetic neurotransmission is **adrenergic.**
   4. **Inability to ejaculate.** Loss of sympathetic function to the reproductive tract results in inability to ejaculate or in retrograde ejaculation into the bladder. It is unclear what effect such a loss produces in the female.

C. **Parasympathetic innervation** arises from spinal segments S2–S4 and represents the sacral portion of the parasympathetic (craniosacral) division of the autonomic nervous system. Sacral parasympathetics innervate the bladder and reproductive tract as well as the lower gastrointestinal tract from the splenic flexure to the pectinate line of the anal canal (see Figures 29-4 and 29-14).
   1. **Presynaptic parasympathetic neurons** leave the sacral spinal nerves near the anterior sacral foramina to **form pelvic splanchnic nerves (nervi erigentes, pelvic nerves).**
      a. These nerves contribute to the **lateral pelvic plexus** and run in the rectal adventitia (see Figure 25-16).
         (1) Some fibers travel retrograde along the rectum to innervate the sigmoid colon.
         (2) Other fibers course in retrograde fashion in the hypogastric plexus as far as the branches of the inferior mesenteric artery, which they follow to the descending bowel.
         (3) Other fibers run to prostatic and uterovaginal plexuses.

b. The long **preganglionic neurons** of the pelvic splanchnic nerves synapse with postganglionic fibers in intramural plexuses, using **cholinergic neurotransmission.**
2. **The postsynaptic parasympathetic neurons** lie in the walls of the structures they innervate (see Figure 25-16).
    a. Parasympathetic postganglionic neurons synapse with effectors, such as smooth muscle cells and secretory cells.
    b. Parasympathetic postsynaptic neurons are also **cholinergic.**
3. **Clinical considerations**
    a. **Aganglionosis** (Hirschsprung's disease) is a genetic defect in which parasympathetic ganglia do not develop in the walls of the descending colon.
        (1) Initiation of propulsive function within the sigmoid colon is lost (functional blockage), with ensuing megacolon. Stretching of the colon may produce ischemia with clinical signs of peritonitis.
        (2) **Toxic megacolon** is usually treated surgically, with resection of the aganglionic portion.
    b. **Impotence.** Loss of parasympathetic function to the reproductive tract results in inability to produce clitoral or penile erection.
    c. **Neurogenic bladder.** Loss of parasympathetic innervation to the urinary bladder results in a grossly dilated atonic bladder that retains urine and requires catheterization.

D. Afferent nerves from the pelvic viscera travel either along sympathetic pathways to spinal levels L1–L2 or along parasympathetic pathways to spinal levels S2–S4. The patterns of referred pain from the pelvic viscera are specific (Table 29-1).

VII. FUNCTIONAL CONSIDERATIONS

A. **Precoitus.** Sexual activity of variable duration and intensity is usually required to establish a receptive psychologic attitude before congress occurs.
1. **Mechanism of erection**
    a. **Tumescence.** When an individual becomes sexually aroused, parasympathetic volleys travel by way of the pelvic nerves to the cavernous plexuses, which supply the penis or clitoris.
        (1) Parasympathetic outflow causes smooth muscle relaxation in the arterioles supplying the erectile tissue. These vessels contain longitudinal ridges that partially occlude the lumina when the circular muscularis is slightly tensed.
        (2) The corpora have a spongy structure with irregular vascular spaces that are essentially collapsed in the flaccid penis or clitoris and vestibular bulbs. With increased blood flow, the corpora fill, expand, and become tumescent.
    b. **Turgor.** Erection is a vascular event that involves not only increased inflow but also reduced efflux.
        (1) As the erectile tissue expands, the peripheral veins are compressed against the enveloping tunica albuginea, which effectively impedes drainage of blood from the cavernous sinuses. Consequently, the penis or clitoris enlarges, elongates, and becomes rigid.
        (2) Although the turgor of the corpora cavernosa in the male is extreme, the corpus spongiosum and its extension, the glans penis, do not become nearly as firm.
            (a) The thinner and more elastic tunica albuginea surrounding the corpus spongiosum in the male allows for a more flexible vaginal penetration and offers some resiliency when the glans penis contacts the cervix or wall of the posterior fornix during coitus.
            (b) The less turgid corpus spongiosum allows the seminal bolus to dilate the urethra so ejaculation can occur.

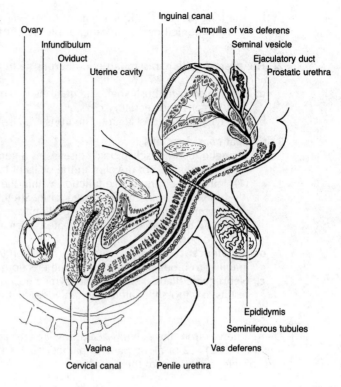

FIGURE 29-15. The second (penetration) phase of coitus (after Da Vinci).

2. **Male emission and secretion**
   a. With sexual stimulation, the male genital ducts become increasingly motile, producing an **emission** (movement of sperm from the epididymis to the ampullary portion of the vas by slow peristalsis).
      (1) If restraint of ejaculation is prolonged in spite of repeated sexual stimuli, the muscular walls of the engorged ampullary portion of the vas deferens may spasm, with the resulting pain referred to the perineum.
      (2) The bulbourethral (Cowper's) glands and the urethral glands (of Littré) secrete mucus.
   b. Emission and secretion are involuntary phenomena, regulated by the parasympathetic nervous system.
3. **Female secretion.** Similar stimulation results in **secretion** from cervical glands and vaginal walls, erection of the clitoris with accompanying enlargement of the vestibular bulb, swelling of the vaginal walls, and secretion from vestibular glands.
   a. When the vestibular bulbs expand, they dilate laterally, causing the overlying labia minor to part.
   b. The cervical glands secrete profusely into the upper reaches of the vagina.
   c. The vestibular glands also secrete mucinous material at the introitus, facilitating the entry of the penis.
   d. Increased vascular flow to the vagina produces a fluid transudate across the vaginal walls.

B. **Second (penetration) phase of coitus.** Penetration of the vagina by the erect penis can be accomplished in varied postures (Figure 29-15).
   1. **Penetration.** The postures of coitus are particularly significant in the event of repeated failure to effect conception or the need to make adjustments for a shorter vagina, an unusually long penis, retroversion of the cervix, or obesity.
      a. Full penetration of the erect penis elevates the mobile cervix and creates a pocket in the posterior fornix so that the average 9-cm vagina accommodates a substantial portion of the average 13-cm penis.

b. At the time of penetration and thereafter, male secretions produced by the bulbourethral glands and the glands of the penile urethra are added to the vestibular and cervical secretions.
c. Coital movements accentuate stimulation of the penis, clitoris, and vaginal walls while emotional responses may intensify and glandular secretions may be augmented.

2. **Emission.** Semen is moved by emission into the prostatic and penile urethras. Activated by the fructose in seminal fluid, spermatozoa become motile.

C. Third (orgasmic) phase of coitus. When a particular degree of stimulation is reached and when the partners so desire, the physical and emotional process of orgasm may occur.

1. **In the male,** orgasm usually is concomitant with ejaculation.
    a. Sympathetic outflow produces spasmodic and rhythmic contraction of the vas deferens, seminal vesicles, prostate gland, and urethra, expelling the seminal bolus. Although the onset of ejaculation may be controlled, once started, ejaculation cannot be interrupted.
    b. Ejaculation is accompanied by spasmodic contractions of the bulbocavernosus and ischiocavernosus muscles as well as the muscles of the urogenital diaphragm, pelvic floor, and external anal sphincter, all innervated by the pudendal nerve.
    c. During coitus, the ejaculate (25 ml) is discharged into the upper part of the vagina where, depending on the coital posture, a seminal pool may form. Fibrinogen in the seminal fluid causes a temporary thickening of the seminal pool, which tends to retain the ejaculate within the vagina and promote seminal access to the external cervical os.

2. **In the female,** orgasm is concomitant with contractile spasms of the uterus, apparently initiated by sympathetic nerves of the uterovaginal plexus.
    a. The uterine contractions tend to aspirate ejaculate into the cervical canal.
    b. Additionally, orgasm may involve contraction of the bulbocavernosus and ischiocavernosus muscles as well as the muscles of the urogenital diaphragm, pelvic floor, and external anal sphincters.

D. Fourth (postcoital resolution) phase of coitus

1. **Resolution**
    a. Because the sympathetic outflow to the genital areas is dominant during the orgasmic phase in both sexes, there is less parasympathetic activity. This situation increases the tone in the periarterial muscle, thereby reducing blood flow to the erectile tissues of the penis, clitoris, and vestibular bulbs. The erectile tissue quickly becomes flaccid.
    b. Usually, but not always, the male (and perhaps the female) enters into a refractory period of variable duration during which a second erection is not possible. This inability to respond is probably attributable to the predominance of sympathetic innervation during orgasm and high circulating epinephrine levels after orgasm that override parasympathetic effects.

2. **Conception.** Without barrier contraception, the sperm cells enter the cervical canal. Within approximately 30–60 minutes at a swim rate of 2 mm per minute, sperm traverse the uterine cavity and enter the ostia of the uterine tubes.
    a. Once in the uterine tube, the sperm become capacitated (attain the ability to penetrate an ovum) by tubal secretions.
    b. Somewhere in the distal one third of the uterine tube, an ovum, if present, may be fertilized by penetration of one sperm cell.

E. Implantation

1. **Uterine tube transit.** Three to four days after fertilization, the zygote reaches the uterus.

FIGURE 29-16. Fetal development at approximately 35 weeks. Note the changed relationships within the abdominal cavity.

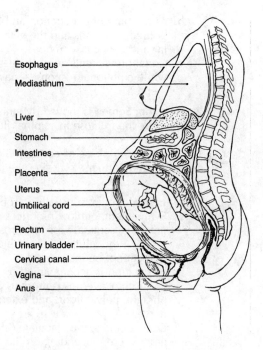

2. **Uterine implantation.** At this time, if the endometrium is in the secretory phase and no intrauterine device is in place, implantation may occur.
    a. With the development of the chorionic membranes, human chorionic gonadotropin maintains the corpus luteum so that the endometrium remains in the secretory stage.
    b. Development and growth of the fetus require 38–40 weeks (Figure 29-16).

PART IIIC PELVIS

STUDY QUESTIONS

DIRECTIONS: Each of the numbered items or incomplete statements in this section is followed by answers or by completions of the statement. Select the ONE lettered answer or completion that is BEST in each case.

Questions 1–4

During a prenatal examination, a 23-year-old woman (gravida 1, para 1) in her third trimester tells her obstetrician that she is bothered by hemorrhoids.

1. Hemorrhoids frequently develop during the later stages of pregnancy, in part because of enlargement of normal anastomotic connections between the middle rectal (hemorrhoidal) vein and which of the following veins?

(A) External pudendal
(B) Internal pudendal
(C) Obturator
(D) Umbilical
(E) Uterine

2. The obstetrician notes external hemorrhoids and, knowing that they will probably resolve after delivery, prescribes a topical ointment for palliation. External hemorrhoids develop in branches of the hemorrhoidal vein and characteristically are

(A) larger than internal hemorrhoids
(B) superior to the pectinate line
(C) painful
(D) prone to profuse bleeding

3. The obstetrician determines the position of the fetus and estimates the size, comparing the anticipated birth weight with the pelvic structure of the mother. Distinguishing features of the female pelvis include all of the following EXCEPT

(A) a pelvic outlet that is larger than that in the male
(B) a narrow subpubic angle between inferior pubic rami compared to a wide angle in the male
(C) an oval pelvic inlet compared to a heart-shaped inlet in the male
(D) a shallow false pelvis with flaring of the ilia compared to the deep false pelvis with more vertical ilia in the male

4. The obstetrician approximates the diagonal conjugate diameter of the pelvic inlet along an index finger and the transverse diameter of the pelvic outlet against a fist. He concludes that cesarean section is likely, and orders a radiographic evaluation of the pelvis to obtain exact measurements. The dimension of the birth canal that usually is most critical in determining whether delivery will proceed vaginally is the

(A) anteroposterior diameter (from the inferior margin of the pubic symphysis to the sacrococcygeal joint)
(B) diagonal conjugate (between the inferior margin of the pubic symphysis and the sacral promontory)
(C) obstetric conjugate diameter (from the widest point of the pubic symphysis to the sacral promontory)
(D) transverse midplane diameter (between the ischial spines)
(E) transverse outlet diameter (between the ischial tuberosities)

(end of group question)

5. Which of the following statements about the bulbourethral glands (of Cowper) is true?

(A) They are homologous with the paraurethral glands (of Skene)
(B) They contribute fructose to seminal fluid
(C) They drain into the membranous urethra
(D) They secrete prior to and during ejaculation

Questions 6–11

A 36-year-old man has occasional dull, throbbing pain associated with the right testis and scrotum. Physical examination reveals a varicocele of the pampiniform plexus. The examining physician remarks to the patient that he probably has had the condition for most of his adult life and that it should not be a bother to him. The patient states emphatically that it has arisen within the last few months.

6. Factors that might be considered by the examining physician include the fact that varicocele of the pampiniform plexus on the right side is associated with

(A) a long, redundant mesorchium
(B) hydrocele processus vaginalis
(C) situs inversus abdominis
(D) testicular torsion

7. The physician is inclined to pass off the complaint, but because the patient continues to complain of pain, he schedules the patient for surgery to ligate the testicular vein. In addition to the vas deferens, structures normally found within the internal spermatic fascia of the spermatic cord include all of the following EXCEPT the

(A) deferential artery
(B) genitofemoral nerve
(C) pampiniform venous plexus
(D) testicular artery
(E) testicular nerves

8. Early in the operation, an incision is made above and parallel to the inguinal ligament. Exploration of the iliac fossa reveals a dilated right testicular vein. The surgeon decides to ligate the testicular vein just cranial to the deep inguinal ring. This procedure would likely

(A) abolish the cremasteric reflex
(B) eliminate pressure by the column of venous blood
(C) reduce blood flow through the testes
(D) reduce countercurrent heat exchange
(E) reduce sperm viability

9. The incision is extended to mobilize the ascending colon. In the vicinity of the lower pole of the right kidney, the surgeon observes a retroperitoneal mass that is encroaching on the right testicular vein and impeding the venous return from the right testis. The right testicular vein normally drains into the

(A) hepatic portal vein
(B) inferior vena cava
(C) internal iliac vein
(D) right renal vein
(E) right suprarenal vein

10. Because the tumor appears to be of renal origin, a computed tomography scan, an arteriogram, and an intravenous pyelogram should be performed before attempting its removal. A needle biopsy of the mass is performed, however, and then the testicular vein is ligated as planned to relieve the pain associated with the varicocele. After ligation of the testicular vein, venous blood from the testes will return via all of the following collateral pathways EXCEPT the

(A) cremaster veins
(B) external pudendal veins
(C) deep dorsal veins of the penis
(D) posterior scrotal veins
(E) veins of the vas deferens

11. Pathologic examination reveals that the kidney mass is benign, but excision is recommended at subsequent admission. During a postsurgical physical examination, the surgeon tests the cremaster reflex of the patient by scratching the inner thigh. Which of the following statements correctly pertains to the cremaster reflex?

(A) Its afferent limb is mediated by the obturator nerve
(B) It tests the integrity of the S2–S4 spinal levels
(C) Its efferent limb causes the scrotum to contract
(D) Its efferent limb is mediated by the femoral branch of the genitofemoral nerve

(end of group question)

Questions 12 and 13

A fit and trim 66-year-old man complains of slow initiation and increased frequency of urination.

12. The examining physician suspects benign prostatic hypertrophy, which results in obstruction of the prostatic urethra by enlargement of the

(A) anterior lobe
(B) entire prostate gland
(C) median lobe
(D) posterior lobe

13. Because the differential diagnosis also includes prostate cancer, a rectal examination is performed to determine the size and consistency of the prostate. All of the following may be palpated by digital examination per rectum in the male EXCEPT

(A) disease of a seminal vesicle
(B) evidence of urinary retention in the bladder
(C) the lateral lobes of the prostate gland
(D) the median lobe of the prostate gland

(end of group question)

14. Which of the following statements correctly pertains to the ischiocavernosus muscle in both the male and the female?

(A) Its contraction is the principal mechanism of erection
(B) It inserts into the central tendon of the perineum
(C) It lies in the superficial perineal space
(D) It receives its motor innervation from the nervi erigentes

Questions 15 and 16

At a checkup with her gynecologist, a 43-year-old woman (gravida 3, para 2) comments on occasional involuntary urine discharge. The physician commences examination of the vulva.

15. All of the following statements describing the vestibule in the female perineum are true EXCEPT

(A) It is bordered by the labia minora
(B) It receives both the urethra and the vagina
(C) It receives drainage from the greater vestibular and paraurethral glands
(D) It receives sensory innervation from the nervi erigentes

16. The examiner notes a small urethrocele protruding into the anterior vaginal wall. This finding confirms the diagnosis of stress incontinence. Which of the following statements correctly characterizes the external urethral sphincter?

(A) Its anatomic structure is identical in males and females
(B) It is a portion of the pelvic diaphragm
(C) It is under involuntary control
(D) It receives innervation by the pudendal nerve

(end of group question)

Questions 17–19

A 27-year-old woman with intense lower right quadrant pain is examined in the emergency room. Pain is elicited by abdominal palpation, becoming intense as the examiner nears the right lower quadrant. During bimanual vaginal examination, the pain increases when the uterus is palpated. The examiner notes that an IUD is in place.

17. All of the following pelvic structures refer pain to the L1–L2 dermatomes EXCEPT the

(A) ovaries
(B) paraurethral glands (of Skene)
(C) urinary bladder
(D) uterine body
(E) uterine tubes

18. The patient is referred for vaginal sonography to rule out a luteal cyst or ectopic pregnancy. The sonic probe is placed in the anterior vaginal fornix and aimed anteriorly. Assuming normal anatomy, structures that can be visualized anteriorly, anterosuperiorly, anterolaterally, and anteroinferiorly include all of the following EXCEPT the

(A) body of the uterus
(B) cervix
(C) left and right ovaries
(D) pubic symphysis
(E) urinary bladder

19. Salpingitis is confirmed on the right side. The IUD is removed and the patient is treated with antibiotics. The patient may be at higher risk for a future ectopic tubal implantation because of possible damage to the tubal mucosa. Viable fertilization of the ovum normally occurs in the

(A) ampulla of the oviduct
(B) body of the uterus
(C) fundus of the uterus
(D) peritoneal cavity
(E) primary follicle of the ovary

(end of group question)

DIRECTIONS: Each of the numbered items or incomplete statements in this section is negatively phrased, as indicated by a capitalized word such as NOT, LEAST, or EXCEPT. Select the ONE lettered answer or completion that is BEST in each case.

20. All of the following structures play a significant role in fecal continence EXCEPT the

(A) external anal sphincter
(B) internal anal sphincter
(C) ischiorectal fat
(D) puborectalis muscle

21. All of the following statements correctly pertain to the ischioanal fossa EXCEPT

(A) It communicates with the superficial perineal space
(B) It contains inferior rectal (hemorrhoidal) vessels and nerves
(C) It extends between the pelvic and urogenital diaphragms
(D) It is continuous with the superficial (fatty) layer of superficial fascia of the perineum and abdominal wall
(E) It is filled with adipose tissue traversed by irregular connective tissue septa

22. The male superficial perineal space includes all of the following structures EXCEPT the

(A) bulbospongiosus muscle
(B) bulbourethral glands
(C) corpus spongiosum
(D) ischiocavernosus muscle
(E) penile urethra

Study Questions   Part IIIC Pelvis | **483**

DIRECTIONS: Each set of matching questions in this section consists of a list of four to twenty-six lettered options (some of which may be in figures) followed by several numbered items. For each numbered item, select the ONE lettered option that is most closely associated with it. To avoid spending too much time on matching sets with large numbers of options, it is generally advisable to begin each set by reading the list of options. Then, for each item in the set, try to generate the correct answer and locate it in the option list, rather than evaluating each option individually. Each lettered option may be selected once, more than once, or not at all.

Questions 23–26

(A) Parasympathetic division
(B) Sympathetic division
(C) Both
(D) Neither

For each pelvic function, select the appropriate division of the autonomic nervous system that is most likely responsible.

23. Sensation from the glans penis

24. Clitoral erection

25. Male ejaculation

26. Detrusor reflex

Questions 27–29

(A) Prostate gland
(B) Seminal vesicles
(C) Both
(D) Neither

For each characteristic of the male reproductive tract, select the structure that is most likely associated with it.

27. Produces seminal fluid

28. Site of spermatozoa maturation and storage

29. Homologous to the paraurethral glands

Questions 30–32

(A) Labia majora
(B) Vagina
(C) Vestibular bulbs
(D) Vestibule

For each male pelvic structure, select the most appropriate female homologue.

30. Scrotum

31. Penile (cavernous) urethra

32. Corpus spongiosum

ANSWERS AND EXPLANATIONS

1. The answer is B [Chapter 27 IV B 2 a; Chapter 29 I C 2 b]. The middle rectal vein anastomoses with the superior rectal branch of the inferior mesenteric vein and with the inferior rectal branch of the internal pudendal vein. Venous compression caused by the fetus promotes development of alternative drainage pathways though existing anastomoses. Similar anastomotic connections usually occur between the superior, middle, and inferior rectal arteries.

2. The answer is C [Chapter 28 III C 2 c]. The portion of the anal canal inferior to the inferior pectinate line is innervated somatically by the inferior rectal branches of the pudendal nerve. Hemorrhoids in this region are painful and demand early attention. Internal hemorrhoids, above the pectinate line, are large and painless and tend to bleed profusely, frequently causing anemia.

3. The answer is B [Chapter 27 I B 3 b; Table 27-1]. The female pelvis, in comparison with that of the male, is larger and has a shallower false pelvis, a more oval inlet, a larger outlet, and a wider inferior pubic angle.

4. The answer is D [Chapter 27 I B 2 a (2) (a), c (2)]. The transverse midplane diameter may be estimated from the estimated transverse diameter and measured radiographically. This is the most critical of the pelvic dimensions in determining craniopelvic disproportion. The obstetric conjugate can be estimated from the diagonal conjugate.

5. The answer is D [Chapter 28 IV D 2 c; V A 1 b (3) (d); Chapter 29 VII B 1 b]. The bulbourethral glands, lying in the deep perineal space, discharge into the penile urethra before and during ejaculation. The prostate gland is homologous to the paraurethral glands in the female. The fructose component of seminal fluid is a product of the seminal vesicles.

6. The answer is C [Chapter 28 IV A 3 b (3) (b); Chapter 29 IV A 3 b (1)]. Varicocele of the pampiniform plexus on the right side is uncommon because this condition usually results from compression of the testicular vein by a full sigmoid colon. In individuals with situs inversus, in whom the sigmoid colon is on the right side, the varicocele also would be on the right side.

7. The answer is B [Chapter 28 IV A 3 b (2) (b), (3)]. The testicular neurovascular bundle, the ductus deferens, and a remnant of the processus vaginalis lie within the internal spermatic fascia of the spermatic cord. The genital branch of the genitofemoral nerve lies in the cremaster layer beneath the external spermatic fascia.

8. The answer is B [Chapter 29 IV A 3 b (1)]. Testicular vein ligation at the inguinal ligament eliminates the hydrostatic pressure produced by the column of venous blood in the abdomen. Blood return from the testis is through anastomotic pathways. Varicocele produces venous thrombosis and stasis, which interferes with the countercurrent heat exchange mechanism and thereby raises scrotal temperature and decreases sperm viability. Ligation reverses this effect.

9. The answer is B [Chapter 28 IV A 3 b (2)]. The right testicular vein normally drains into the inferior vena cava. The left testicular vein joins the left renal vein.

10. The answer is C [Chapter 28 IV A 3 b (3)]. The collateral venous drainage of the testis is primarily along the veins of the vas deferens, but anastomoses occur with the external pudendal, posterior scrotal, and cremaster veins (branches of the veins of the anterior abdominal wall). The deep dorsal vein of the penis, draining the deep structures of the penis, does not offer the possibility for anastomotic connections.

11. The answer is D [Chapter 28 IV A 3 b (2)]. The cremaster reflex (elevation of the testis within the scrotum when the inner thigh is stroked) tests the integrity of the L1–L2 spinal levels. The afferent limb is by the femoral branch of the genitofemoral nerve; the efferent limb by the femoral branch of the genitofemoral nerve. Cold causes contraction of the smooth muscle of the dartos layer to reduce heat loss.

12. The answer is C [Chapter 29 IV E 1 c (3) (a)]. The median (middle) lobe is most commonly involved in proliferative hyperplasia (benign prostatic hypertrophy), which obstructs the prostatic urethra with retention of urine in the bladder. The posterior lobes are typically involved in malignant transformation. The anterior and lateral lobes are synonymous and are not prone to malignant transformation.

13. The answer is D [Chapter 29 IV F 1]. The lateral and (more importantly) the posterior lobes of the prostate gland, as well as a distended urinary bladder and swollen seminal vesicles, are palpable per rectum. The median lobe of the prostate, which surrounds the urethra, cannot be palpated directly. A hard, nodular prostate may indicate carcinomatous involvement, whereras a large, firm prostate is suggestive of benign prostatic hypertrophy. A large, soft prostate is associated with prostatitis.

14. The answer is C [Chapter 28 IV C 2 c; Chapter 29 VII A 1 b]. The ischiocavernosus muscle lies in the superficial pouch. It contracts periodically under voluntary control, being innervated by the perineal branch of the pudendal nerve. It inserts just distal to the point at which the crura join to form the body of the penis or, in the female, into the margin of the pubic arch and into each crus of the clitoris. The bulbocavernosus muscles insert into the central tendon.

15. The answer is D [Chapter 28 V A 1 b (3); F 1 a (2)]. The vestibule is innervated by perineal branches of the pudendal nerve.

16. The answer is D [Chapter 28 IV D 2 b (2); V D 2 b (1) (a)]. The external urethral sphincter, that portion of the deep transverse perineal muscle adjacent to the urethra, is innervated by the perineal branch of the pudendal nerve and is, therefore, under voluntary control. In the male, the external urethral sphincter completely surrounds the membranous urethra. In the female, because the urethra is embedded in the anterior vaginal wall, this sphincter is incomplete and less effective. The external urethral sphincter is part of the urogenital diaphragm.

17. The answer is B [Chapter 29 VI D; Table 29-1]. The paraurethral and bulbourethral glands are perineal structures with pain mediated directly by the pudendal nerve. The cervix refers pain to dermatomes S2–S4. No visceral structures refer pain to dermatomes L3–S1 because there are no white rami or pelvic splanchnic nerves associated with these spinal nerves.

18. The answer is B [Chapter 29 V B 2; C 2 d (1); D 2 a; Figure 29-9]. The cervix lies posterosuperior to the anterior fornix. With the probe in the anterior fornix, the urinary bladder, the anteflexed-anteverted uterus, the ovaries, and the pubic symphysis lie in the anterior sonic field.

19. The answer is A [Chapter 29 V C 1 e (3); VII D 2 b]. Fertilization usually occurs in the ampullary portion of the uterine tube. During the subsequent four days, the zygote moves toward the uterus while undergoing numerous divisions that enable implantation of the resultant morula. Fertilization may occur in the ovary (rarely) or peritoneal cavity, which may result in ectopic implantation. Although fertilization may occur in the proximal portions of the uterine tube or even uterus, the zygote is expelled before it reaches the implantable morula stage.

20. The answer is B [Chapter 27 II A 2 b (3); Chapter 28 III A 2 a, B 3 a (1), (2); E 1]. Prolonged contraction of the puborectalis portion of the pubococcygeus muscle establishes the carrying angle, which precludes emptying of the rectum. The external anal sphincter can be voluntarily contracted upon the approach of a peristaltic wave, but only for about 15 seconds. Compression of the anal canal by the ischioanal fat as a result of gluteus maximus contraction assists fecal continence. The internal anal sphincter relaxes upon the approach of a peristaltic wave and, as such, is not a major factor in fecal continence.

21. The answer is A [Chapter 28 III A 1, 2; VI B 2 e]. The fat that fills the ischioanal fossa is a continuation of the superficial layer of superficial fascia of the perineum and abdominal wall. Within this layer are the inferior rectal (hemorrhoidal) neurovascular bundles. The anterior horns of the ischioanal fossa extend between the pelvic and urogenital diaphragms. The deep layer of superficial fascia (Colles' f.) attaches to the posterior edge of the urogenital diaphragm, preventing communication between the ischioanal fossa and the superficial perineal space.

22. The answer is B [Chapter 28 IV C 2, D 2 c]. The superficial perineal space contains the root of the penis (i.e., the bulb of the corpus spongiosum and crura of the corpora cavernosa) and the associated bulbospongiosus and ischiocavernosus muscles. The bulbourethral glands lie in the deep perineal space.

23–26. The answers are 23-D, 24-A, 25-B, 26-A [Chapter 28 III B 3 a (2); IV F 1; Chapter 29 II B 4 b, 5 b; VI A–C; VII A 1 a, C 1 a]. The pudendal nerve mediates sensation from the external genitalia and provides motor control for the perineal musculature, including the external anal and urethral sphincters. Erection in both men and women is mediated by parasympathetic nerves, which control the blood flow to the erectile tissues. Ejaculation in males is a sympathetic event, as are the uterine contractions that accompany female orgasm. The detrusor (bladder-emptying) reflex is mediated by visceral afferents and parasympathetic efferents traveling along the nervi erigentes.

27–29. The answers are 27-C, 28-D, 29-A [Chapter 29 II C 1 a (1) (b) (i); III B–D; IV B 1 a; D, E]. Both the seminal vesicles and the prostate gland contribute significantly to the seminal fluid. The epididymis is the site of sperm storage and maturation. The prostate is homologous to the paraurethral glands (of Skene) in the female.

30–32. The answers are 30-A, 31-D, 32-C [Chapter 28 II B, C; IV A 1 b (2); V A 1; C 2 b]. Both the scrotum in men and the labia majora in women develop from the labial-scrotal folds. In men, fusion of the urogenital (urethral) groove and vestibular bulbs forms the penile urethra and corpus spongiosum, respectively. In women, the urogenital groove remains as the vestibule flanked by the vestibular bulbs underlying the labia minora.

SECTION IV
HEAD AND NECK

PART IVA SOMATIC NECK AND NEUROCRANIUM

Chapter 30
Posterior Cervical Triangle

I. CERVICAL FASCIA

A. **Introduction.** The head and neck divides developmentally, anatomically, functionally, and medically into an anterior **visceral portion** and a posterior **somatic portion.**

   1. **The somatic portion** comprises the neurocranium and the cervical vertebral column with the associated musculature.

   2. **The visceral portion** comprises derivatives of the upper end of the primitive gut and respiratory tract with its associated branchial (gill) structures.

B. **Divisions**

   1. **The superficial fascia** in the head and neck contains loose connective tissue.
      a. The **cleavage lines** (Langer's lines) run horizontally around the neck. A transverse incision scar is almost invisible.
      b. The more superficial portions contain variable amounts of adipose tissue.
      c. The deeper portions enclose the voluntary muscles of facial expression. This group is represented by the **platysma muscle** in the neck.
      d. In the head and face, the superficial fascia frequently is continuous with and inseparable from the deep fascia overlying bone.

   2. **The deep cervical fascia** surrounds the neck and gives off septa that separate the visceral and somatic portions. This fascial layer can be divided into four parts.
      a. **Superficial (investing) layer**
         (1) This layer ensheathes the neck beneath superficial fascia.
            (a) **Posteriorly,** it attaches to the ligamentum nuchae.
            (b) **Superiorly,** it attaches along the mandible, mastoid process, and ligamentum nuchae.
            (c) **Inferiorly,** it finds attachment along the acromion, the clavicle, and the manubrium sterni.
         (2) This layer splits several times to individually ensheathe the trapezius, omohyoid, and sternomastoid muscles (see Figure 30-1).
            (a) Anteriorly between the sternomastoid muscles, the split layers occasionally fail to reunite, producing the small **suprasternal space** (of Burns) that contains the anterior jugular veins and an occasional lymph node.
            (b) Superiorly, this layer splits to ensheathe the parotid gland and the submandibular glands.
      b. **The prevertebral layer** encircles the vertebral column and its associated muscles, defining the somatic portion of the neck (Figure 30-1).
         (1) After originating from the cervical spinous processes, it passes between the trapezius and the intrinsic muscles of the cervical spine, to insert onto the transverse processes of the cervical vertebrae. Between the spinous processes and transverse processes, this fascial layer encloses and ensheathes the semispinalis and splenius muscles.
         (2) In the posterior triangle, the prevertebral layer is apposed to the superficial layer of deep cervical fascia.
            (a) The spinal accessory nerve (CN XI) and lymphatics lie between these two layers.
            (b) Low in the posterior triangle, the prevertebral layer invests the scalene muscles, leaving a space between them and the superficial layer of the

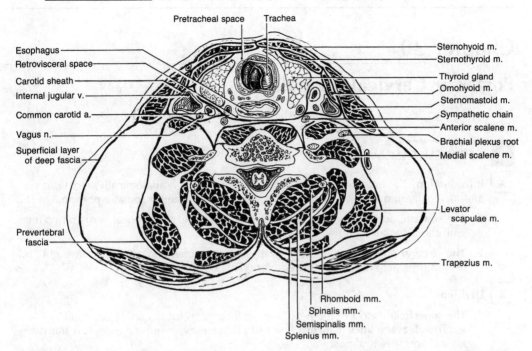

FIGURE 30-1. Deep cervical fascia. The deep layer of deep cervical fascia invests the trapezius and sternomastoid muscles. The pretracheal layer invests the visceral structures. The carotid sheath contains the carotid arteries, internal jugular vein, and vagus nerve. The prevertebral layer invests the musculature of the somatic neck (vertebral muscles).

deep cervical fascia that contains the subclavian and external jugular veins, the transverse cervical and suprascapular arteries, and the inferior belly of the omohyoid muscle.
   - **(c)** The **axillary sheath** is formed by an extension of the prevertebral fascia about the roots of the brachial plexus as they pass between the anterior and middle scalene muscles to reach the upper limb.
- **(3)** Anteriorly, the prevertebral fascia attaches to the transverse processes of the cervical vertebrae. It splits into two leaves.
   - **(a)** The deeper leaf inserts onto the anterior aspect of the cervical vertebrae and intervertebral disks to enclose the longus coli muscles. The more superficial (alar) leaf jumps to the contralateral transverse process (see Figure 30-1).
   - **(b)** The compartment between these two layers of prevertebral fascia is called the **"danger space."**
- **(4)** Inferiorly, the prevertebral fascia extends through the posterior mediastinum into the abdomen.
- **c.** The **pretracheal layer** is a middle fascial layer that originates as septa from the superficial layer of the deep fascia. This layer invests the infrahyoid (strap or ribbon) muscles and the trachea (see Figure 30-1).
   - **(1)** Superiorly, it attaches to the cricoid cartilage. Inferiorly, it continues into the middle mediastinum, where it fuses with the fibrous pericardium.
   - **(2)** Anteriorly, it splits to surround and support the thyroid gland.
- **d.** The **carotid sheath** is a fascial condensation about the carotid arteries, the internal jugular vein, and the vagus nerve (see Figure 30-1).
   - **(1)** The pretracheal fascia and the superficial layer of the deep cervical fascia both send septa to the carotid sheath.
   - **(2)** The sympathetic chain and superior cervical ganglion lie posterior to the carotid sheath, anterior to the prevertebral fascia.

3. **Cervical spaces** are formed by the planes of deep fascia. Extension of infection tends to be limited by these fascial planes but may track up and down the neck within the resultant spaces.
   a. The **visceral compartment** is that region bounded by the pretracheal and prevertebral fascial planes. It is subdivided into two spaces.
      (1) The **pretracheal space** is deep to the pretracheal fascia and surrounds the trachea and thyroid gland, but is anterior to the esophagus (see Figure 30-1). It descends into the superior mediastinum. Infection within this space may come to the surface in the suprasternal space or may track into the superior mediastinum.
      (2) The **retropharyngeal (retrovisceral) space** is posterior to the oropharynx and esophagus and anterior to the prevertebral fascia (see Figure 30-1). This space descends into the posterior mediastinum. Infection in the retrovisceral space may track deeply into the thoracic mediastinum.
   b. The **somatic compartment** is that region enclosed by the prevertebral fascia (see Figure 30-1). It is associated with the **"danger space."** Infection within this space can extend caudally from the base of the skull into the abdomen.

II. SOMATIC NECK

A. **The posterior triangle** contains the structures of the somatic neck.
   1. **Boundaries** are the midline posteriorly, the sternomastoid muscle anteriorly, and the clavicle inferiorly.
   2. **Structures** contained in the posterior triangle include muscles and nerves associated with the cervical vertebral column. Many structures are en route to the upper limb.

B. **Cervical vertebrae** (see Chapter 11 III A 1)

C. **Musculature**
   1. **The superficial muscles** of the posterior cervical triangle are enclosed in the superficial layer of the deep cervical fascia and are associated with the pectoral girdle.
      a. **Sternomastoid (sternocleidomastoid) muscle** (Figure 30-2; see Table 6-1)
         (1) **Attachments.** It has a narrow tendinous head from the manubrium sterni and a flat muscular head from the medial one third of the clavicle. It inserts onto the lateral aspect of the mastoid process of the skull.
         (2) **Action.** It rotates and laterally flexes the head (e.g., it apposes the side of the head to the ipsilateral shoulder).
         (3) **Innervation.** The sternomastoid muscle is innervated by the **spinal accessory nerve (CN XI).**
      b. **Trapezius muscle** (see Figure 30-2 and Table 6-1)
         (1) **Attachments.** It arises from the superior nuchal line of the occiput and the external occipital protuberance as well as from the ligamentum nuchae and the spinous processes of vertebrae C7–T12. Its insertion into the pectoral girdle is extensive.
         (2) **Actions**
            (a) The more **superior portions** insert into the distal one third of the clavicle, elevating and rotating the scapula. When the shoulder is fixed, elevation and rotation of the head occur.
            (b) The **middle portion** inserts into the acromion process and the spine of the scapula, retracting the scapula.
            (c) The more **inferior portion** inserts onto the medial part of the spine of the scapula, depressing and rotating the scapula.
         (3) **Innervation.** The trapezius is innervated by the **spinal accessory nerve (CN XI)** as well as by twigs from the ventral rami of **spinal nerves C3–C5.**

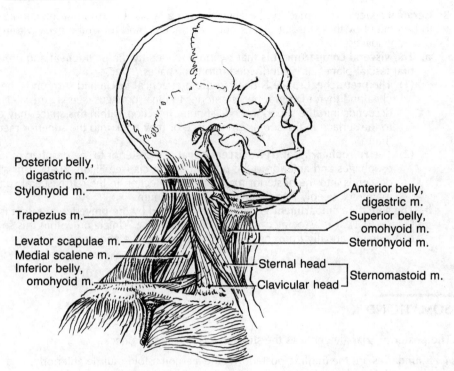

FIGURE 30-2. Superficial cervical musculature.

2. **The intrinsic muscles** lie beneath the prevertebral fascia.
   a. The **lateral group,** including the three **scalene muscles** (anterior, middle, and posterior) and the **levator scapulae,** have clinical importance because of their respiratory functions (Figure 30-3; see Chapter 12 II B 1 and Table 12-3).
   b. The **anterior group,** including the longus colli, longus capitis, rectus capitis anterior, and rectus capitis lateralis muscles, flexes the cervical vertebral column and head (Figure 30-4; see Chapter 12 II C and Table 12-3).
   c. The **posterior group** consists of the spinotransverse, spinospinalis, and transversospinalis groups. This muscle group extends the neck and is innervated by dorsal primary rami of spinal nerves (Figure 30-5; see Chapter 12 I B and Table 12-1).
      (1) **Spinotransverse group.** The splenius muscles constitute the external layer of the intrinsic muscles of the back. Acting bilaterally, this group extends the head and neck. Acting singly, they turn the head toward the ipsilateral shoulder.
      (2) **Sacrospinalis group.** The longissimus capitis and longissimus cervicis muscles, constituting the intermediate layer of intrinsic musculature, run between transverse processes. They act as extensors of the head and neck.
      (3) **Transversospinalis group.** The semispinalis muscles, which form the deep layer of the intrinsic musculature, run between the transverse processes of the thoracic and cervical vertebrae. They act as extensors of the head and neck.
   d. The **suboccipital group** constitutes the deepest posterior layer of intrinsic muscle. This group acts at the atlanto-occipital and atlantoaxial joints to extend and rotate the head (Figure 30-6; see Chapter 12 I B 4).
      (1) The **suboccipital triangle** on each side is bounded by the rectus capitis posterior major and the two obliquus muscles.
      (2) The **suboccipital group** consists of the rectus capitis posterior major, rectus capitis posterior minor, obliquus capitis inferior, obliquus capitis superior muscles (see Figure 30-6 and Table 12-2).

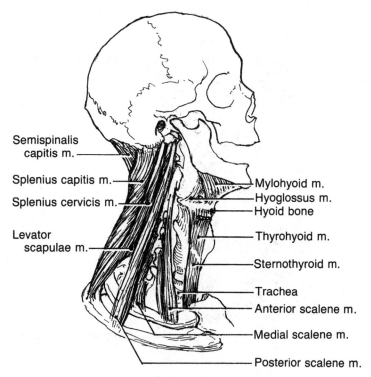

FIGURE 30-3. Lateral cervical musculature.

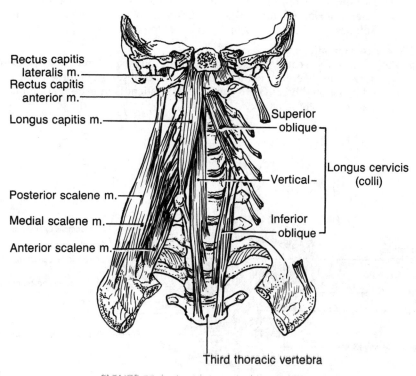

FIGURE 30-4. Anterior cervical musculature.

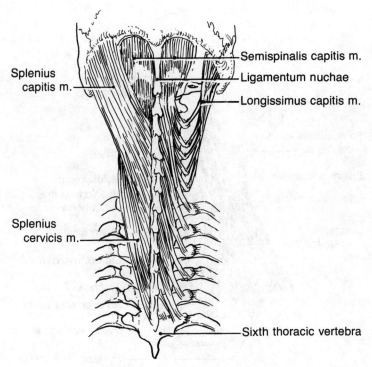

FIGURE 30-5. Posterior cervical musculature. *(Left)* Spinotransverse group; *(right)* transversospinalis group.

e. **Summary of movements**
   (1) **Flexion:** 40°/**extension:** 90°, primarily at the atlanto-occipital joint
   (2) **Lateral flexion (abduction):** ~40°, primarily between vertebrae C2 and C7
   (3) **Rotation:** ±45°, half of which occurs at the atlantoaxial joint
3. **Clinical considerations**
   a. **Respiration.** The **scalene muscles** and the **sternomastoid muscle** assist extreme inspiratory effort by elevating the first and second ribs. Use of these muscles accompanied by labored breathing is a sign that a patient is in extreme respiratory distress, as in an asthma attack.

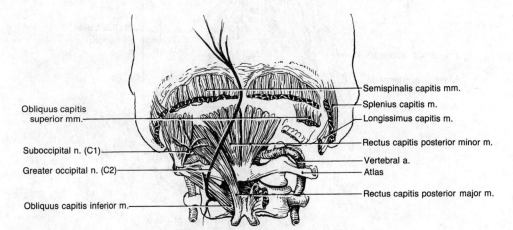

FIGURE 30-6. Suboccipital triangle and musculature.

b. Scalene syndrome. A cervical rib or spasm of the scalene muscles may compress the subclavian artery and brachial plexus where they pass over the first rib. The results are ischemia of the limb and pain along the distribution of the affected nerves.

c. Torticollis

(1) **Spasmodic torticollis** occurs in patients with **sternomastoid** spasm resulting from a viral infection. Full recovery is expected within a few weeks.

(2) **Congenital torticollis** is produced by fibrosis of the sternomastoid, usually following hyperextension injury that occurred during difficult parturition.

d. Central venous lines. The relations of the scalenus anterior are important when inserting a central venous line into the subclavian vein.

(1) The needle must be guided medially, approximating the long axis of the clavicle, to reach the posterior surface where the vein runs over the first rib.

(2) Passing the large bore needle directly perpendicular to the clavicle increases the chance of puncturing the subclavian artery and the cervical pleura.

D. Innervation

1. **The dorsal primary rami** of the cervical nerves innervate the intrinsic musculature of the back and the dorsal portions of the cervical dermatomes.

    a. The **suboccipital nerve** (C1) supplies the suboccipital muscles and, unlike the other spinal nerves, has no sensory root (see Figure 30-6).

    b. The **greater occipital nerve** (C2) pierces the semispinalis and trapezius. It supplies sensory innervation to the posterior regions of the scalp (see Figure 30-6).

    c. The **third (least) occipital nerve** (C3) innervates the posteroinferior portion of the scalp (not shown).

2. **The ventral primary rami** of the cervical nerves pass along a shallow groove on the superior surface of the cervical transverse processes and come to lie in the space between the anterior and middle scalene muscles.

    a. The **ventral primary rami** of nerves C1–C4 intermingle in a predictable pattern to form the **cervical plexus,** which gives rise to cutaneous nerves, muscular branches, and communicating branches (Figure 30-7).

    (1) **Cutaneous nerves** emerge into the posterior triangle at the posterior border of the sternomastoid muscle.

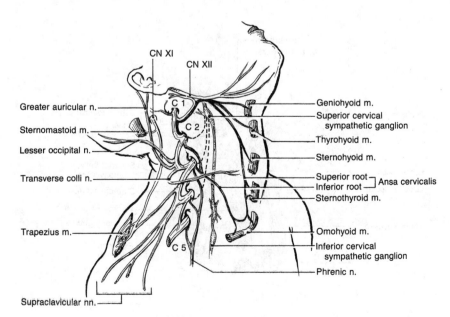

FIGURE 30-7. Cervical plexus.

(a) The **lesser occipital nerve** (C2) ascends to the scalp posterior to the auricle to provide sensation to the inferolateral parts of the scalp (see Figures 30-7 and 30-8).
(b) The **great auricular nerve** (C2–C3) passes vertically by the angle of the jaw before bifurcating about the auricle to provide sensation to the lateral regions of the scalp (see Figures 30-7 and 30-8).
(c) The **transverse cervical (colli) nerve** (C2–C3) crosses the sternomastoid horizontally to mediate sensation from the anterior triangle of the neck (see Figures 30-7 and 30-8).
(d) **Supraclavicular nerves** (C3–C4) supply the dermatomes of the shoulder and upper chest (see Figure 30-7).
(2) **Muscular branches**
(a) Small branches pass from the cervical plexus to muscles arising from the ventral and lateral aspects of the cervical vertebral column as well as to portions of the trapezius, sternomastoid, and diaphragm.
(b) The **ansa cervicalis** is an anastomotic loop superficial to the carotid sheath that innervates the strap muscles (see Chapter 34 VI B 2).
b. The **ventral primary rami** of nerves C4–C8 contribute to the brachial plexus, which is primarily involved with innervation of the upper extremity. Several branches of the brachial plexus course through the posterior triangle.
(1) The **dorsal scapular nerve** (C5, posterior) pierces the scalenus medius muscle, runs posteriorly on the deep surface of the levator scapulae muscle, and descends along the medial border of the scapula to supply the rhomboid muscles (see Figure 6-6).
(2) The **suprascapular nerve** (C5–C6, posterior) runs along the base of the posterior triangle just superior to the clavicle (see Figures 6-6 and 30-9).
(3) The **phrenic nerve** (C3–C5) travels beneath the prevertebral fascia to reach the superior mediastinum and innervate the diaphragm (see Figures 30-7 and 30-9).

3. **Spinal accessory nerve (CN XI)** (Figure 30-8; see Figure 30-9)
   a. **Composition and course**
      (1) Motor rootlets from C1–C5 unite within the vertebral canal and pass superiorly through the foramen magnum into the cranial cavity and exit through the jugular foramen.

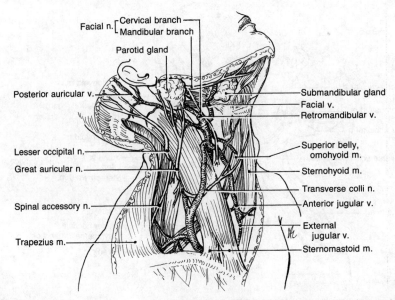

FIGURE 30-8. Superficial neurovascular structures of the posterior cervical triangle.

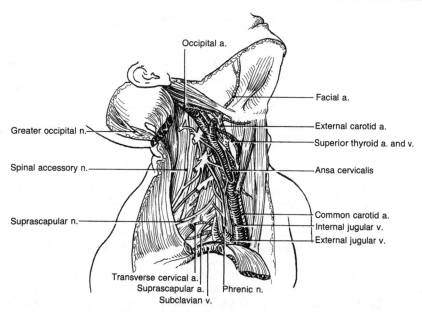

FIGURE 30-9. Deep neurovascular structures of the posterior cervical triangle.

- (2) As this nerve transits the jugular foramen, it is joined for a short distance by a component of the vagus nerve (the so-called **cranial accessory nerve**) before the latter splits off to rejoin the vagus nerve to be distributed as the recurrent laryngeal nerve.
- (3) From the jugular foramen, the spinal accessory nerve courses posteriorly and usually (~75%) passes lateral to the jugular vein.

  b. **Distribution**
  - (1) **From the jugular foramen,** it descends across the medial surface of the stylohyoid and digastric muscles and the deep surface of the **sternomastoid** muscle, to which it sends branches.
  - (2) **In the posterior triangle,** it emerges at the posterior border of the sternomastoid muscle and passes superficial to the levator scapulae to reach and innervate the **trapezius.**
    - (a) Its position approximates the perpendicular bisector of a line joining the mastoid process to the angle of the mandible.
    - (b) It passes deep to the trapezius muscle, to which it sends branches. In addition, the trapezius muscle is variably innervated by twigs of the cervical plexus, some of which join the spinal accessory nerve (see Figure 30-9).

  c. **Clinical considerations**
  - (1) The spinal accessory nerve must be avoided when exploring the posterior triangle. Paralysis of the spinal accessory nerve results in the inability to shrug the shoulder.
  - (2) Swollen lymph nodes in the posterior triangle may irritate this nerve, producing **spasmodic torticollis.**

4. **Cervical sympathetic nerves**
   a. **Presynaptic pathway.** The cervical sympathetic chain runs cranially in the somatic neck posterior to the carotid sheath (see Figure 30-7).
   - (1) **Presynaptic neurons** lie in the lateral cell column of the upper thoracic segments of the spinal cord. These nerve fibers reach the sympathetic chain by way of white rami communicantes associated with the upper thoracic spinal nerves.
   - (2) **Cervical ganglia.** Preganglionic sympathetic fibers run up the chain from upper thoracic levels to synapse in the superior, middle, and inferior cervical ganglia, by **cholinergic transmission.**

(a) The **superior cervical ganglion** is impressively large and sits just below the skull.
(b) The **middle cervical ganglion** may be absent.
(c) The **inferior cervical ganglion** may be conjoined with the first thoracic ganglion as a **stellate ganglion.**

b. **Postsynaptic pathways** lead to the peripheral nerves, the heart, and the head.
   (1) **Postsynaptic neurons** lie in ganglia of the cervical sympathetic chain. The axons are unmyelinated and use **adrenergic neurotransmission.**
   (2) **Gray rami communicantes** carry postganglionic (postsynaptic) nerve fibers from the cervical ganglia to the cervical spinal nerves. These unmyelinated nerve fibers mediate peripheral secretomotor, vasomotor, and pilomotor functions.
   (3) Postganglionic parasympathetic fibers from the **superior cervical ganglion** form a **perivascular plexus** around the internal carotid artery to reach adrenergic effectors in the head.
      (a) **Distribution.** These neurons innervate the pupillary dilators of the iris, the palpebral muscle, and the orbitalis muscle, as well as mediate secretomotor, vasomotor, and pilomotor functions.
      (b) **Lesions** of the cervical sympathetic chain, therefore, produce Horner's syndrome (meiosis, pseudoptosis, enophthalmos, anhidrosis, and flushing).
   (4) **Cardiac accelerator nerves** (cervical splanchnic nerves) carry sympathetic innervation to the heart.

E. **Vasculature**

1. **Arterial supply** is from the external carotid artery and the subclavian artery (Figure 30-9).
   a. The **occipital artery** contributes a profuse blood supply to the scalp (see Figure 34-3).
      (1) **Course.** This large vessel is a branch of the external carotid artery.
         (a) It is crossed near its origin by the hypoglossal nerve. It then passes along the inferior border of the posterior belly of the digastric muscle, deep to the sternomastoid muscle.
         (b) It grooves the medial surface of the mastoid bone, where it may be torn in a fracture of the base of the skull, producing a pathognomonic hematoma—Battle's sign.
      (2) **Distribution.** It emerges from behind the sternomastoid muscle at the apex of the posterior triangle.
   b. The **subclavian artery** arises from the brachiocephalic artery on the right and from the aorta on the left (see Figure 30-9). It gives off three major branches.
      (1) The **vertebral artery** courses through the transverse foramina of vertebrae C6–C1 and the foramen magnum to supply the brain (see Figure 30-6). It also gives off twigs to each cervical vertebrae as well as to the cervical spinal cord.
      (2) **Thyrocervical artery (trunk)** (see Figure 30-9)
         (a) The **inferior thyroid artery** ascends toward the inferior pole of the thyroid gland and gives off the **ascending cervical artery,** supplying deep muscles of the neck and the **inferior laryngeal artery,** which accompanies the recurrent laryngeal nerve.
         (b) The **transverse cervical artery** (when present) passes anterior and lateral to the anterior scalene muscle and across the base of the posterior triangle to reach the edge of the levator scapulae, which divides it into a **superficial branch** and a **deep branch.**
            (i) **Variation.** In about 50% of individuals, there is no transverse cervical artery and the **superficial cervical artery** as well as a **dorsal (descending) scapular artery** arise separately from either the thyrocervical trunk or the subclavian artery.
            (ii) The **superficial branch** of the transverse cervical artery or the **superficial cervical artery** (if present) accompanies the spinal accessory

nerve on the deep surface of the trapezius muscle and anastomoses with the occipital artery.
- (iii) **The deep branch** (dorsal or descending scapular branch) accompanies the dorsal scapular nerve deep to the rhomboids and anastomoses with the circumflex scapular and subscapular arteries.
- (c) The **suprascapular artery** runs laterally with the suprascapular nerve to the shoulder.

2. **Venous return**
   a. The **external jugular vein** is formed at the angle of the jaw by the union of the posterior division of the **retromandibular vein** and the **posterior auricular vein** (see Figure 30-8).
      (1) **Course.** Superficial and variable, the **external jugular vein** passes in the space between the pretracheal and prevertebral fascial layers. At the base of the posterior triangle, it passes posterior to the middle of the clavicle, where it joins the subclavian vein.
      (2) **Distribution.** It receives four tributaries:
         (a) The **suprascapular vein,** which accompanies the like-named artery
         (b) The **superficial cervical vein,** which accompanies the like-named artery
         (c) The **anterior jugular vein** from the anterior triangle
         (d) The **posterior jugular vein** from the apex of the posterior triangle
      (3) **Clinical considerations.** The external jugular vein may be demonstrated by obstructing its drainage with light finger pressure or by having the patient perform the Valsalva maneuver.
         (a) Careful observation of the external jugular usually reveals venous pressure waves.
         (b) The level of the column of blood in the external jugular vein is a useful indication of right atrial pressure.
   b. The **internal jugular vein** drains the head and anterior triangle. It runs within the carotid sheath and receives no branches from the posterior triangle (see Figure 34-4).

3. **Lymphatic drainage**
   a. The **superficial cervical lymph nodes** lie along the external jugular vein.
   b. The **occipital nodes** at the apex of the posterior triangle and the retroauricular nodes over the mastoid process drain the posterior regions of the scalp through the superficial cervical nodes or through the deep cervical nodes (see Figure 34-4).
   c. The numerous **deep cervical nodes** are accessible along the internal jugular vein (see Figure 34-4).
      (1) Most structures of the visceral and somatic portions of the head drain through the deep cervical nodes.
      (2) No physical examination is complete without palpation of the lymph nodes of the neck. A lump in the posterior triangle is a common presentation of malignancy within the lymphatic drainage.
   d. The **thoracic duct** passes anterior to the insertion of the anterior scalene to drain into one of the great veins at the root of the neck on the left side.
   e. The **bronchomediastinal trunks** also pass into the base of the neck and enter the great veins on their respective sides. Tumors of the abdomen and thorax may, therefore, involve the nodes about the subclavian vein.
   f. The **left and right subclavian trunks** in the base of the neck receive drainage from the upper extremity and breast by way of the axillary nodes.

Chapter 31

Neurocranium

I. INTRODUCTION

A. **The skull** is the skeleton of the head, including the mandible (see Figure 31-3).

B. **The cranium** is the portion of the skull without the mandible. The cranium may be divided into an anterior (visceral) portion and a posterior (somatic) portion.

   1. **The neurocranium** is the somatic portion of the cranium. It consists of eight bones (see section III A 1).
      a. **The calvaria** is the vault of the neurocranium. It covers and protects the cerebral hemispheres.
         (1) The brain is protected by meninges, which attach to the calvaria and to the brain (see Figures 31-1 and 31-9).
         (2) The brain is suspended in a pool of cerebrospinal fluid (CSF) confined to the subarachnoid space between arachnoid and pial meningeal layers (see Figure 32-9).
      b. **Cranial nerves** pierce the base of the neurocranium to reach the face and the visceral neck. Conversely, many blood vessels in the face and the visceral neck traverse the bones of the neurocranium to reach the inner aspect of the cranial vault.
      c. **The scalp** is skin and fascia that covers the neurocranium.
   2. **The facial cranium** consists of derivatives of the upper end of the primitive gut and its associated branchial (gill) structures. It consists of 16 bones (see Chapter 35 II B).

II. SCALP

A. **Composition.** The scalp consists of five layers: **s**kin, **c**onnective tissue, **a**poneurosis, **l**oose connective tissue, and **p**eriosteum of the calvaria (mnemonic: scalp)(Figure 31-1).

   1. **The skin** of the scalp has the greatest concentration of hair and sebaceous glands.
      a. Distribution of scalp hair is determined by gender. Androgens produce variable loss of hair in a wide sagittal band from the forehead to the lambdoid suture (male-pattern baldness).
      b. The scalp is a particularly common site for sebaceous cysts, formed by obstructed sebaceous glands.
   2. **The subcutaneous tissue** of the scalp is dense connective tissue that binds the skin strongly to the underlying epicranial aponeurosis.
      a. The vasculature of the scalp runs primarily in the subcutaneous layer. It is rich and widely anastomotic.
      b. **Wounds of the scalp bleed profusely but heal well.**
      c. Swelling from bleeding or edema within this layer is limited by the density of the connective tissue and, therefore, appears as a firm tender lump.
   3. **The occipitofrontalis muscle** and **the epicranial aponeurosis** (galea aponeurotica) constitute the musculotendinous layer of the scalp (see Figure 36-1).
      a. Several flat muscles insert into the epicranial aponeurosis.
         (1) **The occipitofrontalis muscle** consists of the frontalis and occipitalis muscles, together with the common tendon.
            (a) The paired **frontalis muscles** have no direct bony attachment. They arise above the eyebrows within the dense superficial fascia, which is in turn bound by fascial septa to the supraorbital ridges of the frontal bones.

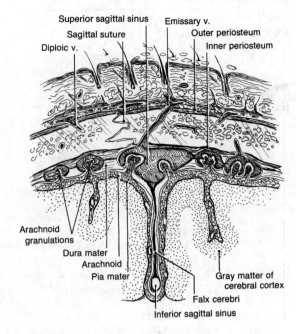

FIGURE 31-1. Scalp, cranium, and meninges in coronal section.

They insert into the epicranial aponeurosis. Contraction of these muscles raises the eyebrows, as in the expression of surprise.
- **(b)** The paired **occipitalis muscles** arise from the highest nuchal line of the occipital bone and the mastoid process of the temporal bone. They insert into the epicranial aponeurosis. Contraction of these muscles draws the scalp posteriorly.
- **(2)** **The anterior and superior auricularis muscles** (but not the posterior auricularis muscle) converge from a broad origin on the epicranial aponeurosis to insert at the base of the auricle.
- **(3)** All scalp muscles are derivatives of primitive second branchiomeric arch musculature and are innervated by the **facial nerve** (CN VII).
- **b.** **The epicranial aponeurosis** is interposed between the frontalis and occipitalis portions of the occipitofrontalis muscle (see Figure 36-1). These muscles place the aponeurosis under tension so that deep transverse lacerations of the scalp gape widely.

4. **Loose connective tissue** between the epicranial aponeurosis and the periosteum permits considerable movement of the scalp and forms the **subaponeurotic space.** This layer contains a rich network of deep arteries and veins.
   - **a. Extracranial hematoma,** the result of bleeding in the subaponeurotic space, can extend over the cranium. It can extend posteriorly, to the highest nuchal line; anteriorly, into the eyelids to produce the "black eye"; and laterally, to the temporal line.
     - **(1)** It frequently occurs in conjunction with normal childbirth, the lumpy clot being resorbed within a few weeks.
     - **(2)** In conjunction with a depressed cranial fracture, the pressure from arterial bleeding into the closed subaponeurotic space can exacerbate the depressed fracture, compressing the brain or even driving bone fragments into the brain.
   - **b. Scalp flaps.** The loose connective tissue layer provides the plane of separation in any injury that tears the scalp from the calvaria or for the surgeon elevating the scalp from the periosteum.
5. **The periosteum** fuses firmly with bone at the sutures and with the periosteum of the adjacent bone, thus limiting the subperiosteal space.

B. Vasculature

1. **Arterial supply** (see Figures 34-3 and 36-5)
   a. The **external carotid artery** provides blood to the scalp through its **occipital, posterior auricular,** and **superficial temporal branches.**
   b. The **internal carotid artery** supplies the anterior scalp by way of the **supraorbital** and **supratrochlear branches** of the **ophthalmic artery.**

2. **Venous return**
   a. **Superficial veins** in the subcutaneous layer generally parallel the arteries (see Figure 36-5).
   b. **Deep veins** in the subaponeurotic space communicate with the diploic veins of the cranium and the dural sinuses within the cranial vault by way of **emissary veins.** Because of the potential for hematogenous spread of infection, the subaponeurotic space has been termed the **"danger space."**

C. Innervation

1. **Somatic innervation** (see Figure 30-6)
   a. The **greater occipital nerve** (from the C2 posterior primary ramus) supplies the posterior regions of the scalp.
   b. The **lesser occipital nerve** (from C2–C3 anterior primary rami) supplies the scalp posterior to the ear.
   c. The **least (third) occipital nerve** (from the C3 posterior primary ramus, not shown) innervates the posterior–inferior portion of the scalp.
   d. The **great auricular nerve** (from C2–C3 anterior primary rami) supplies regions immediately anterior and posterior to the ear.

2. **Branchiomeric innervation.** The anterior half of the scalp (forehead) is innervated by branches of the trigeminal nerve (CN V), which is derived from the nerves to the primitive gills.
   a. **Sensory** (see Figure 36-4)
      (1) The **supraorbital and supratrochlear nerves** arise from the ophthalmic nerve (from CN V1) and supply the forehead.
      (2) The **zygomaticotemporal nerve** (from CN V2) supplies the lateral region of the forehead.
      (3) The **auriculotemporal nerve** (from CN V3) supplies the regions anterior and superior to the ear.
   b. **Motor.** The **facial nerve** (CN VII) innervates the **occipitalis, frontalis,** and **auricular muscles** (see Figure 36-2).

III. BONES AND ARTICULATIONS

A. Basic structure

1. **Composition**
   a. **Three unpaired midline endochondral bones** form the **base of the neurocranium:** the **ethmoid bone** anteriorly, the **sphenoid bone** centrally, and the **occipital bone** posteriorly (see Figures 31-6 and 31-7).
   b. **Paired lateral dermal (intramembranous) bones** form the **calvarium:** the fused **frontal bone** anterolaterally, the **temporal bones** laterally, and the **parietal bones** superiorly and posterolaterally (see Figure 31-3).

2. **Articulations.** Some cranial bones (e.g., the frontal) fuse early in life (synchondroses). Most interdigitate at unyielding **sutures.**

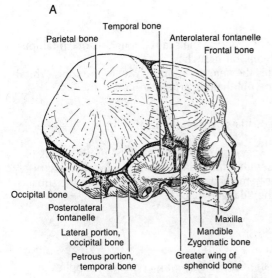
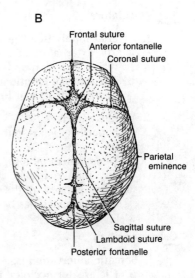

FIGURE 31-2. Skull of the newborn. *(A)* Lateral aspect. Note the relative sizes of the neurocranium and facial cranium, the fontanelles, the absence of the mastoid process, and the shape of the mandible. *(B)* Superior aspect. Note the shape of the anterior and posterior fontanelles.

   a. **In children,** the sutures do not interdigitate extensively, if at all, permitting centrifugal growth of the flat bones (Figure 31-2).
   b. **In adults,** the boundaries of the various bones have little significance but provide useful landmarks.
      (1) The **sagittal (parietal) suture** lies in the midline between left and right parietal bones (see Figure 31-2B). The frontal suture also separates the left and right frontal bones in children. The frontal bones usually are fused in adults.
      (2) The **coronal (frontoparietal) suture** extends laterally between the frontal bone and the left and right parietal bones.
      (3) The **lambdoid (occipitoparietal) suture** extends laterally between the occipital bone and the left and right parietal bones. It may contain one or more small lambdoid bones (see Figures 31-2B and 31-3).
3. **Development and growth**
   a. **Squamous bones** form by intramembranous ossification early in fetal life.
      (1) These plates "float" within the periosteal membranes on the surface of the developing brain and do not articulate with each other.
      (2) During passage through the birth canal, the bony plates ride over one another to some extent, facilitating parturition.
   b. **In infancy,** the bony plates resolve into compact inner and outer **tables** separated by cancellous **diploë**. The bones either fuse with each other at synchondroses or form sutures. In an infant, there are gaps (**fontanelles**) in the calvaria at the rounded corners of the parietal bones (see Figure 31-2).
      (1) The **anterior fontanelle** is diamond-shaped.
         (a) The shape of the anterior fontanelle, which distinguishes it from the posterior fontanelle, enables the obstetrician to determine the orientation of the fetal head by vaginal examination during labor.
         (b) The anterior fontanelle permits access to intracranial veins and enables the perinatologist to monitor intracranial pressure.
      (2) The **posterior fontanelle** is triangular.
      (3) There are also paired **anterolateral fontanelles** at the pterion and **posterolateral fontanelles.**
4. **The marrow cavity** of the diploë contains veins.
   a. **Diploic veins** drain into four main trunks: occipital, posterior temporal, anterior temporal, and frontal diploic veins.

b. Some diploic veins form communicating channels (**emissary veins**) between the superficial veins and the dural sinuses.

B. External surface

1. **The base** of the neurocranium slopes upward from the basal part of the occiput posteriorly, to the ethmoid anteriorly (see Figures 31-4 and 31-6). The oblique surface of the sphenoid and ethmoid bones forms the juncture between the neurocranium and the facial cranium.

2. **Lateral aspect** (Figure 31-3)
   a. **Frontal bone landmarks** (see Figures 31-3 and 35-2)
      (1) The **glabella** is the smooth prominence immediately superior to the bridge of the nose. It marks the **nasion.**
      (2) The **superciliary crest** is a prominent ridge on either side of the midline just superolateral to the glabella.
      (3) A **supraorbital notch** (occasionally foramen) is the point at which the supraorbital artery and nerve leave the orbit and run onto the forehead.
      (4) The **supraorbital margin** continues laterally as the **supraorbital ridge,** which is usually more pronounced in males.
   b. **Temporal bone landmarks** (see Figure 31-3)
      (1) The **articular tubercle** lies just anterior to the condylar notch at the temporal root of the zygomatic frontal process.
         (a) The **condylar notch** and **articular tubercle** form a sigmoidal articular surface for the condylar process of the mandible.
         (b) An **articular disk** is interposed between the bones of the **temporomandibular joint.**
      (2) The **tympanic plate** intervenes vertically between the **condylar notch** and the **external auditory meatus.**
      (3) The **supramastoid crest** is a continuance of the zygomatic arch superior to the external auditory meatus and mastoid process.
         (a) A small **suprameatal spine** lies at the posterosuperior edge of the external auditory meatus. This spine and the supramastoid crest bind the **small suprameatal triangle,** a surgical landmark for the **mastoid antrum.**

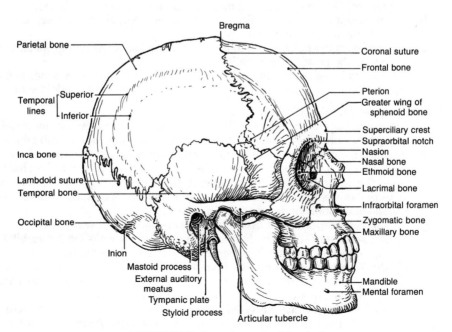

FIGURE 31-3. Lateral aspect of the skull.

(b) The **supramastoid crest** passes along the lateral aspect of the neurocranium as the **temporal line** from which the temporalis muscle originates.
(4) The long **styloid process** projects from the base of the cranium inferior to the external auditory meatus.
  (a) It gives rise to several muscles of the visceral neck, as well as to the stylohyoid and stylomandibular ligaments.
  (b) It is a remnant of the second branchial arch.
(5) The **mastoid process** lies posterolateral to the styloid process.
  (a) This stout process is not present at birth but develops as a result of traction from the sternomastoid muscles. The splenius capitis and longissimus capitis muscles insert onto its posterolateral aspect.
  (b) **Mastoid air cells** fill the mastoid process and communicate with the middle ear cavity through the **mastoid antrum.**
  (c) Foramina for a few emissary veins are posterior to the mastoid process.
c. The **greater wing of the sphenoid bone** fills the small gap between the squamous portion of the temporal bone and the frontal bone (see Figure 31-3).
  (1) The **pterion** is at the H-shaped suture where the wing of the sphenoid bone joins the parietal, frontal, and temporal bones. It lies two finger-breadths above the midpoint of the zygomatic arch. In infants, it is the location of the **anterolateral fontanelle.**
  (2) The **pterygoid processes** arise from the inferior surface of the greater wing of the sphenoid bone.

3. **Posterior aspect**
   a. The **external occipital protuberance (inion)** and the vertical **external occipital crest,** extending in the sagittal plane to the foramen magnum, mark the attachment of the ligamentum nuchae to the occipital bone (see Figures 31-3 and 31-4).
   b. The **superior nuchal line** connects the inion with the mastoid process and gives origin to the trapezius muscle.
   c. The **inferior nuchal line** divides the insertions of the splenius capitis muscle from the insertion of the posterior rectus capitis muscle.
   d. The **highest nuchal line** provides the origin for the occipital belly of the occipitofrontalis muscle.

4. **Superior aspect**
   a. **Sutures** include the **coronal, sagittal,** and **lambdoid** (see section III A 2)
   b. The **bregma** is the point at which the sagittal and coronal sutures meet. It is the location of the anterior fontanelle.
   c. The **lambda** is the point at which the sagittal and lambdoid sutures meet. It is the location of the posterior fontanelle.
   d. **Sutural (lambdoid, wormian, or Inca) bones** are small squamous bones that occasionally form within sutures. They are most common in the lambdoid suture.

5. **Inferior aspect** (Figure 31-4; see Figure 31-5)
   a. **Anteriorly,** the midline bones of the base of the neurocranium roof the nasal cavity, and the corrugated orbital plates of the **frontal bones** roof the orbital cavities.
   b. **Laterally,** the petrous portions of the **temporal bones** consist of two columns of dense bone that abut the anterolateral edges of the basal part of the occipital bone. The greater wings of the unpaired **sphenoid bone,** together with the squamous portions of the temporal bones, form the infratemporal fossae on each side.
     (1) **Sphenoid bone landmarks** (see Figure 31-5)
       (a) The inferior orbital fissure lies between the greater wing of the sphenoid and the maxilla. The inferior orbital fissure transmits the **infraorbital branch of the maxillary nerve** (CN V2) and the infraorbital vessels.
       (b) **Pterygoid processes** arise from the inferior surface of the greater wing of the sphenoid bone. They form the lateral boundaries of the **choanae,** which demarcate the nasal cavity from the nasopharynx.
         (i) The **medial pterygoid plate** gives rise to the superior constrictor of the pharynx and terminates in the **pterygoid hamulus,** which functions as a trochlea for the tendon of the tensor veli palatini muscle.

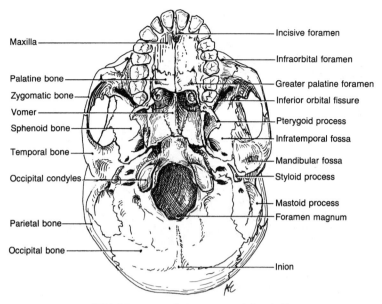

FIGURE 31-4. Inferior aspect of the skull.

        (ii) The **lateral pterygoid plate** gives origin to the pterygoid muscles (lateral and medial).
        (iii) A **scaphoid fossa** lies posteriorly, between the medial and lateral pterygoid plates.
    (c) The **pterygopalatine fossa** is a small, inverted pyramidal space deep within the infratemporal fossa through which arteries and nerves gain access to the nasal cavity, the palate, the orbit, and the face (see Figure 36-9).
        (i) The **pterygomaxillary fissure** opens into the pterygopalatine fossa.
        (ii) It is bounded by the **maxillary tuberosity** anteriorly, by the **pterygoid process** of the sphenoid bone posteriorly, and by the **infratemporal crest** of the sphenoid bone superiorly.
        (iii) Medially, the thin perpendicular plate of the palatine bone separates it from the nasal cavity.
    (d) **Three foramina** lie parallel to the posterior border of the sphenoid bone.
        (i) The **pterygoid (vidian) canal** at the base of the **medial pterygoid plate** carries the **vidian nerve (nerve of the pterygoid canal)** to the **pterygopalatine fossa.** The vidian nerve is formed by conjoining of the **greater superficial petrosal nerve** and the **deep petrosal nerve.**
        (ii) The **foramen ovale** transmits the **mandibular division of the trigeminal nerve** (CN V3), together with the accessory meningeal artery, the lesser superficial petrosal nerve, a recurrent meningeal branch of the trigeminal nerve, and an emissary vein.
        (iii) The **foramen spinosum** (posterolaterally) transmits the **middle meningeal artery.** Fracture of the skull that passes through the foramen spinosum may result in extracranial hemorrhage or intracranial bleeding (epidural hematoma).
(2) **Temporal bone landmarks.** The petrous portion of the temporal bone and the adjacent edges of the squamous portions of the temporal bone contain three nearly parallel rows of numerous foramina (Figure 31-5).
    (a) The **lateral row** lies along the anterior edge of the petrous portion of the temporal bone.
        (i) The **middle ear cleft** continues anteriorly as the **tympanic (auditory, eustachian) tube.** The bony portion emerges from the anteromedial

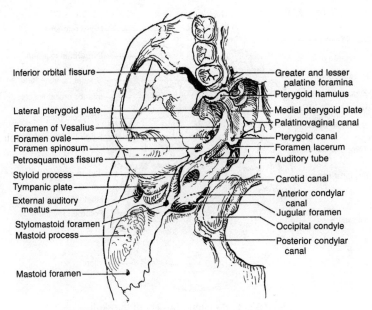

FIGURE 31-5. Landmarks of the infratemporal fossa.

end of the tympanic plate. It is continued by a cartilaginous portion, which runs in a groove along the anterolateral margin of the petrous portion of the temporal bone.
  (ii) The **tympanic plate** forms the sheer vertical anterior wall of the external auditory meatus.
  (iii) The **mastoid process** is grooved on its medial side for the digastric muscle and for the occipital artery.
(b) The **intermediate row** lies in the center of the petrous portion of the temporal bone.
  (i) The **petrosquamous (tympanosquamosal) fissure** lies between the petrous and squamous portions of the temporal bone. Its medial portion, the **petrotympanic fissure,** transmits the **chorda tympani nerve.**
  (ii) The **stylomastoid foramen** lies between the mastoid and styloid processes. The termination of the **facial canal,** it transmits the **facial nerve** (CN VII).
  (iii) The **styloid process** is slender, pointed, and of variable length. The shaft gives origin to the stylohyoid, stylopharyngeus, and styloglossus muscles. The tip gives rise to the stylomandibular and stylohyoid ligaments.
(c) The **medial row** lies along the posterior edge of the petrous portion of the temporal bone.
  (i) The **foramen lacerum** results from superior and inferior defects in the carotid canal at the ragged apex of the temporal bone. As a result, the petrous temporal bone fails to meet the sphenoid bone. This foramen is covered by a layer of fibrocartilage and periosteum. With the exception of the greater superficial petrosal nerve, no structures pass directly through it.
  (ii) The **carotid canal** curves sharply anterior along the petrous temporal bone. Defects in the roof and floor of the canal form the foramen lacerum.
  (iii) The **jugular foramen** runs posteriorly into the cranial cavity. It transmits the internal jugular vein and the glossopharyngeus (CN IX), vagus (CN X), and spinal accessory (CN XI) nerves.

c. **Posteriorly,** the **occipital bone** consists of two parts (see Figure 31-4).
  (1) The **Y-shaped basilar part (basi-occiput)** forms the anterior boundary of the **foramen magnum.** Paired **occipital condyles** articulate with the atlas.
    (a) The **hypoglossal (anterior condylar) canal,** emerging beneath the anterolateral edge of each condyle, transmits the **hypoglossal nerve** (CN XII).
    (b) A **posterior condylar canal** at the posterolateral end of each condyle transmits an emissary vein.
  (2) The **squamous part (occipital squama)** has an inferior triangular midline notch that contributes the posterior boundary of the **foramen magnum.**
    (a) The **foramen magnum** contains the initial segment of the spinal cord as well as the roots of the spinal accessory nerve and the vertebral arteries.
    (b) The large **mastoid foramen** just posterior to the mastoid process transmits an emissary vein.

C. **Internal surface** (Figure 31-6; see Figure 31-7)
  1. **Calvarium.** The **cranial sutures** become obliterated with old age.
    a. A **frontal crest** provides attachment for the falx cerebri, a midsagittal fold of dura mater.
      (1) The frontal crest diminishes and becomes the **sagittal groove,** which deepens posteriorly to accommodate the **sagittal sinus.**
      (2) On either side of midline are **granular lacunae.**
        (a) They accommodate the venous lacunae, into which CSF drains from arachnoid granulations.
        (b) They become larger and deeper with age.
    b. The **intracranial surface** of the temporal bone is grooved and occasionally tunneled by the **middle meningeal vessels** (see Figure 31-6). A blow to the side of the head, which either fractures the skull or tears the periosteum at this point, produces rapid intracranial hemorrhage (**epidural hematoma**).
    c. The **internal occipital crest,** below the confluence of sinuses, provides attachment for the falx cerebelli.
      (1) The **sagittal groove** becomes a raised vertical gutter along the occipital bone.
      (2) **Sulci for the transverse sinuses** converge posteriorly from both sides in the transverse plane.

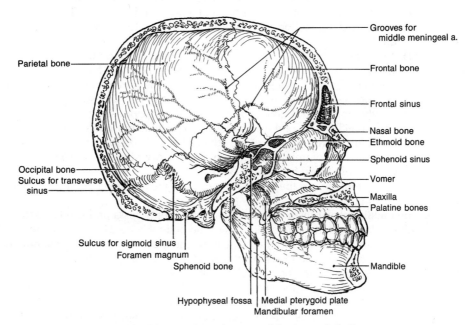

FIGURE 31-6. Lateral aspect of the internal skull.

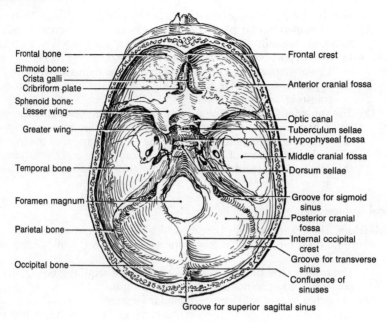

FIGURE 31-7. Floor of the neurocranium.

(3) At the **internal occipital protuberance,** the convergence of the gutters for the transverse sinuses and the superior sagittal sinus marks the **confluence of sinuses** (torculus of Herophilus).

2. **Floor of the cranial cavity.** Medially, the floor declines in a continuous convex curve, interrupted only by the **tuberculum sellae** and the **dorsum sellae** of the sphenoid bone. Laterally, the **cranial fossae** (anterior, middle, and posterior) descend in steps (Figure 31-7).
   a. The **anterior cranial fossa** contains the frontal lobes of the brain.
      (1) The **frontal bone** bounds this fossa anteriorly and laterally. The **orbital plates** of the frontal bone form the floor of the anterior fossa and the roof of the orbit.
      (2) The **ethmoid bone** is surrounded by the frontal bone on three sides.
         (a) In the midline, the **crista galli** provides attachment for the falx cerebri.
         (b) Between the crista galli and the crest of the frontal bone is a small **foramen cecum,** transmitting an emissary vein that passes from the veins of the frontal sinus and nasal cavity to the superior sagittal sinus.
         (c) On either side of the crista galli lie the **olfactory bulbs** of the brain. The thin **cribriform plate** of the ethmoid bone is perforated on each side by 15 to 20 foramina through which **olfactory nerves** pass from the nasal mucosa to each olfactory bulb.
      (3) The smooth arc of the **lesser wing of the sphenoid bone** demarcates the sharp posterior margin of the anterior fossa.
   b. The **middle cranial fossa** is deeper than the anterior fossa and is related to the pituitary gland and the temporal lobes of the brain (Figure 31-8; see Figure 31-7).
      (1) The **anterior wall** is formed largely by the greater wing of the **sphenoid bone.**
         (a) Anteriorly, the **lesser wing** meets the orbital plate. Medially, the lesser wing forms the anterior clinoid process.
         (b) Laterally, the **greater wing** fuses with frontal, parietal, and temporal bones to close the pterion.
         (c) The space between the greater and lesser wings is the **superior orbital fissure.**

FIGURE 31-8. Foramina of the middle cranial fossa.

- **(2)** The **optic groove** (**sulcus chiasmatis**) indents the sphenoid bone between the **optic foramen (canal)** and the tuberculum sellae. The term sulcus chiasmatis aptly describes its contents as the optic chiasm.
- **(3)** The **optic canal** contains the optic nerve, the ophthalmic artery, and the central vein of the retina.
- **(4)** The **sella turcica** is named for its fanciful resemblance to a premedieval Turkish chair.
  - **(a)** The **tuberculum sellae** rises from the body of the sphenoid bone, separating the hypophyseal fossa from the optic groove.
  - **(b)** The **hypophyseal (pituitary) fossa,** a midline depression, contains the **pituitary gland (hypophysis).**
  - **(c)** The **dorsum sellae** is a vertical plate that extends between the posterior clinoid processes to form the posterior wall of the sella.
- **(5)** The **posterior clinoid processes** are continuations of the dorsolateral corners of the vertical dorsum sellae.
- **(6)** **Foramina** form an arc around the medial edge of the greater wing of the sphenoid bone (Table 31-1; see Figure 31-8).
  - **(a)** The **superior orbital fissure** passes between the wings of the sphenoid into the orbital cavity.
    - **(i)** The **oculomotor** (CN III), **trochlear** (CN IV), **ophthalmic division of the trigeminal** (CN V1), and **abducens** (CN VI) **nerves** all pass through the medial end of this fissure to gain access to the orbit.
    - **(ii)** The **superior ophthalmic veins** leave the orbit through this fissure.
  - **(b)** The **foramen rotundum** passes into the pterygopalatine fossa. It contains the maxillary division of the trigeminal nerve (CN V2).
  - **(c)** The **foramen ovale** passes into the infratemporal fossa and transmits the **mandibular division of the trigeminal nerve** (CN V3) to the infratemporal fossa. Other transmitting structures include an accessory meningeal artery, the lesser superficial petrosal nerve, a recurrent meningeal branch of the trigeminal nerve, and an emissary vein.
  - **(d)** The **foramen spinosum** transmits the **middle meningeal artery.**
  - **(e)** Other small, variable foramina lie along this arc (e.g., the emissary foramina, the foramen of Vesalius, and the innominate canal).

TABLE 31-1. Neurocranial Foramina

Foramen	Location	Contents
Anterior cranial fossa		
Olfactory foramina	Ethmoid bone	Olfactory nn.
Foramen cecum	Between ethmoid and frontal bones	Emissary v.
Middle cranial fossa		
Optic canal	Sphenoid bone	CN II, ophthalmic a., central vein of retina
Superior orbital fissure	Sphenoid bone	CN III, CN IV, CN V1, CN VI, superior ophthalmic v.
Foramen rotundum	Sphenoid bone	CN V2
Foramen ovale	Sphenoid bone	CN V3
Foramen spinosum	Sphenoid bone	Middle meningeal a.
Foramen lacerum	Between sphenoid and temporal bones	Carotid artery (longitudinally), gr. superficial petrosal n.
Hiatus of the facial canal	Temporal bone	Gr. superficial petrosal n.
Posterior cranial fossa		
Internal auditory meatus	Temporal bone	CN VII, CN VIII
Jugular foramen	Between temporal and occipital bones	CN IX, CN X, CN XI, internal jugular v.
Hypoglossal canal	Occipital bone	CN XII
Posterior condylar canal	Occipital bone	Emissary v.

(7) The **carotid canal** passes through the body of the petrous portion of the temporal bone to emerge lateral to the dorsum sellae.
    (a) The **foramen lacerum** is formed by superior and inferior defects in the carotid canal (i.e., the petrous temporal bone fails to meet the sphenoid bone anteriorly). In the floor of the middle cranial fossa, the foramen lacerum is covered by a layer of fibrocartilage and periosteum.
    (b) Other than the internal carotid artery within the carotid canal, only the greater superficial petrosal nerve passes (obliquely) through the foramen lacerum.

(8) The **posterior margin of the middle cranial fossa** is formed by the superior margin of the petrous temporal bone lateral to the posterior clinoid process.
    (a) The **roof** of the middle ear cavity is the thin **tegmen tympani**. A severe middle ear infection may erode into the cranial cavity.
    (b) The **anterior semicircular canal** is accommodated by the **arcuate eminence** of the petrous portion of the temporal bone. The superior edge of the petrous portion of the temporal bone is sharp and grooved for the **superior petrosal sinus**.
    (c) At the apex of the petrous portion of the temporal bone is a depression for the **trigeminal (semilunar) ganglion**.

  c. The **posterior cranial fossa,** the deepest of the fossae, is formed by the temporal and occipital bones and contains the cerebellum and the brain stem (see Figure 31-7).
    (1) The **clivus** of the basal portion of the occipital bone dives from the dorsum sellae to the foramen magnum.

(2) The **transverse** and **sigmoid venous sinuses** deeply groove the walls of this fossa. The sigmoid sinus joins the **inferior petrosal sinus** to become the **superior bulb of the internal jugular vein** as it exits the cranial cavity through the **jugular foramen.**

(3) Several foramina pass through the basal walls of this fossa (see Table 31-1).

   (a) The **internal auditory meatus** emerges through the posteromedial surface of the petrous portion of the temporal bone, transmitting the **facial nerve** (CN VII) and the **vestibulocochlear nerve** (CN VIII).

   (b) The **jugular foramen** transmits the internal jugular vein through its posterior part. The **glossopharyngeus** (CN IX), **vagus** (CN X), and **spinal accessory** (CN XI) **nerves** pass through the jugular foramen anterior to the **internal jugular vein.**

   (c) The **anterior condylar (hypoglossal) canal** transmits the **hypoglossal nerve (CN XII).**

   (d) The **posterior condylar canal** transmits an **emissary vein** from the sigmoid sinus to the veins of the pericranium.

   (e) At the foramen magnum, the **brain stem** is continuous with the spinal cord. The spinal roots of the **spinal accessory nerve** (CN XI) and the **vertebral arteries** pass into the cranial cavity through this foramen.

D. Clinical considerations

1. **Fracture of the cranium** may involve the scalp, dura, cranial nerves, and cerebral tissue.

   a. **Locations.** The prominence of the parietal bones makes them most susceptible to fracture. The occipital bones are less often affected. A blow to the side of the head, which either fractures the temporal bone or tears the periosteum at this point, produces rapid intracranial hemorrhage (**epidural hematoma**).

   b. **Injury associated with fracture.** Bone fragments may tear the venous sinuses, dura, or a meningeal vessel or even lacerate the brain.

   c. **Fractures of the base of the skull** frequently cross foramina (weaker portions of the bone) and jeopardize the contained structures (see Table 31-1).

2. **Space-occupying lesions** (depressed skull fractures, tumors, hematomas, and even swelling of cerebral tissue) encroach on the space occupied by the incompressible brain within the rigid neurocranium.

   a. **Temporal lobe herniation.** Regions of the temporal lobes may be forced through the tentorial notch, causing oculomotor nerve dysfunction.

   b. **Medullary herniation.** The brain stem may be forced through the foramen magnum, compressing nerves and arteries. The resultant medullary ischemia rapidly results in death.

   c. **Nerve compression.** The bony confines of a foramen or canal cannot accommodate nerve swelling. Inflammatory edema produces nerve compression and dysfunction (e.g., Bell's palsy).

IV. CRANIAL MENINGES AND VENOUS SINUSES

A. Structure.
The central nervous system is enclosed in three layers of meninges (see Figures 31-1 and 32-9).

1. **The dura mater** is a tough outer protective layer (pachymeninx) consisting of two adherent fibrous membranes.

   a. The **outer dural layer** is the **periosteum** of the cranial vault. It adheres tightly to the inner lamina of the cranium.

   (1) Although the **middle meningeal artery** and its major branches run between the periosteal and meningeal dural layers, the small branches are within the periosteal layer.

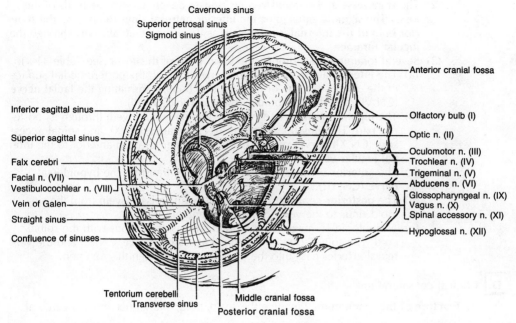

FIGURE 31-9. Dural folds, venous sinuses, and cranial nerves. The cranial nerves pass through the meninges in numeric order.

- (2) The **epidural space** is a potential space between the inner lamina of the cranium and the dura mater.
- b. The **inner dural layer** is the **true dura mater.** It adheres in most places to the outer layer, which explains the nomenclature confusion, given that this structure was named about 2300 years ago.
  - (1) **Folds of the inner dural layer** (back-to-back layers) extend between the major divisions of the brain (Figure 31-9).
    - (a) The **falx cerebri** lies in the median longitudinal fissure between cerebral hemispheres.
      - (i) It attaches to the crista galli and the frontal crest anteriorly and joins the tentorium cerebelli posteriorly.
      - (ii) Superiorly, the two layers separate and attach to the lateral edges of the sagittal groove, forming the **superior sagittal sinus.**
    - (b) The small **falx cerebelli** lies in the midsagittal plane, attaching to the internal occipital crest.
    - (c) The **tentorium cerebelli** lies in a transverse plane between the cerebellum and cerebrum. It attaches around the occipital bone and along the ridge of the petrous portion of the temporal bone and continues medial to the posterior clinoid process.
      - (i) The **tentorial notch** (incisura tentorii) in the midline fits snugly about the midbrain and separates the posterior cranial fossa from the rest of the cranial cavity.
      - (ii) The **transverse sinus** is formed by separation of the two dural layers and their attachment to the lateral calvarial walls.
      - (iii) The **superior petrosal sinus** is formed by separation of the two dural layers and their attachment anteriorly along the ridge of the petrous portion of the temporal bone.
      - (iv) The **trigeminal cave** (of Meckel) for the trigeminal ganglion is a dural evagination at the apex of the petrous portion of the temporal bone.
    - (d) The **diaphragma sellae** covers the hypophyseal fossa with the exception of a small hiatus through which the infundibular stalk and hypophyseal portal veins pass.

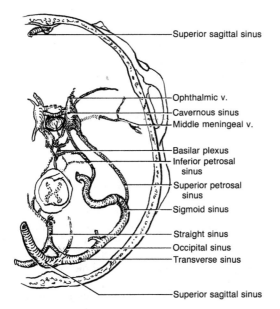

FIGURE 31-10. Venous sinuses of the neurocranium.

(2) **Dural venous sinuses** provide the principal blood return from the brain.
   (a) **Structure** The triangular dural sinuses are bounded by two leaves of meningeal dura with a base of periosteal dura (see Figure 31-1). The sinuses are lined by endothelium.
      (i) The triangular shape prevents collapse.
      (ii) A sinus can be torn by a bone fragment from a compound depressed cranial fracture or accidental surgical incision. Aspiration of air into the venous system, if sufficient to fill the right ventricle with froth, is fatal.
   (b) **Venous drainage pathway** (Figure 31-10; see Figure 31-9)
      (i) The **superior sagittal sinus** lies in the root of the falx cerebri. It begins at the crista galli and runs posteriorly, receiving the superior superficial cerebral veins. Anteriorly, it communicates with the veins of the nasal cavity through the foramen cecum. Posteriorly, it drains predominantly into the right transverse sinus.
      (ii) The **inferior sagittal sinus** lies in the inferior free edge of the falx cerebri. It drains the deeper regions of the cerebrum.
      (iii) The **straight sinus** begins at the apex of the tentorium cerebelli by the joining of the inferior sagittal sinus with the **cerebral vein** that drains the thalamus and posterior aspect of the midbrain. It runs along the intersection of the falx cerebri and the tentorium cerebelli to reach the left transverse sinus.
      (iv) The confluence of sinuses (torculus of Herophilus) formed by the joining of the superior sagittal, straight, and occipital sinuses drains into the left and right transverse sinuses at the internal occipital protuberance.
      (v) The **transverse sinuses** drain from the confluence along the edge of the tentorium cerebelli on both sides. The **right transverse sinus** receives drainage predominantly from the superior sagittal sinus. The **left transverse sinus** receives drainage predominantly from the straight and occipital sinuses.
      (vi) The **superior petrosal sinus** enters the left or right transverse sinus at the point that the transverse sinus becomes the sigmoid sinus. It drains several meningeal veins.
      (vii) The sigmoid sinus continues the path of the transverse sinus at the petrous portion of the temporal bone. Each turns inferiorly and

passes through the **jugular foramen** to become the **internal jugular vein.**

(3) **Cavernous sinus.** Other venous sinuses (e.g., inferior petrosal, sphenoparietal, and basilar) form between the meningeal and periosteal dural layers unrelated to dural folds, but only the **cavernous sinus** is particularly important (see Figure 31-10).

  (a) It lies lateral to the body of the sphenoid bone, extending from the superior orbital fissure to the petrous portion of the temporal bone (see Figures 31-10 and 31-11).

  (b) **Venous flow**

  (i) It receives the **ophthalmic vein,** the **central vein of the retina,** and the **middle and inferior cerebral veins,** as well as emissary veins from the deep face and visceral neck. It drains into the superior and inferior petrosal sinuses.

  (ii) Slow blood flow makes infection in this sinus difficult to treat.

  (c) **Contents.** The carotid artery and cranial nerve VI run through this sinus medially within sheaths of endothelium. Cranial nerves III, IV, V1, and V2 lie laterally along the wall of the cavernous sinus (Figure 31-11).

  (i) **Pathways for spread of infection.** Because the ophthalmic veins drain a small area of the face through connections with the angular facial vein, and emissary veins communicate with the visceral region of the neck, infections in these areas may spread into the cavernous sinus.

  (ii) **Cavernous sinus thrombosis** may result from infection. Symptoms and signs include pain behind the eye, cranial nerve palsies affecting eye movement (ophthalmoplegia), and edema of the optic disk (papilledema) from venous engorgement of the retina with retinal blindness. Thrombosis of the internal carotid artery may result in stroke.

(4) The **spinal dura** is the continuation of the meningeal layer of dura over the spinal cord (see Figure 13-2). This layer begins at the foramen magnum, where the dura mater is no longer fused with the periosteum of the cranial vault.

c. **Cranial nerves** pass through the dura in numeric order from anterior to posterior (see Figure 31-9). This orderly numeric relationship does not hold for the foramina of the bony skull.

d. The **epidural space** is a potential space between the inner lamina of the cranium and the periosteal dura. Bleeding within this space, usually a torn middle meningeal artery, results in a high-pressure hemorrhage **(epidural hematoma).**

e. The **subdural space** is a potential space between the true dura and arachnoid layer.

(1) It normally contains no more than a thin film of serous secretion.

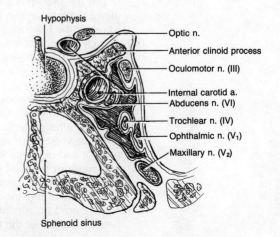

FIGURE 31-11. Contents of the cavernous sinus.

(2) Bleeding within this space, usually from a torn vein, results in **subdural hematoma,** which is a low-pressure bleed.
2. The **arachnoid layer** lies deep to the dura. It consists of a fine fibrous membrane that is closely apposed, but not attached, to the dura. It surrounds the brain, but does not dip into the sulci.
   a. **Arachnoid trabeculae** (fine strands of arachnoid tissue) bridge the subarachnoid space to connect with the underlying pia mater.
   b. The **subarachnoid space,** between the arachnoid layer and the pia mater, contains cerebrospinal fluid (CSF).
      (1) CSF is produced by **choroid plexuses** that project into the ventricles of the brain.
         (a) CSF flows through the ventricular system of the brain, entering the cisterna magna of the subarachnoid space through three foramina in the roof of the fourth ventricle.
         (b) CSF flows through the subarachnoid space of the spinal cord and brain, suspending the brain and functioning as a hydraulic shock absorber.
      (2) CSF returns to the venous system through the **arachnoid villi** (see Figure 31-1). These small, papillary tufts of arachnoid project through the inner layer of dura into the venous sinuses (primarily the superior sagittal sinus).
         (a) CSF pressure usually is greater than venous pressure, so CSF drains into the venous system.
         (b) When venous pressure momentarily becomes greater than CSF pressure, CSF flow stops and the minute tubules are compressed, preventing backflow of blood into the CSF.
      (3) **Major distributing arteries to the brain** run in the **subarachnoid space.**
         (a) A vascular bleed from a major artery (i.e., a ruptured cerebral aneurysm) will disseminate widely in the subarachnoid space.
         (b) Blood in a lumbar tap of CSF is evidence of arterial bleeding.
   c. The arachnoid layer of the brain is continuous with that of the spinal cord.
3. **The pia mater** is a delicate layer attached to brain tissue that intimately follows the surface of the brain through all of its convolutions.
   a. It contains small blood vessels and ensheathes them as they enter the brain parenchyma.
   b. **Contusion of the brain,** resulting from blunt trauma, produces pial and subpial hemorrhage. The latter is contained by the pia.
   c. The pial layer of the brain is continuous with that of the spinal cord.

B. Clinical considerations
   1. **Intracranial bleeding**
      a. **Elevated intracranial pressure** related to a space-occupying hematoma causes headache, nausea, depressed consciousness, and edema of the optic disk.
      b. **Tentorial herniation** (squeezing the temporal lobe through the tentorial notch) compresses the oculomotor nerve to extraocular and intraocular muscles. An early sign is loss of pupillary constriction on the side of the lesion, then bilateral dilation as continuing herniation involves the contralateral nerve. For this reason, pupillary response to light is monitored in individuals who have sustained head injury.
      c. **Compression of the medulla** causes anoxia to the vital centers of the brain stem and subsequent death.
   2. **Epidural hematoma**
      a. **Etiology.** Temporal or parietal bone fracture is the most common cause of a torn meningeal artery. **Arterial bleeding** is rapid and at arterial pressure. Symptoms of brain compression generally occur within 2 to 3 hours.
      b. **Symptoms.** Head trauma may produce a transient loss of consciousness (concussion), followed by a lucid interval that lasts from a few minutes to several hours. Individuals often recover from the initial concussion only to die hours later from an undiagnosed epidural hematoma—the "talk-and-die syndrome."

c. **Emergency craniotomy** is indicated to drain the hematoma, thereby relieving the pressure on the brain, and to effect hemostasis.
3. **Subdural hematoma**
   a. **Etiology.** Large **cerebral veins** are relatively unsupported and vulnerable as they pierce the arachnoid layer and cross the subdural space.
      (1) A blow to the head or even a sudden jarring with resultant shear forces can tear these veins.
      (2) Bleeding extends beneath the dura and superficial to the arachnoid. Blood seldom appears in the CSF if the arachnoid is intact.
   b. **Symptoms.** The venous ooze is often slow and at low pressure. Because CSF production is a function of intracranial pressure, the effects of the subdural hematoma are offset in part by a reduction in CSF production. This compensation explains the range of symptoms from various levels of alertness to transient loss of consciousness.
4. **Subarachnoid hemorrhage**
   a. **Etiology.** Rupture of a **cerebral artery,** which has been weakened by atheroma or idiopathic (Berry's) aneurysm, may be rapidly fatal.
   b. **Symptoms.** Onset of sudden severe headache correlates with strenuous activity. Blood is present in a lumbar tap.
   c. The key to avoiding rupture of cerebral aneurysm is early diagnosis of intense, unexplained headache that may be accompanied by cranial nerve dysfunctions. Surgery after rupture is usually futile.
5. **Pial hemorrhage**
   a. **Etiology. Contusion** of the brain, resulting from blunt trauma, produces **pial** and **subpial hemorrhage.** The latter is contained by the pia.
      (1) Cerebral trauma that causes the brain to strike the cranial surfaces may bruise the small vessels of the pia and the brain tissue.
      (2) Bruising occurs not only directly under the blow (coup), but often also at the opposite side of the brain (contrecoup) as a result of rebound.
   b. **Symptoms.** The individual may lose consciousness (concussion) at the time of the event. The aftermath of hemorrhage is atrophy, degeneration, sclerosis, and corresponding loss of function, as illustrated by the "punch-drunk" boxer.
6. **Pain and headache**
   a. The brain is not directly sensitive to pain or pressure.
   b. The cerebral vasculature is sympathetically innervated. The dura receives afferent fibers from the trigeminal, glossopharyngeal, and vagus nerves, as well as from the dorsal primary ramus of the first cervical spinal nerve. Some forms of **headache,** especially migraine, appear to be related to vascular innervation.

Chapter 32
Brain

I. INTRODUCTION

A. **The brain** is the enlarged, convoluted, and highly developed rostral portion of the central nervous system (CNS) (see Figure 32-1).

1. **Composition.** The average adult human brain weighs about 1400 grams, approximately 2% of body weight.
   a. The **brain stem** (see Figures 32-4 and 32-5) consists of midbrain, pons, and medulla. It is continuous with the **spinal cord** at the foramen magnum.
      (1) In the posterior cranial fossa, the medulla and pons lie on the basilar portion of the occipital bone.
      (2) The brain stem passes through the **tentorial notch** (incisura tentorii) of the tentorium cerebelli to reach the middle cranial fossa where the midbrain lies on the body of the sphenoid bone.
   b. The **cerebellum** (see Figure 32-4) extends dorsally from the pons and fills the posterior cranial fossa.
   c. The **cerebrum** occupies the greater portion of the middle and anterior cranial fossae (see Figure 32-1).
      (1) Its surface is convoluted, which greatly increases its area.
      (2) It is divided into lobes named by the overlying bone.
2. **Substructure.** Over 10^{12} **neurons** interconnect to form specific networks. Neurons are separated from each other by **glial cells.**
   a. **Neurons.** Some neurons synapse with several thousand other neurons; others are more specific and synapse with very few neurons. Neurons communicate with each other by the release of neurotransmitters at points of synaptic contact.
      (1) **Gray matter** contains neuron cell bodies separated by glial cells. **White matter** consists of pathways (tracts) containing the axon processes of neurons wrapped in glial membranes. Unlike the spinal cord, the gray matter lies peripherally and the white matter is central.
      (2) In the adult, neurons may die if subjected to damage or ischemia. Brain tissue does not regenerate, so function may be lost or altered permanently.
   b. **Glial cells** lie in both the white and gray matter.
      (1) One type of glial cell insulates neurons in both gray and white matter by forming **myelin sheaths** about axons and by intercalating between neuron cell bodies.
      (2) Other types provide the selective **blood–brain barrier,** which prevents or restricts substances in the blood from entering the brain.
3. **Meninges.** The gelatinous brain is protected by the bony skull and is invested by a succession of three meningeal membranes—the dura mater, arachnoid layer, and pia mater (see Chapter 31 IV).

B. **Cranial nerves.** Generally arising from the ventral and lateral surfaces of the brain stem, 12 (13, depending on the reference) pairs of cranial nerves exit the cranium through foramina and fissures in the cranial floor (see Figure 32-5). Cranial nerves may be sensory, motor, or mixed.

1. **Sensory pathways**
   a. **First-order peripheral sensory neurons** convey afferent information to the spinal cord. These neurons have cell bodies in a **dorsal root ganglion** (or equivalent) outside the CNS.
   b. **Second-order relay nucleus** are located in nuclei within gray matter. Each nucleus is concerned with one sensory modality. Second-order neurons from these

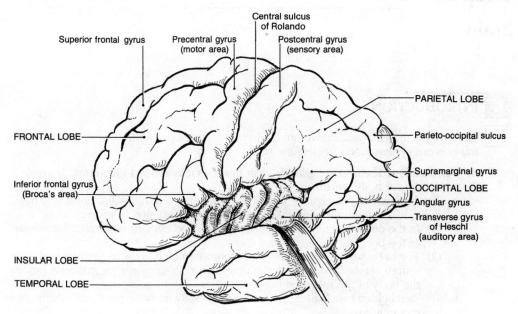

FIGURE 32-1. Cerebrum, lateral aspect of the left (dominant) cerebral hemisphere. Spearing of lateral fissure by traction on the temporal lobe reveals the underlying insular lobe.

nuclei relay the specific afferent sensation along ascending tracts to higher centers, such as the thalamic nuclei.
   c. **Third-order neurons** project from the **thalamus** to the cerebral cortex.
2. **Motor pathways**
   a. **Upper motor neurons** in the cerebral cortex act on lower motor neurons in the brain stem and spinal cord.
   b. **Lower motor neurons** in motor nuclei within the ventral regions of the brain stem contact effectors, such as skeletal muscle.

II. CEREBRUM

A. **Cerebral hemispheres** form most of the brain (Figure 32-1). Fully two thirds of synapses of the brain are within the cerebrum. These hemispheres and **basal ganglia** form the **telencephalon (forebrain).**

1. **Structure.** The cerebral hemispheres are separated by the **longitudinal cerebral fissure,** into which the **falx cerebri** projects (see Figure 32-3).
   a. **Development.** Originally related to the olfactory bulbs, the two hemispheres expanded superiorly and posteriorly during evolutionary development.
   b. **Divisions**
      (1) **Gray matter (cerebral cortex)** consists of nerve cells (neurons) arranged in **cortical columns.** With complex circuitry, these columns function as units.
         (a) **Input** to the cortical columns is principally (99%) from cortical association areas and the thalamus, with little input from subcortical areas. **Thalamic radiations** provide close functional association between the cerebral hemispheres and the thalamus.
         (b) **Output** from each cortical column is to other cortical areas, the thalamus, and the basal ganglia, as well as to the cerebellum. **Pyramidal cells** of cortical columns in the **primary motor area** project to lower motor neurons in the cranial nerve nuclei and spinal cord.

(2) **White matter** lies deeper and consists of tracts (such as **thalamic radiations, internal capsule,** and **corpus callosum**) or fascicles of axons from input and output neurons.
  c. The **surface of the cerebral cortex** is folded into **gyri** (ridges) separated by **sulci** (grooves).
    (1) **Gyri** increase the surface area, thereby accommodating more functional columns. Although the pattern of gyri and sulci is variable, the location of some ridges and grooves is remarkably consistent, providing convenient landmarks.
    (2) **Principal sulci** (see Figure 32-1)
      (a) The **lateral (sylvian) fissure** is a deep groove between the frontal and temporal lobes.
        (i) The **parietal and frontal lobes** lie superior to this fissure.
        (ii) The **temporal lobe** lies inferior to this fissure.
      (b) A **central sulcus** (fissure of Rolando) lies in a coronal plane. From the posterior end of the sylvian fissure, it extends over the lateral and onto the medial surface within the longitudinal fissure. It defines the boundary between frontal and parietal lobes.
      (c) The **parieto-occipital fissure** more or less defines the boundary between the parietal and occipital lobes, especially on the medial surface.
      (d) The **calcarine fissure,** discernible only in the medial aspect, bisects the occipital lobe transversely.

2. **Functions of the hemispheres.** Subdivision of the cerebral lobes by overlying bones is useful because it closely approximates division of function. These cortical subdivisions contain neural circuits for the processing of specific kinds of sensory or motor information.
  a. **Dominance.** Each cerebral hemisphere deals with the contralateral side of the body. Usually one hemisphere (the left in 98% of individuals) is "dominant." Some higher functions are specific to one or the other hemisphere in a predictable way.
    (1) The **left hemisphere** in most individuals is concerned with verbal, calculating, and analytical thinking, as well as interpretation of speech, stereognosis, and motor function to the right hand. **Lesions** affecting this hemisphere may result in paralysis of the right side, with possible loss of speech and associated functions, such as agnosias and aphasias.
    (2) The **right hemisphere** in most individuals is the seat of nonverbal, spatial, temporal, and synthetic function, appreciation of art and music, and motor function to the left hand. **Lesions** affecting this hemisphere may produce loss of visual–spatial awareness such that paralysis of the left side is compounded by a tendency to ignore the paralyzed part.
  b. **Primary cortical areas,** which usually are next to major sulci, are regions most directly related to specific functions.
    (1) **Cortical motor areas** provide upper motor neurons that synapse on lower motor neurons.
    (2) **Cortical sensory areas** receive third-order thalamic relay neurons that project modality-specific sensation.
  c. **Secondary cortical (association) areas,** adjacent to primary areas, are concerned with higher levels of organization and integration.
    (1) An association area usually lies between two primary areas.
    (2) Output from association areas is to other association areas, primary areas, and thalamic nuclei.

3. **Functions of the lobes**
  a. The **frontal lobe** (anterior to the central sulcus and lateral fissure) serves motor function, speech, cognition, and high levels of affective behavior.
    (1) The **primary motor area (precentral gyrus)** of the frontal lobe contains upper motor neuron cell bodies (pyramidal cells) that form the primary motor pathways.

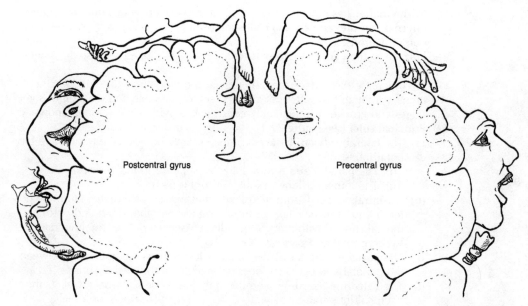

FIGURE 32-2. Somatotopic organization of the cerebral cortex. *(A)* Coronal section through postcentral gyrus. Relative sizes of cortical areas associated with various regions of the body are depicted by the sensory homunculus. *(B)* Coronal section through precentral gyrus. Relative sizes of regions that innervate muscles of various regions of the body are depicted by the motor homunculus.

        (a) Input. Connections from the **ventral lateral** and **ventral anterior nuclei of the thalamus** convey **modulating influence** from the basal ganglia and from the cerebellum to this motor area.
        (b) Output. The upper motor neurons project down the **pyramidal tracts** and cross to reach contralateral lower motor neurons in the brain stem and spinal cord.
        (c) The cortical columns are somatotopically arranged, so areas of the body may be mapped to corresponding points on the surface of the cortex, producing the **motor homunculus** (Figure 32-2B).
        (d) Damage to specific areas of precentral gyrus correlate with loss of specific muscle functions on the opposite side of the body **(contralateral spastic upper-motor neuron paralysis).**
    (2) The **supplementary motor area** of the medial surface of the frontal lobe **(prefrontal gyrus)** is somehow involved in the integration of voluntary movement.
        (a) At the inferior end of the left supplementary motor area is the **speech area** of Broca (see Figure 32-1). This area coordinates the muscles used in speech.
        (b) Damage in this area results in the inability to say what is thought **(motor aphasia).**
  b. The **parietal lobe** (between the central sulcus and the parieto-occipital fissure) is involved with somatosensory processing.
    (1) The primary sensory area **(postcentral gyrus)** of each parietal lobe receives general sensation from the contralateral side of the body through relays in the ventral posterolateral and posteromedial nuclei of the thalamus.
        (a) The cortical columns are also somatotopically arranged so that areas of the body may be mapped to corresponding points on the surface of the cortex, producing the **sensory homunculus** (see Figure 32-2A).
        (b) Damage to specific areas of postcentral gyrus correlates with loss of sensation on the opposite side of the body.

- (2) **Sensory association areas** posterior to the postcentral gyrus of the parietal lobe integrate tactile and visual stimuli to conceptualize shape and form or to evoke memory.
- (3) The **supramarginal gyrus** of the parietal lobe is involved in higher perceptual mechanisms for touch (see Figure 32-1). Lesions in this area produce **tactile agnosia** (inability to identify objects by feel).
- (4) The **angular gyrus** is involved in comprehension of written and visual objects (see Figure 32-1).
- c. The **occipital lobe** (posterior to the parieto-occipital sulcus) contains the visual cortex (see Figure 32-1).
  - (1) **Primary visual areas** (striate cortex) lie on either side of the calcarine sulcus in this lobe.
    - (a) **Retinal pathways** project to the lateral geniculate bodies of the thalamus in such a manner that the left visual field for each eye projects to the right lateral geniculate body. Lesions of the optic nerve produce **total blindness in the ipsilateral eye** (see Figure 32-6).
    - (b) **Optic radiations** project posteriorly from the lateral geniculate body of the thalamus to the occipital cortex. Lesions of the optic tract, optic radiations, or optic cortex produce **contralateral homonymous visual field defects** (see Figure 32-6).
  - (2) **Visual association areas** (peristriate cortex) of each occipital lobe surround the primary visual cortex in concentric bands. It is in this region that dots are resolved into lines and lines are recognized as shapes.
    - (a) Injury to these areas produces defective spatial orientation and visual disorganization in contralateral homonymous visual fields.
    - (b) Lesions between the angular gyrus and the occipital cortex interrupt the interpretation of written language (**alexia** or **visual agnosia**).
- d. The **temporal lobe** lies inferior to the lateral sulcus and is involved in memory and audition.
  - (1) The **primary auditory area** (**transverse gyrus** of Heschl) lies within the temporal surface of the lateral fissure at the caudal end of the superior temporal gyrus (see Figure 32-1).
    - (a) Sensory neurons from the cochlear nerve project to the cortex through relays in the cochlear nuclei, the inferior colliculus, and the medial geniculate body of the thalamus.
    - (b) The main element of auditory processing is appreciation and interpretation of language, which is subserved by the dominant hemisphere.
  - (2) The **auditory association area** (Wernicke's area) is similarly involved in language comprehension. A patient with a lesion in this area cannot understand what he or she hears (**sensory aphasia**).
- e. The **central (insular) lobe** lies at the base of the lateral fissure and involves taste (see Figure 32-1).
- f. The **limbic lobe (cingulate gyrus** and **angular gyrus)**, deep in the sagittal fissure (see Figure 32-4), is concerned with olfaction, regulation of the viscera, primitive emotions, and behavioral activity.

4. **Projections of the cerebral cortex**
   a. **Commissural fibers** pass between hemispheres above the thalamus as the **corpus callosum** (Figure 32-3).
      (1) The corpus callosum contains more than 10^6 crossing axons. These transcerebral connections are important because the hemispheres have different capabilities.
      (2) Cutting the corpus callosum results in a **split-brain individual** in whom information cannot be relayed to the contralateral side for specialized association (e.g., a common object placed in one hand cannot be recognized or matched with a similar object in the other hand).
   b. The **primary motor areas** project to lower motor neurons by way of the **internal capsule** to **corticobulbar pathways** and **pyramidal pathways**.

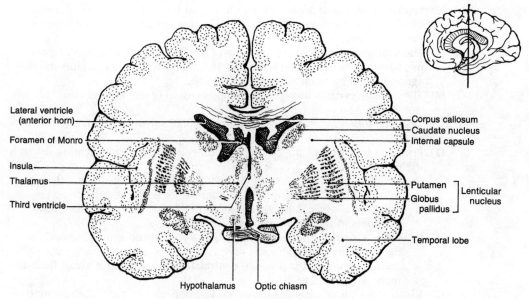

FIGURE 32-3. Coronal section through the cerebrum passes through the basal ganglia, just anterior to the thalamus *(see insert)*.

        (1) The **internal capsule** (see Figure 32-3), composed of numerous connecting axon pathways, flanks the thalamus and hypothalamus (diencephalic structures).
        (2) Corticobulbar pathways project to the lower motor neurons in the brain stem nuclei.
        (3) Pyramidal pathways project to lower motor neurons in the gray matter of the spinal cord.
    c. Extrapyramidal pathways influence lower motor neurons less directly through relays in the brain stem.
  5. Stroke. Primary motor areas project through the **internal capsule** (between the caudate nucleus and thalamus medially and the lenticular nucleus laterally) before passing into the pyramidal tracts. Small lesions in this region caused by thrombosis or hemorrhage of the striate arteries have devastating effects.

B. **Basal ganglia** are large nuclei deep within the base of the cerebral hemispheres (see Figure 32-3).
  1. Structure. They consist of the **caudate nucleus** and **lenticular nucleus** (comprising the **putamen** and **globus pallidus**). The internal capsule separates the caudate nucleus from the lenticular nucleus.
  2. Projections. Many areas of the cerebral cortex and thalamus project to the basal ganglia. The outflow from the basal ganglia is primarily to the thalamus, from which they project back to the motor areas of the cerebral cortex.
  3. Functions. The basal ganglia are processing stations in a feedback circuit that modulates motor outflow from the cortex and smooths voluntary actions.
  4. Dysfunction. Impairment of the basal ganglia produces release dyskinesias, such as tremors, choreas, ballism, and athetosis. All release dyskinesias result from removal of normal inhibitory influence.

III. BRAIN STEM

A. Thalamus (diencephalon) (see Figure 32-3)

1. **The thalamus** occupies most of the rostral portion of the brain stem to either side of the third ventricle.
   a. **Composition.** With over 25 separate nuclei, it serves as a major synaptic relay station.
   b. **Input.** All afferent modalities (except olfaction and pheromonal stimuli) synapse in the thalamus. The basal ganglia also project to it.
   c. **Output.** Third-order neurons project to the primary sensory areas of the cortex as well as to accessory and association areas.
   d. **Function.** As well as a relay point, this massive and complex nucleus is also concerned with crude appreciation of subconscious sensations and vague levels of awareness, including sleep and level of consciousness. Rhythms established by certain thalamic nuclei form the major contributions to the electroencephalogram (EEG).
   e. **Thalamic dysfunction (thalamic syndrome),** usually the result of tumor or thrombosis of the posterior choroidal artery, lowers the threshold for pain, temperature, and tactile sensation to a point where non-noxious stimuli produce marked unpleasant effects.

2. **The subthalamus** contains the **subthalamic nuclei.**
   a. These nuclei modulate motor outflow of the cortex and smooth voluntary actions.
   b. Damage to these nuclei results in **ballism** (lively jerking or shaking movements).

3. **The hypothalamus** lies below the thalamus.
   a. **Autonomic function.** It modulates visceral activities, such as thermoregulation, appetite, thirst, sexual impulses, and emotion through neural mechanisms mediated by the autonomic nervous system.
   b. **Endocrine function.** It also modulates and regulates the release of hormones from the **pituitary gland (hypophysis),** which is suspended from the hypothalamus by the thin **infundibular stalk** (see Figure 32-4).
      (1) The **neurohypophysis,** the posterior section of the pituitary, is an extension of the hypothalamus and releases **vasopressin** (antidiuretic hormone) and **oxytocin** (milk letdown factor) into the blood.
      (2) The **adenohypophysis,** the anterior section, is a pharyngeal derivative (Rathke's pouch) that secretes six trophic hormones.
         (a) Hormone release under hypothalamic control regulates the major endocrine axes, including growth hormone, lactogenic hormone, adrenocorticotropic hormone, thyroid-stimulating hormone, follicle-stimulating hormone, and luteinizing hormone.
         (b) It is prone to the formation of benign tumors, which erode the walls of the pituitary fossa and may impinge on the optic chiasm, causing visual field loss.
   c. **Pain modulation.** Pain-blunting endorphins and enkephalins are released by the hypothalamus in response to trauma or stress.

B. Midbrain (mesencephalon)

1. **Composition.** This transitional region lies between the tentorium cerebelli and the pons. The gray matter begins to become internalized and the white matter becomes peripheral (Figure 32-4).
   a. Ventrally, the midbrain contains the **substantia nigra.**
   b. The roof (tectum) contains the **superior and inferior colliculi.**

2. **Function**
   a. The **substantia nigra** is associated with modulation of movement.
      (1) It connects with the basal ganglia by dopaminergic neurons.
      (2) **Reduced dopamine production** in the substantia nigra results in **paralysis**

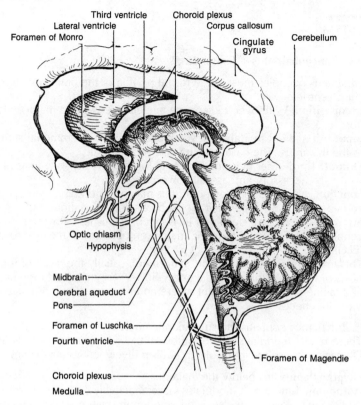

FIGURE 32-4. Sagittal section through the brain, brain stem, and cerebellum.

   agitans (Parkinson's disease), characterized by rigidity, tremors at rest, and difficulty in initiating or ceasing movements (bradykinesia).
   b. **Reflex centers**
      (1) The **superior colliculus** receives direct input from the retina and relayed input from the visual cortex. This information is used to adjust the iris to ambient light conditions and to orient the eyes toward visual stimuli.
      (2) The **inferior colliculus** receives auditory input. This information is used to orient the eyes and head toward auditory stimuli.
   c. It contains the nuclei of the **oculomotor nerve** (CN III) and **trochlear nerve** (CN IV).

C. **Hindbrain (metencephalon and myelencephalon)**

1. **Metencephalon**
   a. The **cerebellum** is a dorsal expansion of the brain stem (see Figures 32-4 and 32-5). Expression of its function is ipsilateral.
      (1) It is concerned with the coordinated contraction and relaxation of agonistic and antagonistic muscle groups to produce smooth and precise movement.
      (2) **Cerebellar dysfunction** produces jerky uncoordinated movements **(ataxia).**
   b. The **pons** is a dilated section of the brain stem so named because the fibers bridge from one side of the cerebellum to the other (see Figure 32-5).
      (1) It carries ascending tracts to the cerebellum from peripheral receptors as well as descending tracts from the cerebral cortex. It also carries tracts from the cerebellum to the thalamus for relay to the cerebral cortex.
      (2) It contains the nuclei of the **trigeminal nerve** (CN V), **abducens nerve** (CN VI), and **facial nerve** (CN VII).

2. **Myelencephalon**
    a. The **medulla** lies caudal to the pons (see Figures 32-4 and 32-5).
    b. Contents
        (1) The medulla contains the nuclei of the **hypoglossal nerve (CN XII), glossopharyngeal nerve (CN IX),** and **vagus nerve (CN X).**
        (2) **Ascending sensory pathways** to the thalamus are relayed to the cerebral hemispheres.
        (3) **Descending motor pathways** from the cortex project to the brain stem nuclei and to the gray matter of the spinal cord.

IV. CRANIAL NUCLEI AND ASSOCIATED NERVES

A. Introduction

1. **Classification.** Cranial nerves are classified according to three criteria (Figure 32-5; see Table 32-1).
    a. **General or special.** This classification signifies a modality as unique or apart from the general grouping.
    b. **Somatic or visceral. Somatic nerves** are associated with parietal (body wall) structures. **Visceral nerves** are associated with splanchnic structures.
    c. **Afferent or efferent.** Nerves are further classified as sensory or motor.
2. **Organization**
    a. **In the spinal cord,** the motor and sensory areas have specific regions (see Figure 13-2).

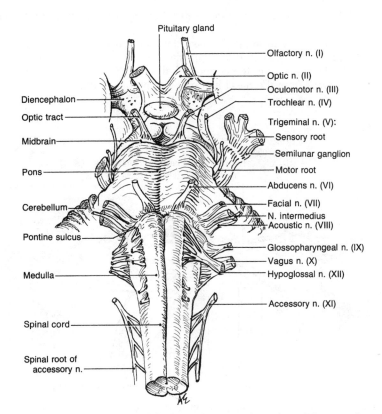

FIGURE 32-5. Anterior aspect of the brain stem reveals origins of the cranial nerves.

(1) **Somatic sensory neurons** relay to the dorsal horn.
(2) **Motor nuclei** are located in the ventral (somatic motor) and lateral (visceral motor) horns.

b. **In the brain stem,** the basic arrangement is modified by the large size of the fourth ventricle. It essentially separates the dorsal horns as though the walls never came together in the dorsal midline.
(1) **Sensory nuclei** lie in two dorsal columns. **Somatic sensory nuclei** are dorsolateral and **visceral sensory nuclei** are dorsomedial (see Figure 32-7).
(2) **Motor nuclei** are arranged in three longitudinal columns that extend through the ventral aspect of the brain stem (see Figure 32-8). **Somatic motor nuclei** are most ventromedial, **parasympathetic nuclei** are intermediate, and **branchiomotor nuclei** are ventrolateral.

B. **Special somatic afferent nuclei.** Special somatic nerves convey information related to the special senses.

1. **Optic nerves (CN II)** mediate **vision.**
   a. **Visual pathway** (Figure 32-6; see Table 32-1)
      (1) **The retina** of each eye receives an inverted and reversed image of the visual fields projected by the lens.
         (a) True optic nerves are represented in the retina by the **bipolar cells,** which contact the **rod and cone receptor cells.**
         (b) Ganglion cells of the retina, the optic "nerve," the optic tracts, and the lateral geniculate bodies are all diencephalic structures.
         (c) Lesions of the retina cause **blindness** in the affected area.
      (2) **Optic nerves** leave the retina at the **optic disk** (blind spot) and course through the **optic canal** to the optic chiasm.
         (a) These tracts are complete with dura, arachnoid, subarachnoid space, and pia.
         (b) Lesions of the optic nerve distal to the chiasm cause **total ipsilateral blindness.**
      (3) At the **optic chiasm,** pathways from both retinas are sorted so that the **same visual fields** project to the opposite sides of the brain.
         (a) Pathways from the temporal half of each retina remain ipsilateral. Pathways from the nasal half cross in the optic chiasm.

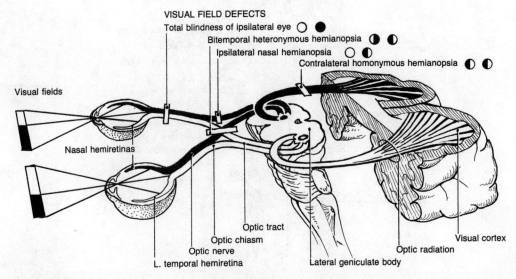

FIGURE 32-6. Optic pathways. The optic nerves, optic tracts, and optic projections are shown with typical lesions and resultant extent of blindness *(shading in diagrams of visual field defects).*

(b) **Hemianopsias.** The optic chiasm sits above the pituitary fossa, so that a pituitary tumor may press on the optic chiasm.
 (i) **Bitemporal heteronymous hemianopsia (tunnel vision)** results from medial compression of the optic chiasm, interrupting input from the nasal retina of both eyes with loss of both temporal visual fields.
 (ii) **Ipsilateral nasal hemianopsia** results from lateral compression of uncrossed nerves, interrupting input from the temporal retina of one eye with loss of one nasal visual field.
 (4) **Optic tracts** pass posteriorly from the optic chiasm, conveying common visual fields to the **lateral geniculate body of the thalamus.**
 (5) **Optic radiations** from each lateral geniculate body project to the primary visual area of each occipital lobe. Lesions of the optic tracts, optic radiations, or occipital cortex result in **cortical blindness** with **contralateral homonymous hemianopsia** (loss of the contralateral half of the visual fields of both eyes).
 b. **Reflex pathways.** The optic nerve also provides connections with midbrain nuclei involved with pupillary reflexes.
 2. **Vestibular and cochlear nuclei** (see Figure 32-7 and Table 32-1) in the hindbrain receive the neurons of the **vestibulocochlear nerve (CN VIII).** This nerve reaches the inner ear through the **internal acoustic meatus.**
  a. **Vestibular nuclei** mediate **balance** and **equilibrium.**
   (1) Through connections to the nuclei of the external ocular muscles, they enable the eye to fix on an object when the head is moving.
   (2) These nuclei also project to the cervical spinal cord and enable the head to remain stable when the body moves.
   (3) Abnormal ocular movements (**nystagmus**) are associated with vestibular dysfunction.
  b. **Cochlear nuclei** relay **acoustic information** to the **medial geniculate body** of the thalamus for subsequent relay to the primary auditory cortex.

C. **General somatic afferent nuclei.** Three **trigeminal nuclei,** which tend to be the most dorsolateral column of sensory nuclei, receive somatic afferent sensation (Figure 32-7). The associated cranial nerves are equivalent to dorsal roots.
 1. **The trigeminal nerve (CN V)** conveys general sensation from the anterior scalp, face, nose, mouth, pharynx, and most of the pinna (see Figure 36-4 and Table 32-1).
  a. **The trigeminal (gasserian, semilunar) ganglion** contains the cell bodies of most of the afferent neurons of the **trigeminal nerve (CN V).**
  b. **Pain, touch, and temperature** are associated with specific brain stem nuclei where second-order neurons are located (see Figure 32-7).
   (1) **The mesencephalic nucleus of V** is most rostral and mediates proprioception (afferent limb of the **jaw-jerk reflex**). This nucleus appears to contain first-order sensory neurons, the only exception to the rule that the first-order neurons are located in dorsal root ganglia or cranial equivalents.
   (2) **The principal nucleus of V** in the pons receives fibers carrying light touch and two-point discrimination.
   (3) **The spinal nucleus of V,** which extends into the cervical spinal cord, receives fibers conveying pain, coarse touch, and temperature sensation.
 2. **The facial nerve (CN VII)** mediates general sensation from a small region of the auricle (see Table 32-1).
  a. The **geniculate ganglion** contains the cell bodies of the first-order afferent neurons.
  b. Fibers synapse in the **spinal nucleus of V** (see Figure 32-7).
 3. **The glossopharyngeal nerve (CN IX)** is responsible for a small area of somatic sensation within the external auditory meatus (see Table 32-1).
  a. The **superior glossopharyngeal (jugular) ganglion** contains the cell bodies of the first-order neurons.
  b. Fibers synapse in the **spinal nucleus of V** (see Figure 32-7).

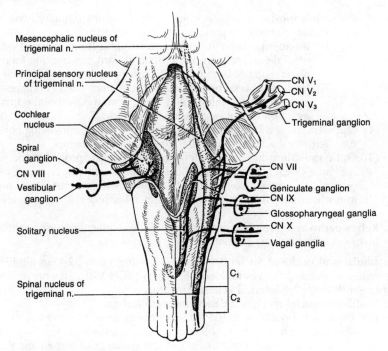

FIGURE 32-7. Sensory nuclei of the brain stem, anterior aspect. Sensory components of the cranial nerves terminate in sensory nuclei.

4. **The vagus nerve (CN X)** is responsible for a small area of somatic sensation on the posterior surface of the ear and a small area within the external auditory meatus (see Table 32-1).
   a. The **superior vagal (jugular) ganglion** contains the cell bodies of the first-order neurons.
   b. Fibers synapse in the **spinal nucleus of V** (see Figure 32-7).

D. **General visceral afferent nucleus.** The **solitary nucleus** (nucleus solitarius), which lies dorsomedially in the brain stem, receives afferent sensation from the cranial, cervical, thoracic, and abdominal viscera.

1. **The glossopharyngeal nerve (CN IX),** with cell bodies in the **inferior glossopharyngeal (petrosal) ganglion,** carries sensation from the pharynx as well as reflex information from the carotid complex (see Figures 32-7 and 37-8 and Table 32-1). These neurons synapse in the **nucleus solitarius** to mediate the **gag reflex** and **carotid reflex** (see Figure 32-7).

2. **The vagus nerve (CN X),** with cell bodies in the **inferior vagal (nodose) ganglion,** carries sensation from the larynx as well as reflex information (see Figures 32-7 and 37-13 and Table 32-1). These neurons synapse in the **nucleus solitarius** to mediate visceral reflexes **(cough reflex, aortic reflex, respiratory reflexes).**

E. **Special visceral afferent nuclei**

1. **Pheromonal stimulus** is mediated by the **terminal nerve (CN "0")** that is closely associated with, but not part of, the olfactory nerve (see Table 32-1). It is a primitive system of which little is known.
   a. **Non-olfactory pheromonal** receptors lie in the **vomeronasal organ** (of Jacobson) in the nasal septum just superior to the nasal vestibule.
   b. Second-order cell bodies lie in the **ganglion terminale** (accessory olfactory bulb), adjacent to each olfactory bulb. They are thought to project to hypothalamic nuclei for relay to the limbic lobe.

2. **Olfaction** is conveyed by **olfactory nerves (CN I)** that arise in close association with the cerebrum (see Table 32-1).
   a. **Olfactory nerves** (approximately 20) pass from the nasal mucosa through the **cribriform plate** to the olfactory bulbs.
      (1) These delicate nerves frequently are damaged by inflammatory processes or severed in head injury (even without fracture of the ethmoid bone), causing **anosmia**.
      (2) These nerves may provide the route of entry for encephalitis or meningitis.
   b. The **olfactory bulbs** lie on the cribriform plates of the ethmoid bone.
   c. An **olfactory tract** extends from each olfactory bulb to the base of the frontal area of the brain.

3. **Taste** from the tongue and epiglottis is conveyed to the s**olitary nucleus** (see Figure 32-7). From the solitary nucleus, second-order neurons convey this information to the thalamus.
   a. **Taste from the anterior two-thirds of the tongue** finds its way through the lingual nerve, to the **chorda tympani** to the **facial nerve (CN VII)**. Cell bodies of the afferent neurons are located in the **geniculate ganglion** (see Figure 36-3 and Table 32-1).
   b. **Taste from the posterior surface of the tongue** is carried via the **glossopharyngeal nerve (CN IX)**. Cell bodies of these afferent neurons are located in the **inferior glossopharyngeal (petrosal) ganglion** (see Figure 37-7 and Table 32-1).
   c. **Taste from the epiglottis** is carried by the **vagus nerve (CN X)**. Cell bodies of these afferent neurons are located in the **inferior vagal (nodose) ganglion** (see Figure 32-7 and 37-13 and Table 32-1).

F. **Special visceral efferent nuclei. Branchiomeric nerves** arising from the **most lateral column** of motor nuclei innervate muscles derived from gill arch derivatives.

1. **The trigeminal motor nucleus** in the midpons gives rise to the branchiomotor component of the **trigeminal nerve (CN V)**.
   a. It is the most rostral special visceral nucleus (see Figure 32-8).
   b. Branchiomotor axons pass through the trigeminal ganglion and exit the cranium through the foramen ovale to join the **mandibular nerve (CN V$_3$)**.
   c. It innervates the ipsilateral **muscles of mastication** and several other muscles (see Figure 36-4 and Table 32-1).

2. **The facial motor nucleus** gives rise to the branchiomotor component of the **facial nerve (CN VII)**.
   a. This nucleus lies just caudal to the trigeminal motor nucleus (see Figure 32-8).
   b. The facial nerve innervates the ipsilateral **muscles of facial expression** (see Figure 36-2 and Table 32-1).

3. **The nucleus ambiguus** (see Figure 32-8) gives rise to the branchiomotor component of the **glossopharyngeal (CN IX)** and **vagus (CN X) nerves.**
   a. The **glossopharyngeal nerve (CN IX)** innervates the ipsilateral stylopharyngeus muscle only (see Figure 37-8 and Table 32-1).
   b. The **vagus nerve proper (CN X)** is that portion of the vagus nerve that develops in association with the fourth branchial arch. It innervates the ipsilateral pharyngeal muscles by the **pharyngeal plexus.** It also innervates the ipsilateral cricothyroid muscle by the **superficial (external) branch of the superior laryngeal nerve** (see Figure 37-13 and Table 32-1).
   c. The **recurrent laryngeal (cranial accessory) nerve (CN X)** is that portion of the vagus nerve that develops in association with the sixth branchial arch. It briefly joins the spinal accessory nerve within the jugular foramen (which explains its alternative name) and then separates to rejoin the vagus nerve. As the **recurrent (inferior) laryngeal branch of the vagus nerve,** it innervates all of the ipsilateral **muscles of the larynx** but the cricothyroid muscle (see Figure 37-13 and Table 32-1).

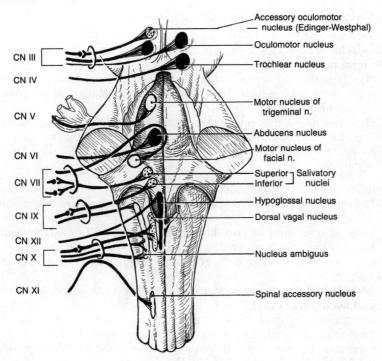

FIGURE 32-8. Motor nuclei of the brain stem, anterior aspect. Somatic motor *(shaded)*, branchiomeric motor *(unshaded)*, and parasympathetic *(stippled)* nuclei contribute axons to motor components of various cranial nerves.

4. **The lateral cell column of the upper five or six cervical segments** forms the **spinal accessory nerve (CN XI)** from a series of rootlets (Figure 32-8).
   a. This nerve passes into the cranial cavity through the foramen magnum and joins the cranial accessory portion of the vagus nerve within the jugular canal. After passing through the jugular foramen, the two nerves separate.
   b. The spinal accessory nerve innervates the ipsilateral **sternomastoid** and **trapezius muscles** (see Table 32-1).

G. **General visceral efferent nuclei.** These nerves arise from the **intermediate column** of motor nuclei.

1. **The accessory oculomotor nucleus** (of Edinger-Westphal) is the most rostral parasympathetic nucleus (see Figure 32-8 and Table 32-1). It gives rise to the parasympathetic component of the **oculomotor nerve (CN III).**
   a. **Preganglionic parasympathetic neurons** from this nucleus travel along the oculomotor nerve (CN III) to the ciliary ganglion (see Figure 33-4A).
   b. **Postganglionic axons of the ciliary ganglion** course to the **iridial sphincter** and **ciliary muscles** of the lens, controlling pupillary constriction and accommodation to near vision. Injury to the oculomotor nerve is indicated by a dilated pupil that is unresponsive to light.

2. **The superior salivatory nucleus** (see Figure 32-8 and Table 32-1) gives rise to the parasympathetic secretomotor component of the **facial nerve (CN VII).**
   a. **Preganglionic axons** travel with the **nervus intermedius** of the facial nerve to the **pterygopalatine ganglion** (via the greater superficial petrosal and vidian nerves) and the **submandibular ganglion** (via the chorda tympani and lingual nerve).
   b. **Postganglionic axons of the pterygopalatine ganglion** supply the lacrimal, nasal, and palatine glands (see Figure 36-3).

c. **Postganglionic axons of the submandibular ganglion** supply the submandibular gland, sublingual gland, and glands of the anterior tongue (see Figure 36-3).

3. **The inferior salivatory nucleus** (see Figure 32-8 and Table 32-1) gives rise to the parasympathetic secretomotor fibers of the **glossopharyngeal nerve (CN IX).**
   a. **Preganglionic axons** travel with the glossopharyngeal nerve, then with the tympanic nerve (of Jacobson) and the lesser superficial petrosal nerve to reach the otic ganglion (see Figure 37-8).
   b. **Postganglionic axons of the otic ganglion** then supply the parotid gland (see Figure 37-8).

4. **The dorsal vagal nucleus,** the most caudal of the cranial parasympathetic nuclei (see Figure 32-8 and Table 32-1), gives rise to the parasympathetic component of the **vagus nerve (CN X).**
   a. **Preganglionic parasympathetic fibers** are long and synapse in small ganglia associated with the viscera of the thorax and much of the abdomen.
   b. **Postganglionic neurons from small parasympathetic ganglia** are secretomotor and stimulate visceral muscular activity, but slow cardiac muscle activity.

H. General somatic efferent nuclei. **Somatic nerves** arising from the **most medial column** of motor nuclei are equivalent to motor roots and innervate muscle derived from embryologic somites.

1. **Nuclei of extraocular muscles**
   a. The **oculomotor nucleus** (see Figure 32-8 and Table 32-1) provides the somatic component of the **oculomotor nerve (CN III).**
      (1) Axons from this nucleus innervate five extraocular muscles (levator palpebrae superioris, superior rectus, medial rectus, inferior rectus, and inferior oblique).
      (2) Lesion or compression within the tentorial notch produces **lateral strabismus** (wall eyes) with nearly complete **ophthalmoplegia** (ocular paralysis) of the eye as well as **ptosis** (drooping) of the eyelid. **Mydriasis** (a dilated pupil that does not respond to light) is an associated finding.
   b. The **trochlear nucleus** (see Figure 32-8 and Table 32-1) gives rise to the **trochlear nerve (CN IV).**
      (1) Axons from this nucleus innervate the superior oblique muscle (see Figure 33-4B).
      (2) Lesions produce **diplopia** when looking down and out.
   c. The **abducens nucleus** (see Figure 32-8 and Table 32-1) gives rise to the **abducens nerve (CN VI).**
      (1) Axons from this nucleus innervate the lateral rectus muscle (see Figure 33-4B).
      (2) Lesions produce **medial strabismus (crossed eyes).**
   d. A **medial longitudinal fasciculus** connects the extraocular nuclei, coordinating conjugate eye movements.
   e. **Courses**
      (1) The oculomotor and trochlear nerves pass through the tentorial notch before entering the dura.
      (2) The oculomotor, trochlear, and abducens nerves pass through the **cavernous sinus** to access the **superior orbital fissure** through which they reach the orbit.

2. **The hypoglossal nucleus** (see Figure 32-8 and Table 32-1) gives rise to the **hypoglossal nerve (CN XII).**
   a. This nerve, formed by a series of rootlets that emerge from the anterolateral sulcus of the medulla, leaves the cranium through the **anterior condylar (hypoglossal) canal.**
   b. It innervates the ipsilateral muscles of the tongue (see Figure 34-5).
   c. Lesions result in **deviation of the tongue** toward the side from which the lesion protrudes (Table 32-1).

Table 32-1. Summary of Cranial Nerves

Functional Component	Branches	Cell Bodies	Cranial Foramen	Distribution	Associated Nucleus
Terminal nerve (CN 0)					
SVA		Accessory olfactory bulb	Cribriform plate	Pheromonal receptors of vomeronasal organ	Hypothalamic nuclei
Olfactory nerve (CN I)					
SVA		Olfactory epithelium	Cribriform plate	Olfactory epithelium	Olfactory bulb and tract
Optic nerve (CN II)					
SSA		Bipolar cells of retina	Optic foramen	Rods and cones of nasal and temporal retina	Ganglion cells of retina to lateral geniculate body of thalamus
Oculomotor nerve (CN III)					
GSE	Superior	Oculomotor nucleus	Superior orbital fissure	Levator palpebrae superioris muscle and superior rectus muscle	
	Inferior	Oculomotor nucleus	Superior orbital fissure	Medial rectus, inferior rectus, and inferior oblique muscles	
GVE	Ciliary	Accessory oculomotor nucleus and ciliary ganglion	Superior orbital fissure	Ciliary muscle (accommodation) and iris (constriction)	
Trochlear nerve (CN IV)					
GSE		Trochlear nucleus	Superior orbital fissure	Superior oblique muscle	
Trigeminal nerve (CN V)					
GSA	Ophthalmic Lacrimal Nasociliary Supratrochlear Supraorbital	Semilunar ganglion	Superior orbital fissure	Forehead, mucous membranes of nasal cavity and sinuses; corneal blink reflex	Trigeminal sensory nucleus
GSA	Maxillary Infraorbital Nasopalatine Descending palatine Superior alveolar Zygomaticofacial Zygomaticotemporal	Semilunar ganglion	Foramen rotundum	Midface, upper teeth, mucous membranes of nasal cavity, maxillary sinus, and hard palate; sneeze reflex	Trigeminal sensory nucleus
GSA	Mandibular Inferior alveolar Lingual Mental Mylohyoid Buccal Auriculotemporal	Semilunar ganglion	Foramen ovale	Mandible, lower teeth, cheeks, and anterior two thirds of tongue; jaw-jerk reflex	Trigeminal sensory nucleus

(continued)

Table 32-1. (continued)

Functional Component	Branches	Cell Bodies	Cranial Foramen	Distribution	Associated Nucleus
SVE	Motor root of mandibular n. Muscular	Motor nucleus of CN V	Foramen ovale	Muscles of mastication, tensor tympani muscle, tensor veli palatini muscle; motor limb of jaw-jerk reflex	
	Mylohyoid			Mylohyoid muscle, and anterior belly of digastric muscle	
Abducens nerve (CN VI)					
GSE		Abducens nucleus	Superior orbital fissure	Lateral rectus muscle	
Facial nerve (CN VII)					
GSA	Auricular	Geniculate ganglion	Facial canal	External ear	Spinal nucleus of CN V
GVA	Greater superficial petrosal	Geniculate ganglionfacial	Hiatus of the canal	Nasal mucosa	Nucleus solitarius
SVA	Chorda tympani via lingual branch of V3	Geniculate ganglion	Petrotympanic fissure, middle ear, facial canal, and internal auditory meatus	Taste, anterior two thirds of tongue	Nucleus solitarius
GVE	Greater superficial petrosal (to nerve of pterygoid canal)	Superior salivatory nucleus and pterygopalatine ganglion	Hiatus of the facial canal and pterygoid canal	Secretomotor to lacrimal, nasal, and palatine glands	
	Chorda tympani	Superior salivatory nucleus and submandibular ganglion	Facial canal, middle ear, petrotympanic fissure, and lingual nerve (V)	Secretomotor to submandibular, and intrinsic lingual glands	
SVE	Temporal Zygomatic Buccal Mandibular Cervical	Facial nucleus	Stylomastoid foramen	Muscles of facial expression, stapedius muscle, posterior belly of digastric muscle, and stylohyoid muscle (efferent limb of corneal blink reflex)	
Vestibulocochlear nerve (CN VIII)					
SSA	Vestibular	Vestibular ganglion	Internal auditory meatus	Utricle, saccule, and semicircular canals	Vestibular nuclei
	Cochlear	Spiral ganglion	Internal auditory meatus	Spiral organ to medial geniculate body to temporal lobe	Cochlea nuclei
Glossopharyngeal nerve (CN IX)					
GSA	Tympanic	Jugular ganglion (of IX)	Tympanic canal	External auditory meatus and middle ear	Spinal nucleus of CN V
GVA	Lingual	Petrosal ganglion	Jugular foramen	Sensation, posterior third of tongue	Nucleus solitarius
	Pharyngeal	Petrosal ganglion	Jugular foramen	Nasopharynx and oropharynx (gag reflex)	Nucleus solitarius
	Carotid	Petrosal ganglion	Jugular foramen	Carotid sinus and body (baroreflex and chemoreflex)	Nucleus solitarius

(continued)

Table 32-1. *(continued)*

Functional Component	Branches	Cell Bodies	Cranial Foramen	Distribution	Associated Nucleus
SVA	Lingual	Petrosal ganglion	Jugular foramen	Taste, posterior third of tongue	Nucleus solitarius
GVE	Tympanic and lesser superficial petrosal	Inferior salivatory nucleus and otic ganglion	Tympanic canal, middle ear, lesser superficial petrosal, and foramen ovale	Secretomotor to parotid gland	
SVE	Stylo-pharyngeal	Nucleus ambiguus	Jugular foramen	Stylopharyngeus muscle	
Vagus nerve (CN X)					
GSA	Auricular	Jugular ganglion (of X)	Tympanic canal	External auditory meatus	Spinal nucleus of CN V
GVA	Aortic plexus	Nodose ganglion	Jugular for.	Aortic body (chemoreceptor reflex)	Nucleus solitarius
	Superior laryngeal, internal br.	Nodose ganglion	Jugular for.	Superior larynx (cough reflex)	Nucleus solitarius
	Inferior laryngeal	Nodose ganglion	Jugular for.	Inferior larynx	Nucleus solitarius
	Vagal	Nodose ganglion	Jugular for.	Thoracic and upper abdominal viscera	Nucleus solitarius
SVA	Superior laryngeal, internal br.	Nodose ganglion	Jugular for.	Taste from epiglottis	Nucleus solitarius
GVE	Vagus	Dorsal motor nucleus of CN X and distal para-sympathetic ganglia	Jugular for.	Visceral smooth muscle and gland control	
SVE	Pharyngeal	Nucleus ambiguus	Jugular for.	Palate muscles (except tensor veli palatini) and pharyngeal muscles (except stylopharyngeus)	
	Superior laryngeal, external br.	Nucleus ambiguus	Jugular for.	Cricothyroid muscle	
	Inferior laryngeal	Nucleus ambiguus	Jugular for.	Laryngeal muscles (except cricothyroid muscle)	
Spinal accessory nerve (CN XI)					
SSE	Sternomastoid Trapezius	Lateral column of upper cervical cord	Vertebral canal, foramen magnum, and jugular foramen	Sternomastoid muscle and trapezius muscle	
Hypoglossal nerve (CN XII)					
GSE		Hypoglossal nucleus	Hypoglossal (anterior condylar) canal	Intrinsic tongue muscles, genioglossus muscle, hypoglossus muscle, and styloglossus muscle	

SVA, special visceral afferent; SSA, special somatic afferent; GSE, general somatic efferent; GVE, general visceral efferent; GSA, general somatic afferent; SVE, special visceral efferent; GVA, general visceral afferent.

V. VENTRICULAR SYSTEM

A. Structure. The ventricles represent the rostral end of the **central canal** of the neural tube. The terminal expansion of the neural tube divides into two lateral ventricles, which drain sequentially through the midline third ventricle, the cerebral aqueduct, and the midline fourth ventricle (see Figure 32-9).

1. **The lateral ventricles** are cavities within each cerebral hemisphere (see Figure 32-3).
   a. **Horns** extend into the frontal, temporal, and occipital lobes.
   b. The **interventricular foramen** (of Monro) opens from each lateral ventricle into the third ventricle.
2. **The third ventricle,** a cleft-like cavity between the thalami, is continuous with the **cerebral aqueduct** or **iter** posteriorly (see Figure 32-4).
3. **The iter or cerebral aqueduct** (of Sylvius) in the midbrain connects the third and fourth ventricles (see Figure 32-4). It is prone to obstruction, which produces **hydrocephalus.**
4. **The fourth ventricle** lies in the pons beneath the cerebellum (see Figure 32-4).
   a. Posteriorly, it extends into the **central canal** of the spinal cord.
   b. It communicates with the **cisterna magna (cerebellomedullary cistern)** of the subarachnoid space (located between the medulla and the cerebellum) through three openings in the roof of the fourth ventricle: two **lateral foramina** (of Luschka) and, somewhat caudally, the **median foramen** (of Magendie).

B. Cerebrospinal fluid (CSF) is clear, low in protein, and normally has few lymphocytes.

1. **Within the ventricles,** CSF is formed at the choroid plexuses.
   a. **Choroid plexus** occurs along the medial wall of the body and inferior horns of the lateral ventricles, along the roof of the third ventricle, and on the caudal region of the fourth ventricle roof (see Figure 32-4).
      (1) **Structure.** Nervous tissue around the ventricles is deficient where the pia mater and ependyma fuse. Overlying pia and vascular mesenchyme invades this tissue to form vascular **choroid plexuses.**
      (2) **CSF is produced** by ultrafiltration, resulting from the pressure differential between arterial and venous blood. Elevation in either venous or intracranial pressure slows or stops CSF production.
   b. **CSF flows** through the ventricles and enters subarachnoid space, where it returns to the venous system (see Figure 32-4).
2. **In the subarachnoid space,** CSF bathes and suspends the spinal cord and brain (Figure 32-9).
   a. CSF functions as a hydraulic shock absorber.
   b. **Arachnoid villi** project into the superior sagittal sinus, returning CSF to the venous system (see Figures 31-1 and 32-9).
      (1) **Villus structure.** Small, papillary tufts of arachnoid project through the inner layer of dura mater into the venous sinuses.
      (2) **CSF resorption.** CSF pressure usually is greater than venous pressure, so CSF drains into the venous system. When venous pressure becomes momentarily greater than CSF pressure, CSF flow stops. The minute tubules compress, preventing backflow of blood into the CSF.
3. **Clinical considerations**
   a. **Lumbar puncture.** CSF samples may be drawn at the L4–L5 intervertebral level and then analyzed. An increased number of white blood cells is indicative of bacterial meningitis. Blood is indicative of subarachnoid (arterial) hemorrhage.
   b. **Cranial fractures** with leakage of CSF. Fracture of the ethmoid bone may allow CSF to drain through the nose **(rhinorrhea).** Fracture of the base of the skull may allow CSF to drain from the ear **(otorrhea).**
   c. **Hydrocephalus.** Flow of CSF may be interrupted by congenital malformations, usually within the iter, or by scarring of arachnoid villi after meningitis.

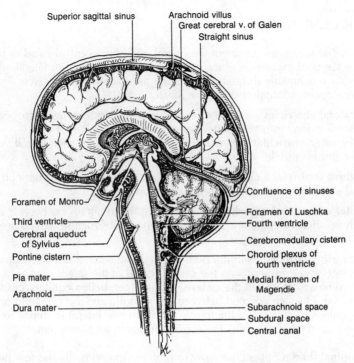

FIGURE 32-9. Ventricular system of the brain. Cerebrospinal fluid (CSF) is secreted by the choroid plexuses of the lateral, third, and fourth ventricles. CSF enters the subarachnoid space through three foramina in the roof of the fourth ventricle. The brain is suspended in CSF, which is resorbed into the venous system at the arachnoid villi.

   (1) In children, the neurocranium enlarges and the fontanelles fail to close. In adults, the symptoms are similar to those associated with chronic subdural hematoma.
   (2) In both children and adults, treatment involves placement of a shunt to drain the CSF into the venous system or the peritoneal cavity.
  d. **Pneumocephalus** (air in the subarachnoid space) may occur as a complication of a lumbar tap or as a result of cranial fracture. It produces severe headache.

VI. VASCULATURE

A. Introduction. The brain is a highly metabolic organ with imperative vascular needs. Although it represents but 2% of total body weight, the brain uses about 20% of the cardiac output and requires sufficient and constant blood circulation. **Ischemia** rapidly leads to loss of consciousness, followed by convulsions. Irreversible brain damage begins after 4 minutes and brain death occurs after 8 minutes of blood deprivation.

B. Arterial supply

  1. **Carotid arteries** (see Figure 32-10)
     a. In the **superior mediastinum,** the **right common carotid artery** originates from the brachiocephalic artery. The **left common carotid artery** arises directly from the arch of the aorta.
     b. In the **neck,** each common carotid bifurcates into an **external carotid artery,** which supplies the face and neurocranium, and an **internal carotid artery,** which primarily supplies the brain.

c. The **internal carotid artery** enters the cranium through the carotid canal.
   (1) **Intracranial course**
      (a) **Within the carotid canal,** a plexiform sheath of **postganglionic sympathetic fibers** surrounds the artery.
      (b) **At the apex of the petrous bone,** it turns up and medially along the body of the sphenoid, where it turns anteriorly again within the cavernous sinus.
      (c) **On leaving the cavernous sinus,** it arches backward beneath the anterior clinoid process and the optic nerve.
   (2) **Major branches**
      (a) An **ophthalmic branch** accompanies the optic nerve and gives off the **central retinal artery** that enters the optic nerve.
      (b) The **posterior communicating branch** passes posteriorly above the oculomotor nerve to contribute to the circle of Willis.
      (c) The **anterior cerebral artery,** formed by bifurcation of the internal carotid artery, passes dorsally around the corpus callosum at the base of the sagittal sulcus in company with its contralateral counterpart.
         (i) They supply the medial aspect of the frontal and parietal cerebral cortex.
         (ii) The two anterior cerebral arteries are connected by a small **anterior communicating branch** that contributes to the circle of Willis.
      (d) The **middle cerebral artery,** formed by bifurcation of the internal carotid artery, runs superoposteriorly in the lateral fissure.
         (i) It supplies most of the lateral aspect of the cerebral cortex.
         (ii) Lenticulostriate arterial branches supply the basal ganglia, thalamus, and internal capsule. These thin-walled vessels are common sites of hemorrhage or thrombosis, which causes contralateral paralysis, termed **stroke** (arch. apoplexy) given its sudden and devastating effects.

2. **Vertebral arteries** arise from the subclavian arteries (Figure 32-10).
   a. They pass through the **transverse foramina** of vertebrae C6–C1. Above the atlas, they pierce the dura mater and arachnoid at the atlanto-occipital membrane.
   b. They pass through the **foramen magnum** to enter the cranial cavity and ascend on the clivus of the basilar portion of the occipital bone, giving off several branches.
      (1) **Posterior–inferior cerebellar arteries** pass laterally over the inferior surface of the cerebellum. These vessels, in turn, give off **posterior spinal arteries.**
      (2) The **anterior spinal artery** forms by union of a small branch from each vertebral artery. It runs down the midline in the anterior median fissure of the medulla.
   c. The **basilar artery** forms at the pontomedullary junction by the joining of the two vertebral arteries. It has numerous branches.
      (1) The **anterior–inferior cerebellar artery** passes close to the abducens nerve, the vestibulocochlear nerve, and the glossopharyngeal nerve.
      (2) **Pontine branches** are small and numerous.
      (3) **Superior cerebellar arteries** arise just caudal to the anterior border of the pons.
      (4) **Posterior cerebral arteries** form by the bifurcation of the basilar artery at the anterior border of the pons.
         (a) Each of these arteries runs laterally along the pons to supply the tentorial surface of the occipital lobe.
         (b) The oculomotor nerve (CN III) emerges between the superior cerebellar and posterior cerebral arteries. Aneurysm of either of these vessels often is discovered because of signs of oculomotor nerve compression.
         (c) The **posterior communicating branch** passes anteriorly above the oculomotor nerve to contribute to the circle of Willis.

3. **Arterial anastomotic circle** (of Willis) (see Figure 32-10)
   a. **Structure**
      (1) Paired **anterior cerebral arteries** from the internal carotid arteries

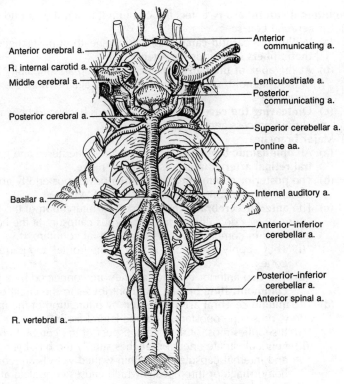

FIGURE 32-10. Internal carotid and vertebral arteries supply blood to the brain.

- (2) The **anterior communicating branch** between the left and right anterior cerebral arteries
- (3) Paired **posterior communicating branches** between the internal carotid arteries and the posterior cerebral arteries
- (4) Paired **posterior cerebral arteries** from the basilar artery
- b. **Function**
  - (1) It provides anastomotic connection between the carotid and vertebral arteries on the ventral surface of the brain stem.
  - (2) The arterial anastomotic circle is complete in 90% of cases, although the communicating arteries often are small and of little functional significance.
  - (3) The vessels are prone to **Berry's aneurysm** formation (see section VI B 4 a).
4. **Clinical correlation.** The cerebral arteries pass dorsally around the sides of the brain within deep fissures. All of the major arteries lie in the subarachnoid space.
   - a. **Subarachnoid hemorrhage** occurs on rupture of an atheromatous artery or a berry aneurysm. Many victims experience a stepwise progression of disease before a crippling attack, including headache, transient episodes, dizziness, and weakness. Aneurysms are best treated surgically before rupture. Surgical intervention after rupture usually is futile.
   - b. **Cerebral hemorrhage (stroke)** results from rupture of cerebral vessels within the brain parenchyma or from penetrating trauma.
   - c. **Occlusion.** Atherosclerosis causes progressive narrowing of arteries.
     - (1) Symptoms include transient episodes, dizziness, weakness, loss of memory—all of which depend on the region affected.
     - (2) Angiography and PET scans are used to localize the site and extent of blockage.
     - (3) Some blockages are alleviated by endarterectomy or anastomosis to an alternative supply, such as the superficial temporal artery.

C. Venous return (see Figure 31-9)
   1. **Distally,** cerebral veins generally parallel the terminal portion of the arterial tree in the pial layer. Large collecting veins run in the subarachnoid layer.
   2. **Proximally,** cerebral veins pierce the arachnoid layer and traverse the subdural space to gain access to the **dural venous sinuses.** These triangular sinuses are bounded by two leaves of meningeal (true) dura; the base is periosteal dura (see Figure 31-1). Endothelium lines the sinuses.
   3. **Venous drainage pathway** [see Chapter 31 IV A 1 b (1) (b)]
   4. **Clinical considerations**
      a. **Venous hemorrhage.** As the larger cerebral veins pierce the arachnoid layer and cross the subdural space to gain access to the dural sinuses, they receive little support and may be torn, resulting in **subdural hematoma** (see Chapter 31 IV B 3).
      b. **Air embolism** [see Chapter 31 IV A 1 b (2) (a) (ii)]

Chapter 33
Orbit, Eye, and Ear

I. ORBIT

A. **The bony orbit (orbita, orbital cavity)** is a shell of bone that nearly surrounds and protects the eyeball.

1. **Structure.** It is a pyramidal cavity with the axis directed about 23° from the sagittal plane (see Figure 33-3).
   a. The **orbital roof (superior wall)** is formed by the frontal bone with a minor contribution from the lesser wing of the sphenoid bone (Figure 33-1).
      (1) Laterally, the **roof** contains a depression for the lacrimal gland, the **lacrimal fossa.**
      (2) Anteriorly, the frontal bone contains **frontal sinuses.** At the supraorbital margin, the **supraorbital notch** or **foramen** transmits the supraorbital branch of the frontal nerve.
      (3) Superiorly, the orbital roof is also the floor of the anterior cranial fossa.
   b. The **medial wall** (in a sagittal plane) consists of the **ethmoid bone** and the small **lacrimal bone** (see Figure 33-1).
      (1) The **lamina papyracea,** a paper-thin portion of the ethmoid bone, separates the orbit from the underlying ethmoid air cells.
      (2) The **anterior and posterior ethmoid foramina** transmit neurovascular structures from the orbit to the nasal cavity and paranasal sinuses.
      (3) The **fossa of the lacrimal sac (lacrimal groove)** in the lacrimal bone continues as the **nasolacrimal canal.** It contains the nasolacrimal duct and terminates in the **inferior nasal meatus.**
      (4) **Medial wall fracture**
         (a) Because the orbital contents are incompressible, "orbital compression injury" from catching a ball in the orbit transmits forces that may fracture orbital walls. The ethmoidal lamina papyracea bone is weak and prone to fracture.
         (b) This injury provides communication between the nasal cavity and orbit by way of the ethmoid sinuses, resulting in **orbital emphysema.**
   c. The **lateral wall** (about 45° from the sagittal plane) is formed by the **zygomatic bone** and the **greater wing of the sphenoid bone** (see Figure 33-1).
   d. The **orbital floor** is formed by the **maxilla,** with a minute contribution from the **palatine bone** (see Figure 33-1).
      (1) Traversing the floor is the **infraorbital groove,** which transmits the infraorbital branch of the maxillary nerve. This groove becomes the **infraorbital canal** just before reaching the infraorbital margin and terminates as the **infraorbital foramen.**
      (2) The floor of the orbit is also the roof of the maxillary sinus.
      (3) **Orbital floor fracture**
         (a) The infraorbital canal is weak and prone to fracture.
         (b) This injury provides communication between the nasal cavity and orbit by way of the maxillary sinus, resulting in **orbital emphysema.**

2. **Foramina and fissures.** At the apex of the bony orbit are three apertures (see Figure 33-1).
   a. The **optic canal** passes through the lesser wing of the sphenoid bone and connects the orbit with the middle cranial fossa (see Figure 33-1).
      (1) The **optic nerve** passes through this canal to reach the eye.
      (2) The **ophthalmic artery** leaves the internal carotid artery near the optic nerve. After giving off the central artery of the retina that enters the optic nerve, the ophthalmic artery follows the optic nerve through the optic canal into the bony orbit.

FIGURE 33-1. Bones of the right orbit.

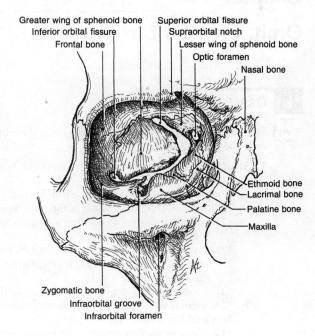

b. The **superior orbital fissure** (between the greater and lesser wings of the sphenoid bone) connects the orbit with the middle cranial fossa (see Figure 33-1). It transmits several neurovascular structures.
  (1) The **oculomotor nerve** (CN III) innervates several extraocular muscles (see Figure 33-4A). A parasympathetic component innervates the ciliary muscle and iridial constrictor.
  (2) The **trochlear nerve** (CN IV) innervates the superior oblique muscle (see Figure 33-4B).
  (3) The **ophthalmic division of the trigeminal nerve** (CN V_1) provides sensory innervation to the supraorbital region, forehead, and anterior scalp, and to the skin of the nose medial to the eye.
  (4) The **abducens nerve** (CN VI) innervates the lateral rectus muscle (see Figure 33-4B).
  (5) The **main (superior branch) of the ophthalmic vein** drains the eyeball and orbital contents primarily to the **cavernous sinus.**
c. The **inferior orbital fissure** connects the orbit with the pterygopalatine fossa (see Figure 33-1). It is spanned in large part by the involuntary **orbitalis muscle** (Müller's first).
  (1) The **infraorbital nerve** (a branch of the maxillary division of the trigeminal nerve) lies in the infraorbital groove and passes into the face through the infraorbital foramen.
  (2) The **infraorbital artery** lies in the infraorbital groove.
  (3) The **inferior ophthalmic vein** drains the eyeball and orbital contents variably to the cavernous sinus and pterygoid venous plexus.
d. The **superior and inferior orbital fissures** are continuous around the medial orbital surface of the greater wing of the sphenoid bone.

B. Eyelids (palpebrae)
  1. **Overview.** The eyelids protect the anterior aspect of the eye.
    a. They enclose a potential space, the **conjunctival sac,** which opens into the skin of the face at the palpebral fissure.
    b. The **palpebral fissure** is bounded by the upper and lower palpebral margins.
    c. The **palpebral margins** meet at the **medial canthus** and **lateral canthus.**

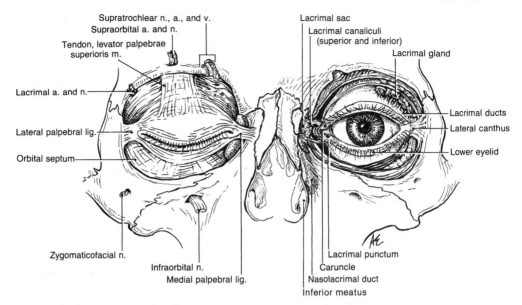

FIGURE 33-2. Palpebrae and lacrimal apparatus. Eyelid structure, insertion of the levator palpebrae superioris muscle, and associated neurovascular structures of the right eye; the lacrimal apparatus of the left eye.

2. **Structure.** The upper eyelid is larger and more mobile than the lower. Each lid contains a **tarsal plate** of dense connective tissue, which maintains the shape of the lid and to some extent protects the underlying eyeball (Figure 33-2).
   a. The **superior and inferior tarsal plates** merge to form the **medial** and **lateral palpebral ligaments,** which insert into the respective orbital margins.
   b. The **orbital septum** is a fascial sheet that attaches the tarsal plates to the superior and inferior orbital margins.
   c. The **levator palpebrae superioris muscle** attaches to the tarsal plate of the superior eyelid (see Figure 33-2).
   d. The **subcutaneous tissue** of the eyelids is loose and may swell significantly with collection of fluid (puffy eyelids) or extravasated blood (black eye).

3. **Musculature**
   a. The **palpebral portion of the orbicularis oculi muscle,** a muscle of facial expression, runs between the skin of the eyelid and the tarsal plate (see Figures 33-2 and 36-1).
      (1) Its fibers insert into the skin.
      (2) Contraction produces the blink. Muscle fibers of the palpebral portion move toward geodesic lines (the shortest distance between two points over the surface of a sphere) to close the palpebral fissure.
      (3) It is innervated by the **facial nerve** (CN VII).
         (a) This nerve forms the **efferent limb of the corneal blink reflex.**
         (b) Paralysis of the palpebral portion, such as occurs in **facial nerve (Bell's) palsy,** results in an inability to close the eyelids. When the eyes cannot blink, the conjunctiva dries and the cornea may ulcerate.
   b. The **levator palpebrae superioris** is an extraocular muscle (see Figures 33-2 and 33-3A).
      (1) It arises from the apex of the orbit and widens into an aponeurosis that inserts into the superior tarsal plate.
      (2) It draws the lid upward when the eyeball is elevated.
      (3) It is innervated by the **superior division of the oculomotor nerve** (CN III). Paralysis produces **ptosis** or **blepharoptosis,** with inability to lift the lid when the gaze is directed upward.

c. The **superior tarsal muscle** (Müller's third) consists of smooth (involuntary) muscle.
   (1) It runs from the superior border of the tarsal plate of the upper lid to the tendon of the levator palpebrae superioris muscle.
   (2) Contraction accentuates the opening of the palpebral fissure, as in "wide-eyed fear." Lack of muscle tone because of decreased sympathetic activity or decreased circulating levels of epinephrine results in difficulty keeping the eyes open ("tired eyes").
   (3) This muscle is innervated by **sympathetic nerves** that have preganglionic cell bodies in the **upper thoracic levels of the spinal cord** and postsynaptic cell bodies in the **superior cervical ganglion.**
   (4) A lesion involving the cranial sympathetic pathway **(Horner's syndrome)** causes paralysis of this muscle with a slight drooping of the lid **(pseudoptosis).**
4. The **palpebral margins** contain numerous glands and a double or triple row of hairs. The medial ends tend to be hairless.
   a. **Tarsal (meibomian) glands** lubricate the lids. They occasionally become infected **(acute meibomianitis).**
   b. **Sebaceous glands** are associated with the hair follicles. Infection of a sebaceous gland produces the **sty.**
   c. **Sweat glands** are also present.
5. **Conjunctiva** lines the inner surfaces of the eyelid and the anterior eyeball, forming the **conjunctival sac** (see Figure 33-6).
   a. Palpebral conjunctiva reflects off the base of each lid onto the eyeball as the bulbar conjunctiva. This reflection forms the **superior** and **inferior conjunctival fornices.**
      (1) **Lacrimal ducts** drain into the superior fornix.
      (2) **Palpebral conjunctiva** is richly supplied with blood. It normally is pink, but becomes pallid in anemia.
   b. The **bulbar (corneal) conjunctiva** protects the cornea and anterior sclera.
      (1) It is transparent and contains many small blood vessels. Vasodilation in response to allergy, irritation, or inflammation produces the **bloodshot eye** and **pinkeye.**
      (2) Innervation by the **ophthalmic division of the trigeminal nerve** (CN V_1) provides the **afferent limb of the blink reflex.**
         (a) Any foreign contact with the conjunctiva causes blinking. Interconnections in the brain stem produce reflex blinking in both eyes simultaneously—the **consensual blink reflex.**
         (b) The **blink reflex** provides a mechanism for testing the afferent portion of the trigeminal nerve and the efferent portion of the facial nerve.
            (i) If both eyes blink when the right eye is touched but not when the left eye is touched, a problem involves the left trigeminal nerve.
            (ii) If only the right eye blinks when either eye is touched, the left facial nerve is affected.
6. A **lacrimal gland** lies in the superior lateral region of each orbit (see Figure 33-2).
   a. Each **lacrimal gland** has an intimate relationship with the tendon of the levator palpebrae superioris muscle. The major portion of this gland lies above the levator palpebrae superioris; a smaller portion lies beneath it. Thus, movement of the eyelids tends to milk the gland, lubricating and moistening the conjunctiva.
   b. **Lacrimal ductules,** about a dozen, drain the lacrimal gland into the lateral region of the superior conjunctival fornix.
   c. **Parasympathetic neurons** mediate lacrimation by the facial nerve (CN VII). With cholinergic stimulation, tears flush across the eyeball and are drawn by capillary action along the palpebral fissure.
7. **Lacrimal apparatus**
   a. The **lacus lacrimalis** is a triangular area at the medial canthus enclosed by the **plica semilunaris,** a small fold of conjunctiva that is a remnant of the nictitating membrane.

b. A **small lacrimal papilla** with an apical **punctum** borders the lacus lacrimalis of each lid and forms the orifice of the lacrimal canaliculus (see Figure 33-2).
c. The **upper and lower lacrimal canaliculi** drain behind the medial palpebral ligament and join to drain into the lacrimal sac.
d. The **lacrimal sac** is the blind upper end of the **nasolacrimal duct.** It rests in the **lacrimal groove** of the lacrimal and maxillary bones (see Figure 33-2).
   (1) The **lacrimal fascia** passes around the sac from the crest on the frontal process of the maxilla to the lacrimal crest.
   (2) The medial attachments of the palpebral portion of the orbicularis oculi muscle split around the lacrimal sac. Those fibers posterior to the sac are referred to as the **lacrimal portion** (Horner's muscle). Blinking compresses the lacrimal sac. With muscle relaxation, fluid is aspirated into the lacrimal canaliculi from the lacus.
   (3) The **nasolacrimal duct** drains into the **inferior meatus** of the nasal cavity.
      (a) A small flap of mucosa at its nasal orifice usually prevents the percolation of air into the conjunctival sac.
      (b) If the duct is blocked, the lacrimal fluid cannot drain into the nose and so runs out onto the cheek (epiphora).
   (4) **Lacrimation,** in the early stages, involves "sniffling" as the lacrimal fluid drains into the nose. When the lacrimal canaliculi become overwhelmed, lacrimal fluid overflows the palpebrae and tears flow down the cheeks.

C. **Extraocular musculature.** The orbital cavity contains seven voluntary muscles of somatic origin, six of which move the eyeball.
1. **Ocular axes.** The eyeballs, suspended within the orbit by a fibrous capsule, move about three mutually perpendicular axes.
   a. **Elevation and depression** occur about a **transverse axis** through the equator of the eyeball.
   b. **Abduction and adduction** occur about a **vertical axis** through the poles.
   c. **Intorsion and extorsion** occur about an **anteroposterior axis** through the pupil. **Intorsion** is a medial rotation of the upper portion of the eye toward the midline; **extorsion** is the opposite.
2. **The annulus tendineus** (of Zinn) gives rise to four rectus muscles at the apex of the orbit (see Figure 33-3).
   a. Contents
      (1) The **optic nerve** and the **ophthalmic artery** pass through the annulus as they leave the optic canal to enter the orbital cavity.
      (2) The superior and inferior divisions of the **oculomotor nerve, abducens nerve,** and **nasociliary branch** of the ophthalmic nerve pass through the annulus as they leave the superior orbital fissure to enter the orbital cavity.
   b. **Relations.** The **ophthalmic vein** and **trochlear nerve,** as well as the frontal and lacrimal branches of the **ophthalmic nerve,** pass superior to the annulus tendineus.
3. **Four rectus muscles** diverge from the annulus tendineus to insert into the sclera of the anterior portion of the eyeball (see Table 33-1).
   a. The **superior rectus muscle** inserts onto the anterosuperior aspect of the eyeball (Figure 33-3). It primarily **elevates the eye** and is innervated by the superior branch of the **oculomotor nerve** (CN III).
   b. The **inferior rectus muscle** inserts onto the anteroinferior aspect of the eyeball (see Figure 33-3B). It **depresses the eye** and is innervated by the inferior branch of the **oculomotor nerve** (CN III).
   c. The **lateral rectus muscle** inserts onto the anterolateral aspect of the eyeball (see Figure 33-3). It **abducts the eye** and is innervated by the **abducens nerve** (CN VI).
   d. The **medial rectus muscle** inserts onto the anteromedial aspect of the eyeball (see Figure 33-3). It **adducts the eye** and is innervated by the superior branch of the **oculomotor nerve** (CN III).
4. **Two oblique muscles** have different origins, but insert into the sclera of the posterior portion of the eyeball (see Table 33-1).

Table 33-1. Extraocular Musculature

Muscle	Origin	Insertion	Primary Action	Secondary Actions (Normally cancel out)	Innervation
Superior rectus	Annulus tendineus	Anterosuperior aspect	Elevation	Adduction and intorsion	Oculomotor (CN III)
Inferior oblique	Lacrimal bone	Posterolateral inferior quadrant	Elevation	Abduction and extorsion	Oculomotor (CN III)
Inferior rectus	Annulus tendineus	Anteroinferior aspect	Depression	Adduction and extorsion	Oculomotor (CN III)
Superior oblique	Sphenoid bone	Posterolateral superior quadrant	Depression	Abduction and intorsion	Trochlear (CN IV)
Medial rectus	Annulus tendineus	Anteromedial aspect	Adduction		Oculomotor (CN III)
Lateral rectus	Annulus tendineus	Anterolateral aspect	Abduction		Abducens (CN VI)
Levator palpebrae superioris	Sphenoid bone	Tarsal plate of upper eyelid	Raises upper eyelid		Oculomotor (CN III)

    a. **Superior oblique muscle** has a unique course (see Figure 33-3). It runs anteriorly from the sphenoid bone superomedial to the annulus tendineus. Its tendon passes through a fibrous trochlea (L. pulley) attached to the frontal bone at the anterosuperior aspect of the medial orbital margin. The tendon then reverses to insert onto the posterolateral quadrant of the superior surface of the eyeball. It draws the posterior portion of the eyeball upward to **depress the eye** and is innervated by the **trochlear nerve** (CN IV).
    b. **Inferior oblique muscle** originates from the anteroinferior aspect of the medial orbital margin. It passes posterolaterally to insert onto the posterolateral quadrant of

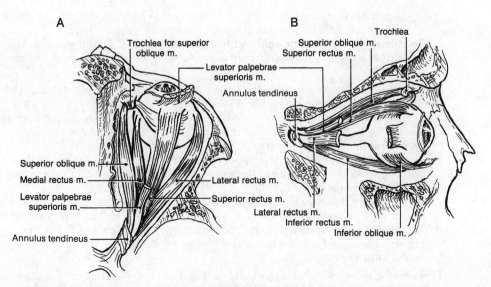

FIGURE 33-3. Extrinsic muscles of the right eye. *(A)* Viewed from above; *(B)* viewed from the right side.

the inferior surface of the eyeball (see Figure 33-3B). It seems to represent the distal portion of a primitive muscle that paralleled the superior oblique muscle. It draws the posterior portion of the eyeball downward to **elevate the eye** and is innervated by the inferior branch of the **oculomotor nerve** (CN III).

5. **The levator palpebrae superioris muscle** originates superior to the annulus tendineus and inserts into the tarsal plate of the superior eyelid (see Figure 33-3B). It is innervated by the superior branch of the **oculomotor nerve** (CN III) and draws the eyelid upward when the eyeball is elevated.

6. **Group actions**
   a. **Primary actions** are straightforward (Table 33-1).
      (1) **Abduction/adduction** occurs about the vertical axis.
         (a) **Abduction** is by the **lateral rectus muscle.**
         (b) **Adduction** is by the **medial rectus muscle.**
      (2) **Elevation/depression** occurs about the **transverse axis.**
         (a) **Elevation.** The **superior rectus** muscle draws upward on the anterior superior quadrant. The **inferior oblique muscle** draws downward on the posterior inferior quadrant of the eyeball.
         (b) **Depression.** The **inferior rectus** muscle draws downward on the anterior–inferior quadrant. The **superior oblique muscle** draws upward on the posterior–superior quadrant of the eyeball.
   b. **Secondary actions** are complex because the bony orbit is directed somewhat laterally (23°) and the pupil normally is directed anteriorly (0°).
      (1) **Abduction/adduction.** The lines of action of the extraocular muscles above and below the eye pass medial to the vertical axis. Therefore, they draw the points of attachment medially about the vertical axis.
         (a) The **superior and inferior recti** draw the anterior portion of the eye medially, thereby adducting the eye.
         (b) The **superior and inferior oblique muscles** draw the posterior portion of the eyeball medially, thereby abducting the eye.
      (2) **Intorsion/extorsion.** Similarly, the lines of action of the extraocular muscles above and below the eye pass medial to the anteroposterior axis. Therefore, they draw the points of attachment medially about the anteroposterior axis.
         (a) The **superior rectus and superior oblique muscles** secondarily rotate the superior portion of the eyeball medially **(intorsion).**
         (b) The **inferior rectus and inferior oblique muscles** secondarily rotate the inferior portion of the eyeball medially (**extorsion,** because the top of the eye rotates laterally).
      (3) **Cancellation of secondary actions**
         (a) Secondary actions of extraocular muscles tend to balance or cancel, leaving the primary actions (see Table 33-1).
         (b) Paralysis of one or more extraocular muscles produces imbalance of the secondary actions. Complex forms of double vision occur when gaze is directed in certain directions.

7. **The fascia bulbi** (Tenon's capsule) forms a connective tissue socket in which the eyeball is suspended.
   a. Extraocular muscles pass through Tenon's capsule to insert into the sclera.
   b. Sheaths of the medial and lateral rectus tendons thicken, forming the **medial and lateral check ligaments,** which check the action of opposite muscles.
   c. These fascial condensations continue around the inferior surface of the eyeball to form a fascial hammock that suspends the eyeball between the medial and lateral margins of the bony orbit.

D. Innervation

1. **Nerves of the extraocular musculature**
   a. The **oculomotor nerve** (CN III) contains somatic and autonomic components (Figure 33-4A).
      (1) Axons from the **oculomotor nucleus** innervate five extraocular muscles.

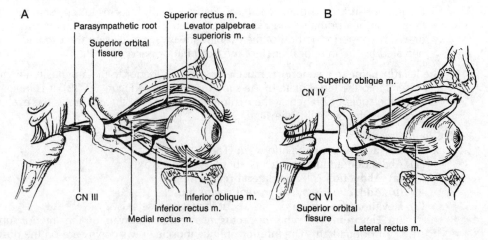

FIGURE 33-4. Innervation of the extrinsic muscles of the eye. *(A)* Oculomotor nerve; *(B)* trochlear and abducens nerves.

- (a) The **superior division** innervates the levator palpebrae superioris and the superior rectus, and usually the medial rectus muscles.
- (b) The **inferior division** innervates the inferior rectus and the inferior oblique muscles.
- (2) Axons from the **accessory oculomotor nucleus** (of Edinger-Westphal) contribute the parasympathetic component of the oculomotor nerve (see Figure 32-7).
  - (a) **Preganglionic parasympathetic neurons** leave the oculomotor nerve through the **parasympathetic (motor) root** to reach the ciliary ganglion.
  - (b) **Postganglionic axons of the ciliary ganglion** course through the **short ciliary nerves** to the iridial sphincter for pupillary constriction and to the ciliary muscle for accommodation to near vision.
- (3) **Lesions** produce **lateral strabismus** and nearly complete **ophthalmoplegia** of the eye as well as **ptosis** of the eye lid. In addition, the pupil is dilated **(mydriasis)**, and there is loss of accommodation.

b. The **trochlear nerve** (CN IV) conveys motor innervation (see Figure 33-4B).
- (1) Axons arise from the **trochlear nucleus** and innervate the **superior oblique muscle.**
- (2) Lesions produce **diplopia** when looking down. Individuals with diplopia usually experience difficulty and apprehension on descending stairs.

c. The **abducens nerve** (CN VI) provides motor innervation (see Figure 33-4B).
- (1) Axons arise from the **abducens nucleus** and innervate the **lateral rectus muscle.**
- (2) Lesions produce **medial strabismus** (crossed eyes). Diplopia is minimal when looking toward the side opposite the lesion.

2. **The ophthalmic division of the trigeminal nerve (CN V_1)** provides sensory branches within and about the orbit.
  a. The **lacrimal nerve** runs superolaterally within the orbit (Figure 33-5; see Figure 36-4).
   - (1) It supplies the lateral portion of the eyelids and palpebral conjunctiva.
   - (2) Proximally, it contains no secretomotor fibers to the lacrimal gland. Distally, it receives postsynaptic parasympathetic secretomotor fibers from the pterygopalatine ganglion through the zygomatic branch of the maxillary nerve.
  b. The **frontal nerve,** the largest branch of the ophthalmic nerve, bifurcates (see Figures 33-5 and 36-4).
   - (1) The **supraorbital nerve** runs inferior to the orbital roof. It passes through the **supraorbital notch** or foramen to supply the skin of the eyelid, forehead, and scalp.

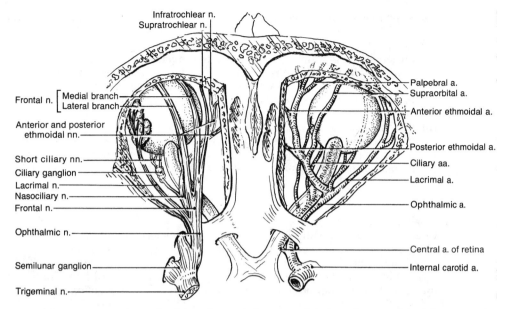

FIGURE 33-5. Nerves and vessels of the orbit, including the ophthalmic division of the left trigeminal nerve and the right ophthalmic artery.

- (2) The **supratrochlear nerve** runs more medially and supplies the skin of the medial portions of the eyelids and the central portion of the forehead.
- c. The **nasociliary nerve** runs superomedially within the orbit and becomes the **anterior ethmoidal nerve** (see Figures 33-5 and 36-4).
  - (1) A **sensory root (ramus communicans)** connects with the **ciliary ganglion.** Sensation from the cornea passing along the **short ciliary nerves** gains access to the nasociliary nerve by this ramus. (Sensory pathways synapse in the trigeminal, not the ciliary, ganglion). They contribute to the afferent limb of the **blink reflex.**
  - (2) **Long ciliary nerves** also convey corneal sensation directly from the eyeball. They contribute to the afferent limb of the **blink reflex.**
  - (3) The **posterior ethmoidal nerve** passes through the **posterior ethmoidal foramen** to supply the mucosa of the sphenoidal and ethmoidal air sinuses as well as a portion of the nasal mucosa.
  - (4) The **infratrochlear branch** supplies skin of the medial portion of the eyelids and palpebral conjunctiva.
  - (5) The **anterior ethmoidal nerve,** the continuation of the nasociliary nerve, passes through the **anterior ethmoidal foramen** and descends through the ethmoid bone.
    - (a) The **internal nasal branch** supplies the mucosa of the ethmoidal and frontal air sinuses as well as portions of the nasal mucosa.
    - (b) The **external nasal branch** supplies the skin at the tip of the nose and about the external nares.
3. **The optic nerves** are cerebral tracts with complete meningeal sheaths of dura, arachnoid, and pia.
   a. These nerves pass through the **optic foramen** in company with the ophthalmic artery (see Figure 33-5).
   b. The **central artery of the retina,** a branch of the ophthalmic artery, runs within the optic nerve.
4. **Autonomic innervation to the lacrimal gland**
   a. **Sympathetic innervation** follows the perivascular plexuses to reach the lacrimal gland. The sympathetic role in lacrimation is not defined.

b. **Parasympathetic innervation** mediates lacrimation by the **facial nerve** (see Figure 36-3). With cholinergic stimulation, tears flush across the eyeball, drawn by capillary action along the palpebral fissure.
   (1) **Preganglionic pathway. Presynaptic parasympathetic neurons** arise from the **superior salivatory nucleus.** These axons travel with the **nervus intermedius** of the facial nerve, the **greater superficial petrosal nerve,** and the **nerve of the vidian (pterygoid) canal** to the **pterygopalatine ganglion** in the pterygopalatine fossa.
   (2) **Postganglionic pathway.** In the pterygopalatine fossa, postsynaptic axons from the **pterygopalatine ganglion** travel along the **maxillary nerve** (CN V_2), its **zygomaticofacial branch,** and a communicating branch to the **lacrimal branch of the ophthalmic nerve** (CN V_1) along which it travels to reach the lacrimal gland.

E. **Vasculature**
   1. **Ophthalmic artery**
      a. **Course.** This branch of the internal carotid artery enters the orbit through the **optic canal** beneath the optic nerve. Within the orbit, it winds to the medial surface of the optic nerve in company with the nasociliary nerve (see Figure 33-5).
      b. **Intraorbital branches.** It supplies the eyeball through the **central artery of the retina** as well as by **anterior** and **posterior ciliary branches.** In addition, it sends twigs to the extraocular muscles.
      c. **Extraorbital branches** correspond to the named branches of the ophthalmic division of the trigeminal nerve: **anterior ethmoidal, posterior ethmoidal, supratrochlear, supraorbital,** and **lacrimal arteries.**
   2. **Ophthalmic vein**
      a. The **superior ophthalmic vein** drains the superior regions of the orbit, eyelids, and forehead, as well as receiving **vorticose veins** from the eyeball.
         (1) It anastomoses with the **angular vein,** a branch of the facial vein, and receives most of the return of the inferior ophthalmic vein.
         (2) It leaves the orbit through the **superior orbital fissure** to drain into the **cavernous sinus.**
      b. The **inferior ophthalmic vein** drains the inferior regions of the orbit and eyelids as well as receiving **vorticose veins** from the eyeball.
         (1) It anastomoses with the **angular vein** and the **superior ophthalmic vein.**
         (2) A major trunk usually anastomoses the superior ophthalmic vein, draining to the **cavernous sinus.** A lesser trunk usually leaves the orbit through the inferior orbital fissure, draining to the **pterygoid venous plexus.**
      c. **Inflammatory thrombosis of the cavernous sinus.** Anastomoses between the angular and ophthalmic veins may result in infectious spread from periorbital and perinasal regions to the cavernous sinus.
         (1) **Retinal sequelae.** Cavernous sinus thrombosis interferes with venous drainage of the retina, resulting progressively in **engorgement of the retinal arteries,** retinal ischemia, and blindness.
         (2) **Neural sequelae.** Because the ophthalmic, oculomotor, trochlear, and abducens nerves also pass through the cavernous sinus, local infection may result in ophthalmoplegia, mydriasis, and sensory deficits in the periorbital region (see Figure 31-11).
         (3) **Cerebrovascular sequelae.** Because the carotid artery passes through the cavernous sinus, inflammation may produce **carotid arteritis** with thrombus formation. When dislodged, thrombi produce showers of cerebral thromboemboli with stroke.

II. EYEBALL (BULBUS OCULI)

A. **Overview.** The eyeball is formed as an outgrowth of the brain, the **optic vesicle.** The anterior half of the optic vesicle involutes against the posterior half, forming the **optic cup.** Two primitive layers are formed by occlusion of the optic vesicle.

1. **Two layers of the retina** of the adult eye represent an extension of the brain. The retina transduces electromagnetic radiation into electrical impulses.
2. Other parts of the eyeball, such as the **lens,** are derived from ectoderm. The lens inverts and focuses the visual fields onto the retina.

B. **Ocular tunics.** The eyeball is a durable structure with a tough fibrous coat and a fluid-filled cavity that maintains the shape. It supports the lens in such a way that the focal point lies consistently on the retinal surface.
1. **Fibrous tunic**
    a. The **sclera,** the white of the eye, accounts for about five sixths of the eyeball (Figure 33-6).
        (1) It is composed of dense connective tissue that counteracts hydraulic pressure within the chambers to maintain the shape of the eye.
        (2) It provides insertion for extraocular muscles as well as zonular fibers that suspend the lens.
        (3) Anteriorly, the sclera is covered by bulbar conjunctiva, which is transparent and contains many small blood vessels and nerve endings.
            (a) Conjunctiva is innervated by the ophthalmic division of the **trigeminal nerve** (CN V_1), which mediates the **afferent limb of the blink reflex.**
            (b) **Inflammation of conjunctiva** (conjunctivitis) causes vascular engorgement (bloodshot eye or pinkeye). Although the usual cause is infection or allergy, occasionally the pathologic process is within the eyeball (iritis, glaucoma). This situation is often accompanied by pain or impaired vision.
        (4) The sclera joins the cornea at the **limbus.**
    b. The **cornea** is transparent and accounts for the anterior one sixth of the eyeball (see Figure 33-6).
        (1) It consists of dense collagen fibers, the extreme regularity of which results in a **crystalloid structure** that is transparent to light.
            (a) The cornea is the **principal refractor** of the eye (more than the lens). Its small radius of curvature and the separation of media with different refractive indices (air and aqueous humor) strongly bends light waves.

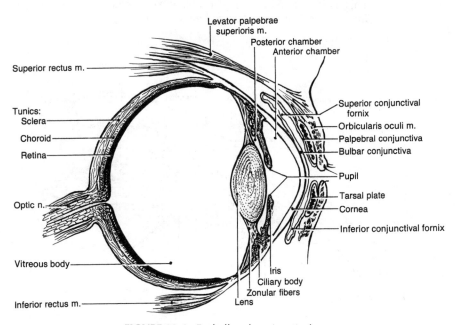

FIGURE 33-6. Eyeball and conjunctival sac.

(b) The shape of the cornea and the anteroposterior diameter of the eye determine the focal point.
  (i) Insufficient corneal refraction that cannot be corrected by the lens results in a focal point behind the retina **(hyperopia or hypermetropia)**. The result is farsightedness, because distant objects can be focused on the retina by accommodation of the lens.
  (ii) Excessive corneal refraction results in a focal point in front of the retina **(myopia)**. The result is nearsightedness, because close objects focus on the retina without accommodation by the lens.
  (iii) Irregularities in corneal shape produce variation in the focal points **(astigmatism)**.
(2) Corneal epithelium, a continuation of the conjunctiva, is transparent and contains small blood vessels and nerve endings.
(3) Injury or inflammation (keratitis) heals but with a loss of the high degree of organization and, consequently, loss of transparency. Corneal transplantation **(keratoplasty)** is effective treatment in individuals with corneal scars. Some focal problems can be treated by surgically changing the corneal shape **(radial keratotomy)**.

2. Vascular tunic
  a. The **choroid layer** consists primarily of blood vessels supplied by the **short ciliary arteries** and drained by the **vorticose veins** (see Figure 33-6).
  b. The **ciliary body** is the anterior continuation of the choroid layer. The epithelium of the highly vascular ciliary body secretes **aqueous humor** into the posterior chamber.
    (1) The ciliary body suspends the lens by a multitude of **zonular fibers** (of Zinn), which insert into the capsule of the lens.
    (2) The **ciliary muscle** runs from the base of the ciliary body to the scleral spur at the limbus (Figure 33-7). It controls tension on the **zonular fibers** and, thus, the shape of the lens.
      (a) Far accommodation results from ciliary muscle relaxation. The normal elastic tension exerted on the capsule of the lens through the zonular fibers tends to flatten the lens, resulting in minimal refraction of light rays.
      (b) Near accommodation results from muscle contraction, which draws the ciliary body anteriorly and thereby releases tension on the zonular fibers. This movement allows the elastic lens to become more spherical, increasing refraction and converging light.

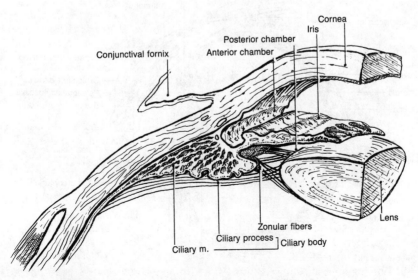

FIGURE 33-7. Limbus of the eyeball, ciliary body, and lens.

(c) **Control of accommodation** is by complex central pathways and mediated by parasympathetic nerves of the accessory oculomotor nucleus (of Edinger-Westphal).
(3) The **iris** arises from the anterior portion of the ciliary body (see Figures 33-6 and 33-7).
   (a) It divides the space between the cornea and lens into **anterior** and **posterior chambers.**
   (b) It contains variable amounts of pigment.
      (i) Heavy pigmentation in the anterior layer results in brown eyes.
      (ii) If pigment is limited to the posterior layer, the iris appears blue or gray because light refracts within the unpigmented anterior layer.
   (c) The iris contains primitive myoepithelial cells surrounding a central aperture, the **pupil.**
      (i) **The sphincter pupillae** consist of myoepithelial cells arranged in a circle under parasympathetic control.
      (ii) **The dilator pupillae** consist of radially arranged myoepithelial cells under sympathetic control.
   (d) The size of the pupil is a reflex response to the intensity of light reaching the retina.
(4) **Autonomic innervation** of the iris and ciliary body
   (a) **Parasympathetic pathways**
      (i) **Presynaptic parasympathetic nerves** from the **accessory oculomotor nucleus** (of Edinger-Westphal) travel to the orbit along the oculomotor nerve (CN III), leave through the parasympathetic root, and synapse in the **ciliary ganglion.**
      (ii) **Postsynaptic parasympathetic fibers** from the ciliary ganglion travel along the short ciliary nerves to reach the eyeball, ciliary body, and iridial sphincter.
   (b) **Sympathetic pathways**
      (i) **Preganglionic sympathetic neurons** lie in the upper thoracic segments of the spinal cord and then course superiorly in the sympathetic chain to reach the **superior cervical ganglion.**
      (ii) **Postsynaptic sympathetic fibers** then course in perivascular plexuses to reach the iridial dilator.
   (c) **Clinical considerations**
      (i) Injury to the oculomotor nerve releases parasympathetic influence and produces a dilated pupil **(mydriasis).**
      (ii) Injury to the upper thoracic spinal cord or to the cervical sympathetic chain **(Horner's syndrome)** releases sympathetic influence and produces a constricted pupil **(miosis).**

3. **Retina** (see Figure 33-6).
   a. The **neural retina** (the anterior primitive layer) consists of four layers of nerve and supporting cells through which light rays must pass to reach the photosensitive cells.
      (1) The **first layer** consists of **nerve axons** that collect at the **macula** or **optic disk** and pass through the **cribriform plate** of the sclera to form the optic nerve. The result is a blind spot.
      (2) The **second layer** is formed by the so-called **ganglion cells** and is equivalent to a brain stem nucleus.
      (3) The **third layer** consists of **bipolar cells,** equivalent to cells of dorsal root ganglia.
      (4) The **fourth layer** contains the light-sensitive **rods and cones.** At the **fovea centralis,** a great concentration of cones maximizes visual acuity in the center of the visual field.
   b. The **pigmented retina** (the posterior primitive layer) absorbs light that passes completely through the anterior layer, preventing backscatter (blurring of vision).

c. Clinical considerations
   (1) Retinal detachment is the result of separation of the anterior and posterior layers, re-establishing the primitive optic vesicle.
   (a) Separation usually starts anteriorly and is easily missed until it is well advanced.
   (b) Predisposition to detachment is bilateral. Care must be taken to protect the other eye.
   (c) The layers may be reattached by coagulation procedures.
   (2) Funduscopic examination
   (a) Retinal blood flow
      (i) The central artery of the retina enters the eyeball with the optic nerve and branches over the retina as end-arteries (terminal branches that have no anastomoses). These vessels are the only arteries in the body that can be examined visually for signs of systemic disease, such as hypertension and diabetes.
      (ii) Obstruction to venous return, such as by cavernous sinus thrombosis, is indicated by retinal engorgement.
   (b) Papilledema. Because the optic nerve is a portion of the central nervous system and is surrounded by layers of meninges, elevated cerebrospinal fluid pressure produces edema of the optic disk, which can be detected by ophthalmoscopic examination.

C. Chambers of the eye

1. **Anterior cavity** (see Figure 33-6). The **anterior chamber** lies between the cornea and the iris; the **posterior chamber** lies between the iris and the lens.
   a. These chambers contain thin, watery **aqueous humor.**
      (1) Aqueous humor is secreted by the **ciliary process** into the posterior chamber.
      (2) It passes through the pupil into the anterior chamber.
      (3) It drains into the venous system through **sinus venosum sclerae (Schlemm's canal)** at the angle of the anterior chamber.
   b. If drainage is impaired, intraocular pressure increases and retinal blood flow is impaired, producing retinal ischemia (**glaucoma**) and blindness.

2. **The posterior cavity** or **vitreous chamber,** behind the lens and ciliary body, contains the **vitreous body,** a transparent and semigelatinous material. Osmotic pressure produced by the composition of the vitreous body provides hydraulic pressure that maintains the shape of the eyeball.

3. **The lens** separates the aqueous humor from the vitreous body (see Figures 33-6 and 33-7).
   a. **Composition.** It consists of highly ordered connective tissue cells. The high degree of order confers transparency.
      (1) It is enclosed in an elastic capsule into which the zonular fibers insert. The lens itself is deformable and elastic. Variable tension from the ciliary muscle through the **zonular fibers** alters the shape of the lens.
      (2) **Cataract** is the most common cause of blindness. Progressive opacities result from degenerative changes and disorganize the regularity of the connective tissue fibers. When a cataract is sufficiently large to impair vision, the lens can be removed surgically.
   b. **Refractivity.** Its refractive index is slightly different from the aqueous and vitreous humors, providing some degree of refraction.
      (1) Tension exerted on the lens by the zonular fibers adjusts the shape, thereby altering the refracting power.
      (2) With aging, the lens tends to harden and lose intrinsic elasticity. It then cannot be deformed sufficiently to accommodate near and far extremes (**presbyopia**), both of which may require correction with bifocal lenses.

III. EAR

A. External ear

1. **The auricles** (pinnas), lying on either side of the head and directed slightly forward, concentrate sound waves and facilitate stereophonic localization of the source (see Figure 33-8).
   a. **Structure.** A single convoluted plate of **elastic cartilage** forms the skeleton of the ear. The skin of the auricle attaches to the underlying perichondrium.
   b. The **helix** is the posterior free rolled edge of the pinna.
      (1) It begins as the **crus** in the **concha,** superior to the external auditory meatus.
      (2) A small **superior (darwinian) tubercle** is a vestige of the point of the ear.
      (3) The helix ends as a flabby lobule or **ear lobe.**
   c. The **antihelix,** a second ridge, runs approximately parallel to the helix, dividing the auricle into an outer **scaphoid fossa** and the deeper **concha.**
      (1) It begins anterosuperiorly as two crura that define the **triangular fossa.**
      (2) Inferiorly, the antihelix terminates as the **antitragus.**
   d. The **tragus** is a projection from the anterior portion of the external ear.
      (1) Directed posteriorly, it partially covers and protects the external auditory meatus.
      (2) The **intertragic incisure** separates the tragus from the antitragus.

2. **The external auditory (acoustic) meatus** extends into the temporal bone from the concha to the tympanic membrane (Figure 33-8).
   a. **Development.** It is a remnant of the first branchial (gill) groove.
   b. **Structure.** It is somewhat S-shaped, with cartilaginous and bony segments.
      (1) The **external one third** is formed by a continuation of the elastic cartilage of the concha. It is directed upward and backward.
         (a) The **skin** lining the external meatus contains sebaceous glands, associated with hair follicles, and ceruminous glands.
         (b) **Cerumen (ear wax)** consists of secretions of both glands along with desquamated cells and dust. Excessive accumulations of ear wax can clog the external auditory meatus.

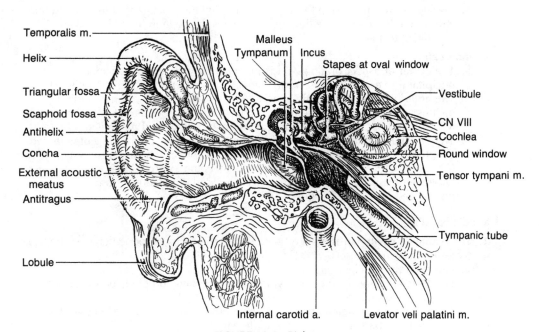

FIGURE 33-8. Right ear.

(2) The **internal two thirds** of the external meatus are within a bony canal. This segment is directed slightly anteriorly and downward.
  (a) The boundary between the external third and internal two thirds is marked by a slight ridge anteroinferiorly. Foreign objects may become impacted by this constriction.
  (b) The epithelium contains fewer glands and is devoid of hairs.
  (c) **Chronic infection** (e.g., **swimmer's ear**) may involve the underlying bone, producing bony hyperplasia and narrowing of the meatus.
  c. **Function.** The shape of the concha and external auditory meatus amplifies sound waves by 5–10 decibels.
3. **The tympanum** (tympanic membrane or ear drum) separates the external ear from the internal ear (see Figures 33-8 and 33-9).
  a. **Structure.** It consists of a sheet of fibrous tissue covered on both sides by epithelium. Circular and radial arrangements of fibers keep the membrane moderately tense and, thus, receptive to sonic vibrations.
  b. **Orientation.** It tilts across the external auditory meatus so that the anteroinferior quadrant is deeper than the posterosuperior quadrant. It is slightly concave.
    (1) The tympanum attaches to the malleolus along the length of the manubrium (see Figure 33-9).
    (2) The **umbo,** the central portion of the concavity, marks the tympanic attachment to the manubrium of the malleus.
    (3) Superior to the attachment of the malleus, the tympanic membrane appears less tense and more vascular, **pars flaccida.** The region below the pars flaccida is the **pars tensa.**
  c. **Function.** Pressure changes associated with sonic waves are transduced into mechanical vibrations at the tympanum.
    (1) Movement of air molecules against the tympanum by brownian movement is just below threshold in a young individual with acute hearing.
    (2) The faintest sound audible to the normal human ear corresponds to an excess pressure of about 10^{-10} atmospheres, which produces a tympanic excursion of about 5 nanometers.
    (3) The distance through which the tympanum moves in response to normal (60 decibels) conversation is only tens of nanometers.
  d. **Myringotomy,** an incision through the anteroinferior quadrant of the tympanum with insertion of an indwelling tube, provides drainage in individuals with chronic otitis media.
4. **Innervation**
  a. The **posterior aspect** of the pinna is innervated by the **great auricular** and **lesser occipital nerves,** which arise from spinal levels C2 and C3 (see Figure 30-7).
  b. The **anterior aspect** of the pinna and a variable part of the external auditory meatus are innervated by the auriculotemporal branch of the **mandibular nerve** (CN V3) and by the posterior auricular branch of the **facial nerve** (CN VII).
  c. The **external auditory meatus** is also innervated by twigs from the **glossopharyngeal** (CN IX) and **vagus** (CN X) **nerves,** which explains why a patient may gag (CN IX) or cough (CN X) when an insect enters the external auditory meatus or when cerumen is removed by curettage. In elderly individuals, cardiac arrest may be induced (carotid reflex of CN IX).

B. **Middle ear** (Figure 33-9)
  1. **The tympanic cavity (antrum)** is a space within the temporal bone between the squamous and petrous portions.
    a. **Lateral wall** (see Figure 33-9)
      (1) **Major features** include the **tympanic membrane** and the lateral wall of the epitympanic recess.
      (2) The **chorda tympani** arises from the facial nerve near the stylomastoid foramen and passes through the middle ear.
        (a) It enters the tympanic cavity at the posterior canaliculus of the chorda tympani (iter chordae posterius) in the posterior wall.

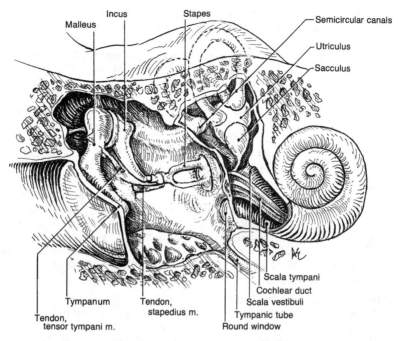

FIGURE 33-9. Right middle ear and inner ear.

- **(b)** In the middle ear, it passes deep to the manubrium of the malleus.
- **(c)** It leaves the tympanic cavity through the anterior canaliculus (iter chordae anterius) in the petrotympanic fissure.
- **(d)** This nerve conveys taste from the anterior two thirds of the tongue as well as parasympathetic presynaptic secretomotor fibers to the submandibular ganglion.
- **b.** The **roof (tegmen tympani)** separates the epitympanic recess from the overlying middle cranial fossa. The **epitympanic recess**, a superior extension of the antrum, contains the head of the malleus and the body of the incus (see Figure 33-9).
- **c.** The **floor** of the tympanic cavity is a plate of bone separating the middle ear cavity from the jugular canal.
- **d.** The **posterior wall** has numerous communications with **mastoid air cells** (see Figure 33-9).
    - **(1)** Infection of the middle ear may spread into these spaces **(mastoiditis).** Treatment often is difficult, requiring mastoidectomy.
    - **(2)** A bony pyramidal eminence projecting from the posterior wall contains the **stapedius muscle,** which inserts onto the stapes.
- **e.** The **anterior wall** separates the tympanic cavity from the carotid canal (see Figure 33-9).
    - **(1)** The **auditory (pharyngotympanic, tympanic, eustachian) tube** opens into the anterior wall at the **tympanic orifice.**
        - **(a)** It connects the middle ear with the nasopharynx and aids in equalizing the pressure across the tympanum. A pressure differential of 100–150 mm Hg will rupture the tympanum.
        - **(b)** It is a pathway for the spread of infection from the nasopharynx to the middle ear.
        - **(c)** It is a remnant of the first branchial pouch.
    - **(2)** The canal of the **tensor tympani muscle** (just medial to the auditory tube) opens into the anterior wall just superior to the tympanic orifice. The contained muscle inserts onto the malleus.

f. The **medial wall** is the most complex (see Figure 33-9).
   (1) A **superior prominence** marks the position of the lateral semicircular canal of the inner ear.
   (2) The **prominence of the facial (fallopian) canal** marks the course of the horizontal portion of the facial nerve. Occasionally, only a layer of periosteum lies between the middle ear and the nerve. **Otitis media** (middle ear infection) may involve the underlying facial nerve.
   (3) The **oval window** (fenestra vestibuli) receives the footplate of the stapes and transmits sonic vibrations of the ossicles to the perilymph of the scala vestibuli.
   (4) The **tympanic bulla** (promontory) is formed by the basal turn of the cochlea.
      (a) The **tympanic plexus** (of Jacobson) passes across this promontory and contains sensory contributions from the glossopharyngeal and vagus nerves, which distribute to the tympanum and external auditory meatus.
      (b) Anteriorly, the **lesser superficial petrosal nerve** forms from this plexus and conveys presynaptic parasympathetic secretomotor fibers of glossopharyngeal nerve (CN IX) origin to the **otic ganglion.**
   (5) The **round window** (fenestra tympani) is covered by an elastic membrane. It accommodates the pressure waves transmitted to the perilymph of the scala tympani.

2. Three auditory ossicles bridge the tympanic cavity.
   a. The **malleus** (L. handle) is the most lateral bone, attaching to the tympanum and articulating with the incus (see Figure 33-9).
      (1) The **manubrium** of the malleus is attached along its length to the tympanum.
      (2) The **head** projects into the epitympanic recess and articulates with the incus at the saddle-shaped **incudomalleolar joint.** The head is stabilized by the **superior ligament** to the tegmentum tympani.
      (3) An **anterior process** provides attachment for the **anterior ligament,** which passes through the **petrotympanic fissure** and seems to be developmentally continuous with the **sphenomandibular ligament.** Both ligaments are remnants of Meckel's cartilage and, thus, are first branchial arch derivatives.
      (4) The **tensor tympani muscle** originates within the like-named canal in the anterior wall and inserts into the neck of the malleus (Table 33-2).
         (a) As a first arch derivative, it is innervated by the motor division of the **mandibular nerve** (CN V).
         (b) **During chewing,** contraction mediated by the **mandibular nerve** pulls the tympanic membrane inward and thereby tenses the tympanum. This change dampens the vibration of the malleus in response to noise produced by chewing. Mandibular nerve injury, however, appears to produce no deficit comparable to the hyperacusis associated with facial nerve injury.
      (5) The **chorda tympani nerve** passes deep to the manubrium of the malleus.
   b. The **incus** (L. anvil) lies medially, primarily in the epitympanic recess (see Figure 33-9).

Table 33-2. Musculature of the Middle Ear

Muscle	Origin	Insertion	Primary Action	Innervation
Tensor tympani	Canal of the tensor tympani	Handle of malleus	Dampen tampanum	Mandibular branch of trigeminal nerve (CN V_3)
Stapedius muscle	Posterior wall of middle ear cavity	Neck of stapes	Dampen stapes	Stapedius br. of facial nerve (CN VII)

(1) The body articulates with the head of the malleus at the **incudomalleal joint.**
(2) A short posterior crus provides the attachment for the posterior ligament, which runs to the posterior wall.
(3) A longer, descending crus articulates with the stapes at the **incudostapedial joint.**
(4) Like the malleus, the incus is a first branchial arch derivative.
c. The **stapes** (L. stirrup) is most medial of the ossicles (see Figure 33-9).
(1) It articulates with the descending crus of the incus at the **incudostapedial joint.**
(2) The **body** bifurcates into two limbs, which end in a single footplate.
(3) The **footplate** inserts into the oval window, and the articulation is maintained by an **annular ligament. Otosclerosis** at the edge of the oval window impedes movement and is the most common cause of adult deafness.
(4) The **stapedius muscle** originates within the pyramidal eminence on the posterior wall of the antrum and inserts into the head of the stapes (see Table 33-2).
(a) It is innervated by the **facial nerve** (CN VII).
(b) **Reflex contraction** of this muscle damps the vibrations of the stapes. Paralysis of this muscle, as a result of facial nerve palsy, produces **hyperacusis,** whereby normal sounds are perceived as annoyingly loud.
(5) The stapes and stapedius muscle derive from the second branchial arch.
3. **Function.** Auditory ossicles pivot about axes defined by supporting ligaments. They transmit sonic vibrations between the outer ear and the inner ear and amplify the force.
a. **Mechanical advantage.** The area ratio of the tympanum to the oval window is about 18:1. The inferior crus of the incus is not as long as the handle of the malleus, however, so the excursion of the stapes is only about one third of the tympanum. The result is about a sixfold net mechanical advantage.
b. This **amplification of force** compensates for the fivefold difference in impedance between the air on one side of the tympanum and the more dense perilymph on the other side of the oval window.

C. Inner ear. This division lies within the petrous portion of the temporal bone.
1. **Structure** (see Figure 33-9).
a. The **osseous labyrinth,** a series of bony canals, contains perilymph that suspends the membranous labyrinth.
b. The **membranous labyrinth,** a system of continuous membranous canals, is filled with endolymph and contains the sensory organs.
2. **Vestibular apparatus**
a. **Labyrinth** of the vestibular portion (Figure 33-10)
(1) The **vestibule** is a chamber in the osseous labyrinth situated behind the oval window.
(a) From this chamber radiate three bony **semicircular canals** and one of the chambers **(scala vestibuli)** of the spiral cochlea.
(b) Within these chambers and canals lie the comparable portions of membranous labyrinth.
(2) The **utricle and saccule** are dilations of the membranous labyrinth within the vestibule.
(a) Within each dilation is a sensory **macula** that projects into the endolymph.
(b) The cytoarchitecture of the macula is such that sensory nerve endings deform in response to static gravity and inertia as well as to vibration.
b. **Semicircular canals** (see Figure 33-10)
(1) **Three bony canals** are arranged in mutually perpendicular planes.
(a) The **anterior (superior) semicircular canal** projects vertically with the long axis directed anteromedially at about 45°.
(b) The **lateral semicircular canal** is nearly horizontal and projects slightly into the middle ear cavity.

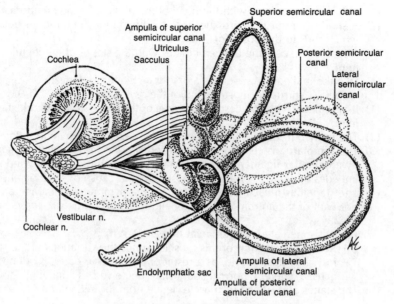

FIGURE 33-10. Right membranous labyrinth.

- (c) The **posterior semicircular canal** projects vertically with the long axis directed posterolaterally at about 45°.
- (d) The **anterior semicircular canal** is parallel to the contralateral posterior semicircular canal.
- (2) **Membranous semicircular canals** are suspended in **perilymph.**
  - (a) Dilation of the membranous labyrinth at one end of each semicircular canal contains a **crista ampullaris.**
  - (b) The cytoarchitecture of the crista ampullaris is such that sensory nerve endings are stimulated on structural deformation caused by rotational inertia on the enclosed endolymph.
- c. The **vestibular portion of the vestibulocochlear nerve** (CN VIII) arises from the maculae and cristae ampullaris.

3. **Cochlear apparatus** (Figure 33-11)
   a. The **bony cochlea** consists of two adjacent ducts that are less than semicircular in cross section and spiral two and three-quarters turns about a central **modiolus.**
      (1) The **scala vestibuli** begins in the vestibule and receives the vibrations transmitted to the perilymph at the oval window.
      (2) The **lower scala tympani** connects with the scala vestibuli through the **helicotrema** at the apex of the cochlea and terminates at the round window, at which the sound waves are dissipated.
   b. The **membranous cochlear duct** (scala media) is wedged distally between the scala vestibuli and scala tympani as far as the helicotrema.
      (1) This duct contains the **spiral organ** (of Corti), which is suspended in endolymph and responds to vibrations in that fluid.
      (2) Given the cytoarchitecture of the **spiral organ,** specific portions resonate harmonically with each audible frequency.
         (a) Its width is greater toward the apex of the cochlea than at the base; thus, the lower frequencies resonate near the helicotrema and the higher frequencies near the oval window.
         (b) Sensory cells detect the regions of resonant vibration.
   c. The **cochlear nerves** leave the hollow **modiolus** to form the acoustic division of the **vestibulocochlear nerve** (CN VIII).

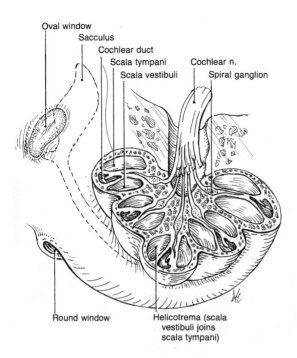

FIGURE 33-11. Right cochlea.

(1) Sensory neurons from specific portions of the spiral organ project to the medial geniculate body and then to the auditory cortex, where tone is perceived.
(2) Slight differences in the quality, amplitude, and phasing of the sound result in stereophonic localization of the source.

PART IVA SOMATIC NECK AND NEUROCRANIUM

STUDY QUESTIONS

DIRECTIONS: Each of the numbered items or incomplete statements in this section is followed by answers or by completions of the statement. Select the ONE lettered answer or completion that is BEST in each case.

Questions 1–5

An 84-year-old man, febrile and gasping for air, is brought to the emergency room by ambulance. The emergency medical technician reports "impending respiratory arrest" with no obvious trauma. The airway is intubated but with considerable difficulty because of inflamed and severely edematous pharyngeal and supraglottic tissues. No foreign material is noted in the airway. Radiographic evaluation of the neck and chest reveals infection in the retrovisceral space, as evident by extensive extravasation opening fascial planes.

1. The pretracheal and retrovisceral spaces are defined by prevertebral fascia. All of the following structures are enclosed by pretracheal fascia EXCEPT the

(A) infrahyoid (ribbon) muscles
(B) muscles of the cervical vertebral column
(C) thyroid gland
(D) trachea and esophagus

2. Because of the extent of infection, surgical drainage of the retrovisceral space is recommended. An incision is made posterior and parallel to the sternomastoid muscle. After anterior retraction, the carotid sheath is located. All of the following neurovascular structures are contained within the carotid sheath EXCEPT the

(A) common carotid artery
(B) internal carotid artery
(C) internal jugular vein
(D) sympathetic chain
(E) vagus nerve

3. Anterior retraction of the carotid sheath preserves the innervation of the strap muscles. These infrahyoid muscles are innervated by all of the following EXCEPT the

(A) hypoglossal nerve
(B) inferior ramus of the ansa cervicalis
(C) superior ramus of the ansa cervicalis
(D) ventral primary rami of spinal nerves C1–C3

4. When branches of the cervical plexus are encountered, most are preserved, but one nerve is sectioned. A motor deficit is unlikely because all of the following nerves of the cervical plexus have sensory function EXCEPT the

(A) greater occipital
(B) lesser occipital
(C) posterior auricular
(D) suboccipital
(E) transverse cervical

5. After entering the retrovisceral space, the surgeon clears all accumulated pus and closes the incision, leaving a drain in place. With aggressive intravenous antibiotic therapy, the patient is expected to recover fully. The retrovisceral space is continuous with which of the following spaces?

(A) "Danger space"
(B) Pretracheal space
(C) Space of the carotid sheath
(D) Submandibular space
(end of group question)

6. A 37-year-old woman with severe headache, stiff neck, and fever is admitted to the hospital with a likely diagnosis of meningitis. Infection may spread from the nasal cavity to the meninges along the olfactory nerves. Olfactory fibers pass from the olfactory mucosa to the olfactory bulb via the

(A) anterior and posterior ethmoid foramina
(B) cribriform plate
(C) hiatus semilunaris
(D) nasociliary nerve
(E) sphenopalatine foramen

Questions 7–10

A 57-year-old woman is involved in a minor automobile accident in which her head is shaken, but not bruised. While exchanging information with the other driver, she briefly loses consciousness. She is transported to the emergency room for evaluation, during which she becomes progressively groggy and then seems to recover.
Radiographs show no evidence of fracture, but a mass is observed in the left hemisphere. A CT scan confirms the finding is a subdural hematoma with compensation.

7. Compensation for subdural hematoma results from decreased cerebrospinal fluid (CSF) production. CSF is produced in all of the following regions EXCEPT the

(A) cisterna magna
(B) fourth ventricle
(C) lateral ventricles
(D) third ventricle

8. Cerebrospinal fluid enters the subarachnoid space at the

(A) arachnoid villi
(B) choroid plexus
(C) foramina of Luschka and Magendie
(D) foramina of Monro
(E) iter

9. Cerebrospinal fluid enters the venous system

(A) at arachnoid villi
(B) at the cisterna magna
(C) through the subarachnoid veins
(D) via capillaries in the ependyma
(E) at the end of the medullary canal of the spinal cord

10. The neurosurgeon recommends craniotomy for drainage of the hematoma and to attempt to control bleeding. A subdural hematoma is usually caused by

(A) fracture of the diploic space
(B) laceration of the superficial temporal artery
(C) leak from a cerebral vein
(D) rupture of a cerebral artery
(E) tearing of a meningeal artery
(end of group question)

Questions 11 and 12

A 73-year-old woman is evaluated for a vision problem. The differential diagnosis includes a cerebrovascular accident in the right visual cortex.

11. The sign providing the diagnosis of a cerebral vascular accident in the primary visual area of the right cerebral hemisphere is

(A) bitemporal heteronymous hemianopsia
(B) contralateral homonymous hemianopsia
(C) ipsilateral homonymous hemianopsia
(D) ipsilateral nasal hemianopsia
(E) total blindness of the right eye

12. The artery that primarily supplies the right visual cortex is the

(A) anterior cerebral
(B) middle cerebral
(C) ophthalmic
(D) basilar
(E) right vertebral
(end of group question)

Questions 13–17

A 17-year-old male with blurred vision, lethargy, and severe headache is seen in the emergency room. Tests of ocular movement indicate possible early stage right ophthalmoplegia.

13. Directing the eye outward involves the use of more than one extraocular muscle. All of the following nerves mediate this movement EXCEPT the

(A) abducens nerve
(B) inferior division of the oculomotor nerve
(C) superior division of the oculomotor nerve
(D) trochlear nerve

14. Palsy of the right abducens nerve usually results in diplopia, which can be minimized by directing the gaze

(A) downward to the right
(B) laterally to the left
(C) laterally to the right
(D) upward to the right

15. Further testing reveals that the right pupil is dilated and the pupillary reflex is diminished. The cell bodies of the neurons responsible for pupillary dilation are located in the

(A) accessory oculomotor nucleus
(B) ciliary body
(C) ciliary ganglion
(D) pterygopalatine ganglion
(E) superior cervical ganglion

16. During the physical examination, the emergency room physician notes a large, indurated pustule on the right side of the nose. In addition, the right retina appears somewhat engorged. Retinal engorgement indicates blockage of venous return. Drainage of the eye and orbit is primarily to the cavernous sinus through the

(A) annulus of Zinn
(B) cribriform plate
(C) inferior orbital fissure
(D) orbital canal
(E) superior orbital fissure

17. The physical findings are consistent with a diagnosis of cavernous sinus thrombosis. High-dose intravenous antibiotic therapy is recommended. All of the following nerves pass through the cavernous sinus EXCEPT the

(A) abducens
(B) mandibular
(C) oculomotor
(D) ophthalmic
(E) trochlear
(end of group question)

DIRECTIONS: Each of the numbered items or incomplete statements in this section is negatively phrased, as indicated by a capitalized word such as NOT, LEAST, or EXCEPT. Select the ONE lettered answer or completion that is BEST in each case.

---

18. Initial examination of a 17-year-old bicycle accident victim, who was not wearing a helmet, reveals neurologic deficits involving cranial nerves. Radiographs demonstrate a basal skull fracture that extends through the jugular foramen. All of the following structures might be involved by such a cranial fracture EXCEPT the

(A) cranial accessory nerve
(B) glossopharyngeal nerve
(C) hypoglossal nerve
(D) spinal accessory nerve
(E) vagus nerve

19. All of the following statements correctly describe the precentral gyrus of the brain EXCEPT

(A) It is the primary motor area
(B) It projects to brain stem nuclei and the gray matter of the spinal cord
(C) It receives direct sensory input
(D) It receives input from the basal ganglia and cerebellum

20. All of the following statements about the lacrimation process are correct EXCEPT

(A) blinking of the eyelids expresses small amounts of lacrimal fluid from the lacrimal gland
(B) lacrimal secretion is controlled by parasympathetic nerves
(C) lacrimal canaliculi drain through the ampulla into the lacrimal sac
(D) blinking causes the lacrimal sac to aspirate the lacrimal fluid from the lacus lacrimalis
(E) the nasolacrimal duct ends in the hiatus semilunaris

21. All of the following nerves provide sensory innervation to the external ear EXCEPT the

(A) cochlear nerve
(B) facial nerve
(C) glossopharyngeal nerve
(D) mandibular division of the trigeminal nerve
(E) vagus nerve

22. Loud, low-pitched sounds and vibrations cause all of the following EXCEPT

(A) bulging of the round window membrane into the middle ear cavity
(B) greater movement at the stapes than at the malleus
(C) maximal vibration in the apical portion of Corti's spiral organ
(D) reflex activity along a branch of CN V
(E) reflex activity along a branch of CN VII

DIRECTIONS: Each set of matching questions in this section consists of a list of four to twenty-six lettered options (some of which may be in figures) followed by several numbered items. For each numbered item, select the ONE lettered option that is most closely associated with it. To avoid spending too much time on matching sets with large numbers of options, it is generally advisable to begin each set by reading the list of options. Then, for each item in the set, try to generate the correct answer and locate it in the option list, rather than evaluating each option individually. Each lettered option may be selected once, more than once, or not at all.

Questions 23–24

(A) Right CN V injury
(B) Right CN VII injury
(C) Both
(D) Neither

For each case history, select the nerve damage with which it is most likely to be associated.

23. After an injury to the face, stimulation of the right cornea results in blinking of the left eye, but not the right eye.

24. After a patient sustains a fracture that separated the facial skull from the neurocranium (Le Fort III), stimulation of the right cornea fails to produce blinking in either eye. Stimulation of the left cornea produces blinking in the left eye only.

Questions 25–32

(A) Anterior condylar canal
(B) Foramen lacerum
(C) Foramen magnum
(D) Foramen ovale
(E) Foramen rotundum
(F) Foramen spinosum
(G) Hiatus of the facial canal
(H) Inferior orbital fissure
(I) Optic canal
(J) Posterior condylar canal
(K) Superior orbital fissure
(L) Stylomastoid foramen

For each structure, select a foramen, fissure, or canal through which it passes.

25. Facial nerve

26. Middle meningeal artery

27. Hypoglossal nerve

28. Motor division of the trigeminal nerve

29. Abducens nerve

30. Greater superficial petrosal nerve

31. Ophthalmic artery

32. Spinal accessory nerve

ANSWERS AND EXPLANATIONS

1. The answer is B [Chapter 30 I B 2 b; Figure 30-1]. The prevertebral fascia encloses and invests the somatic musculature of the neck. The sternomastoid and trapezius muscles are associated with the superficial layer of the deep cervical fascia, whereas the strap muscles lie between that layer and the pretracheal fascia. The visceral structures of the neck are bounded by pretracheal fascia.

2. The answer is D [Chapter 30 I B 2]. The sympathetic chain and the phrenic nerve lie posterior to the carotid sheath. Because the carotid arteries, jugular vein, and vagus nerve are contained within a connective tissue sheath, the structure and contents can be safely retracted.

3. The answer is A [Chapter 30 II D 2 a (2). The superior ramus of the ansa cervicalis from spinal nerve C1 runs with (but is not a part of) the hypoglossal nerve to innervate the geniohyoid, thyrohyoid, and sternohyoid muscles. The inferior ramus of the ansa cervicalis from spinal nerves C2 and C3 innervates the omohyoid and sternothyroid muscles. The two limbs of the ansa communicate and provide cross innervation to some of these muscles, such as the sternothyroid. The hypoglossal nerve innervates all of the intrinsic and some of the extrinsic muscles of the tongue.

4. The answer is D [Chapter 30 II D 1 a, b; 2 a (1), (2)]. The suboccipital nerve (C1, posterior) supplies the muscles of the occipital triangle and has no sensory function. All of the other named branches of the cervical plexus (C2–C4, anterior), with the exception of the ansa cervicalis, are sensory nerves.

5. The answer is D [Chapter 30 I B]. The retrovisceral space continues superiorly as the retropharyngeal space, which extends anteriorly to join the submandibular space. Thus, oral infection may track to the mediastinum. The pretracheal space is separate, so local infection is retained within the infrahyoid neck. The carotid sheath is an enclosed space. The "danger space," a somatic space defined by the prevertebral fascia, extends from the base of the skull through the posterior mediastinum.

6. The answer is B [Chapter 31 III C 2 a (2) (c); Chapter 32 IV E 2; V B 2 b]. The cribriform plate of the ethmoid bone provides passageways for the olfactory nerves from the olfactory mucosa of the superior nasal meatus to the olfactory bulb of the brain. Infection may track along these nerves, thereby spreading to the meninges. Fracture of the ethmoid bone may result in leaking of cerebrospinal fluid through the nose. The anterior and posterior ethmoid nerves, branches of the nasociliary nerve, pass through the similarly named foramina. The medial and lateral nasopalatine nerves (branches of the maxillary nerve) pass through the sphenopalatine foramen. The frontal and maxillary sinuses drain into the hiatus semilunaris.

7. The answer is A [Chapter 31 IV A 2 b (1); B 3 b; Chapter 32 V B 1, 2]. Cerebrospinal fluid (CSF) is produced entirely within the ventricular system of the brain, by the choroid plexuses on the roof of the lateral ventricles, the third ventricle, and the fourth ventricle. The cisterna magna is a widening of the subarachnoid space outside of the brain stem.

8. The answer is C [Chapter 31 IV A 2 b (1) (a); Chapter 32 V B]. Cerebrospinal fluid (CSF) passes from the lateral ventricles through the foramina of Monro, into the third ventricle, and through the iter into the fourth ventricle. It leaves the ventricular system of the brain and enters the cisterna magna of the subarachnoid space through the foramina of Luschka and Magendie in the roof of the fourth ventricle.

9. The answer is A [Chapter 31 IV A 2 b (1) (c); Chapter 32 V B 2; Figure 32-9]. Cerebrospinal fluid circulates through the subarachnoid space, providing fluid protection for the brain, spinal cord, and spinal roots. It enters the superior sagittal sinus via arachnoid villi.

10. The answer is C [Chapter 31 IV B 3]. Cerebral veins are most vulnerable to tearing as they pass between the arachnoid and dura mater to drain into the venous sinuses. Injury at this site results in extravasation of blood in the subdural space (subdural hematoma). Tearing of a meningeal artery produces an epidural hematoma, whereas the rupture of a cerebral artery results in bleeding into the subarachnoid space.

11. The answer is B [Chapter 32 II A 3 c (2) (b)]. Pathways for visual fields are sorted out in the optic chiasm, so the left visual fields from each eye are collected together for projection to the right visual cortex. Because the optic projections to the visual cortex contain fibers that convey information from the contralateral visual fields, loss of the primary cortical visual area on one side produces loss of the contralateral visual field in each eye—contralateral homonymous hemianopsia.

12. The answer is D [Chapter 32 VI B 1 c, 2 c (4)]. The posterior cerebral artery is a branch of the basilar artery, the continuation of the joined vertebral arteries. The basilar artery supplies not only the brain stem but also the occipital lobe of the cerebrum.

13. The answer is C [Chapter 33; Table 33-1]. The lateral rectus muscle, the principal abductor of the eye, is innervated by the abducens nerve (CN VI). The superior oblique muscle, a depressor and abductor of the eye, is innervated by the trochlear nerve (CN IV). The inferior oblique muscle, an elevator and abductor of the eye, the medial rectus muscle (an adductor of the eye), and the inferior rectus muscle (also an elevator of the eye) are innervated by the inferior division of the oculomotor nerve (CN III). The superior rectus, an elevator of the eye, and the levator palpebrae superioris are both innervated by the superior division of the oculomotor nerve.

14. The answer is B [Chapter 32 IV H 1 c; Chapter 33 I D 1 c]. Because the abducens nerve innervates the lateral rectus muscle, abducens palsy results in a medial strabismus. Diplopia is minimized when the gaze is directed toward the opposite side.

15. The answer is E [Chapter 33 I D 4]. The cell bodies of the neurons that innervate the dilator pupillae muscle in the superior cervical ganglion receive presynaptic stimulation from neurons located in the uppermost thoracic levels of the spinal cord. Generally, the sympathetic neurons follow perivascular pathways to the site of innervation.

16. The answer is E [Chapter 31 IV A 1 b (3); Chapter 33 I E 2 a, b]. The superior ophthalmic vein usually receives an anastomotic connection from the inferior ophthalmic vein before passing superior to the annulus fibrosus (of Zinn) to reach the superior orbital fissure and cavernous sinus.

17. The answer is B [Chapter 31 IV A 1 b (3) (c); Figure 31-11]. The oculomotor, trochlear, abducens, and ophthalmic nerves run through the cavernous sinus to reach the orbit. Usually, the maxillary nerve is included. The mandibular nerve, which passes inferiorly from the trigeminal ganglion to the foramen ovale, misses the cavernous sinus. The carotid artery also lies in the medial wall of this sinus.

18. The answer is C [Chapter 31 III B 5 b (2) (c)]. In addition to the jugular vein, the jugular foramen transmits the glossopharyngeal nerve (CN IX), the vagus nerve (CN X) with its cranial accessory component, and the spinal accessory nerve (CN XI). The hypoglossal nerve leaves the cranium through the anterior condylar (hypoglossal) canal.

19. The answer is C [Chapter 32 II A 3 a (1), B 2]. This primary motor area in the frontal lobe receives input from the cortical sensory areas as well as modulating influences from the basal ganglia and cerebellum. Its somatotopically organized pyramidal cells project to the lower motor neurons of the brain stem and spinal cord. The motor area receives no direct sensory projections, however; most input is relayed through thalamic nuclei.

20. The answer is E [Chapter 33 I B 6, 7]. The nasolacrimal duct enters the inferior meatus of the nose. The hiatus semilunaris in the middle meatus receives drainage from the frontal, ethmoidal, and maxillary sinuses.

21. The answer is A [Chapter 33 III A 4]. The anterior aspect of the pinna is innervated by the auriculotemporal branch of the mandibular division of the trigeminal nerve. The posterior aspect is innervated by the auricular branch of the facial nerve. The external auditory meatus is innervated by branches from both the glossopharyngeal and vagus nerves. The cochlear division of the vestibulocochlear nerve (CN VIII) conveys audition from the inner ear.

22. The answer is B [Chapter 33 III B 1 f (5), 2 a (4), c (4); C 3 b (2) (a)]. Low-pitched sounds cause harmonic resonance in the apical portion of the spiral organ. Masticatory vibrations cause reflex contraction of the tensor tympani muscle (innervated by CN V). Loud sounds result in reflex contraction of the stapedius muscle (innervated by CN VII). Every action of the stapes at the oval window produces an opposite action at the round window. The malleus always has a greater range of excursion than the stapes, but the stapes always has greater mechanical advantage.

23–24. The answers are: 23-B, 24-C [Chapter 33 I B 3 a, 5 b (2)]. The trigeminal nerve (CN V) is the afferent limb of the blink reflex. Complex neural pathways in the brain stimulate both ipsilateral and contralateral facial nerve nuclei so that a bilateral (consensual) blink response is mediated by the facial nerves, forming the efferent limb of the blink reflex. The results of the tests may be deduced from this basis.

25–32. The answers are: 25-L, 26-F, 27-A, 28-D, 29-K, 30-G, 31-I, 32-C [Chapter 31 III C 2]. The facial nerve enters the temporal bone through the internal acoustic meatus to gain access to the facial canal, which terminates at the stylomastoid foramen. The middle meningeal artery transits the foramen spinosum. The hypoglossal nerve passes through the anterior condylar (hypoglossal) canal. The mandibular branch carries the motor division of the trigeminal nerve. It leaves the cranium through the foramen ovale. The nerves to the extraocular muscles, including the abducens (CN VI), enter the bony orbit through the superior orbital fissure along with the ophthalmic branch of the trigeminal nerve (CN V) and the ophthalmic vein. The greater superficial petrosal nerve leaves the facial nerve by way of the hiatus of the facial canal before entering the pterygoid (vidian) canal. The ophthalmic artery enters the orbit through the optic canal, along with the optic nerve. The cervical roots of the spinal accessory nerve enter the cranial cavity by passing upward through the foramen magnum. This nerve then leaves the cranial cavity via the jugular foramen.

PART IVB FACIAL CRANIUM AND VISCERAL NECK

Chapter 34
Anterior Cervical Triangle

I. CERVICAL FASCIA

A. Introduction. The head and neck may be divided developmentally, anatomically, functionally, and medically into an anterior **visceral portion** and a posterior **somatic portion.** The course of the **sternomastoid muscle** defines an **anterior triangle** and a **posterior triangle** (see Figure 30-2).

1. **Somatic cervical structures** lie in the **posterior triangle.** They include the neurocranium and the cervical vertebral column with associated musculature.
2. **Visceral cervical structures** lie in the **anterior triangle.** They include derivatives of the upper end of the primitive gut and associated branchial (gill) structure.

B. Fascia of the anterior triangle

1. **Superficial fascia** in this region contains loose connective tissue (see Chapter 30 I B 1).
2. **Deep cervical fascia** surrounds the neck and gives off septa, which separate the visceral and somatic portions (see Chapter 30 I B 2).
3. **Cervical spaces** are formed by the planes of deep fascia. Extension of infection tends to be limited by these fascial planes but may track up and down the neck within the resultant spaces (see Chapter 30 I B 3).

II. HYOID BONE

A. Characteristics (see Figure 34-2)

1. This U-shaped, free-floating bone consists of a median body, paired lesser horns (cornua) laterally, and paired greater horns posteriorly.
2. It does not articulate with any other bone. Muscles provide support and stability.
3. It is tethered by the **stylohyoid ligament,** which runs between the styloid process and the lesser horn on each side (see Figure 37-7).

B. Development. The lesser horns of the hyoid bone and stylohyoid ligament are second arch derivatives. The body and greater horn come from the third arch.

C. Muscle attachments

1. **Suprahyoid muscles** run between the hyoid bone and the mandible (see Figures 34-1, 34-2, 34-5, and 37-7). They stabilize and draw the hyoid bone anterosuperiorly, superiorly, and posterosuperiorly during the early phases of deglutition.
   a. Along the greater horn, the **middle constrictor** attaches posteriorly and the **hyoglossus muscle** attaches superiorly.
   b. At the lesser horn, the **digastric** and **stylohyoid** muscles attach superiorly.
   c. Along the body, the **geniohyoid** and **mylohyoid muscles** attach superiorly.

2. **Infrahyoid muscles** run from the hyoid bone to the thyroid cartilage, sternum, or scapula (see Figures 34-1 and 34-2). They stabilize and draw the hyoid bone inferiorly during the final phase of deglutition.
   a. **Along the greater horn,** the **thyrohyoid muscle** attaches inferiorly.
   b. **Along the body,** the **omohyoid** and **sternohyoid muscles** attach inferiorly.

III. MUSCULATURE

A. **Boundaries of the anterior triangle** are defined by the sternomastoid muscle laterally, the anterior midline, and the lower border of the mandible superiorly (Figure 34-1).

B. **Subdivisions**
   1. **The digastric muscle** defines the **digastric** and **submental triangles** (see Figure 34-1).
      a. The **digastric triangle** is bounded by the lower edge of the mandible and the anterior and posterior bellies of the **digastric muscle** that radiate from the hyoid bone (anteriorly to the chin and posteriorly to the mastoid process).
      b. The **submental triangles** are formed by the left and right anterior bellies of the digastric muscles, the hyoid bone, and the midline raphe.
   2. **The omohyoid muscle** further divides the anterior triangle into a **carotid triangle** posteriorly and a **muscular triangle** anteriorly (see Figure 34-1).
      a. The **carotid triangle** is bounded by the superior belly of the omohyoid muscle anteroinferiorly, the posterior belly of the digastric muscle anterosuperiorly, and the sternomastoid muscle posteriorly.

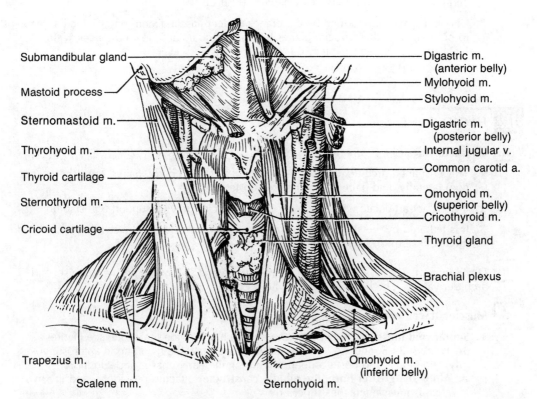

FIGURE 34-1. Musculature of the anterior cervical triangle. The hyoid bone divides the cervical musculature into suprahyoid and infrahyoid groups.

Table 34-1. Anterior Triangle Musculature

Muscle	Origin	Insertion	Primary Action	Innervation
Infrahyoid Muscles				
Sternohyoid	Sternum	Hyoid bone	Lowers hyoid	Superior ramus of ansa cervicalis (C1)
Thyrohyoid	Hyoid bone	Thyroid cartilage	Raises larynx	Superior ramus of ansa cervicalis (C1)
Sternothyroid	Sternum	Thyroid cartilage	Lowers larynx	Inferior ramus of ansa cervicalis (C2–C3)
Omohyoid	Scapular notch	Hyoid bone	Lowers hyoid	Inferior ramus of ansa cervicalis (C2–C3)
Suprahyoid Muscles				
Digastric				
Posterior belly	Mastoid process of temporal bone	Anterior belly through hyoid trochlea at lesser horn	Elevates hyoid bone	Facial n. (CN VII)
Anterior belly	Posterior belly through hyoid trochlea	Mandible	Elevates hyoid bone	Mylohyoid branch of inferior alveolar n. (CN V)
Mylohyoid	Body of mandible	Midline raphe	Elevates hyoid bone, floor of mouth and tongue	Mylohyoid branch of inferior alveolar n. (CN V)
Stylohyoid	Styloid process	Lesser horn of hyoid	Elevates and retracts hyoid bone	Facial n. (CN VII)
Geniohyoid	Anterior mandible	Hyoid bone	Protracts hyoid	Superior ramus of ansa cervicalis (C1)

    b. The **muscular triangle** is defined by the superior belly of the omohyoid, the sternomastoid muscle, and the midline.

C. Infrahyoid (strap or ribbon) muscles (see Figure 34-1 and Table 34-1)
    1. Organization. These muscles are named for their attachments.
        a. The **sternohyoid muscle,** running between the sternum and the hyoid bone, is the most superficial strap muscle.
        b. The **thyrohyoid muscle** is short, running between the greater horn of the hyoid bone and the oblique line of the thyroid cartilage.
        c. The **sternothyroid muscle** continues the path of the thyrohyoid muscle from the oblique line of the thyroid cartilage to the sternum. It lies deep to the sternohyoid muscle.
        d. The **omohyoid muscle** has two bellies.
             (1) The **superior belly** (omos, G. shoulder) originates from the **hyoid bone** and descends vertically to the central tendon.
            (2) The **central tendon** of the omohyoid muscle is held in position deep to the

sternomastoid muscle at the level of the cricoid cartilage by an investing septa from the deep cervical fascia. This trochlea allows a directional change from vertical to nearly horizontal.

   (3) The **inferior belly** runs nearly horizontally from the central tendon, passes deep to the sternomastoid muscles, and enters the posterior triangle to attach in the scapular notch of the scapular.

2. **Group actions** (see Table 34-1)
   a. **On the hyoid bone.** The sternohyoid and omohyoid muscles lower the hyoid bone and larynx during deglutition (swallowing) and phonation.
   b. **On the thyroid cartilage.** The thyrohyoid raises the larynx during deglutition and phonation; the sternothyroid lowers the larynx during deglutition (swallowing) and phonation. Acting together, they stabilize the larynx.
3. **Innervation.** The infrahyoid muscles are innervated by the **ansa cervicalis (C1–C3)**.

D. **Suprahyoid muscles** (Table 34-1; see Figures 34-1 and 34-2)
1. **Organization**
   a. The **digastric muscle** has two bellies that radiate anteriorly and posteriorly from a central tendon attached to the lesser horn of the hyoid bone.
      (1) The **posterior belly** originates on the mastoid process of the temporal bone and courses toward the lesser horn of the hyoid bone.
      (2) The **intermediate tendon** passes through a connective tissue trochlear that binds it loosely to the body of the hyoid bone near the lesser horn.
      (3) The **anterior belly** courses to the digastric fossa on the anteroinferior border of the mandible.
   b. The **stylohyoid muscle** runs from the styloid process to the lesser horn of the hyoid bone. The tendon of insertion splits for the passage of the central tendon of the digastric muscle.
   c. The **mylohyoid muscle** arises from the mylohyoid line on the medial surface of the body of the mandible.
      (1) Fibers run anteroinferiorly to meet the corresponding fibers from the opposite side in a midline raphe that is attached to the hyoid bone posteriorly.

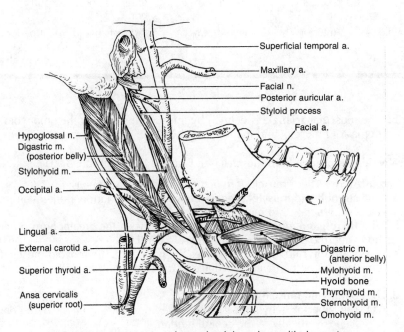

FIGURE 34-2. Nerves and vessels of the submandibular region.

(2) This muscle forms a sling beneath the mandible and is considered the **floor of the mouth.**
d. The **geniohyoid muscle** runs from the inferior genial tubercle of the anterior mandible to the anterosuperior aspect of the hyoid bone. It lies within the oral cavity, deep to the mylohyoid muscle.

2. **Group actions.** Suprahyoid muscles raise and stabilize the hyoid bone and larynx during deglutition and phonation (see Table 34-1).
   a. The anterior and posterior bellies of the **digastric muscle** elevate the hyoid bone.
   b. The **stylohyoid muscle** raises and retracts the hyoid.
   c. The **mylohyoid muscle** raises the hyoid, thereby elevating the floor of the mouth.
   d. The **geniohyoid muscle** draws the hyoid forward.
   e. Together, this muscle group stabilizes the hyoid bone and larynx during deglutition and phonation (see Table 34-1).

3. **Innervation** (Figure 34-2; see Table 34-1)
   a. The **mylohyoid nerve,** a branch of the inferior alveolar branch of the mandibular division of the **trigeminal nerve,** innervates the **anterior belly of the digastric muscle** and the **mylohyoid muscle** (see Figure 36-4).
   b. The **digastric branch of the facial nerve** innervates the **posterior belly of the digastric muscle** and the **stylohyoid muscle** (see Figure 34-2).
   c. Fibers from **spinal nerve C1** that continue with the hypoglossal nerve (after the superior ramus of the ansa cervicalis separates) innervate the **geniohyoid muscle.**

IV. THYROID GLAND AND PARATHYROID GLANDS

A. Thyroid gland

1. **Structure.** This essential endocrine gland consists of two (sometimes three) lobes (see Figure 34-1).
   a. The pear-shaped **left and right lobes** extend around the trachea and esophagus as far as the carotid sheath.
      (1) The thyroid is partially covered by the sternohyoid and sternothyroid muscles.
      (2) Superiorly, each lobe is confined by the insertion of the sternothyroid muscle onto the thyroid cartilage. Inferiorly, each lobe extends to the level of the fifth tracheal ring.
   b. An **isthmus** connects the left and right lobes and lies anterior to the tracheal rings just caudal to the cricoid cartilage, although it may be absent (10%).
   c. A **pyramidal lobe** occasionally (43%) extends superiorly from the middle of the isthmus.

2. **Development**
   a. It develops from a diverticulum in the posterior segment of the tongue, marked by the **foramen cecum.**
      (1) The thyroid primordium migrates caudally in the neck anterior to the hyoid bone to reach its final position anterior to the trachea. As it descends, it maintains a connection to the foramen cecum, the **thyroglossal duct.**
         (a) The median **pyramidal lobe,** when present, is a remnant of the thyroglossal duct.
         (b) The apex of a pyramidal lobe may be joined to the hyoid bone by a fibrous band that occasionally contains some muscle fibers (levator glandulae thyroideae).
      (2) Although the thyroglossal duct normally disappears, **rests** of functional thyroid tissue may remain along the track of the thyroglossal duct as **accessory thyroid glands.** Thyroid rests may develop into benign thyroglossal **cysts,** which often require surgical removal.
   b. The **parafollicular cells,** which secrete calcitonin, arise from the fourth branchial pouches.

3. **Support**
   a. **Pretracheal fascia** splits at the posterior border of the left and right lobes (see Figure 30-1).
      (1) The **anterior layer** binds tightly to the larynx.
      (2) The **posterior layer** extends around the esophagus.
      (3) Branches of the inferior thyroid artery and the recurrent laryngeal nerve lie between the anterior and posterior fascial layers.
   b. The thyroid gland is palpable as it moves up and down during swallowing.

4. **Function.** The follicular cells of the thyroid gland produce **thyroxine,** which regulates metabolism. Parafollicular cells produce **calcitonin,** which regulates calcium balance.

5. **Clinical considerations**
   a. **Goiter.** An enlarged thyroid gland may extend superiorly in the carotid triangle and inferiorly into the thoracic inlet. Because the gland lies deep to the sternothyroid muscle, pressure on the trachea at the thoracic inlet may cause shortness of breath and a wheeze, especially when raising the arms above the head.
   b. **Tracheostomy**
      (1) In an emergency, tracheostomy is accomplished through the cricothyroid membrane, a relatively avascular region.
      (2) In the operating room, the surgeon may identify and ligate bleeding vessels or even divide the thyroid isthmus, placing the tracheal stoma just above the jugular notch.
   c. **Thyroidectomy** (partial or total) is common but postoperative thyroid hormone replacement therapy is required. If total thyroidectomy is indicated, care must be taken to locate radiographically and excise any thyroid rests (accessory thyroid tissue) along the path of the thyroglossal duct.

B. Parathyroid glands

1. **Structure.** They are small and brownish pink because of their vascularity. They can be exasperatingly difficult to locate surgically.

2. **Development**
   a. The **superior (IV) parathyroid glands** are fourth branchial pouch derivatives. They lie posterior to the apex of each lobe of the thyroid gland.
   b. The **inferior (III) parathyroid glands** are third branchial pouch derivatives. They are related to the base of each lobe. Occasionally, however, they accompany the thymus (also a third branchial pouch derivative) into the anterior mediastinum. Locating a displaced parathyroid gland is frustrating if it develops a tumor necessitating excision.

3. **Support.** The parathyroid glands are supported by the thyroid parenchyma.

4. **Function.** They secrete **parathyroid hormone** (PTH), which controls calcium metabolism. As such, the parathyroids are essential glands and removal necessitates exogenous PTH administration.

5. **Total parathyroidectomy** (usually inadvertent) is followed by a gradual decrease in blood calcium levels, which can produce fatal tetany. These glands are extremely hardy and continue to function if located and transplanted from an excised thyroid into the sternomastoid muscle.

C. Thyroid vasculature

1. **Arterial supply** is from the **superior thyroid artery** (a branch of the external carotid artery) and the **inferior thyroid artery** (a branch of the thyrocervical trunk off the subclavian artery). A median **thyroidea ima** may arise (4%) from the brachiocephalic artery. These vessels anastomose freely over the thyroid gland (see Figure 34-4).

2. **Venous return.** A venous plexus on the surface of the gland and over the trachea drain the superior, middle, and inferior thyroid veins.

V. VASCULATURE

A. Carotid sheath

1. **Composition.** Formed by septa from the superficial layer of deep cervical fascia, the carotid sheath contains the **carotid arteries, internal jugular vein,** and the **vagus nerve** (see Figure 30-1).

2. **Location.** For the most part, the carotid sheath lies under cover of the sternomastoid muscle and anterior to the transverse processes of the cervical vertebrae.
   a. A strong **carotid pulse** is palpable in the carotid triangle inferior to the midpoint of the thyroid cartilage (Adam's apple) by gently pressing the common carotid artery against the underlying vertebra.
   b. In the elderly, atheromatous plaques may be dislodged by carotid palpation. It is wise, therefore, to take the carotid pulse on the right side, because a stroke induced in the nondominant cerebral hemisphere would be less devastating.

B. Arterial supply

1. **The common carotid arteries** supply most of the heart and neck.
   a. The **left common carotid artery** arises from the aortic arch.
   b. The **right common carotid artery** arises with the right subclavian artery from the brachiocephalic artery, a remnant of the artery of the right fourth branchial arch.
   c. The **carotid bifurcation** (Figure 34-3) occurs at the level of the superior edge of the thyroid cartilage (C4). Two receptors at the carotid bifurcation provide feedback to the vasomotor and respiratory centers of the brain.
      (1) The **carotid sinus** is a fusiform dilation that functions as a baroreceptor to monitor blood pressure. It initiates the **cardiac reflex.**

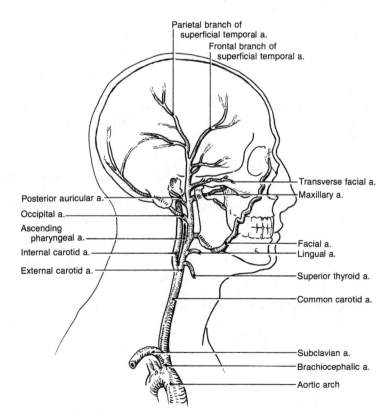

FIGURE 34-3. Branches of the carotid arteries in the neck.

- **(a)** The **afferent limb of the carotid reflex** is mediated by the carotid branch of the glossopharyngeal nerve (CN IX). This branch is part of the post-trematic division of the nerve of the third branchial arch. The information is relayed to the vasomotor center of the brain stem.
- **(b)** The **efferent limb of the carotid reflex** is mediated by the vagus nerve (CN X). It slows heart rate and diminishes cardiac stroke volume.

(2) The **carotid body** is a highly vascular area between the bifurcating vessels. It functions as a chemoreceptor sensitive to blood oxygen concentration (pO_2) and, to lesser extent, blood pH and partial pressure of carbon dioxide (pCO_2).
- **(a)** The **afferent limb of the respiratory reflex** is conveyed by the carotid nerve, a branch of the glossopharyngeal nerve (CN IX), and relayed to the respiratory center of the brain stem. Low pO_2 or high pCO_2 levels result in increased respiration rates.
- **(b)** The **efferent limb of the respiratory reflex** is by numerous nerves that control the muscles of respiration.
- **(c)** Carotid body tumors are difficult to excise because of their location.

(3) Palpating a **carotid pulse** requires a gentle touch and efforts to avoid the immediate vicinity of the carotid bifurcation, staying inferior to the midpoint of the thyroid cartilage. Even slight external pressure can slow the heart rate and give a false pulse count.

d. Carotid hemostasis may be achieved in an emergency by common carotid artery compression at the level of the cricoid cartilage.

2. The internal carotid artery ascends to the **carotid canal** in the base of the cranium without giving off branches in the neck.
   a. Course. The styloid process and its associated muscles interpose between the internal and external carotid arteries (see Figure 34-2).
   b. The **ophthalmic branch** supplies the forehead and anastomoses with the facial and superficial temporal arteries.

3. The external carotid artery supplies most of the face and visceral neck with eight major branches, four of which are given off in the neck (see Figure 34-3).
   a. The **occipital artery** arises **posteriorly** from the external carotid artery.
   - **(1)** Deep to the mastoid origin of the digastric muscle, this artery continues posteriorly, grooving the medial surface of the mastoid process, to emerge into the posterior cervical triangle.
   - **(2)** It gives off two branches to the sternomastoid muscle. The upper branch is associated with the spinal accessory nerve; the lower with the hypoglossal nerve.

   b. The **posterior auricular artery** also arises **posteriorly** from the external carotid artery just above the origin of the occipital artery.
   - **(1)** It ascends between the superior surface of the posterior belly of the digastric muscle and the parotid gland.
   - **(2)** It grooves the base of the skull between the mastoid process and the auricle.

   c. The **ascending pharyngeal artery** is a long slender branch that arises close to the bifurcation of the common carotid artery.
   - **(1)** It runs cranially between the internal carotid and the pharyngeal constrictors.
   - **(2)** It anastomoses freely with the ascending palatine branch of the facial artery.
   - **(3)** Superiorly, it passes over the free edge of the superior constrictor to supply the submucosal regions of the nasopharynx.

   d. The **superior thyroid artery** arises **anteriorly** at the tip of the greater horn of the hyoid bone. Occasionally, it arises with the lingual artery from a common trunk.
   - **(1)** It runs along the posterior border of the thyrohyoid muscle. At this point, it gives off a **superior laryngeal branch** that passes beneath the muscle and pierces the thyrohyoid membrane in company with the internal branch of the superior laryngeal nerve.
   - **(2)** The main trunk continues inferiorly along the medial edge of the upper lobe of the thyroid gland, supplying its anterior surface. Near the isthmus, it anastomoses with its contralateral counterpart. Because the superior thyroid artery

may lie close to the external laryngeal branch of the superior laryngeal nerve, care is needed when ligating this vessel to avoid cricothyroid paralysis.
    e. **Other branches** include the lingual, facial, superficial temporal, and maxillary arteries (see Chapter 36 VI and 37 III).
  4. **The thyrocervical artery (trunk),** arising from the subclavian artery, supplies the inferior cervical regions and portions of the shoulder (see Figure 6-5).
    a. **The inferior thyroid artery** ascends toward the inferior pole of the thyroid gland.
      (1) The **ascending cervical artery** supplies the deep muscles of the neck.
      (2) The **inferior laryngeal artery** approaches the thyroid gland posteriorly at about the level of the isthmus. It then reverses toward the lower pole and ascends again parallel to the recurrent laryngeal nerve. It supplies the posterior portions of that gland and the larynx. When the surgeon is ligating this vessel, care is needed to avoid the recurrent laryngeal nerve.
    b. The **suprascapular artery** supplies the posterior scapular region.
    c. The **transverse cervical artery,** when present, crosses the posterior triangle and bifurcates into superficial and deep branches.

C. Venous return (Figure 34-4)
  1. **The external jugular vein** is superficial and variable. It may be demonstrated (when present) by obstructing its drainage by light finger pressure or by having the patient perform the Valsalva maneuver. Careful observation of this vessel reveals venous pressure waves. To the experienced clinician, the level of the column of blood in the external jugular veins is a useful indication of right atrial pressure.
    a. **Course and variability.** It usually (35%) is formed at the angle of the jaw by the union of the posterior division of the **retromandibular vein** and the **posterior auricular vein.** It is absent in about 10% of the population.
    b. **Tributaries**
      (1) The **suprascapular vein** drains the base of the neck and accompanies the like-named artery.
      (2) The **superficial cervical vein** drains the anterior region of the neck and also accompanies the like-named artery.
      (3) The **anterior jugular vein** (when present) drains the muscular triangle.
      (4) The **posterior jugular vein** drains the apex of the posterior triangle.

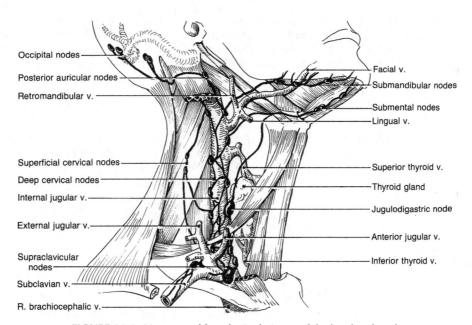

FIGURE 34-4. Venous and lymphatic drainage of the head and neck.

c. **Valves.** This vein variably has two pairs of downward-directed **valves.** The inferior valve is at the entrance to the subclavian vein; the superior valve is about 4 cm above the clavicle. For this reason as well as its occasional absence, the external jugular vein is not preferred for the insertion of central venous lines.

2. **The internal jugular vein** is formed by the sigmoid sinus as it exits the skull through the jugular foramen in company with the glossopharyngeal, vagus, and spinal accessory nerves (see Figure 34-4).
   a. **Course and tributaries.** As it descends through the neck in the carotid sheath, this vein winds laterally.
      (1) It usually receives the **facial vein,** one or more **lingual veins,** the **superior and middle thyroid veins,** as well as veins from the pharyngeal venous plexus.
      (2) Anastomoses usually occur with the anterior jugular and external jugular veins.
   b. It comes together with the subclavian vein on each side behind the sternoclavicular joint to form the **brachiocephalic vein.**
   c. In the root of the neck, the internal jugular veins lie anterior to the corresponding arteries immediately posterior to the sternomastoid muscle. This location, as well as the absence of valves in these veins, facilitates insertion of central venous lines.

D. **Lymphatic drainage** (see Figure 34-4)
   1. **Cervical nodes.** Lymph from the face and anterior region of the neck drains by two pathways. **Superficial cervical nodes** parallel the external jugular vein. **Deep cervical nodes** parallel the internal jugular vein.
      a. **Superficial (external jugular) nodes** receive drainage principally from the **occipital nodes,** the **retroauricular nodes,** and the **parotid nodes,** with some drainage from the anterior cervical nodes.
      b. **Deep (internal jugular) nodes** generally receive drainage from the superficial face and deep structures via the **submandibular, submental,** and **jugulodigastric nodes.**
         (1) The **jugulodigastric node** lies just below the posterior belly of the digastric muscle. It drains the tonsillar region and is therefore valuable in the diagnosis of pharyngeal inflammation.
         (2) **Anastomoses** are abundant between the various nodal regions as well as between the superficial and deep cervical lymphatics.
      c. **Jugular trunks** form from the deep cervical lymphatics in the root of the neck.
         (1) The **left jugular trunk** enters the thoracic duct just before it enters the venous system. This trunk also receives drainage from the **signal node** (of Virchow) that lies just superior to the middle one third of the clavicle close to the termination of the thoracic duct. Pathologic changes in this node signal tumor growth below the diaphragm.
         (2) The **right jugular trunk** terminates at the junction of the internal jugular vein and the subclavian vein.
   2. **The thoracic duct** enters the neck behind the left subclavian artery. From the thorax, it ascends into the neck. About the C6 vertebral level, it crosses anterior to the left vertebral artery and the left thyrocervical trunk. It drains into the venous system, where the internal jugular vein and the subclavian vein join to form the brachiocephalic vein.

VI. INNERVATION

A. **Hypoglossal nerve (CN XII)** (Figure 34-5)
   1. **Course and composition.** This nerve exits the cranium through the **anterior condylar (hypoglossal) canal.**

Anterior Cervical Triangle | 583

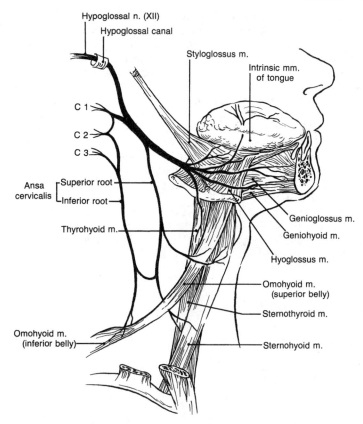

FIGURE 34-5. Hypoglossal nerve and upper cervical plexus.

   a. **Just inferior to the hypoglossal foramen,** it winds over the surface of the inferior vagal ganglion and passes between the internal carotid artery and the internal jugular vein.
   b. **At the angle of the mandible,** it courses laterally, turning anteriorly to pass superficial to the occipital artery.
   c. **At the level of the greater horn of the hyoid bone,** it crosses the external carotid artery, the loop of the lingual artery, and the hyoglossus muscle to reach the root of the tongue.
2. **Distribution.** The purely motor hypoglossal nerve innervates all the **intrinsic muscles of the tongue** and three extrinsic muscles of somatic origin, the **styloglossus, hyoglossus,** and **genioglossus muscles.**
   a. **Rule:** The hypoglossal nerve innervates every muscle that ends in the suffix "-glossus." **Exception:** The palatoglossus muscle is innervated by the pharyngeal branch of the vagus nerve.
   b. **Hypoglossal nerve injury.** Deviation of the tongue toward the affected side on protrusion is the result of unopposed action of the unaffected (contralateral) genioglossus muscle.

B. **Cervical plexus (C1–C3)**

1. **Sensory innervation.** The **transverse cervical (colli) nerve** (C2–C3) innervates the skin of the anterior triangle.
2. **Motor innervation.** The **ansa cervicalis** (C1–C3) innervates the strap muscles as well as one suprahyoid muscle.
   a. **Composition.** The ansa cervicalis receives contribution from spinal nerves C1–C3.

- (1) Fibers from spinal nerve C1 join and travel with the hypoglossal nerve. Most soon leave as the **superior root of the ansa cervicalis.**
- (2) Fibers from C2–C3 form the **inferior root of the ansa cervicalis.**
- (3) The superior and inferior roots of the ansa cervicalis anastomose to form a loop on the surface of the carotid sheath, the derivation of the term ansa.

b. **Distribution**
- (1) The **superior ramus** of the ansa cervicalis is formed by nerve fibers from C1 that run alongside the hypoglossal nerve and then leave the hypoglossal nerve (derivation of the former name **descendens hypoglossi**).
  - (a) These fibers innervate the **sternohyoid muscle, sternothyroid muscle,** and **superior belly of the omohyoid muscle.**
  - (b) Some nerve fibers from C1 that accompany the hypoglossal nerve extend beyond the descendens hypoglossi to reach the **thyrohyoid** and **geniohyoid muscles.**
- (2) The **inferior ramus of the ansa cervicalis** is formed by nerve fibers from C2–C3 (derivation of the former name **descends cervicalis**). These fibers innervate the inferior belly of the omohyoid muscle and a portion of the sternothyroid muscle.

Chapter 35
Facial Skeleton

I. INTRODUCTION

A. **Developmental considerations.** The facial cranium develops in conjunction with sensory organs for vision and smell, nasal passages for respiration, and oral stoma for taste and ingestion.

1. **Visceral derivatives.** Because the facial skeleton and anterior skeletal structures of the neck derive from the primitive gill arch system (Table 35-1), they exhibit branchiomeric segmentation (G. branchia, gill + meros, part).
2. **Innervation**
   a. **Special sensory nerves** are related to organs of special senses: the **olfactory nerve** (CN I), the **optic nerve** (CN II), and the **vestibulocochlear nerve** (CN VIII).
   b. **Somatic motor nerves** are related to muscles with somatic segmentation.
      (1) **Extraocular muscles** derive from cephalic somites and are innervated by the **oculomotor nerve** (CN III), the **trochlear nerve** (CN IV), and the **abducens nerve** (CN VI).
      (2) **Muscles of the tongue** mostly derive from cephalic somites and are innervated by the **hypoglossal nerve** (CN XII).
   c. **Branchiomeric nerves** are associated with muscles of branchiomeric segmentation and are related to the gill (branchial) arches and clefts (see Figure 35-1A). Each nerve divides into two branches, one anterior to the gill cleft (pre-trematic; from G. trema, hole) and the other posterior (post-trematic).
      (1) **Pretrematic branches** are sensory only.
      (2) **Post-trematic branches** contain both sensory and motor nerves.
      (3) The pretrematic branch of one arch usually joins the post-trematic branch of the preceding branchial arch, so each arch has dual innervation. For example, the chorda tympani branch of the facial nerve joins the lingual branch of the trigeminal nerve.

B. **Organization.** In most vertebrates, a series of branchial (gill) clefts develop in the pharyngeal wall. Cephalad and caudal to each branchial cleft are branchial arches composed of bone or cartilage, muscles, blood vessels, and nerves. In humans, remnants of these gill arches are incorporated into other structures to serve other functions.

1. **Branchial skeletal derivatives** arise from each arch (see Table 35-1).
2. **Branchiomeric muscle derivatives** maintain relationships with the bones associated with the arch, but some muscle migration occurs (see Table 35-1).
3. **Branchiomeric nerve derivatives** contain afferent and efferent fibers that run from the brain stem to the region of the cleft and divide characteristically to form pretrematic and post-trematic branches (Figure 35-1B; see Table 35-1).
   a. The **trigeminal nerve (CN V)** derives from the **first arch**. The **maxillary nerve** (pretrematic branch of the trigeminal nerve) is totally sensory and the **mandibular nerve** (the post-trematic branch) is mixed.
   b. The **facial nerve (CN VII)** derives from the **second arch**. The **chorda tympani** (sensory pretrematic branch) joins the lingual branch of the mandibular nerve (post-trematic nerve of the first arch). The **facial nerve proper** (the post-trematic branch) is mixed.
   c. The **glossopharyngeal nerve (CN IX)** derives from the **third arch**. The **tympanic nerve** (the sensory pretrematic branch) joins the acoustic branch of the facial nerve (the post-trematic branch of the third arch) to innervate the external auditory meatus. The **glossopharyngeal nerve proper** (the post-trematic branch) is mixed.
   d. The **vagus nerve (CN X)** derives in part from the **fourth arch**. The **pharyngeal**

Table 35-1. Branchial Arch Derivatives

Branchial Arch	Bone	Muscle	Innervation
First arch	Mandible Sphenomandibular lig. Malleus Incus	Masticatory mm.* Tensor tympani m. Tensor palatini m. Mylohyoid m. Digastric ant. belly m.	Trigeminal n. (CN V) Pre: Maxillary n. (S) Post: Mandibular n. (Sensory + Motor)
Second arch	Hyoid lesser horn Stylohyoid lig. Styloid process Stapes	Mm. of facial expression† Stylohyoid m. Digastric post. belly m. Stapedius m.	Facial n. (CN VII) Pre: Chorda tympani n. (S) Post: Facial n. proper (S + M)
Third arch	Hyoid body and greater horn	Stylopharyngeus m.	Glossopharyngeal n. (CN IX) Pre: Tympanic br. (S) Post: Glossopharyngeal n. (S + M)
Fourth arch	Laryngeal cartilages	Pharyngeal mm.‡ Cricothyroid m.	Vagus n. (CN X) Pre: Pharyngeal br. (S) Post: Superior laryngeal br. (S + M)
Sixth arch	Laryngeal cartilages	Laryngeal mm.§	Vagus n. (CN X) Pre: ? Post: Recurrent laryngeal br. (S + M)

* Masseter, medial pterygoid, lateral pterygoid, temporalis.
† Occipitofrontalis, corrugator cupercillii, procerus, orbicularis oculi, compressor naris, nasalis, levator labii superioris, levator anguli oris, zygomaticus minor, zygomaticus major, risorius, depressor anguli oris, depressor labii inferioris, orbicularis oris, mentalis, platysma, buccinator, anterior auricular, superior auricular, posterior auricular.
‡ Superior constrictor, middle constrictor, inferior constrictor, levator palatini, palatoglossus, palatopharyngeus, salpingopharyngeus.
§ Posterior cricoarytenoid, lateral cricoarytenoid, transverse arytenoid, thyroarytenoid (vocalis).

branch (the sensory pretrematic branch) joins the pharyngeal branch of the glossopharyngeal nerve (the post-trematic branch of the third arch). The **superior laryngeal branch** (the post-trematic branch) is mixed.

e. The **vagus nerve (CN X)** also derives in part from the **sixth arch.** Because the fifth arch never develops in humans, there apparently are no fifth arch components. Also, there is no identifiable pretrematic branch of that portion of the vagus nerve associated with the sixth arch. The **recurrent (inferior laryngeal) branch** of the vagus nerve is the mixed, post-trematic nerve of the sixth arch.

II. FACIAL CRANIUM

A. **Regions.** The facial cranium is bounded posteriorly by the sphenoid bone and superiorly by the floor of the anterior cranial fossa. It is the external counterpart of the step between the anterior and middle cranial fossae.

1. **In the upper face,** the orbital margin is bounded by the frontal, zygomatic, and maxillary bones. The sphenoid, palatine, ethmoid, lacrimal, and nasal bones also contribute to the orbital walls (Figure 35-2; see Figure 33-1).

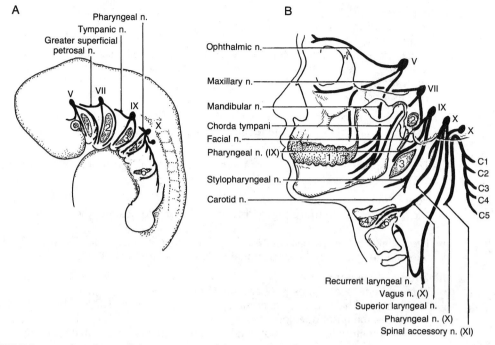

FIGURE 35-1. Developmental organization of the cranial nerves. *(A)* In early development, each branchiomeric nerve has a sensory pre-trematic division and a mixed post-trematic division that innervate adjacent sides of each gill cleft. *(B)* The same pattern can be discerned in the adult.

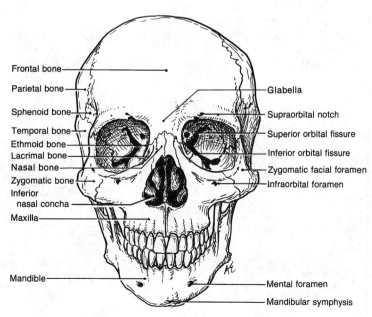

FIGURE 35-2. Facial cranium.

2. **In the midface,** on either side of the sagittal plane, a pyramidal stack of hollow bones surrounds the nasal cavity (see Figure 35-2).
   a. Its **apex** is the ethmoid bone.
   b. Its **floor** (the hard palate) is formed by the maxilla and palatine bones. In between, the vomer contributes to the nasal septum.
   c. The **walls** on either side are the maxilla, palatine, inferior nasal concha, lacrimal, and nasal bones.
3. **In the lower face,** the mandible surrounds the floor of the mouth and pharynx (see Figure 35-2). It articulates with the temporal bone of the neurocranium to complete the skull.
4. **The lateral aspect** displays features of the frontal, temporal, sphenoid, and zygomatic bones (see Chapter 31 B 2).
5. **Severe fracture** may separate the visceral face from the neurocranium (Le Fort type III), allowing the face to move posteriorly and inferiorly to obstruct the airway.

B. Bones. The facial cranium consists of 16 bones, including the mandible. The frontal, ethmoid, sphenoid, temporal, and basi-occipital bones are shared with the neurocranium.
1. **Unpaired midline bones**
   a. **Ethmoid bone** (see Figures 31-7 and 35-3B).
      (1) **Location.** The ethmoid sits at the apex of the pyramidal stack of bones that defines the nasal opening.
         (a) It contributes to the roof, walls, septum, sinuses, and conchae of the nasal cavity (Figure 35-3).
         (b) It contributes to the medial wall of the orbit (see Figure 33-1).
         (c) It contributes to the floor of the anterior cranial fossa.
      (2) The **cribriform plate** forms the roof of the nasal cavity, separating it from the anterior cranial fossa (see Figure 33-3B).
      (3) **Ethmoid labyrinths** (see Figure 35-3A)
         (a) Thin, scroll-shaped superior and middle conchae divide the lateral wall into a **sphenoethmoidal recess,** a **superior meatus,** and a **middle meatus.**
         (b) The ethmoid bone contains about a dozen **ethmoid sinuses** (air cells), which open medially into the nasal cavity.
         (c) The **lateral plates of the ethmoid labyrinth** form the **lamina papyracea** of the medial orbital wall.
      (4) The **vertical plate,** in the midline, extends above the cribriform plate as the

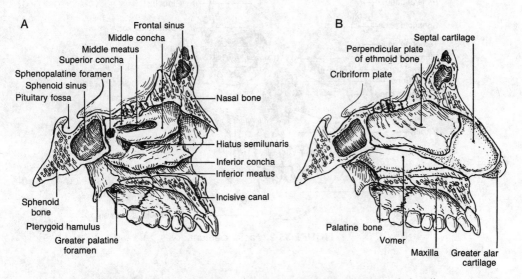

FIGURE 35-3. Bones of the nasal cavity. *(A)* Lateral wall; *(B)* medial wall.

crista galli, and articulates with the vomer below to form a portion of the bony **nasal septum** (see Figure 35-3B).

(5) **Fracture**

(a) **Injury to the cribriform plate** may produce rhinorrhea (discharge of cerebrospinal fluid through the nose). It also allows pathogens to gain entry into the cranial cavity, with resultant meningitis. Tearing of the olfactory nerves results in anosmia (loss of olfaction).

(b) **Fracture of the lamina papyracea** breaches the separation of nasal and orbital cavities, which leads to orbital emphysema and momentary exophthalmos when blowing the nose.

b. **Vomer** (see Figures 31-4 and 35-3B)

(1) Anteriorly, this thin midline bone is grooved for articulation with the septal cartilage. Posteriorly, it splits into two alae to accommodate the sphenoidal crest.

(2) It usually deviates to one side. Simple preliminary observation of the nasal septum can save time when a nasogastric or nasotracheal tube must be inserted.

c. **Sphenoid bone** (see Figures 31-4 and 35-3A)

(1) It lies between the left and right middle cranial fossae and separates the cranial cavity from the nasal cavity, the orbit, and the infratemporal fossa.

(2) The **body** abuts the posterior surface of the ethmoid bone. The **median crest** on its inferior surface contributes to the bony nasal septum.

(a) **Sphenoid sinuses,** within the body of the sphenoid bone bilaterally, open into the sphenoethmoidal recess above the superior concha.

(b) The **septum** between the left and right sphenoid sinuses usually deviates to one side, an important consideration in trans-sphenoidal hypophysectomy.

(3) **Greater and lesser sphenoidal wings** reach laterally and articulate with the orbital plates of the frontal bone.

(a) The **lesser wing** contributes to the floor of the anterior cranial fossa (see Figure 31-7).

(b) The **vertical surface of the greater wing** comprises the anterior wall and contributes to the floor of the middle cranial fossa as well as the posterior walls of the orbital cavities (see Figure 31-4).

(c) **Pterygoid processes** project perpendicularly from the infratemporal skull surface (see Figure 31-4).

(i) These processes broaden into medial and lateral pterygoid plates for attachment of pterygoid muscles.

(ii) Between the two plates, a triangular defect (the scaphoid fossa) is roofed by the pyramidal process of the palatine bone.

(4) The **pterygopalatine fossa** is a gap between the pterygoid process of the greater wing of the sphenoid bone and the maxilla. It is an inverted four-sided pyramid (Figures 35-4; see Chapter 36 IV B and Figure 36-9).

2. **Fused lateral bones**

a. **Frontal bone** (see Figures 31-3 and 35-2)

(1) The **squamous portion** forms the forehead and turns sharply posteriorly at the orbital margins to form the **orbital plates.** These plates roof the orbit and form the floor of the anterior cranial fossa (see Figure 31-7).

(a) The **ethmoid notch** between the left and right orbital plates receives the cribriform plate of the ethmoid bone. On each side in the frontal bone is a tiny canal, which transmits the anterior ethmoid neurovascular bundle from the orbit to the superior surface of the cribriform plate and on to the nasal cavity.

(b) The **supraorbital notch or foramen,** in the center of the supraorbital margin, transmits the supraorbital neurovascular bundle to the forehead (see Figure 35-2).

(c) The **nasal notch,** anteriorly in the midline, articulates with the nasal bone and the frontal process of the maxilla. It is the location of the **nasion.**

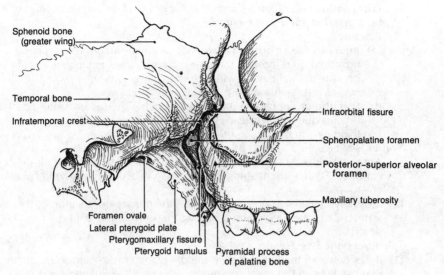

FIGURE 35-4. Infratemporal and pterygopalatine fossae.

- **(2) Frontal sinuses** (air cells) lie anteriorly between the bony tables of the frontal bone (see Figure 35-3A).
- **b. Maxillae** fuse to form the largest bone of the facial cranium (see Figure 35-2).
  - **(1) Location.** Each maxilla lies below the frontal bone and ethmoid labyrinth. It is approximately pyramidal and from it extend several processes.
  - **(2) Structure**
    - **(a)** The **superior (orbital) surfaces** of the maxilla contain the **infraorbital canals,** which terminate as the **infraorbital foramina.** Each foramen transmits an infraorbital neurovascular bundle.
    - **(b) Zygomatic processes** articulate with the zygomatic bone and mark the lateral boundary of the anterior face and the infratemporal fossa.
    - **(c) Frontal processes** extend around the anterior ends of the ethmoid bone to articulate with the frontal bone.
      - **(i)** Each forms a portion of the medial orbital margin (see Figure 33-1).
      - **(ii)** A bony canal in each frontal process of the maxilla transmits the nasolacrimal duct, which drains the lacrimal sac into the inferior meatus of the nasal cavity.
    - **(d) Alveolar processes** (superior alveolar margins) support the teeth of the upper jaw (see Figure 35-6A).
      - **(i)** Each process has sockets for eight teeth.
      - **(ii)** The posterior end of each process continues a little way beyond the third molar as the maxillary tubercle (see Figure 35-4).
    - **(e) Horizontal palatine processes,** on the inner surface above the alveolar margin, form the anterior floor of the nasal cavity and the roof of the oral cavity (see Figure 35-3B).
      - **(i)** They fuse at the midline to form the anterior four fifths of the **hard palate** (see Figure 35-6A).
      - **(ii)** Each process is pierced in the anterior midline by the **incisive foramen (canal),** which transmits nasopalatine neurovascular bundles.
    - **(f) Maxillary tuberosities** mark the posterior end of the maxilla. Each **pterygopalatine fossa** (the narrow space behind the maxilla) is closed anteriorly by the maxillary tuberosity, medially by the perpendicular plate of the palatine bone, and posteriorly by the pterygoid process of the sphenoid bone (see Figure 35-4).
  - **(3) Large maxillary sinuses** lie within the maxillary bone on each side.

- **(4) Fracture**
  - **(a) Fracture of the orbital floor,** usually in the infraorbital groove (often from catching a ball in the orbit), produces communication between the nasal and orbital cavities. This breach predisposes to orbital emphysema, momentary exophthalmos when blowing the nose, and enophthalmos as a result of herniation of periorbital fat into the maxillary sinus.
  - **(b) Collapse of the thin anterior wall** of the maxillary sinus (the result of a fist or a ball) may denervate the anterior maxillary teeth because the anterior–superior alveolar nerves run in tunnels through this bone.
  - **(c) Transverse fracture** may occur at the level of the nasal floor (Le Fort type I fracture).
  - **(d) Vertical fracture** separates the maxilla from the frontal and sphenoid bones (Le Fort type III fracture), detaching the facial cranium from the neurocranium.
- **c. Mandible.** This U-shaped bone forms the lower jaw and bounds the floor of the mouth. Technically, it is part of the skull and not the facial cranium (see Figures 31-3 and 35-2).
  - **(1) Rami** constitute the two posterior vertical portions of the mandible that turn sharply upward from the body of the jaw at the angle.
    - **(a) External surface** (Figure 35-5A)
      - **(i)** The **condyloid process** has a rounded head (condyle) for articulation with the temporal bone at the temporomandibular joint.
      - **(ii)** The **neck** is narrow, lying between the head (condyle) and the ramus. It receives the lateral pterygoid muscle.
      - **(iii)** The **coronoid process** arises from the anterior border of the ramus and provides insertion for the temporalis muscle.
      - **(iv)** The **mandibular incisure (notch)** lies between the coronoid and condyloid processes.
      - **(v)** At the **angle of the mandible,** a series of oblique ridges provides insertion for the masseter muscle, and the **stylomandibular ligament** attaches to the mandible.
      - **(vi)** The ramus continues into the base, anteriorly, as the oblique line.
    - **(b) Internal surface** (see Figure 35-5B)
      - **(i)** The **mandibular foramen** opens into the mandibular canal, which carries the inferior alveolar branch of the trigeminal nerve and the inferior alveolar artery. The small triangular **lingula** guards the anterior margin of this foramen and provides attachments for the **sphenomandibular ligament** from which the mandible pivots.

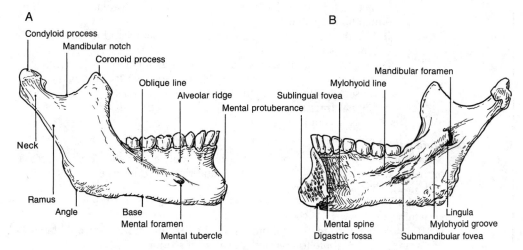

FIGURE 35-5. Mandible. *(A)* External aspect; *(B)* internal aspect.

(ii) The **mylohyoid groove,** immediately inferior to the lingula and running just inferior to the mylohyoid line, carries the mylohyoid branch of the inferior alveolar nerve.

(iii) A series of oblique ridges near the **angle of the mandible** provides insertion for the medial pterygoid muscle.

Periosteum is not attached to the bone at this site because of the paucity of Sharpey's fibers.

The medial pterygoid and masseter muscles form the **mandibular sling** at the angle of the ramus.

(2) **The body** constitutes the horizontal portion of the mandible. Its thick, rounded inferior margin is the base.

(a) **External surface** (see Figure 35-5A)
  (i) The **mandibular symphysis** marks the fusion of the two mandibular anlage. At birth, the halves of the body are joined by dense, fibrous tissue. Bony fusion, occurring after the age of 2 years, is evident by a faint midline ridge, which widens inferiorly as the **mental tubercles.**
  (ii) **Mental foramina,** located on each side inferiorly to the second premolar tooth, transmit the mental artery and the mental branch of the inferior alveolar nerve.

(b) **Internal surface** (see Figure 35-5B)
  (i) The **mylohyoid line** (into which the mylohyoid muscle inserts) separates the sublingual fossa (for the sublingual gland) from the submandibular fossa (for the submandibular gland).
  (ii) **Genial tubercles** for the genioglossus muscles (superior) and geniohyoid muscles (inferior) are just above the anterior end of the mylohyoid line.

(c) **Alveolar ridges** run along the superior margin of the body.
  (i) On each side, these ridges have sockets for eight teeth (see Figure 35-6B).
  (ii) Ridges form in response to the presence of teeth and transmit masticatory forces to the jaw. In the edentulous individual, the alveolar ridges atrophy.

(3) **Fracture.** The shape of the mandible makes it particularly susceptible to fracture, second in frequency only to the nasal bone.

(a) **Subcondylar fracture** produces mandibular instability through interference with protrusion of the jaw. Contraction of the lateral pterygoid muscle on the injured side draws the proximal fragment forward, but not the rest of the ramus.

(b) **Fractures of the body** may result in malocclusion of the teeth anterior to the fracture line as well as damage to the inferior alveolar nerve within the mandibular canal.
  (i) "Favorable" fractures are parallel to the ramus and result in less displacement.
  (ii) "Unfavorable" fractures slant toward the angle of the jaw with greater displacement because the anterior digastric and masseter muscles pull the fragments in opposite directions.

(c) **Symphyseal fracture** produces malocclusion between left and right sides as well as loss of jaw stability.

3. **Separate lateral bones**
   a. **Nasal bones** form the bridge of the nose and articulate with the frontal processes of the maxilla as well as with the ethmoid and frontal bones (see Figures 31-3 and 35-2). They are a common site of fracture.
   b. **Lacrimal bones** not only are the smallest bones of the facial cranium, but also are very thin (see Figures 31-3 and 35-2).
      (1) They lie between the frontal process of the maxilla and the ethmoid labyrinth, and contribute to the medial wall of the orbit.
      (2) Each lacrimal bone bears a groove (lacrimal fossa) for the lacrimal sac.

c. **Inferior nasal conchae,** separate bones, lie in the lateral wall of each nasal cavity, curling above the inferior meatus (see Figure 35-3A).
    (1) They articulate with the maxillary, ethmoid, and palatine bones.
    (2) They form the inferior boundary of the hiatus semilunaris, into which the frontal, anterior, and middle ethmoidal sinuses, as well as the maxillary sinus, drain.
d. **Palatine bones** have both vertical and horizontal processes (see Figure 35-3B).
    (1) The **perpendicular plate** is flush with the posteromedial edge of the maxilla and forms the lateral walls of the posterior portion of the nasal cavity. Posteriorly, nerves and vessels within the laterally situated perpendicular plate pierce the adjacent hard palate through the greater and lesser palatine foramina.
    (2) The **horizontal palatine process** parallels the palatine process of the maxilla and forms the posterior one fifth of the hard palate (see Figure 35-6A).
        (a) Horizontal processes of the maxilla and palatine bones articulate to form the **hard palate.** Medially, the horizontal plate of the palatine bone articulates with the vomer, completing the nasal septum.
        (b) The posterior edge of the hard palate is sharp and concave on either side of the nasal spine.
    (3) The **sphenopalatine notch,** superiorly, incompletely articulates with the sphenoid bone to form the **sphenopalatine foramen,** which transmits the sphenopalatine neurovascular bundle (see Figure 35-3).
e. **Zygomatic bones** are laterally placed (see Figures 31-3 and 35-2).
    (1) **Bone landmarks**
        (a) The **orbital margin** turns sharply inferior along the frontal process of the zygomatic bone.
        (b) The **temporal process** of the zygomatic bone projects posteriorly to fuse with the zygomatic frontal process of the temporal bone, forming the zygomatic arch.
        (c) The **zygomatic arch** bridges the temporal fossa and conceals the base of the cranium from lateral view. It transmits forces from the facial cranium to the neurocranium.
    (2) **Fractures** result from trauma to the cheek bone with loss of stability in the zygomatic arch. Because the zygomaticotemporal and zygomaticofacial branches of the maxillary nerve pass through foramina in the zygomatic bone, this injury may be associated with paraesthesias.

III. TEETH

A. Introduction

1. **In children,** there are **20 deciduous (milk) teeth**—five on each side of the jaw, including:
    a. **Two incisors,** which come in at approximately 6 and 8 months
    b. **One canine,** which comes in at approximately 10 months
    c. **Two premolars,** which come in during the second year

2. **In the adult,** there are **32 permanent teeth**—eight on each side of each jaw, including:
    a. **Two incisors,** which erupt in the sixth or seventh years
    b. **One canine,** which erupts in the tenth year
    c. **Two premolars** (bicuspid teeth), which erupt in the ninth and eleventh years
    d. **Three molars** (tricuspid teeth), the first of which comes in during the sixth year, followed by the second molar in the early teen years and the third molar (wisdom tooth) in the late teen years or even early twenties. Some individuals have supernumerary, especially mandibular, third molars.

B. **Tooth structure.** The basic structural element is dentine, a yellowish substance that is nurtured through the fine dental tubules of odontoblasts lining the central pulp space.

1. **The root** is embedded in the alveolar part of the mandible or maxilla.
   a. It is covered with **cementum,** which is connected to the bone of the socket by a layer of modified periosteum, the **periodontal ligament.**
   b. The number of roots (from one to three) depends on the type of tooth.

2. **The crown** projects into the oral cavity, composed of one or more cusps.
   a. It is covered with **enamel,** a hard, white crystalline substance formed before the tooth erupts. After eruption, this inert material changes little. Its only change is to adsorb fluoride ions, which reduce its solubility in the acid metabolites of oral bacteria.
   b. Enamel, like bone, is stained by tetracycline antibiotics, but only while it is being formed. Therefore, tetracycline should not be given to children.

3. **The neck,** an intermediate cervical part, is related to the gingiva (gum).

C. **Organization** of the teeth is according to position, shape, and number of roots.

1. **Classification.** Each alveolar margin bears 16 teeth classified according to the shape of the crown (Figure 35-6).
   a. **Incisors** (4) have a thin cutting edge.
   b. **Canines or cuspids** (2) have a single prominent cone or cusp. They are usually longer than the incisors or premolars, but length of this tooth varies considerably among individuals.
   c. **Premolar or bicuspids** (4) have a crown divided by a sagittal groove into two cusps.
   d. **Molars or tricuspids** (6) of adults have two, three, or, occasionally, more cusps.
      (1) Many years of grinding an unrefined diet may flatten the molar surfaces completely.
      (2) The upper molars have three roots, and the lower molars have two roots.

2. **Sockets.** The bone of the socket has a thin cortex, the lamina dura, separated from the adjacent labial and lingual cortices by a variable amount of trabeculated bone.
   a. The labial wall of the socket is particularly thin over the incisor teeth. It is best to break this surface when removing an incisor tooth.
   b. When removing molars, the lingual route is easier.

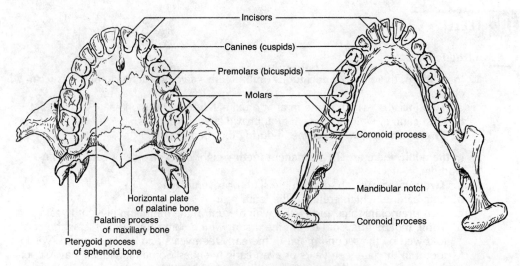

FIGURE 35-6. Teeth. *(A)* Maxilla; *(B)* mandible.

D. Innervation is predominantly by divisions of the trigeminal nerve.
1. **Maxillary teeth** are innervated by the anterior, middle, and posterosuperior alveolar branches of the maxillary nerve (CN V_2).
2. **Mandibular teeth** are innervated by the inferior alveolar branch of the mandibular nerve (CN V_3).

E. Clinical considerations
1. **Advanced infection of the root** involves the periodontal ligament and erodes through the lamina dura (radiographic evidence of chronic infection).
2. **Abscess involving the mandibular teeth** may spread through the lower jaw to emerge on the face or in the floor of the mouth. The mandibular nerve also innervates a portion of the ear, and pain from an infected lower tooth may be referred to the ear.
3. **Abscess involving the maxillary teeth** may spread through the upper jaw to emerge on the face or in the roof of the mouth. The relations of the upper teeth may influence the mode of presentation.
   a. **Infection of a second incisor** may spread along the palate.
   b. **Infection of a canine** may point (track superficially) in the face, resulting in thrombosis of the angular facial vein and spread along the superior ophthalmic vein to the cavernous sinus.
   c. Because the roots of the maxillary molars project into the maxillary sinus, infection of an upper tooth may produce symptoms of sinusitis with pain referred to the distribution of the maxillary nerve.
   d. Similarly, sinus infection may irritate the nerves to these teeth, causing toothache.

Chapter 36
Face

I. INTRODUCTION

A. Developmental considerations. The face comprises derivatives of the primitive gut (nose and mouth), branchiomeric (gill) structures, and special sensory receptors (eye and ear)(see Chapter 35 I A, B).

B. Facial features. The orifices and their associated superficial structures are the features of the face (see Figure 36-1). Variations in the proportion of each feature determines an individual's facial identity.

1. **Facial openings.** At the palpebral fissures, nostrils, and oral fissure, the skin of the face is continuous with the mucous membrane that lines these structures.
   a. **Palpebral fissures** are bounded superiorly and inferiorly by eyelid margins (see Figure 33-2).
      (1) **Eyelids** protect the anterior surface of each eye.
      (2) They enclose a potential space, the **conjunctival sac,** which opens into the skin of the face at the palpebral fissure.
      (3) The **palpebral portion of the orbicularis oculi muscle** runs beneath the skin of each eyelid. Contraction of this muscle blinks the eyelid.
      (4) Palpebral margins meet at **medial** and **lateral canthi.**
   b. **External nares (nostrils)** are supported by the nose, a pyramidal anterior extension of the nasal cavity (see Figure 36-1).
      (1) Its skeleton is both bony and cartilaginous.
      (2) The frontal process of the maxilla and the nasal bones articulate with the nasal notch of the frontal bone to form the arch of the nose.
      (3) The **septal cartilage** is a quadrangular plate of fibrocartilage thickened around its margin (see Figure 35-3).
         (a) It lies approximately in the midline sagittal plane.
         (b) Its posterior edge meets the septal plate of the ethmoid bone and the vomer.
         (c) The anterior edge emerges from behind the nasal bones as the ridge of the nose.
         (d) It articulates with the lateral and alar plates of the nasal cartilage to complete the cartilaginous portion of the anterior nose.
      (4) The external nares are associated with muscles, but the human nares cannot be closed, only pinched.
   c. The **oral fissure** is bounded by the superior and inferior labial folds that meet laterally at commissures (see Figure 36-1).
      (1) The **labial margin** has prominent superficial vascular papillae (thelia) that provide a **vermilion border.**
      (2) Thin median **frenula** extend from the inner surface of each gum to the associated lip.
      (3) Between the labial skin and labial mucosa are pea-sized **labial salivary glands** and the **orbicularis oris muscle.**
   d. **External ear** (see Figure 33-8).
      (1) **Auricles or pinnae,** lying on either side of the head and directed slightly forward, concentrate sound waves and enable stereophonic localization of the source.
      (2) The **external auditory (acoustic) meatus** extends from the concha to the tympanic membrane.
      (3) The **tympanum** (tympanic membrane or ear drum) separates the external ear from the internal ear and is receptive to sonic vibrations.

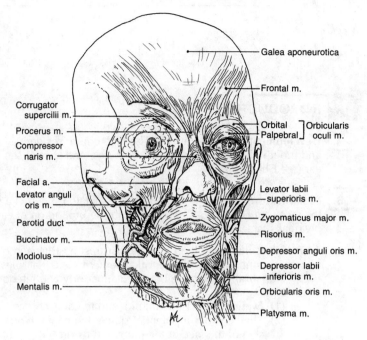

FIGURE 36-1. Muscles of facial expression. *Left side,* superficial muscles; *right side,* deeper muscles.

2. **Facial musculature.** Around each orifice, a system of subcutaneous muscles forms sphincters and dilators, the **muscles of facial expression** (see Figure 36-1). In addition, the jaw receives a group of deep muscles, the **muscles of mastication** (see Figure 36-4).
3. **Innervation**
   a. **General principle.** Cranial nerves are associated with muscles related to the branchial (gill) arches (see Chapter 35 I and Figure 35-1B).
   b. **Sensory** stimuli are mediated predominantly by the **trigeminal nerve.**
   c. **Motor control**
      (1) **Muscles of facial expression** are innervated by the **facial nerve** (see Figure 36-9).
      (2) **Muscles of mastication** are innervated by the **mandibular branch** (CN V_3) of the trigeminal nerve (see Figure 36-10).

II. PAROTID GLAND

A. **Relations.** The parotid gland is wrapped around the ramus of the mandible, forming lateral and retromandibular portions (see Figure 36-5).

1. **Laterally,** the parotid lies superficial to the masseter muscle.
   a. An anterior extension (facial process) surrounds the parotid duct. The facial process may form a separate **accessory parotid gland.**
   b. The firm cord-like **parotid duct** runs anteriorly from the facial process of the gland.
   c. The **facial nerve** forms a plexus of terminal branches within the parotid gland. Each branch is separated from the parotid parenchyma by fascial invaginations.
2. **Medially,** the retromandibular portion extends between the mandible and the medial pterygoid muscle just superficial to the styloid process and the external carotid artery.

B. Drainage. The **parotid duct** (of Stensen) drains the serous saliva from the parotid gland into the oral cavity.

1. **In the cheek,** the duct runs anteriorly from the lateral portion of the gland. At the anterior edge of the masseter muscle, it dives medially and passes through the buccinator muscle.
2. **In the mouth,** it opens into the vestibule opposite the second maxillary molar. The oblique course of the duct between the mucous membrane and the buccinator muscle allows it to be compressed and therefore act as a valve.

C. Innervation

1. **Parasympathetic neurons** from the **glossopharyngeal nerve (CN IX)** innervate the parenchyma of the parotid gland and control secretory activity (see Figure 36-7).
   a. **Presynaptic neurons** lie in the **inferior salivatory nucleus** of the midbrain and run in the **glossopharyngeal nerve** (CN IX).
      (1) In the jugular canal, the secretomotor fibers leave in the **tympanic branch (nerve of Jacobson)**.
      (2) In the middle ear, the tympanic nerve forms the **tympanic plexus** (of Jacobson) on the medial wall. The **lesser superficial petrosal nerve** leaves this plexus through a canal of the same name in the anterior wall and gains access to the middle cranial fossa (see Figure 36-7).
      (3) In the middle cranial fossa, the **lesser superficial petrosal nerve** courses briefly across the temporal bone and exits the cranium by the foramen ovale or the innominate foramen to access the infratemporal fossa.
      (4) In the infratemporal fossa, the fibers synapse in the **otic ganglion,** which lies just distal to the foramen ovale and adjacent to the mandibular division of the trigeminal ganglion (see Figure 36-7).
   b. The **postsynaptic secretomotor fibers** leave the **otic ganglion** and gain access to the **auriculotemporal branch** of the mandibular nerve, which carries them to the parotid gland (see Figure 36-10).
      (1) This nerve also carries sympathetic sudomotor fibers to the sweat glands of the scalp.
      (2) If the nerve is severed, autonomic fibers can regenerate into each other's pathways and innervate the wrong gland. The anticipation or taste of food then produces facial sweating instead of salivation (Fray's syndrome).
2. **Sympathetic pathways** innervate only the vasculature of the parotid gland. They regulate blood flow through the parotid and thereby the consistency of the saliva.
   a. **Presynaptic neurons** reside in the lateral cell columns of the upper thoracic segments of the spinal cord. Their axons reach the sympathetic chain through **white rami communicantes** and course cranially within the cervical sympathetic chain to the superior cervical ganglion.
   b. **Postsynaptic neurons** reside in the **superior cervical ganglion.** Their axons form a plexus about the internal carotid artery and likely reach the parotid gland along twigs from the deep petrosal nerve.

D. Tumors

1. **Tumors** usually arise in the lateral portion of the gland.
2. Surgical intervention requires dissecting glandular parenchyma away from the nerves. The surgeon is presented with a challenge, because iatrogenic facial nerve injury has major visible effects.

III. MUSCLES OF FACIAL EXPRESSION

A. Basic principles

1. **Development.** The muscles of facial expression (Table 36-1) derive from a single plate of muscle associated with the second branchial arch that subsequently migrates over the skull.

2. **Functions.** Most muscles of facial expression comprise sphincters and dilators that adjust the facial openings. All are instrumental in the display of human emotion.
3. **Innervation.** The **facial nerve (CN VII),** the nerve of the second branchial arch, innervates these muscles.

B. **Oral musculature** (see Figure 36-1 and Table 36-1)
1. **Sphincter.** The **orbicularis oris muscle** is the sphincter of the mouth (Figure 36-1).
   a. **Orientation.** This muscle lies circumferential to the margin of the lips. At the lateral periphery, some fascicles pass into the **modiolus labii** (L. hub of the lip), a small mass of connective tissue on either side of the mouth that acts as a common point of insertion for several muscles of facial expression.
   b. **Function.** The **orbicularis oris** purses or puckers the lips and closes the oral cavity.
2. **Dilators**
   a. These muscles of the mouth are numerous and quite small, and have either cutaneous or bony origins. Their insertions blend with the muscular substance of the lips. Many of these muscles insert into the **modiolus,** an arrangement that provides an extensive range of facial expression.
   b. **Layers** (see Figure 36-1)
      (1) The **subcutaneous layer** consists of the **risorius muscle** (from L. risus, to laugh). Small and variable, it arises from the parotid fascia and runs horizontally to insert at the angle of the mouth.
      (2) **Superficial layer**
         (a) The **zygomatic muscles** form an inverted "V" as they diverge from their origins at the orbital margins. The **zygomaticus major** inserts on the angle of the mouth; the **zygomaticus minor** inserts onto the upper lip.
         (b) The **levator labii superioris alaeque nasi** inserts on the upper lip laterally and on the major alar cartilage medially.
      (3) **Middle layer**
         (a) The triangular **depressor anguli oris (triangularis),** arising from the oblique line on the body of the mandible, and the **levator anguli oris,** arising from the incisive fossa of the maxilla, converge at the modiolus.
         (b) The quadrangular **depressor labii inferioris,** arising from the oblique line of the body of the mandible, and the **levator labii superioris,** arising from the orbital margin, insert onto the lower and upper lips, respectively.
      (4) **Deep layer**
         (a) The **mentalis muscle** runs between the incisive fossa of the mandible and the lower lip near the midline.
         (b) The **buccinator muscle** (L. trumpeter) arises from the maxilla, mandible, and pterygomaxillary raphe.
            (i) Posterior to the third molar teeth, it interdigitates with fibers of the superior pharyngeal constrictor in the **pterygomandibular raphe.** Anteriorly, its fibers pass directly into the lips.
            (ii) Most important, the buccinator draws the cheeks against the molar teeth to retain food on the occlusal surfaces during mastication.

C. **Nasal musculature** (see Figure 36-1 and Table 36-1)
1. **Sphincter.** The **compressor naris muscle** (the transverse part of the **nasalis muscle**) acts as the nasal sphincter. Its function is rudimentary in humans, merely pinching the nostrils, but impressive in diving mammals.
2. **Dilators.** The **dilator naris** (the alar part of the **nasalis muscle**) and the nasal part of the **levator labii superioris alaeque nasi** are the nasal dilators. Both insert onto the alar cartilage and flare the nostrils.

D. **Orbital musculature**
1. **Sphincters** (see Figure 36-1 and Table 36-1)
   a. The **orbitalis oculi,** the orbital sphincter, has two parts.

(1) The thick **orbital portion (pars orbitalis)** encircles the eyes in continuous loops.
   (a) It arises from the medial orbital margin and the medial palpebral ligament.
   (b) When it contracts, the eye closes tightly **(the wink)**.
(2) The **palpebral portion (pars palpebrae)** crosses the eyelids beneath the skin.
   (a) Its fibers are arranged so that, on shortening, they move toward geodesic lines (the shortest distance between two points on the surface of a sphere)

Table 36-1. Muscles of Facial Expression

Muscle	Origin	Insertion	Primary Action	Innervation
Oral Musculature				
Orbicularis oris	Maxilla, mandible, and pterygomandibular raphe	Skin of lips	Purses lips	Buccal and mandibular brs. of facial n. (CN VII)
Buccinator	Maxilla, mandible, and pterygomandibular raphe	Skin of lips	Compresses cheeks	Buccal br. of facial n. (CN VII)
Zygomaticus major	Zygomatic bone	Modiolus	Draws angle of mouth upward	Zygomatic br. of facial n. (CN VII)
Zygomaticus minor	Zygomatic bone	Upper lip	Raises upper lip	Zygomatic br. of facial n. (CN VII)
Levator anguli oris	Maxilla	Modiolus	Draws angle of mouth upward	Buccal br. of facial n. (CN VII)
Levator labii superioris	Maxilla	Upper lip	Raises upper lip	Buccal br. of facial n. (CN VII)
Depressor anguli oris	Mandible	Modiolus	Draws angle of mouth downward	Mandibular br. of facial n. (CN VII)
Depressor labii inferioris	Mandible	Lower lip	Lowers lower lip	Mandibular br. of facial n. (CN VII)
Mentalis	Mandible	Lower lip	Lowers lower lip	Mandible br. of facial n. (CN VII)
Platysma	Skin of upper thorax and shoulder	Mandible and modiolus	Tenses skin over anterior neck, draws angle of mouth downward	Cervical br. of facial n. (CN VII)
Orbital Musculature				
Orbicularis oculi				
Orbital portion	Orbital margin of eye	Palpebral skin	Eye wink	Zygomatic br. of facial n. (CN VII)
Palpebral portion	Lateral palpebral ligament	Medial palpebral ligament	Eye blink and compresses lacrimal sac	Zygomatic br. of facial n. (CN VII)
Occipitofrontalis				
Occipitalis	Nuchal line	Epicranial aponeurosis	Draws scalp posteriorly	Auricular br. of facial n. (CN VII)
Frontalis	Epicranial aponeurosis	Skin at the supraciliary crest	Draws scalp anteriorly	Temporal br. of facial n. (CN VII)
Corrugator supercilii	Frontal bone	Skin of medial end of upper eyelid	Draws eyebrows inferomedially	Temporal br. of facial n. (CN VII)

to close the palpebral fissure. Contraction of this muscle portion produces **the blink.**
- **(b)** The origin of the palpebral portion encircles the lacrimal sac. Those fibers posterior to the sac may be referred to as the **lacrimal portion** (Horner's muscle). Thus, blinking compresses the lacrimal sac, so that on relaxation, fluid is aspirated from the medial corner of the eye.
- **(3) Innervation.** As a muscle of facial expression, it is innervated by the **facial nerve** (CN VII).
  - **(a)** This nerve forms the efferent limb of the **blink reflex.**
  - **(b)** Paralysis of the palpebral portion of this muscle results in inability to close the eyelids [see Chapter 33 I B 3 a (3) (b)].
- **b.** Although not sphincters in the true sense, the small **procerus** and **corrugator supercilii muscles** pull the eyebrows downward and inward to shield the eyes from bright light or in an expression of concentration.

2. **Dilators** (see Figures 33-2 and 36-1)
- **a.** The **levator palpebrae superioris muscle** acts directly on the upper eyelid to open the palpebral fissure. As an extraocular muscle, it is innervated by the oculomotor nerve.
- **b.** The **occipitofrontalis muscle** appears to widen the eyes by raising the eyebrows, as in an expression of surprise or fear.

E. **Auricular musculature** has little action in humans. The **anterior and superior auricular muscles** converge on the auricle from the epicranial aponeurosis. The **posterior auricular muscle** arises from the mastoid process and inserts on the posterior aspect of the pinna.

F. **Platysma.** The extent of this muscle is variable (see Figure 36-1). It migrates over the anterior surface of the neck and may extend onto the upper thoracic wall. Contraction tenses the skin of the anterior regions of the neck, which is of value for men while shaving.

IV. INFRATEMPORAL AND PTERYGOPALATINE FOSSAE

A. **Infratemporal fossa** (see Figures 31-5 and 35-4)

1. **Boundaries**
    - **a.** The **roof** is formed by the infratemporal surface of the greater wing of the sphenoid bone.
        - **(1)** It is pierced by the foramen ovale (transmitting the mandibular division of the trigeminal nerve) and foramen spinosum (transmitting the middle meningeal artery).
        - **(2)** The sphenoid bone bears a pterygoid process that expands inferiorly into two pterygoid plates.
            - **(a)** The **medial pterygoid plate** provides attachment for the superior constrictor muscle.
            - **(b)** The **lateral pterygoid plate** provides attachment for the medial and lateral pterygoid muscles.
    - **b.** The **anterior wall** is formed by the maxilla.
    - **c.** The **medial wall** is formed by the lateral pterygoid plate of the sphenoid bone.
    - **d.** The **posterior wall** is formed by the temporal bone.
    - **e.** The **lateral boundary** is formed by the ramus of the mandible, which articulates with the temporal bone at the temporomandibular joint.

2. **Contents**
    - **a.** The **parotid gland** curves around the posterior border of the mandible into the infratemporal fossa.
    - **b.** The **carotid sheath** passes between the parotid gland and the intrinsic pharyngeal musculature.

c. The **muscles of mastication,** maxillary artery and its branches, pterygoid venous plexus, and branches of the mandibular division of the trigeminal nerve lie in this fossa.

B. Pterygopalatine fossa

1. **Structure.** This narrow fossa approximates an inverted four-sided pyramid (see Figures 35-4 and 36-2).
    a. **Laterally,** it is open through the pterygomaxillary fissure, communicating with the infratemporal fossa.
    b. The **posterior wall** is formed by the pterygoid process of the sphenoid bone and contains three foramina (Figure 36-2).
        (1) The pterygoid (vidian) canal traverses the root of the pterygoid process. It contains the vidian nerve, which consists of presynaptic parasympathetic neurons of the greater superficial petrosal branch of the facial nerve that synapse in the pterygopalatine ganglion. It also contains postsynaptic sympathetic fibers from the carotid perivascular plexus by way of the deep petrosal nerve.
        (2) The **foramen rotundum** pierces the greater wing of the sphenoid bone just anterior to the root of the pterygoid process, connecting with the middle cranial fossa and transmitting the maxillary division of the trigeminal nerve (CN V_2).
        (3) The **pterygovaginal canal** contains vessels that communicate with the pharyngeal spaces.
    c. The **medial (deep) wall** is formed by the **perpendicular plate of the palatine bone.** It contains the **sphenopalatine foramen** that communicates with the nasal cavity and transmits the **sphenopalatine neurovascular bundle.**
    d. The **anterior wall** is formed by the posterior surface of the maxilla. It contains two fissures and a group of small foramina (see Figure 36-2).
        (1) The **inferior orbital fissure** is formed where the maxilla and the greater wing

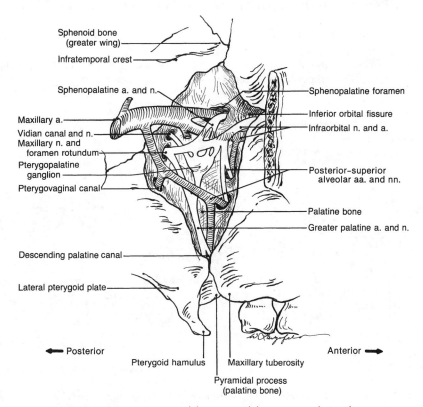

FIGURE 36-2. Contents and foramina of the pterygopalatine fossa.

of the sphenoid fail to fuse, forming a communication with the orbit. The **infraorbital branches** of the maxillary nerve and artery reach the orbit and face through this hiatus.
- (2) The **superior orbital fissure** results from a separation between the greater and lesser wings of the sphenoid bone, forming a communication between the middle cranial fossa and the orbit. It transmits the oculomotor nerve (CN III), trochlear nerve (CN IV), ophthalmic division of the trigeminal nerve (CN V_1), and the abducens nerve (CN VI), as well as the ophthalmic vein.
- (3) **Posterior superior alveolar foramina** (usually two or three) transmit posterior–superior alveolar neurovascular bundles to the maxillary molars.
- e. The **roof** (base), formed by the greater wing of the sphenoid bone, contains no foramina.
- f. **Inferiorly,** the pterygopalatine fossa narrows, forming the palatine canal by the approximation of the maxillary process, the palatine bone, and the pterygoid process of the sphenoid bone. It communicates with the oral cavity, transmitting the palatine neurovascular bundle (see Figure 36-2).
2. **Contents.** This fossa has been called the "neurovascular junction" of the deep face (see Figure 36-2).
    - a. **Vascular structures.** The **maxillary artery** enters this fossa laterally through the pterygomaxillary fissure.
    - b. **Nerves.** The **maxillary and vidian nerves** enter through the posterior wall by the foramen rotundum and vidian (pterygoid) canal, respectively. The **pterygopalatine ganglion** lies within this fossa.
    - c. Neurovascular bundles leave the pterygopalatine fossa through foramina or fissures in the medial wall (sphenopalatine foramen), in the anterior wall (inferior orbital fissure, posterior superior alveolar foramina), and inferiorly at the apex (palatine canal).

V. TEMPOROMANDIBULAR JOINT AND MUSCLES OF MASTICATION

A. Temporomandibular joint
1. **Structure** (Figure 36-3)
    a. The **condyle (head)** of the mandible, the articular tubercle, and the mandibular fossa of the temporal bone join to form this joint (see Figure 36-3A).

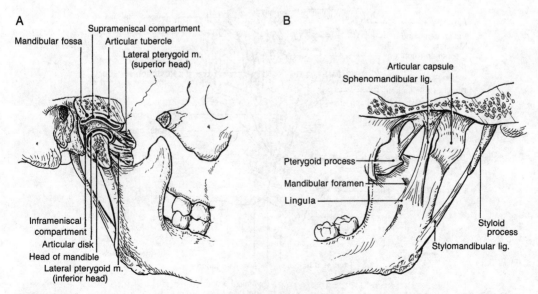

FIGURE 36-3. Temporomandibular joint. *(A)* Lateral aspect with the joint capsule opened to show the articular disk; *(B)* medial aspect.

b. A **fibrocartilaginous, biconcave articular disk** is interposed between the rounded condyle of the mandible and the sigmoid articular surface of the temporal bone (see Figure 36-3A). This disk divides the joint into a suprameniscal compartment and an inframeniscal compartment. Specific movements occur in each compartment.

c. Ligamentous support (see Figure 36-3B)
  (1) A fibrous **articular capsule** spans the joint from temporal bone to the articular disk before continuing to the neck of the mandible.
  (2) The **sphenomandibular ligament,** medial to the joint, runs from the spine of the sphenoid bone to the lingula of the mandible. Its attachment defines the resultant axis of rotation of the temporomandibular joint.
  (3) The **stylomandibular ligament,** posterior to the joint, runs from the tip of the styloid process to the angle of the mandible.
  (4) The **temporomandibular ligament** runs from the lateral zygomatic process, anterior to the articular capsule, to the mandibular neck.

2. Movements associated with opening and closing the mouth
  a. Elevation/depression occurs in the **inframeniscal compartment** as the condyle **rotates** on the articular disk about a transverse axis through the mandibular condyles.
  b. Protrusion/retraction occurs in the **suprameniscal compartment** as the articular disk **glides** anteriorly and posteriorly along the temporal bone between the mandibular fossa and articular tubercle.
  c. The compound **horizontal axis of rotation** (produced by rotation with anterior–posterior gliding at the temporomandibular joint) passes through the **lingula** of the ramus, adjacent to the mandibular foramen.
    (1) As the mouth opens, the condyle glides anteriorly in the suprameniscal compartment with the articular disk. This action is accomplished by the inferior and superior heads of the lateral pterygoid muscle. Concomitantly, the mandible rotates in the inframeniscal compartment.
    (2) The **resultant axis of rotation** passes through the lingulae where the jaw is suspended by the **sphenomandibular ligament.** The neurovascular bundle enters the mandibular foramen adjacent to the lingula—the point of minimal movement.
  d. Mandibular dislocation (subluxation)
    (1) If the mandible is sufficiently depressed and protrusion occurs beyond the apex of the articular tubercle, spasm of the temporalis muscle may sublux the head under the zygoma, locking the jaw.
    (2) The physician or dentist must reduce the dislocation.
      (a) With a rolled gauze sponge between the molars, the clinician first pushes downward on the mandibular alveolar ridges to overcome spasm of the temporalis, masseter, and medial pterygoid muscles.
      (b) The clinician then pushes down and back to ease the condyle over the articular tubercle. As the jaw snaps back into the mandibular fossa, because of muscular spasm, he or she should beware of the forceful apposition of the molars.

B. Muscles of mastication (Table 36-2)
  1. The temporalis muscle occupies the temporal fossa (Figure 36-4A; see Table 36-2).
    a. Fibers arise from the lateral aspect of the cranium, pass deep to the zygomatic arch, and converge onto the coronoid process of the mandible.
    b. The anterior and medial fascicles **elevate the jaw;** the posterior fascicles **retract the jaw.**
  2. The masseter lies deep to the external lobe of the parotid gland and completely covers the outer surface of the mandibular ramus (see Figure 36-4A and Table 36-2). A **buccal fat pad** separates the masseter from the buccinator muscle.
    a. The masseter arises from the medial and lateral surfaces of the zygomatic arch and inserts into the periosteum at the angle of the mandible and along the mandibular ramus.
    b. It is a strong **elevator of the jaw.**

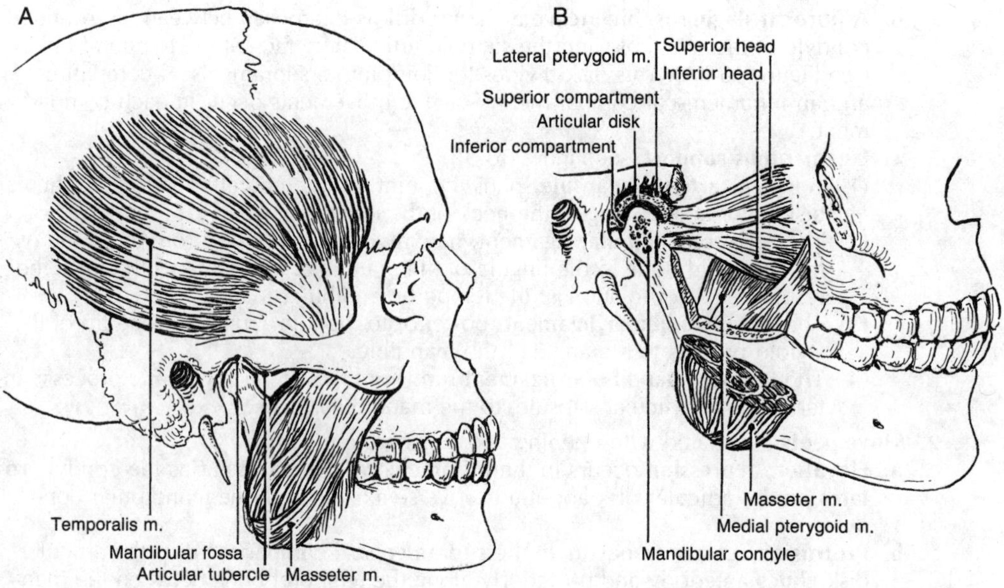

FIGURE 36-4. Muscles of mastication. *(A)* Temporalis and masseter muscles; *(B)* medial pterygoid and lateral pterygoid muscles.

3. **The lateral pterygoid muscle** lies in a horizontal plane (see Figure 36-4B and Table 36-2). The medial surface is adjacent to the mandibular nerve, the maxillary artery, and the pterygoid venous plexus.
   a. **Divisions** (see Figure 36-4B)
      (1) The **superior head** arises from the infratemporal surface of the sphenoid bone and inserts mainly into the articular capsule and disk of the temporomandibular joint.
      (2) The **inferior head** arises from the lateral surface of the lateral pterygoid plate and inserts into the condylar process of the mandible.
   b. This muscle **opens the jaw** by pulling the condyle and the articular disk anteriorly.
      (1) Together, the left and right lateral pterygoid muscles protrude the jaw.
      (2) Unilaterally, they swing the jaw to the opposite side and thus effect the grinding movements of mastication.
4. **The medial pterygoid muscle** parallels the masseter (see Figure 36-4B and Table 36-2). It is separated from the mandible by the internal lobe of the parotid gland, the lingual nerve, the chorda tympani nerve, the inferior alveolar nerve, and the inferior alveolar artery.
   a. The deep part arises from the medial surface of the lateral pterygoid plate and the pyramidal process of the palatine bone. The smaller superficial part arises from the tubercle of the maxilla. It inserts into the periosteum at the angle of the mandible.
   b. A strong **elevator of the mandible,** this thick quadrangular muscle resembles its counterpart—the masseter—with which it forms the **masseteric (mandibular) sling.** This sling supports and elevates the angle of the mandible but, lacking Sharpey's fibers, is not attached to the bone.
5. **Accessory muscles of mastication** (see Figures 34-1 and 34-2)
   a. Although not grouped with the muscles of mastication, the anterior belly of the **digastric muscle,** the **mylohyoid muscle,** and the **geniohyoid muscle** insert onto the mandible and act to produce motion at the temporomandibular joint, generally depressing (opening) the jaw.
   b. **Gravity** plays an important role in jaw opening.

Table 36-2. Muscles of Mastication

Muscle	Origin	Insertion	Primary Action	Innervation
Temporalis	Temporal and parietal bones	Coronoid process of mandible	Elevates and retracts jaw	Mandibular n. (CN V_3)
Masseter	Zygomatic arch	Angle of mandible	Elevates jaw	Mandibular n. (CN V_3)
Lateral pterygoid				
Superior head	Sphenoid bone	Temporomandibular disk	Draws articular disk forward	Mandibular n. (CN V_3)
Inferior head	Lateral pterygoid plate	Neck of mandible	Bilaterally, protracts jaw Unilaterally, abducts jaw	Mandibular n. (CN V_3)
Medial pterygoid	Medial and lateral pterygoid plates	Angle of the mandible	Elevates jaw	Mandibular n. (CN V_3)

VI. VASCULATURE

A. Arterial supply

1. **The internal carotid artery** gives rise to the ophthalmic artery. Its supratrochlear and supraorbital branches supply the forehead and anterior scalp (see Figures 33-5 and 34-3). It anastomoses with the superficial temporal, facial, and posterior auricular branches of the external carotid artery.

2. **The external carotid artery** gives rise to four major branches that supply the face (Figure 36-5).

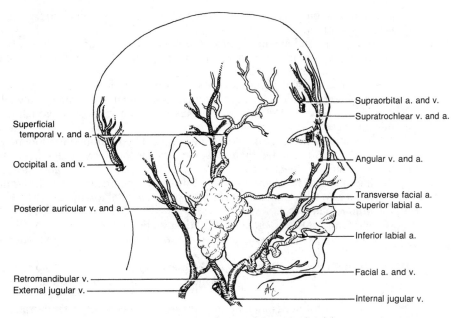

FIGURE 36-5. Vasculature of the superficial face.

a. The **facial artery** arises from the external carotid anteriorly opposite the angle of the mandible between the lingual and maxillary arteries (see Figures 34-3 and 36-5).
   (1) **Beneath the mandible,** this artery continues cranially before turning anteriorly behind the angle of the mandible and then inferiorly. It gives off three major submandibular branches.
      (a) The **tonsillar branch,** usually the largest blood vessel supplying the tonsil, arises at the apex of this loop (see Figure 37-6). It is separated from the palatine tonsil by the styloglossus muscle.
      (b) The **ascending palatine branch** is small and parallels the ascending pharyngeal artery with which it anastomoses.
      (c) The **submental branch** arises as the facial artery emerges between the mandible and the submandibular gland. It runs forward to the chin along the insertion of the mylohyoid muscle.
   (2) **In the face,** the facial artery continues over the body of the mandible, in a sinuous course, to enter the face (see Figure 36-5). It gives rise to three major branches.
      (a) The **inferior labial** and **superior labial branches** arise lateral to the mouth, and each anastomoses with contralateral counterparts (see Figure 36-5).
      (b) The **angular artery,** lateral to the nose, continues the facial artery. It anastomoses with the infraorbital branch of the maxillary artery and the supraorbital branch of the ophthalmic artery (see Figure 35-5).
   (3) A **facial pulse** is palpable where the facial artery passes over the body of the inferior edge of the mandible.
b. The **posterior auricular artery** arises posteriorly from the external carotid artery (see Figure 36-5). It ascends between the superior surface of the posterior belly of the digastric muscle and the parotid gland. It grooves the base of the skull between the mastoid process and the auricle.
c. The **superficial temporal artery** ascends anterior to the external auditory meatus and supplies the lateral and central portions of the scalp (see Figure 36-5).
   (1) It gives off a **transverse facial branch** before crossing the zygomatic arch just anterior to the external auditory meatus.
   (2) It anastomoses freely with contralateral counterparts, with supraorbital and supratrochlear branches of the ophthalmic artery, and with the posterior auricular artery.
   (3) The **temporal pulse** is palpable where the superficial temporal artery courses across calvarium.
d. The **maxillary (internal maxillary) artery** runs anteriorly out of the retromandibular portion of the parotid gland to supply the deep face as well as portions of the nasal and oral cavities (Figure 36-6).
   (1) The first (mandibular) portion of the maxillary artery lies on the deep surface of the mandibular ramus, where it crosses superficial to the inferior alveolar and lingual nerves (see Figure 36-6). Most of its branches pass through foramina.
      (a) The **middle meningeal artery** enters the cranium through the foramen spinosum surrounded by a plexus of postganglionic sympathetic fibers.
      (b) The **inferior alveolar artery** runs with the inferior alveolar nerve through the mandibular foramen into the lower jaw and accompanies its terminal branches. It exits into the face through the mental foramen.
      (c) Minor branches
         (i) Tiny tympanic branches and deep auricular branches supply the ear.
         (ii) An accessory meningeal artery runs counter to the mandibular nerve through the foramen ovale.
   (2) The **second (pterygoid) portion of the maxillary artery** passes deep to the inferior head of the lateral pterygoid muscle (see Figure 36-6). Its branches do not pass through foramina.
      (a) **Muscular branches** include the **masseteric, buccal, deep temporal,** and **pterygoid** arteries.

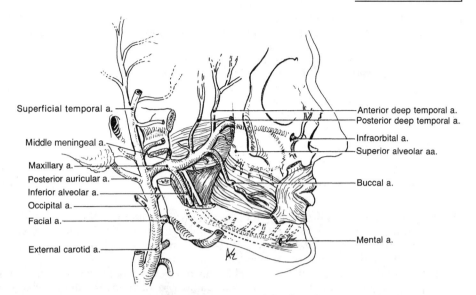

FIGURE 36-6. Arteries of the deep face.

      (b) These vessels accompany similarly named branches of the anterior division of the mandibular nerve.
   (3) The **third (pterygopalatine or terminal) portion of the maxillary artery** dives into the pterygomaxillary fissure to enter the pterygopalatine fossa (see Figure 36-2). Branches that arise within this fossa accompany branches of the maxillary nerve to the nose, palate, and upper jaw. Its branches pass through foramina.
      (a) The **pharyngeal branch** courses posteriorly through the pterygovaginal canal.
      (b) **Posterior superior alveolar arteries** enter the maxillary tuberosity through several foramina to supply the upper molars.
      (c) The **sphenopalatine artery** enters the nasal cavity through the **sphenopalatine foramen,** supplying the medial wall of the nasal cavity by way of the **nasopalatine artery** and the lateral walls by way of the **lateral nasal arteries.**
      (d) The **descending palatine artery** exits through the **pterygopalatine canal** at the apex (inferiorly) to supply the hard and soft portions of the palate.
      (e) The **infraorbital artery** accompanies the infraorbital nerve through the **inferior orbital fissure.** It supplies the anterior maxillary teeth before exiting through the **infraorbital foramen,** supplying facial structures, and anastomosing with branches of the facial artery.

B. Venous return
   1. Ophthalmic veins
      a. The **superior ophthalmic vein** collects blood from the forehead through the supratrochlear and supraorbital veins, the orbital contents, and eye by the vorticose veins. It passes through the superior orbital fissure to drain into the cavernous sinus, which in turn drains into the sigmoid sinus, becoming the **internal jugular vein** (see Figure 31-10).
      b. The **inferior ophthalmic vein** collects blood from the orbital contents and eye through the vorticose veins. It usually joins the superior ophthalmic vein. It also anastomoses through the inferior orbital fissure with the pterygoid venous plexus, which in turn drains through the maxillary and retromandibular veins to the internal jugular vein (see Figure 36-5).
      c. Clinical correlations

(1) Cavernous sinus thrombosis. Superficial infection in the danger zone about the nose may spread along anastomotic pathways between the facial and superior ophthalmic veins to reach the cavernous sinus. Infection in this sinus is life-threatening [see Chapter 31 IV A 1 b (3) (c)].

(2) Pteryoid plexus thrombosis. Superficial infection in the danger zone about the nose may also spread along deep anastomotic pathways between the facial vein and the inferior ophthalmic vein to the pterygoid plexus and from there into cranial, orbital, and pharyngeal regions.

2. **The facial vein** usually drains into the internal jugular vein, but it also connects with the external jugular vein through the posterior jugular vein (see Figures 34-4 and 36-5).

3. **The superficial temporal vein,** together with the maxillary vein, forms the retromandibular vein, which drains into the internal jugular vein (see Figure 36-5).

4. **The pterygoid venous plexus** lies in close association with the lateral pterygoid muscle and the middle portion of the maxillary artery. It drains into the maxillary vein and receives drainage from the veins of the orbit, nasal cavities, paranasal sinuses, oral cavity, and structures of the infratemporal fossa.
   a. Vessels that drain into the pterygoid plexus have anastomotic connections with superficial veins of the face by numerous pathways, including the **deep facial (buccal) vein,** the **inferior ophthalmic vein,** and the **pharyngeal plexus.**
   b. It anastomoses with the cavernous sinus by way of the sphenoid emissary vein (of the foramen of Vesalius).
   c. Infection in the danger zone about the nose may spread along these anastomotic pathways to the pterygoid plexus and from there into cranial, orbital, and pharyngeal regions.

5. **The maxillary vein** drains the deep face. Together with the superficial temporal vein, it forms the retromandibular vein, which drains into the external jugular vein.

C. **Lymphatic drainage.** The lymph vessels of the superficial and deep face generally parallel the veins and drain to the deep cervical nodes (see Figure 34-4).

VII. INNERVATION

A. **The facial nerve (CN VII)** is the principal motor nerve of the superficial face. It represents the nerve of the second branchial arch.

1. **Course and distribution** (see Table 32-1)
   a. **In the posterior cranial fossa,** it enters the petrous portion of the temporal bone through the internal auditory meatus (Figure 36-7). It accompanies the vestibulocochlear nerve (CN VIII).
   b. **Within the temporal bone,** the facial nerve follows the facial canal (see Figure 36-7).
      (1) The **facial canal** (fallopian canal) begins at the apex of the internal auditory meatus and runs laterally for a short distance.
      (2) At the **genu** (L. knee), the facial canal turns sharply posteriorly in the medial wall of the middle ear cavity. It then runs inferiorly in the posterior wall of the middle ear to reach the stylomastoid foramen.
         (a) The **geniculate ganglion** lies at the genu (see Figures 36-7 and 36-8). It is the equivalent of a dorsal root ganglion, containing the cell bodies of the general and special sensory neurons of the facial nerve.
         (b) The **greater superficial petrosal nerve** leaves the facial nerve at the genu and passes anteriorly to enter the middle cranial fossa through the hiatus of the facial canal (see Figures 36-7 and 36-8).
            (i) This nerve conveys preganglionic parasympathetic secretomotor fibers.

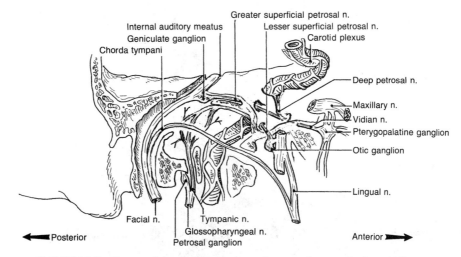

FIGURE 36-7. Course of the facial and glossopharyngeal nerves in the middle ear.

- (ii) It joins with the sympathetic deep petrosal nerve (carrying postganglionic sympathetic fibers from the carotid perivascular plexus) to form the **nerve of the pterygoid canal (vidian nerve).** This nerve continues through the base of the pterygoid process to reach the pterygopalatine fossa (see Figure 36-2).
- (iii) Parasympathetic nerves synapse in the **pterygopalatine (sphenopalatine) ganglion.**
- (iv) Postsynaptic fibers from the pterygopalatine ganglion merge with various branches of the maxillary nerve to reach the lacrimal glands of the palate and glands of the nasal mucosa.
- (3) The **nerve to the stapedius muscle** leaves the facial nerve in the posterior portion of the facial canal.
- (4) The **chorda tympani nerve** (the pretrematic branch of the nerve of the second branchial arch) leaves just as the facial nerve exits the skull (see Figures 36-7 and 36-8).
  - (a) **Course**
    - (i) Just within the stylomastoid foramen, this nerve runs forward within the **posterior canaliculus** of the chorda tympani to reach the middle ear cavity.
    - (ii) Within the middle ear, it runs anteriorly between the incus and the handle of the malleus and re-enters the bone through the **anterior canaliculus** of the chorda tympani.
    - (iii) In the infratemporal fossa, it exits the skull at the **petrosquamous portion of the petrotympanic fissure** just medial to the mandibular condyle.
    - (iv) It courses anteriorly and joins the lingual nerve (the post-trematic branch of the nerve of the first branchial arch).
  - (b) **Composition**
    - (i) **Secretomotor function.** The chorda tympani nerve conveys parasympathetic presynaptic fibers to the submandibular ganglion. Postsynaptic fibers return to the lingual nerve to reach the submandibular and sublingual salivary glands (Figure 36-8).
    - (ii) **Special sensory function.** It conveys sensory fibers for taste from the anterior two thirds of the tongue. These fibers have their cell bodies in the geniculate ganglion.
- (5) The **stylomastoid foramen** marks the termination of the facial canal.
- c. **In the infratemporal fossa,** the facial nerve gives off numerous branches (see Figure 36-8).

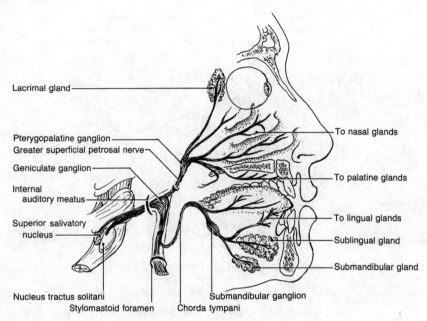

FIGURE 36-8. Parasympathetic and sensory components of the facial nerve (CN VII). Autonomic portions run to the pterygopalatine and submandibular ganglia. Sensory portion conveys taste from the anterior part of the tongue. Cell bodies of sensory neurons are located in the geniculate ganglion.

  (1) As it exits the skull, it gives off a **digastric branch** to the posterior belly of that muscle and **a stylohyoid branch** to that muscle.
  (2) Almost immediately, it gives rise to the **posterior auricular nerve** that divides into a small **auricular branch** (to the posterior auricular muscle) and a larger **occipital branch** (to the occipital belly of the occipitofrontal muscle). The posterior auricular nerve also carries sensory fibers from the external ear.
  d. **In the face,** the facial nerve passes anterolaterally through the parotid gland (Figure 36-9).

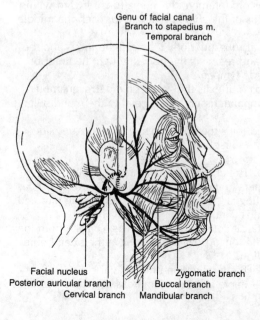

FIGURE 36-9. Branchiomotor component of the facial nerve (CN VII) innervates the muscles of facial expression (posterior belly of the digastric, the stylohyoid, and the stapedius muscles).

(1) A **plexus of branches** passes through the parotid gland, each separated from the parotid parenchyma by a fascial invagination.
(2) **Terminal branches** emerge from the parotid onto the face. These superficial branches are easily damaged by facial trauma.
   (a) **Temporal branches** supply muscles superior to the eyes, including the anterior and superior auricular muscles, the frontal belly of the occipitofrontalis, the corrugator supercilii, and the orbicularis oculi.
   (b) The **zygomatic branch** passes to the orbicularis oculi. It is the principal mediator of the efferent limb of the **corneal blink reflex.**
   (c) **Buccal branches** supply muscles of the cheek (including the buccinator muscle), upper lip, and nose.
   (d) The **mandibular branch** loops down over the mandible. It supplies the risorius muscles and the depressors of the lower lip.
   (e) The **cervical branch** innervates the platysma muscle.

2. **Injury**
   a. **Lesions involving the terminal branches** of the facial nerve produce imbalance of the muscles of facial expression, with unilateral expressionless drooping of the face.
      (1) Zygomatic branch injury paralyzes the orbicularis oculi with epiphora and possible conjunctival and corneal ulceration. The corneal blink reflex is absent.
      (2) Buccal branch injury paralyzes the buccinator, which results in accumulation of food in the cheeks and difficulty in masticating food.
      (3) Lesions involving the buccal or mandibular branches paralyze the orbicularis oris, resulting in drooling and difficulty masticating food.
   b. **Lesion at the sternomastoid foramen** produces complete paralysis of the muscles of facial expression with all the above signs and symptoms.
   c. **Inflammation of the nerve within the facial canal** is the most common cause of Bell's palsy, for which no effective treatment is known. Most patients recover spontaneously, often with little or no nerve damage. **Otitis media** may also involve the facial nerve.
      (1) Such inflammation is associated with the signs and symptoms as described for lesions involving the terminal branches.
      (2) Involvement of the chorda tympani results in loss of taste from the anterior two thirds of the tongue and reduced salivary secretion on the ipsilateral side.
      (3) Involvement of the nerve to the stapedius muscle produces **hyperacusis.**
   d. **Lesions in the internal auditory meatus** often are the result of **acoustic neuroma** or **fracture.**
      (1) They produce the signs and symptoms as described for inflammation of the nerve within the facial canal.
      (2) Involvement of the greater superficial petrosal nerve results in loss of lacrimation and secretion from nasal and palatine glands on the affected side.
      (3) **Syndrome of crocodile tears.** If the secretomotor fibers re-grow along each other's pathways and innervate the wrong gland, the anticipation of food produces lacrimation instead of salivation and vice versa.
      (4) **Hearing and balance dysfunction** often accompany lesions within the internal acoustic meatus.

B. **The trigeminal nerve (CN V)** is the principal sensory nerve of the superficial face. It represents the nerve of the first branchial arch. A coronal line drawn over the scalp and connecting the tragi of the ears marks the boundary between branchiomeric innervation (anteriorly) and somatic innervation (posteriorly). It also conveys motor innervation to the muscles of mastication (see Table 36-2).

1. **Main trunk**
   a. **Distribution** (Figure 36-10)
      (1) **Intracranial course.** The mixed trigeminal nerve originates from several nuclei in the pons. It courses anteriorly to enter a dural evagination, the trigeminal cave (of Meckel) in the middle cranial fossa. It is here that the trigeminal

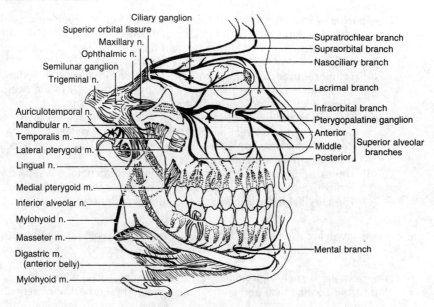

FIGURE 36-10. Trigeminal nerve (CN V). Three divisions provide sensory innervation to the face. The motor root of the mandibular division innervates the muscles of mastication, the tensor tympani muscles, and the tensor veli palatini muscle.

(semilunar) ganglion lies, containing the cell bodies of most of the sensory neurons of the trigeminal nerve.

  (2) **Extracranial course.** The trigeminal nerve (as the name implies) gives off three divisions. The cutaneous supply to the face conforms to the boundaries of these three regions.

  b. **Trigeminal neuralgia (tic douloureux)** produces excruciating facial pain in response to touching a trigger area. So severe and unremitting is the pain that some patients have become suicidal. The pain is often along the distribution of the mandibular nerve. The trigger area is often maxillary.

2. **Ophthalmic division (CN V$_1$)**
    a. **Course.** It exits the middle cranial fossa through the **superior orbital fissure** and exits the orbit by way of the **supraorbital notch** (sometimes supraorbital foramen)(see Figures 33-5 and 36-10 and Table 32-1).
    b. **Distribution.** It mediates general sensation from the forehead and nose (see Figure 36-10).
        (1) The **lacrimal nerve** runs superolaterally within the orbit.
            (a) It supplies the lateral portion of the eyelids.
            (b) Distally, it receives postsynaptic parasympathetic secretomotor fibers from the pterygopalatine ganglion by way of the zygomaticofacial branch of the maxillary nerve.
        (2) The **frontal nerve,** the largest branch of the ophthalmic, bifurcates.
            (a) The **supraorbital nerve** runs superior to the orbital axis. It passes through the **supraorbital notch** or foramen to supply the skin of the eyelid, forehead, and scalp.
            (b) The **supratrochlear nerve** runs more medially to supply the skin of the medial portions of the eyelids and the central portion of the forehead.
        (3) The **nasociliary nerve** runs superomedially within the orbit.
            (a) **Long ciliary nerves** convey conjunctival sensation directly from the eyeball.
            (b) **Short ciliary nerves** along with the **sensory root (ramus) of the ciliary ganglion** provide another sensory pathway from the conjunctiva to the nasociliary nerve.

(c) The **posterior ethmoidal nerve** passes through the **posterior ethmoidal foramen** to supply the mucosa of the sphenoidal and ethmoidal air sinuses and a portion of the nasal mucosa.
(d) The **infratrochlear nerve** supplies the medial portion of the eyelids.
(e) The **anterior ethmoidal nerve,** the continuation of the nasociliary nerve, passes through the **anterior ethmoidal foramen** and descends through the ethmoid bone.
  (i) The **internal nasal branch** supplies the mucosa of the ethmoidal and frontal air sinuses and portions of the nasal mucosa.
  (ii) The **external nasal branch** supplies the skin at the tip of the nose and about the external nares.
 c. **Clinical considerations**
  (1) The **afferent limb of the corneal blink reflex** is provided by the nasociliary branches of the ophthalmic nerve that convey general sensation from the conjunctiva.
  (2) **Inflammation of the cavernous sinus** or a **lesion at the superior orbital fissure** results in paresthesias of the forehead and nose, with loss of the corneal blink reflex.

3. **Maxillary division (CN V$_2$)**
 a. **Characteristics.** This nerve mediates general sensation form the skin between the lower eyelid and the upper lip as well as from the maxillary teeth (see Figure 36-10 and Table 32-1). It represents the pretrematic division of the trigeminal nerve and is entirely sensory.
 b. **Course.** It enters the pterygopalatine fossa through the foramen rotundum. Although supported by fatty tissue, it appears to hang freely in the pterygopalatine fossa until it enters the inferior orbital fissure (see Figure 36-2).
 c. **Distribution** (see Figure 36-10)
  (1) **Communicating branches** from the pterygopalatine ganglion carry postsynaptic secretomotor fibers to the infraorbital, sphenopalatine, and palatine branches of the maxillary nerve (see Figure 36-2). The communicating branches appear to suspend the pterygopalatine ganglion from the maxillary nerve.
  (2) The **zygomatic nerve** frequently arises from the maxillary nerve in the pterygopalatine fossa. It passes into the orbit through the **inferior orbital fissure** and divides.
   (a) The **zygomaticofacial nerve** traverses the lateral wall of the orbit through a small foramen to supply small areas of skin lateral to the eye. As part of the pathway for lacrimation, this branch also conveys parasympathetic secretomotor neurons to the lacrimal branch of the ophthalmic division whereby they reach the lacrimal gland.
   (b) The **zygomaticotemporal nerve** traverses the lateral wall of the orbit through a small foramen to supply the posterolateral zygomatic area.
  (3) **Posterior–superior alveolar nerves** (two or three) run along the posterior surface of the maxillary tuberosity and pass through small canals in the bone to supply the maxillary molars (see Figure 36-2).
  (4) The **sphenopalatine (posterior–superior nasal) nerve** passes through the **sphenopalatine foramen** into the nasal cavity, where it divides into two groups (see Figure 36-2). As part of the pathway for **nasal secretion,** this branch also conveys postsynaptic parasympathetic secretomotor neurons to glands in the nasal mucosa.
   (a) The **medial group** arches over the roof of the nasal cavity to reach the nasal septum. The longest of this group is the **nasopalatine nerve,** which descends along the nasal septum to reach the incisive foramen through which it passes to innervate the anterior portion of the oral palate (see Figure 37-1B).
   (b) The **lateral group** innervates the posterior ends of the nasal conchae as well as the posterolateral nasal walls (see Figure 37-1A).
  (5) The **palatine nerves** descend in the pterygopalatine fossa and the

pterygopalatine canal (see Figure 36-9). As part of the pathway for **palatine secretion,** this branch also conveys postsynaptic parasympathetic secretomotor neurons to the glands of the palate.
- (a) The **greater palatine nerve** passes through the **greater palatine foramen.** It then runs anteriorly along the inner aspect of the alveolar margin in a groove on the surface of the hard palate. It innervates the mucosa of the central and posterior regions of the hard palate.
- (b) The **lesser palatine nerve** emerges from the lesser palatine foramen and runs posteriorly. It innervates the soft palate, the tonsils, and the palatoglossal folds.

(6) The **main trunk of the maxillary nerve** continues through the **inferior orbital fissure** to become the **infraorbital nerve.** In the floor of the orbit, it lies in the infraorbital groove (see Figure 36-10).
- (a) This trunk gives off nerve branches that run in canals in the anterior wall of the maxilla.
  - (i) The **middle superior alveolar nerves** are inconsistent. They innervate the maxillary bicuspids and cuspids.
  - (ii) The **anterior–superior alveolar nerves** innervate the maxillary incisors and give off branches to the lateral wall of the nasal cavity as well as to the maxillary sinus.
- (b) The **infraorbital nerve** traverses the orbital margin through a bony canal, the **infraorbital foramen.** It terminates in a spray of small branches that innervate the lower eyelid, lateral region of the nose, nasal vestibule, and superior lip.

d. **Clinical considerations**
  (1) The **afferent limb of the sneeze reflex** is provided by the branches of the maxillary nerve that convey general sensation from the nasal vestibule. The sneeze reflex is tested by noting a patient's aversion as a swab contacts the vibrissae at the external nares.
  (2) **Lesion of the maxillary nerve** at the foramen rotundum, within the pterygopalatine fossa, or along the infraorbital canal results in paresthesias of the midface and maxillary teeth, with loss of the sneeze reflex.
  (3) **Anesthesia of the maxillary nerve**
   - (a) **Lateral approach.** A needle is inserted through the mandibular notch until it strikes the pterygoid process. The tip is then guided anteriorly along the lateral pterygoid plate until it slips into the pterygopalatine fossa. The result is anesthesia of the entire maxillary division of the trigeminal nerve.
   - (b) **Inferior approach.** A curved needle is passed into the greater palatine foramen and upward along the pterygopalatine canal to reach the pterygopalatine fossa. The result is anesthesia of the entire maxillary division of the trigeminal nerve.
   - (c) **Anterior approach.** A needle may be inserted into the infraorbital foramen, reaching the nerve within the infraorbital groove. The result is anesthesia of the anterior branches of the infraorbital nerve, including the anterior and middle superior alveolar nerves to the incisor, cuspid, and bicuspid teeth.

4. **Mandibular division (CN V$_3$)**
  a. **Characteristics.** This nerve represents the post-trematic branch of the trigeminal nerve. As a mixed nerve, it provides motor innervation to the muscles of mastication and mediates general sensation from the jaw and lateral side of the face, the supratentorial meninges, the mucous membranes of the inferior oral cavity, and the mandibular teeth (see Table 32-1).
  b. **Course** (see Figure 36-10)
   (1) It leaves the cranium through the **foramen ovale.**
   (2) The sensory cell bodies are located in the **semilunar (trigeminal, gasserian) ganglion.**
  c. **Distribution**

(1) After passing through the foramen ovale, where the large **sensory root** is joined by a small **motor root** (see Figure 36-10), it gives off two branches (one sensory, one motor). It then divides into anterior and posterior trunks.
  (a) A **meningeal branch** (sensory) enters the cranium through the foramen spinosum along with the middle meningeal artery.
  (b) The **medial pterygoid nerve** (motor) emerges from the medial aspect of the mandibular nerve to supply the **medial pterygoid, tensor veli palatini,** and **tensor tympani muscles.**
(2) The **anterior trunk** is mainly motor. It passes inferior to the lateral pterygoid muscle and gives off several branches.
  (a) Motor. The branches to the **masseter, lateral pterygoid,** and **temporalis muscles** emerge from the lateral aspect of the mandibular nerve. These branches also convey the efferent limb of the jaw-jerk reflex.
  (b) Sensory. The **buccal nerve** is the terminal sensory branch. It courses deep to the mandibular coronoid process to innervate the cheek and superior buccal gingiva (but not the buccinator muscle).
(3) The **posterior trunk** is mixed and gives off several branches.
  (a) The **auriculotemporal nerve** (sensory) is formed by the union of two branches of the posterior trunk. These branches pass to either side of the middle meningeal artery before converging.
   (i) This nerve runs between the neck of the mandible and the external auditory meatus to reach the temple, where it has a cutaneous distribution.
   (ii) Postsynaptic parasympathetic secretomotor fibers from the otic ganglion pass along this nerve to innervate the parotid gland.
  (b) The **lingual nerve** (sensory) is joined by the **chorda tympani nerve** (the pretrematic branch of the facial nerve) at the inferior border of the lateral pterygoid muscle.
   (i) Course (see Figures 36-10 and 37-4A).
     In the infratemporal fossa, this nerve descends beyond the lateral border of the lateral pterygoid muscle onto the lateral surface of the medial pterygoid muscle.
     Behind the ramus of the mandible, it lies on the lateral surface of the styloglossus muscle below the attachment of the superior constrictor. At the posterior end of the inferior alveolar margin, it lies near the third molar.
     In the floor of the mouth, it lies on the lateral surface of the hyoglossus muscle and passes superficial to the submandibular duct. It splays out toward the tip of the tongue.
   (ii) General sensation from the anterior two thirds of the tongue is conveyed by the lingual nerve with cell bodies located in the **semilunar (trigeminal) ganglion.**
   (iii) Taste from the anterior two thirds of the tongue is carried distally in the lingual nerve and then proximally in the **chorda tympani nerve.** The cell bodies are in the **geniculate ganglion.**
   (iv) Presynaptic parasympathetic secretomotor fibers join the lingual nerve from the chorda tympani nerve and leave the lingual nerve by a few short communicating branches that synapse in the **submandibular ganglion** (see Figure 36-8).
  (c) The **inferior alveolar nerve** (mixed) runs inferior to the lingual nerve to enter the **mandibular foramen** on the inner surface of the mandibular ramus (see Figure 36-10).
   (i) Motor. The **mylohyoid branch** arises immediately before the mandibular foramen and passes along the inner surface of the mandible below the mylohyoid line. The only motor branch of the posterior trunk, it innervates the **mylohyoid muscle** and the **anterior belly of the digastric muscle.**
   (ii) Sensory. Within the mandibular canal, the inferior alveolar nerve conveys general sensation from the mandibular molars and premolars. It bifurcates at the mental foramen.

The **incisive branch** innervates the lower incisors and canine teeth.

The cutaneous **mental nerve** exits into the face through the mental foramen and provides sensation to the anterior aspects of the chin and lower lip.

d. **Clinical considerations**
   (1) **Lesion at the foramen ovale** results in paresthesias along the mandible, mandibular teeth, and side of the face; paralysis of the muscles of mastication; and loss of the jaw-jerk reflex.
   (2) **Anesthesia of the mandibular division**
      (a) A needle is inserted through the mandibular notch until it strikes the lateral pterygoid plate. It is then walked posteriorly along the pterygoid plate to reach the vicinity of the foramen ovale, where the anesthestic agent is administered. The result is anesthesia of the entire **mandibular nerve.**
      (b) The **inferior alveolar nerve** is anesthetized intraorally by inserting the needle lateral to the pterygomandibular raphe. The tip is then walked posteriorly along the medial aspect of the ramus of the mandible close to the mandibular foramen, where the anesthetic agent is injected.
      (c) The **mental** and **incisive nerves** are anesthetized by injection of anesthetic directly into the mental foramen.
   (3) **Fracture of the mandible** may injure or sever the inferior alveolar nerve.

Chapter 37
Cranial and Cervical Viscera

I. INTRODUCTION

A. Developmental considerations. The cranial end of the primitive gut develops into two systems: the respiratory tract and the gastrointestinal tract. Despite the functional dichotomy of these two systems, the anatomic division is incomplete between the palate and the larynx.

B. Anatomic divisions
1. **The nasal cavity** serves primarily in respiration and olfaction.
2. **The oral cavity** functions primarily in ingestion and taste, but it may also function in respiration and articulation of speech.
3. **The pharynx** serves both respiration and ingestive functions with a minor taste function.
4. **The larynx** functions in respiration and phonation.

II. NASAL CAVITY

A. Basic structure. The nasal cavity is a pyramidal bony chamber that lies beneath the anterior cranial fossa. It is capped by the cartilaginous nose (see Figure 35-3B).

1. **The nares** (nostrils) open into the nasal cavity at the vestibule.
2. **The choanae** open into the nasopharynx, posteriorly (see Figure 37-3).
3. **The nasal septum** extends medially from the roof to the floor and has both bony and cartilaginous portions (see Figure 35-3B). It usually deviates to one side of the midline.
4. **Conchae (turbinate bones)** project scroll-like into the nasal cavity from the lateral walls (see Figure 35-3A). The **superior concha** and **middle concha** project from the ethmoid bone. The **inferior concha** is a separate bone. Conchae divide the nasal cavity into **recesses** or **meatuses**.
    a. The **sphenoethmoidal recess** is a narrow space above the small superior concha lined with olfactory epithelium. The ostium of the sphenoid sinus enters posteriorly.
    b. The **superior meatus** below the superior concha receives several ducts from the posterior ethmoid sinuses.
    c. The **middle meatus** below the middle concha has a prominent bulge, the **ethmoid bulla,** which is the medial wall of the middle ethmoid sinuses.
        (1) The **hiatus semilunaris** (a slitlike groove) curves anteriorly, beneath the ethmoid bulla (see Figure 35-3A).
        (2) The **infundibulum of the frontal sinus** as well as numerous ducts from the anterior and middle ethmoid sinuses enter the anterosuperior end of the hiatus semilunaris.
        (3) The **ostium of the maxillary sinus** opens into the posteroinferior end of the hiatus semilunaris.
    d. The **inferior meatus** below the inferior concha receives the nasolacrimal duct anteriorly (see Figure 35-3A). A small fold of mucous membrane at the nasolacrimal orifice acts as a valve.

e. The **antrum** is a shallow depression, just anterior to the middle meatus, that is limited superiorly by a small rudimentary concha **(agger nasi)**.
   (1) Inferiorly, the antrum is continuous with the **vestibule** of the nose, just inside the nares.
   (2) The **nasovomeral organ** (of Jacobson) lies in the nasal septum at the junction between the antrum and vestibule. Its receptors mediate nonsmell pheromonal information.

B. Mucous membranes

1. **Vestibular mucosa.** The vestibule is lined by skin that bears short thick hairs, **vibrissae**.
2. **Olfactory mucosa.** The mucous membrane lining the sphenoethmoid recess in the apex of the nasal cavity contains olfactory cells.
3. **Nasal mucosa.** Elsewhere, the mucosa contains numerous mucus-secreting glands and ciliated cells.
   a. The layer of mucosa covering the middle and inferior concha traps particulate matter and moistens the inspired air.
   b. Mucus is swept backward by cilia into the nasopharynx and is swallowed.
      (1) Cilia are affected by low temperatures and cease to beat at about 10°C, explaining, in part, why the nose "runs" in cold weather.
      (2) Upper respiratory tract infections and cigarette smoke also inhibit ciliary action.

C. Vasculature

1. **Arterial supply**
   a. **Distribution** is from branches of the internal and external carotid arteries (Figure 37-1).
      (1) The **anterior ethmoidal artery,** a branch of the ophthalmic artery, supplies the anterior regions of the nasal cavity. The **posterior ethmoidal artery** supplies the superior region.
      (2) The **sphenopalatine artery,** a branch of the maxillary artery, divides into lateral and septal branches.
      (3) The **greater palatine artery,** also a branch of the maxillary artery, supplies an anteroinferior region by passing upward through the incisive foramen.

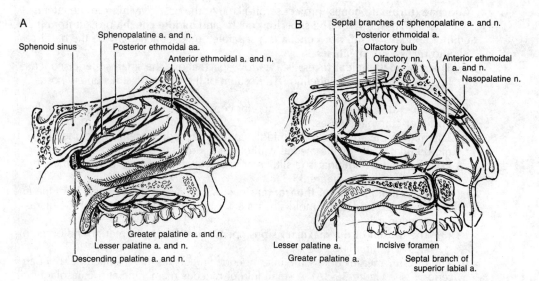

FIGURE 37-1. Neurovascular structures of the nasal cavity. *(A)* Lateral wall; *(B)* nasal septum.

(4) The **facial artery,** a branch of the external carotid, supplies the inferior and interior portions through its superior labial and ascending palatine branches.
- b. **Epistaxis** is associated with trauma, hypertension, infection, and clotting disorders.
  - (1) Treatment depends on the cause, but topical pressure, vasoconstrictors, and cautery are usually effective.
  - (2) Ligation is impractical because the branches are deep, although limited success has been reported with external carotid ligation.

2. **Venous return**
   - a. Nasal veins tend to drain into the **pterygoid plexus.**
   - b. The **submucous venous plexus** functions as a heat exchanger, warming inspired air and conserving heat loss in expired air. Vasomotor control of this plexus is a delicately balanced system capable of accommodating changes in humidity, temperature, and air flow.
   - c. **Clinical considerations.** Vasodilation and venous engorgement are accompanied by some mucosal edema, which may occlude the nasal airway. Disturbances in control are sometimes the result of allergic responses (e.g., pollen and house dust), but often they are of uncertain etiology. Relief produced by vasoconstrictor agents is temporary because tolerance quickly develops and rebound vasodilation occurs.

D. Innervation. Like the arterial supply, nerves of the nasal cavity are divided into an anterior ethmoidal (ophthalmic) area and a posterior sphenopalatine (maxillary) area.

1. **The terminal nerve (CN "0")** is closely associated with, but not part of, the olfactory nerve.
   - a. This nerve mediates nonsmell pheromonal information from the **vomeronasal organ** (of Jacobson) in the nasal septum just superior to the nasal vestibule (see Table 32-1).

Table 37-1. Palatine Musculature

Muscle	Origin	Insertion	Primary Action	Innervation
Intrinsic Musculature				
Tensor veli palatini	Lateral surface of auditory tube	Velum via the hamulus process	Tenses velum	Mandibular n. (CN V_3)
Levator veli palatini	Temporal bone and medial surface of auditory tube	Velum	Raises velum	Pharyngeal branch of vagus n. (CN X)
Uvulae	Spine of hard palate	Uvula	Tenses velum	Pharyngeal branch of vagus nerve (CN X)
Extrinsic Musculature				
Palatoglossus	Velum	Tongue	Raises posterior portion of tongue	Pharyngeal branch of vagus n. (CN X)
Palatopharyngeus	Velum	Pharynx	Raises larynx	Pharyngeal branch of vagus n. (CN X)

b. The cell bodies are located in the **ganglion terminale** (accessory olfactory bulb), next to each olfactory bulb.
2. **Olfactory nerves (CN I)**
   a. The mucous membrane lining the **sphenoethmoid recess** in the apex of the nasal cavity contains olfactory cells.
   b. Approximately 20 **olfactory nerves** convey special visceral sensation from the olfactory cells through the cribriform plate of the ethmoid bone to synapse in the olfactory bulbs (see Figure 37-1B and Table 32-1).
   c. These nerves have small meningeal sheaths. Infection may track to the meninges through the cribriform plate.
   d. Olfactory nerves are delicate. Ethmoid fracture may result in **anosmia** (olfactory anesthesia).
3. **Ophthalmic branches (CN V_1)** (see Table 32-1)
   a. The **anterior ethmoid nerve,** a branch of the nasociliary nerve, mediates general sensation from the anterior half of the nasal cavity (see Figures 36-10 and 37-1B).
   b. The **posterior ethmoid nerve** mediates general sensation from the superior half. It is also a branch of the nasociliary nerve.
4. **Maxillary branches (CN V_2)**
   a. The **sphenopalatine (posterior nasal) nerve** passes through the **sphenopalatine foramen.** It divides into lateral and septal (nasopalatine) branches and conveys sensation from the posterior and central regions of the nasal cavity (see Figures 36-10 and 37-1).
   b. The **anterior–superior alveolar branch** supplies a portion of the vestibule and mediates the afferent limb of the **sneeze reflex.**
   c. Branches from the **intraorbital branch** of the maxillary nerve supply the external nares.

E. **Paranasal sinuses**
1. **Frontal sinuses** develop as superior extensions of an anterior ethmoid sinus (see Figure 35-3A).
   a. The left and right frontal sinuses usually are asymmetric. Often one sinus extends across the midline to the opposite side.
   b. They usually drain by the **frontonasal duct** (infundibulum), which empties into the **hiatus semilunaris** of the middle meatus.
2. **The ethmoid labyrinth** consists of 6–18 thin-walled air cells between each lateral nasal wall and the medial orbital wall.
   a. Ethmoid sinuses open medially above and below the middle concha.
      (1) **Posterior ethmoid sinuses** (1–7) drain into the superior meatus.
      (2) **Middle ethmoid sinuses** (average 3), which open into the hiatus semilunaris of the middle meatus, form the **ethmoid bulla.** Occasionally (~10%), an ethmoid air cell extends into the middle concha.
      (3) **Anterior ethmoid sinuses** (2–11) drain into the frontal recess of the middle meatus, into the frontal infundibulum of the frontal sinuses, or into the hiatus semilunaris of the middle meatus. Usually (~89%), an anterior ethmoid air cell extends into the agger nasi.
   b. Infection of these sinuses may erode through the lamina papyracea into the orbit.
3. **Two sphenoid sinuses,** divided by a thin septum, fill the body of the sphenoid bone (see Figure 35-3A).
   a. **Medially,** the septum between the left and right sphenoid sinuses rarely lies in the midline (see Figure 31-11).
   b. **Anteriorly,** each sinus opens by a small ostium high on the anterior wall, so drainage into the sphenoethmoid recess cannot occur by gravity with the head held erect.
   c. **Lateral** to the sphenoid sinus are the cavernosus sinus, optic nerve, internal carotid artery, maxillary nerve, and pterygoid (vidian) nerve. **Chronic sphenoid sinusitis** may erode the thin walls and involve these structures.

d. **Superiorly,** the pituitary gland lies in the pituitary fossa. This gland may be reached surgically through the nose and sphenoid sinus.
   (1) This trans-sphenoidal approach follows the septum of the nose through the body of the sphenoid.
   (2) Care to avoid the cavernous sinus and internal carotid artery reduces the risk of intractable venous ooze or potentially catastrophic arterial hemorrhage.
e. **Posteriorly,** the sphenoid sinus may extend to the clivus.

4. **The maxillary sinus** lies in the body of the maxilla just lateral to the nasal cavity.
   a. Each drains into the **hiatus semilunaris** of the middle meatus. The slitlike ostium is high on the medial surface, so it does not drain by gravity with the head held erect (see Figure 35-3A).
      (1) It is closely related to the orifices of the frontal and ethmoid sinuses at the hiatus semilunaris. Infection in any one of these channels quickly spreads to the maxillary sinus.
      (2) The maxillary sinus is particularly prone to infection, which tends to be complicated by the collection of a stagnant pool of mucus. Surgical drainage of this pool involves piercing the bony nasal wall of the sinus beneath the inferior concha.
   b. Much of the medial wall is composed of cartilage.
   c. The floor is formed by the alveolar process. Only a thin layer of bone separates the sinus cavity from the roots of the teeth.
   d. The infraorbital canal projects from the roof.
   e. The sinuses are lined by mucoperiosteum, which is thinner and less richly supplied with blood vessels and glands than the mucosa of the nasal cavity. Cilia sweep mucus toward the ostia. In mucosal infections, ciliary action is inhibited and mucus collects.
   f. **Maxillary sinusitis** mimics the clinical sign of maxillary tooth abscess.
      (1) Most chronic cases (as high as 89%) are related to an infected tooth that projects into the sinus.
      (2) Infection may also spread from the maxillary sinus to the upper teeth.
      (3) Sinusitis may also affect the infraorbital nerve if the layer of covering bone is especially thin or absent.

III. ORAL CAVITY

A. **Basic structure.** This cavity extends from the oral fissure to the palatoglossal arches, marking the oropharyngeal isthmus.

1. **The roof of the mouth** is formed by the **hard** and **soft palates** (Figure 37-2; see Figure 37-5).

2. **The walls** of the oral cavity are the cheeks (see Figure 37-2). The maxillary and mandibular alveolar ridges with associated dentition subdivide the oral cavity into outer **vestibule** and inner **oral cavity proper.**

3. **The floor** of the oral cavity is formed and supported inferiorly by the **mylohyoid muscle** (see Figure 34-1).

4. **The oropharyngeal isthmus** is flanked by the palatoglossal arches, which comprise the **palatoglossal** and **palatopharyngeal folds** (see Figure 37-5). Between these folds is the tonsillar **fossa,** containing the **palatine tonsil.**

B. **Palate**

1. **The hard palate** forms the anterior three fifths of the palate and consists of the horizontal processes of the maxillary and palatine bones (see Figure 35-3 and 37-5).
   a. It separates the respiratory and digestive tracts. Because it occurs only in mammals, it can be surmised that its primary function is to isolate the mouth in suckling.

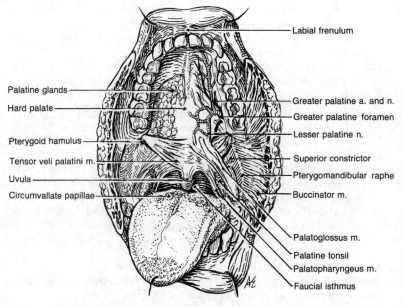

FIGURE 37-2. Oral cavity.

   b. **Palatine mucous glands** are abundant on the oral side of the hard palate (see Figure 37-2).
2. **The soft palate (velum)** contains muscle fibers, glands, lymphoid tissue, and an aponeurosis (see Figure 37-2).
   a. The mucous membrane of the palate is stratified squamous epithelium.
   b. The **uvula** hangs from the posterior border of the soft palate in the midline.
   c. The **aponeurosis of the tensor veli palatini muscle** continues the plane of the hard palate and forms a structural support for the soft palate.
3. **Musculature**
   a. **Intrinsic musculature** (see Table 37-1)
      (1) **Tensor veli palatini muscle** (Figure 37-3; see Figure 37-2)
         (a) It arises from the lateral (membranous) surface of the auditory tube and from the scaphoid fossa at the base of the pterygoid process. It descends along the lateral surface of the superior constrictor muscle. The tendon of tensor veli palatini passes around the **hamulus** of the medial pterygoid plate, pierces the buccinator, which also arises from the pterygoid plate, and spreads out within the soft palate to become the palatine aponeurosis.
         (b) Contraction of this muscle tightens the aponeurosis, thereby tensing the palate as well as opening the auditory tube.
         (c) Developmentally, it is a first arch muscle appropriated by the developing pharynx. Therefore, it is innervated by the **mandibular nerve** (CN V_3).
      (2) **Levator veli palatini muscle** (see Figure 37-3)
         (a) It arises from the petrous portion of the temporal bone and the medial (cartilaginous) surface of the auditory tube. It follows the auditory tube over the superior constrictor muscle and inferiorly along its inner surface. It inserts perpendicularly into the tensor aponeurosis of the soft palate.
         (b) It elevates the palate to establish the nasopharyngeal seal and opens the auditory tube.
         (c) It is innervated by the pharyngeal branch of the **vagus nerve** (CN X).
      (3) **Uvulae muscles** (see Figure 37-3)
         (a) This pair descends from the spine of the hard palate into the uvula on each side.

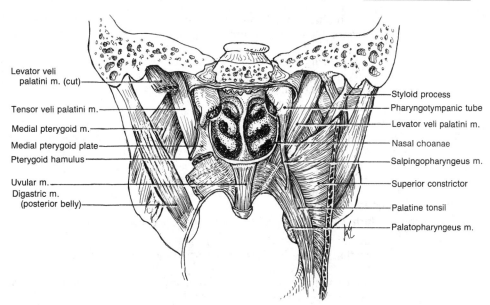

FIGURE 37-3. Musculature of the soft palate *(posterior view).*

        (b) These muscles contract when the palate is raised.
           (i) They help establish the nasopharyngeal seal by tensing the posterior margin of the soft palate.
           (ii) When one muscle contracts alone, it draws the uvula to the same side. The integrity of innervation by the pharyngeal branch of the **vagus nerve** (CN X) is observed when the examining physician has the patient say "ahhhhh."
   b. Extrinsic musculature (Table 37-1)
     (1) The **palatoglossus muscle** arches upward from the side of the tongue to the inferior surface of the soft palate (see Figures 37-2 and 37-6).
       (a) It underlies the **anterior faucial pillar or arch (palatoglossal fold).**
       (b) It acts as a sphincter (in conjunction with the base of the tongue) between the oral cavity and the pharynx.
       (c) It is innervated by the **vagus nerve** (CN X) through the pharyngeal plexus.
     (2) The **palatopharyngeus muscle** arises laterally from the palate and inserts into the pharyngeal musculature (see Figures 37-2 and 37-6).
       (a) It underlies the **posterior faucial pillar (palatopharyngeal fold)** and inserts onto the border of the thyroid cartilage.
       (b) It also acts as a sphincter (in conjunction with the base of the tongue) between the oral cavity and the pharynx. It helps to raise the larynx in swallowing (deglutition).
       (c) It is innervated by the **vagus nerve** (CN X) through the pharyngeal plexus.
   c. Combined muscle function (see Table 37-1)
     (1) Intrinsic muscle function. Closure of the nasopharyngeal isthmus by elevation of the palate (establishment of the **velopharyngeal seal**) is important in swallowing, speaking, and blowing.
       (a) During quiet nasal respiration, the soft palate hangs vertically.
       (b) In swallowing, the soft palate is raised to prevent food from entering the nose.
       (c) In blowing or in the production of the explosive consonants, the escape of air through the nose is prevented so pressure can build up within the mouth.
     (2) Extrinsic muscle function. Closure of the oropharyngeal isthmus by

contraction of the palatoglossus and palatopharyngeus muscles against the base of the tongue forms the **oropharyngeal seal,** which is important in swallowing.

   d. **Cleft palate,** a developmental anomaly, prevents the establishment of the velopharyngeal seal. Without surgical correction, the individual experiences difficulty with speech and deglutition.

4. **Vasculature**
   a. **Arterial supply**
      (1) The **greater** and **lesser palatine branches** of the descending palatine artery supply the hard and soft portions, respectively (see Figures 37-1B and 37-2).
      (2) The **ascending palatine branch** of the facial artery supplies the lateral sections (see Figure 37-6).
      (3) The **tonsillar branch** of the facial artery also supplies the lateral sections (see Figure 37-6).
   b. **Venous return** is to the pterygoid plexus and a peritonsillar plexus.
5. **Innervation** (see Figure 37-1 and Table 32-1)
   a. **General sensation for the hard palate** is carried by branches of the maxillary nerve (CN V_2) in the pterygopalatine fossa.
      (1) The **greater palatine nerve** reaches the hard palate through the **greater palatine foramen** (see Figures 37-1A and 37-2).
      (2) The **nasopalatine nerve** reaches anterior portions of the hard palate through the **incisive foramen** (see Figure 37-1B).
   b. **General sensation for the soft palate** is carried by the **lesser palatine nerve,** a branch of the maxillary nerve (see Figures 37-2 and 37-6). The **glossopharyngeal nerve** (CN IX) mediates the afferent limb of the **gag reflex** (see Figure 37-8).
   c. **Motor innervation.** Most muscles of the palate are innervated by the pharyngeal branch of the **vagus nerve** (see Figure 37-13). The tensor palatini muscle is innervated by the **mandibular nerve** (see Table 37-1).

C. Walls of the oral cavity

1. **The cheeks** or fleshy walls of this cavity comprise a layer of muscle between the skin of the face and the mucous membrane of the vestibule (see Figure 37-2).
   a. The **lips (labial folds),** separated by the **oral fissure** anteriorly, contain the orbicularis oris muscle and some pea-sized labial salivary glands.
   b. The **buccinator muscle** lies within the cheek (see Table 36-1).
      (1) This muscle arises from the posterior ends of the alveolar margins and the anterior aspect of the pterygomandibular raphe. It is pierced by the tendon of the tensor veli palatini muscle as it approaches the hamulus and by the parotid duct at the anterior border of the masseter muscle. Fibers pass around the cheeks and into the lips.
      (2) It compresses the vestibule and, together with the tongue, keeps food on the cusps of the molars for mastication.
      (3) It is innervated by the buccal branch of the **facial nerve** (see Figure 36-9).
2. **The alveolar margins** and associated **dentition** form the inner wall of the **vestibule** (see Figure 37-2).
   a. These margins are covered by **gingivae (gums).**
   b. In a young person, the gums surround the cervical portion of the teeth. Some gum recession occurs with age.
   c. Teeth and their sockets are discussed in Chapter 35 (III).
   d. In the anterior midline, small folds of mucosa are related to the inner and outer surfaces of the gums: **frenulum labia** of the upper and lower lips and the **frenulum linguae** of the tongue.

D. The tongue is a muscular organ concerned with ingestion, swallowing, speech, taste, and general sensation.

1. **Development.** The tongue develops from second and third branchial arch derivatives.

- **a.** Its oral and pharyngeal portions have different origins, different mucosa, and different innervation.
- **b.** A V-shaped sulcus lies between the second and third arch derivatives, separating the oral and pharyngeal parts. The apex of the sulcus points posteriorly and marks the **foramen cecum** from which the **thyroglossal duct** originated.

2. **Oral and pharyngeal portions** (see Figure 37-2)
    - **a.** The **oral (presulcal) portion** of the tongue has a tip, a dorsal surface, and a ventral surface.
        - **(1)** The **dorsal lingual mucosa** contains **lingual papillae** that increase the surface area available for taste receptors.
            - **(a)** They vary from the large, flat-topped vallate (circumvallate) papillae through the fungiform to the conical filiform.
            - **(b)** The **vallate papillae** are confined to the area immediately next to the lingual sulcus.
            - **(c)** Small transverse mucosal folds at the lateral edges of the tongue create foliate papillae.
        - **(2)** The **ventral lingual mucosa** is smooth and highly vascular.
            - **(a)** Two fringed folds of mucosa, the **plica fimbriata**, extend from the floor of the mouth almost to the tip of the tongue.
            - **(b)** A median **frenulum** also extends the length of the ventral surface.
                - **(i)** At the root of the tongue, the frenulum is related to the paired **sublingual papillae** for the ducts (of Wharton) of the **submandibular salivary glands.**
                - **(ii)** The associated **sublingual glands** create **sublingual folds** around the root of the tongue.
            - **(c)** Superficial vasculature of the **ventral surface** facilitates rapid sublingual absorption of substances, such as nitroglycerin and nicotine.
    - **b.** The **pharyngeal (postsulcal) portion** of the tongue begins at the row of vallate papillae (see Figures 37-2 and 37-5).
        - **(1)** The boundaries between the oral (anterior two thirds) and pharyngeal (posterior one third) portions of the tongue are blurred by the extension of third arch primordia superficially across the V-shaped sulcus. The area marked by the vallate papillae (in front of the sulcus) actually falls in the posterior one third.
        - **(2)** The lumpy surface of the mucosa covers lymphoid tissue of the **lingual tonsil.**

3. **The muscular root of the tongue** lies deep to the mandible and above the hyoid bone.

Table 37-2. Lingual Musculature

Muscle	Origin	Insertion	Primary Action	Innervation
Hyoglossus	Hyoid bone superolaterally	Tongue laterally	Depresses sides of tongue	Hypoglossal n. [CN XII]
Genioglossus	Anterior mandible	Tongue	Protracts tongue	Hypoglossal n. [CN XII]
Styloglossus	Styloid process of temporal bone	Tongue	Retracts tongue	Hypoglossal n. [CN XII]
Palatoglossus	Velum	Tongue	Elevates tongue	Pharyngeal br. of vagus n. (CN X)
Intrinsic muscles	Tongue	Tongue	Change shape of tongue	Hypoglossal n. [CN XII]

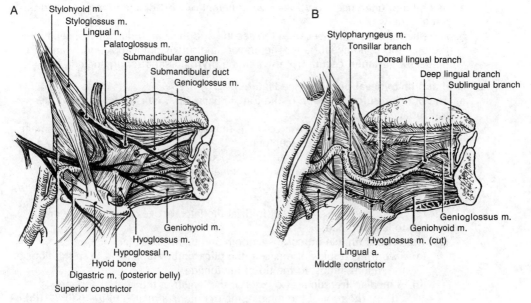

FIGURE 37-4. Tongue. *(A)* Musculature; *(B)* deep neurovascular structures.

   a. **Extrinsic muscles** (Figure 37-4; see Table 37-2).
      (1) **Hyoglossus muscle** (see Figure 37-4A)
         (a) This muscle originates along the ramus of the hyoid bone and inserts into the tongue by interdigitating with the styloglossus muscle and the lateral intrinsic musculature of the tongue.
         (b) **Relations** superficial to the hyoglossus muscle
            (i) The hypoglossal nerve (CN XII) passes near its inferior border.
            (ii) The submandibular duct passes superior to the muscle to empty into the oral vestibule.
            (iii) The lingual nerve, which winds around the submandibular duct, crosses first on its lateral aspect and then on its medial aspect.
            (iv) A portion of the submandibular gland lies superficial to this muscle.
         (c) **Relations** deep to the hyoglossus muscle
            (i) The lingual artery, which runs its tortuous course deep to the hyoglossus muscle and superficial to the genioglossus muscle, gives off dorsal and deep branches and the sublingual artery.
            (ii) The terminal portion of the deep branch of the glossopharyngeal nerve passes posteriorly to this muscle to enter the posterior portion of the tongue.
         (d) The hyoglossus draws the side of the tongue inferiorly.
         (e) It is innervated by the **hypoglossal nerve** (CN XII).
      (2) **Genioglossus muscle** (see Figure 37-4B)
         (a) It arises from the superior genial tubercle, just superior to the geniohyoid, and passes posteriorly deep to the hyoglossus muscle to insert into the tongue. Some fibers continue beyond the tongue into the middle constrictor muscle.
         (b) This muscle pulls the base of the tongue anteriorly in protrusion.
            (i) Acting singly, each pushes the protruded tongue to the opposite side.
            (ii) Integrity of innervation by the **hypoglossal nerve** (CN XII) is checked by the examining physician when he or she has the patient stick out the tongue.
      (3) **Styloglossus muscle** (see Figure 37-4A)
         (a) It arises from the styloid process and interdigitates with the hyoglossus muscle and the lateral intrinsic musculature of the tongue.

(b) It draws the posterior portion of the tongue superiorly and posteriorly. It is the principal retractor of the tongue.
(c) It is innervated by the **hypoglossal nerve.**
(4) **Palatoglossus muscle** (see Figures 37-2 and 37-4A)
(a) It originates on the hard palate, lies within the palatoglossal arch at the oropharyngeal isthmus, and inserts into the tongue.
(b) It acts more on the palate than on the tongue, establishing the palatoglossal sphincter.
(c) Unlike the other muscles of the tongue, it is innervated by the **vagus nerve** (CN X).
b. **Intrinsic muscles of the tongue** are arranged in longitudinal, transverse, and vertical groups (Table 37-2).
(1) They arise and insert within the substance of the tongue. The two halves of the tongue are separated by a fibrous septum into which some of these muscles insert.
(2) They change the shape of the tongue.
(3) All are striated muscles of somatic origin innervated by the **hypoglossal nerve** (CN XII).

4. **Vasculature**
   a. **Arterial supply** (see Figure 37-4B)
      (1) The **lingual artery,** which provides the principal blood supply, usually arises from the external carotid artery slightly cranial to the superior thyroid artery, but may arise from a common trunk with the facial artery. It courses deep to the posterior belly of the digastric muscle and continues deep to the hyoglossus muscle to enter the tongue through the root.
      (2) **Lingual branches**
         (a) The **tonsillar branch** supplies the palatine tonsil, lateral pharyngeal wall, and pharyngeal portions of the tongue (see Figures 37-4B and 37-6).
         (b) The **dorsal lingual branch** supplies the posterior portions of the tongue and oral cavity by looping deep to the hyoglossus muscle.
         (c) The **sublingual branch** passes along the genioglossus muscle to supply the anterior portion of the tongue and the oral cavity beneath the tongue.
         (d) The **deep lingual branch,** the terminal branch, supplies the deep portions of the central part of the tongue.
   b. **Venous return.** The **lingual vein** drains most of the tongue.
      (1) It is formed by like-named veins, which drain along the arterial branches.
      (2) The **lingual vein** usually drains into the **retromandibular vein.** Occasionally, however, it drains into the external jugular vein.
   c. **Lymphatic drainage** is along three routes from different regions of the tongue. Drainage of the lips and tongue is an important consideration in the spread of tobacco-associated carcinoma of these structures.
      (1) The **posterior one third** of the tongue drains unilaterally into the deep cervical jugulodigastric nodes that lie at the point where the posterior belly of the digastric muscle crosses the internal jugular vein.
      (2) The **marginal portions** of the middle one third drain unilaterally through the mylohyoid muscle to the submandibular nodes, which in turn drain to the jugulo-omohyoid group of deep cervical nodes.
      (3) The **central portions** of the middle one third drain bilaterally through the mylohyoid muscle to the left and right submandibular nodes, which in turn drain to the respective jugulo-omohyoid group of deep cervical nodes.
      (4) The **tip of the tongue** (anterior one third) drains bilaterally through the mylohyoid muscle to the left and right submental nodes and the respective jugulo-omohyoid group of deep cervical nodes.

5. **Innervation**
   a. The **lingual nerve (CN V),** the post-trematic nerve of the first branchial nerve, supplies general branchiomeric sensation for the anterior two thirds of the tongue (see Figure 37-4A and Table 32-1).
   b. The **chorda tympani (CN VII),** the pretrematic branch of the second branchial

nerve, carries taste and presynaptic secretomotor fibers for the intrinsic glands (of Nuhn) of the anterior two thirds of the tongue (see Figures 36-7 and 36-8). It also carries presynaptic secretomotor fibers for the submandibular gland and the sublingual gland.
    c. The **glossopharyngeal nerve (CN IX)**, the third branchial nerve, supplies general branchiomeric sensation (gag reflex) and taste for the posterior one third of the tongue (see Figures 37-6 and 37-8).
    d. The **hypoglossal nerve (CN XII)** supplies general somatic motor innervation to the tongue (see Figures 34-5 and 37-4A). This nerve innervates the muscles of the tongue (hyoglossus, genioglossus, and styloglossus muscles, as well as the intrinsic muscles of the tongue), with the exception of the palatoglossus.

E. **Submandibular and sublingual salivary glands**
   1. **Location and drainage**
      a. The **submandibular gland** folds around the posterior edge of the mylohyoid muscle, so a portion lies within the floor of the oral cavity and a portion is external to the oral cavity (see Figure 34-1).
         (1) This gland is ensheathed by the investing layer of deep cervical fascia.
         (2) The **submandibular duct** (of Wharton) runs from the anterior end of the deep portion, superficial to the hyoglossus muscle. It empties into the oral cavity at a small papilla at the side of the frenulum of the tongue.
         (3) The **lingual nerve** winds around the duct lateromedially.
      b. The **sublingual gland** surrounds the terminal portion of the submandibular duct.
         (1) It empties directly into the floor of the mouth by 10 to 20 short ductules.
         (2) A few ductules join the submandibular duct. Occasionally, the anterior sublingual ducts join to form a discrete **sublingual (Bartholin's) duct,** which usually joins the submandibular duct.
   2. **Innervation.** These glands are innervated by parasympathetic secretomotor fibers from the **facial nerve** (see Figure 36-3).
      a. **Presynaptic parasympathetic fibers** arise from the **superior salivatory nucleus,** run with the facial nerve, and then join the lingual nerve by way of the chorda tympani. The fibers synapse in the **submandibular ganglion,** which is attached to the lingual nerve by a few short connecting branches.
      b. **Postsynaptic parasympathetic fibers** leave the submandibular ganglion by the lingual nerve to reach the submandibular and sublingual glands as well as the intrinsic glands of the anterior tongue.

IV. PHARYNX

A. **Basic structure.** The pharynx lies deep to the rami of the mandible and extends to the base of the cranium. It is part of both the respiratory and the digestive tracts.
   1. **Divisions.** The upper end is divided in the plane of the palate into the **nasopharynx** and the **oropharynx**. It continues inferiorly as the **hypopharynx.**
   2. **Relations.** On either side of the pharynx, the muscles of mastication pass through the infratemporal fossa from the cranium to the mandible (see Figure 37-3). Posteriorly, the pharynx is apposed to the prevertebral fascia of the somatic neck (Figure 37-5), defining the **retrovisceral space.**

B. **The nasopharynx** opens anteriorly into the nasal cavity.
   1. **The nasal choanae** mark the beginning of the nasopharynx. The floor is formed by the soft palate. The posterior and superior aspects are related to the basilar occipital bone and the arch of the atlas (see Figure 37-5).
   2. **The auditory (pharyngotympanic, eustachian) tube** opens in the lateral wall of the nasopharynx at the level of the inferior meatus of the nasal cavity (see Figure 37-5).

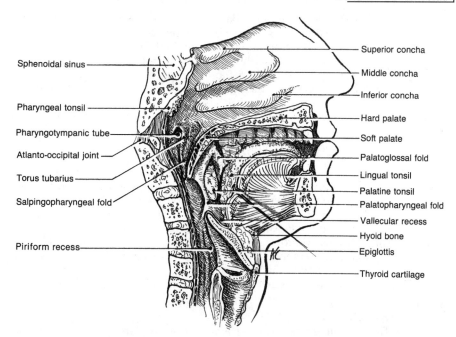

FIGURE 37-5. Nasopharynx and oropharynx.

    a. The cartilaginous wall of the tube raises a tubal elevation (**torus tubarius**).
    b. A small aggregation of lymphoid tissue forms the **tubal tonsil** in this region. Hypertrophy or edema of this tonsil may occlude the auditory tube with accumulation of secretions in the middle ear.
    c. Infection in the nasopharynx may track along the auditory tube to produce **otitis media**.
    d. The auditory tube is a remnant of the first branchial pouch.
  3. **The salpingopharyngeus muscle** originates from the end of the auditory tube and inserts into the musculature of the pharynx (see Figure 37-6).
    a. The **salpingopharyngeal fold** overlies this muscle.
    b. It raises the pharynx during deglutition.
    c. It is innervated by the pharyngeal branch of the **vagus nerve** (CN X).
  4. **The pharyngeal tonsil (adenoid)** consists of lymphoid tissue embedded in the posterior wall of the nasopharynx (see Figure 37-5). Hypertrophy of this tonsil (swollen adenoids) may interfere with nasal respiration and phonation.
  5. **The nasopharyngeal isthmus** is the terminal section of the nasopharynx.
    a. The pharynx narrows where the soft palate (velum) establishes the **velopharyngeal seal.**
    b. When the pharyngeal musculature is drawn superiorly in deglutition, a slightly posterior (Passavant's) ridge helps close this seal.

C. The **oropharynx** opens anteriorly into the **oral cavity** (see Figure 37-5).
  1. The **faucial pillars** (faucial arches or folds) bound the **oropharyngeal isthmus** (see Figure 37-6).
    a. The **palatoglossal folds** overlie the **palatoglossus muscles** anteriorly on the lateral wall of the oropharynx.
    b. The **palatopharyngeal folds** overlie the **palatopharyngeus muscles** posteriorly on the lateral wall of the oropharynx.
  2. **Tonsillar fossae** lie between the diverging fauces on each side.
    a. Each triangular fossa contains a mass of lymphoid tissue, the **palatine tonsil** (see Figure 37-5).

(1) This tonsil extends from the base of the tongue to the edge of the soft palate.
(2) The medial surface has a superior intratonsillar cleft and 12 to 15 tonsillar crypts that extend deep into the lymphoid tissue.
(3) The lateral surface has a fibrous **tonsillar capsule.**
   (a) This capsule limits the peritonsillar space.
   (b) It is easily separated from the pharyngeal wall, except at the root of the tongue. The tonsillar arteries enter the gland at this point.
   b. Tonsillar fossae lie between the palatine muscles and are related to the space between the superior and middle pharyngeal constrictors (see Figure 37-6).
   (1) Posteriorly, the gap is limited by the origin of the middle constrictor and the stylohyoid ligaments from which it arises.
   (2) Anteriorly, the hyoglossus muscle ascends to the lateral surface of the tongue.
   (3) The **stylopharyngeus muscle** carries the **glossopharyngeal nerve** through the gap between the superior and middle constrictors.
      (a) Therefore, both structures are related to the tonsil, laterally. The nerve is at risk during tonsillectomy.
      (b) If the styloid process is unusually long (in 4% of individuals), a swollen tonsil may compress the glossopharyngeal nerve against the styloid, with pain referred to the pharynx and ear.
   c. The tonsillar fossa represents the second branchial pouch.

3. **Tonsillar vasculature**
   a. **Arterial supply** (Figure 37-6)
      (1) The **ascending pharyngeal artery** is a branch of the external carotid artery.
      (2) The **ascending palatine branch** arises from the facial artery.
      (3) A **tonsillar branch** arises from the lingual artery.
      (4) The **tonsillar branch** of the facial artery is the primary source to the tonsil.
      (5) A **lesser palatine branch** arises from the maxillary artery in the pterygopalatine fossa.
   b. **Venous return**
      (1) A **peritonsillar venous plexus** is drained by veins that parallel the arterial branches.
      (2) Principal drainage is by the **tonsillar branch of the lingual vein.**
      (3) Any vein that drains the tonsil may be large, and as such is a source of profuse venous hemorrhage after tonsillectomy.

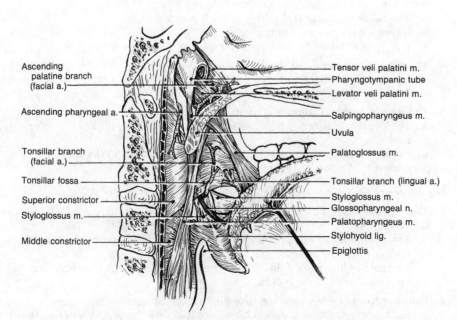

FIGURE 37-6. Internal pharyngeal musculature and vasculature.

4. **Waldeyer's ring** is a ring of lymphoid tissue extending around the pharynx (see Figure 37-5). The component parts hypertrophy and regress at different stages in life.
   a. The **pharyngeal tonsil (adenoid),** which lies where the nasopharynx contacts the sphenoid bone, hypertrophies in the toddler but regresses by 8 years of age.
   b. **Tubal tonsils** are accumulations of lymphoid tissue about the opening of each pharyngotympanic tube. Hypertrophy may constrict or occlude the auditory tube and predispose the child to recurrent otitis media.
   c. **Palatine tonsils,** lying in the tonsillar crypts, hypertrophy during childhood but regress by puberty.
   d. **Lingual tonsils,** at the base of the tongue, enlarge at the time of puberty and regress very little during adult life.

5. **Tonsillectomy**
   a. Because the palatine tonsils usually regress, the treatment of choice for **tonsillitis** currently tends to be conservative (antibiotic) therapy while awaiting regression. This move away from routine surgical intervention has resulted in more post-adolescent individuals being hospitalized for tonsillectomy.
   b. Tonsillectomy is indicated in persistent infection, particularly if complicated by sinusitis, otitis media, or direct spread into the loose tissue of the pharynx (**quinsy**).

D. The **hypopharynx** is the inferior extension of the oropharynx.

1. **The lingual tonsil** and base of the tongue form the anterior wall (see Figure 37-5).

2. **The epiglottis** marks the upper surface of the larynx and guards the opening into the larynx (see Figure 37-5).
   a. A **median glossoepiglottic fold** extends from the base of the tongue to the epiglottis, with a vallecular recess to each side (see Figure 37-9A).
   b. **Lateral glossoepiglottic (pharyngoepiglottic) folds** run from the anterolateral walls of the hypopharynx to the base of the epiglottis and define the inferolateral boundaries of the **vallecular recesses** (see Figure 37-9A).
   c. The **vallecular recesses** are the remnants of the third branchial pouches.

3. **The piriform recesses** of the hypopharynx extend inferiorly beneath the lateral glossoepiglottic folds on either side of the larynx (see Figure 37-9A).
   a. These areas dilate considerably. Swallowed food and liquid are diverted by the epiglottis into these recesses to either side of the larynx.
   b. Swallowed foreign bodies may lodge in these recesses.

E. **Musculature** (Table 37-3)

1. **Superior constrictor**
   a. **Course** (Figure 37-7; see Figure 37-6)
      (1) **Anteriorly,** this muscle is attached to the pterygoid plate, the pterygomandibular raphe (which it shares with the buccinator muscle), and the posterior portions of the maxillary and mandibular alveolar processes.
      (2) **Posteriorly,** it attaches to the pharyngeal tubercle of the basioccipital bone as well as to the posterior pharyngeal raphe.
      (3) The **lateral superior edge** is free and does not meet the cranium (see Figures 37-4 and 37-7).
         (a) The **auditory tube** passes through this hiatus to open into the nasopharynx.
         (b) The **tensor veli palatini muscle** descends vertically from the auditory tube, lateral to this hiatus, to reach the hamulus. It then turns medially through the hiatus to insert into the soft palate.
         (c) The **levator veli palatini muscle** remains deep to the superior constrictor.
   b. The superior constrictor raises the oropharynx to establish Passavant's ridge and begins the stripping action that propels the bolus into the hypopharynx.
   c. It is innervated by the **vagus nerve** (CN X) through the pharyngeal plexus.

Table 37-3. Pharyngeal Musculature

Muscle	Origin	Insertion	Primary Action	Innervation
Superior constrictor	Pterygoid plate, pterygomandibular raphe, mandible	Occipital bone, pharyngeal raphe	Constricts upper part of pharynx in deglutition	Pharyngeal br. of vagus n. (CN X)
Middle constrictor	Hyoid bone	Pharyngeal raphe	Constricts middle part of pharynx in deglutition	Pharyngeal br. of vagus n. (CN X)
Inferior constrictor	Thyroid cartilage	Pharyngeal raphe	Constricts lower part of pharynx in deglutition	Pharyngeal br. of vagus nerve (CN X)
Palato-pharyngeus	Velum	Pharynx and thyroid cartilage	Raises pharynx and larynx	Pharyngeal br. of vagus n. (CN X)
Salpingo-pharyngeus	Auditory tube	Pharynx	Raises pharynx and larynx	Pharyngeal br. of vagus n. (CN X)
Stylopharyngeus	Styloid process of temporal bone	Pharynx and thyroid cartilage	Raises pharynx	Glossopharyngeal n. (CN IX)

2. **Middle constrictor**
   a. **Course** (see Figures 37-6 and 37-7)
      (1) Anteriorly, this fan-shaped muscle arises from the stylohyoid ligament and the greater and lesser horns of the hyoid bone.
      (2) Its fibers pass posteriorly, external to those of the superior constrictor, to insert into the pharyngeal raphe.
      (3) A gap between the lowest origin of the superior constrictor and the uppermost origin of the middle constrictor provides passage for the stylopharyngeus muscle, the pharyngeal branch of the glossopharyngeal nerve, and the tonsillar branch of the facial artery.
   b. It strips the bolus from oropharynx into hypopharynx.
   c. It is innervated by the **vagus nerve** (CN X) through the pharyngeal plexus.

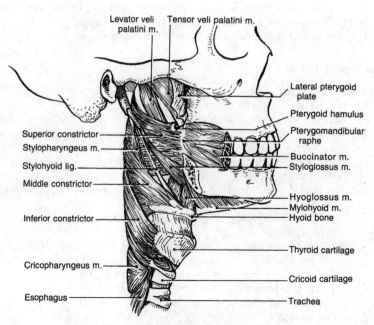

FIGURE 37-7. External pharyngeal musculature.

3. **Inferior constrictor**
   a. **Course** (see Figure 37-7)
      (1) This muscle, which is fan-shaped superiorly and tubular inferiorly, arises from the oblique line on the lamina of the thyroid cartilage and from the cricoid cartilage.
      (2) The fibers pass almost horizontally, external to the middle constrictor, to insert into the pharyngeal raphe.
   b. The cricoid portion of this muscle (the **cricopharyngeus muscle**) functions as a sphincter at the superior end of the esophagus.
   c. It is innervated by the **vagus nerve** (CN X) through the pharyngeal plexus.
4. **The palatopharyngeus muscle** underlies the posterior faucial pillar (palatopharyngeal fold) [see section III B 3 b (2)].
5. **The salpingopharyngeus muscle** underlies the salpingopharyngeal fold (see Figure 37-6).
   a. It originates from the end of the auditory tube at the torus tubarius and inserts into the musculature of the middle part of the pharynx.
   b. It raises the pharynx and larynx during swallowing.
   c. It is innervated by the **vagus nerve** (CN X) through the pharyngeal plexus.
6. **Stylopharyngeus muscle**
   a. It originates from the styloid process and passes through the hiatus between the superior and middle constrictors to interdigitate with the pharyngeal musculature. Some fascicles insert onto the posterolateral border of the thyroid cartilage (see Figure 37-7).
   b. It raises the pharynx during deglutition.
   c. It is the only muscle innervated by the **glossopharyngeal nerve** (CN IX).

F. **Innervation**

1. **The glossopharyngeal nerve (CN IX)** comprises the pretrematic and post-trematic portions of the nerve of the third branchial arch (see Figure 35-1 and Table 32-1).
   a. **Proximally,** this nerve passes out of the cranium in a deep groove on the medial margin of the jugular foramen (Figure 37-8).
      (1) The **pretrematic tympanic branch** conveys sensory neurons from the ear and

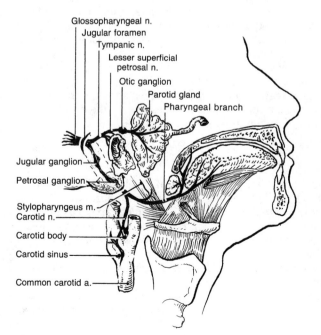

FIGURE 37-8. Glossopharyngeal nerve.

parasympathetic neurons continue along the lesser superficial petrosal nerve to the otic ganglion for parotid gland secretion.
- (2) A small **superior (jugular) ganglion** and an **inferior (petrosal) ganglion** just below the jugular foramen contain cell bodies for the sensory neurons.
- b. **Distally,** it passes laterally between the internal jugular vein and the internal carotid artery. It then winds around the stylopharyngeus muscle to enter the pharynx between the superior and middle constrictors (see Figure 37-8).
  - (1) The **pharyngeal branch** (also pretrematic) supplies branchiomeric sensation from the posterior one third of the tongue and the walls of the pharynx.
    - (a) This branch supplies the afferent limb of the **gag reflex.**
      - (i) The integrity of the glossopharyngeal nerve is tested by the examining physician with a wooden tongue depressor.
      - (ii) Stimulation in the external auditory meatus (such as by an insect or curettage) may initiate a gag reflex.
    - (b) This branch also supplies taste sensation from the posterior one third of the tongue.
  - (2) The post-trematic motor branch supplies the **stylopharyngeus muscle.**
    - (a) The stylopharyngeus seems to be the only branchiomeric muscle supplied by the glossopharyngeal nerve, although some sources claim it also supplies the middle constrictor.
    - (b) The post-trematic sensory branch is associated with the carotid body and sinus and mediates the carotid reflex.
2. **The vagus nerve (CN X)** comprises pretrematic and post-trematic portions of the nerves of the fourth and sixth branchial arches (see Figure 35-1).
   - a. The **pharyngeal branch** supplies most of the branchiomeric musculature of the pharynx (see Figure 37-13 and Table 32-1), with the exception of the tensor veli palatini (CN V_3) and stylopharyngeus (CN VII) muscles.
   - b. The only vagal sensory function in the pharynx is mediation of special visceral sensation from the epiglottic taste buds.

G. Deglutition

1. **The oral phase** is voluntary.
   - a. **Elevation of the hyoid bone** by the **digastric** and **mylohyoid muscles** raises the tongue to the roof of the mouth.
   - b. **Movement of the bolus into the pharynx**
     - (1) The **intrinsic tongue muscles** press the tip of the tongue against the maxillary incisors and the hard palate to squeeze the food toward the pharynx.
     - (2) The **palatoglossus muscle** elevates the base of the tongue to squeeze the food through the fauces into the pharynx.
     - (3) The **styloglossus muscle** draws the base of the tongue posteriorly, propelling the bolus into the oropharynx.
   - c. **Establishing the oropharyngeal seal.** Together with the elevated and retracted tongue, the **palatoglossus** and **palatopharyngeus muscles** narrow the oropharyngeal isthmus behind the bolus, forming the **oropharyngeal seal.**
2. **The pharyngeal phase** is also voluntary, but once started, it cannot be comfortably interrupted.
   - a. **Establishment of velopharyngeal seal**
     - (1) The soft palate is elevated by the action of the **levator veli palatini** and **tensor veli palatini muscles** to seal off the nasopharyngeal isthmus.
     - (2) The **superior constrictor, palatopharyngeus,** and **salpingopharyngeus muscles** draw the upper portion of the pharynx upward over the bolus, accentuating Passavant's ridge and thereby reinforcing the velopharyngeal seal.
   - b. **Elevation of the hyoid bone**
     - (1) The concomitant contraction of the **stylohyoid and digastric muscles** draws the hyoid bone cranially.
     - (2) This action also draws the attached larynx cranially under the tongue so that the **epiglottis** assumes a more horizontal posture. Thus, the voice box (larynx) rises to close itself against the lid (epiglottis).

c. **Glottal closure.** The arytenoid cartilages tilt forward by contraction of the thyroarytenoid muscle and approximate to assist in closing off the larynx. This mechanism is sufficient to prevent the inhalation of food if the epiglottis is excised.

d. **Stripping action.** Sequential contraction of the superior, middle, and inferior pharyngeal constrictors propels the bolus through the piriform recesses to either side of the larynx.

3. **The esophageal phase** is involuntary.
   a. **Stripping action.** The inferior constrictor and the cricopharyngeus muscles squeeze the bolus into the esophagus; peristalsis moves the bolus toward the stomach.
   b. **Re-establishment of the respiratory tract.** The lingual and pharyngeal musculature relax, breaking the oropharyngeal and nasopharyngeal seals.
   c. **Repositioning of the larynx.** The infrahyoid muscles contract to draw the larynx inferiorly.
      (1) Contraction of the hyoglossus and genioglossus muscles returns the tongue to the floor of the oral cavity.
      (2) With depression of the hyoid bone, the epiglottis assumes a more vertical position, opening the larynx for respiration.

V. LARYNX

A. Superficial structure

1. **Laryngeal ossicles.** The hyoid bone, epiglottis, thyroid cartilage, and cricoid cartilage are midline supporting structures (see Figure 37-10). The arytenoid, corniculate, and cuneiform cartilages are paired laterally.

2. **Internal structure**
   a. **Hypopharyngeal recesses**
      (1) **Vallecular recesses** beyond the tongue are limited inferolaterally by the **lateral glossoepiglottic folds** and separated by the **median glossoepiglottic fold** (Figure 37-9A).
      (2) **Piriform fossae (recesses)** lie on either side of the laryngeal opening inferior to the lateral glossoepiglottic folds (see Figure 37-9A).

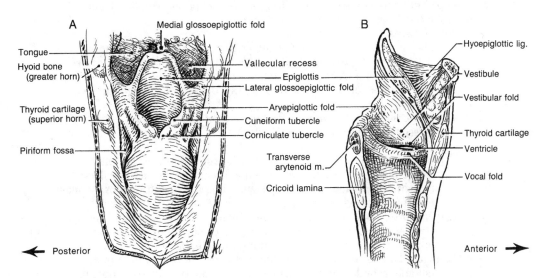

FIGURE 37-9. Hypopharynx and larynx. *(A)* Posterior aspect; *(B)* lateral aspect.

b. The fibroelastic **quadrangular membrane** on each side runs between the arytenoid cartilages and the epiglottis.
      (1) The free upper border of this membrane forms the **aryepiglottic fold** (see Figure 37-9).
         (a) This fold forms the wall separating the larynx from the **piriform recesses** of the hypopharynx.
         (b) A small, rod-like **cuneiform cartilage** lies within the free edge of each fold, forming the cuneiform **tubercle.**
      (2) The free lower border of this membrane forms the horizontal **vestibular fold** or **false vocal cord** (see Figure 37-9B).
      (3) The **laryngeal vestibule** lies between the aryepiglottic fold and the vestibular fold.
   c. Inferior to the vestibular (false vocal) fold is the fusiform **laryngeal sinus** or **ventricle** (see Figure 37-9B) that represents the fourth branchial pouch. A **saccule** of variable size invaginates into the laryngeal wall from the ventricle.
   d. **True vocal folds** (cords) define the **rima glottidis** (see Figure 37-9B).
3. **Muscles** run between the laryngeal cartilages, producing abduction, adduction, and tensing of the vocal cords.
4. **Function.** The larynx is a compound sphincter. It closes the airway during swallowing and during the Valsalva maneuver (as in coughing, urination, and defecation). With fine motor control, it constricts the airway for phonation.

B. Laryngeal cartilages (ossicles)

1. **The hyoid bone** is U-shaped with a median **body,** paired **lesser horns** (**cornua**) laterally, and paired **greater horns** posteriorly (see Figure 37-7).
   a. The **stylohyoid ligament** runs between the styloid process and the lesser horn (see Figure 37-7).
   b. Numerous muscles attach along the outer surface of the greater horn and body of the hyoid (see Figure 37-2).
   c. The **thyrohyoid membrane** runs in a broad sheet from the medial surface of the hyoid bone to the upper border of the thyroid cartilage (see Figure 37-10).
      (1) This membrane thickens anteriorly to form the **medial thyrohyoid ligament** as well as laterally to form the **lateral thyrohyoid ligament.**
      (2) It is pierced superolaterally by the **internal branch of the superior laryngeal neurovascular bundle.**
2. **The epiglottis** projects posterosuperiorly into the hypopharynx (see Figure 37-9) and is stabilized by several ligaments.
   a. Attachments
      (1) The superior surface of the epiglottis has a **medial glossoepiglottic fold** and a pair of **lateral glossoepiglottic folds,** defining the **vallecular recesses.**
      (2) At its base, it attaches to the anterior inner surface of the thyroid cartilage.
      (3) The anterior surface is linked to the hyoid by the **hyoepiglottic ligament.**
   b. Innervation
      (1) **Branchiomeric sensation** from the upper epiglottic surface is carried by the **glossopharyngeal nerve** (see Figure 37-8), which provides the afferent limb of the **gag reflex.**
      (2) Branchiomeric sensation from the lower surface of the epiglottis and supraglottic larynx is carried by the internal branch of the **superior laryngeal branch of the vagus nerve** (see Figure 37-13), which provides the afferent limb of the **cough reflex.**
      (3) **Epiglottic taste buds** are innervated by the **internal branch of the superior laryngeal nerve,** which arises from the vagus nerve and represents the pretrematic division of the nerve of the fourth branchial arch.
   c. **Function.** During the pharyngeal phase of deglutition, elevation of the hyoid bone draws the attached larynx cranially so the **epiglottis** assumes a more horizontal posture.

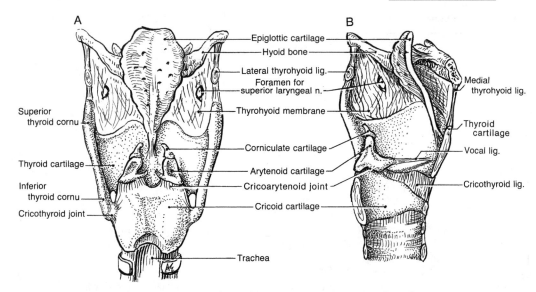

FIGURE 37-10. Laryngeal ossicles. *(A)* Posterior aspect; *(B)* lateral aspect.

3. **The thyroid cartilage** comprises two laminae fused in the anterior midline (Figure 37-10).
   a. The upper border has a characteristic **median thyroid notch.**
      (1) Below the notch is a **laryngeal prominence.**
      (2) In males, the laryngeal prominence (Adam's apple) is larger than in females, which results in longer vocal cords that vibrate at a lower frequency.
   b. The outer surface has an **oblique line** along which the sternothyroid muscle inserts, separating the origins of the thyrohyoid muscle and the inferior constrictor.
   c. The open posterolateral borders are drawn into **superior horns** and **inferior horns** (see Figure 37-10A).
      (1) From the **superior horns,** a thickened portion of the thyrohyoid membrane forms the **lateral thyrohyoid ligament.**
      (2) At the **inferior horns,** the thyroid cartilage articulates with the cricoid cartilage at the **cricothyroid joint.**
      (3) Each open posterior border receives the insertions of the stylopharyngeus and palatopharyngeus muscles.
4. **The cricoid cartilage** (G. krikos, ring) has a broad posterior **lamina** and narrow anterior **arch** (see Figure 37-10).
   a. Posterolaterally, this cartilage articulates with the **inferior horns** of the thyroid to form the **cricothyroid joints** (see Figure 37-10A).
   b. The **lamina** provides muscle attachments.
   c. **Arytenoid cartilages** articulate with the upper posterolateral borders of the cricoid lamina.
5. **The arytenoid cartilages** are pyramidal (see Figure 37-10).
   a. The **base** forms a shallow ball-and-socket articulation with the upper border of the cricoid lamina.
   b. The **apex** is surmounted by the small, nonfunctional corniculate cartilage.
   c. The **muscular process** receives the posterior and lateral cricoarytenoid muscles as well as the transverse arytenoid muscle.
   d. The **vocal process** receives the vocal ligament.

C. Laryngeal movements

1. **Articulations.** Movements of the larynx occur primarily at the cricothyroid and cricoarytenoid joints.

Table 37-4. Laryngeal Musculature

Muscle	Origin	Insertion	Primary Action	Innervation
Cricothyroid	Cricoid cartilage anteriorly	Thyroid cartilage laterally	Tenses vocal cords	Superior laryngeal br. of vagus n. (CN X)
Posterior cricoarytenoid	Cricoid cartilage posteriorly	Muscular process of arytenoid cartilage	Abducts vocal cords	Inferior laryngeal br. of vagus n. (CN X)
Lateral cricoarytenoid	Cricoid cartilage laterally	Muscular process of arytenoid cartilage	Adducts vocal cords	Inferior laryngeal br. of vagus n. (CN X)
Transverse arytenoid	Muscular process of arytenoid cartilage	Muscular process of arytenoid cartilage	Adducts vocal cords	Inferior laryngeal br. of vagus n. (CN X)
Oblique arytenoid	Muscular process of arytenoid cartilage	Aryepiglottic fold	Constricts rima glottidis	Inferior laryngeal br. of vagus n. (CN X)
Thyroarytenoid	Thyroid cartilage anteriorly	Vocal process of arytenoid cartilage	Adducts and relaxes vocal folds	Inferior laryngeal br. of vagus n. (CN X)
Vocalis portion	Thyroid cartilage anteriorly	Vocal process of arytenoid cartilage	Tenses vocal folds to raise pitch	Inferior laryngeal br. of vagus n. (CN X)

    a. Cricothyroid joints occur between the inferior horns of the thyroid cartilage and the posterior lateral walls of the cricoid cartilage (see Figure 37-10A).
        (1) A common transverse axis through these joints permits one degree of freedom.
        (2) Movement changes the distance between the anterior point of the thyroid cartilage and the vocal process of each arytenoid cartilage, thereby altering tension on the vocal cords. When the distance between the thyroid and cricoid cartilages is reduced anteriorly, the vocal cords are stretched and tensed.
    b. Cricoarytenoid joints occur between the superior surface of the cricoid lamina and the bases of the arytenoid cartilages (see Figure 37-10).
        (1) These joints have three degrees of freedom.
            (a) Sliding medially and laterally occurs along the cricoid lamina in abduction and adduction.
            (b) Rotation medially and laterally occurs about a vertical axis.
            (c) Tilting forward and backward slightly occurs about a transverse axis.
        (2) Movements are conjoined so that the abduction always occurs with external rotation, and adduction with internal rotation.
2. Intrinsic musculature (see Table 37-4)
    a. Cricothyroid muscle (Figure 37-11B; see Figures 34-1 and 37-12)
        (1) This muscle, crossing the cricothyroid joint, runs between the external surfaces of the thyroid and cricoid arches.
        (2) In contraction, it approximates the anterior edges of the cricoid and thyroid cartilages, thereby stretching and tensing the vocal folds.
    b. Posterior cricoarytenoid muscle (see Figure 37-11A)
        (1) Fibers converge to the muscular process of the arytenoid cartilage from a broad origin on the posterior lamina of the cricoid cartilage.

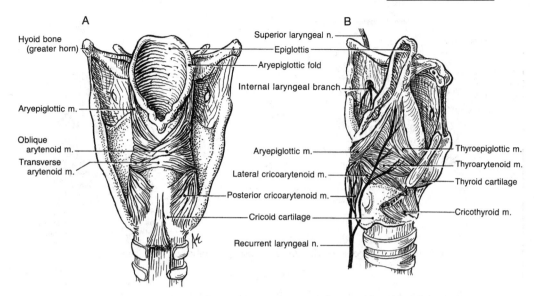

FIGURE 37-11. Laryngeal musculature. *(A)* Posterior aspect; *(B)* lateral aspect.

      (2) It is the only dilator muscle of the rima glottidis.
         (a) Horizontal fibers rotate the arytenoid cartilage laterally, thereby swinging the vocal process outward to abduct the vocal folds.
         (b) Vertical fibers tilt the arytenoid cartilage backward, thereby tensing the vocal cords.
  c. Transverse arytenoid and **oblique arytenoid muscles** (see Figures 37-11A and 37-12B)
      (1) These fibers run between the medial surfaces of the arytenoid cartilage and fibers from the oblique arytenoid muscles, which continue to the epiglottis as the aryepiglotticus muscle.
      (2) These muscles adduct the arytenoid cartilage, thereby adducting the vocal cords.
  d. Lateral cricoarytenoid muscle (see Figures 37-11B and 37-12B)
      (1) It arises from the outer surface of the cricoid arch and inserts onto the muscular process of the arytenoid cartilage.
      (2) It rotates the arytenoid cartilage medially and tilts it forward, thereby adducting the vocal cords and reducing the tension on the cords.
  e. Thyroarytenoid muscle (Figure 37-12B; see Figure 37-11B)
      (1) This muscle passes obliquely from the thyroid cartilage to the aryepiglottic folds and arytenoid cartilage superficial to the quadrangular membrane.
         (a) Superiorly, it is the **thyroepiglottic muscle.**
         (b) Inferomedially, it is the **thyroarytenoid muscle proper.**
         (c) The most medial portion constitutes the **vocalis muscle.**
      (2) Actions
         (a) The **thyroepiglotticus muscle** widens the laryngeal inlet.
         (b) The **thyroarytenoid proper** internally rotates the arytenoid cartilage and thereby adducts the vocal folds.
         (c) At the same time, the **vocalis portion** draws the arytenoid cartilage forward, slackening the vocal cords.
            (i) Although the vocalis muscle slackens the posterior portion of the cord, it tenses the anterior portion and, thus, raises the pitch of the voice.
            (ii) Both thyroarytenoid proper and vocalis muscles contract during deglutition to close the **rima glottidis.**

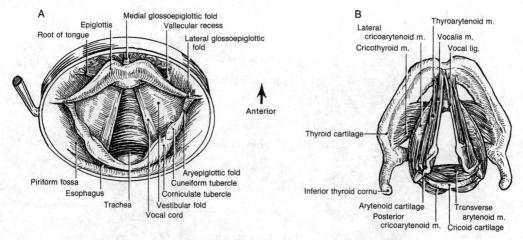

FIGURE 37-12. Vocal cords. *(A)* Laryngoscopic view; *(B)* vocalis muscle.

3. **Group actions** (Table 37-4)
   a. **Abduction of the vocal folds** is accomplished by the **posterior cricoarytenoid muscle,** acting about vertical axes through the cricoarytenoid joints.
   b. **Adduction of the vocal folds**
      (1) The **lateral cricoarytenoid** is the principal adductor, producing rotation about the vertical axes through the cricoarytenoid joints.
      (2) The **transverse arytenoid** and **oblique arytenoid muscles** slide the arytenoid cartilage closer to the midline, acting at the saddle-shaped cricoarytenoid joint.
      (3) The **thyroarytenoid muscle** internally rotates the arytenoid cartilage and thereby adducts the vocal folds weakly.
   c. **Tensing the focal folds** is accomplished by the **cricothyroid muscle** and the **thyroarytenoid muscle,** especially its medial **vocalis** portion, acting about a transverse axis through the cricothyroid joint.
4. **Group innervation.** The cricothyroid muscle is innervated by the **superior laryngeal branch** of the vagus nerve (CN X). All other intrinsic laryngeal muscles are innervated by the **recurrent (inferior) laryngeal branch** of the vagus nerve (see Figure 37-11B).

D. **Vasculature**

1. **Arterial supply**
   a. The **superior thyroid artery** arises as the first branch of the external carotid artery. The external branch supplies the thyroid gland. The **superior laryngeal branch** passes through the thyrohyoid membrane to the larynx (see Figure 30-9).
   b. The **inferior thyroid arteries** arise from the thyrocervical trunk, which supplies the thyroid gland before reaching the laryngeal muscles.
2. **Venous return** parallels the arteries (see Figure 34-4).

E. **Innervation** is by the **vagus nerve.**

1. **Composition and course.** The **vagus nerve (CN X;** see Table 32-1) has fibers that belong to all the functional categories except somatic efferent. The vagal trunk passes down the neck within the carotid sheath between the internal jugular vein and the carotid arteries.
2. **Distribution.** The majority of vagal fibers are general visceral afferents and efferents associated with thoracic and abdominal viscera.
   a. The **auricular branch** from the external auditory meatus and a **meningeal branch**

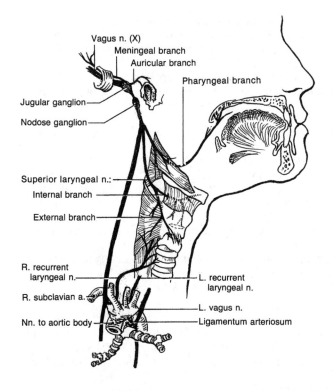

FIGURE 37-13. Vagus nerve.

provide the majority of the cell bodies in the **superior (jugular) ganglion** (Figure 37-13).
- b. The **pharyngeal branch** passes between the carotid arteries and onto the pharynx in the interval between the superior and middle constrictors to bringing motor fibers to the **pharyngeal plexus** (see Figure 37-13).
- c. The **superior laryngeal nerve** passes deep to the carotid arteries onto the middle constrictor, where it divides into motor and sensory branches (see Figure 37-13).
  - (1) The **internal branch** pierces the thyroid membrane and is sensory only (see Figure 37-11B).
    - (a) It provides mucosal sensation between the inferior surface of the epiglottis and the vocal folds and also conveys taste fibers from the epiglottis.
    - (b) It provides the afferent limb of the **cough reflex.**
    - (c) Its cell bodies lie in the **inferior (nodose) ganglion.**
    - (d) It represents pretrematic and post-trematic divisions of the nerve of the fourth branchial arch.
  - (2) The **external branch** descends on the surface of the inferior constrictor to innervate only the **cricothyroid muscle** (see Figure 37-13).
    - (a) This nerve has a posterior relation to the upper pole of the thyroid gland and must be avoided when attempting to clamp the upper vascular peduncle in thyroidectomy.
    - (b) This nerve represents the post-trematic division of the nerve of the fourth branchial arch.
- d. **Inferior (recurrent) laryngeal nerves**
  - (1) **During development,** the recurrent laryngeal nerve is associated with the sixth branchial arch as the post-trematic division.
    - (a) **On the right side,** the fifth and sixth arches degenerate, allowing this nerve to pass around the subclavian artery, which is associated with the fourth branchial arch.
    - (b) **On the left side,** this nerve is trapped below the **ligamentum arteriosum,** the remnant of the artery of the sixth branchial arch.
  - (2) **Course** (see Figures 37-11B and 37-13)

- **(a) In the lower neck,** it ascends between the esophagus and the trachea.
- **(b) In the middle of the neck,** it runs behind the posterior surface of the thyroid gland, lying among the terminal branches of the inferior thyroid artery. This nerve is at risk during thyroidectomy.
- **(c)** At the level of the inferior border of the inferior constrictor (cricopharyngeus muscle), it dives into the larynx and ascends posterior to the articulation between the thyroid and cricoid cartilages.

(3) **Distribution**
- **(a)** It conveys **sensation** to the laryngeal and tracheal mucosa below the vocal folds.
    - **(i)** It provides the afferent limb of a **cough reflex.**
    - **(ii)** Its cell bodies lie in the **inferior (nodose) ganglion.**
- **(b)** It provides motor innervation to the **intrinsic muscles of the larynx** (except the cricothyroid muscle).
    - **(i)** Because the vocal folds have several adductors and only one abductor, laryngospasm apposes the vocal folds. This situation requires immediate intubation of the patient.
    - **(ii)** When paralyzed, the vocal fold initially adducts (owing to elastic elements in the muscle) and then adopts a middle position as atrophy occurs. This lack of tension produces hoarseness.

F. Phonation and speech

1. **Phonation** is accomplished by adducting the vocal cords sufficiently (lateral cricoarytenoid and transverse arytenoid muscles) to vibrate in a stream of exhaled air.
    - **a.** Tensing the cords (cricothyroid muscle) strengthens the vibrations and raises the pitch.
    - **b.** Effectively shortening the cords (vocalis muscle) also raises the pitch.
    - **c.** Sonic vibrations are transmitted to pharyngeal, oral, and nasal passages, as well as to the paranasal sinuses, producing resonance.

2. **Articulation** is accomplished by varying the resonance characteristics of the nasopharynx, oropharynx, and oral cavity to produce the vowel sounds. Interrupting the resonance produces consonant sounds.
    - **a.** Lingual and palatine musculature changes the shape of the oropharynx and intermittently limits access to the nasopharynx, altering resonant characteristics.
    - **b.** Mandibular, lingual, and facial musculature about the lips changes the shape of the oral cavity, also varying the resonant characteristics.
    - **c.** In combination, the tongue, palate, teeth, and lips interrupt resonance.

PART IVB   FACIAL CRANIUM AND VISCERAL NECK

STUDY QUESTIONS

DIRECTIONS: Each of the numbered items or incomplete statements in this section is followed by answers or by completions of the statement. Select the ONE lettered answer or completion that is BEST in each case.

1. The buccal area has a dual nerve supply. Which of the following statements concerning the two buccal nerves is true?

   (A) The fibers of one buccal nerve are associated with the geniculate ganglion, whereas those of the other are not associated with any ganglion
   (B) The fibers of one buccal nerve are associated with the submandibular ganglion, whereas those of the other are associated with the pterygopalatine ganglion
   (C) The fibers of one buccal nerve are associated with the trigeminal ganglion, whereas those of the other are associated with the geniculate ganglion
   (D) The fibers of one buccal nerve are associated with the trigeminal ganglion, whereas those of the other are associated with the pterygopalatine ganglion
   (E) The fibers of one buccal nerve are associated with the trigeminal ganglion, whereas those of the other are not associated with a ganglion

2. Anesthesia of the maxillary premolar teeth can be effected by infiltrating the nerve as it leaves the

   (A) foramen rotundum
   (B) greater palatine foramen
   (C) incisive canal
   (D) infraorbital foramen
   (E) lesser palatine foramen

3. A clinical manifestation of a fracture that passes through the left stylomastoid foramen and injures the contained nerve is

   (A) facial palsy
   (B) hyperacusis in the left ear
   (C) loss of lacrimation on the left side
   (D) loss of left parotid gland secretion

4. When cell bodies in the left trigeminal ganglion are damaged by a viral infection, the signs and symptoms include

   (A) inability to elicit a bilateral corneal blink reflex from the right side
   (B) left facial paralysis
   (C) loss of sensation of pain from the anterior two thirds of the tongue
   (D) loss of taste from the anterior two thirds of the tongue
   (E) weakness or paralysis of the left muscles of mastication

5. The articular disk, or meniscus, of the temporomandibular joint is correctly characterized by

   (A) a fibrocartilage composition
   (B) an attachment from the medial pterygoid muscle
   (C) an attachment for the temporalis muscle
   (D) separating medial and lateral joint compartments

6. Right-sided effects associated with the administration of local anesthetic into the right greater palatine canal far enough to reach the ganglion situated superiorly to the canal include

   (A) dry eye from loss of secretion of the lacrimal glands
   (B) dry mouth from loss of secretion of the parotid gland
   (C) dry mouth from loss of secretion of the submandibular and sublingual glands
   (D) loss of taste from the posterior one third of the tongue
   (E) loss of sensation from the anterior two thirds of the tongue

7. A physical examination reveals moderate swelling of the deep cervical nodes. Which of the following regions is not likely involved in the pathologic process responsible for such lymphadenopathy?

(A) Nasal sinuses
(B) Occipital region
(C) Parotid gland
(D) Tongue
(E) Tonsils

8. An elderly man has difficulty manipulating food within the oral cavity before swallowing. Physical examination reveals his tongue deviates to the left on protrusion. A CT scan demonstrates an extramedullary mass in the vicinity of the anterior condylar canal. Which of the following muscles remains functional with paralysis of the hypoglossal nerve?

(A) Genioglossus
(B) Hyoglossus
(C) Palatoglossus
(D) Styloglossus

9. During a physical examination of a patient with severe facial ache, an otolaryngologist notes drainage of purulent discharge into the left middle meatus when the patient is lying on the right side. The most likely source of the infection is the

(A) anterior ethmoid sinuses
(B) frontal sinus
(C) maxillary sinus
(D) nasolacrimal duct
(E) sphenoid sinus

10. Which of the following statements about the inferior laryngeal nerve is correct?

(A) It has an internal branch that conveys sensation from the larynx superior to the vocal cords
(B) It produces muscle contraction that lengthens the (true) vocal folds
(C) It conveys taste from the epiglottis
(D) It innervates all of the laryngeal musculature by an external branch except the cricothyroid muscle

DIRECTIONS: Each of the numbered items or incomplete statements in this section is negatively phrased, as indicated by a capitalized word such as NOT, LEAST, or EXCEPT. Select the ONE lettered answer or completion that is BEST in each case.

11. The tissues of the hard and soft palates receive an autonomic neural innervation that is described by all of the following statements EXCEPT

(A) preganglionic parasympathetic fibers travel along the greater superficial petrosal nerve, a branch of CN VII
(B) postganglionic sympathetic fibers arrive via the deep petrosal nerve
(C) the lesser superficial petrosal nerve contributes to the nerve of the pterygoid canal
(D) the greater and lesser palatine nerves pass through the pterygopalatine canal
(E) the anterior portion of the hard palate is supplied by the nasopalatine nerve, which enters the nose through the sphenopalatine foramen

12. All of the following statements correctly describe the epiglottis EXCEPT

(A) during swallowing, the epiglottis becomes horizontal to close the laryngeal aditus
(B) it contains taste buds innervated by the vagus nerve
(C) it is connected to the root of the tongue
(D) the piriform recesses lie on either side of it in the hypopharynx
(E) two lateral aryepiglottic folds connect it to the laryngeal cartilages

13. All of the following statements correctly pertain to the carotid sinus EXCEPT

(A) it is found in the bifurcation of the common carotid artery
(B) it is innervated by the vagus nerve
(C) it is a pressure receptor
(D) it regulates cardiac rate
(E) it regulates respiratory rate

14. An infection that spreads into the left pterygopalatine fossa from the left pterygoid venous plexus may subsequently track directly to all of the following spaces EXCEPT the

(A) left nasal cavity
(B) left orbit
(C) left maxillary sinus
(D) middle cranial fossa
(E) oral cavity

15. Stenosis of the first and second cervical intervertebral foramina results in weakness of all of the following muscles EXCEPT the

(A) geniohyoid
(B) mylohyoid
(C) omohyoid
(D) sternohyoid
(E) thyrohyoid

DIRECTIONS: Each set of matching questions in this section consists of a list of four to twenty-six lettered options (some of which may be in figures) followed by several numbered items. For each numbered item, select the ONE lettered option that is most closely associated with it. To avoid spending too much time on matching sets with large numbers of options, it is generally advisable to begin each set by reading the list of options. Then, for each item in the set, try to generate the correct answer and locate it in the option list, rather than evaluating each option individually. Each lettered option may be selected once, more than once, or not at all.

Questions 16–21

(A) CN V_1
(B) CN V_2
(C) CN V_3
(D) CN IX
(E) CN X

For each reflex, select the nerve that mediates its afferent limb.

16. Carotid

17. Corneal blink

18. Cough

19. Gag

20. Jaw-jerk

21. Sneeze

Questions 22–26

(A) First
(B) Second
(C) Third
(D) Fourth
(E) Sixth

For each muscle, select the branchial arch from which it is derived.

22. Orbicularis oris

23. Stylopharyngeus

24. Stapedius

25. Vocalis (thyroarytenoid)

26. Lateral pterygoid

Questions 27–29

(A) Anterior condylar canal
(B) Foramen ovale
(C) Jugular foramen
(D) Posterior condylar canal
(E) Stylomastoid foramen

To exit the cranial cavity, each cranial nerve passes through a foramen or fissure that may be affected by a cranial fracture. For each clinical manifestation of cranial nerve injury, select the foramen or fissure that is most likely involved in a cranial fracture.

27. The uvula deviates to the right when the palate is raised

28. The left eye cannot blink

29. The jaw deviates to the left upon protrusion

Questions 30–32

(A) Ciliary ganglion
(B) Geniculate ganglion
(C) Otic ganglion
(D) Trigeminal ganglion
(E) Submandibular ganglion

For each group of cell bodies, select the ganglion with which it is most appropriately associated.

30. Cell bodies for taste afferents

31. Cell bodies for the jaw-jerk reflex

32. Cell bodies for conjunctival sensation

ANSWERS AND EXPLANATIONS

1. **The answer is E** [Chapter 36 VII A 1 d (2) (c), B 4 b (2) (b); Table 32-1]. The buccal branch of the trigeminal nerve conveys sensation to the cheek and has its cell bodies located in the trigeminal (semilunar) ganglion. The buccal branch of the facial nerve provides motor innervation to the muscles of facial expression, including the buccinator muscle. These fibers originate in the facial nucleus of the brain stem and are not associated with a ganglion.

2. **The answer is A** [Chapter 36 VII B 3 C (6), d (3)]. The maxillary premolar teeth are innervated by the middle superior alveolar nerves. The maxillary premolars can be anesthetized effectively by infiltrating the maxillary nerve as it enters the pterygopalatine fossa through the foramen rotundum.

3. **The answer is A** [Chapter 36 VII A 1 b, 2 a, b]. The portion of the facial nerve that leaves the stylomastoid foramen provides motor innervation to the muscles of facial expression, so injury in this area causes facial paralysis only. The branch of the facial nerve to the stapedius muscle comes off higher in the facial canal, and the nerve to the lacrimal gland arises at the genu. The parotid gland is innervated by the tympanic branch of the glossopharyngeal nerve.

4. **The answer is C** [Chapter 32 IV C 1 a; Chapter 36 VII A 1 b (4) (b) (ii), 2 a, B 1]. The semilunar or trigeminal ganglion, the equivalent of a dorsal root ganglion, contains the cell bodies of the sensory neurons of the trigeminal nerve (CN V). Selective damage to these neurons results in loss of sensation from the face and anterior two thirds of the tongue. The motor neurons that innervate the muscles of mastication lie in the motor nucleus of V. The motor division of the facial nerve supplies the muscles of facial expression and taste from the anterior two thirds of the tongue.

5. **The answer is A** [Chapter 36 V A 1 b, 2 a–c]. This fibrocartilaginous articular disk separates the superior and inferior compartments. It receives the superior head of the lateral pterygoid muscle—the jaw protractor muscle. The inferior head of the lateral pterygoid muscle inserts primarily onto the mandibular condyle and neck. The medial pterygoid muscle inserts onto the mandibular at the angle and the temporalis muscle inserts onto the coronoid process.

6. **The answer is A** [Chapter 36 VII A 1 b, B 3 d (3)]. Injection of anesthetic into the pterygopalatine fossa, by the inferior approach through the palatine canal or by the lateral approach through the pterygomaxillary fissure, anesthetizes the maxillary division of the trigeminal nerve (CN V_2) as well as the nerves associated with the pterygopalatine ganglion. The postsynaptic parasympathetic neurons of the pterygopalatine ganglion control secretion of the nasal and oral mucosa as well as lacrimation. Branches of neither the mandibular nerve (sensation from the anterior one third of the tongue) nor the glossopharyngeal nerve (sensation and taste from the posterior two thirds of the tongue) would be affected.

7. **The answer is B** [Chapter 34 V D 1 a, b]. The lymphatic drainage of the anteroinferior portion of the face, the nasal cavities, and the anterior portion of the oral cavity, including the anterior margin of the tonsils, tongue, gingivae, and teeth, is through the submandibular lymph nodes to the deep cervical nodes. The lymphatic drainage of the occipital region, the external ear, the parotid gland, and the anterosuperior portion of the face is toward the superficial cervical lymph nodes.

8. **The answer is C** [Chapter 34 VI B 2; Chapter 37 III D 5 d]. The hypoglossal nerve (CN XII) innervates the intrinsic muscles of the tongue as well as the styloglossus, genioglossus, and hyoglossus muscles. The palatoglossus muscle is innervated by the branchiomeric motor division of the vagus nerve (CN X). Thus, a brain stem lesion in the vicinity of the hypoglossal nucleus would not affect the palatoglossus innervation.

9. The answer is C [Chapter 37 II A 4 c]. The maxillary, frontal, and anterior ethmoid sinuses, as well as the middle ethmoid sinus, drain into the hiatus semilunaris within the middle nasal meatus. Because the ostium of the maxillary sinus is high, it does not drain by gravity except when lying on the opposite side. The sphenoid and posterior ethmoid sinuses drain into the sphenoethmoid recess. The nasolacrimal duct drains into the inferior meatus.

10. The answer is D [Chapter 37 V E 2 d (3)]. The recurrent (inferior) laryngeal nerve, a branch of the vagus nerve (CN X), innervates all but one of the muscles of the larynx. The cricothyroid muscle, which lengthens the vocal cords, is innervated by the external branch of the superior laryngeal nerve. Taste from the epiglottic taste buds is conveyed by the superior laryngeal branch.

11. The answer is C [Chapter 36 II C 1 a; VII A 1 b]. The lesser superficial petrosal nerve, a continuation of the tympanic branch of the glossopharyngeal nerve, conveys parasympathetic preganglionic fibers to the otic ganglion for parotid secretion. The greater superficial petrosal nerve, a branch of the facial nerve, and the deep petrosal nerve convey the parasympathetic presynaptic and sympathetic postsynaptic fibers, respectively, to the pterygopalatine fossa.

12. The answer is D [Chapter 37 IV D 2, 3; V B 2]. The piriform recesses lie inferior to the epiglottis and caudal to the lateral glossoepiglottic folds, which connect the epiglottis to the tongue. The vallecular recesses lie to either side of the epiglottis.

13. The answer is B [Chapter 37 IV F 1 b (2) (b)]. The carotid body and sinus are located at the bifurcation of the common carotid artery. The carotid body (a chemoreceptor) monitors the partial pressure of the dissolved oxygen in the blood; the carotid sinus, a pressure receptor, monitors blood pressure. Both are innervated by the carotid branch of the glossopharyngeal nerve, which functions as the afferent limb of a reflex that controls the heart and respiratory rates.

14. The answer is C [Chapter 36 IV B]. The branches of the nerves and vessels of the pterygopalatine fossa reach the nose, eye, and mouth through several foramina. The pterygopalatine fossa communicates with the middle cranial fossa via the foramen rotundum and by the vidian canal; with the nasal cavity via the sphenopalatine foramen; with the orbital cavity via the inferior orbital fissure; and with the oral cavity via the palatine canal. The maxillary sinus communicates with the nasal cavity; any involvement by this spread of infection would be by an indirect route.

15. The answer is B [Chapter 34 VI B 2]. The superior ramus of the ansa cervicalis, arising from C1 and accompanying the hyoglossal nerve, innervates the geniohyoid and thyrohyoid muscles. The inferior ramus, arising from C1 and C3, innervates the omohyoid and sternothyroid muscles. The mylohyoid, along with the anterior belly of the digastric muscle, are innervated by the mylohyoid nerve, a branch of the mandibular nerve.

16–21. The answers are: 16-D, 17-A, 18-E, 19-D, 20-C, 21-B
[Chapter 36 VII B 2 c (2), 3 d (1), 4 b (2) (a); Chapter 34 V B 1 c (1) (a); Chapter 37 IV F 1 b (1) (a), (2) (b); V E 2 c (1) (b)]. The trigeminal nerve mediates three reflexes. The ophthalmic division provides the afferent limb of the corneal blink reflex; the maxillary division, the afferent limb of the sneeze reflex; and the mandibular division, both limbs of the jaw-jerk reflex. The glossopharyngeal nerve provides the afferent limbs of the gag and carotid reflexes. The vagus nerve provides the afferent limb of the cough reflex.

22–26. The answers are: 22-B, 23-C, 24-B, 25-E, 26-A [Chapter 35 I B 1–3]. The muscles of mastication, including the lateral pterygoid muscle, as well as the tensors tympani and veli palatine, mylohyoid, and posterior belly of the digastric, are derived from the first branchial arch and are innervated by the mandibular division of the trigeminal nerve (CN V). The muscles of facial expression, including the orbicularis oris, as well as the stapedius, stylohyoid, and posterior belly of the digastric are second branchial arch derivatives and thus innervated by the facial nerve (CN VII). The stylopharyngeus is the only muscle innervated by the glossopharyngeal nerve (CN IX), the nerve of the third branchial arch. The cricothyroid muscle, innervated by the superior laryngeal nerve (a branch of the vagus) is a fourth arch derivative. The vocalis (thyroarytenoid) muscle and the intrinsic laryngeal musculature (except the cricothyroid muscle) are sixth branchial arch derivatives innervated by the recurrent laryngeal nerve (a branch of the vagus).

27–29. The answers are: 27-C, 28-E, 29-B
[Chapter 36 VII A 2 b, B 4 d; Chapter 37 III B 3 a (3); D 3 a; Table 32-1]. The glossopharyngeal (CN IX), vagus (CN X), and spinal accessory (CN XI) nerves exit the cranial cavity via the jugular foramen. Injury to the vagus nerve at this point results in deviation of the uvula away from the injured side, because of the unopposed action of the contralateral levator veli palatini muscle as well as hoarseness from paralysis of the intrinsic laryngeal musculature. A fracture involving the stylomastoid foramen can injure the facial nerve with paralysis of the muscles of facial expression on the injured side. Injury to the mandibular division of the trigeminal nerve as it passes through the foramen ovale would paralyze the muscles of mastication with deviation of the jaw toward the injured side due to the unopposed action of the lateral pterygoid muscle. The hypoglossal nerve passes through the anterior condylar canal; a large emissary vein is the only structure passing through the posterior condylar canal.

30–32. The answers are: 30-B, 31-D, 32-D
[Table 32-1]. Taste afferents, which run from the anterior two thirds of the chorda tympani, have their cell bodies in the geniculate ganglion. The trigeminal ganglion contains afferent cell bodies that innervate the face and, therefore, provide the afferent limbs of the blink, sneeze, and jaw-jerk reflexes. The ciliary pterygopalatine, submandibular, and otic ganglia contain only postsynaptic neurons of parasympathetic motor pathways.

Index

Note: Page numbers in *italic* indicate illustrations; those followed by (t) indicate tables. Q and A denote questions and answers.

A

Abdominal aorta, 314, 395
Abdominal cavity, 329
Abdominal esophagus, *344,* 344–346
Abdominal expiration, 256, 318
Abdominal hernias, 318, 322
Abdominal incisions, 315
Abdominal muscle, 317
Abdominal reflex, 325, 398
Abdominal rigidity, 399Q, 404E
Abdominal wall, 401Q, 405E
  arterial supply, *319,* 319–320
    lymphatic drainage, 320
    venous return, 320
  clinical considerations, 321–322
    access to peritoneal cavity, 321
    hernias, 322
  definition, 313
  fixation of, 318
  fossae, 340–341
  inguinal region, 322–323
    canal, *323,* 323–324
    femoral ring, 327–328
    nerves of, 325–326
    spermatic cord, 324–325, *325*
    triangle, *326,* 326–327
  innervation
    intercostal nerves, 320–321
    lumbar nerves, 321
  internal structures, 313–314
  lymphatic channels, 396–397
  musculature
    actions, 316t, 318
    organization, 315, *316,* 316t, 317
    rectus sheath, *316,* 318
  nerves, 397–398
  skin and fasciae
    deep fascia, 315
    integument, 314–315
    superficial fascia, 315
    transversalis fascia, 315
  structure, *393,* 393–394
  subdivisions, 313, *314*
  surface landmarks, 313, *314*
  vessels, *390,* 395–396
Abdominocentesis, 330
Abducens nerve, 526, 533, 544, 547, 550, 568Q, 571E, 585
  palsy of, 566Q, 570E
Abducens nucleus, 533, 550
Abduction/adduction, 5, 19, 46Q, 49E
  of carpometacarpal joint, 103
  of fifth digit, 115
  of glenohumeral joint, 122Q, 127E
  of hip joint, 166, 170, 171t, 172, 172t
  of metacarpophalangeal joint, 104, 113
  of radiocarpal joint, 87
  of thumb, 114
  of ulnocarpal joint, 88
Abductor digiti minimi, 228
Abductor hallucis brevis, 227
Abductor lurch, 417
Abductor pollicis longus muscles, 124Q, 126Q, 128E, 129E
Abscess, of tooth, 595
Accessory breasts, 13
Accessory muscles of respiration, 247
Accessory oculomotor nucleus, 550
Accommodation, control of, 555
Acetabular fossa, 163
Acetabular labrum, 163
Acetabular notch, 163
Acetabulum, *162,* 163, 165, 407
Acetylcholine, 32, 33
Achalasia, 297
Achilles' tendon, 202
Acoustic information, 529
Acoustic neuroma, 613
Acquired umbilical hernia, 322
Acromioclavicular joint, 56
  movement, 58
  structure, 58
  support, 58
Acromioclavicular ligament, 58
Acromioclavicular subluxation, 58
Acromion process, 57
Acute pancreatitis, 362
Addison's disease, 392
Adductor brevis muscle, 189–190, 197
Adductor canal, 194–195
Adductor hallucis muscle, 227
Adductor longus muscle, 189, 197
Adductor magnus muscle, 190
Adductor pollicis muscle, 125Q, 126Q, 129E
Adductor tubercle, 179
Adenohypophysis, 525
Adrenal arteries, 392
Adrenal cortex, 391
Adrenal glands, 403Q, 406E
  clinical considerations, 392
  location, 391
  structure, 391–392
  vasculature, 392
Adrenal medulla, 33, 383, 392
Adrenal vein, 392
Adrenalectomy, 392
Adrenergic neurotransmission, 498
Adrenergic sympathetic activity, 286
Adrenergic system, 32
Afferent innervation, 352
Afferent limb, 484
  of cremaster reflex, 398
Afferent pathways, 33

Aganglionosis, 382, 475
Air embolism, 541
Airway, 269
Alae, 408
Alar plates, 407, 409
Aldosterone, 391–392
Alexia, 523
Allantoic stalk, 332
Allantois, 453
Alpha motoneurons, 151
Alveolar artery, 608
Alveolar foramina, 604
Alveolar nerve, 617–618, *618*
Alveolar processes, 590
Alveolar respiration, 269
Alveolar ridges, 592
Alveolar-capillary block, 270
Amphiarthrosis, 18, 48Q, 50–51E, 131, 132, 139, 165
Ampulla, 458, 480, 482Q, 485E
Anal canal, 375, 419, 480
Anal columns, 422
Anal fistulas, 423
Anal glands, 422
Anal hemorrhoids, 424
Anal incontinence, 424
Anal membranes, 419
Anal sinuses, 422
Anal sphincter, 422, 423–424, 425, 482Q, 485E
Anal triangle, 419
Anal valves, 422
Anastomoses, 277, 582
  arterial, 38
  lymphatic, 416
Anatomic dead space, 257
Anatomic principal, 7
Anatomic snuffbox, 106
Anatomic terminology
  adjectives, 3–4, *4*
  movements, *5,* 5–6
  planes, *4,* 4–5
  position, 3, *4*
  vocabulary, 6–7
Anconeus muscle, 77
Anesthesia
  caudal, 155, 471
  epidural, 149, 155, 472
  spinal, 26, 155, 472
Aneurysms, 195, 396
Angina pectoris, 277, 306Q, 310E
Anginal pain, 306Q, 310E
  intestinal, 376
Angular vein, 552
Ankle and tarsal joints, 162, *202, 204*
  conjoined movements of foot, 206
  distal tibiofibular, *202,* 203
  subtalar, 204–205
  talocrural, 203–204, *204*
  transverse, 205–206

653

Ankle-jerk reflex, 221
Annular ligaments, 75, 110
Annular pancreas, 363
Annulus fibrosus, 139, 273, 278
Annulus tendineus, 547
Anococcygeal raphe, 412
Ansa cervicalis, 496, 576, 583, 584
Antagonistic muscles, 21
Antebrachial cutaneous nerve, 70, 102
Antebrachial fascia, 93–94, 105, 107
Antebrachial vein, 100
Antebrachium, 56
Anterior (ventral), 3
Anterolateral fontanelles, 504, 506
Anteroposterior axes, 5
Antihelix, 557
Antrum of ear, 558–559, 620
Anus, 422
Aorta, 253, 313, 390
　abdominal, 314
　ascending, 294
　descending, 294
Aorta valve, 283
Aortic arch, 294
Aortic body, 269
Aortic hiatus, 298, 394, 395
Aortic insufficiency, 283
Aortic nodes, 396
Aortic plexus, 474
Aortic pulse, 313, 396
Aortic semilunar valve sounds, 289
Aortic sinuses, 283
Aortic stenosis, 253, 283
Aortic vestibule, *281, 282,* 282–283
Aorticorenal ganglia, 383, 390
Aorticorenal plexus, 457
Apex of heart, 305Q, 310E
Aphasia, motor, 522
Apical pulse, 273
Aponeurosis, 20, 106
Appendices epiploicae, 368
Appendicitis, 370
Appendicular artery, 370
Appendicular pain, 370
Appendix, 334, 403Q, 406E
　vermiform, *369,* 369–370
Appendix testis, 455
Aqueous humor, 556
Arachnoid, 25, 149–151, *150*
Arachnoid trabeculae, 517
Arachnoid villi, 565Q, 569E
Arch of Riolan, 377
Arcuate artery, 212
Arcuate eminence, 512
Arcuate ligament, 393, 433
Areola, 13
Arm, 56
　movement of, 60, *63,* 63–65, 64*t*
Arterial anastomoses, 38
Arterial circulation, 274, 275–276
Arterial flow to forearm, 123Q, 127E
Arterial pulse, 38
Arterial system, 37–38
　arterioles, *36,* 38
　clinical considerations, 38
　conducting arteries, 37
　distributing arteries, 37–38
Arteries, 35. *(See also specific arteries)*
Arterioles, *36,* 38
Arteriovenous anastomoses, 41
Arteriovenous shunt, 270
Arteritis, carotid, 552
Arthritis
　degenerative, 18, 182
　osteo-, 167, 184

Articular capsule, 605
Articular cartilage, 17, 18
Articular disk, 505, 645Q, 649E
Articular processes, 132, 133
Articular tubercle, 505
Articulations, 17–19, *18*
　with forearm, 73
　of shoulder girdle, 58–60
　speech, 644
Aryepiglottic fold, 638
Arytenoid cartilages, 639
Arytenoid muscle, 641, 642
Ascending aorta, 294
Ascending cervical artery, 498
Ascending colon, 314, 334, 367–372
Ascending mesocolon, 333
Ascites, 43–44, 330
Aspiration pneumonia, 264
Asthma, 269
Ataxia, 526
Atebrachial cutaneous nerve, 102
Atelectasis, 266
Atlantoaxial joint, 134, 158Q, 160E
Atlanto-occipital joint, 134, 158Q, 160E
Atlas, 156Q, 159E
Atrial diastole, 287
Atrial septal defects, 280, 292
Atrial systole, 286–287
Atrioventricular bundle, 274, 284
Atrioventricular insufficiency, 289
Atrioventricular nodal artery, 275
Atrioventricular node, 278, 283–284
Atrioventricular stenosis, 289
Atrium
　left, 274
　right, 274
Atrophy, 21
Attachments sites, 20
Auditory association area, 523
Auditory meatus, 505, 558, 597
Auditory ossicles, *559,* 560–561
Auditory tube, 507, 559, 630–631
Auricles, 557, 597
Auricular artery, 580, 608
Auricular muscle, 503, 602
Auricular nerve, 503
Auricular surface, 407
Auricular vein, 499
Auricularis muscles, 502
Auriculotemporal nerve, 503, 617
Auscultation, 243
Autonomic motor system, 23
Autonomic nervous system (ANS), 24, 300–301
　composition, *31,* 31–32
　divisions, 32–33
　visceral efferent pathway, *31,* 32
Avascular necrosis, 86, 87, 234Q, 238E
Axillary artery, 67–68
Axillary lymph nodes, 13, 80
Axillary nerve, 82–83, *83,* 125Q, 129E
Axillary pulse, 67
Axillary syndrome, 71
Axillary tail of breast, 13
Axillary vasculature
　axilla, 65
　vascular supply, 65
Axillary vein, 68, 122Q, 127E, 320
Axis, 5
　of rotation, 19
Axon, 23
　myelinated, 32
　unmyelinated, 32
Azygos vein, 271, 293, 298, 345, 379

B

Back, soft tissues of
　anterior vertebral muscles, 144, 146, 147*t*, 148
　functional considerations, 148
　posterior vertebral muscles, 143, *144*
　　innervation, 144
　　spinotransverse group, 143, *144,* 145*t*
　　suboccipital muscles, 144, 146*t*
　　transversospinalis group, 143–144, *144,* 145*t*
Back pain, 409
Ballism, 525
Bartholin's cysts, 436, 442
Bartholin's duct, 630
Bartholin's glands, 436
Basal ganglia, 520
Basicranial anastomotic circle, 539–540
Basilar artery, 539, 565Q, 570E
Basilic vein, 68, 80, 100, 117
Batson's plexus, 26
Bell's palsy, 545
Benign prostatic hypertrophy, 460
Berry's aneurysm, 540
Biceps brachii, 63, 76, 77
Bicipital aponeurosis, 94
Bicornuate uterus, 473
Bicuspid teeth, 594
Bilateral axes, 5
Bile, storage of, 359
Bile canaliculi, 356
Bile concentration, 359
Bile duct, 356–357
Bile ductules, 356
Bile peritonitis, 359
Bile pigments, 356
Bile salts, 356
Bile secretion, 356
Biliary obstruction, 356, 360
Biliary overload, 356
Biliary pain, 358, 360
Bipennate muscles, 20
Birth canal, *408,* 409–410, 410*t,* 479Q, 484E
Bitemporal heteronymous hemianopsia, 529
Bladder, neurogenic, 475
Blepharoptosis, 545
Blindness, 528
　cortical, 529
Blinking, 546, 551, 553, 567Q, 571E, 602, 613, 615
Blood clot, 46Q, 50E
Blood pressure, 38
Blood-air barrier, 269
Bloodless line of kidney, 348
Bloodshot eye, 546
Body of uterus, 481Q, 485E
Bones, 15. *(See also* Fractures; *specific)*
　healing, 17
Bony cochlea, 562
Bony orbit, 543–544, *544,* 548
Borborygmi, 314, 366
Bowel, 314
Bowel sounds, 314
Brachial artery, 67, 123Q, 127E
Brachial cutaneous nerve, 82
Brachial plexus, 122Q, 127E
　development, 68
　lesions, 71
　subdivisions, 69–71
Brachial veins, 68
Brachialis muscle, 76
Brachiocephalic artery, 294

Brachiocephalic veins, 293–294, *294*, 582
Brachioradialis muscle, 76, 77, 124Q, 128E
Brachium, 56
Brain, 23
  cerebrum
    basal ganglia, 524
    cerebral hemispheres, *520*, 520–524
    composition, 519
  cranial meninges, 25
  cranial nuclei and associated nerves, *527*, 527–533, 534–536*t*
  functions, 25
  meninges, 519
    cranial nerves, 519–520
  morphology, 24–25
  stem, 24, 152, 513, 519, 525
    hindbrain (metencephalon and myelencephalon), 526–527
    midbrain (mesencephalon), 525–526, *526*
    thalamus (diencephalon), *524*, 525
  substructure, 519
  vasculature, 538–541
  ventricular system
    cerebrospinal fluid, 537–538
    structure, 537, *538*
Branchiomeric nerves, 531–532, 585
Branchiomeric sensation, 638
Breast, *12*, 12–13
  lobes, 45Q, 49E
  lymphatic drainage from, 45Q, 49E
Bregma, 506
Bronchial arteries, 298, 345
Bronchial tree
  clinical considerations, 266
  lobar bronchi, *263*, 266
  main stem bronchi, *265*, 265–266
  segmental bronchi, *263*, 266
  trachea, 265, *265*
Bronchiectasis, 266, 270
Bronchomediastinal lymph nodes, 268, 271, 300
Bronchomediastinal trunks, 499
Bronchopulmonary segments, 264, 264*t*, 305Q, 309–310E
Bronchoscopy, 305Q, 309E
Brown-Sequard syndrome, 153
Bruit, 38
Buccal area, 645Q, 649E
Buccal nerve, 617
Buccinator muscle, 600, 626
Buccopharyngeal membrane, 330
Bucket-handle effect, 255
Bulbar arteries, 434
Bulbar conjunctiva, 546
Bulbocavernosus muscles, 431–432, 433*t*, 437
Bulbospongiosus muscle, 426, 431, 439
Bulbourethral (Cowper's) glands, 428, 433, 455, 480Q, 482Q, 484E, 486E
Bursa, *12*, 59
  midpalmar, 95
  subacromial, 59
  thenar, 95
Bursitis, 73

C

Calcaneal spur, 203
Calcanean tendon, 202
Calcaneocuboid joint, 203, 206
Calcaneofibular ligament, 204
Calcaneonavicular joint, 205
Calcaneonavicular ligament, 202, 236Q, 239–240E
Calcaneus, 202
Calcarine fissure, 521
Calcification, connective tissue, 17
Calcitonin, 578
Callus formation, 17
Calvaria, 501, 509
Camper's fascia, 11, 47Q, 50E, 315
Cancer. (*See* Carcinoma)
Canine teeth, 593, 594
Canthi, 597
Canthus, 544
Capillaries, 35, *36*, 38–39
  lymph, 42–43
Capillary beds
  capillaries, *36*, 38–39
  clinical considerations, 40
  sinusoids, 39–40
Capitate bone, 93
Capitulum of radius, 73
Capsular tear, 59
Capsule, internal, 521
Caput medusae, 320, 379
Carcinoma
  esophageal, 402Q, 405E
  lung, 265
  lymphatic metastatic, 402Q, 406E
  ovarian, 467
  pancreatic, 348, 363
  prostate, 481Q, 485E
  rectal, 482
Cardia of stomach, 346
Cardiac accelerator nerves, 498
Cardiac auscultation areas, 289, *289*
Cardiac chambers
  conducting system, 283–285, *284*
  left atrium, 281–282
  left ventricle, 282
  right atrium, *279*, 279–280
  right ventricle, 280–281
Cardiac control, 36
Cardiac dynamics, 35
  cycle, 286, *287*
    atrial diastole, 287
    atrial systole, 286–287
    auscultation areas, 289, *289*
    conduction velocity, 288
    ejection volume, 288
    valve locations, 288, *289*
    valvular defects and disease, 289–290
    ventricular diastole, 288
    ventricular systole, 288
  fetal and early postnatal circulation, *290*, 290–292
Cardiac failure, 290
Cardiac incisure of stomach, 263
Cardiac innervation, motor control
  parasympathetic division, *285*, 285–286
  sympathetic division, 286
Cardiac ligament of stomach, 272
Cardiac notch of stomach, 260
Cardiac plexus, 301, 303, 308Q, 311E
Cardiac reflex, 579–580
Cardiac sphincter, 296, 298, 347
Cardiac structure, external, *272*, 273–275
Cardiac wall
  endocardium, 278
  epicardium, 278
  fibrous ring, 278–279
  myocardium, 278
Carotid arteries, 294, 503, 538, 579, 607, *607*, 647Q, 650E
Carotid arteritis, 552
Carotid bifurcation, 579
Carotid body, 580
  and sinus, 269
Carotid canal, 508, 512, 580
Carotid hemostasis, 580
Carotid pulse, 579
Carotid sheath, 564Q, 569E, 602
Carotid sinus, 579, 646Q, 650E
Carotid triangle, 574
Carpal bones, 85–87
Carpal radiate ligament, 93
Carpal tunnel syndrome, 86, 93, *94*, 95, 102, 109, 110, 112, 119
Carpometacarpal joints, 103–104, *104*
Carpus, 56
Cartilage, 17
Cataract, 556
Cauda equina, 26, 149, 153
Caudal anesthesia, 137, 155, 471
Caudal pancreatic arteries, 361, 362, 375
Caudate lobe of liver, 339
Caudate nucleus, 524
Caval hiatus, 394
Cavernous plexuses, 474
Cavernous sinus, 516, 552, 566Q, 570E
Cavernous sinus thrombosis, 609–610
Cecal arteries, 369
Cecal fossae, 368–369
Cecum, 334, 368
Celiac artery, 330, 343, 347, 375–376
Celiac ganglia, 297, 348, 352, 383
Celiac plexus, 348, 381
Celiac trunk, 403Q, 406E
Celiac veins, 378
Cementum, 594
Central, 4
Central gray area, 151
Central lobe, 523
Central nervous system (CNS), 23
  brain
    cranial meninges, 25
    functions, 25
    morphology, 24–25
  spinal cord
    meninges, 26
    morphology, 25–26
Central neurons, 30
Central sensory neuron, 48Q, 51E
Central space, 109
Central sulcus, 521
Central venous lines, 495
Cephalic vein, 68, 80, 99–100, 117, 122Q, 127E
Cerebellum, 24, 519, 526
Cerebral cortex, projections of, 523
Cerebral hemorrhage, 540
Cerebral vascular accident, 565Q, 570E
Cerebral vein, 565Q, 569–570E
Cerebrospinal fluid (CSF), 25, 150, 537–538, 565Q, 569E
Cerebrum, 24, 519
  basal ganglia, 524
  cerebral hemispheres, *520*, 520–524
Cerumen, 557
Cervical and thoracic splanchnic nerves, 306Q, 310E
Cervical artery, 581
Cervical curvature, 131
Cervical enlargement, 25
Cervical ganglia, 497, 566Q, 570E
Cervical intervertebral foramina, 647Q, 650E

Cervical lymph nodes, 499, 646Q, 649E
Cervical nerve, 583
Cervical nerves, 28
Cervical pleura, 259
Cervical plexus, 564Q, 569E, 583–584
Cervical rib, 246
Cervical spaces, 573
Cervical splanchnic nerves, 286, 301
Cervical sympathetic nerves, 497–498
Cervical triangle
 fascia, 573
 divisions, 489–491, *490*
 hyoid bone, 573–574, *576*
 innervation, 582–584, *583*
 musculature, *574*, 574–577
 parathyroid glands, 578
 somatic neck
  musculature, 491–492, *492, 493, 494*, 494–495
  triangle, 491
  vertebrae, 491
 thyroid gland, *574*, 577–578
 thyroid vasculature, 578
 vasculature, *579*, 579–582, *581*
Cervical vein, 581
Cervical vertebrae, 133–135, *134*
Cervical vertebral column, 564Q, 569E
Cervix, 468
Cesarean section, 479Q, 484E
Check ligament, 549
Cheeks, 626
Chemical reflex, 269
Choanae, 506, 619
Cholecystitis, 360, 403Q, 406E
 with jaundice, 360
 without jaundice, 360
Cholecystokinin, 351, 359, 360
Choleliths, 359–360
Cholinergic nerves, 33
Cholinergic neurotransmission, 382
Cholinergic transmission, 382, 474, 497
Chorda tympani nerve, 508, 558, 560, 585, 611, 629–630
Chordae tendineae, 280, 282
Choroid layer, *553*, 554
Choroid plexus, 537
Chronic back pain, 139
Chronic sphenoid sinusitis, 622
Chyle, 42
Chylothorax, 43, 262, 300
Chyluria, 43
Chyme, 349
Ciliary body, 554
Ciliary muscle, *554*, 554–555
Ciliary nerve, 551, 614
Circulation, 35
Circulatory system, 35, *36*
 arterial
  arterioles, *36*, 38
  clinical considerations, 38
  conducting arteries, 37
  distributing arteries, 37–38
 capillary beds
  capillaries, *36*, 38–39
  clinical considerations, 40
  sinusoids, 39–40
 heart
  cardiac pump, 35–37, *36*
  coronary circulation, 36
  fetal circulation, 36–37
 lymphatic
  clinical considerations, 43–44
  composition, 41
  function, 41–42

lymph flow, 43
structure, 42–43
venous
 arteriovenous anastomoses, 41
 capillary beds, 40
 portal, 41
 veins, *36*, 40–41
Circumcision, 426
Circumduction, 6, 19, 46Q, 49E, 166
 of hip joint, 173
 of metacarpophalangeal (MP) joint, 104
 of radiocarpal joint, 87
 of ulnocarpal joint, 88
Circumflex artery, 276
Cisterna chyli, 42, 299, 380, 396
Cisterna magna, 565Q, 569E
Clavicle, 58, 243
 fracture, 56–57
 location, 56
 ossification, 56
Clavipectoral nodes, 117–118
Claw-hand, 121
Cleavage (Langer's) lines, 315
Cleft palate, 626
Clinical anatomy, 3
Clitoral erection, 483Q, 486E
Clitoris, 421, *436*, 436–437, *438*
Clivus, 512
Cloacal folds, 419
Cloacal membrane, 331, 419
Coarctation of aorta, 253, 396
Coccygeal cornua, 138
Coccygeal ligament, 149, 151, 156Q, 159E
Coccygeal plexus, 417
Coccygeal vertebrae, 138
Coccygeus muscle, 411
Coccygodynia, 138, 409
Coccyx, 158Q, 160E, 409
Cochlear apparatus, 562–563, *563*
Cochlear nerve, 562, 567Q, 570–571E
Cochlear nuclei, 529
Coelom, 259
Coelomic cavity (coelom), 331
Colic artery, 371, 372, 376, 401Q, 405E
Colic flexure, 371
Colic nodes, 371
Colic vessels, 402Q, 405E
Collagenous fibers, 17
Collagenous matrix, 15
Collateral ligament, 204, *204*
 injury, 186
Collateral ligaments, 204, *204*
Collecting ducts, 386, 389
Colles' fascia, 11, 430, 431, 437–438
Colles' fracture, 85
Colliculus of brain, 526
Colliculus seminalis, 452
Colon, 314
Commissure, 435
Commissurotomy, 307Q, 311E
Common fibular nerve, 213, 236Q, 240E
Common hepatic artery, 347
Common iliac arteries, 314, 396, *413*, 413–414
Common iliac nodes, 396
Compressor naris muscle, 600
Concentration, bile, 359
Conception, 477
Concha, 557
Conchae, 619
Conduction velocity, 288
Condylar canal, 509, 567Q, 571E
Condylar notch, 505

Condyle, 179, *180*, 181, 604
Condyle osteotomy, 123Q, 127E
Congenital absence, 467
Congenital anomalies of penis, 421
Congenital hypertrophic pyloric stenosis, 347
Congenital torticollis, 495
Congenital umbilical hernia, 322
Congestive heart failure, 271, 290, 304Q, 309E
Conjoined movements of foot, 206
Conjoined tendon, 324
Conjunctiva, 546
 inflammation of, 553
Conjunctival sac, 544, 597
Connective tissue, 47Q, 50E
 calcification, 17
Conn's syndrome, 392
Conoid ligament, 58
Continence
 fecal, 412
 urinary, 412, 484–485
Contraction, 20
Contralateral homonymous hemianopsia, 529, 565Q, 570E
Contralateral homonymous visual field defects, 523, *528*
Contusion of brain, 517
Conus medullaris, 25, 149
Coracoacromial arch, 59
Coracoacromial ligament, 57, 59, 122Q, 124Q, 127E, 128E
Coracobrachialis muscle, 63
Coracoclavicular ligament, 58
Coracohumeral ligament, 59
Coracoid process, 57, 58
Cornea, *553*, 553–554, 567Q, 571E
Corneal blink reflex, 647Q, 650E
Corneal epithelium, 554
Corneal refraction, 554
Corneal transplantation, 554
Coronal planes, 5
Coronal suture, 504
Coronary, 235Q, 239E
Coronary (atrioventricular) sulcus, 273
Coronary bypass, 253
Coronary circulation, 36, *275*, 275–277
 arterial, 275
 venous return, 277–279
Coronary ligaments, 331
Coronary obstruction, 277
Coronary perfusion, 277
Coronary sinus, 277–278, 279
Coronary sulcus, 275
Coronoid fossa, 73
Coronoid process, 73
Corpora albicantia, 466
Corpora cavernosa, 426, 427, 431, 437, 438
Corpora lutea, 466
Corpus callosum, 521, 523
Corpus spongiosum, 426, 427, 428, 452, 483Q, 486E
Corrugation of skin over thenar eminence, 115
Corrugator supercilii muscle, 602
Cortex of adrenal gland, 386
Cortical blindness, 529
Cortical motor area, 521
Cortical sensory area, 521
Corticobulbar pathways, 524
Corticobulbar tract, 152
Corticospinal tract, 152
Costal angle, 245
Costal articulations, 246
Costal cartilages, 245–246, 246

Costal fractures, 252
Costal groove, 245
Costal pleura, 259, 260
Costal processes, 133
Costal ventilation, 257
Costocervical artery, 67
Costochondritis, 246
Costomediastinal recess, 261
Cothyroid muscle, 646Q, 650E
Cough reflex, 643, 647Q, 650E
Counter current heat exchanger, 457
Coxal bone, 407–408, *408*
Cranial and cervical viscera
  basic structure, 619
  innervation, 621t, 621–622
  larynx, 619
    innervation, 642–644
    laryngeal cartilages (ossicles), 638–639
    laryngeal movements, 639–642
    phonation and speech, 644
    superficial structure, *637*, 637–638, *639*
    vasculature, 642
  mucous membranes, 620
  nasal cavity, 619–620
  oral cavity, 619, 623, *624*
    basic structure, 623
    palate, 623–626, *631*
    submandibular duct, 630
    submandibular gland, 630
    tongue, 626–630
    walls of, 626
  paranasal sinuses, 622–623
  pharynx, 619
    basic structure, 630
    deglutition, 636–637
    hypopharynx, 633
    innervation, *635*, 635–636
    musculature, *632*, 633–635, 634t
    nasopharynx, 630–631
    oropharynx, *631*, 631–633
    vasculature, *620*, 620–621
Cranial fossa, 510, 610
Cranial fracture, 537, 566Q, 570E
Cranial nerves, 23, 501, 516, 519–520
Cranial nuclei and associated nerves, *527*, 527–533, 534–536t
Cranial parasympathetic outflow, 33
Cranial sutures, 48Q, 50–51E
Craniotomy, 565Q, 569–570E
  emergency, 518
Cranium, 501
Cremaster artery, 456
Cremaster muscle, 317, 429
Cremaster reflex, 197, 326, 399Q, 404E, 429, 480Q, 484E
  afferent limb of, 398
  efferent limb of, 398
Cribriform fascia, 195
Cribriform plate, 510, 565Q, 569E, 588, 589
Cricoarytenoid joints, 640
Cricoarytenoid muscle, 640–641, *641*, *641*, 642
Cricoid cartilage, 639
Cricopharyngeal muscle, 296
Cricopharyngeus muscle, 297, 635
Cricothyroid joints, 639, 640
Cricothyroid muscle, 640, 642, *642*, 643
Crista ampullaris, 562
Crista galli, 510, 589
Crista terminalis, 279, 280
Crista urethralis, 452
Crocodile tears, syndrome of, 613

Crown of tooth, 594
Cruciate anastomosis, 195
Cruciate ligament, 110, 185–186
  injury, 186
Cruciform ligament, 141
Crural compartments
  fascia, 206
  musculature, 206–208
Crural fascia, 209
Crus of ear, 557
Crutch palsy, 84
Cruveilhier's fascia, 11, 429–430, 437
Cryptorchidism, 323
Cubital vein, 80, 100, 122Q, 127E
Cuboid bone, 203, 223
Culdocentesis, 473
Cuneocuboid joint, 223–224
Cuneonavicular joint, 223–224
Cupula, 263
Cushing's syndrome, 392
Cusp of aortic valve, 282
Cuspid teeth, 594
Cutaneous nerves, 495
Cyanosis, 262
Cyst, 461
Cystic artery, 376
Cystic veins, 359
Cystocele, 453, 473
Cystohepatic triangle, 358, 359
Cystoscopy, transurethral, 486

D

Danger space, 503
De Quervain's syndrome, 88
Deep circumflex iliac artery, 319–320
Deep dorsal fascia, 105
Deep fascia, 11–12, 47Q, 50E, 206
Deep femoral artery, 195
Deep fibular nerve, 213
Deep lymphatics, 320
Deep palmar fascia, 108–109
Deep plantar artery, 212
Deep posterior compartment syndrome, 216, 221
Deferential artery, 415, 458
Deferentitis, 461
Degenerative arthritis, 18, 182
Deglutition, 636–637
Deltoid muscle, 63
Deltopectoral triangle, 80
Dendrites, 23
Dens odontoid process, 134
Denticulate ligament, 151
Dentition, 626
Depression, 6
Depressor anguli oris, 600
Depressor labii inferioris, 600
Dermatome, 10, 27, *28*, 69, 161, 253
Dermis, 9
Descending aorta, 294
Descending artery, 308Q, 311E
Descending colon, 314, 372–374
Descending duodenum, 351
Descending mesocolon, 333
Detrusor reflex, 483Q, 484, 485–486, 486E
Developmental anomalies, 391
Dextrocardia, 292
Diagnostic nerve conduction studies, 21
Diagonal conjugate, 409
Diaphragm, 243, 305Q, 309E, *393*, 393–394, 403Q, 406E
  action, 251
  apertures, 251

development, 251
innervation, 251
structure, 250–251
urogenital, 412
Diaphragma sellae, 514
Diaphragmatic pleura, 259, 260
Diaphragmatic ventilation, 256, 257
Diaphysis, 15
Diarthrodial joints, 18–19, *19*
Diarthroses, 18, 48Q, 50–51E, 132
Diastole, 36
Diastolic pressure, 38, 288
Didelphia, 473
Digastric muscle, 573, 574, 576, 577, 606
Digastric triangle, 574
Digestion, 351
Digital arteries, 124Q, 129E
Digital palmar prehension, 121
Digital prehension, 121
Dilator naris, 600
Dilator pupillae, 555
Diploë, 504
Diploic veins, 504
Diplopia, 533, 550, 566Q, 570E
Disk degeneration, 139
Disk herniation, 140
Dislocation
  lunate, 86–88, 93
  mandibular, 605
  patellar, 183
  shoulder, 59–60
Displacement of lunate bone, 119
Distal, 4
Distal carpal row, 93, *94*
Distal interphalangeal joints, 103, *104*, 104–105, 223, 228
Distal phalanx, 234Q, 239E
Distal radioulnar joint, 85, *86*, 87
Distal tarsal group, 223
Distal tibiofibular joint, 201, *202*, 203
Distal tubule, 389
Diverticulitis, 375, 403Q, 406E
Diverticulosis, 375
Dorsal and ventral mesenteries, 331
Dorsal artery of clitoris, 441
Dorsal calcaneonavicular ligaments, 205
Dorsal digital arteries, 212
Dorsal digital branch of ulnar nerve, 125Q, 129E
Dorsal foot, 235Q, 239E
Dorsal intercarpal ligaments, 93
Dorsal interossei, 227, 228
Dorsal interosseous muscles, 112–113, 126Q, 129E
Dorsal mesentery, 332
Dorsal mesoduodenum, 332
Dorsal mesogastrium, 332
Dorsal mesointestine, 332–333
Dorsal pancreatic artery, 362
Dorsal pedal artery, *211*, 211–212, 219
Dorsal primary rami, 29, 46Q, 49E, 144, 156Q, 159E, 254, *495*, 495–498, *496*, *497*
Dorsal root ganglia, 27, 48Q, 51E, 152, 153, 253, 301, 519
Dorsal roots, 27
Dorsal scapular artery, 498
Dorsal scapular nerve, 496
Dorsal subaponeurotic space, 105
Dorsal subcutaneous space, 105, 117
Dorsal vagal nucleus, 533

Dorsal vein, 441
Dorsalis pedis artery, 235Q, 239E
Dorsiflexion, 5
Dorsum sellae, 510, 511
Drawer sign, 186
Ductal variations, 361
Ducts
  lymph, 42
  thoracic, 43
Ductus arteriosus, 37, 291
  closure of, 291
Ductus venosus, 37, 290, 292, 308Q, 311E
  closure of, 292
Duodenal cap, 350
Duodenal papilla, 351
Duodenal stenosis, 334
Duodenal ulcers, 352
Duodenojejunal flexure, 351
Duodenum, *350,* 350–352
Dupuytren contracture, 109, 121
Dura mater, 25, 149, 156Q, 159E, 513–517
Dural sac, 150
Dural venous sinus, *515,* 515–516, 541
Dysphagia, 297, 298, 345
Dyspnea, 262

E

Ear
  external, *557,* 557–558, 567Q, 570–571E, 597
  inner, *559,* 561–563, *562*
  middle, 558–561, *559*
Ear lobe, 557
Ear wax, 557
Ectopic adrenal tissue, 392
Ectopic implantation, 467–468
Ectopic kidneys, 391
Ectopic ovaries, 467
Ectopic pancreatic tissue, 363
Ectopic pregnancy, 481Q, 485E
Edema, 40, 42, 43
  exudative, 40
  pulmonary, 44
  transudative, 40
Effective strength of muscle, 21
Efferent ductules, 456
Efferent limb, 484
  of cremaster reflex, 398
Effusion, pulmonary, 44
Ejaculation, 435, 459
Ejaculatory duct, 452, 454, 459
Ejection volume, 288
Elastic cartilage, 17
Elastic fibers, 17
Elbow
  articulations, 74–76
    joint, *74,* 74–75
    joint stability, 75–76
  bones, 73
    fractures of, 75–76
    humerus, 73
      articulations with forearm, 73
      characteristics, 73
    radius, 73–74
    ulna, 73–74
  brachial innervation
    axillary (humeral circumflex) nerve, 82–83, *83*
    medial brachial cutaneous nerve, 82
    median nerve, 82
    musculocutaneous nerve, *81,* 81–82
    radial nerve, *83,* 83–84

ulnar nerve, 82
brachial vasculature
  artery, 78–80, *79*
  lymphatic drainage, 80
  venous return, 80
muscle function
  clinical considerations, 78
  group actions, 76–77, 78*t*
  group innervations, 77, 78*t*
  movement, 76, *76,* 77, 78*t*
muscle function at
  clinical considerations, 78
  group actions, 76–77, 78*t*
  group innervations, 77, 78*t*
  movement, 76, *76,* 77, 78*t*
Electrocardiogram (ECG), 284–285, *287*
Electromyography, 22
Elephantiasis, 44
Elevation, 6
Embryonic disk, 330
Embryonic mesenteries, 331
Emphysema, 256
  orbital, 543
Enamel, 594
End-arteries, 38, 46Q, 50E
Endocardium, 278
Endochondral bones, 503
Endometrium, 469
Endopelvic fascia, 458, *464,* 464–465, 483
Enteric nervous system, 33
Enteric reflexes, 366
Enteric tube, 330
Enterogastric reflex, 349
Enterogastrone, 351
Epicardium, 278
Epicondylar fracture, 102
Epicondyle, 179
Epicranial aponeurosis, 501, 502
Epidermis, 9, 314
Epididymis, 456, 457–458
Epididymitis, 461
Epidural anesthesia, 26, 149, 155, 472
Epidural hematoma, 25, 509, 513, 516, 517–518
Epidural space, 25, 26, 149, 155, 156Q, 159E, 514, 516
Epigastric artery, 319
Epigastric herniation, 322
Epigastric region, 313
Epigastric veins, 320, 379
Epiglottic taste buds, 638
Epiglottis, 633, 638, *639,* 646Q, 650E
Epinephrine, 33, 383, 392
Epiphyseal disk, 15, 46Q, 49E
Epiphyseal plate, 57, 163
  fracture through, 164
Epiphyses, 15
Epiploic appendages, 368
Episiotomy, 442–443
Epispadias, 421
Epistaxis, 621
Epitympanic recess, 559
Epoöphoron, 455, 463
Erection, 435
Esophageal arteries, 345
Esophageal carcinoma, 402Q, 405E
Esophageal hiatus, 394
Esophageal plexus, 297, 301, 303, 345
Esophageal varices, 296–297, 345, 348, 379
Esophageal venous plexus, 296, 345
Esophagectomy, 402Q, 405E
Esophagitis, 298, 345
  regurgitative, 345

Esophagus, *294,* 296–298
  lower, *344,* 344–346
Estrogen, 466
Ethmoid bone, 503, 510, 543, *588,* 588–589
Ethmoid bulla, 619, 622
Ethmoid labyrinths, 588
Ethmoid nerve, 622
Ethmoid notch, 589
Ethmoid sinus, 588, 622
Ethmoidal artery, 620
Ethmoidal foramen, 551
Ethmoidal nerve, 551, 615
Eustachian tube, 507
Eversion, 6
Eversion sprains, 204
Excitation-contraction coupling, 21
Expiration, abdominal, 318
Expiratory reserve volume, 256, 257
Extension, 46Q, 49E
  of forearm, 77
  of IP joints, 113
  of thumb, 114
Extensor aponeurosis, 105, 106
Extensor digitorum, 107
Extensor digitorum brevis, 210, *210*
Extensor digitorum longus, 209–210
Extensor hallucis brevis, 210
Extensor hallucis longus, 210
Extensor indicis proprius, 107
Extensor pollicis brevis, 107
Extensor pollicis longus, 106, 107, 124Q, 128E
External rotation, 19
Extracapsular fractures, 164
Extracranial hematoma, 502
Extrahepatic occlusion, 355
Extrahepatic variation, 355
Extraocular muscles, 566Q, 570E, 585
  nuclei of, 533
Extrapyramidal pathways, 152, 524
Extrinsic elastic recoil, 246, 255–256
Exudative edema, 40
Eyeball, 552–553
  chambers of eye, 556
  ocular tunics, *553,* 553–556
Eyelids, 544–547, 597

F

Face
  developmental considerations, 597
  features, 597–598, *598*
  infratemporal fossa, 602–603
  innervation, 610–618, *611, 612, 614*
  muscles of facial expression, *598,* 599–602, 601*t*
  parotid gland, 598–599, *607*
  pterygopalatine fossa, *603,* 603–604
  temporomandibular joint, *604,* 604–605
  vasculature
    arterial supply, *607,* 607–609, *609*
    lymphatic drainage, 610
    venous return, 609–610
Facial artery, *607,* 608, 621
Facial cranium, 501
  bones, *588,* 588–593, *590, 591*
Facial (fallopian) canal, 560
Facial motor nucleus, 531
Facial nerve, 502, 503, 526, 529, 545, 552, 567Q, 571E, 577, 585, 598, 602, 610, *611, 612,* 613, 626
Facial palsy, 645Q, 649E
Facial skeleton
  cranium, regions, 586, *587,* 588

developmental considerations, 585, 586t
organization, 585–586, 586, 586t
Facial vein, 610
Falciform ligament, 331, 353
False ribs, 245
False vocal cord, 638
Falx cerebri, 514
Falx inguinalis, 317, 324, 326
Fascia
  Camper's, 11, 47Q, 50E
  clinical considerations, 12
  Colles,' 11
  Cruveilhier's, 11
  deep, 11–12, 47Q, 50E
  divisions, 11, 11–12
  membranous, 47Q, 50E
  Scarpa's, 11, 47Q, 50E
  superficial, 11
Fascia bulbi, 549
Fascia lata, 188
Fasciae latae, 206
Fascial divisions, 11, 11–12
Fascial septa, 189
Fatty layer, 48Q, 50E
Fecal continence, 412, 422
Fecal incontinence, 482Q, 485E
Female reproductive tract
  development of, 455
  endopelvic fascia, 464, 464–465
  external genitalia, 421
  mesometrium, 462, 462–464, 463
  ovaries, 465, 465–467
  parietal peritoneum, 461–462
  secretion, 476
  urethra, 452–453
  uterine tubes, 467–468
  uterus, 454, 468–472
  vagina, 472–473
Feminization of males, 392
Femoral artery, 174, 174–175, 193, 194, 194–195, 320, 414
Femoral canal, 193
Femoral circumflex artery, 195
Femoral cutaneous nerve, 199, 233Q, 238E, 398
Femoral hernias, 193, 327
Femoral neck, 164
Femoral nerve, 192, 196, 197, 220, 236Q, 240E, 398
Femoral pulse, 194
Femoral ring, 193, 327–328
Femoral shaft, 179
  fractures of, 164
Femoral sheath, 193
Femoral triangle, 190, 192–193, 194
Femoral vein, 193, 196, 220, 320
Femoropatellar joint, 181, 182–183
Femorotibial joint, 181, 183–184
Femur, 162, 163–164
Fetal circulation, 36–37, 290, 290–292
Fibroblasts, 11
Fibrocartilage, 17, 131
  composition, 645Q, 649E
Fibrocartilaginous, biconcave articular disk, 605
Fibrous capsule of knee, 184
Fibrous flexor sheaths, 110–111
Fibrous pericardium, 271
Fibrous ring, 278–279
Fibula, 162, 180, 201, 202
  fractures of, 204
Fibular artery, 218–219, 219
Fibular nerve, 198
Fibular notch, 203

Fifth lumbar nerve, 156Q, 159E
Fight, fright, and flight responses, 32
Filum terminale, 26, 149, 151
First dorsal metatarsal artery, 212
First-order neuron, 152, 153
Fistula, portacaval, 379
Flail chest, 257
Flat bones, 15
Flexion/extension, 5, 18, 46Q, 49E
  of carpometacarpal joints, 103
  of fifth digit, 115
  of forearm, 77
  of hip joint, 166, 169–170, 170t, 171t
  of metacarpophalangeal joint, 104
  of metacarpophalangeal joints, 113
  of radiocarpal joint, 87
  of ulnocarpal joint, 88
Flexor carpi radialis, 95, 96
Flexor carpi ulnaris, 95, 96
Flexor digiti minimi, 228
Flexor digitorum brevis, 227, 228
Flexor digitorum longus, 215–216, 225, 227, 228, 234Q, 239E
Flexor digitorum profundus, 94, 95, 96, 109–110, 123Q, 128E
Flexor digitorum superficialis, 94, 95
Flexor forearm and anterior wrist
  antebrachial vasculature, 97, 98, 99
  lymphatic drainage, 100
  radial artery, 97, 98, 99
  venous return, 99–100
  bones of
    carpal, 93, 94
    midcarpal joint, 93, 94
  compartment
    antebrachial fascia, 93–94
    musculature, 94–96, 97t
  innervation of compartment
    lateral antebrachial cutaneous nerve, 102
    medial antebrachial cutaneous nerve, 102
    median nerve, 100–102
    ulnar nerve, 102
Flexor hallucis brevis, 227, 228
Flexor hallucis longus, 215, 216–217, 225, 228, 235Q, 239E
Flexor pollicis brevis, 125Q, 129E
Flexor pollicis longus, 94, 95, 109, 110
Flexor retinaculum, 94, 107, 110, 215, 217
Floating ribs, 245
Fluid, exchange of, 39
Fontanelles, 504
Foot, 161, 162, 202, 223
  bones of, 201–203
  dorsum of
    crural (deep) fascia, 209
    musculature, 207, 209–210, 210
  eversion of, 206
  inversion of, 206
Foot drop, 213, 417
Foramen, 543
Foramen cecum, 510, 577, 627
Foramen lacerum, 508, 512, 567Q, 571E
Foramen magnum, 509, 539, 567Q, 571E
Foramen ovale, 37, 280, 290–291, 291, 507, 511, 567Q, 571E, 616, 647Q, 651E
  closure of, 291
Foramen rotundum, 511, 567Q, 571E, 603, 645Q, 649E

Foramen spinosum, 507, 511, 567Q, 571E
Foramina, 511
  of Luschka and Magendie, 565Q, 569E
Force-area relation, 21
Forced costal expiration, 256
Forced costal inspiration, 256
Forearm and posterior wrist, 56
  articulations with, 73
  bones of
    articulations, 87–88
    bony landmarks, 85
    carpal, 85–87
    radius, 85
    ulna, 85
  extensor compartment
    antebrachial fascia, 88
    musculature, 88–90, 91t
    innervation of extensor compartment
      nerve injury, 92
      radial nerve, 90, 92, 92
Foregut
  abdominal esophagus, 344, 344–346
  biliary tree and gallbladder, 356–360, 357
  duodenum, 350, 350–352
  liver, 352–356, 353
  pain pathways, 384, 384t
  pancreas, 360–363
  spleen, 363–364
  stomach, 346, 346–350
Foreskin, 426, 437
Fornices, 472
Fornix, 469, 472
Fossa navicularis, 419, 428, 436, 455
Fossa ovalis, 37, 280, 281, 281, 291
Fourchette (frenulum of labia), 436
Fovea centralis, 555
Foveal notch, 163
Fractures, 16–17
  of clavicle, 56–57
  Colles', 85
  of elbow, 75–76
  epicondylar, 102
  extracapsular, 164
  healing of, 46Q, 49E
  intracapsular, 164
  intracondylar, 181
  of metacarpal shafts, 103
  midhumeral, 78
  midradial, 96
  pelvic, 452
  proximal phalangeal, 103, 223
  radial, 78
  scaphoid, 86
  supracondylar, 78, 102, 181
  tarsal, 203
Frenulum, 426, 436, 597, 627
Frenulum labia, 626
Frenulum linguae, 626
Friction rub, 260–261, 273
Frontal bone, 503, 506, 510, 587, 589–590
  landmarks, 505
Frontal crest, 509
Frontal lobe, 521–522
Frontal nerve, 550, 614
Frontal processes, 590
Frontal sinus, 543, 590, 619, 622
Frontalis muscle, 501, 503
Frontonasal duct, 622
Fulcrums, 18–19, 19
Functional anatomy, 3
Functional pulmonary stenosis, 292

Fundus of uterus, 468
Funicular process, 322
Fusiform, 20

G

Gag reflex, 626, 636, 647Q, 650E
Gallaudet's fascia, 430
Gallbladder, 313, 358–360
Gallstones, 359–360
Gamma motoneurons, 151
Ganglia, 26
  paravertebral, 33
  prevertebral, 33
  sympathetic, 33
Ganglion terminale, 622
Gartner's cyst, 455
Gartner's ducts, 464
Gas, exchange of, 39
Gaseous diffusion, 270
Gastric arteries, 345, 347, 375, 376, 400Q, 404E
Gastric ulcer, 400Q, 404E
Gastric vein, 345
Gastritis, 349
Gastrocnemius muscle, 215
Gastrocolic ligament, 333, 336, 372
Gastroduodenal artery, 362, 376, 400Q, 404E
Gastrohepatic ligament, 331, 353, 400Q, 404E
Gastrointestinal tract
  developmental considerations, 343
  divisions, 343–344
    abdominal esophagus, 344, 344–346
    anal canal, 375
    ascending colon, 367–372
    descending (left) colon, 372–374
    duodenum, 350, 350–352
    gallbladder, 358–360
    jejunum and ileum, 364t, 364–367, 365
    liver, 352–356, 353
    pancreas, 360–363
    rectum, 375
    sigmoid colon, 374–375
    spleen, 363–364
    stomach, 346, 346–350
  innervation
    autonomic nervous system, 380–383, 381
    visceral afferent nerves, 383–384, 384t
  vasculature
    arterial supply to, 375–377
    lymphatic drainage of, 379–380
    venous return, 378–379
Gastro-omental artery, 347, 375
Gastrophrenic ligament, 332, 346
Gastroschisis, 337
Gastrosplenic ligament, 332
General visceral efferent nuclei, 532–533
Genial tubercles, 592
Geniculate anastomosis, 195, 218
Geniculate ganglion, 610, 611, 612, 648Q, 651E
Genioglossus muscle, 628, 628
Geniohyoid muscle, 573, 577, 584, 606
Genital duct, 453

Genital swelling, 419
Genital tubercle, 419
Genitofemoral nerve, 197, 321, 326, 398, 399Q, 404E, 429, 435, 442, 480Q, 484E
Genu valgum, 179
Genu varum, 179
Germinal epithelium, 466
Gingivae, 626
Glabella, 505
Glands, 23
Glans clitoridis, 437
Glans penis, 419, 426, 455, 483Q, 486E
Glaucoma, 556
Glenohumeral joint, 57, 59, 124Q, 128E
  adduction of, 122Q, 127E
  bursae, 59
  dynamic stability, 59
  ligamentous support, 59
  movement, 59
  shoulder dislocation, 59–60
  structure, 59
Glenohumeral ligaments, 59
Glenoid fossa, 57
Glenoid labrum, 57
Glial cells, 519
Globus pallidus, 524
Glossopharyngeal nerve, 269, 527, 529, 530, 531, 533, 558, 585, 599, 626, 630, 632, 635, 635–636
Glottal closure, 637
Glucagon, 362
Glucocorticoids, 392
Gluteal artery, 414
Gluteal nerves, 416, 417
Gluteal region
  basic principles, 161
  innervation, 166, 175–176, 176
  muscle function at hip joint
    group actions, 169–170, 170t, 171t, 172t, 172–173, 173t
    organization, 167–169, 168, 170t
  organization, 161–162
  pelvic articulations
    greater sciatic foramen, 165
    hip (coxal) joint, 162, 165–167
    lesser sciatic foramen, 165
    obturator foramen, 165
    pubic symphysis, 165
    sacroiliac joint, 164–165
  pelvic bones
    bony landmarks, 162–163
    femur, 162, 163–164
    girdle, 163
  vasculature, 173–175, 174
Gluteus maximus muscle, 188–189
Gluteus medius, 233Q, 238E
Goiter, 578
Gomphosis, 18
Gonadal arteries, 390, 401Q, 405E
Gonads, 322
Gracilis muscle, 189, 197
Granular lacunae, 509
Gravity, effects of, on ventilation, 256
Gray matter, 24–25, 25, 151, 151, 519, 520
  of cerebral cortex, 48Q, 51E
Gray rami communicantes, 29, 302, 498
Gray ramus communicans, 32
Great cardiac vein, 306Q, 310E
Great pancreatic artery, 361, 362

Great saphenous vein, 195–196, 212, 219, 235Q, 239E
Greater occipital nerve, 495
Greater omentum, 332, 333, 334, 346
Greater palatine artery, 620
Greater sac, 338–339, 340
Greater sciatic foramen, 165, 407, 414
Greater sciatic notch, 407
Greater splanchnic nerves, 343, 345, 352, 360
Greater trochanter, 164, 235Q, 239E
Gubernaculum, 322, 325, 465
Gubernaculum testis, 420
Gynecomastia, 13, 392
Gyri, 521

H

Hamate bone, 93
Hamstring group, 190, 191
Hamstring reflex, 199
Hamulus, 93
Hand, 56
  articulations
    carpometacarpal joints, 103–104, 104
    interphalangeal (IP) joints, 104, 104–105
    MP joints, 104, 104
  blood supply of, 124Q, 129E
  bones
    metacarpals, 103
    phalanges, 103
  dorsal musculature
    extrinsic muscles, 105–107
    fascia of hand, 105
  function, 121
  innervation
    median nerve, 118–120, 119
    radial nerve, 118
    ulnar nerve, 120, 120–121
  palmar musculature
    extrinsic muscles, 109–112
    fascia of palm, 107–109
    intrinsic, 112–116, 115t
  vasculature
    lymphatic drainage, 117–118
    radial artery, 116, 116–117
    venous return, 117
Hard palate, 593, 623–624
Hartmann's pouch, 358
Haustra, 368
Hearing and balance dysfunction, 613
Heart, 274–292
  cardiac pump, 35–37, 36
  coronary circulation, 36
  fetal circulation, 36–37
Heart block, 285
Heartburn, 298, 345
Heel-strike, 231
Helicotrema, 562
Helix, 557
Hematemesis, 379
Hematoma, 40
  epidural, 25, 516, 517–518
  extracranial, 502
  subdural, 25, 517, 518, 541
Hematopoiesis, 364
Hemianopsias, 529
Hemiazygos vein, 298–299, 299, 345, 348
Hemigastrectomy, 350
Hemodynamics, 37, 40–41
Hemorrhage
  cerebral, 540

pial, 518
subarachnoid, 518, 540
subpial, 518
venous, 541
Hemorrhagic pancreatitis, 361
Hemorrhoidal arteries, 434
Hemorrhoidal vein, 479Q, 484E
Hemorrhoids, 379, 424, 479Q, 480, 482, 484E
Hemothorax, 252, 258, 262
Hepatic artery, *346*, 354–355, *355*, 356, 376, 400Q, 404E
　aberrant left, 348, 375
Hepatic blood flow, 290
Hepatic capsule, 354
Hepatic diverticulum, 352
Hepatic ducts, 356
Hepatic flexure, 372
Hepatic pedicle, 339, 356
Hepatic plexus, 381
Hepatic portal system, 41, 343
Hepatic portal vein, 320, 343, 352, 355, 378, *378*
Hepatic resection, 356
Hepatic tree, 356–358, *357*
Hepatic veins, 353, 355, 378–379
Hepatoduodenal ligament, 331, 339, 350
Hepatomegaly, 313
Hepatopancreatic ampulla, 351, 357, 361
Hepatopancreatic sphincter, 351, 357
Hepatorenal recess, 339
Hernias, 473
　abdominal, 318, 322
　femoral, 193, 327
　hiatus, 394
　inguinal, 326–327, 402Q, 405–406E
　medullary, 513
　omental, 339
　paraduodenal, 337
　paraesophageal, 345
　temporal lobe, 513
　umbilical, 334, 337
Herniated vertebral disks, 221
Hiatus hernia, 345–346, 394
Hiatus of facial canal, 567Q, 571E
Hiatus semilunaris, 619, 622, 623
Hiccups, 251
Higher order neurons, 153
Hilar nodes, 267, 300
Hilus, 391
Hindbrain, 526–527
Hindgut, 331
　anal canal, 375
　descending (left) colon, 372–374
　pain pathways, 384, 384*t*
　rectum, 375
　sigmoid colon, 374–375
Hip disease, 231–232
Hip joint, 161, 162, *162*, 163, 165–167
Hirschsprung's disease, 375, 382
Horizontal incisions, 321
Horizontal palatine processes, 590
Horner's syndrome, 265, 546, 555
Horseshoe abscess, 112
Horseshoe kidney, 391
Hour glass stomach, 345
Housemaid's knee, 182
Human anatomy
　organization, 3
　origins, 3
Humeral abduction, 125Q, 129E
Humeral circumflex artery, 67
Humeroradial joint, 73, 74
　movement, 75
　support, 75
Humeroulnar joint, 73
　movement, 74
　support, 74
　variation, 74–75
Humerus, 56, 57–58
　articulations with forearm, 73
　characteristics, 73
Hunter's canal, 194–195
Hyaline cartilage, 17
Hyaline membrane disease, 270
Hydrocele processus vaginalis, 429
Hydrocephalus, 537
Hydrochloric acid, 349
Hydronephrosis, 391
Hydrostatic-osmotic pressure, 43
Hydrothorax, 262
Hymen, 419, 436
Hyoglossus muscle, 628, *628*
Hyoid bone, 573–574, 575, *576*, 636, 638
Hyperacusis, 561
Hypermetropia, 554
Hyperopia, 554
Hyperplasia, 392
Hypertension, systemic, 389
Hypertrophy, muscular, 21
　benign prostatic, 460
Hypochondriac regions, 313
Hypogastric nerve, 484
Hypogastric nerves, 474
Hypoglossal canal, 509
Hypoglossal nerve, 509, 513, 527, 533, 564Q, 566Q, 568Q, 569E, 570E, 571E, 582–583, *583*, 585, 628, 629, 630
Hypoglossal nucleus, 533
Hypopharyngeal recesses, 637–638
Hypopharynx, 630, 633
Hypophyseal fossa, 511
Hypophyseal portal system, 41
Hypospadias, 421
Hypothalamus, 525
Hypothenar compartment, 109
Hypothenar muscles, *115,* 115*t,* 115–116

I

Icterus, 356
Ileal diverticulitis, 367
Ileal diverticulum, 330, 365, 367, 402Q, 406E
Ileal papilla, 368
Ileocecal fossa, 340
Ileocecal valve, 368, 403Q, 406E
Ileocolic artery, 369, 371
Ileojejunal resection, 367
Ileum, 364*t,* 364–367, *365,* 399Q, 404E
Iliac arteries, 314, 390, 414
Iliac circumflex arteries, 414
Iliac crest, 158Q, 160E, 315, 395, 407
Iliac fossa, 407, 480Q, 484E
Iliac nodes, 424
Iliac spines, 158Q, 160E
Iliac vein, 415
Iliococcygeus muscle, 412
Iliocostalis muscles, 143
Iliofemoral ligament, 166
Iliohypogastric nerve, 29
Iliohypogastric nerves, 29, 321, 325, 397, *397*
Ilioinguinal nerve, 29, 325
Ilioinguinal nerves, 29, 321, 325, 398, 435, 442
Iliolumbar artery, 414
Iliolumbar ligament, 409
Iliopsoas muscle, 164, 192, 393*t,* 394–395
Iliotibial tract, 188–189
Iliotrochanteric band, 166
Ilium, *162,* 163, 407
Impotence, 475
Inca bone, 506
Incisors, 593, 594
Incontinence
　anal, 424
　fecal, 482Q, 485E
　sphincteric, 422
　stress, 481Q, 485E
　urinary stress, 440
Incudomalleal joint, 561
Incudomalleolar joint, 560
Incudostapedial joint, 561
Incus, *559,* 560–561
Indirect inguinal hernia, 402Q, 405–406E, 429
Infarct, 38, 277
Infection of subaponeurotic space, 105
Inferior articular processes, 133
Inferior duodenum, 351
Inferior epigastric arteries, 414
Inferior extensor retinaculum, 209
Inferior gluteal artery, 174, *174*
Inferior gluteal nerve, *169, 176,* 177
Inferior gluteal neurovascular bundle, 407
Inferior hypogastric plexuses, 474
Inferior oblique capitis muscle, 144
Inferior peroneal retinaculum, 209
Inferior thoracic aperture, 243
Inferior thyroid artery, 498
Inferior vena cava, 279, 339
Infracolic compartments, 339, *340*
Infraglenoid tubercle, 57
Infrahepatic recess, 339
Infrahyoid muscles, 564Q, 569E, 574, *574,* 575*t,* 575–576
Inframeniscal compartment, 605
Infraorbital artery, 544, 609
Infraorbital canals, 590
Infraorbital foramen, 543, 609
Infraorbital foramina, 590
Infraorbital groove, 543
Infraorbital nerve, 544, 616
Infrapatellar bursa, 182
Infrapatellar bursitis, 182
Infrapiriform portion of greater sciatic foramen, 414
Infrapiriform recess, 407
Infrasigmoid fossa, 340
Infraspinatus muscle, 125Q, 129E
Infraspinous fossa, 57
Infratemporal crest, 507
Infratemporal fossa, 611–612
Infratrochlear nerve, 615
Infundibulopelvic ligament, 465
Infundibulum of heart, 281
Inguinal canal, *323,* 323–324, 420
Inguinal hernias, 326–327
Inguinal ligament, 193
Inguinal nodes, 396
Inguinal triangle (of Hesselbach), *326,* 326–327
Innermost intercostal muscles, 250
Innervation, *635,* 635–636
Inspiratory reserve volume, 256, 257
Insulin, 362
Integument
　cleavage lines, 9–10, *10*

clinical significance, 10
divisions, 9
functions of, 9
innervation, 10
surface area, 9
Intercondylar eminence, 181, 183
Intercostal arteries, 253, 298, 319
Intercostal muscles, 249t, 249–250, 256
Intercostal nerve block, 254
Intercostal nerves, 29, 304Q, 309E, 394
Intercostal neurovascular bundle, 245
Intercostal nodes, 299
Intercostal space, 304Q, 309E
Intercuneiform joint, 223–224
Intermediate bursa, 112
Intermuscular septum, 189, 206, 215
Internal, 4
Internal os, 468
Internal rotation, 19
Interosseous ligaments, 203, 409
Interosseous membrane, 75, 206
Interosseous talocalcaneal ligament, 205
Interphalangeal articulations, 224
Interphalangeal joints, 104, 104–105
Intersigmoid recess, 374
Interspinales muscles, 144
Interspinous ligaments, 140
Interstitial cells, 454
Intertransversarii muscles, 144
Intertrochanteric crest, 164
Intertubercular groove, 57
Intertubercular plane, 313
Interventricular artery, 276
Interventricular septum, 281, 282
Interventricular sulcus, 273–274
Intervertebral arteries, 133, 154
Intervertebral disks, 131, 132, 149
Intervertebral foramina, 133
Intervertebral ligaments, 131
Intervertebral veins, 154
Intestinal angina, 376
Intestinal lymph trunk, 372
Intorsion/extorsion, 6
Intracapsular femoral neck fracture, 234Q, 238E
Intracapsular fractures, 164
Intracardiac injection, 272–273
Intracondylar fracture, 181
Intracranial bleeding, 517
Intrahepatic perfusion, 355–356
Intramembranous bone, 503
Intrascalene block, 69
Intravenous line, placement of, 123Q, 127E
Intravenous pyelography (IVP), 389
Intrinsic elastic recoil (pulmonary compliance), 256
Intussusception, 368, 403Q, 406E
Inversion, 6
Inversion sprains, 204
Involuntary muscles, 23
Ipsilateral nasal hemianopsia, 529
Ipsilateral spinal cord crush injury, 153
Iris, 555
Ischemia, 538
Ischial ramus, 163, 407
Ischial spine, 163, 407
Ischial tuberosity, 163, 407
Ischioanal fat, 422
Ischioanal fossa, *421*, 421–422, 482Q, 485E
Ischiocavernosus muscles, 431, 437, 438, 481Q, 485E
Ischiofemoral ligament, 167

Ischium, *162*, 163, 407–408, *408*
Isotonic contraction, 20, 288

J

Jaundice, 363
 cholecystitis with, 360
 cholecystitis without, 360
Jaw-jerk reflex, 647Q, 650E
Jejunum, 364t, 364–367, *365*
Joint, 18
Jugular foramen, 508, 513, 516, 647Q, 651E
Jugular vein, 499, 516, 581, *582*, 609
Jundice, 356

K

Keratitis, 554
Keratoplasty, 554
Kidney stones, 389
Kidneys, 314, 385
 ectopic, 391
 external structure, 386, *386*
 functions, 388–389
 horseshoe, 391
 innervation, 388
 internal structure, 386–387
 lobated, 391
 polycystic, 391
 vasculature, *387*, 387–388
Knee
 bones at, 179
  distal femur, 179, *180*, 181
  patella, 181–182, *190*
  proximal tibia, *180*, 181
 dynamic action, *187*, 187–188
 ligamentous support
  collateral ligaments, 184–186
  fibrous capsule, 184
  instability, 186–187
  patellar retinaculum, 184–186
  stability, 186
  synovial capsule, 184
 structure, 182–184
Knee extension, 234Q, 238E
Knee-jerk reflex, 196, 233Q, 238E
Kyphosis, 131

L

Labia major, 421, 435, 483Q, 486E
Labia minor, 421, 435–436
Labial salivary glands, 597
Labii superioris alaeque nasi, 600
Labioscrotal folds, 420, 435
Labyrinth, 561
Lacrimal apparatus, 546–547
Lacrimal bones, 543, 592
Lacrimal canaliculi, 547
Lacrimal ductules, 546
Lacrimal fascia, 547
Lacrimal fossa, 543
Lacrimal glands, 546, 551–552, 645Q, 649E
Lacrimal groove, 543
Lacrimal nerve, 550, 614
Lacrimal papilla, 547
Lacrimal portion, 547
Lacrimal sac, 547
Lacrimation, 547
Lactiferous ducts, 13
Lacunar ligament, 193
Lambda, 506

Lambdoid bone, 506
Lambdoid suture, 504
Lamina of dorsal horn, 152, 153
Lamina papyracea, 588, 589
Langer's lines, 45Q, 49E
 basis for, 9
 direction of, 9–10
Laryngeal artery, 498
Laryngeal musculature, 646Q, 650E
Laryngeal nerves, 295, 531, 643
Laryngeal ossicles, 637
Laryngeal prominence, 639
Laryngeal sinus, 638
Laryngeal vestibule, 638
Larynx, 619
Lateral, 4
Lateral horn, 28
Lateral regions, 313
Lateral rotation, 6
Latissimus dorsi muscle, 63, 395
Law of descent, 26
Least splanchnic nerve, 388, 390
Left atrium, 274, 281–282
Left A-V valve, 282
Left coronary artery, *275*, 276
Left hemisphere, 521
Left pericardiacophrenic artery, 306Q, 310E
Left ventricle, 274, 282–283
Left-to-right shunt, 280
Leg, 161, 162, *180*, *202*
 bones of, 201
 innervation
  course and composition, 212
  cutaneous sensory, 220
  distribution, 212–213, *213*
  injury to common fibular nerve, 213
  motor, 221
 musculature
  fascia, 215
  muscles of, 215–218, *216*, 217t
 vasculature
  arteries, 211–212, 218–219, *219*
  veins, 212, 219–220
Lens, 553
Lenticular nucleus, 524
Leptomeninges, 149
Leptomeninx, 25
Lesions, 521
Lesser omentum, 331, 353
Lesser sac, 334
Lesser sciatic foramen, 165, 408
Lesser sciatic notch, 407, 408
Lesser trochanter, 164
Levator anguli oris, 600
Levator ani muscle, 411, 422, 432
Levator labii superioris, 600
Levator palpebrae superioris, 545, 602
Levator scapulae, 146
Levator veli palatini muscle, 624, 633
Ligamenta flava, 149
Ligaments, 19
 function, 19
 torn, 19, 46Q, 49E
Ligamentum arteriosum, 37, 295, 298, 643
Ligamentum flavum, 140, 157Q, 159–160E
Ligamentum teres, 163, 167
Ligamentum venosum, 37
Ligamentum venosus, 292
Limbic lobe, 523
Linea alba, 313, 315, 317, 321
Linea aspera, 179
Linea semilunaris, 313, 317
Linea terminalis, 407, 409

Lingual artery, 629
Lingual nerve, 617, 629
Lingual papillae, 627
Lingual tonsils, 627, 633
Lingual veins, 582, 629
Lingula, 263, 591
Lips, 626
Liver, 352–356, *353*
Lobar bronchus, 263
Lobated kidneys, 391
Long bones, 15
Longissimus muscle, 143
Longitudinal fasciculus, 533
Longitudinal ligament, 140, 141, 156Q, 159E
Longitudinal plantar arch, 225
Longus colli muscle, 146
Lordosis, 131–132
Lower motor neuron, 48Q, 51E
Lumbar cistern, 26, 150
Lumbar curvature, 131
Lumbar herniation, 317
Lumbar nerves, 28, 394
Lumbar plexus, 176, *176*, 321
Lumbar puncture, 150, 155, 537
Lumbar region, 158Q, 160E
Lumbar rib, 246
Lumbar splanchnic nerves, 344, 388, 390, 397, 400Q, 405E, 474, 482
Lumbar tap, 26
Lumbar trigone, 395
Lumbar vertebrae, 136
Lumbocostal triangle, 394
Lumbodorsal fascia, 395
Lumbosacral enlargement, 25
Lumbosacral plexus, 166, 175–176, *176*, 397, *416*, 416–417
Lumbosacral trunk, 176, *176*, 398, 416
Lumbrical muscles, 113, 126Q, 129E, 227, 228
Lumen, 296
Lunate, 86, 93, 123Q, 128E
  dislocation of, 86–88, 93
Lungs
  carcinoma of, 265
  clinical considerations, 264–265
  development, 263
  differences, 263
  external structure, *263*, 263–264
  innervation
    course and composition, *268*, 268–269
    regulation of respiration, 269
  vasculature
    arterial supply, *265,* 266–267
    lymphatics, 267–268
    venous return, 267
Luteal cyst, 481Q, 485E
Lymph, 42
Lymph capillaries, 42–43
Lymph ducts, 42
Lymph nodes, 42, 43, 300
  aortic, 396
  atrioventricular, 278, 283–284
  aupratrochlear, 80, 100
  axillary, 13, 80
  bronchomediastinal, 268, 271, 300
  cervical, 499, 646Q, 649E
  clavipectoral, 117–118
  colic, 371
  common iliac, 396
  hilar, 267, 300
  illiac, 424
  inguinal, 396
  intercostal, 299

mesenteric, 371, 402Q, 406E
occipital, 499
para-aortic, 300, 457, 470
preaortic, 300
pulmonary, 267
supraclavicular, 13, 117–118
supratrochlear, 80
Lymph trunk, 371
Lymphatic anastomoses, 416
Lymphatic drainage, 68, 80, 352
Lymphatic metastatic carcinoma, 402Q, 406E
Lymphatic system, 35
  clinical considerations, 43–44
  composition, 41
  function, 41–42
  lymph flow, 43
  structure, 42–43
Lymphatic vessels, 35, 41–42, 42
Lymphatics, 43
Lymphedema, 44

M

Macula, 561
Male reproductive tract
  development of, *454*, 454–455
  ejaculation, 483Q, 486E
  emission and secretion, 476
  epididymis, 457–458
  external genitalia, 419–421
  testes, 455–457
Male urethra, *483,* 487
Malignant metastases, 350
Malleolar artery, 218
Malleolus, 181, 201, 235Q, 239E
Malleus, *559,* 560, 567Q, 570–571E
Mammary glands, 9, 243, 306Q, 310E
  anomalies, 13
  blood supply, 13
  lymphatic drainage, 13
  structural considerations, *12,* 12–13
Mammary papilla, 13
Mandible, 591–592, *592*
  dislocation, 605
  fracture of, 618
Mandibular foramen, 591
Mandibular nerve, 531, 560, 566Q, 570E, 585, 618, 624, 626
Mandibular sling, 592
Mandibular symphysis, 592
Mandibular teeth, 595
Manubriocostal joints, 244
Manubriosternal synchondrosis, 244
Manubrium, 58, 244–245, *245,* 560
Marginal artery, 371, 372, 374, 377, *377*
Masculinization of females, 392
Masseter muscle, 605
Mastication, 531
  muscles of, 598, 605–606, 607t
Mastoid air cells, 506, 559
Mastoid antrum, 505, 506
Mastoid foramen, 509
Mastoid process, 506, 508
Mastoiditis, 559
Maxilla, 543, *587,* 590–591
Maxillary artery, 604, 608–609, *609, 609*
Maxillary nerve, 552, 585, 604, 616
Maxillary sinus, 590, 619, 623, 646Q, 647Q, 650E
Maxillary sinusitis, 623
Maxillary teeth, 595
Maxillary tuberosities, 590
Maxillary tuberosity, 507

Maxillary vein, 610
Maximum isometric force, 21
Meatus, 557
Mechanical advantage, 21
Meckel's diverticulum, 337, 365, 367, 402Q, 406E
Medial, 4
Medial rotation, 5, 63–64
Median, 4
Median crest, 589
Median nerve, 70, 82, 94, 96, 113, 114, 118–120, *119,* 123Q, 126Q, 127E, 128E, 129E
Median sacral crest, 408
Mediastinal flutter, 262
Mediastinal pleura, 259, 260
Mediastinal shift, 262
Mediastinum, 259
  anterior, 293
  innervation, 300–303
  middle, 271
  parasympathetic pathways, 301
  posterior, *295,* 295–298
  superior, 293–295, 299
  sympathetic pathways, 301
  vasculature, 298–300
  visceral sensation, 301–302
Medulla oblongata, 149
Medullary arteries, 154
Medullary herniation, 513
Membranous cochlear duct, 562
Membranous fascia, 47Q, 50E
Membranous semicircular canal, 562
Membranous urethra, 485
Mendelson's syndrome, 264
Meningeal artery, 507, 513, 567Q, 571E, 608
Meningeal ramus, 29
Meningitis, 565Q, 569E
Meniscal injury, 184
Meniscal tears, 234Q, 238E
Menisci, 183–184
Meniscofemoral ligament, 185
Meniscus, 183, 645Q, 649E
  injury, 186
Mental foramina, 592
Mentalis muscle, 600
Mesenteric artery, 330, 343, 344, 362, 371, *373,* 374, 376–377, 401Q, 403Q, 405E, 406E
Mesenteric artery syndrome, 351
Mesenteric ganglia, 383
Mesenteric infarction, 376
Mesenteric nodes, 371, 402Q, 406E
Mesenteric vein, 352, 371, 374, 378
Mesenteries, 261, 329, 413, 462
  dorsal, 332
  embryonic, 331
Mesoappendix, 370
Mesometrium, *462,* 462–464, *463,* 468, 469
Mesonephric duct, 453, 457, 458
Mesonephric ducts, 454, 455
Mesonephros, 453
Mesosalpinx, 462, *463*
Mesothelioma, 261
Mesovarium, 462, 465
Metacarpal shafts, fractures of, 103
Metacarpals, 56, 103
Metacarpophalangeal joint, 103, 104, *104*
Metanephric diverticula, 453, 454
Metanephric duct, 453
Metanephros, 453
Metaphyses, 15
Metastases, malignant, 350

Index

Metatarsal bones, 201–203, 223
Metatarsal shafts, fractures of, 223
Metatarsophalangeal joints, 223, 224, 228
Metencephalon, 526
Micropenis, 421
Microvilli, 367
Micturition, 485–486
Midbrain, 525–526, *526*
Midclavicular planes, 313
Middle meatus, 588
Midgut, 330, 343–344
  ascending colon, 367–372, *371*
  jejunum and ileum, 364t, 364–367, *365*
  transverse colon, *371*, 372
Midgut pain pathways, 384, 384t
Midhumeral fractures, 78
Midinguinal planes, 313
Midline sternotomy, 271
Midpalmar, 95
Midpalmar bursa, 95, 110
Midpalmar space, 109
Midradial fracture, 96
Midsagittal plane, 4
Mineralocorticoids, 391–392
Minimum isometric force, 21
Minor calyx, 386, 389
Mitral insufficiency, 282
Mitral stenosis, 282, 307Q, 311E
Mitral valve, 307Q, 310–311E
Mitral valve insufficiency, 289
Mitral valve sounds, 289
Modiolus, 562, 600
Modiolus labii, 600
Molar teeth, 593, 594
Morton's metatarsalgia, 224
Morton's toe, 232
Motor aphasia, 522
Motor deficit, 564Q, 569E
Motor homunculus, 522, *522*
Motor innervation, 46Q, 49E
Motor tracts, 151–152, *152*
Movement, 20
  initiation of, 21
Multifidus muscles, 143
Multipennate, 20
Murmurs, 289
Muscle tone, 22
Muscles. *(See also specific muscle)*
  action, 20–21
  antagonistic, 21
  composition, 20
  force-generating capacity of, 21
  function, 21
  nervous control of musculoskeletal movement, 21–22
  strength of, 47Q, 50E
  synergistic, 21
  voluntary, 23
Muscle-splitting approach, 321
Musculature, *632*, 633–635, 634t
Musculocutaneous nerve, 70, 76, 81, 124Q, 128E
Musculophrenic artery, 253, 319
Musculoskeletal system
  basic functions, 15
  composition, 15
  functional and clinical considerations, 16–17
  skeleton
    articulations, 17–19, *18*
    bone, 15–17
    cartilage, 17
    types, 15
Myasthenia gravis, 293

Mydriasis, 550, 555
Myelencephalon, 527
Myelin, 23
Myelin sheaths, 519
Myelinated axon, 32
Mylohyoid, 647Q, 650E
Mylohyoid groove, 592
Mylohyoid line, 592
Mylohyoid muscle, 573, 576, 577, 606
Mylohyoid nerve, 577
Myocardium, 278
Myometrium, 469
Myopia, 554
Myotomes, 10, 28, 69, 161, 254
Myringotomy, 558

N

Nares, 597, 619
Nasal bones, 592
Nasal choanae, 630
Nasal conchae, 593
Nasal mucosa, 620
Nasal notch, 589
Nasal septum, 589, 619
Nasalis muscle, 600
Nasion, 505, 589
Nasociliary branch, 547
Nasociliary nerve, 551, 614
Nasolacrimal duct, 547, 567Q, 570E
Nasopalatine artery, 609
Nasopalatine nerve, 615, 626
Nasopharyngeal isthmus, 631
Nasopharynx, 630–631
Nasovomeral organ, 620
Navicular bone, 202
Necrosis, avascular, 86, 87
Negative-pressure devices, 258
Nephrolith, 391, 400Q, 405E
Nephron, 386, 388
Nephroptosis, 385
Nerve bundle, 26
Nerve compression, 140, 513
Nerve entrapment, 233Q, 238E
Nerve palsy, 118
Nerve root compression, 71
Nerves. *(See also specific nerve)*
  cranial, 23
  intercostal, 29
  spinal, 23
Nervi erigentes, 382, 481Q, 485E
Nervous system
  autonomic, 300–301
  subdivisions, 23–24, *24*
Neural arch, 149
Neurocranium, 501, *502*, 567Q, 571E
  bones and articulations
    basic structure, 503–505, *505*, *509*, *510*
    clinical considerations, 513
    external surface, *505*, 505–509
    internal surface, *509*, 509–513, *510*
  cranial meninges and venous sinuses
    clinical considerations, 517
    structure, 513–517, *514*
  scalp
    composition, 501–502, *502*
    innervation, 503
    vasculature, 503
Neurogenic bladder, 475
Neurohypophysis, 525
Neurology, 10
Neurons, 23
  brain, 519, 520
  central, 30

peripheral, 30
postganglionic, 32
Neurovascular bundle, 252
Neurovascular groove, 252
Neurovascular structures, 293–295
Nipple, 10, 13
Nomina Anatomica, 6
Norepinephrine, 33, 383
Notochord, 139
Nuclei, cranial nerve, 527–533
Nuclei of lateral horn, 151
Nuclei of ventral horn, 151
Nucleus ambiguus, 531
Nucleus pulposus, 139, 157Q, 159–160E
Nucleus pulposus extrusion, 156Q, 159E

O

Oblique aponeurosis, 399Q, 404E
Oblique muscle, 317, 318, 395, 401Q, 405E, 547–549, 548t, 550
Oblique popliteal ligament, 186
Oblique sinus, 272
Oblique vein, 278
Obliterated umbilical arteries, 399Q, 404E
Obstetric conjugate, 410
Obstruction
  of airway, 269
  biliary, 360
Obturator artery, 174, *174*, 193–194, 414
  aberrant, 193
Obturator externus muscle, 408, 422
Obturator foramen, 163, 165
Obturator internus muscle, 408, 422
Obturator nerve, 197, *197*, 236Q, 240E, 398
Obturator veins, 415
Occipital artery, 498, 580
Occipital bone, 503
Occipital lobe, 523
Occipital nerve, 503, 558
Occipital nodes, 499
Occipital region, 646Q, 649E
Occipitalis muscles, 502, 503
Occipitofrontalis muscle, 501, 602
Occlusion, 540
Ocular axes, 547
Oculomotor nerve, 526, 533, 544, 547, 549–550, *550*, 566Q, 570E, 585
Oculomotor nucleus, 533
Olecranon bursa, 73
Olecranon fossa, 73
Olecranon process, 73
Olfaction, 531
Olfactory bulbs, 510
Olfactory mucosa, 620
Olfactory nerves, 510, 622
Omental bursa, 334, 339
Omental foramen, 331, 334
Omental herniation, 339
Omohyoid muscle, 574, 575, 584
Omphalocele, 337
Omphalomesenteric duct, 330, 334
Ophthalmic artery, 503, 543, 547, 552, 568Q, 571E
Ophthalmic branches, 622
Ophthalmic nerve, 547, 552
Ophthalmic vein, 544, 547, 552, 609–610
Ophthalmoplegia, 533, 550
Opposition, 6

of carpometacarpal joints, 103
of fifth digit, 115
of thumb, 114
Optic canal, 511, 567Q, 571E
Optic cup, 552
Optic foramen, 511
Optic groove, 511
Optic nerve, 528, 528–529, 534–535*t*, 543, 547, 551, 585
Optic radiations, 523, *528*, 529
Optic tracts, 529
Optic vesicle, 552
Oral cavity, 619, 623, *624*
  basic structure, 623
  palate, 623–626, *631*
  submandibular duct, 630
  submandibular gland, 630
  tongue, 626–630
  walls of, 626
Oral fissure, 597, 626
Orbicularis oculi muscle, 545
Orbicularis oris, 597, 600, 647Q, 650E
Orbit
  bony, 543–544, *544, 548*
  extraocular musculature, 547–549, 548*t*
  eyeball, 552–553
    chambers of eye, 556
    ocular tunics, *553*, 553–556
  eyelids, 544–547
  innervation, 549–552, *550*
  vasculature, 552
Orbital emphysema, 543
Orbital fissure, 510, 566Q, 567Q, 570E, 571E, 603, 604, 609
Orbital floor, 543
Orbital plates, 589
Orbital roof, 543
Orbital septum, 545
Orbitalis oculi, 600
Orchitis, 461
Oropharyngeal isthmus, 623, 631
Oropharyngeal seal, 626, 636
Oropharynx, 630, *631*, 631–633
Osteoarthritis, 167, 184
Osteocytes, 15
Osteophytes, 139
Osteoporosis, 164
Ostium, 455
Otic ganglion, 560, 599
Otitis media, 560, 613, 631
Otorrhea, 537
Otosclerosis, 561
Oval window, 560
Ovarian artery, 463, 466
Ovarian fimbria, 465
Ovarian fossae, 465
Ovarian ligament, 421, 465
Ovarian torsion, 466
Ovarian tumors, 467
Ovarian veins, 463, 466
Ovaries, 314, 421, *454,* 455, *465,* 465–467
  definitive, 323
  ectopic, 467
Oxytocin, 525

P

Pachymeninx, 149
Pachymenix, 25
Pain
  anginal, 306Q, 310E
  appendicular, 370
  biliary, 358, 360
  chronic back, 139
  referred, 34, 297, 345, 352, 360, 362
  visceral, 368
Pain afferents, 301
Paired motions, 18
Palate, 623–626, *631*, 646Q, 650E
Palatine artery, 609
Palatine bones, 543, 593, 603
Palatine nerve, 615–616, 616, 626
Palatine secretion, 616
Palatine tonsils, 633
Palatoglossal folds, 623, 631
Palatoglossus muscles, 625, 629, 631, 636, 646Q, 649E
Palatopharyngeal folds, 623, 631
Palatopharyngeal muscles, 625, 631, 635, 636
Palmar, 3
Palmar aponeurosis, 94, *95,* 108
Palmar bursae, 111
Palmar cutaneous branch, 119–120
Palmar interosseous muscles, 113
Palmar muscles, *112,* 112–113, 115*t*
Palmar prehension, 121
Palmaris brevis muscle, 107
Palmaris longus, 95, 96, 108, 109
Palpebral conjunctiva, 546
Palpebral fissures, 544, 597
Palpebral ligament, 545
Palpebral margins, 544, 546
Palsies
  abducens nerve, 566Q, 570E
  Bell's, 545
  crutch, 84
  facial, 645Q, 649E
  nerve, 118
  Saturday night, 71, 84
Pampiniform plexus, 429, 456
Pancoast tumor, 265, 308Q, 311E
Pancreas, 360–363, 403Q, 406E
Pancreatic artery, 375
Pancreatic carcinoma, 348, 363
Pancreatic duct, 351
Pancreatic islets, 361, 362
Pancreatic pain, 362–363
Pancreatic sphincter, 357
Pancreatic veins, 362
Pancreaticoduodenal artery, 352, 362
Pancreatitis
  acute, 362
  hemorrhagic, 361
Panniculus adiposus, 11
Papillary choledochus, 357
Papillary muscles, 306Q, 310E
Papillary sphincter, 357
Papilledema, 556
Para-aortic lymph nodes, 300, 457, 470
Paracentesis, 341
Paracolic gutter, 371, 373
Paracolic recesses, 340
Paraduodenal fossae, 340
Paraduodenal hernias, 337
Paraesophageal hernia, 345
Parafollicular cells, 577
Paralysis, 153
Paralysis agitans, 525–526
Paralytic ileus, 367
Paramedian incisions, 321
Paramesonephric ducts, 453, 455
Parametrium, 464
Paranasal sinuses, 622–623
Pararenal fat, 385
Parasagittal planes, 4
Parasympathetic ganglia, 33
Parasympathetic innervation, 352
Parasympathetic nerves, 268
Parasympathetic neurons, 546
Parasympathetic system, 24
Parathyroid glands, 578
Parathyroid hormone, 578
Parathyroidectomy, total, 578
Paraurethral glands, 436, 481Q, 485E
Paravertebral ganglia, 32–33, 33
Paresis, 124Q, 128E
Parietal bone, 503
Parietal fascia, 412–413
Parietal lobe, 521, 522–523
Parietal peritoneum, 329, 331, 341, 393, 413, 461–462
Parietal pleura, 393
  clinical considerations, 260–261
  divisions, 259
  innervation, 260
  lines of reflection, 260, *260*
  mesenteries, 261
  pleural space, 261
  visceral, 261
Parietal serous pericardium, 271
Parieto-occipital fissure, 521
Parkinson's disease, 525–526
Paroöphoron, *454,* 455, 463
Parotid duct, 598, 599
Parotid gland, 598–599, 602, *607*
Pars flaccida, 558
Pars orbitalis, 601
Pars palpebrae, 601
Pars tensa, 558
Patella, 181–182, 183, *190*
Patellar reflex, 196
Patellar retinaculum, 182, 183, 186
Patellar tendon, 181, 186
Patent ductus arteriosus, 292
Patent urachus, 332, 486
Pectinate line, 422
Pectineal ligament, 193
Pectineus muscle, 189
Pectoral girdle, 56
  bones of, *56,* 56–58
  movement of, 60, 61*t,* 62
Pectoral muscles, 256
Pectoral nerve, 70
Pectoralis major, 247
Pectoralis minor, 122Q, 127E, 247
Pedicles, 133
Pelvic brim, 401Q, 405E
Pelvic cavity, 329, 410
Pelvic floor, 411
Pelvic fractures, 452
Pelvic girdle, 162, *162,* 163
Pelvic inflammatory disease (PID), 467
Pelvic inlet, 409, 479Q, 484E
Pelvic ligaments, 413
Pelvic outlet, 410, 479Q, 484E
Pelvic splanchnic nerves, 344, 390, 416, 482
Pelvic viscera
  development of urogenital ducts
    early stages, 453
    of female reproductive tract, *454,* 455
    indifferent stage, 453–454, *454*
    of male reproductive tract, *454,* 454–455
  female reproductive tract
    endopelvic fascia, *464,* 464–465
    mesometrium, *462,* 462–464, *463*
    ovaries, *465,* 465–467
    parietal peritoneum, 461–462
    uterine tubes, 467–468
    uterus, *454,* 468–472
    vagina, 472–473
  functional considerations
    fourth phase of coitus, 477
    implantation, 477–478, *478*

precoitus, 475–476
second phase of coitus, *476*, 476–510
third phase of coitus, 477
innervation of
  parasympathetic, *486, 506*, 509–510
  pelvic plexus (lateral pelvic plexus), 473–474
  sympathetic, *486, 506*, 509
male reproductive tract
  clinical considerations, 461
  epididymis, 457–458
  prostate gland, *458, 459*, 460–461
  seminal vesicles, *458*, 459–460
  testes, 455–457
  vas deferens, *458*, 458–459
pelvic portion of urinary system
  ureters, 482
  urethra, *483*, 487–488
  urinary bladder, *481*, 482–487, *483*, 485t
rectum
  clinical considerations, 482
  external structure, 480, *481*
  innervation, 482
  internal structure, 480
  vasculature, 480, *481*, 482
Pelvis, 161
  bones and joints
    birth canal, *408*, 409–410, 410t
    general structure, 407–409, *408*
    characterization of, 410
    differences between male and female, 410, 411t
  fascia and peritoneum
    parietal, 412–413
    peritoneum, 413
  female, 479Q, 484E
  lumbosacral plexus, *416*, 416–417
  musculature, organization, 411t, 411–412
  vasculature
    arterial supply, *413*, 413–415
    lymphatic drainage, 415–416
    venous return, 415
Penile lymphatics, 434–435
Penile urethra, 419, 426, 428, 452, 454, 455, 483Q, 486E
  perforations of, 432
Penis, *425*, 425–428
Pennate, 20
Pepsin, 349
Peptic ulcer, 349, 367, 400Q, 404E
Percussion, 243, 307Q, 311E
Perforating veins, 196
Perfusion, pulmonary, 270
Perfusion-ventilation ratio, 304Q, 309E
Pericardial cavity
  characteristics, 271, *272*
  clinical considerations, 272–273
  sinuses, 272
  structure, 271
Pericardial fluid, 271
Pericardial sinuses, 272
Pericardial tamponade, 272
Pericardiocentesis, 272–273
Pericarditis, 273
Pericardium, 393
Perilymph, 562
Perineal body, 419
Perineal muscle, 485
Perineal nerve, 417
Perineum
  anal triangle
    canal, 422–424
    fecal continence, 425

innervation, 424–425
ischioanal (ischiorectal) fossa, *421*, 421–422
vasculature, 424
development of external genitalia
  female, 419–421
  indifferent stage, 419
  male, 419–421
  divisions, 419, *420*, 420t
female urogenital triangle
  clinical considerations, 442–443
  deep perineal space (pouch), *439*, 439–440
  external genitalia, 435–437, *436*
  fascia, 437–438
  innervation, 442
  superficial perineal space (pouch), *438*, 438–439
  vasculature, 440–442, *441*
  general structure, 419
male urogenital triangle
  deep perineal pouch, *432*, 432–433
  external genitalia, *425*, 425–429, *426*
  fascia, 429–430, *430*
  functional considerations, 435
  innervation, 435
  superficial perineal pouch, *430*, 431–432
  vasculature, *434*, 434–435
Periodontal ligament, 594
Periosteal dura, 25
Periosteum, 502, 513
Peripheral, 4
Peripheral nervous system, 23
  definitions, 26
  spinal nerves, 28–29
  spinal roots, *27*, 27–28, *28*
Peripheral sensory neuron, 30, 48Q, 51E
Perirenal fat, 385, 386
Peristalsis, 366
Peristaltic activity, 391
Peristaltic rush, 366
Peristaltic waves, 349, 366
Peritoneal cavity, 259, 401Q, 405E
  abdominal, 329
  development of, and viscera
    early differentiation, *330*, 330–333, *331, 332*
    rotation of foregut, *333*, 333–334
    rotation of midgut, 334–337, *335, 336*
  gender differences, 329
  pelvic, 329
  peritoneum, 329–330
  relationships, definitive positions, 339–340, 340t
  subdivisions, 338–341, *339*
  clinical considerations, 341
Peritoneal fluid, 329–330
Peritoneal implantation, 467
Peritoneal structures, 338
Peritoneal subdivisions, 338–341, *339*
Peritoneum, 329, 331, 399Q, 404E, 413
Peritonitis, 341
  pain secondary to, 363
Peritonsillar venous plexus, 632
Periumbilical veins, 379
Periureteral sheath, 389, *390*
Perivascular plexus, 498
Peroneal nerve, 220, 417
Peroneus brevis muscle, 234Q, 238–239E
Peroneus longus muscle, 225

Petrosal nerve, 552, 560, 599, 646Q, 650E
Petrosal sinus, 512, 513, 514, 515
Petrosquamous fissure, 508
Petrotympanic fissure, 508
Phalanges, 56, 103, 201–203, 223
Phallus, 419
Pharyngeal artery, 580
Pharyngeal tonsil, 631, 633
Pharynx, 619
  basic structure, 630
  deglutition, 636–637
  hypopharynx, 633
  innervation, *635*, 635–636
  musculature, *632*, 633–635, 634t
  nasopharynx, 630–631
  oropharynx, *631*, 631–633
Pheochromocytoma, 392
Pheromonal stimulus, 530–531
Phlebotomy, 80, 100
Phonation, 644
Phrenic arteries, 403Q, 406E
Phrenic nerves, 69, 271, 304Q, 306Q, 309E, 310E, 394, 496
  lesion, 258
Pia mater, 25, 151, 157Q, 160E, 517
Pial hemorrhage, 518
Pinkeye, 546
Pinnae, 597
Piriform recesses, 633, 638, 646Q, 650E
Piriformis muscle, 164, 233Q, 238E, 407
Pisiform bone, 87, 93
Pituitary gland, 525
Plantar, 3
Plantar aponeurosis, 225
Plantar arches and ligaments, 224–225, 229
Plantar arterial arch, 212
Plantar artery, 229
Plantar calcaneonavicular ligament, 205
Plantar digital arteries, 229
Plantar fascia, 225
Plantar fasciitis, 225
Plantar flexion, 5
Plantar foot
  ambulation
    normal gait, 231
    pathologic gait, 231–232
  articulations, 223–225
  bones, 223
  innervation
    cutaneous sensory, 229
    motor, 230, *230*
  musculature
    extrinsic muscles, 225
    fascia, 225
    intrinsic muscles, *226*, 226–228, 227t
  vasculature
    arteries, 229
    veins, 229
Plantar interosseous, 228, 234Q, 239E
Plantar ligaments, 224
Plantar nerve, 230
Plantaris muscle, 215, 234Q, 238E
Plastic surgery, 10
Platysma muscle, 602
Pleura, 331
Pleural cavities, 259
  functional considerations, 261–262
  mesenteries (pulmonary ligaments), 261
  parietal pleura, 259
    clinical considerations, 260–261
    divisions, 259

innervation, 260
lines of reflection, 260, *260*
pleural space, 261
visceral pleura, 261
Pleural fluid, 261
Pleural recesses, 261
Pleural space, 259, 261
Pleural taps, 252
Pleurisy, 304Q, 309E
Plexus, 26
Plica fimbriata, 627
Plicae circulares, 367
Pneumocephalus, 538
Pneumothorax, 258, 261–262
Point of maximal impulse (PMI), 273
Polycystic kidneys, 391
Polymastia, 13
Polythelia, 13
Pons, 526
Popliteal artery, 195, 218, 234Q, 239E
Popliteal fossa, 195
Popliteal lesion, 221
Popliteal pulse, 195
Popliteal vein, 220
Popliteus muscle, 215, 234Q, 238E
Portacaval fistula, 379
Portal hypertension, 296, 345, 379, 482
Portal-systemic anastomoses, 348
Positive-pressure devices, 258
Postcentral gyrus, 522
Posterolateral disk prolapse, 140
Posterolateral fontanelles, 504
Postganglionic axons, 533, 550
Postganglionic neuron, 32
Postganglionic sympathetic fibers, 539
Postnatal circulation, *290,* 290–292
Postsynaptic neurons, 32–33
Postsynaptic parasympathetic, 32–33
Pouch of Morison, 339, 400Q, 404E
Practical nomenclature, 6–7
Preaortic nodes, 300
Precentral gyrus, 566Q, 570E
Precoitus, 475–476
Prefrontal gyrus, 522
Preganglionic axons, 533
Preganglionic neurons, 32, 151
Preganglionic parasympathetic neurons, 550
Preganglionic sympathetic fibers, 301
Pregnancy, changes in pubic symphysis, 165
Prehension
digital, 121
digital palmar, 121
palmar, 121
Premature ventricular systole, 285
Premolar teeth, 593, 594, 645Q, 649E
Prepatellar bursitis, 182
Prepuce, 419–420, 426, 436, 437
Presbyopia, 556
Pressure reflex, 269
Presynaptic parasympathetic fibers, 32–33
Presynaptic pathway, 32–33
Pretracheal space, 564Q, 569E
Prevertebral ganglia, 33
Primary auditory area, 523
Primary intestinal loop, 334
Primary motor area, 520, 521–522, 523
Primary pancreatic duct, 361
Primary rami, 408
Primary visual area, 523
Primitive gut, 330
Primitive yolk sac, 330
Procerus muscle, 602

Processus vaginalis, 324, 420, 421, 429
Proctodeum, 331
Progesterone, 466
Prolapse of rectum, 482
Pronation, 6
of forearm, 77
Pronator quadratus, 95
Pronator teres, 95
Prone, 4
Pronephros, 453
Proper hepatic artery, 347, 355
Prostate gland, 455, 458, *458, 459,* 460–461, 481Q, 483Q, 485E, 486E
carcinoma of, 481Q, 485E
Prostatectomy, 461
Prostatic concretions, 461
Prostatic ducts, 460
Prostatic urethra, 459, 481Q, *483,* 485E, 487
Prostatic utricle, 452, 455, 459
Prostatic venous plexus, 434, 461, 484
Prostatitis, 461
Protraction, 6
Proximal, 4
Proximal carpal row, 93
Proximal interphalangeal joints, 103, *104,* 104–105, 223, 228
Proximal phalangeal fractures, 103, 223
Proximal radioulnar joint, 73
movement, 75
support, 75
Proximal (superior) tibiofibular joint, 181
Proximal tarsal group, 223
Proximal tibiofibular joint, 201
Proximal tubule, 386, 389
Pseudoptosis, 546
Pterion, 506
Pterygoid, 647Q, 650E
Pterygoid hamulus, 506
Pterygoid muscles, 606, 607t
Pterygoid plate, 507, 602
Pterygoid plexus, 621
Pterygoid plexus thrombosis, 610
Pterygoid process, 507
Pterygoid processes, 506, 589
Pterygoid venous plexus, 610, 647Q, 650E
Pterygoid (vidian) canal, 507
Pterygomandibular raphe, 600
Pterygomaxillary fissure, 507
Pterygopalatine fossa, 507, 589, 590, *590,* 647Q, 650E
Pterygopalatine (sphenopalatine) ganglion, 604, 611
Pterygovaginal canal, 603
Ptosis, 533, 545
Pubic crest, 313, 315
Pubic (hypogastric) region, 313
Pubic ramus, 408
Pubic symphysis, 163, 165, 408
Pubic tubercle, 315
Pubis, *162,* 163, 407, 408
Pubocervical (vesicovaginal) ligaments, 465
Pubococcygeus muscle, 412, 422
Pubofemoral ligament, 167
Puboprostaticus muscle, 412
Puborectalis muscle, 412, 425, 480, 485, 486

Pubovaginalis muscle, 412
Pudendal artery, 195, 414
Pudendal block, 417
Pudendal canal, 412, 417, 424, 441
Pudendal cleft, 435
Pudendal nerve, 417, 422, 424, 435, 442, 473, 481Q, 485, 485E
Pudendal neurovascular bundle, 407, 408
Pudendal vein, 479Q, 484E
Pudendal veins, 441
Pudendal vessels, 422
Pulled elbow, 75
Pulmonary arteries, 271
Pulmonary circulation, 35, *36*
Pulmonary edema, 44, 270
Pulmonary effusion, 44
Pulmonary embolus, 307Q, 311E
Pulmonary fibrosis, 270
Pulmonary lymph nodes, 267
Pulmonary plexuses, 268
Pulmonary semilunar valve sounds, 289
Pulmonary sinuses, 281
Pulmonary valve, 281
Pulmonary veins, 271
Pulse, 124Q, 128E
axillary, 67
Pulse pressure, 288
Pulse rate, 38
Pump-handle effect, 246, 255
Punctum, 547
Pupil, 555
Pupillary dilation, 566Q, 570E
Pupillary reflex, 566Q, 570E
Putamen, 524
Pyloric antrum, 347, 400Q, 404E
Pyloric canal, 347
Pyloric sphincter, 347
Pyloric stenosis, 334
Pylorus, 334, 347
Pyothorax, 262
Pyramidal cells, 520
Pyramidal pathways, 523, 524
Pyramidal tracts, 522
Pyramidalis muscle, 317

Q

Quadrangular space, 82–83
Quadratus lumborum muscle, 148, 256, *393,* 393t, 394
Quadratus plantae, 227, 228

R

Radial artery, 97, *98,* 99, 116
Radial bursa, 95, 110, 112
Radial collateral ligament, 75
Radial keratotomy, 554
Radial nerve, 90, 92, *92,* 118
Radial notch, 73
Radial pulse, 99, 116, 117
Radial tuberosity, 74
Radicular arteries, 154
Radiocarpal joint, *86,* 87–88
Radioulnar joint, 74
Radius, 85
articulations, 73
characteristics, 73
Rami, 591–592
Rami communicantes, 29
Reanastomosis, 402Q, 405E
Rectal artery, 376, 415, 480
Rectal folds, 480

Rectal nerve, 417
Rectal plexus, 415, 474
Rectal sling, 412, 480
Rectal vein, 480
Rectocele, 473
Rectosigmoid artery, 374, 376
Rectosigmoid junction, 377
Rectouterine, 462
Rectouterine folds, 464
Rectouterine pouch, 469, 472
Rectovaginal fistulas, 473
Rectovaginal pouch, 462
Rectovaginal septum, 469, 472
Rectovesical pouch, 482
Rectovesical septum, 482
Rectum, 375, 403Q, 406E
  carcinoma of, 482
  clinical considerations, 482
  external structure, 480, *481*
  innervation, 482
  internal structure, 480
  prolapse of, 482
  vasculature, 480, *481*, 482
Rectus abdominis muscle, 317, 318, 400Q, 404E
Rectus capitis muscle, 144, 146
Rectus femoris muscle, 183, 189, 190
Rectus muscles of eye, 547, 550
Recurrent patellar dislocation, 183
Reference landmarks, 10
Referred cardiac pain, 286
Referred esophageal pain, 297, 345
Referred pain, 34, 352, 360, 362
Referred small bowel pain, 366
Reflex
  abdominal, 325, 398
  ankle-jerk, 221
  blink, 551
  cardiac, 579–580
  chemical, 269
  cremaster, 197, 326, 429, 480Q, 484E
  enteric, 366
  enterogastric, 349
  hamstring, 199
  knee-jerk, 233Q, 238E
  knee-jerk (patellar), 196
  patellar, 196
  pressure, 269
  respiratory, 580
  sneeze, 622
  stretch, 269
Reflex control, 22
Reflex sympathetic dystrophy (RSD), 38
Refractivity, 556
Regional anatomy, 3
Regurgitative esophagitis, 298, 345, 346
Remodeling, 16, 17
Renal arteries, 386, 390
Renal calculus, 400Q, 405E
Renal capsule, 386
Renal colic, 401Q, 405E
Renal corpuscle, 386, 389
Renal fascia, 385, 401Q, 405E
Renal pelvis, 386, 389
Renal sinus, 386
Renal vein, 386
Reposition of thumb, 114, 125Q, 129E
Respiration, 269
Respiratory distress, 305Q, 309E
Respiratory function test, 305Q, 309E
Respiratory reflex, 580

Respiratory system, functional anatomy of
  basic principles, 269
  pulmonary function, 269–270
Rete testis, 456
Retina, 528, *553*, 555–556
Retinacula, 12, 166
Retinal blood flow, 556
Retinal detachment, 556
Retinal engorgement, 566Q, 570E
Retinal pathways, 523
Retraction, 6
Retrocecal fossa, 340
Retroduodenal arteries, 351
Retromandibular vein, 499, 581
Retroperitoneal mass, 480Q, 484E
Retroperitoneal structure, 480
Retroperitoneal structures, 338
Retropubic space, 483
Retrovisceral space, 564Q, 569E, 630
Rhinorrhea, 537
Rib cage, 245
  movement, 246–247
    diaphragm, 250–251
  organization, 243–246
Ribs, *245*, 245–246
Right atrium, 274, *279*, 279–280
Right A-V (tricuspid) valve, 280
Right coronary artery, 275, 308Q, 311E
Right hemisphere, 521
Right ventricle, 274, 280–281
Right ventricular hypertrophy, 292
Right-sided effects, 645Q, 649E
Right-to-left shunt, 281, 292
Right-ventricular heart failure, 289
Rima glottidis, 638, 641
Rods and cones, 555
Roof (tegmen tympani), 559
Root, 594
Rootlets, 27
Rotation, 166
  in hip joint, 172–173, 173*t*
  medial, 63–64
Rotation effect, 255
Rotator cuff, 59, 63
  tears, 59
Round window, 560
Runners stitch, 354

S

Saccule, 561
Sacral artery, 414
Sacral cornua, 137, 158Q, 160E, 409
Sacral curvature, 131
Sacral foramina, 156Q, 159E, 408
Sacral hiatus, 137, 408–409, *416*
Sacral nerves, 28
Sacral parasympathetic outflow, 33
Sacral plexus, 176, *176*, 416
Sacral region, 158Q, 160E
Sacral vertebrae, 136–138, *137*
Sacrococcygeal joint, 138
Sacroiliac joint, 138, 163, 164–165, 407, 409
Sacrospinalis group, 143, *144*, 145*t*
Sacrospinous ligament, 165, 407, 408, 409, 411
Sacrotuberous ligament, 407, 408, 409, 422
Sacrum, 408–409, *416*
Saddle block, 435, 442
Sagittal groove, 509
Sagittal sinus, 509, 514, 515
Sagittal suture, 504

Salpingitis, 467, 482Q, 485E
Salpingopharyngeal fold, 631
Salpingopharyngeus muscle, 631, 635
Saphenous hiatus, 195
Saphenous nerve, 196
Saphenous vein, 212, 219, 235Q, 239E
Sartorius muscle, 189, 191
Saturday night palsy, 71, 84
Scala tympani, 562
Scala vestibuli, 561, 562
Scalene muscles, 146, 247, 256, 494
Scalene syndrome, 495
Scalp, 501
  composition, 501–502, *502*
  innervation, 503
  vasculature, 503
Scaphoid, 85–86, 93, 124Q, 128E
  fracture, 86
Scaphoid fossa, 507, 557
Scapula, 243
  articulations, 57
  characteristics, 57
Scapulothoracic joint
  movement, 58–59
  structure, 58
Scarpa's fascia, 11, 47Q, 50E, 315
Sciatic foramen, 414, 424, 434, 441
Sciatic nerve, 177, *198*, 198–199, 407
Sciatic notch, 163
Sciatica, 138, 198, 407
Sclera, *553*, 553
Scoliosis, 132
Scrotal septum, 420
Scrotum, 420, *425*, *427*, 428, 483Q, 486E
Sebaceous glands, 546
Second digit, motor function of, 123Q, 128E
Secondarily retroperitoneal structures, 338
Secondary cortical (association) area, 521
Second-order neuron, 152, 153
Secretin, 350
Segmental innervation, 10
Segmental pulmonary resection, 264
Segmentation, 10
Sella turcica, 511
Semicircular canals, 561–562, 562
Semilunar herniation, 322
Semilunar insufficiency, 290
Semilunar stenosis, 290
Seminal fluid, 459
Seminal vesicles, 454, *458*, 459–460, 483, 483Q, 486E
Seminiferous cords, 454
Seminiferous tubules, 456
Semispinalis muscle, 143
Sensory association areas, 523
Sensory homunculus, 522, *522*
Sensory pathways, 352
Sensory tracts, *152*, 152–153
Septal cartilage, 597
Septal defects, 292
Septate uterus, 473
Septum, nasal, 589
Septum primum, 280
Septum secundum, 280
Septum secundum defects, 280
Serous fluid, 329
Sesamoid bones, 15–16, 20, 87, 223
Shin splints, 217
Shoulder dislocation, 59–60
Shoulder girdle, articulations of, 58–60

Index

Shoulder joint
  axillary vasculature
    axilla, 65
    vascular supply, 65–68, *66*
  muscle function at, 60, 61*t, 62, 63,* 63–65, 64*t*
  combined movements of shoulder and arm, 65
  movement of arm, 60, *63,* 63–65, 64*t*
  movement of pectoral girdle, 60, 61*t, 62*
Shoulder region
  basic organization, 55
  basic principles, 55
  brachial plexus
    development, 68
    lesions, 71
    subdivisions, 69–71
  regions of upper extremity, 55–56
Sigmoid arteries, 374
Sigmoid colon, 314, 374–375
Sigmoid mesocolon, 333, 374
Sigmoidal arteries, 376
Sign of benediction, 119
Sinoatrial nodal branch, 275
Sinoatrial node, 278, 283
Sinovaginal bulbs, 455
Sinus thrombosis, 566Q, 570E
Sinus venarum, 279
Sinus venosum sclerae, 556
Sinuses, pericardial, 272
Sinusoids, 39–40
Situs inversus abdominis, 480Q, 484E
Situs inversus viscerum, 335, 337
Skeleton
  articulations, 17–19, *18*
  bone, 15
  cartilage, 17
Skull, 501, *502*
Sliding hiatus hernia, 345–346
Sneeze reflex, 616, 622, 647Q, 650E
Soft palate, 624, *624,* 626
Soleus muscles, 215
Somatic cell bodies, 27–28
Somatic cervical structures, 573
Somatic dermatome, 34
Somatic efferent nuclei, 533
Somatic nerves, 397, 527
Somatic nervous system, 23–24, 29, *30*
  afferent nerves, 29–30
    classification, 29–30
    pathway, 30
  efferent nerves, 30, *30*
    classification, 30
    pathway, 30–31
    reflex arcs, 31
  visceral afferent nerves, 31
Somatic sensory system, 23
Somatic wall, 259
Somatopleure, 330
Spasmodic torticollis, 495, 497
Special visceral efferent nuclei, 531–532
Speech area, 522
Spermatic cord, 317, 324–325, *325, 425,* 428–429, 458
Spermatozoa, 460
Sphenoethmoid recess, 588, 619, 622
Sphenoid bone, 503, 506, 510, 543, 589
  landmarks, 506
Sphenoid sinuses, 589, 622
Sphenoidal wings, 589
Sphenomandibular ligament, 591, 605
Sphenopalatine artery, 620

Sphenopalatine foramen, 593, 603, 609, 615, 622
Sphenopalatine nerve, 615
Sphenopalatine neurovascular bundle, 603
Sphenopalatine notch, 593
Sphincter, anal, 481Q, 485E, 600
Sphincter choledochus, 357
Sphincter pupillae, 555
Sphincteric incontinence, 422
Spinal accessory nerve, 496, 496–497, *497,* 513, 532, 568Q, 571E
Spinal anesthesia, 26, 151, 155, 472
Spinal arachnoid, 26
Spinal cord, 23, 519
  clinical considerations
    caudal anesthesia, 155
    epidural anesthesia, 155
    intrathecal anesthesia, 155
    lumbar puncture, 155
  general organization
    characteristics, 149
    clinical considerations, 153
    external structure, 149
    internal structure, *151,* 151–153
    spinal meninges, 149–151, *150*
    vertebral canal, 149
  meninges, 26
  morphology, 25–26
  nerve roots
    primary rami, 154
    spinal, *150,* 153–154
    termination of, 157Q, 159E
    transection, 153
  vasculature
    arterial supply, 154
    venous return, 154
Spinal dura, 516
Spinal dura mater, 26
Spinal meninges, 26, 149–151, *150*
Spinal nerves, 23, 28–29, 133, 149, 254, 399Q, 404E
Spinal pia mater, 26
Spinal roots, 26–28, *27, 28*
Spinal segment, 27, 156Q, 159E
Spinal tap, 151
Spinalis muscle, 143
Spinocerebellar pathways, 153
Spinoreticulothalamic tract, 152–153
Spinous processes, 133, 158Q, 160E, 244
Spiral organ, 562
Splanchnic nerves, 33, 297, 301, 302, 329, 344, 348, 382, 399Q, 404E
Splanchnopleure, 330
Spleen, 314, 334, 363–364
Splenectomy, 364
Splenic artery, 348, 363, 375
Splenic flexure, 372
Splenic laceration, 364
Splenic vein, 363, 378
Splenius capitis muscle, 143
Splenius cervicis muscle, 143
Splenomegaly, 363
Splenorenal anastomosis, 379
Splenorenal ligament, 332, 334, 346, 363
Split-brain individual, 523
Spondylolisthesis, 137–138
Spondylolysis, 137
Sprains, 19
Squamous bones, 504
Stapedius muscle, 559, 561, 611, 647Q, 650E
Stapes, 561, 567Q, 570–571E
Stenosis, 377, 396, 647Q, 650E
  congenital hypertrophic pyloric, 347

Sternoclavicular joint, 56, 244
  movement, 58
  structure, 58
  support, 58
Sternocleidomastoid muscle, 256
Sternocostal angle, 272
Sternocostal triangle, 394
Sternohyoid muscle, 574, 575, 584
Sternomastoid muscles, 247, 494, 532, 564Q, 569E, 573
Sternopericardial ligaments, 271, 293
Sternothyroid muscle, 575, 584
Sternotomy, 244, 306Q, 310E
  midline, 271
Sternum, 244–245
Stiff knee, 182
Stomach, 313, *346,* 346–350
  empty, 314
Strabismus, 533, 550
  lateral, 550
  medial, 550
Straight sinus, 515
Stress incontinence, 481Q, 485E
Stretch marks, 315
Stretch reflexes, 269
Striae, 315
Stroke, 524, 540
Student's elbow, 73
Styloglossus muscle, *628,* 628–629
Stylohyoid ligament, 573, 638
Stylohyoid muscle, 573, 576, 577
Styloid process, 85, 201, 506, 508
Stylomandibular ligament, 591, 605
Stylomastoid foramen, 508, 567Q, 571E, 611, 645Q, 647Q, 649E, 651E
Stylopharyngeus muscle, 632, 635, 636, 647Q, 650E
Subacromial bursa, 59, 122Q, 127E
Subacromial bursitis, 59
Subacute tenosynovitis, 106
Subaponeurotic space, 225, 502
  infection of, 105
Subarachnoid hemorrhage, 518, 540
Subarachnoid space, 25, 26, 150, 517, 537–538, 565Q, 569E
Subclavian arteries, 65–67, 294, *294,* 498
Subclavian trunks, 499
Subclavian vein, 68
Subclavius nerve, 70
Subcondylar fracture, 592
Subcostal nerve, 29
Subcostal plane, 313
Subcutaneous bursae, 182
Subcutaneous fascia, 107
Subdeltoid bursa, 59
Subdural hematoma, 25, 517, 518, 541, 565Q, 569E
Subdural space, 25, 26, 46Q, 49E, 149, 150, 516
Subhepatic recess, 339
Sublingual duct, 630
Sublingual gland, 627, 630
Submandibular duct, 630
Submandibular ganglion, 630
Submandibular gland, 627, 630
Submandibular space, 564Q, 569E
Submental triangle, 574
Suboccipital muscles, 144, 146*t*
Suboccipital nerve, 495, 564Q, 569E
Subphrenic recess, 339
Subpial hemorrhage, 518
Subscapular artery, 67
Subscapularis muscle, 125Q, 129E
Substantia nigra, 525
Subtalar joint, 204–205

Subthalamic nuclei, 525
Subthalamus, 525
Sucking pneumothorax, 262
Sulcus tali, 202
Summation effect, 249
Superciliary crest, 505
Superficial cervical artery, 498
Superficial epigastric arteries, 195, 320
Superficial fascia, 206, 225
Superficial fibular nerve, 213
Superficial inguinal ring, 399Q, 404E
Superficial lymphatics, 320
Superficial perineal space, 481Q, 482Q, 485E, 486E
Superficial petrosal nerve, 568Q, 571E
Superior duodenum, 339, 350
Superior epigastric artery, 253
Superior extensor retinaculum, 209
Superior gluteal artery, 173–174, *174*, 414
Superior gluteal nerve, *169, 176*, 176–177
Superior gluteal neurovascular bundle, 407
Superior hypogastric plexus, 474
Superior meatus, 588
Superior oblique capitis muscle, 144
Superior peroneal retinaculum, 209
Superior ramus, 163
Superior thoracic aperture (thoracic inlet), 243
Superior vena cava, 279, 293
Supernumerary nipples, 13
Supination, 6
 of forearm, 77
Supine, 4
Supplementary motor area, 522
Supraclavicular nodes, 13, 117–118
Supracolic compartment, 339, *340*
Supracondylar fracture, 78, 102, 181, 234Q, 239E
Supracondylar ridges, 73
Supraduodenal artery, 351
Supraglenoid tubercle, 57
Suprahyoid muscles, 573, *574*, 576
Supramarginal gyrus, 523
Supramastoid crest, 505, 506
Suprameatal spine, 505
Suprameniscal compartment, 605
Supraorbital nerve, 503, 550
Supraorbital notch, 505, 543, 550, 589, 614
Supraorbital ridge, 505
Suprapatellar bursa, 182, 184
Suprapatellar bursitis, 182
Suprapiriform portion of greater sciatic foramen, 414
Suprapiriform recess, 407
Suprarenal arteries, 392
Suprascapular artery, 66, 499, 581
Suprascapular nerve, 70, 496
Suprascapular vein, 581
Supraspinatus muscle, 125Q, 129E
Supraspinatus tendinitis, 59
Supraspinous fossa, 57
Supraspinous ligament, 140
Supratrochlear lymph nodes, 80, 100
Supratrochlear nerve, 503, 551
Supratrochlear notch, 614
Supraventricular crest, 280
Sural nerve, 220
Suspensory ligaments, 13
Sustentaculum tali, 202
Sutural bone, 506
Suture, 17, *18*
Sweat glands, 9, 546
Swimmer's ear, 558
Sylvian fissure, 521
Sympathectomy, 383
Sympathetic chain, 564Q, 569E
Sympathetic ganglia, 33, 474
Sympathetic postganglionic fibers, 32–33
Sympathetic postganglionic neurons, 47Q, 50E
Sympathetic preganglionic fibers, 32
Sympathetic system, 24
Sympathetic trunk, 302–303, 397
Symphyseal fracture, 592
Symphyses, 132
Synarthroses, 17
Synarthrosis, 48Q, 50–51E
Syndesmosis, 17, *18*, 48Q, 50–51E
Synergistic muscles, 21
Synovial capsule, 18
Synovial cavity, 18
Synovial fluid, 18
Synovial joints, 132
Synovial membrane, 18
Synovial tendon sheaths, 12, 105, 110
Syringomyelia, 153
Systemic anatomy, 3
Systemic circulation, 35
Systemic hypertension, 389
Systole, 36
Systolic pressure, 38, 288

T

Tabes dorsalis, 153
Tactile agnosia, 523
Talocalcaneal joint, 205
Talocalcaneal ligament, 204, 205
Talocalcaneonavicular joint, 204–205, 205
Talocrural joint, 201, 203–204, *204*
Talofibular ligaments, 203, 204
Talonavicular joint, 205
Talonavicular ligament, 205
Talotibial ligament, 204
Talus, 201–202
Tarsal bones, 201–203
 fractures, 203
Tarsal glands, 546
Tarsal joints, 206
Tarsal muscle, 546
Tarsal plate, 545
Tarsometatarsal joint, 223–224, 234Q, 239E
Tarsus, 162, *202, 204*
Taste, 531
Teeth, 593–595, *594*
Tegmen tympani, 512
Temporal artery, 608
Temporal bone, 503, 506, 610–611
 landmarks, 505, 507–509
Temporal lobe, 521, 523
 herniation, 513
Temporal process, 593
Temporal vein, 610
Temporalis muscle, 605
Temporomandibular joint, 505, *604*, 604–605, 645Q, 649E
Temporomandibular ligament, 605
Tendinous inscriptions, 317
Tendons, 20
Tennis elbow, 75
Tenosynovitis, 112
 subacute, 106
Tension pneumothorax, 262
Tensor fasciae latae muscle, 188
Tensor tympani muscle, 559, 560, 560*t*
Tensor veli palatini muscle, 624, 633
Tentorial herniation, 517
Tentorial notch, 514, 519
Tentorium cerebelli, 514
Teres major muscle, 122Q, 127E
Teres minor muscle, 125Q, 129E
Terminal ileum, 334
Terminal nerve, 621
Testes, 420, 454, 455–457
 definitive, 322–323
Testicular arteries, 429, 456
Testicular (internal spermatic) veins, 456
Testicular neurovascular bundle, 323
Testicular torsion, 457
Testicular vein, 480Q, 484E
Tetralogy of Fallot, 281, 292
Thalamic radiations, 521
Thalamic syndrome, 525
Thalamus, 152, 520, *524*, 525
Thebesian veins, 306Q, 310E
Thenar bursa, 95, 109–110
Thenar compartment, 109
Thenar muscles, 113–114, *115*
Thenar space, 109
Thermoregulation, 9
Thigh, 162, *162*, 163, *180*
 fascia and muscles of
  fascia lata, 188–189
  fascial septa, 189
  musculature, 189–192, *190*, 191*t*, 192*t*
 innervation of
  anterior compartment, 196–197, *197*
  posterior compartment, *197*, 197–199
 vasculature of
  arterial supply, 193–195
  venous return of, 195–196
Third (least) occipital nerve, 495
Third-order neuron, 152
Thoracentesis, 262, 304Q, 305Q, 309E
Thoracic, duct, 42–43
Thoracic age, musculature
 extrinsic inspiratory, 247, *248*, 248*t*
 intrinsic (body wall), 247, 249, *249*, 249*t*
Thoracic artery, 66, 253, 319
Thoracic cage
 innervation, 253–254
 respiratory mechanics
  approximate ventilation capacities, 257
  clinical considerations, 257–258
  costal ventilation, *255*, 255–256
  diaphragmatic ventilation, 256
  respiration, 254
 rib, 243–247, *244*, *245*, *247*
 vasculature, 252–253
Thoracic cavity, 259
Thoracic curvature, 131
Thoracic duct, 42–43, 43, 47Q, 50E, 299–300, *300*, 396, 499, 582
Thoracic nerves, 28, 253
Thoracic outlet syndrome, 246
Thoracic region, 158Q, 160E
Thoracic splanchnic nerves, 286, 301, 397
Thoracic vertebrae, *135*, 135–136, 157Q, 159E, 243–244
Thoracodorsal nerve injury, 71
Thoracolumbar outflow, 32
Thorax, 243, 299

Thrombosis, 377
Thymectomy, 293
Thymus gland, 293
Thyroarytenoid muscle, 642
Thyrocervical artery, 66–67, 498–499, 581
Thyroepiglotticus muscle, 641
Thyroglossal duct, 577, 627
Thyrohyoid muscle, 574, 575, 584
Thyroid arteries, 345, 580–581
Thyroid cartilage, 639
Thyroid gland, *574*, 577–578
Thyroidea ima, 578
Thyroidectomy, 578
Thyroxine, 578
Tibia, 162, *180*, 201, *202*
  fractures of, 204
Tibial artery, 195, 211, *211*, 215, 218, *219*, 235Q, 239E
Tibial condyle, 181
Tibial nerve, 198, 215, 221, 236Q, 240E, 417
Tibial pulse, 215
Tibial tubercle, 201
Tibial tuberosity, 181
Tibial vein, 219
Tibialis anterior muscle, 209
Tibialis posterior muscle, 215, 217, 225
Tibiofibular ligaments, 203
Tibionavicular ligament, 204
Tic douloureux, 614
Tidal volume, 255, 257
Tongue, 626–630
Tonsillar capsule, 632
Tonsillar fossae, 631
Tonsillectomy, 633
Tonsillitis, 633
Torticollis, 495
Torus tubarius, 631
Total ipsilateral blindness, 523, 528, *528*
Total parathyroidectomy, 578
Toxic megacolon, 475
Trabeculae carneae, 280
Trachea, 265, *265*
Tracheobronchial nodes, 268
Tracheostomy, 578
Tragus, 557
Transitional epithelium, 484
Transperitoneal anastomotic connections, 352
Transperitoneal veins, 379
Transudative edema, 40
Transurethral cystoscopy, 486
Transversalis fascia, 429
Transverse abdominal muscles, 317
Transverse abdominis muscles, 401Q, 405E
Transverse carpal arch, 93, 94
Transverse carpal ligament, 93, 94, 108, 124Q, 128–129E
Transverse cervical artery, 66–67, 498
Transverse colon, 334, *371*, 372, 402Q, 405E
Transverse crural septum, 215
Transverse foramen, 134, *135*
Transverse foramina, 157Q, 159E, 539
Transverse humeral ligament, 57
Transverse incisions, 321
Transverse mesocolon, 333
Transverse metacarpal arch, 104, *104*
Transverse midplane diameter, 479Q, 484E
Transverse perineal ligament, 433
Transverse plantar arches, 224
Transverse processes, 133, 243
Transverse sinus, 514, 515
Transverse tarsal joint, 205–206
Transverse thoracic muscles, 250, 256
Transverse thoracis, 304Q, 309E
Trapezium, 93
Trapezius muscle, 497, 532
Trapezoid bone, 93
Trapezoid ligament, 58
Trendelenburg lurch, 233Q, 238E
Triangular disk, 85
Triangular fossa, 557
Triceps muscle, 63
Triceps brachii muscle, 76, 77
Triceps surae muscle, 215
Tricuspid valve, 594
  insufficiency, 289
  sounds, 289
Trigeminal cave, 514
Trigeminal ganglion, 512, 645Q, 648Q, 649E, 651E
Trigeminal motor nucleus, 531
Trigeminal nerve, 526, 529, 544, 550, 553, 568Q, 571E, 577, 585, 613–618, *614*
Trigeminal neuralgia, 614
Trigeminal nuclei, 529
Triglycerides, 41
Triquetral bone, 87, 93
Trochlea, 73, 74
Trochlea tali, 202
Trochlear nerve, 526, 533, 544, 547, 550, 585
Trochlear notch, 73
Trochlear nucleus, 533, 550
Tropocollagen, 11
True ribs, 245
True vocal folds, 638
Tubal branches, 415
Tubal implantation, 467–468
Tubal tonsils, 631, 633
Tubercle, 245, 246, 638
Tuberculum sellae, 510, 511
Tumescence, 475
Tumors. (*See* Carcinoma)
Tunica albuginea, 427, 456
Tunica vaginalis, 420, 421, *427*, 429, 456
Tunnel vision, 529
Turgor, 475
Tympanic bulla, 560
Tympanic cavity, 558–559
Tympanic membrane, 558
Tympanic nerve, 585
Tympanic orifice, 559
Tympanic plate, 505, 508
Tympanic plexus, 560, 599
Tympanic tube, 507
Tympanosquamosal fissure, 508
Tympanum, 558, 597

U

Ulcers
  duodenal, 352
  peptic, 349, 400Q, 404E
Ulna, 85
  articulations, 73
  characteristics, 73
Ulnar artery, *98*, 99
Ulnar bursa, 112, 125Q, 129E
  inflammation, 119
Ulnar collateral ligament, 74
Ulnar groove, 82
Ulnar head of pronator teres, 124Q, 128E
Ulnar nerve, 82, 96, 102, 113, 114, *120*, 120–121, 123Q, 126Q, 128E, 129E
Ulnar notch, 85
Ulnar pulse, 117
Ulnocarpal joint, 85, *86*, 88
Umbilical arteries, 483, 484
  closure of, 291–292
Umbilical artery, 415
Umbilical folds, 331–332, 399Q, 404E
Umbilical hernia, 322, 334, 337
  acquired, 322
  congenital, 322
Umbilical ligaments, 483
Umbilical region, 313
Umbilical ring, 334
Umbilical vein, 292, 331
  closure of, 292
Umbilicus, 10
Umbo, 558
Unicornuate uterus, 473
Unipennate muscles, 20
Unmyelinated axon, 32
Upper motor neuron, 48Q, 51E
Urachal cyst, 332, 486
Urachus, 332, 453, 483
Ureterovesical valves, 484
Ureters
  clinical considerations, 391
  course and relations, 389
  innervation, 390–391
  vasculature, 390
Urethra, 453, 455, *483*, 487–488
  perforation of, 433
Urethral crest, 452
Urethral folds, 436, 455
Urethral glands, 428, 453
Urethral meatus, 419, 428, 455
Urethral orifice, 453
Urethral sphincter, 433, 485
Urethrocele, 453, 481Q, 485E
Urethrovaginal fistulas, 473
Urinary bladder, 314, 453–454, 468, *481*, 482–487, *483*, 485t
  capacity, 485
Urinary continence, 412, 484–485
Urinary stress incontinence, 440
Urinary tract infection, 461
Urination, 391
  frequency of, 481Q, 485E
Urogenital diaphragm, 412, 419, 432, 433, 468, 483
Urogenital groove, 419
Urogenital hiatus, 432
Urogenital membranes, 419
Urogenital sinus, 419, 436
Urogenital triangle, 419
Uterine arteries, 415, 466, 469, 472
Uterine cavity, 468
Uterine implantation, 478, *478*
Uterine neurovascular bundle, 463
Uterine prolapse, 471
Uterine tube transit, 477
Uterine tubes, 463, 467–468
Uterine veins, 469
Uterorectal pouch, 455
Uterosacral ligaments, 464
Uterosalpingography, 468
Uterovaginal plexus, 470, *471*, 473, 474
Uterovesical pouch, 455, 462
Uterus, 314, *454*, 468–472
Uterus duplex, 473
Utricle, 561

Uvula, 485, 647Q, 651E
Uvulae muscles, 624

V

Vagal trunks, 345
Vagina, 455, 472–473, 483Q, 486E
Vaginal anomalies, 473
Vaginal epithelium, 472
Vaginal examinations, 473
Vaginal introitus, 436
Vaginal venous plexus, 472
Vagotomy
  selective, 382
  total, 382
Vagus nerves, 269, *294*, 294–295, *295*, 301, 303, 305Q, 309E, 343, 344, 345, 352, 366, 372, 382, 527, 530, 531, 558, 585, 586, 624, 625, 626, 629, 633, 634, 635, 636, 642, 646Q, 650E
Vallate papillae, 627
Valsalva fixation, 322
Valsalva maneuver, 318
Valve locations, 288, *289*
Valvular defects and disease, 289
Valvular stenosis, 289
Varices, esophageal, 345
Varicocele, 457, 480Q, 484E
Varicose veins, 196
Vas deferens, 323, 429, 454, 457, *458*, 458–459, 459
Vascular malformations, 35
Vasculature, 343
Vasectomy, 458
Vasopressin, 525
Vastus intermedius muscles, 183
Vastus lateralis muscle, 183
Vastus medialis longus muscle, 189
Vastus medialis muscle, 183
Vastus medialis obliquus muscle, 189
Veins, 35. *(See also specific veins)*
Velopharyngeal seal, 625, 631
Vena cava, 396, 480Q, 484E
Venae cava, 271
Venae comitantes, 100, 196
Venipuncture, 41, 100, 123Q, 127E, 196
Venous anastomoses, 415
Venous hemorrhage, 541
Venous homograph, 306Q, 310E
Venous plexus, 415
Venous pump, 196
Venous return, 277–279
Venous sinuses, 25
Venous system, 565Q, 569E
  arteriovenous anastomoses, 41
  capillary beds, 40
  portal, 41
  veins, *36*, 40–41
Ventilation, 269
  costal, 257
  diaphragmatic, 257
  effects of gravity on, 256
Ventilation capacity, 304Q, 309E
Ventilation-perfusion ratio, 270
Ventral gray matter, 254
Ventral herniation, 321
Ventral horn, 27, 48Q, 51E
Ventral lingual mucosa, 627

Ventral mesogastrium, 353
Ventral primary rami, 29, 254, 320–321, 495, 496
Ventral roots, 27–28, 253
Ventricle
  left, 274
  right, 274
Ventricular diastole, 288
Ventricular septal defects, 281, 292
Ventricular systole, 288
Vermiform appendix, *369*, 369–370
Vertebra, 131
Vertebra prominens, 135, 244
Vertebral arteries, 65, 135, 154, 498, 539, *540*
Vertebral body, 149
Vertebral column
  composition, 131
  intervertebral disks
    clinical considerations, 139
    external structure, 138–139, *139*
    functions, 139
    internal structure, 139
  ligaments
    anterior longitudinal, 140
    cruciform, 141
    ligamentum flavum, 140
    posterior longitudinal, 141
    supraspinous, 140
  regional modifications of characteristics, 133–135, *134*
  clinical considerations, 138
  coccygeal, 138
  lumbar, 136
  sacral, 136–138, *137*
  thoracic, *135*, 135–136
  spinal curvatures
    abnormal, 131–132
    normal, 131
  structure of typical vertebra
    arch, 132–133, *135*
    body, 132, *135*
    foramina, 133, *135*
    processes, 133, *135*
Vertebral foramen, 133, *135*
Vertebral venous plexus, 149, 154, 156Q, 159E
Vertebrate body, 259
Vertical axis, 5
Vertical incisions, 321, 322
Vesical arteries, 390, 484
Vesical artery, 415
Vesical plexus, 474
Vesical venous plexus, 437
Vesicouterine pouch, 469, 483
Vesicovaginal fistulas, 473
Vesicovaginal septum, 483
Vesiculitis, 461
Vessels, lymphatic, 42
Vestibular bulbs, 437, 438, 483Q, 486E
Vestibular fold, 638
Vestibular glands, 436, 439
Vestibular mucosa, 620
Vestibule, 436, 483Q, 486E, 561
Vestibule sinus, 421, 436
Vestibulocochlear nerve, 562, 585
Vicar's knee, 182
Vidian nerve, 604
Vincula, 110
Visceral afferent (GVA), 33–34

Visceral cell bodies, 28
Visceral cervical structures, 573
Visceral efferent model, 531–532
Visceral nerve, 527
Visceral nervous system, 24
  autoimmune composition, *31*, 31–32
  divisions, 32–33
  visceral efferent pathway, *31*, 32
  visceral afferent (GVA), 33–34
Visceral origin, pain of, 46Q, 50E
Visceral pain, 368
Visceral peritoneum, 329, 331, 341, 413, 462, 469
Visceral pleura, 261
Visceral sensory system, 23
Visceral serous pericardium (epicardium), 271
Visual agnosia, 523
Visual association area, 523
Visual cortex, 565Q, 570E
Vital capacity, 257
  reduction of, 269
Vitreous chamber, 556
Vocalis muscle, 641, 647Q, 650E
Volar carpal ligament, 94, 108
Volkmann's ischemic contracture, 80
Voluntary muscles, 23
Volvulus, 367
Vomer, 589
Vomeronasal organ, 621
Vorticose veins, 552, 554
Vulva, 435–436, *436*

W

Waldeyer's ring, 633
Water on the knee, 182
Weakness of pronation, 124Q, 128E
White matter, 25, *151*, 151–153, *152*, 519, 521
White rami communicantes, 29, 32, 33, 46Q, 50E, 301, 302
White ramus communicans, 47Q, 50E
Winged scapula, 69
Wolffian duct, 453, 457, 458
Wolff-Parkinson-White syndrome, 285
Wormian bone, 506
Wrist, 56
Wrist articulations, 87–88
Wrist drop, 84, 92, 118

X

Xiphisternal synchondrosis, 244–245, 245
Xiphoid process, 245, 313

Z

Zonular fibers, 554
Zygomatic arch, 593
Zygomatic bones, 543, 593
Zygomatic muscles, 600
Zygomatic nerve, 615
Zygomatic processes, 590
Zygomaticofacial nerve, 615
Zygomaticotemporal nerve, 503, 615